Advances in Intelligent Systems and Computing

Volume 1048

The series "Advances in Intelligent Systems and Computing" contains publications on theory, applications, and design methods of Intelligent Systems and Intelligent Computing. Virtually all disciplines such as engineering, natural sciences, computer and information science, ICT, economics, business, e-commerce, environment, healthcare, life science are covered. The list of topics spans all the areas of modern intelligent systems and computing such as: computational intelligence, soft computing including neural networks, fuzzy systems, evolutionary computing and the fusion of these paradigms, social intelligence, ambient intelligence, computational neuroscience, artificial life, virtual worlds and society, cognitive science and systems, Perception and Vision, DNA and immune based systems, self-organizing and adaptive systems, e-Learning and teaching, human-centered and human-centric computing, recommender systems, intelligent control, robotics and mechatronics including human-machine teaming, knowledge-based paradigms, learning paradigms, machine ethics, intelligent data analysis, knowledge management, intelligent agents, intelligent decision making and support, intelligent network security, trust management, interactive entertainment, Web intelligence and multimedia.

The publications within "Advances in Intelligent Systems and Computing" are primarily proceedings of important conferences, symposia and congresses. They cover significant recent developments in the field, both of a foundational and applicable character. An important characteristic feature of the series is the short publication time and world-wide distribution. This permits a rapid and broad dissemination of research results.

** Indexing: The books of this series are submitted to ISI Proceedings, EI-Compendex, DBLP, SCOPUS, Google Scholar and Springerlink **

More information about this series at http://www.springer.com/series/11156

Kedar Nath Das · Jagdish Chand Bansal ·
Kusum Deep · Atulya K. Nagar ·
Ponnambalam Pathipooranam ·
Rani Chinnappa Naidu

Editors

Soft Computing for Problem Solving

SocProS 2018, Volume 1

 Springer

Editors
Kedar Nath Das
Department of Mathematics
National Institute of Technology Silchar
Silchar, Assam, India

Jagdish Chand Bansal
Department of Mathematics
South Asian University
New Delhi, Delhi, India

Kusum Deep
Department of Mathematics
Indian Institute of Technology Roorkee
Roorkee, Uttarakhand, India

Atulya K. Nagar
Department of Mathematics
Faculty of Science
Liverpool Hope University
Liverpool, UK

Ponnambalam Pathipooranam
School of Electrical Engineering
VIT University
Vellore, Tamil Nadu, India

Rani Chinnappa Naidu
School of Electrical Engineering
VIT University
Vellore, Tamil Nadu, India

ISSN 2194-5357   ISSN 2194-5365   (electronic)
Advances in Intelligent Systems and Computing
ISBN 978-981-15-0034-3   ISBN 978-981-15-0035-0   (eBook)
https://doi.org/10.1007/978-981-15-0035-0

This Springer imprint is published by the registered company Springer Nature Singapore Pte Ltd.
The registered company address is: 152 Beach Road, #21-01/04 Gateway East, Singapore 189721, Singapore

Preface

SocProS stands for *Soft Computing for Problem Solving*. It is an Eight years old series of International Conferences held annually under the joint collaboration among a group of faculty members from the institutes of repute like NIT Silchar, IIT Roorkee, South Asian University Delhi, Liverpool Hope University, UK and VIT Vellore.

For the first time, SocProS was held at IE(I), RLC, Roorkee, India during Dec 20-22, 2011, with General Chairs as Prof Kusum Deep, Indian Institute of Technology Roorkee and Prof Atulya K. Nagar, Liverpool Hope University, UK. The second SocProS was held at JKLU, Jaipur, India during Dec 28–20, 2012. Similarly, the third SocProS was held at the Greater Noida Extension Centre of IIT Roorkee during December 26–28, 2013, fourth SocProS was held at NIT Silchar, Assam during December 27–29, 2014, Fifth SocProS was held at Saharanpur Campus of IIT Roorkee, during December 18–20, 2015, Sixth SocProS was held at Thapar University, Patiala, Punjab, during December 23–24, 2016, Seventh SocProS was held at IIT Bhubaneswar, Odisha, During December 23–24, 2017, Now the name 'SocProS' became a brand name which has already established its benchmark in last eight years through its successful milestones every time in attracting many participants from all over the world like UK, US, Korea, France, Dubai, South Africa etc.

This time, the Eighth SocProS has been held at VIT Vellore, India during Dec 17–19, 2018. Like earlier SocProS conferences, the focus of SocProS 2018 lies in Soft Computing and its applications to solve real life problems occurring in different domains in the field of medical and health care, supply chain management, signal processing and multimedia, industrial optimization, image processing, cryptanalysis etc. SocProS 2018 attracted a wide spectrum of thought-provoking research papers on various aspects of Soft Computing with umpteen applications, theories and techniques. A total 176 quality research papers are selected for publication in the form of proceedings in its Volume 1 and Volume 2.

We are sure that the research findings in the novel papers contained in this proceeding will be much fruitful and may inspire more and more researchers to work in the field of *soft computing*. The topics that are presented in this proceedings are Fuzzy logic & Fuzzy controller, Artificial Neural Network, Face Recognition & Classification, Feature Extraction, Machine learning, Reinforcement learning, Deep learning, Supervised learning, Different optimization techniques like Spider-Monkey Optimization, Particle Swarm Optimization, Meta heuristic Optimization, Artificial Bee Colony Optimization, Walk Grey Wolf Optimization, Algorithms like Flower Pollination Algorithm, Parallel Random Forest Algorithm, C-mode Clustering Algorithm, Crow Search Algorithm, Genetic Algorithm, Artificial Bee Colony Algorithm, Adaptive Multi-Swarm Bat Algorithm etc. Therefore this proceeding must provide an excellent platform to explore the assorted soft computing techniques to the readers.

The editors would like to express their sincere gratitude to its Patron, Plenary Speakers, Invited Speakers, Reviewers, Programme Committee Members, International Advisory Committee, and Local Organizing Committee; without whose support the quality and standards of the Conference could not be maintained. Special thanks to Springer and its team for this valuable publication.

Over and above, we would like to express our deepest sense of gratitude to 'VIT Vellore' for hosting this conference. Also, sincere thanks to all sponsors of SocProS' 2018.

Silchar, India	Kedar Nath Das
New Delhi, India	Jagdish Chand Bansal
Roorkee, India	Kusum Deep
Liverpool, UK	Atulya K. Nagar
Vellore, India	Ponnambalam Pathipooranam
Vellore, India	Rani Chinnappa Naidu

About This Book

The proceedings of SocProS 2018 will serve as an academic bonanza for scientists and researchers working in the field of Soft Computing. This book contains theoretical as well as practical aspects using fuzzy logic, neural networks, evolutionary algorithms, swarm intelligence algorithms, etc. with many applications under the umbrella of 'Soft Computing'. This book is beneficial for the young as well as experienced researchers dealing across complex and intricate real world problems for which finding a solution by traditional methods is a difficult task.

The different application areas covered in the proceedings are: Image Processing, Cryptanalysis, Industrial Optimization, Supply Chain Management, Newly Proposed Nature Inspired Algorithms, Signal Processing, Problems related to Medical and Health Care, Networking Optimization Problems etc. This will surely helpfully for the researchers/scientists working in similar fields of optimization.

Contents

Analysis of Fractional-Order Deterministic HIV/AIDS Model During Drug Therapy Treatment 1
Ajoy Dutta, Asish Adak and Praveen Kumar Gupta

Load Bearing Capacity for a Ferrofluid Squeeze Film in Double Layered Porous Rough Conical Plates 9
Yogini D. Vashi, Rakesh M. Patel and Gunamani B. Deheri

Effect of Slip Velocity on a Ferrofluid-Based Longitudinally Rough Porous Plane Slider Bearing 27
Mohmmadraiyan M. Munshi, A. R. Patel and G. M. Deheri

Intelligent Controller Based Solar Photovoltaic with Battery Storage System for Conditioning the Electrical Power 43
Ravi Dharavath and I. Jacob Raglend

Autonomous Vehicle for Obstacle Detection and Avoidance Using Reinforcement Learning 55
C. S. Arvind and J. Senthilnath

Evaluation of Deep Learning Model with Optimizing and Satisficing Metrics for Lung Segmentation 67
Usma Niyaz, Abhishek Singh Sambyal and Devanand Padha

Improved Flower Pollination Algorithm for Linear Antenna Design Problems 79
Rohit Salgotra, Urvinder Singh, Sriparna Saha and Atulya K. Nagar

Texture-Based Fuzzy Connectedness Algorithm for Fetal Ultrasound Image Segmentation for Biometric Measurements 91
S. Jayanthi Sree and C. Vasanthanayaki

THD Analysis of Flying-Capacitor Multilevel Converter Using Fuzzy Logic Controller 105
M. Priya, Ponnambalam Pathipooranam and K. Muralikumar

Adaptive Neuro-Fuzzy Inference System for Predicting Strength of High-Performance Concrete 119
L. V. Prasad Meesaraganda, Nilarghya Sarkar and Nilanjan Tarafder

Optimization of Target Oriented Network Intelligence Collection for the Social Web by Using k-Beam Search 135
Aditya Shaha and B. K. Tripathy

Compressed Air Energy Storage Driven by Wind Power Plant for Water Desalination Through Reverse Osmosis Process 145
M. B. Hemanth Kumar and B. Saravanan

Fully Fuzzy Semi-linear Dynamical System Solved by Fuzzy Laplace Transform Under Modified Hukuhara Derivative 155
Purnima Pandit and Payal Singh

Comparison of Performance of Four-Element Microstrip Array Antenna Using Electromagnetic Bandgap Structures 181
K. Prahlada Rao, R. M. Vani and P. V. Hunagund

Model Development for Strength Properties of Laterized Concrete Using Artificial Neural Network Principles 197
P. O. Awoyera, J. O. Akinmusuru, A. Shiva Krishna, R. Gobinath, B. Arunkumar and G. Sangeetha

Wind Power Forecasting Using Parallel Random Forest Algorithm 209
V. Anantha Natarajan and N. Sandhya Kumari

Leukemia Cell Segmentation from Microscopic Blood Smear Image Using C-Mode 225
Neha Singh and B. K. Tripathy

Implementation of Exploration in TONIC Using Non-stationary Volatile Multi-arm Bandits 239
Aditya Shaha, Dhruv Arya and B. K. Tripathy

Fuzzy Logic-Based Model for Predicting Surface Roughness of Friction Drilled Holes 251
N. Narayana Moorthy and T. C. Kanish

Face Recognition and Classification Using GoogleNET Architecture 261
R. Anand, T. Shanthi, M. S. Nithish and S. Lakshman

Enhancing Public Health Surveillance by Measurement of Similarity Using Rough Sets and GIS 271
Priyansh Jain, Harshal Varday, K. Sharmila Banu and B. K. Tripathy

Data Analytics Implemented over E-commerce Data to Evaluate Performance of Supervised Learning Approaches in Relation to Customer Behavior 285
Kailash Hambarde, Gökhan Silahtaroğlu, Santosh Khamitkar, Parag Bhalchandra, Husen Shaikh, Govind Kulkarni, Pritam Tamsekar and Pranita Samale

Optimal Renewable Energy Resource Based Distributed Generation Allocation in a Radial Distribution System 295
Kola Sampangi Sambaiah and T. Jayabarathi

PV Module Temperature Estimation by Using ANFIS 311
Challa Babu and Ponnambalam Pathipooranam

Modified Artificial Potential Field Approaches for Mobile Robot Navigation in Unknown Environments 319
Ngangbam Herojit Singh, Salam Shuleenda Devi and Khelchandra Thongam

Analysis of BASNs Battery Performance at Different Temperature Conditions Using Artificial Neural Networks (ANN) 329
B. Banuselvasaraswathy, R. Vimalathithan and T. Chinnadurai

ASIC Implementation of Fixed-Point Iterative, Parallel, and Pipeline CORDIC Algorithm 341
Grande Naga Jyothi, Kundu Debanjan and Gorantla Anusha

Elephant Herding Optimization Based Neural Network to Predict Elastic Modulus of Concrete 353
B. S. Adarsha, Narayana Harish, Prashanth Janardhan and Sukomal Mandal

Adaptive Sensor Ranking Based on Utility Using Logistic Regression 365
S. Sundar, Cyril Joe Baby, Anirudh Itagi and Siddharth Soni

Detection of Dementia from Brain Tissues Variation in MR Images Using Minimum Cross-Entropy Based Crow Search Algorithm and Structure Tensor Features 377
N. Ahana Priyanka and G. Kavitha

A Hybrid Approach for Intrusion Detection System 391
Neelam Hariyale, Manjari Singh Rathore, Ritu Prasad and Praneet Saurabh

Inspection of Crop-Weed Image Database Using Kapur's Entropy and Spider Monkey Optimization 405
V. Rajinikanth, Nilanjan Dey, Suresh Chandra Satapathy and K. Kamalanand

**Implementation of Fuzzy-Based Multicarrier and Phase Shifted
PWM Symmetrical Cascaded H-Bridge Multilevel Inverter**   415
K. Muralikumar, Ponnambalam Pathipooranam and M. Priya

Derived Shape Features for Brain Hemorrhage Classification   431
Soumi Ray and Vinod Kumar

Prediction of Crime Rate Using Data Clustering Technique   443
A. Anitha

**Identification of Astrocytoma Grade Using Intensity, Texture,
and Shape Based Features**   455
Arkajyoti Mitra, Prasun Chandra Tripathi and Soumen Bag

**Early Prenatal Diagnosis of Down's Syndrome-A Machine
Learning Approach**   467
Esther Hannah, Lilly Raamesh and Sumathi

**Recent Research Advances in Black and White Visual Cryptography
Schemes**   479
T. E. Jisha and Thomas Monoth

**An N-Puzzle Solver Using Tissue P System with Evolutional
Symport/Antiport Rules and Cell Division**   493
Resmi RamachandranPillai and Michael Arock

**Renewable Energy Management and Implementation in Indian
Engineering Institutions—A Case Study**   505
Shekhar Nair, Senthil Prabu Ramalingam,
Prabhakar Karthikeyan Shanmugam and C. Rani

**Inverse Kinematics Analysis of Serial Manipulators Using Genetic
Algorithms**   519
Satyendra Jaladi, T. E. Rao and A. Srinath

Deep Learning for People Counting Model   531
T. Revathi and T. M. Rajalaxmi

**Hybrid Variable Length Partial Pulse Modulation for Visible
Light Communication**   539
Jyothi and Ponnambalam Pathipooranam

**An Efficient Dynamic Background Subtraction Algorithm
for Vehicle Detection Tracking System**   551
Rashmita Khilar, Sarat Kumar Sahoo, C. Rani
and Prabhakar Karthikeyan Shanmugam

PV-Based High-Gain Boost Converter   563
Ritanjali Behera, Sarat Kumar Sahoo, M. Balamurugan,
Prabhakar Karthikeyan Shanmugam and C. Rani

Indoor Object Classification Using Higher Dimensional MPEG Features 573
Dibyendu Roy Chaudhuri, Dhairya Chandra and Ankush Mittal

Lung Nodule Segmentation Using 3-Dimensional Convolutional Neural Networks 585
Subham Kumar and Sundaresan Raman

Improved Performance and Execution Time of Face Recognition Using MRSRC 597
Jitendra Madarkar, Poonam Sharma and Rimjhim Singh

Motion Detection Using a Hybrid Texture-Based Approach 609
Rimjhim Padam Singh, Poonam Sharma and Jitendra Madarkar

Selection of Television Channels for Product Promotion: A Fuzzy-TOPSIS Approach 621
Arshia Kaul, Sugandha Aggarwal and P. C. Jha

Implementation of ACO Tuned Modified PI-like Position and Speed Control of DC Motor: An Application to Electric Vehicle 629
Geetha Mani

A Bidirectional Converter for Integrating DVR-UCap to Improve Voltage Profile Using Fuzzy Logic Controller 647
T. Y. Saravanan and Ponnambalam Pathipooranam

Comparative Study on Histogram Equalization Techniques for Medical Image Enhancement 657
Sakshi Patel, K. P. Bharath, S. Balaji and Rajesh Kumar Muthu

A Fuzzy Multi-criteria Decision Model for Analysis of Socio-ecological Performance Key Factors of Supply Chain 671
Rahul Solanki, Jyoti Dhingra Darbari, Vernika Agarwal and P. C. Jha

A Fuzzy MCDM Model for Facility Location Evaluation Based on Quality of Life 687
Aditi, Arshia Kaul, Jyoti Dhingra Darbari and P. C. Jha

Analytical Structural Model for Implementing Innovation Practices in Sustainable Food Value Chain 699
Rashi Sharma, Jyoti Dhingra Darbari, Venkata S. S. Yadavalli, Vernika Agarwal and P. C. Jha

Mathematical Design and Analysis of Photovoltaic Cell Using MATLAB/Simulink 711
CH Hussaian Basha, C. Rani, R. M. Brisilla and S. Odofin

**Development of Cuckoo Search MPPT Algorithm for Partially
Shaded Solar PV SEPIC Converter**   727
CH Hussaian Basha, Viraj Bansal, C. Rani, R. M. Brisilla and S. Odofin

**An Improved Fuzzy Clustering Segmentation Algorithm Based
on Animal Behavior Global Optimization**   737
A. Absara, S. N. Kumar, A. Lenin Fred, H. Ajay Kumar and V. Suresh

**Fitness-Based Controlled Movements in Artificial Bee Colony
Algorithm**   749
Harish Sharma, Kritika Sharma, Nirmala Sharma, Assif Assad
and Jagdish Chand Bansal

**Herbal Plant Classification and Leaf Disease Identification
Using MPEG-7 Feature Descriptor and Logistic Regression**   761
Ajay Rana and Ankush Mittal

**Simulation of Metaheuristic Intelligence MPPT Techniques for Solar
PV Under Partial Shading Condition**   773
CH Hussaian Basha, C. Rani, R. M. Brisilla and S. Odofin

**Closed Loop Control of Diode Clamped Multilevel Inverter
Using Fuzzy Logic Controller**   787
K. Muralikumar and Ponnambalam Pathipooranam

**Prediction of California Bearing Ratio Using Particle Swarm
Optimization**   795
T. Vamsi Nagaraju, Ch. Durga Prasad and M. Jagapathi Raju

Adaptive Multi-swarm Bat Algorithm (AMBA)   805
Reshu Chaudhary and Hema Banati

**Fuzzy-Based-Cascaded-Multilevel Inverter Topology
with Galvanic Isolation**   823
K. Muralikumar, C. Sivakumar, Ankit Rautela
and Ponnambalam Pathipooranam

**Modeling the Efficacy of Geopolymer Mosquito Repellent Strips
Leachate Distribution Using Meta-heuristic Optimization**   839
D. K. D. B. Rupini and T. Vamsi Nagaraju

**Optimization of Drilling Rig Hydraulics in Drilling Operations
Using Soft Computing Techniques**   849
G. Sangeetha, B. Arun kumar, A. Srinivas, A. Siva Krishna, R. Gobinath
and P. O. Awoyera

**Renewable Energy Harnessing by Implementing a Three-Phase
Multilevel Inverter with Fuzzy Controller**   863
K. Muralikumar and Ponnambalam Pathipooranam

**Design of SVPWM-Based Two-Leg VSI for Solar PV
Grid-Connected Systems** 879
CH Hussaian Basha, V. Govinda Chowdary, C. Rani, R. M. Brisilla
and S. Odofin

**Performance Analysis and Optimization of Process Parameters
in WEDM for Inconel 625 Using TLBO Couple with FIS** 893
Anshuman Kumar, Chinmaya P. Mohanty, R. K. Bhuyan
and Abdul Munaf Shaik

**Application of WDO for Decision-Making in Combined Economic
and Emission Dispatch Problem** 907
V. Udhay Sankar, Bhanutej, C. H. Hussain Basha, Derick Mathew,
C. Rani and K. Busawon

**Application of Wind-Driven Optimization for Decision-Making
in Economic Dispatch Problem** 925
V. Udhay Sankar, Bhanutej, C. H. Hussain Basha, Derick Mathew,
C. Rani and K. Busawon

**Reliability–Redundancy Allocation Using Random Walk Gray
Wolf Optimizer** 941
Shubham Gupta, Kusum Deep and Assif Assad

Optimal Control of Roll Axis of Aircraft Using PID Controller 961
V. Bagyaveereswaran, Subhashini, Abhilash Sahu and R. Anitha

Adaptive Noise Cancellation Using Improved LMS Algorithm 971
Sai Saranya Thunga and Rajesh Kumar Muthu

**Variant Roth-Erev Reinforcement Learning Algorithm-Based Smart
Generator Bidding as Agents in Electricity Market** 981
P. Kiran and K. R. M. Vijaya Chandrakala

**Standalone Solar Photovoltaic Fed Automatic Voltage Regulator
for Voltage Control of Synchronous Generator** 991
Garapati Vinayramsatish, K. R. M. Vijaya Chandrakala
and S. Sampath Kumar

**Optimizing Vertical Air Gap Location Inside the Wall for Energy
Efficient Building Enclosure Design Based on Unsteady Heat
Transfer Characteristics** 1003
Saboor Shaik, Sunnam Nagaraju, Shaik Mohammed Rizvan
and Kiran Kumar Gorantla

Author Index 1011

About the Editors

Dr. Kedar Nath Das is an Assistant Professor at the Department of Mathematics, National Institute of Technology, Silchar, Assam, India. Over the past 10 years, he has made substantial contributions to research on soft computing, and has published several research papers in prominent national and international journals. His chief area of interest is in evolutionary and bio-inspired algorithms for optimization.

Dr. Jagdish Chand Bansal is an Associate Professor at the South Asian University, New Delhi, India and visiting research fellow at Liverpool Hope University, Liverpool, UK. He has an excellent academic record and is a leading researcher in the field of swarm intelligence. Further, he has published numerous research papers in respected international and national journals.

Prof. Kusum Deep is a Professor at the Department of Mathematics, Indian Institute of Technology Roorkee, India. Over the past 25 years, her research has made her a central international figure in the areas of nature-inspired optimization techniques, genetic algorithms and particle swarm optimization.

Prof. Atulya K. Nagar holds the Foundation Chair as Professor of Mathematical Sciences and is Dean of the Faculty of Science at Liverpool Hope University, UK. Prof. Nagar is an internationally respected scholar working at the cutting edge of theoretical computer science, applied mathematical analysis, operations research, and systems engineering. He received a prestigious Commonwealth Fellowship for pursuing his doctorate (DPhil) in Applied Non-Linear Mathematics, which he earned from the University of York (UK) in 1996; and he holds BSc (Hons.), MSc, and MPhil (with Distinction) from the MDS University of Ajmer, India.

Prof. Ponnambalam Pathipooranam is an Associate Professor at the School of Electrical Engineering, VIT University, India. His areas of research interests are Multilevel Converters, Fuzzy controller for multilevel converters, MPC controllers, Thermoelectric Generators for Solar Photo voltaic cells areas in which he is actively publishing. He is having 15 years of teaching experience.

Prof. Rani Chinnappa Naidu received the B.Eng. and M.Tech. degrees from VIT University, Vellore, India, and Ph.D. degree from Northumbria University, Newcastle upon Tyne, UK., all in Electrical Engineering. After that, she joined as a Postdoctoral Researcher in Northumbria Photovoltaic Applications Centre, Northumbria University, UK. She is currently an Associate Professor at VIT University. She is an Senior member in IEEE. She leads an appreciable number of research groups and projects in the areas such as solar photovoltaic, wind energy, power generation dispatch, power system optimization, and artificial intelligence techniques.

Analysis of Fractional-Order Deterministic HIV/AIDS Model During Drug Therapy Treatment

Ajoy Dutta, Asish Adak and Praveen Kumar Gupta

Abstract In this study, we discussed the Caputo sense fractional-order HIV/AIDS model including the drug therapy, and mathematically examined the dynamic behaviour of the model. We have discussed qualitative analysis of the proposed mathematical model and defined the existence and uniqueness conditions. Local stability is also checked for HIV-free equilibrium point. We have given some facts about the growth rate of HIV/AIDS, the source of HIV virus, as well as death rate of $CD4^+$ T cells, which play a vital role in HIV dynamics. The numerical simulations are demonstrated to reveal the analytical results.

Keywords Mathematical modelling · Caputo derivative · HIV/AIDS model · Stability analysis

1 Introduction

At present, more than 50 million citizens worldwide are living with Human Immunodeficiency Virus, and most of the citizens have become resistant to the existing antiretroviral therapies. In the current scenario, there has been a lot of progress in the HIV treatment due to tremendous use of cART (combined antiretroviral therapy) and HAART (highly active antiretroviral therapy). These treatments have yielded considerable improvements in diagnosis and have moderated the rate of infections in citizens who follow their drug treatment. Therefore, scientists need to build up new antiretroviral drugs to fight HIV while it is located in the vital fact that replication of this virus is a very unproductive process [1, 2]. The impact of the traditional drug has to conquer the drug deficiency in lymphoid cells and tissues (CD4 T lymphocytes, host HIV infection). Ho et al. [3] constructed a 'Systems Approach' for HIV treatment through multi-drug-involved nanoparticles. Otunuga [4] has defined and

A. Dutta · A. Adak · P. K. Gupta (✉)
Department of Mathematics, National Institute of Technology Silchar, Silchar 788010, Assam, India
e-mail: pkguptaitbhu@gmail.com

© Springer Nature Singapore Pte Ltd. 2020
K. N. Das et al. (eds.), *Soft Computing for Problem Solving*,
Advances in Intelligent Systems and Computing 1048,
https://doi.org/10.1007/978-981-15-0035-0_1

examined a 2n + 1-dimensional differential equation model with the introducing noise in the transmission rate and treatment of HIV disease.

In the current scenario, fractional calculus is in this phase where numerous models are going to be proposed, described, and applied to real-world problems in the area of physical, biological, engineering sciences and many more branches [5, 6]. However, the researchers have previously accounted for many outstanding results. But, in real situation, many non-local phenomena are unexplored and it will be discovered in the near future. Recently, Pinto and Carvalho [7] analysed the effect of screening and pre-exposure prophylaxis on HIV dynamics in infected citizens, and the said model reported that the fractional derivative order has an influential role during the HIV epidemics. Recently, Pinto and Carvalho [8] studied a mathematical model for HIV infection with fractional-order derivatives where latent T helper cells are incorporated. With these motivations of application of fractional derivatives, we analysed a fractional-order HIV/AIDS dynamical model with the impact of drug therapy.

2 Fractional HIV/AIDS Model

In this part of the study, we constructed a mathematical model which has four compartments: uninfected, HIV-infected CD4$^+$ T cells, virus cells, and drugs concentration as T, I, V, C, respectively. More precisely, we constructed an HIV/AIDS dynamic model that describes the relationship between all the said compartments, and it is formulated by the following fractional-order non-linear system of differential equations in the Caputo sense

$$_cD_t^\alpha T = s - \beta VT + \gamma I - d_1 T - f_1 CT, \tag{1}$$

$$_cD_t^\alpha I = \beta VT - \gamma I - d_2 I - f_2(1 - \eta)CI, \tag{2}$$

$$_cD_t^\alpha V = bd_2 I - d_3 V - f_3(1 - \eta)CV, \tag{3}$$

$$_cD_t^\alpha C = \nu - \delta C, \tag{4}$$

with

$$T(0) = T_0, I(0) = I_0, V(0) = V_0, C(0) = C_0. \tag{5}$$

Here, the inflow rate and the die rate are s, $d_1 T$ of uninfected CD4$^+$ T cells. The uninfected CD4$^+$ T cells converted into infected CD4$^+$ T cells by a virus as βVT, recovered or cured infected CD4$^+$ T cells as a rate γI, obliterated at a rate $f_1 CT$ due to injecting of drugs. Infected cells might be killed because of virion in their nucleus.

The loss rate of an HIV infected cell is considered as $(\gamma + d_2)I$, where $d_2 I$ is the elimination rate of HIV-infected cells, γI is the cure rate of infected cells into the uninfected compartment, and infected cells are destroyed at a rate $f_2(1 - \eta)CV$ due to injecting of drugs. Virions are generated by infected cells at a rate $b\, d_2 I$, decayed at a rate $d_3 V$, and destroyed at a rate $f_3(1 - \eta)CV$ due to injecting of drugs.

3 Analysis of the Model

Before starting the study of stability analysis of the fractional-order system (1)–(5), we begin with some basic results: definition of Caputo fractional-order differentiation, theorems, and lemmas. Let us consider $x(t) = [T(t), I(t), V(t), C(t)]^T$ and $\mathfrak{R}_+^4 = \{x \in \mathfrak{R}^4 : x \geq 0\}$.

Definition 1 (see [9]). Consider that $\alpha > 0$, $t > a$ where $\alpha, a, t \in \mathrm{R}$. Therefore, Caputo fractional operator formula is

$$
_c D_t^\alpha f(t) = \begin{cases} \dfrac{1}{\Gamma(n-\alpha)} \displaystyle\int_a^t \dfrac{f^{(n)}(\tau)}{(t-\tau)^{\alpha+1-n}} d\tau, & (n-1) < \alpha \leq n \\ \dfrac{d^n f(t)}{dt^n}, & \alpha = n \end{cases}
$$

of order α, where $\Gamma(\cdot)$ is the Euler Gamma function.

3.1 Positivity and Boundedness

Assume that $f : \mathrm{R}^n \to \mathrm{R}^n$ for n $= 4$. Consider the fractional-order system

$$
_c D_t^\alpha x(t) = F(x), \ 0 < \alpha \leq 1, \ \text{and}\ x(0) = x_0, \tag{6}
$$

where $F(x) = [f_1, \ f_2, \ f_3, \ f_4]^T$ and $x_0 \in R^n$. For the global existence of the solution for the system (1)–(4) with initial conditions (5), we have to define the subsequent lemmas.

Lemma 2 (see [10]). *If $F(x)$ and $\frac{\partial F}{\partial x}(x)$ are continuous and $\| F(x) \| \leq \lambda + \mu \| x \|$ for all $x \in R^n$, where λ and μ are two positive constants. Then, the system (1)–(4) has the one and only solution on $[0, +\infty)$.*

Theorem 3 (see [10]). *Consider that $f(t) \in C[a, b]$ and Caputo derivative $_c D_t^\alpha f(t) \in C[a, b]$ for $0 < \alpha \leq 1$, so we define*

$$
f(t) = f(a) + \frac{1}{\Gamma(\alpha)} \big(_c D_t^\alpha f\big)(\xi)(t - a)^\alpha, \tag{7}
$$

for $a < \xi < t, \forall t \in (a, b]$.

Lemma 4 (see [10]). *Consider that* $f(t) \in C[a, b]$ *and Caputo derivative* $_cD_t^\alpha f(t) \in C[a, b]$ *for* $0 < \alpha \leq 1$. *If* $_cD_t^\alpha f(t) \geq 0$, $\forall t \in [a, b]$, *then* $f(t)$ *is increasing, moreover* $_cD_t^\alpha f(t) \leq 0$, $\forall t \in [a, b]$, *then* $f(t)$ *is decreasing,* $\forall t \in [a, b]$.

Theorem 5 (see [11]). *Let* $x^* = [T^*, I^*, V^*, C^*]^T$ *is one of the equilibrium points of the fractional-order system* $_cD_t^\alpha x(t) = F(x)$, $0 < \alpha \leq 1$ *and* $x(0) = x_0$. *Then* x^* *is asymptotically stable locally if the spectrum of the Jacobian matrix* $J(x^*)$ *for the system* (1)–(5) *satisfies the following:*

$$\left| \arg\left(Eigenvalues\ J(x^*) \right) \right| > \frac{\alpha\pi}{2},$$

where $J(x^*) = [b_{ij}]_{x=x^*}$, $i, j = 1, 2, 3, 4$, *and* $b_{ij} = \frac{\partial f_i}{\partial x_j}$.

3.2 Equilibrium Points and Reproduction Number

Equilibrium Points. In this subsection, we define all the equilibria for the system (1)–(4) after solving the following non-linear algebraic equations:

$$_cD_t^\alpha T = {}_cD_t^\alpha I = {}_cD_t^\alpha V = {}_cD_t^\alpha C = 0. \tag{8}$$

Then, we get two equilibrium points:

(i)   the first one, infection-free equilibrium point, i.e.

$$E_0 = \left[\frac{s\delta}{d_1\delta + f_1 v}, 0, 0, \frac{v}{\delta} \right]^T. \tag{9}$$

(ii)   the second one, endemic equilibrium point, i.e.

$$E_1 = [T_1, I_1, V_1, C_1]^T, \tag{10}$$

where

$$T_1 = \frac{\left[(d_2 + \gamma)\delta + f_2(1 - \eta)v \right][d_3\delta + f_3(1 - \eta)v]}{d_2 b\beta\,\delta^2},$$

$$I_1 = \frac{sd_2 b\beta\delta^3 - (d_1\delta + f_1 v)\left[(d_2 + \gamma)\delta + f_2(1 - \eta)v \right][d_3\delta + f_3(1 - \eta)v]}{d_2 b\beta\delta^2[d_2\delta + f_2(1 - \eta)v]},$$

$$V_1 = \frac{sd_2 b\beta\delta^3 - (d_1\delta + f_1 v)\left[(d_2 + \gamma)\delta + f_2(1 - \eta)v \right][d_3\delta + f_3(1 - \eta)v]}{\beta\delta[d_2\delta + f_2(1 - \eta)v][d_3\delta + f_3(1 - \eta)v]},$$

and $C_1 = \frac{v}{\delta}$.

Reproduction Number. Afterwards, we calculate the basic reproduction number ($\mathfrak{R}_0$) of system (1)–(4). Biologically, let $\mathfrak{R}_0$ be a symbol of the average number of new infections developed by single infected cell during infection.

$$\mathfrak{R}_0 = \frac{s d_2 b \beta \delta^3}{(d_1 \delta + f_1 v)\left[(d_2 + \gamma)\delta + f_2(1-\eta)v\right]\left[d_3\delta + f_3(1-\eta)v\right]}. \tag{11}$$

3.3 Local Stability

In this subsection, we define the local stability of system (1)–(4) with the help of the Jacobian matrix,

$$J(x)|_{E_0} = \begin{pmatrix} -\beta V - d_1 - f_1 C & \gamma & -\beta T & -f_1 T \\ \beta V & -d_2 - \gamma - f_2(1-\eta)C & \beta T & -f_2(1-\eta)I \\ 0 & b d_2 & -d_3 - f_3(1-\eta)C & -f_3(1-\eta)V \\ 0 & 0 & 0 & -\delta \end{pmatrix}$$

Hence, the associated transcendental equation for the above matrix is

$$\left| J(x)|_{E_0} - \lambda I \right| = 0, \tag{12}$$

where I is the 4×4 identity matrix.

Now, we defined the stability behaviour of the infection-free equilibrium point, i.e. $E_0 = \left[\frac{s\delta}{d_1\delta + f_1 v}, 0, 0, \frac{v}{\delta}\right]^T$ in the following theorem:

Theorem 6 *The infection-free equilibrium point E_0 of fractional-order system (1)–(4) is locally asymptotically stable for $\mathfrak{R}_0 < 1$ if all eigenvalues λ_i of the Jacobian matrix $J(E_0)$ satisfy the condition $|\arg(\lambda_i)| > \alpha \frac{\pi}{2}$.*

Proof The Jacobian matrix $J(E_0)$ for the systems (1)–(4) calculated at the infection-free equilibrium point E_0 is

$$J(E_0) = \begin{pmatrix} -d_1 - f_1\left(\frac{v}{\delta}\right) & \gamma & -\beta\left(\frac{s\delta}{d_1\delta + f_1 v}\right) & -f_1\left(\frac{s\delta}{d_1\delta + f_1 v}\right) \\ 0 & -d_2 - \gamma - f_2(1-\eta)\left(\frac{v}{\delta}\right) & \beta\left(\frac{s\delta}{d_1\delta + f_1 v}\right) & 0 \\ 0 & b d_2 & -d_3 - f_3(1-\eta)\left(\frac{v}{\delta}\right) & 0 \\ 0 & 0 & 0 & -\delta \end{pmatrix}.$$

The characteristic equation of the Jacobian matrix $J(E_0)$ is

$$(\lambda + \delta)\left(\lambda + d_1 + f_1\frac{v}{\delta}\right)(\lambda^2 + a_1\lambda + a_2) = 0 \tag{13}$$

Fig. 1 Stability region of
the fractional-order system
(1)–(4) is enlarged when $0 <
\alpha \leq 1$, where λ is the root of
the characteristic equation

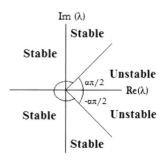

where

$$a_1 = \left(d_2 + d_3 + \gamma + (f_2 + f_3)(1 - \eta)\frac{v}{\delta} \right) > 0$$

and

$$a_2 = \frac{1}{\delta^2}\left[(d_2 + \gamma)\delta + f_2(1 - \eta)v\right][d_3\delta + f_3(1 - \eta)v] - \left(\frac{sd_2b\beta\delta}{d_1\delta + f_1v} \right),$$

or

$$a_2 = \frac{1}{\delta^2}\left[(d_2 + \gamma)\delta + f_2(1 - \eta)v\right][d_3\delta + f_3(1 - \eta)v] \, (1 - \Re_0) \qquad (14)$$

The literature suggests the Routh–Hurwitz stability criterion for fractional-order systems [7, 10], and describes the necessary and sufficient condition $|\arg(\lambda_i)| > \alpha\frac{\pi}{2}$ for various models. According to this criterion, it is clear that all roots of the characteristic Eq. (13) have negative real parts if and only if $a_1 > 0$ and $a_2 > 0$. Equation (14) implies that if $\Re_0 < 1$, then all roots will be negative and for $\Re_0 < 1$ the necessary and sufficient condition will satisfy (Fig. 1).

Hence, the system (1)–(4) is asymptotically stable at E_0 if $\Re_0 < 1$; otherwise, if $\Re_0 > 1$, the characteristic Eq. (13) has given at least one positive eigenvalues, therefore, E_0 is unstable.

4 Numerical Solution and Discussion

In this paper, we presented a numerical solution of the fractional-order HIV/AIDS model (1)–(5). Here, we solve this system using Mathematica 8.0. Consider that $s = 3 \, \text{mm}^{-3}\text{day}^{-1}$, $\beta = 0.000024 \, \text{mm}^3\text{day}^{-1}$, $\gamma = 0.2 \, \text{day}^{-1}$, $b = 1940$, $d_1 = 0.01 \, \text{day}^{-1}$, $d_2 = 0.5 \, \text{day}^{-1}$, $d_3 = 3.4 \, \text{day}^{-1}$ (since removal rate of infected cells will be higher than uninfected cells), $\alpha = 1$, $f_1 = 0.009 \, \text{day}^{-1}$, $f_2 = 2 \times 10^{-10} \, \text{day}^{-1}$ and $f_3 = 10^{-4} \, \text{day}^{-1}$ [10, 11].

In view of the reality that the recovery rate (γ) will also depend on drugs which are given to the patient, we are defining the variation of γ with initial conditions $T(0) = 1000$, $I(0) = 0$, $V(0) = 0.001$ and $C(0) = 0.5$ in Fig. 2.

Fig. 2 Plot of **a** uninfected CD4 + T cells **b** infected CD4 + T cells **c** virus versus time for various values of γ

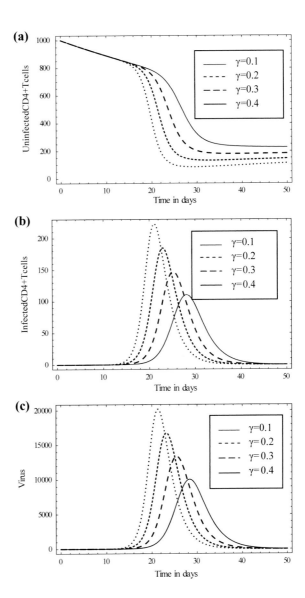

5 Conclusion

In this paper, we presented a Caputo sense fractional-order HIV/AIDS dynamics model with drug therapy. The authors have defined the equilibrium points and reproduction number by Jacobian-based spectral radius method for the proposed model. The recent appearance of fractional differential equations as models makes it necessary to investigate analysis of solution for such equations. So, the authors describe the stability analysis on a proposed fractional order model, and obtained a sufficient condition on the parameters for asymptotically stable infection-free steady state. The numerical solutions have demonstrated the impact of drugs in the patients and it is depicted through Fig. 2 for various values of γ.

Acknowledgements The authors are very thankful to the respected reviewers for their positive comments towards the improvement of the manuscript. This research work is financially supported by TEQIP-III, National Institute of Technology Silchar, Assam-788010 under MHRD, India.

References

1. Perelson, A.S., Kirschner, D.E., Boer, R.: Dynamics of HIV infection of CD4 + T cells. Math. Biosci. **114**, 81–125 (1993)
2. Duncan, S., Dorrell, L.: Promising new drugs and drug targets for HIV treatment. Future Prescr. **14**, 1–4 (2013)
3. Ho, R.J.Y., Yu, J., Li, B., Kraft, J.C., Freeling, J.P., Koehn, J., Shao, J.: Systems approach to targeted and long-acting HIV/AIDS therapy. Drug Deliv. Trans. Res. **5**(6), 531–539 (2015)
4. Otunuga, O.M.: Global stability for a 2n + 1 dimensional HIV/AIDS epidemic model with treatments. Math. Biosci. **299**, 138–152 (2018)
5. Sun, H.G., Zhang, Y., Baleanu, D., Chen, W., Chen, Y.Q.: A new collection of real world applications of fractional calculus in science and engineering. Commun. Nonlinear Sci. Numer. Simul. **64**, 213–231 (2018)
6. Gupta, P.K., Dhar, B.: Dynamical behaviour of fractional order tumor-immune model with targeted chemotherapy treatment. Int. J. Eng. Tech. **7**, 6–9 (2018)
7. Pinto, C.M.A., Carvalho, A.R.M.: The impact of pre-exposure prophylaxis (PrEP) and screening on the dynamics of HIV. J. Comput. Appl. Math. **339**, 231–244 (2018)
8. Pinto, C.M.A., Carvalho, A.R.M.: A latency fractional order model for HIV dynamics. J. Comput. Appl. Math. **312**, 240–256 (2017)
9. Caputo, M.: Linear model of dissipation whose Q is almost frequency independent-II. Geophys. J. R. Astron. Soc. **13**, 529–539 (1967)
10. Dutta, A., Gupta, P.K.: A mathematical model for transmission dynamics of HIV/AIDS with effect of weak CD4 + T cells. Chin. J Phys. **56**(3), 1045–1056 (2018)
11. Gupta, P.K.: Local and global stability of fractional order HIV/AIDS dynamics model. In: Ghosh, D., Giri, D., Mohapatra, R., Savas, E., Sakurai, K., Singh, L. (eds.) Mathematics and Computing. ICMC 2018. CCIS, vol. 834, pp. 141–148. Springer, Singapore (2018)

Load Bearing Capacity for a Ferrofluid Squeeze Film in Double Layered Porous Rough Conical Plates

Yogini D. Vashi, Rakesh M. Patel and Gunamani B. Deheri

Abstract This article goals to determine the enactment of double layered porous rough conical plates with ferrofluid based squeeze film lubrication. The Neuringer–Roseinweig model has been employed for magnetic fluid flow. For the characterization of roughness two different forms of polynomial distribution function have been used and comparison is made between both roughness structure. The stochastic model of Christensen and Tonder regarding transverse roughness has been invoked to develop the associated Reynolds' equation from which the pressure circulation is found. This provides growth to the calculation of load-bearing capacity. From the graphical appearance it is established that from the design point of view roughness pattern G1 is more suitable compared to G2. The results presented here confirm that the introduction of double layered plates results in improved load carrying capacity. This is further enhanced by the ferrofluid lubrication. Further, the roughness affects the bearing system significantly, however, the situation enhanced in the case of negatively skewed roughness. A noticeable fact is that the porosity of the outer layer influences more as compared to the inner layer even in the presence of mild magnetic strength.

Keywords Squeeze film · Conical plates · Roughness · Ferrofluid · Load carrying capacity

Y. D. Vashi (✉)
Department of Applied Sciences and Humanity, Alpha College of Engineering and Technology, Khatraj, Kalol 382721, Gujarat, India
e-mail: yogini.vashi@gmail.com

R. M. Patel
Department of Mathematics, Gujarat Arts and Science College, Ahmedabad 380006, Gujarat, India
e-mail: rmpatel12711@gmail.com

G. B. Deheri
Department of Mathematics, Sardar Patel University, Vallabh Vidyanagar 388120, Gujarat, India
e-mail: gm.deheri@rediffmail.com

© Springer Nature Singapore Pte Ltd. 2020
K. N. Das et al. (eds.), *Soft Computing for Problem Solving*,
Advances in Intelligent Systems and Computing 1048,
https://doi.org/10.1007/978-981-15-0035-0_2

Nomenclature

a	Dimension of bearing (mm)
$\boldsymbol{h}$	Uniform fluid film thickness(mm)
h	Mean film thickness
h_s	Deviation from the mean film thickness
$\dot{h}_0$	Normal velocity of bearing surface
H_1	The thickness of the inner layer of the porous plate (mm)
H_2	The thickness of the outer layer of the porous plate (mm)
$\overline{H}$	Magnetic field vector
p	Pressure distribution(N/m^2)
$\overline{p}$	Non dimensional Pressure distribution
W	Load carrying capacity (N)
$\overline{W}$	Non dimensional Load bearing capacity
η	Dynamic viscosity of fluid (NS/m^2)
μ_0	Permeability of free space (N/A^2)
$\overline{\mu}$	Magnetic susceptibility of magnetic field
σ	Standard deviation (mm)
α	Variance (mm)
ε	Skewness (mm)
ϕ_1	The permeability of the inner layer (m^2)
ϕ_2	The permeability of the outer layer (m^2)
ψ_1	Porosity of inner layer
ψ_2	Porosity of outer layer
α^*	Non dimensional variance
ε^*	Non dimensional skewness
σ^*	Non dimensional standard deviation
ρ	Density of fluid
$\overline{q}$	Velocity of fluid
$\underline{\eta}$	Fluid viscosity
$\overline{M}$	Magnetization vector

1   Introduction

Porous bearing is used very widely in many devices such as Vacuum cleaners, extractor fans, motorcar starters, hair dryer, etc. They are also used in business machines,
farm and construction equipment, and aircraft automotive accessories. In addition,
porous bearing can work hydrodynamically longer without maintenance and more
stable than the equivalent conventional bearing. Also, in these bearings friction is less

as compared to the non-porous bearings. so many researchers have studied the effect of double layered porous bearings of various shapes. Uma Srinivasan [1] worked to study double layered slider bearing with a porous surface. The double layered surface enhanced the bearing's load carrying capacity as well as the friction drag. However, it reduced the friction coefficient. Verma [2] investigated the influence of double-layered porous slider bearing. The study of Rao et al. [3] focused on the relation between Brinkman model and a double layered porous journal bearing's performance. The results suggested that in a double layered bearing, the low permeability layer stuck to the high permeability one, leading to increased bearing capacity and as a result, a decreased friction coefficient. Uma Srinivasan [4] intended to study the impact of time-height of squeeze films on a bearing's load capacity. Various geometrical aspects like circular, elliptical, rectangular, etc. were used for the purpose. It was a comparative study focusing on two-layered porous bearing and conventional bearings. The results suggested that double layered plates enhance a bearing's load carrying capacity. Cusano [5] analyzed an infinitely long two-layered porous bearing.

Conical bearings have been developed for use in agricultural and construction machinery, for the suspension of jolts and insulation of engine vibrations from cabins. Lin et al. [6] studied the behavior of non-Newtonian micropolar fluid squeeze film between conical plates. The non-Newtonian effects of micropolar fluid were found to be better in comparison with the Newtonian case also its effect lengthened the approaching time of squeeze film conical plates. Dinesh Kumar et al. [7] studied the effect of ferrofluid squeeze film for spherical and conical bearings using perturbation analysis. Prakash and Vij [8] analyzed the effect of the shape of the plate and porosity on the performance of squeeze films between porous plates of various shapes.

Practically, a perfectly smooth surface does not exist as all surfaces are rough to some extent. In applied settings, a smooth surface bearing does not provide an optimum idea of performance and bearing life span. Thus, in the recent year studies have focused on correlating surface roughness with the bearing capacity. Christensen and Tonder [9–11] worked on the stochastic surface roughness theory with hydrodynamic lubrication. Many authors have used this technique to understand the impact of surface roughness on performance. Patel and Deheri [12] made a comparative study of different porous configurations and their impact on double layered slider bearing with roughness and magnetic fluid base. The results suggested that Kozeny carman model is more effective than Irmay's model. Deheri et al. [13] made a theoretical study of the influence of squeeze film with a magnetic base on porous rough conical plates. The results showed that an appropriate semi-vertical angle can revert the negative impacts of porosity and standard deviations for negatively skewed roughness. Patel and Deheri [14] deliberated the impact of slip velocity on a squeeze film with ferrofluid in conical plates with longitudinal roughness. It was found that standard deviation and magnetization can substantially neutralize the negative impact created by slip velocity and surface roughness on bearing performance, provided that the negatively skewed roughness was appropriate. Vashi et al. [21] studied the impact of ferrofluid based rough porous parallel plates with couple stress effect.

Various good research articles are available in the literature for the study of squeeze film lubrication of conical bearings and truncated conical bearings. For examples Shimpi and Deheri [15] in porous truncated conical plates, Patel and Deheri [16] in porous conical plates, Vadher et al. [17] in porous rough conical plates, Patel et al. [20] in rough conical bearing with deformation effect.

At present no work has been made to study the influence of surface roughness with two different patterns on ferrofluid based squeeze film in double layered porous conical plates. So in this current study the investigation of Patel and Deheri [13] is extended to the double layered porous conical plates with two different forms of the transverse surface roughness patterns.

2 Analysis

All the traditions of hydrodynamic lubrication are reserved here. The lubricant is an incompressible ferrofluid lubrication, considered for the analysis. Both the porous facings are supposed to be homogeneous and isotropic and porosity is directed by a generalized form of Darcy's law.

Figure 1 displays the geometrical structure of squeeze film lubrication of porous rough conical plates bearing.

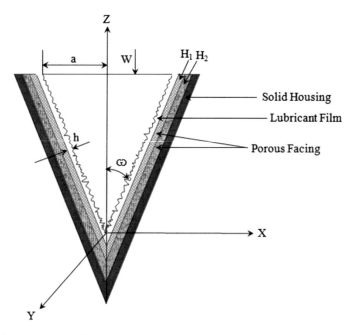

Fig. 1 Physical structure of the bearing system

In the sought of the discussion of Uma Srinivasan [1] the modified Reynolds equation comes out to be

$$\frac{1}{x}\frac{d}{dx}\left(\frac{dp}{dx}\right) = \frac{12\eta\, \dot{h_0}\sin\omega}{h^3\sin^3\omega + 12\phi_1 H_1 + 12\phi_2 H_2} \tag{1}$$

The bearing faces are deliberated to be transversely rough in the context of Christensen and Tonder's [9–11] discussion the lubricant film thickness is taken as

$$h = h + h_s$$

where h is mean film thickness and h_s is the part due to the surface roughness as measured from nominal film thickness. According to Christensen and Tonder [9–11] Stochastic part h_s is defined by the polynomial probability distribution function $f(h_s)_1$ for the domain $-c \leq h_s \leq c$, where c represents the maximum deviation from the mean film thickness.

$$f(h_s)_1 = \begin{cases} \frac{35}{32c^7}\left(c^2 - h_s^2\right)^3, & -c \leq h_s \leq c \\ 0, & \text{elsewhere} \end{cases} \tag{2}$$

Further, a different form of this type of polynomial distribution from Prajapati [18] is

$$f(h_s)_2 = \begin{cases} \frac{15}{16c^5}\left(c^2 - h_s^2\right)^2, & -c \leq h_s \leq c \\ 0, & \text{elsewhere} \end{cases} \tag{3}$$

The measure of the symmetry of the random variable h_s are mean α the standard deviation σ and the parameter ε defined by the relations

$$\alpha = E(h_s) \quad \sigma^2 = E\left[(h_s - \alpha)^2\right] \quad \varepsilon = E\left[(h_s - \alpha)^3\right]$$

where $E(\bullet)$ is the expectancy operator given by the formula

$$E(\bullet) = \int_{-\infty}^{\infty} (\bullet)f(h_s)dh_s \tag{4}$$

The detailed study regarding the roughness model can be observed in Christensen and Tonder [9–11].

Neuringer and Rosensweig [19] established a model to designate the stable flow of magnetic fluid. This model involves the following equations.

Equation of motion:

$$\rho(\bar{q}.\nabla)\bar{q} = -\nabla p + \eta\nabla^2\bar{q} + \mu_0\left(\overline{M}.\nabla\right)\overline{H} \tag{5}$$

Equation of continuity:

$$\nabla.\overline{q} = 0 \tag{6}$$

Maxwell's equations:

$$\nabla \times \overline{H} = 0 \tag{7}$$

$$\nabla.(\overline{H} + \overline{M}) = 0 \tag{8}$$

Equation of Magnetization:

$$\overline{M} = \overline{\mu}\overline{H} \tag{9}$$

The magnetic field is

$$\overline{H} = H(x)(\cos \phi, 0, \sin \phi), \phi = \phi(x, z) \tag{10}$$

and ϕ is given by the equation

$$\nabla \times \overline{H} = 0 \tag{11}$$

is

$$\cot \phi \frac{\partial \phi}{\partial x} + \frac{\partial \phi}{\partial z} = -\frac{1}{H}\frac{dH}{dx} \tag{12}$$

The magnetic field deliberated here is slanting to the lower surface and its magnitude is defined as

$$H^2 = K(a^2 - x^2 \sin^2 \omega) \tag{13}$$

wherin K is a suitably chosen constant depending on the material to produce a field of desired magnetic strength.

Now using the averaging practices of Christensen and Tonder [9–11] and Neuringer and Rosensweig [19] model of magnetic fluid flow Eq. (1) transfer to the form

$$\frac{1}{x}\frac{d}{dx}\left(x\frac{d}{dx}\left(p - 0.5\mu_0\overline{\mu}H^2\right)\right) = \frac{12\eta \dot{h}_0 \sin \omega}{g_1(h)} \tag{14}$$

wherin

$$g_1(h) = h^3 \sin^3 \omega + 3\sigma^2 h \sin \omega + 3\alpha^2 h \sin \omega$$

$$+ 3\alpha h^2 \sin^2 \omega + 3\sigma^2 \alpha + \alpha^3 + \varepsilon + 12\phi_1 H_1 + 12\phi_2 H_2$$

For the different form of polynomial probability distribution function the Eq. (1) transfer to the form

$$\frac{1}{x}\frac{d}{dx}\left(x\frac{d}{dx}(p - 0.5\mu_0\overline{\mu}H^2)\right) = \frac{12\eta\, \dot{h}_0 \sin \omega}{g_2(h)} \tag{15}$$

Where

$$g_2(h) = h^3 \sin^3 \omega + 4\sigma^2 h \sin \omega + 3\alpha^2 h \sin \omega$$
$$+ 2\alpha h^2 \sin^2 \omega + 4\sigma^2 \alpha + \alpha^3 + \varepsilon + 12\phi_1 H_1 + 12\phi_2 H_2$$

The related pressure boundary conditions are

$$p(a\, \mathrm{cosec}\omega) = 0 \, and \, \left(\frac{dp}{dx}\right)_{x=0} = 0 \tag{16}$$

Solving expression (14) with the suitable boundary conditions (16) the appearance for dimensional less pressure established in the film region is found as

$$\overline{p} = \frac{\mu^*(1 - X^2) \sin \omega}{2\pi} + \frac{3(1 - X^2)}{\pi G_1} \tag{17}$$

wherin

$$G_1 = \frac{g_1(h)}{h_0^3} = \sin^3 \omega + 3\sigma^{*2} \sin \omega + 3\alpha^{*2} \sin \omega$$
$$+ 3\alpha^* \sin^2 \omega + 3\sigma^{*2}\alpha^* + \alpha^{*3} + \varepsilon^* + 12\psi_1 + 12\psi_2$$

$$\mu^* = \frac{-\mu_0\overline{\mu}h_0^3 K}{\eta\, \dot{h}_0}, \overline{h} = \frac{h}{h_0} = 1, \alpha^* = \frac{\alpha}{h_0}, \varepsilon^* = \frac{\varepsilon}{h_0^3},$$

$$\sigma^* = \frac{\sigma}{h_0}, \psi_1 = \frac{\phi_1 H_1}{h_0^3},$$

$$\psi_2 = \frac{\phi_2 H_2}{h_0^3}, X = \frac{x}{a\, \mathrm{cosec}\omega}, \overline{p} = \frac{-h_0^3\, \mathrm{cosec}\omega p}{\eta\, \dot{h}_0 a^2\pi}$$

expression for dimensional less pressure for different form of transverse roughness is found as

$$\overline{p} = \frac{\mu^*(1 - X^2) \sin \omega}{2\pi} + \frac{3(1 - X^2)}{\pi G_2} \tag{18}$$

wherin

$$G_2 = \frac{g_2(h)}{h_0^3} = \sin^3 \omega + 4\sigma^{*2} \sin \omega + 3\alpha^{*2}$$
$$+ 2\alpha^* \sin^2 \omega + 4\sigma^{*2}\alpha^* + \alpha^{*3} + \varepsilon^* + 12\psi_1 + 12\psi_2$$

Now the load-bearing capacity of the bearing can be achieved integrating the film pressure over the squeezing film region as follow:

$$W = 2\pi \int_{0}^{a \, cosec\omega} x p(x) dx \qquad (19)$$

Lastly, the expression for $\overline{W}$ is given by

$$\overline{W} = -\frac{h_0^3 W}{\eta \, \overset{\bullet}{h_0} \, a^4 \cos ec^2 \omega} = \frac{3 cosec\omega}{2\pi \, G_1} + \frac{\mu^*}{4\pi} \qquad (20)$$

for the different form of roughness structure expression for $\overline{W}$ is obtained as

$$\overline{W} = \frac{3 cosec\omega}{2\pi \, G_2} + \frac{\mu^*}{4\pi} \qquad (21)$$

3   Result and Discussion

The influence of ferrofluid squeeze film in double layered porous rough conical plate is studied with different forms of transverse surface roughness, also comparison is made between two different structure of transverse surface roughness. In the deficiency of magnetization this investigation diminishes to the work of Uma Srinivasan [1] for smooth bearing. The analytic expression for $\overline{W}$ and $\overline{p}$ for both the roughness structure is given by the Eqs. (17), (18), (20) and (21). It is found that the $\overline{W}$ increases $\frac{\mu^*}{4}$ times as related to the traditional lubrication-based bearing system. Here the solid line represents the roughness structure G_1 and the dotted lines represent the roughness structure G_2. Figures 1, 2, 3, 4, 5 described the profile of $\overline{W}$ with regards to μ^* for different values of σ^*, ω, ε^*, ψ_1 and ψ_2. It is detected that $\overline{W}$ increases as growing the values of μ^* for both the roughness structures. It is clearly seen from Figs. 3 and 4 that influence of G_1 is more related to G_2.

The contrary influence of σ^* on $\overline{W}$ with regards to different values of α^*, ψ_1, ψ_2 and ω can be seen from Figs. 7, 8, 9, 10. From Fig. 8 it is noticed that the adverse influence of σ^* can be registered to be more in the case of G_2.

The Figs. 11, 12, 13, 14, 15, 16, 17 presents the profile of $\overline{W}$ with regards to roughness parameter α^* and ε^* for different values of ψ_1, ψ_2, ω. It is perceived that the negatively rising values of α^* and ε^* rises the load-bearing capacity while

Fig. 2 Profile of $\overline{W}$ with reference to μ^* and σ^*

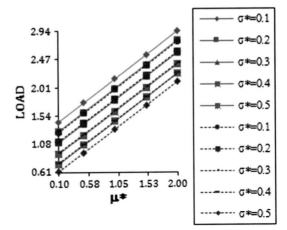

Fig. 3 Profile of $\overline{W}$ with reference to μ^* and ε^*

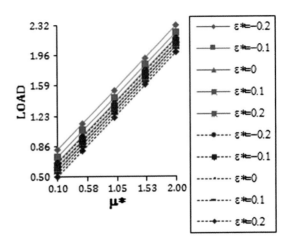

Fig. 4 Profile of $\overline{W}$ with reference to μ^* and ψ_1

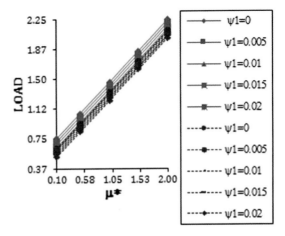

Fig. 5 Profile of $\overline{W}$ with reference to μ^* and ψ_2

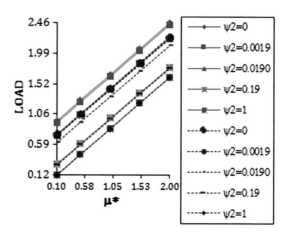

Fig. 6 Profile of $\overline{W}$ with reference to μ^* and ω

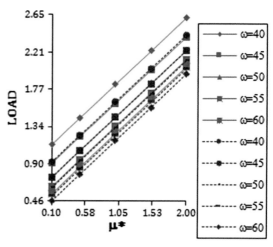

Fig. 7 Change in $\overline{W}$ reference to σ^* and α^*

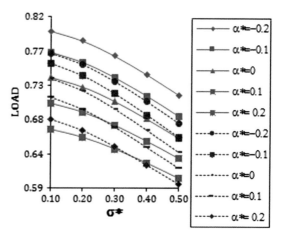

Fig. 8 Change in $\overline{W}$
reference to σ^* and ψ_1

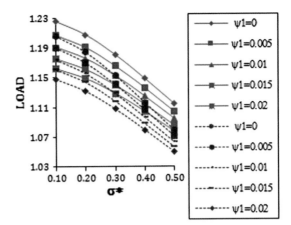

Fig. 9 Change in $\overline{W}$
reference to σ^* and ψ_2

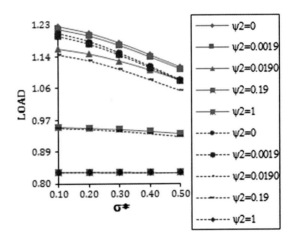

Fig. 10 Change in $\overline{W}$
reference to σ^* and ω

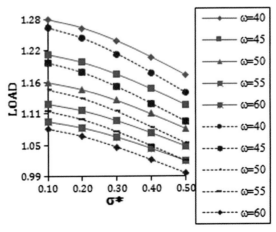

Fig. 11 Profile of $\overline{W}$
reference to α^* and ε^*

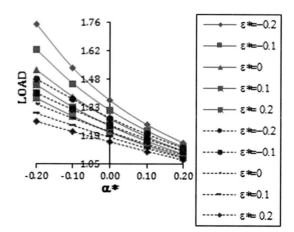

Fig. 12 Profile of $\overline{W}$
reference to α^* and ψ_1

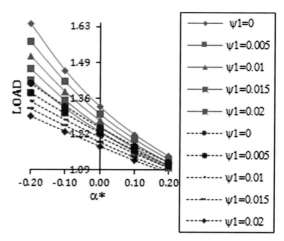

Fig. 13 Profile of $\overline{W}$
reference to α^* and ψ_2

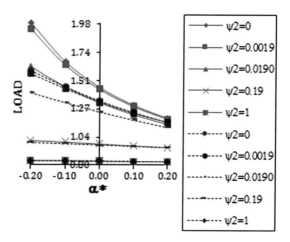

Fig. 14 Profile of $\overline{W}$
reference to α^* and ω

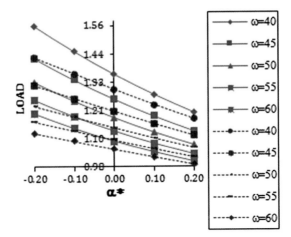

Fig. 15 Profile of $\overline{W}$
reference to ε^* and ψ_1

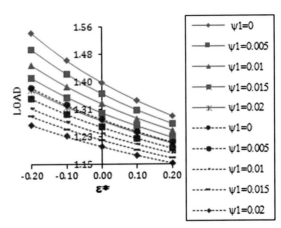

Fig. 16 Profile of $\overline{W}$
reference to ε^* and ψ_2

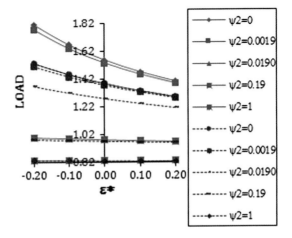

Fig. 17 Profile of $\overline{W}$ reference to ε^* and ω

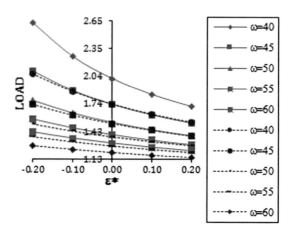

Fig. 18 Profile of $\overline{W}$ reference to ψ_1 and ψ_2

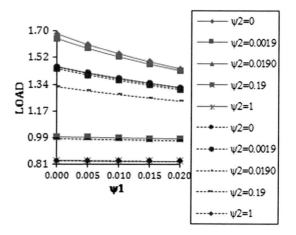

the positively rising values of α^* and ε^* decreases the value of $\overline{W}$. The negatively skewed roughness in conjunction with variance (–ve) may offer necessary help for the ferrofluid lubrication of the bearing system.

Figures 18, 19 shows the distribution of $\overline{W}$ concerned with ψ_1 for different values of ψ_2 and ω. It is mentioned in Fig. 18 that initial influence of second porous layered is virtually minimal for both the roughness structures.

A closed look at the Tables 1, 2, 3 suggests that the first form G_1 of the roughness pattern is found to be more favorable for the adoption in the bearing system. In addition, even G_2 can be taken into consideration when the porosity is at the reduced level and magnetic strength is in force. Although there is atleast 1.8% decrease in load-bearing capacity with regards to G_1.

Fig. 19 Profile of $\overline{W}$ reference to ψ_1 and ω

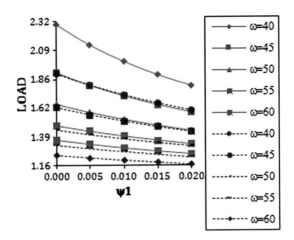

Table 1 Change in $\overline{W}$ with regards to different values of σ^*

		$\overline{W}$ of G_1	$\overline{W}$ of G_2	% increase in $\overline{W}$
$\omega = 45$	$\sigma^* = 0.1$	2.15390788	1.98220325	8.5
$\alpha^* = -0.10$	$\sigma^* = 0.2$	1.99474005	1.82296845	9.3
$\varepsilon^* = -0.10$	$\sigma^* = 0.3$	1.79837823	1.63464754	9.8
$\mu^* = 1$	$\sigma^* = 0.4$	1.61075813	1.462575132	10.2
$\psi_1 = 0.001$	$\sigma^* = 0.5$	1.45201373	1.322623759	9.6
$\psi_2 = 0.019$				

Table 2 Change in $\overline{W}$ with regards to different values of α^*

		$\overline{W}$ of G_1	$\overline{W}$ of G_2	% increase in $\overline{W}$
$\omega = 45$	$\alpha^* = -0.20$	1.62101704	1.40572313	15.7
$\sigma^* = 0.05$	$\alpha^* = -0.10$	1.45201373	1.32262376	9.8
$\varepsilon^* = -0.10$	$\alpha^* = 0$	1.32083663	1.24519182	6.4
$\mu^* = 1$	$\alpha^* = 0.10$	1.21850784	1.175702334	3.4
$\psi_1 = 0.001$	$\alpha^* = 0.20$	1.13831751	1.115138546	1.8
$\psi_2 = 0.019$				

Table 3 Change in $\overline{W}$ with regards to different values of ε^*

		$\overline{W}$ of G_1	$\overline{W}$ of G_2	% increase in $\overline{W}$
$\omega = 45$	$\varepsilon^* = -0.20$	1.53184786	1.37329613	11
$\sigma^* = 0.05$	$\varepsilon^* = -0.10$	1.45201373	1.32262376	9.8
$\alpha^* = -0.10$	$\varepsilon^* = 0$	1.38760648	1.27999338	8.6
$\mu^* = 1$	$\varepsilon^* = 0.10$	1.33454851	1.24363128	7.2
$\psi_1 = 0.001$	$\varepsilon^* = 0.20$	1.29008292	1.212249626	6.6
$\psi_2 = 0.019$				

Fig. 20 Profile of $\overline{W}$
reference to ω and μ^*

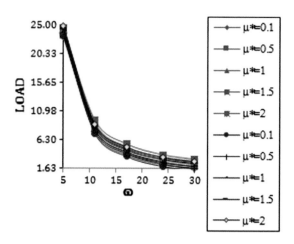

4 Conclusions

The combined influence of surface roughness and ferrofluid lubrication of doubled layered porous rough conical plate is analyzed with two different forms of transverse roughness patterns. From the graphical results our study discovered the following conclusions:

(1) The improved load due to magnetization gets sustained due to double layered.
(2) Double layered porous conical bearing with the roughness pattern G_1 is better than that of the bearing with roughness pattern G_2.
(3) It has been found that the load bearing capacity remains maximum for ω lying between $\frac{\pi}{30}$ to $\frac{\pi}{10}$ approximately (Fig. 20).
(4) The porosity of outer layer is favorable to develop the lubrication performance of the double layered conical bearing. Therefore, when designing the double layered porous bearing the surface porosity should be reduced as far as possible and the roughness needs to be treated carefully from bearing designs point of view. If developed properly this investigation may provide a good opportunity for the industry.

Acknowledgements The authors would like to thank both the reviewers and the editor for their fruitful comments and constructive suggestions for improving the overall presentation and quality of the article.

References

1. Srinivasan, U.: The analysis of double layered porous slider bearing. Wear **42**(2), 205–215 (1977). https://doi.org/10.1016/0043-1648(77)90052-7
2. Verma, P.D.S.: Double layered porous journal bearing. Mech. Mater. **2**(3), 233–238 (1983). https://doi.org/10.1016/0167-6636(83)90017-0

3. Rao, T.V.V.L.N., Rani, A.M.A., Nagrajan, T., Hashim, F.M.: Analysis of journal bearing with double layer porous lubricant film, Influence of surface porous layer configuration. Tribol. Trans. **56**(5), 841–847 (2013). https://doi.org/10.1080/10402004.2013.801100

4. Srinivasan, U.: Load capacity and time height relations for squeeze films between double layered porous plates. Wear **1**(43), 211–225 (1977). https://doi.org/10.1016/0043-1648(77)90115-6

5. Cusano, C.: Analytical investigation of an infinitely long, two layer, porous bearing. wear **22**(1), 59–67 (1972). https://doi.org/10.1016/0043-1648(72)90427-9

6. Lin, J.R., Kuo, C.C., Liao, W.H., Yang, C.B.: Non newtonian micropolar fluid squeeze film between conical plates. Zeitschrift fur Naturforschung J. Phys. Sci. **67**(a), 333–337 (2012)

7. Kumar, D., Sinha, P., Chandra, P.: Ferrofluid squeeze film for spherical and conical bearings. Int. J. Eng. Sci. **30**(5), 645–656 (1992). https://doi.org/10.1016/0020-7225(92)90008-5

8. Prakash, J., Vij S.K.: Load capacity and time height relations for squeeze film between porous plates. Wear **24**(3), 309–322 (1973). https://doi.org/10.1016/0043-1648(73)90161-0

9. Christensen, H., Tonder, K.C.: Tribology of rough surface: stochastic models of hydrodynamic lubrication. SINTEF **10**, 18–69 (1969)

10. Christensen, H., Tonder K.C.: Tribology of rough surfaces: parametric study and comparison of lubrication models. SINTEF **22**, 69–18 (1969b)

11. Christensen, H., Tonder K.C.: The hydrodynamic lubrication of rough bearing surfaces of finite width. In: ASME-ASLE Lubrication Conference, Cincinnati, Ohio, Paper no. 70-Lub-7, October 12–15 (1970)

12. Patel, J.R., Deheri, G.M.: Performance of a magnetic fluid based double layered rough porous slider bearing considering the combined porous structures. Acta Technica Corviniensis-Bull. Eng. Tome **7**, 115–125 (2014)

13. Deheri, G.M., Patel R.M., Patel, H.C.: Magnetic fluid based squeeze film between porous rough conical plates. J. Comput. Methods Sci. Eng. **13**, 419–432 (2013). https://doi.org/10.3233/jcm-130475

14. Patel, J.R., Deheri, G.M.: The effect of slip velocity on the ferrofluid based film in longitudinally rough conical plates. J. Serb. Soc. Comput. Mech. **10**(2), 18–29 (2016)

15. Shimpi, M.E., Deheri, G.M.: Effect of slip velocity and bearing deformation on the performance of a truncated conical plates. Iran. J. Sci. Technol. Trans. Mech. Eng. **38**, 195–206 (2014)

16. Patel, R.M., Deheri, G.M.: Magnetic fluid based squeeze film between porous conical plates. Ind. Lubr. Tribol. **59**(3), 309–322 (2007)

17. Vadher, P.A., Deheri, G.M., Patel, R.M.: Performance of a hydromantic Squeeze films between conducting porous rough conical plates. Mech. Int. J. Theor. Appl. Mech. **45**(6), 767–783 (2010)

18. Prajapati, B.L.: On certain theoretical studies in hydrodynamic and electromagnet hydrodynamic lubrication. Ph.D. thesis, Department of physics, Sardar Patel University, V.V. Nagar (1995)

19. Neuringer, J.L., Rosensweig, R.E.: Magnetic fluids. Phys. Fluids **7**(12), 1927–1937 (1964)

20. Patel, J.R., Shimpi, M.E., Deheri, G.M.: Ferrofluid based squeeze film for a rough conical bearing with deformation effect. Int. Conf. Res. invoat. Sci. Eng. Technol Kalpa Publ Comput **2**, 119–129 (2017)

21. Vashi, Y.D., Patel, R.M., Deheri, G.M.: Ferrofluid based squeeze film lubrication between rough stepped plates with couple stress effect. J. Appl. Fluid Mech. **11**(3), 597–612 (2018)

Effect of Slip Velocity on a Ferrofluid-Based Longitudinally Rough Porous Plane Slider Bearing

Mohmmadraiyan M. Munshi⬛, A. R. Patel and G. M. Deheri

Abstract This paper studies the changes on a Ferrofluid (FF)-based longitudinally rough porous plane slider bearing (PSB) caused by the slip velocity (SV). The impact of magnetic fluid (MF) lubrication has been analyzed by using the Neuringer and Rosensweig model. The changes exerted by longitudinal roughness have been studied using the stochastic averaging model of Christensen and Tonder. The effects of SV are calculated by using the slip model of Beavers–Joseph. The pressure distribution (PD) expression has been calculated by solving the related nondimensional Reynolds' type equation. These calculations have been used to calculate the load carrying capacity (LCC). According to the results, LCC increases due to MF lubricant. The surface roughness, on the other hand, has a negative impact on the performance. The same goes for the SV. However, it has been found that these adversities caused by surface roughness, SV, and porosity can be partially neutralized by the positive impact of magnetization, though it can be said that SV in general decreases the bearing performance.

Keywords Slip velocity · Magnetic fluid · Load carrying capacity

Nomenclature

h	Film thickness (mm)
$\overline{h}$	Mean film thickness (mm)

M. M. Munshi (✉)
Alpha College of Engineering and Technology, Gujarat Technological University, Kalol 382721, Gujarat, India
e-mail: raiyan.munshi@gmail.com

A. R. Patel
Vishwakarma Government Engineering College, Gujarat Technological University, Ahmedabad 382424, Gujarat, India

G. M. Deheri
Sardar Patel University, Vallabh Vidyanagar 388120, Gujarat, India

© Springer Nature Singapore Pte Ltd. 2020
K. N. Das et al. (eds.), *Soft Computing for Problem Solving*,
Advances in Intelligent Systems and Computing 1048,
https://doi.org/10.1007/978-981-15-0035-0_3

h_s	Deviation from mean level
H	Nondimensional film thickness
H_0	Thickness of porous facing (mm)
$\overline{H}$	Magnetic field (Gauss)
H^2	Magnitude of magnetic field (N/A.m)
p	Pressure found in the area covered by the film (N/mm^2)
$\overline{p}$	Anticipated degree of pressure
P	Nondimensional pressure generated due to the film
Q	Integrating constant
$\overline{S}$	Slip parameter
u, v, w	Fluid film velocity components in x, y, z directions, respectively
W	Load capacity (N)
$\overline{W}$	Nondimensional load capacity
α	Variance (mm)
$\overline{\alpha}$	Nondimensional variance
ε	Skewness (mm^3)
$\overline{\varepsilon}$	Skewness in dimensionless form
μ_0	Permeability of the free space (N/A^2)
$\overline{\mu}$	Magnetic susceptibility of particles
μ^*	Dimensionless magnetization parameter
σ	Standard deviation (mm)
$\overline{\sigma}$	Dimensionless standard deviation
ϕ	Permeability of porous facing (m^2)
ψ	Porosity

1 Introduction

The slider bearings are primarily created to aid the transverse load in any given engineering system. The PSB study is a classical one. PSB has a lot of applications in various fields including domestic appliances, automobile transmissions, and clutch plates. Murti [1], Patel and Gupta [2], Tichy and Chen [3], Patel et al. [4] and Patel et al. [5] have also carried out research works on slider bearings.

More and more studies have been recently carried out that analyze the relationship between surface roughness and the associated hydrodynamic lubrication for a variety of bearing systems. The reason for this is that practically, all surfaces contain a certain level of roughness. This may be further exaggerated by some wear and tear. Many researchers have studied the impact of roughness on LCC of a system; Tzeng and Saibel [6], Andharia et al. [7, 8], Chiang et al. [9]. Christensen and Tonder [10–12] gave a general study that analyzed longitudinal and transverse roughness. This method has been further used in many different ways in a number of different investigations [13, 14, 15, 16, 17].

Contemporary researchers are consistently focusing on studying the MF lubrication, in theory as well as in practice [18]. Minute magnetic gains covered with surfactants are suspended and then dispersed into solvents like kerosene, fluorocarbons, hydrocarbons, etc., that are nonconducting and magnetically passive in nature to create MF lubrication. Other researchers including [19–22]; Hamrock [22] have studied hydrodynamic lubrication with a variety of film shapes.

Furthermore, porosity was introduced in an attempt to decrease friction. Porous bearings are used in horsepower motors of hair dryers, record players, vacuum cleaners, tape recorders, sewing machines, water pumps, etc. Morgan and Cameron [23] were the first investigators to study the hydrodynamic lubrication theory of bearings with porous structure.

All the particles undergo a body force when subjected to a magnetic field, resulting in the drag to flow. Therefore, for industrial application, the study of porous metal lubrication with MF is of primary importance [24]. Some researchers [25–28, 29] have also used MF as a lubricant in order to aid the tribological performance of a sliding interface. Lin [30] studied the LCC of a bearing system by replacing the lubricant with an MF. All these studies had similar conclusions mentioning that an MF, when used as a lubricant, enhances the bearing system's performance.

Many studies have attempted to explore the impact of slip on different bearings in a theoretical as well as experimental manner [31–35]. All these investigations have concluded that slip has a substantial impact on the working of any bearing system. Andharia and Deheri [36] concluded that the standard deviation in longitudinal roughness is crucial for increasing the LCC. Thus, it was understood that the study of PSB having MF lubrication should always be assisted with surface roughness for precise results.

In the work mentioned above, no-slip condition has been taken into account. That is why, this study uses MF for lubricant along with the calculation of load in terms of magnetic parameter, roughness parameters, and slip parameter. Due to this reason, the configuration of [36] has been used to investigate the impacts of SV and porosity.

2 Analysis

Figure 1 displays the bearing configuration that is considered to be infinite on the Y-axis. The X-axis represents the uniform velocity U of the slider. The minimum and maximum film thicknesses are represented by h_0 and h_1, respectively. L is the bearing length.

The mentioned field of magnetism is thought to be sloping against the stator, as suggested by [36]. The $h(x)$ is believed to be

$$h(x) = \overline{h}(x) + h_s \tag{1}$$

using the works of Christensen and Tonder [10–12]. Also, the study uses h_s using the probability density function

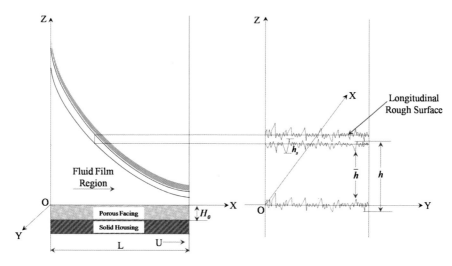

Fig. 1 Physical geometry of the bearing system [36]

$$f(h_s) = \begin{cases} \frac{35}{32c}\left(1 - \frac{h_s^2}{c^2}\right)^3, & -c \le h_s \le c \\ 0 & , \text{ elsewhere} \end{cases} \tag{2}$$

In this case, c is the highest possible deviation from the average width of the film. α, σ, and ε are considered by

$$\alpha = E(h_s), \sigma^2 = E(h_s - \alpha)^2, \varepsilon = E(h_s - \alpha)^3 \tag{3}$$

Here, $E(\cdot)$ represents the anticipated value as provided by

$$E(\cdot) = \int_{-c}^{c} (\cdot) f(h_s) dh_s \tag{4}$$

Neuringer and Rosensweig [26] devised a theory that describes the stable movement of MF. The model was as follows:

Equation of motion

$$\rho(\overline{q}.\nabla)\overline{q} = -\nabla p + \eta\nabla^2\overline{q} + \mu_0(\overline{M}.\nabla)\overline{H} \tag{5}$$

Equation of magnetization

$$\overline{M} = \overline{\mu}\overline{H} \tag{6}$$

Continuity equation

$$\nabla . \overline{q} = 0 \tag{7}$$

The formulae given by Maxwell

$$\nabla \times \overline{H} = 0 \tag{8}$$

and

$$\nabla . (\overline{H} + \overline{M}) = 0 \tag{9}$$

where $\rho, \overline{q}, \overline{M}, p,$ and η are fluid density, fluid velocity, magnetization vector, film pressure, and fluid viscosity, respectively.

Also,

$$\overline{q} = ui + vj + wk \tag{10}$$

Further, the magnetic field's magnitude is given by

$$H^2 = kx(L - x) \tag{11}$$

where k is suitable constant. The general hydrodynamic lubrication assumption modified [13, 16, 36] equation as follows:

$$\frac{d}{dx}\left[h^3\frac{d}{dx}\left(\overline{p} - \frac{1}{2}\mu_0\overline{\mu}H^2\right)\right] = 6\eta U\frac{dh}{dx} \tag{12}$$

Following the stochastically average process discussed in [36], Eq. (12) takes the form:

$$\frac{d}{dx}\left[\frac{1}{E(\overline{h} + h_s)^{-3}}\frac{d}{dx}\left(\overline{p} - \frac{1}{2}\mu_0\overline{\mu}H^2\right)\right] = 6\eta U\frac{d}{dx}\left[\frac{1}{E(\overline{h} + h_s)^{-1}}\right]$$

$$\frac{d}{dx}\left[\frac{1}{m(\overline{h}, \alpha, \sigma, \varepsilon, \phi)}\frac{d}{dx}\left(\overline{p} - \frac{1}{2}\mu_0\overline{\mu}H^2\right)\right] = 6\eta U\frac{d}{dx}\left[\frac{1}{n(\overline{h}, \alpha, \sigma, \varepsilon, \phi)}\right] \tag{13}$$

where

$$m(\overline{h}, \alpha, \sigma, \varepsilon, \phi) = \overline{h}^{-3}\left[1 - 3\alpha\overline{h}^{-1} + 6\overline{h}^{-2}(\sigma^2 + \alpha^2)\right.$$
$$\left. - 10\overline{h}^{-3}(\varepsilon + 3\sigma^2\alpha + \alpha^3 + 12\phi H_0)\right]$$

$$\tag{14}$$

$$n(\bar{h}, \alpha, \sigma, \varepsilon, \phi) = \bar{h}^{-1}\left[1 - \alpha\bar{h}^{-1} + \bar{h}^{-2}(\sigma^2 + \alpha^2)\right.$$
$$\left. - \bar{h}^{-3}(\varepsilon + 3\sigma^2\alpha + \alpha^3 + 12\phi H_0)\right]$$

(15)

The following dimensionless quantities are used

$$X = \frac{x}{L}, \quad M(H, \bar{S}, \bar{\alpha}, \bar{\sigma}, \bar{\varepsilon}, \psi) = h_0^3 m(\bar{h}, \alpha, \sigma, \varepsilon, \phi),$$
$$N(H, \bar{S}, \bar{\alpha}, \bar{\sigma}, \bar{\varepsilon}, \psi) = h_0 n(\bar{h}, \alpha, \sigma, \varepsilon, \phi),$$

$$H = \frac{\bar{h}}{h_0}, \quad \bar{\alpha} = \frac{\alpha}{h_0}, \quad \bar{\sigma} = \frac{\sigma}{h_0}, \quad \bar{\varepsilon} = \frac{\varepsilon}{h_0^3},$$
$$\bar{Q} = \frac{Q}{h_0}, \quad \psi = \frac{\phi H_0}{h_0^3}, \quad \mu^* = \frac{k\mu_0\bar{\mu}h_0^2 L}{2\eta U}, \quad P = \frac{\bar{p}h_0^2}{\eta U L}$$

(16)

The associated boundary conditions are

$$P(0) = 0, \quad P(1) = 0$$

(17)

Solving Eq. (13) with the aid of Eq. (17), the nondimensional type of dispersal of pressure is

$$P = \mu^*(X - X^2) + \int_0^X 6M(H, \bar{S}, \bar{\alpha}, \bar{\sigma}, \bar{\varepsilon}, \psi)\left(\frac{1}{N(H, \bar{S}, \bar{\alpha}, \bar{\sigma}, \bar{\varepsilon}, \psi)} + \bar{Q}\right) dX$$

(18)

where

$$M(H, \bar{S}, \bar{\alpha}, \bar{\sigma}, \bar{\varepsilon}, \psi) = \left[\frac{4 + \bar{S}H}{1 + \bar{S}H}\right]^{-1}\left[1 - 3\bar{\alpha}\left(\frac{4 + \bar{S}H}{1 + \bar{S}H}\right)^{-1/3}\right.$$
$$+ 6(\bar{\sigma}^2 + \bar{\alpha}^2)\left(\frac{4 + \bar{S}H}{1 + \bar{S}H}\right)^{-2/3}$$
$$\left. - 10\left(\frac{4 + \bar{S}H}{1 + \bar{S}H}\right)^{-1}(\bar{\varepsilon} + 3\bar{\sigma}^2\bar{\alpha} + \bar{\alpha}^3 + 12\psi)\right],$$

(19)

$$N(H, \bar{S}, \bar{\alpha}, \bar{\sigma}, \bar{\varepsilon}, \psi) = \left[\frac{4 + \bar{S}H}{1 + \bar{S}H}\right]^{-1/3}\left[1 - \bar{\alpha}\left(\frac{4 + \bar{S}H}{1 + \bar{S}H}\right)^{-1/3}\right.$$

$$+ \left(\overline{\sigma}^2 + \overline{\alpha}^2\right)\left(\frac{4 + \overline{S}H}{1 + \overline{S}H}\right)^{-2/3}$$

$$-\left(\frac{4 + \overline{S}H}{1 + \overline{S}H}\right)^{-1}\left(\overline{\varepsilon} + 3\overline{\sigma}^2\overline{\alpha} + \overline{\alpha}^3 + 12\psi\right) \Bigg] \tag{20}$$

and

$$\overline{Q} = \frac{-\int\limits_{0}^{1} \frac{M(H, \overline{S}, \overline{\alpha}, \overline{\sigma}, \overline{\varepsilon}, \psi)}{N(H, \overline{S}, \overline{\alpha}, \overline{\sigma}, \overline{\varepsilon}, \psi)} dX}{\int\limits_{0}^{1} M(H, \overline{S}, \overline{\alpha}, \overline{\sigma}, \overline{\varepsilon}, \psi) dX} \tag{21}$$

The LCC in dimensionless form is obtained by

$$\overline{W} = \frac{h_0^2}{\eta U L^2} W \tag{22}$$

$$\overline{W} = \frac{\mu^*}{6} + \int \left[\frac{6M\left(H, \overline{S}, \overline{\alpha}, \overline{\sigma}, \overline{\varepsilon}, \psi\right)(1 - X)}{N\left(H, \overline{S}, \overline{\alpha}, \overline{\sigma}, \overline{\varepsilon}, \psi\right)} \right.$$

$$\left. + 6\overline{Q}M\left(H, \overline{S}, \overline{\alpha}, \overline{\sigma}, \overline{\varepsilon}, \psi\right)(1 - X) \right] dX \tag{23}$$

where the LCC is calculated using

$$W = \int\limits_{0}^{1} \overline{p} dx \tag{24}$$

3 Results and Discussion

The study shows that Eqs. (18) and (23) display the dimensionless PD and dimensionless LCC, respectively. Andharia and Deheri [36] suggested the equation of film thickness H in terms of $2 - X$ to study the porous PSB. The results are used to analyze the correlation of SV and longitudinally rough porous PSB with MF.

1/3 rule as given by Simpson having a step size of 0.2 is helpful in calculating Eq. (23) for altering the measure of μ^*, porosity ψ, roughness parameters $\overline{\alpha}, \overline{\sigma}, \overline{\varepsilon}$, and slip parameter $\overline{S}$. Figures 2, 3, 4, 5, 6, 7, 8, 9, 10, 11, 12, 13, 14, 15, and 16 represent these results graphically.

Fig. 2 Trends of $\overline{W}$ concern with μ^* and $\overline{\alpha}$

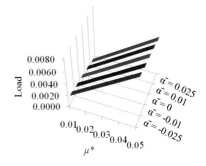

Fig. 3 Trends of $\overline{W}$ concern with μ^* and $\overline{\sigma}$

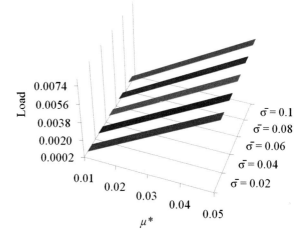

Fig. 4 Trends of $\overline{W}$ concern with μ^* and $\overline{\varepsilon}$

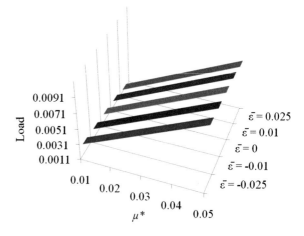

Fig. 5 Trends of $\overline{W}$ concern with μ^* and ψ

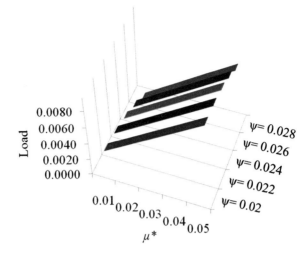

Fig. 6 Trends of $\overline{W}$ concern with μ^* and $\overline{S}$

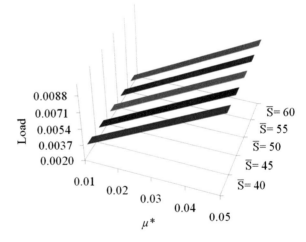

The variation in the LCC with regards to μ^* for different values of $\overline{\alpha}$, $\overline{\sigma}$, $\overline{\varepsilon}$, ψ, and $\overline{S}$ is shown in the Figs. 2, 3, 4, 5, and 6. It can be concluded from these figures that using MF lubrication, substantially increases the bearing performance. The impact of standard deviation due to magnetization of this performance is almost marginal. Figures 7, 8, 9, and 10 suggests that with positive variance, the LCC decreases while with negative variance, it increases. Figures 11, 12, and 13 suggests that LCC is positively related to $\overline{\sigma}$. From Figs. 14 and 15, it can be said that negative skewed roughness has a positive impact on the LCC. The total impact of $\overline{\varepsilon}$(+ve), $\overline{\alpha}$(+ve), and porosity are important as they can severely decrease the LCC. Thus, it can be concluded that SV has a negative impact on bearing performance (Fig. 16).

Fig. 7 Trends of $\overline{W}$ concern with $\overline{\alpha}$ and $\overline{\sigma}$

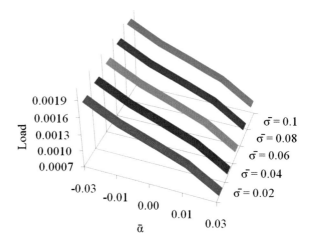

Fig. 8 Trends of $\overline{W}$ concern with $\overline{\alpha}$ and $\overline{\varepsilon}$

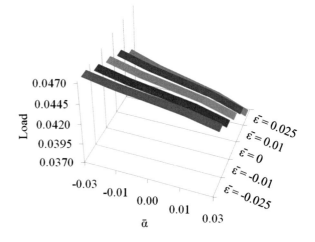

4 Conclusions

It can be said that the attempts made to neutralize the adverse impacts of surface roughness, porosity, and SV with the help of magnetization are considerably limited. Thus, surface roughness of a bearing system should be given some special attention during designing of the system even if the SV is held minimum. However, one thing that remains consistent is that for a better bearing performance, the SV should always be minimum.

Acknowledgements The authors acknowledge with regards the constructive comments, fruitful suggestions and remarks of the reviewer/Editor, leading to an overhauling of the materials presented in the paper.

Fig. 9 Trends of $\overline{W}$ concern with $\overline{\alpha}$ and ψ

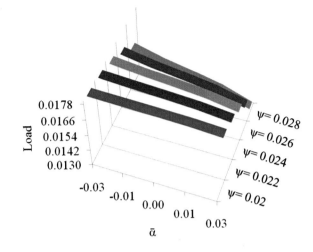

Fig. 10 Trends of $\overline{W}$ concern with $\overline{\alpha}$ and $\overline{S}$

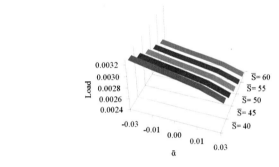

Fig. 11 Trends of $\overline{W}$ concern with $\overline{\sigma}$ and $\overline{\varepsilon}$

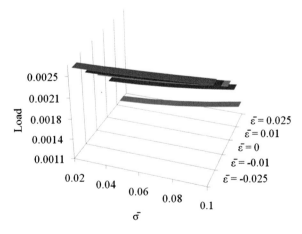

Fig. 12 Trends of $\overline{W}$
concern with $\overline{\sigma}$ and ψ

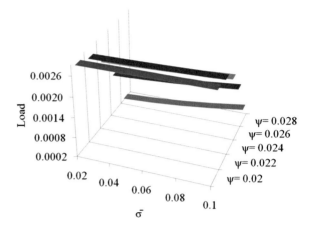

Fig. 13 Trends of $\overline{W}$
concern with $\overline{\sigma}$ and $\overline{S}$

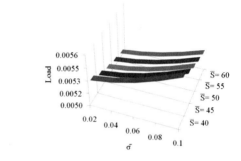

Fig. 14 Trends of $\overline{W}$
concern with $\overline{\varepsilon}$ and ψ

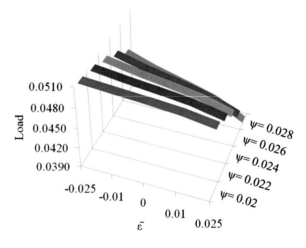

Fig. 15 Trends of $\overline{W}$ concern with $\overline{\varepsilon}$ and $\overline{S}$

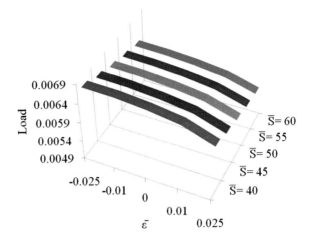

Fig. 16 Trends of $\overline{W}$ concern with ψ and $\overline{S}$

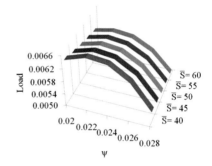

References

1. Murti, P.R.K.: Analysis of porous slider bearings. Wear **28**(1), 131–134 (1974)
2. Patel, K.C., Gupta, J.L.: Hydrodynamic lubrication of a porous slider bearing with slip velocity. Wear **85**(3), 309–317 (1983)
3. Tichy, J.A., Chen, S.H.: Plane slider bearing load due to fluid inertia-experiment and theory. J. Tribol. **107**(1), 32–38 (1985)
4. Patel, S.J., Deheri, G.M., Patel, J.R.: Ferrofluid lubrication of a rough porous hyperbolic slider bearing with slip velocity. Tribol. Ind. **36**(3), 259–268 (2014)
5. Patel, P.A., Deheri, G.M., Patel, A.R.: The performance of an idealized rough porous hydrodynamic plane slider bearing. Int. J. Appl. Math. Sci. **8**(3), 187–196 (2015)
6. Tzeng, S.T., Saibel, E.: Surface roughness effect on slider bearing lubrication. ASLE Trans. **10**(3), 334–338 (1967)
7. Andharia, P.I., Gupta, J.L., Deheri, G.M.: Effect of longitudinal surface roughness on hydrodynamic lubrication of slider bearings. In: Proceedings of Tenth International Conference on Surface Modification Technologies, pp. 872–880. The Institute of Materials, Singapore (1997)
8. Andharia, P.I., Gupta, J.L., Deheri, G.M.: On the shape of the lubricant film for the optimum performance of a longitudinal rough slider bearing. Ind. Lubr. Tribol. **52**(6), 273–276 (2000)
9. Chiang, H.L., Hsu, C.H., Chou, T.L., Hsu, C.H., Lin, J.R.: Surface roughness effects on the dynamic characteristics of finite slider bearings. J. Chung Cheng Inst. Technol. **34**(1), 1–11 (2005)

10. Christensen, H., Tonder, K.C.: Tribology of rough surfaces: stochastic models of hydrodynamic lubrication. SINTEF Report No. 10/69–18 (1969a)
11. Christensen, H., Tonder, K.C.: Tribology of rough surfaces: parametric study and comparison of lubrication model. SINTEF Report No. 22/69–18 (1969b)
12. Christensen, H., Tonder, K.C.: The hydrodynamic lubrication of rough bearing surfaces of finite width. In: ASME-ASLE Lubrication Conference, Cincinnati, Ohio, USA (1970)
13. Deheri, G.M., Andharia, P.I., Patel, R.M.: Longitudinally rough slider bearings with squeeze film formed by a magnetic fluid. Ind. Lubr. Tribol. **56**(3), 177–187 (2004)
14. Patel, J.R., Deheri, G.M.: Magnetic fluid based squeeze film in a rough porous parallel plate slider bearing. Ann. Facul. Eng. Hunedoara-Int. J. Eng. **9**(3), 443–463 (2011)
15. Patel, N.D., Deheri, G.M., Patel, H.C.: Magnetic fluid lubrication of a rough, porous composite slider bearing. Int. J. Surf. Eng. Interdiscip. Mater. Sci. **1**(2), 46–65 (2013)
16. Panchal, G.C., Patel, H.C., Deheri, G.M.: Influence of magnetic fluid through a series of flow factors on the performance of a longitudinally rough finite slider bearing. Global J. Pure Appl. Math. **12**(1), 783–796 (2016)
17. Patel, J.R., Deheri, G.M.: The effect of slip velocity on the ferrofluid based squeeze film in longitudinally rough conical plates. J. Serb. Soc. Comput. Mech. **10**(2), 18–29 (2016)
18. Shukla, J.B., Kumar, D.: A theory for ferromagnetic lubrication. J. Mag. Mag. Mater. **65**(2–3), 375–378 (1987)
19. Bagci, C., Singh, A.P.: Hydrodynamic lubrication of finite slider bearing: effect of one dimensional film shape and their computer aided optimum designs. J. Lubr. Technol. **105**(1), 48–66 (1983)
20. Pinkus, O., Sternlicht, B.: Theory of Hydrodynamic Lubrication. McGraw Hill Book Company, New York (1961)
21. Andharia, P.I., Gupta, J.L., Deheri, G.M.: Effect of surface roughness on hydrodynamic lubrication of slider bearings. Tribol. Trans. **44**(2), 291–297 (2001)
22. Hamrock, B.J.: Fundamentals of Fluid Film Lubrication. McGraw Hill, New York (1994)
23. Morgan, V.T., Cameron, A.: Mechanism of lubrication in porous metal Bearings. In: Proceedings of Conference on Lubrication and Wear. Institution of Mechanical Engineers, London (1957)
24. Patel, N.S., Vakharia, D.P., Deheri, G.M.: A study on the performance of a magnetic fluid based hydrodynamic short porous journal bearing. J. Serb. Soc. Comput. Mech. **6**(2), 28–44 (2012)
25. Bhat, M.V., Deheri, G.M.: Porous composite slider bearing lubricated with magnetic fluid. Jpn. J. Appl. Phys. **30**(10), 2513–2514 (1991)
26. Neuringer, J.L., Rosensweig, R.E.: Magnetic fluids. Phys. Fluids **7**(12), 1927–1937 (1964)
27. Shah, R.C., Bhat, M.V.: Ferrofluid lubrication in porous inclined slider bearing with velocity slip. Int. J. Mech. Sci. **44**(12), 2495–2502 (2002)
28. Snyder, W.T.: The magnetohydrodynamic slider bearing. J. Basic Eng. **84**(1), 197–202 (1962)
29. Shimpi, M.E., Deheri, G.M.: Effect of bearing deformation on the performance of a magnetic fluid-based infinitely rough short porous journal bearing. In: Proceedings of International Conference on Advances in Tribology and Engineering Systems-SPRINGER, Gujarat, India, pp. 19–34, Gujarat Technological University (2013)
30. Lin, J.R.: Dynamic characteristics of magnetic fluid based sliding bearings. Mechanika **19**(5), 554–558 (2013)
31. Sparrow, E.M., Beavers, G.S., Hwang, I.T.: Effect of velocity slip on porous walled squeeze films. J. Lubr. Technol. **94**(3), 260–265 (1972)
32. Deheri, G.M., Patel, R.U.: Effect of slip velocity on the performance of a short bearing lubricated with a magnetic fluid. Acta Polytechn. **53**(6), 890–894 (2013)
33. Shukla, S.D., Deheri, G.M.: Effect of slip velocity on magnetic fluid lubrication of rough porous Rayleigh step bearing. J. Mech. Eng. Sci. **4**, 532–547 (2013)
34. Munshi, M.M., Patel, A.R., Deheri, G.M.: Effect of slip velocity on a magnetic fluid based squeeze film in rotating transversely rough curved porous circular plates. Ind. Eng. Lett. **7**(8), 28–42 (2017)

35. Patel, N.D., Deheri, G.M.: Effect of surface roughness on the performance of a magnetic fluid based parallel plate porous slider bearing with slip velocity. J. Serb. Soc. Comput. Mech. **5**(1), 104–118 (2011)
36. Andharia, P.I., Deheri, G.M.: Performance of a magnetic fluid based longitudinally rough plane slider bearing. Indian Str. Res. J. **4**(4), 1–8 (2014)

Intelligent Controller Based Solar Photovoltaic with Battery Storage System for Conditioning the Electrical Power

Ravi Dharavath and I. Jacob Raglend

Abstract In the current scenario, providing continuous power supply and meeting the peak load demand are becoming the prime challenges in the power sector. The existing nonrenewable energy source causes global warming, environmental pollution, and inability to provide necessary clean power due to the presence of modern technology, population growth, domestic appliances, agriculture sectors, and interconnection of a number of nonlinear loads. This problem can be avoided using hybrid power generation. In this paper, the necessary power and peak demand are provided by inviting the integration of a grid-connected photovoltaic with battery storage hybrid system. The solar photovoltaic power is integrated with the DC link of a syncro converter through the boost converter. The switching control function of the boost converter is operated with an intelligent controller to step up the voltage and track the maximum power from the intermittent nature of solar photovoltaic system. The Radial Basis Function Neural (RBFN) network-based intelligent controller is utilized for tracking the smoothening of the maximum power under dynamic irradiance and temperature conditions. The battery storage system is connected to a DC link through an appropriate DC–DC converter. The battery storage system stabilizes the DC link voltage using a voltage droop control. The performance of the proposed system is simulated in a grid-connected mode under variable load conditions using MATLAB–SIMULINK software.

Keywords Photovoltaic (PV) · Battery storage systems · Radial Basis Function Neural (RBFN) controller · DC–DC converters

R. Dharavath · I. Jacob Raglend (✉)
School of Electrical Engineering, Vellore Institute of Technology, Vellore, Tamilnadu, India
e-mail: jacobraglend.i@vit.ac.in

R. Dharavath
e-mail: rv.dharavath@gmail.com

© Springer Nature Singapore Pte Ltd. 2020
K. N. Das et al. (eds.), *Soft Computing for Problem Solving*,
Advances in Intelligent Systems and Computing 1048,
https://doi.org/10.1007/978-981-15-0035-0_4

43

1 Introduction

The power utilization is increasing every day in the developing country and is playing a major task in economic. The power utilization demand can be fulfilled with renewable and nonrenewable energy sources. But the redundancy in fossil fuel and green economy lead to creating interest in renewable energy generation. The utility sector in India has one national grid installed with a capacity of 33.86 GW as of 30 November 2017. The total installed capacity is 330260.53 GW as of 31 May 2017. The total renewable energy installed in India is with a capacity of 57260.23 GW as of 31 March 2017 [1]. Still, there is a shortage of power during the peak demand in India with a capacity of 1565 MW as of 3 January 2018 [2]. The power and energy shortage in the last 3 years is shown in Table 1. The peak demand has been reducing every year due to the installation of multi-distributed renewable energy sources. The most prominent among the renewable sources is solar energy. The installed solar power in India is 14,771.69 MW as of 31 March 2017. Solar energy is the green economy and free from pollution leading to motivation for installing the solar PV systems.

The solar PV work is based on the principle of the photoelectric effect. It converts solar energy into electrical energy due to an incident of sunlight on the PV array. The solar energy is intermittent in nature, unable to provide a continuous supply during the entire day and night. Hence, solar PV is connected with a storage device and other renewable energy sources [3] to make continuity of supply. The solar PV power is extracted using various kinds of Maximum Power Point Tracking (MPPT) techniques [4, 5]. The solar PV is integrated with battery storage system at the DC link to provide reliable power.

The solar PV system and grid are integrated at the Point of Common Coupling (PCC) through the converter to meet the peak demand and provide continuous power supply. But the intermittent nature of solar PV and dynamic variation in climate cause power quality problem due to voltage and frequency variations [6]. These variations in sources influence the PCC in the grid-connected hybrid system. The frequency variations cause the change in real power flow and the voltage variation affects the reactive power flow. The grid-connected system may include some other small-scale energy sources like wind, diesel generation, and gas engine based generation [7] majorly causing the variations in the supply sources that lead to influencing the sensitive load and finally reducing the life span of the load. Diesel generation causes global warming.

Table 1 Electricity shortage for the last 3 years [1]

S. No.	Years	Shortage of Peak Power (MW)	Shortage of Energy (MU)
1	JAN 2018	1565	12
2	JAN 2017	2,171	17
3	JAN 2016	2,707	54

In this work, the solar PV with battery storage hybrid system is integrated with the grid through the synchro converter meeting the peak demand. The battery, grid synchronization unit provided continues harness power to the sensitive load. The performance of the solar photovoltaic battery storage is implemented in the grid mode using MATLAB–SIMULINK software. The proposed system is simulated under dynamic conditions of the source and load. The following sections deal with the proposed hybrid system.

2 Modules of Proposed Solar PV—Battery Storage, Fuel Cell Based Hybrid System

The modules of integration of solar photovoltaic, battery storage hybrid system are shown in Fig. 1.

The major components of the proposed hybrid system are as follows:

(A) Solar Photovoltaic (PV) system
(B) Boost converter
(C) Battery storage system with bidirectional converter
(D) Synchro converter
(E) Grid synchronization control unit

The solar PV is integrated with the DC link of the synchronous converter and the battery storage system is connected to the same DC link to maintain constant DC link voltage. The synchro converter is placed between the grid and the DC link to make adequate continuous power supply. The AC load-1 is considered as the baseload and the load-2 is considered as the peak load or dynamic load. Under the dynamic

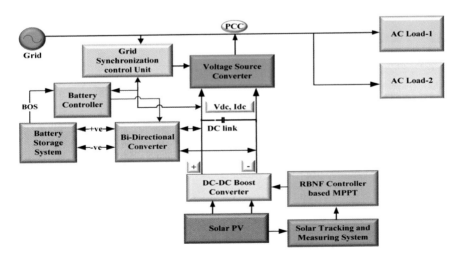

Fig. 1 Grid-connected solar PV with battery storage hybrid system

conditions of the load, the proposed system accommodates continuous power and protects the sensitive load. The descriptions of each module are explained in the following sections.

2.1 Solar PV (Photovoltaic) System

Basically, the solar PV cell is a nonlinear device. The basic solar PV equivalent circuit is shown in Fig. 2a. To build the required voltage and current, these PV cells are framed in a series–parallel configuration in such a way that the solar array is formed. The PV array characteristics are shown in Fig. 2b for dynamic irradiance conditions varying from 100 to 1000 W/m². The PV array model is obtained from the voltage and current relation, Eqs. 1 and 2.

$$V_{PV} == \frac{nKT}{q} \ln\left(\frac{I_{sc}}{I_{pv}} + 1\right) \tag{1}$$

$$I_{pv} = I_{sc} - I_{pvo}\left[\exp\left(\frac{q(V_{pv} + I_{pv}R_s)}{N_s K T n}\right) - 1\right] - \frac{V_{pv} + R_s I_{sc}}{R_{sh}} \tag{2}$$

Power from a photovoltaic system is given by

$$P_{PV} = V_{pv} \times I_{pv} \tag{3}$$

where I_{pv},—solar PV current (A), V_{pv}—PV voltage (V), T—reference PV cell temperature, and K—Boltzmann constant. The maximum power from the solar PV is obtained using a RBFN-based MPPT Controller.

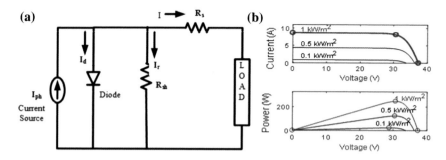

Fig. 2 Solar photovoltaic equivalent circuit [8]

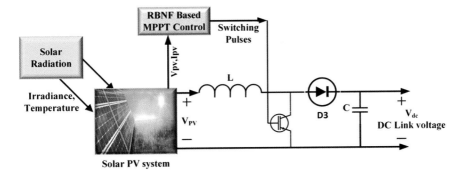

Fig. 3 Boost DC–DC converter with solar PV source

2.2 Boost Converter

The boost converter is connected between the solar PV and the DC link. It is used to step up the solar PV voltage. The basic boost converter is shown in Fig. 3. The boost converter is operated based on the RBFN-based MPPT technique.

The boost converter output voltage is calculated using Eq. 4

$$V_o = \frac{V_{in}}{1 - D} \tag{4}$$

where V_o—boost converter output voltage (V), V_{in}—input voltage (V), D—duty ratio. The boost converter [8] is modelled based on Eqs. 5 and 6.

$$\Delta V_c = \frac{I_o * D}{C * F} \tag{5}$$

$$\Delta I = \frac{V_s * D}{L * F} \tag{6}$$

L (mH) and C (µF) are the boost converter inductor and capacitor which are calculated based on the output voltage and current rating. F—temperature in °C, ΔV_c and ΔI are tolerable voltage and current limits, and V_s—input voltage (V).

2.3 Battery Storage Systems with Bidirectional DC–DC Converter

Battery storage with a bidirectional DC–DC converter provides the energy backup in the proposed hybrid system.

The DC bus voltage of the battery can be regulated with a PI (proportional plus integral) controller either in a charging or discharging mode based on the battery state

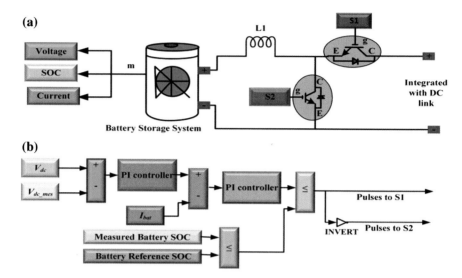

Fig. 4 Battery storage systems. **a** Battery with bidirectional converter. **b** Bidirectional controller

of the charge as shown in Fig. 4. The first section of the PI controller regulates the voltage by taking the reference voltage (V_{dc}) and measuring the actual DC voltage (V_{dc_meas}). The PI charge controller works on the basis of the battery state of the charge. The switching pulses of the charge controller work on the battery depth of the charge.

2.4 Synchro Converter

The proposed synchro converter is a single-phase voltage source inverter. It will convert DC bus voltage into AC with a specified voltage and frequency based on the switching pulses generated from the grid synchronization unit as shown in Fig. 1.

3 Control Algorithm

In the proposed hybrid system, the control techniques are mainly divided into two parts, namely:

1. Radial basis function neural network-based MPPT method
2. Control algorithm for Voltage Source Inverter (Synchro converter)

Each controller description is given in the following sections.

3.1 Radial Basis Function Neural Network Based MPPT

The radial basis function neural network is used in the proposed system to get the smoothening power from the intermittent nature of the solar PV systems within the instant. The convergence of switching pulses is faster and it can direct the signal based on the RBFN training. It will work on two types of training such as supervised and unsupervised training. RBFN mainly contains input, output, and the hidden layer [9, 10]. The layout of RBFN is shown in Fig. 5. The basic MPPT function is analyzed using unsupervised learning and then the weights are updated based on the targeted output. In the input layer, at the node, the net input and output variables are defined using Eqs. 7 and 8.

$$net_i^1 = x_i(n) \tag{7}$$

$$Y_i = f_i^1(net_i^1(N)) \tag{8}$$

where i = 1, 2, 3……n, x_i (n) is the set of input variables and Y_i is the output at the input node. In the hidden layer, the performance of each node is computed using the Gaussian function. At the hidden layer, the net input and output are computed as follows:

$$net_j^2(N) = (X - M_j)^T \sum (X - M_j) \tag{9}$$

$$Y_j^2(N) = f_i^2(net_j^2(N)) = Exp(net_j^2(N)) \tag{10}$$

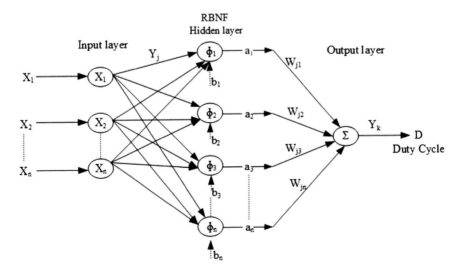

Fig. 5 Layout of RBFN

Here j = 1, 2, 3, …

$$M_{mean} = M_j = [\,m_{1j}\ m_{2j}\ m_{ij}\,]^T \tag{11}$$

$$Standard\ deviation = \sum_j = diag\left[\frac{1}{\sigma_{1j}^2}\ \frac{1}{\sigma_{2j}^2}\ \cdots\ \frac{1}{\sigma_{ij}^2}\right]^T \tag{12}$$

In the output layer, the duty cycle is computed at the output node corresponding to each input set of variables and it can be calculated by the following equations:

$$net_k^3 = \sum W_j Y_j^2(N) \tag{13}$$

$$Y_k^3(N) = f_k^3(net_k^3(N)) = net_k^3(N) = D \tag{14}$$

$$Y_i^1(N) = f_i^1(net_i^1(N)) = net_i^1(N) \tag{15}$$

where i = 1, 2, 3 …

The RBFN-based MPPT technique is computed by the duty cycle within the short duration to get the smoothening output.

3.2 Control Technique for Voltage Source Inverter (Synchro Converter)

The Synchro converter is connected between the DC link and the PCC point of the grid. The Synchro converter controller is shown in Fig. 6. The switching pulses of the synchro converter are controlled using a PWM generator and are regulated by the control of the DC bus voltage and measured AC (Alternating Current) output voltages.

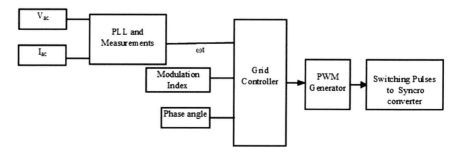

Fig. 6 Control algorithms for synchro converter

Table 2 Solar PV array with boost converter specification

S. No.	Name of the parameter	Rating
	Solar parallel strings	02
	Solar series-connected module per string	08
	Solar PV array open-circuit voltage	37.3 V
	Solar PV short-circuit current	8.66 A
	Maximum voltage at MPP	30.7 V
	Maximum current at MPP	8.15 A
	Capacitor at the solar PV	212.33 μF
	Boost converter switching frequency	5 kHz
	Boost inductor	2 mH
	Boost capacitor	35.211 μF
	Inverter DC link voltage	720 V
	AC load	5 kW, 300 V
	DC load resistor	78 Ω

4 Integration of Solar PV with Battery Storage Hybrid System

The proposed grid connected solar photovoltaic, battery storage is simulated in MAT-LAB–SIMULINK software as shown in Fig. 1. The 4 kW solar PV array is designed with eight series-connected modules per string and two parallel strings. The modules of the PV array parameters and boost converter are mentioned in Table 2. The 5 kHz switching frequency of the boost converter steps up the input natural voltage to 720 V and the designed the inverter DC link bus voltage is 720 V. The inverter and battery bidirectional converters are implemented with IGBT switches. The hybrid system model of a 4 kW array is connected to a 300 V distribution grid through the boost converter and synchro converter. In the first stage, the solar PV voltage can stepped up with the boost converter using an RBFN controller. Then, the PV voltage is synchronized with the DC link voltage. The battery is connected to the DC link using s bidirectional converter and is designed with the 720 V DC bus voltage specification. In the second stage, the synchro converter output voltage is synchronized with a grid voltage of 300 V. The performance results and brief explanation are given in Sect. 5.

5 Simulation Results of Proposed System

The proposed system is simulated under dynamic irradiance and temperature conditions as shown in Fig. 7. At t = 0.2 s, the irradiance is varied from 600 to 800 W/m^2 and then at t = 0.4– 0.6 s, it is varied from 800 to 1000 W/m^2. The temperature is

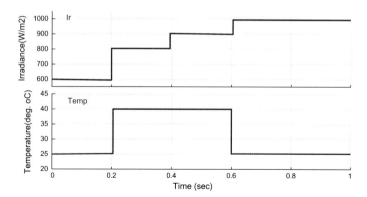

Fig. 7 Dynamic irradiance and temperature

also varied from 25 to 40 °C at t = 0.2 s. Initially, the boost converter and synchro converter are in operation but the output voltage of the synchro converter and the grid voltage are not synchronized till the time 0.04 s. In this duration, the battery is in charging mode. The boost converter output is obtained by regulating the duty cycle using RBFN network-based maximum power tracking method.

The synchronization of the grid voltage and inverter are regulated by the modulation index and phase angle at time t = 0.1 s and the corresponding variation in the load, grid voltage, and synchro converter are shown in Fig. 8. The grid voltage and synchro converter are synchronized even under the dynamic conditions of the load during the interval between t = 0.1 to 0.2 s and t = 0.87–0.93 s. Under dynamic variations in the solar PV, source is balanced by the battery storage system and the DC link voltage is maintained constant as shown in Fig. 9. The DC link bus voltage 720 V reaches steady state at time t = 0.15 s. The battery storage state of the charge is shown in Fig. 10.

The different source and load variations in the power are balanced using the battery storage system and grid control unit. The battery is supporting the DC link voltage to maintain the DC bus voltage constant under different dynamic natures of the sources and load conditions. The proposed hybrid systems are supported to meet the load demand.

6 Conclusion

The proposed grid-integrated solar PV with battery storage hybrid system is performed under different load and source conditions. The dynamic conditions of solar PV source is simulated using an RBFN network controller for obtaining the desired DC link voltage through the boost converter. At the point of common coupling, the voltage is stabilized under the grid-connected mode by the process of grid synchronization through the synchro converter. The solar PV with batteries maintain the

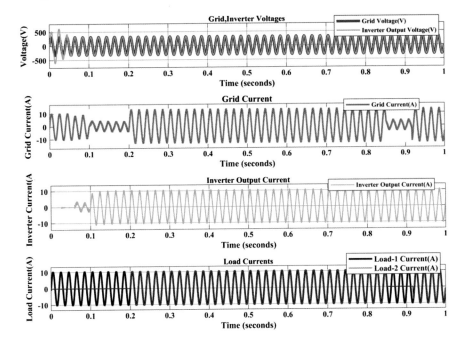

Fig. 8 Grid voltages, current

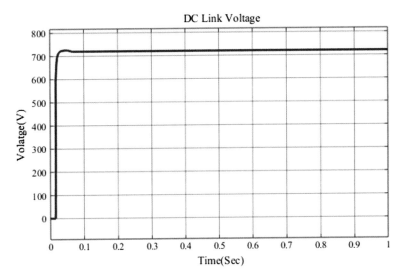

Fig. 9 DC link voltages

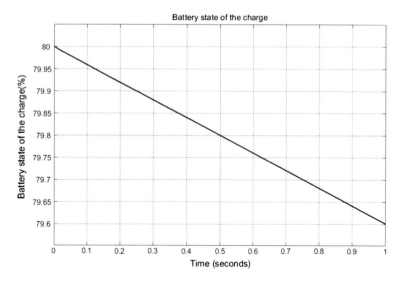

Fig. 10 Battery system voltages and SOC

necessary grid voltage and power. The harness and continuous necessary power are met by maintaining the appropriate DC bus voltage of the inverter and grid voltage.

References

1. https://en.wikipedia.org/wiki/Electricity_sector_in_India#cite_note-capacity-4 (2017)
2. http://vidyutpravah.in/ (2018)
3. Khare, V., Nema, S., Baredar, P.: Solar–wind hybrid renewable energy system: a review. Renew. Sustain. Energy Rev. **58**, 23–33 (2016)
4. Manju, S., Sagar, N.: Progressing towards the development of sustainable energy: a critical review of the current status, applications, developmental barriers and prospects of solar photovoltaic systems in India. Renew. Sustain. Energy Rev. **70**, 298–313 (2017)
5. Saravanan, S., Babu, N.R.: Maximum power point tracking algorithms for photovoltaic system–a review. Renew. Sustain. Energy Rev. **57**, 192–204 (2016)
6. Liang, X.: Emerging power quality challenges due to integration of renewable energy sources. IEEE Trans. Ind. Appl. **53**(2), 855–866 (2017)
7. Chung, Y.H., Kim, H.J., Kim, K.S., Choe, J.W., Choi, J.: Power quality control center for the microgri system. In: IEEE 2nd International Power and Energy Conference(PECon), pp. 942–947 (2008)
8. Sumathi, S., Kumar, L.A., Surekha, P.: Solar PV and Wind Energy Conversion Systems: An Introduction to Theory, Modeling with MATLAB/SIMULINK', and the Role of Soft Computing Techniques. Springer, Berlin (2015)
9. Mellit, A., Kalogirou, S.A.: Artificial intelligence techniques for photovoltaic applications: a review. Prog. Energy Combust. Sci. **34**(5), 574–632 (2008)
10. Reddy, D., Ramasamy, 'S.: Design of RBFN controller based boost type vienna rectifier for grid-tied wind energy conversion system. IEEE Access **6**, 3167–3175 (2018)

Autonomous Vehicle for Obstacle Detection and Avoidance Using Reinforcement Learning

C. S. Arvind and J. Senthilnath

Abstract Obstacle detection and avoidance during navigation of an autonomous vehicle is one of the challenging problems. Different sensors like RGB camera, Radar, and Lidar are presently used to analyze the environment around the vehicle for obstacle detection. Analyzing the environment using supervised learning techniques has proven to be an expensive process due to the training of different obstacle for different scenarios. In order to overcome such difficulty, in this paper Reinforcement Learning (RL) techniques are used to understand the uncertain environment based on sensor information to make the decision. Policy free, model-free Q-learning based RL algorithm with the multilayer perceptron neural network (MLP-NN) is applied and trained to predict optimal vehicle future action based on the current state of the vehicle. Further, the proposed Q-Learning with MLP-NN based approach is compared with the state of the art, namely, Q-learning. A simulated urban area obstacles scenario is considered with the different number of ultrasonic radar sensors in detecting obstacles. The experimental result shows that Q-learning with MLP-NN along with the ultrasonic sensors is proven to be more accurate than conventional Q-learning technique with the ultrasonic sensors. Hence it is demonstrated that combining Q-learning with MLP-NN will improve in predicting obstacles for autonomous vehicle navigation.

Keywords Autonomous vehicle · Ultrasonic radar · Reinforcement learning · Q-learning · Multilayer perceptron neural network

C. S. Arvind (✉)
Department of Computer Science, Dr. Ambedkar Institute of Technology, Bengaluru, India
e-mail: csarvind2000@gmail.co

J. Senthilnath
School of Electrical and Electronic Engineering, Nanyang Technological University, Singapore, Singapore
e-mail: senthil.iiscb@gmail.com

© Springer Nature Singapore Pte Ltd. 2020
K. N. Das et al. (eds.), *Soft Computing for Problem Solving*,
Advances in Intelligent Systems and Computing 1048,
https://doi.org/10.1007/978-981-15-0035-0_5

55

1 Introduction

Technological advancement in advance driver assistance system (ADAS) has empha-sized on passenger and vehicle safety. Autonomous vehicle navigate within the urban situation is way too complex because of static and dynamic obstacles present around the vehicle environment. Detection and avoidance of these obstacles are one of the important aspects of autonomous vehicle navigation. State of the art data analysis using different sensors like camera, radar, and lidar can effectively detect obstacles using computer vision and machine learning in particular reinforcement techniques [1]. At present in ADAS, obstacle detection is conducted by processing radar and visual inputs, which requires a lot of training and ground truth for early detection [2]. To overcome the limitation of the above methods researchers are working an early detection of obstacles using reinforcement technique. It will reduce the effort of generating the ground truth of obstacles manually.

Reinforcement learning (RL) is a machine learning technique in which agent that is an autonomous vehicle will learn its environment based on action, state and reward it obtains from the previous action [3]. There are two modes of learning (i) model-based and (ii) model-free. In the model-based learning, dynamics of the environment is simulated, such that the model learns the transition probability $T(s_1|(s_0, a))$ from the pair of current state s_0 and action 'a' to the next state 's_1'. The main disadvantage of this method is a state space and action space grows, storing this information becomes impractical [4]. To overcome this drawback model-free algorithms rely on trial and error to update its knowledge. As a result, it does not require space to store all the combination of state and actions to learn about the environment.

Madhu et al. [5] has used the Q-Learning algorithm to predict future action to be taken for an autonomous motion of robot based on obstacle detection from a vision-based sensor. Chu et al. [6] has successfully demonstrated using reinforce-ment learning robot will become intelligent and automatically navigate by avoiding collision with the static and dynamic obstacle. The main disadvantage of Q-learning based obstacle detection are (i) learning is slow because of an iterative method in finding optimal Q-value of the next state 's' and greedy action 'a' using off-policy model (ii) Memory requirement is more to store Q-values of all possible situation and action values. To overcome this drawback, Huang et al. [7] has approached robot obstacle avoidance using model-free, off-policy based Q-learning reinforce-ment technique alone with the neural network and has stated, the simulated robot can learn obstacle even in a complex environment. Duguleana et al. [8] has proposed a new path planning algorithm based on Q-learning and artificial neural network for mobile robots and has analyzed the simulated obstacle avoidance using virtual reality for easier visualization and safer testing activities. Xia et al. [9] have also used Q-learning reinforcement learning for obstacle avoidance for the industrial mobile vehicles in an unknown environment using the neural network.

The present research work of obstacle detection has been done on the controlled uncertain environment for the robotic task. There are two major challenges that make autonomous vehicle different from the robotic task (i) For precise vehicle control,

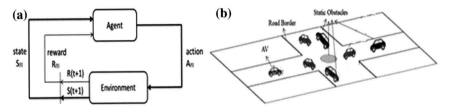

Fig. 1 a Representation of an agent to environment interaction using reinforcement learning. **b** Obstacle detection scenario of the urban situation

action space must be continuous which cannot be dealt with the traditional Q-learning algorithm (ii) the autonomous vehicle must satisfy various constraints like traffic rules and vehicle dynamics.

In this paper, we propose static obstacle detection using reinforcement learning for autonomous vehicle navigation in a simulated environment. The ultrasonic radar sensors distance measure is used to determine the obstacle ahead. The simulation is performed for the dynamic urban scenario. MLP-NN will improve the continuous action space problem by optimally predicting the next action based on vehicle acceleration, heading angle, and distance measure from the ultrasonic sensor. The comparative study is carried out using two reinforcement learning, namely, conventional Q-learning and Q-learning with multilayer perceptron neural network for static obstacle detection. The static obstacle is detected with the number of ultrasonic sensors connected to the vehicle; to determine the accuracy of continuous action state. Also, 'stop-go' collision avoidance mechanism is tested on an autonomous car prototype hardware model which is been developed for this research work.

The paper is organized as follows: Sect. 2 describes the reinforcement methodology for the detection of static obstacles. In Sect. 3 the details of the experiment and the result are presented. Section 4 discusses the conclusion and future work.

2 Methodology

In our research, the reinforcement learning agent is an autonomous vehicle and the surrounding area is its environment. The autonomous vehicle performs the following actions $A(t)$ {Turn Left, Turn Right, Move Forward, Move Reverse}. The environment will return the future state of the vehicle $S(t + 1)$ and reward $R(t + 1)$. Figure 1 represents the agent "vehicle" performing actions and obtaining the future state $S(t + 1)$ and $R(t + 1)$.

Figure 1b represents the actual urban scenario of autonomous vehicle with sensors navigating on narrow roads with vehicles parked on road (static obstacles) and the circle indicated the road junction to help both turning and to maintain one-way circular roadway.

2.1 Q-Learning Algorithm

Detection of the obstacle is based on the distance information from ultrasonic sensors. To get the wide angle information, one or multiple sensors are placed in front of the vehicle. The agent (vehicle) will perform action 'A' like move forward, move reverse, turn left, and turn right. State 'X' signifies the present state of vehicle-based on 'A' action $\{1, ..., N_x\}$. Q-learning is model-off, policy less reinforcement learning algorithm [10]. Initial arbitrary state and action will help in calculating policy (Π). Based on policy (Π) future action is determined using trial and error method. Using reward points 'R' iteratively Q-learning model will understand the environment. Positive reward points are awarded for corrective action. Negative reward points are penalized for the wrong action. Optimal state and action for every action are calculated using,

$$Q(S, A) = R + \gamma (max(Q(S', A')))$$ (1)

where $Q(S, A)$ is the current policy of action 'A' from state 'S'. The reward for the action is 'R', $max(Q(S', A'))$ define the maximum future reward and γ is the discount factor varying from 0 to 1 which determines the significance of future rewards. A factor of 0 will make the agent consider current rewards. While a factor nearing 1 will make it stay for long-term high reward. This will helps in accelerate learning.

Predicted optimal new state and action are stored in Q-table. The agent will navigate using Q-table information in the environment which helps in detecting obstacles and avoiding it. Algorithm 1 explains the steps taken to detect obstacles using Q-learning.

Algorithm1: Obstacle detection using Q-learning

Input: *Action = A {Move forward, Move Reverse, Turn Left, Turn Right}*
            State = X {1.........................Ns}
Output *= Q(X,A) optimal State and Action*
Let γ∈ [0,1] →Discount factor,
Let α∈ {0,1} → learning rate = 0.1
Let R→ Reward
Q-learning Parameters (X, A, R, T, alpha, λ)
Initialize S = Random State, A = Random Action, R = Arbitrarily
            Q: S*A→ R
For *ii to Iteration do*
        Start in state s ∈ S
        while *s is not terminal* **do**
        Calculate Policy (Π(x)) ← arg max_aQ(x,a)
            Action ← Policy (state)

```
        if CollisionwithObstacle == True
                reward ← R(state, action)
                reward = -500
        else
                reward ← R(state, action)
                update(reward)
        endif
            s' ← T(s,a)              // Receive new state
            Q(s',a) ← Equation 1
            s ← s'
    return
    endFor
```

2.2 Q-Learning with MLP-NN

The main drawback of obstacle detection using Q-learning is that it cannot be applied to solve complex problems because of the sparse nature of Q-table to store large data. To solve complex problems like obstacle detection, a multilayer perceptron neural network algorithm [11] combined with Q-learning can predict optimal state value. The Q-learning output is of two dimensions (state and action). Combining with MLP output is reduced to optimal one dimension (new state) from present state, action, reward, and new state input values. Optimal $Q(S, A)$ is predicted using MLP-NN loss function Eq. 2.

$$Loss\big(Q\big(S', A'\big)\big) = 1/n \sum |x_i - Q(S, A)| \tag{2}$$

where x_i is the input (state, reward, action, and new state) and Q(S,A) is the previous iteration Q-value. In our research, we have used single layer MLP with one epoch, with varying hidden layers. Input data is processed using mini-batch processing using RELU activation [12] function. RMSE is minimized using the optimization technique, namely, gradient descent to obtain optimal $Q(S, A)$. The steps taken to detect obstacles using hybrid Q-learning with MLP algorithm is discussed in Algorithm 2.

Algorithm 2: Obstacle detection using Q-learning with MLP-NN

Input: Action = A {Move forward, Move Reverse, Turn Left, Turn Right}
            State = X {1........................Ns}
Output = Q(X,A) optimal State and Action
Let $\gamma \in [0,1]$ →Discount factor,

Let $\alpha \in \{0,1\} \rightarrow$ *learning rate*
Let $R \rightarrow$ *Reward* Let $E \rightarrow$ *Epochs* = 5
Let *minibatch* = 64
Parameters for Qlearning $(X,A,R,T,alpha,\lambda)$
Initialize S = Random State, A = Random Action, R = Arbitrarily
          $Q: S*A \rightarrow R$
For *ii to Epochs* **do**
   For *jj to iteration* **do**
      Start in state s $\in$ S
      while s is not terminal **do**
      Calculate Policy $(\Pi(x) \leftarrow arg\ max_a Q(x,a)$
         action$\leftarrow$ Policy (state)
      if *CollisionwithObstacle == True*
            reward $\leftarrow$ R(state,action)
            reward = -500
   else
            reward $\leftarrow$ R(state,action)
            update(reward)   // *update the reward*
   endif
      $s' \leftarrow T(s,a)$ // *Receive new state*
   For *kk to miniBatch* **do**
      find $max_a(s',a')$ *using* **MLP()**
      $Q(s',a)$ = *append* $max_a(s',a')$
   endFor
   endFor
endFor

3 Simulation Results

Obstacle detection is conducted on a simulation environment, considering the urban scenario using pygame [13]. Tensorflow [14] MLP-NN learning library is used for model building. Figure 2 represents the simulation urban scenario setup, where the blue moving object is an autonomous vehicle with the ultrasonic radar of maximum 30 m range. Red color static objects represent road obstacles. The green color object represents a road circle at the road intersection. Black lines represent the road border/lanes. The velocity of the vehicle is fixed at 10 m/s such that autonomous vehicle moves at the same speed within the environment. Reinforcement learning algorithm will learn about the environment, from sensor input values based on action and rewards. Without rewards it would be difficult to know whether the crashing is bad, not crashing for a long time is good. In our experiment, using trial and error approach a negative reward of -500 is given each time when a car crashes into the obstacle or road borders, a reward of $+5$ for each step without a crash. Different

Fig. 2 Obstacle detection simulation training setup of an urban scenario

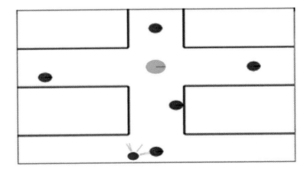

Table 1 Training parameters for Q-learning with MLP-NN model for obstacle detection

Parameters	Value
Number of ultrasonic sensors	3, 5, 7
Ultrasonic sensor placement angle	0°, 45°, 60°, 75°
Neural network hidden units	[256]
Number of epochs	1
Number of iteration per epoch	100,000
Batch size	[40, 100, 400]
Buffer size for Q-values	[10000, 50000]
Learning rate α	0.9
Discount factor γ	0.1

numbers of the ultrasonic sensor with varying hyperparameters of the MLP-NN are simulated.

Table 1 depicts inputs and hyperparameters used in training reinforcement model to detect obstacles using Q-learning and Q-learning with MLP methodology.

Ultrasonic sensors are placed on the front bonnet of the vehicle at the height of 2 feet from the ground. We have experimented with obstacle detection and avoidance using three, five, and seven ultrasonic sensors at different angles. Angles at which ultrasonic sensors are placed on the bonnet of the vehicle are given in Table 1.

3.1 Training Q-Learning with MLP-NN Model

To find optimal obstacle detection, MLP-NN with different values of hyperparameters like the number of hidden units, batch size, learning rate, discount factor, and buffer size is tested. Table 1 shows different hyperparameters values used in building the MLP-NN model. Figure 3 shows the model training loss function value with the number of iteration for [256] hidden units of 40 batch size and learning rate of 0.9. In Fig. 3a red color line represent model built using 7 sensors input is converging

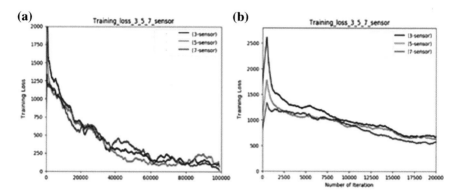

Fig. 3 **a** Moving average MLP-NN training loss for 3 different sensors for [256] hidden layers, with the batch size of 100 for 100,000 iterations. **b** Fast converging red lines of 7 sensors in fewer iteration frames compared to green (5 sensors) and blue (3 sensors) in understanding the environment (20,000 iterations)

Fig. 4 Simulation testing environment setup for evaluating Q-learning and Q-leaning MLP-NN for obstacle detection

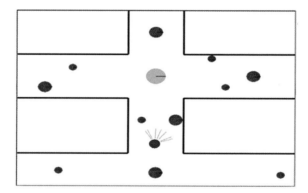

faster within less number of iterations in understanding the environment for obstacle detection compared to 5 and 3 sensors. This figure clearly illustrates that more sensors will help in faster learning about the environment.

3.2 Assessment of Q-Learning and Q-Learning MLP-NN

Evaluation of Q-learning and Q-learning MLP-NN is tested on different testing simulation environment with more static obstacles of different sizes and more in number. Figure 4 shows the simulated testing environment setup evaluating two algorithms for obstacle detection. The red color objects represent static obstacles and the blue color circle object represents autonomous vehicle navigating in the urban scenario.

The performance analysis of the obstacle detection using Q-learning and Q-learning with MLP is evaluated using ROC curve [15–17]. True positive (TP) is the number of correct actions predicted, in our experiment correct action means detecting an obstacle and avoiding it. False positive (FP) is the number of false action by the agent when colliding with the obstacles. In our experiment, false action means obstacle is not detected and vehicle collides with the obstacle. If vehicle collides with road border/lane edge, then those action is considered as false negative (FN). Table 2 shows the performance of Q-learning and Q-learning with MLP. The results depict Q-learning can detect and avoid obstacle with more sensors (7 sensors) with high false positive and false negative. Inferring Q-learning alone cannot learn complex scenarios. Combination of Q-learning with MLP-NN helps in understanding complex scenarios with fewer sensor inputs (3 sensors). In our experiment with 7 sensors inputs, 256 hidden layers of batch size 40 are detecting and avoiding obstacle with an accuracy of 0.9995 F1 scores.

Table 2 Comparative study result of Q-learning and Q-learning with MLP for obstacle detection using different sensor input

Obstacle detection using Q-learning				
Sensors	Input	TP	FP	FN
3	17492	17131	361	399
	Precision	Recall		F1 score
	0.9793	0.9772		0.9782
Sensors	Input	TP	FP	FN
5	17492	17268	224	227
	Precision	Recall		F1 score
	0.9871	0.9870		0.9871
Sensors	Input	TP	FP	FN
7	17492	17330	162	171
	Precision	Recall		F1 score
	0.9907	0.9902		0.9904
Obstacle detection using Q-learning with MLP				
Hidden layer [256], Batch size = 100				
Sensors	Input	TP	FP	FN
3	17492	17451	36	46
	Precision	Recall		F1 score
	0.9979	0.9973		0.9976
5	17492	17462	30	32
	Precision	Recall		F1 score
	0.9982	0.9981		0.9982
7	17492	17486	8	4
	Precision	Recall		F1 score
	0.9995	0.9996		0.9995

3.3   Simulation to Hardware Prototype

We have developed a self-driving car prototype model which can detect an obstacle using an ultrasonic sensor. Figure 5 represents complete hardware design and measurement of self-driving car prototype and customized track to test the obstacle detection with 'STOP-GO' mechanism using ultrasonic radar distance and camera input. The vehicle automatically stops if any obstacle is detected and gradually navigates forward when the distance from the obstacle to the vehicle reaches a safer threshold. For experiment 20 cm is set as safer threshold. In Fig. 5a hardware blueprint of vehicle design is shown with a length of 25 cm and height of 5 cm from the ground. Ultrasonic radar is mounted at 2 cm from the ground level and the camera is placed at 23 cm from ground at a tilted angle of 20°. Figure 5b shows detail hardware circuit connection, where "M" motors of the vehicle are connected to L-298 N H-bridge [18] which act like invertor in controlling the motor spin. Pulse Width Modulation (PWM) of 50 Hz frequency is set to make the spinning of the motor at constant speed. Camera and Ultrasonic radar is connected to Raspberry Pi 3, which is onboard electronic processing unit on the vehicle where all the algorithms are running. Figure 5c

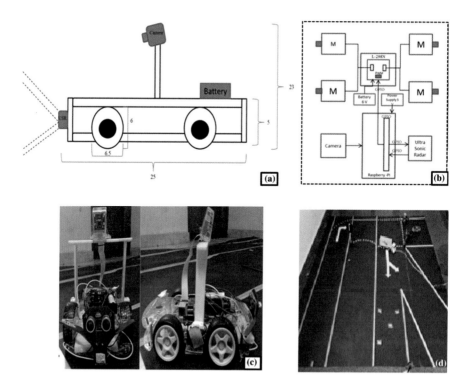

Fig. 5 **a** Hardware design and measurements of self-driving car prototype. **b** Complete hardware design of robotic self-driving car prototype. **c** Self-driving robotic car prototype builds for this research work. **d** Customized test track to test obstacle detection

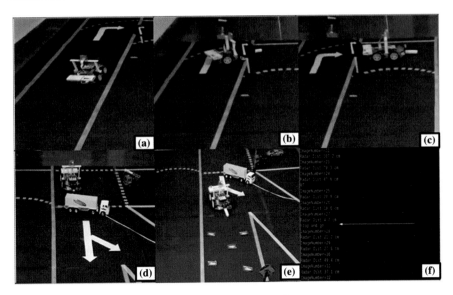

Fig. 6 Experimental result of "stop-go" mechanism on a hardware prototype

built hardware setup of autonomous vehicles used in this work. Figure 5d customized test track setup to test obstacle detection for the urban situation.

Figure 6 shows frame by frame output of obstacle detection and avoidance using the 'stop-go' mechanism on hardware prototype. Figure 6a–c Vehicle is autonomously navigating on test track based on perception algorithms. Figure 6d Vehicle automatically stops as it has detected an obstacle ahead, Figure 6e vehicle autonomously navigate forward when navigation path is free from obstacles Figure 6f shows the experimental result of 'stop –go' mechanism where an obstacle ahead is detected and collision avoidance is mitigated.

4 Conclusions and Future Work

In this paper, we developed a reinforcement learning technique, namely, Q-learning with MLP-NN for obstacle detection and avoidance. The proposed method is proven to be an effective and efficient way of collision avoidance for autonomous vehicle navigation. Simulation experiment results depict Q-learning with MLP-NN is able to understand and detect complex urban scenarios like static obstacles, road border/lane marking better than conventional Q-learning technique. We also developed a hardware prototype and tested 'stop-go' when the sensor detects an obstacle.

Although the results are very promising, some challenges not considered part of this study. For example, including the dynamic moving obstacles on the road.

Detecting and avoiding them can be accomplished using fused inputs from multiple sensors. Learning this complex dynamic environment can be solved using deep reinforcement learning method which of future interest.

References

1. Pendleton, S.D., Andersen, H., Du, X., Shen, X., Meghjani, M., Eng, Y.H., Rus, D., Ang, M.H.: Perception, planning, control, and coordination for autonomous vehicles. Machines **5**, 6 (2017)
2. Garcia, F., Martin, D., De La Escalera, A., Armingol, J.M.: Sensor fusion methodology for vehicle detection. IEEE Trans. Intell. Transp. Syst. Mag. **9**, 123–133 (2017)
3. http://outlace.com/rlpart1.html
4. Qiao, J.-F., Hou, Z.-J., Ruan, X.-G.: Application of reinforcement learning based on neural network to dynamic obstacle avoidance. In: Proceedings of the 2008 IEEE International Conference on Information and Automation, pp. 784–788 (2008)
5. Babu, V.M., Krishna, U.V., Shahensha, S.K.: An autonomous path finding robot using Q-learning. In: IEEE International Conference on Intelligent Systems and Control (2016)
6. Chu, P., Vu, H., Yeo, D., Lee, B., Um, K., Cho, K.: Robot reinforcement learning for automatically avoiding a dynamic obstacle in a virtual environment. In: Park, J., Chao, H.C., Arabnia, H., Yen, N. (eds.) Advanced Multimedia and Ubiquitous Engineering. Lecture Notes in Electrical Engineering, vol. 352. Springer, Berlin, Heidelberg (2015)
7. Huang, B.-Q., Cao, G.-Y., Guo, M.: Reinforcement learning neural network to the problem of autonomous mobile robot obstacle avoidance. In: 2005 International Conference on Machine Learning and Cybernetics, Guangzhou, China, pp. 85–89 (2005)
8. Duguleana, M.: Neural networks based reinforcement learning for mobile robots obstacle avoidance. Expert. Syst. Appl. Int. J. **62**(C), 104–115 (2016)
9. Xia, C., El Kamel, A.: A reinforcement learning method of obstacle avoidance for industrial mobile vehicles in unknown environments using neural network. In: Qi, E., Shen, J., Dou, R. (eds.) Proceedings of the 21st International Conference on Industrial Engineering and Engineering Management 2014. Proceedings of the International Conference on Industrial Engineering and Engineering Management. Atlantis Press, Paris (2015)
10. http://outlace.com/rlpart3.html
11. Thorpe, C.E.: Neural network based autonomous navigation. In: Vision and Navigation. The Carnegie Mellon Navlab, Kluwer (1990)
12. Ide, H., Kurita, T.: Improvement of learning for CNN with ReLU activation by sparse regularization. In: International Joint Conference on Neural Networks (IJCNN) (2017)
13. https://www.pygame.org
14. https://www.tensorflow.org/
15. Qin, Z.-C.: ROC analysis for predictions made by probabilistic classifiers. In: International Conference on Machine Learning and Cybernetics (2005)
16. Senthilnath, J., Kulkarni, S., Benediktsson, J.A., Yang, X.S.: A novel approach for multispectral satellite image classification based on the bat algorithm. IEEE Geosci. Remote Sens. Lett. **13**(4), 599–603 (2016)
17. Senthilnath, J., Bajpai, S., Omkar, S.N., Diwakar, P.G., Mani, V.: An approach to multi-temporal MODIS image analysis using image classification and segmentation. Adv. Space Res. **50**(9), 1274–1287 (2012)
18. Noman, A.T., Chowdhury, M.A.M., Rashid, H.: Design and implementation of microcontroller based assistive robot for person with blind autism and visual impairment. In: 2017 20th International Conference of Computer and Information Technology (ICCIT) (2017)

Evaluation of Deep Learning Model with Optimizing and Satisficing Metrics for Lung Segmentation

Usma Niyaz, Abhishek Singh Sambyal and Devanand Padha

Abstract The segmentation in medical image analysis is a crucial and prerequisite process during the diagnosis of the diseases. The need for segmentation is important to attain the region of interest where the probability of occurrence of an abnormality such as a nodule in the lungs or tumor in the brain is high. In this paper, we have proposed a new architecture called FS-Net which is a convolutional neural network-based model for the segmentation of lungs in CT scan images. It performs encoding of images into the feature maps and then decodes the feature maps into their respective lung masks. We have also trained the state-of-the-art U-Net on the same dataset and compared the results on the basis of optimizing and satisficing metrics. These metrics are useful for the selection of a better model with the maximum score at the satisfying condition. The FS-Net is computationally very efficient and achieves promising dice coefficient and loss score when compared with the U-Net taking one-third of the time.

Keywords FS-Net · U-Net · Data augmentation · Lung segmentation · Neural network · Deep learning

1 Introduction

Deep Convolutional Neural Networks (DCNN) have certainly given the astonishing results in image analysis such as classification [1], segmentation [2], detection, and localization [3]. The rise of the deep learning techniques in medical image analysis for diagnosis of patients has proven to be effectively beneficial to the society by

U. Niyaz (✉) · A. Singh Sambyal · D. Padha
Central University of Jammu, J and K, Jammu, India
e-mail: usmabhatt@gmail.com

A. Singh Sambyal
e-mail: abhishek.sambyal@gmail.com

D. Padha
e-mail: devanand@gmail.com

© Springer Nature Singapore Pte Ltd. 2020
K. N. Das et al. (eds.), *Soft Computing for Problem Solving*,
Advances in Intelligent Systems and Computing 1048,
https://doi.org/10.1007/978-981-15-0035-0_6

67

reducing the efforts of radiologists and pathologists by giving the human-level accuracy. Segmentation in the medical images is a complicated problem but it increases the efficiency of the model by performing analysis only in the region of interest and neglecting the irrelevant information. The segmentation is done in the preprocessing stage, so any mistake committed affects the other stages resulting in undesirable output. The manual segmentation is a slow process so there is a need for computer-aided systems that can do segmentation fast and accurate without human interaction. Automatic processing of medical images without human intervention reduces time, cost, and the human error. The existing computer-aided systems have produced appreciable results in segmentation but deep learning has outperformed the experts.

Traditional machine learning approaches for medical image segmentation including the graph cut approach [4], amplitude segmentation based on histogram features [5] involves the long sequence of algorithms and the filters were manually chosen for the detection of the features such as lines, edges, and curves which was very time consuming and often fails when the images were of low contrast, blur, noisy which increases the efforts of experts. However, the deep learning approaches automatically chooses the best filters for an image and results in most prominent features for the better classification and segmentation.

In the big data domain, where the medical data is enormously increasing, it becomes necessary to utilize such a big amount for useful work. The utilization of deep neural networks to get trained on the large data speeds up the process of the diagnosis of the disease and performs comparatively better than human. However, in medical image segmentation, there is a limitation of less data availability and class imbalance. Therefore, augmentation techniques like rotation, shifting, translation, and scaling are applied to increase the data for the network to get trained well. Due to the ability to process and learn from a large amount of data, deep learning techniques segment the region of interest in an accurate manner. To measure the loss and similarity between the predicted and actual output, two techniques cross-entropy and dice similarity are used for the efficient training of classification and segmentation tasks [6].

In our work, we have proposed a CNN-based architecture that is trained on lung CT scan images for the prediction of lung masks. The model consists of the encoder and decoder networks and the main idea is to get the features of CT scan images in the encoder network which includes the pooling layers that downsample the resolution of images by half and then use the upsampling layers in the decoder network to get the downsampled images into high-resolution images yielding lung masks. The network has all the vital layers of CNN like convolutional layer, max-pooling, ReLU, and dense layer, so this architecture is so-called CNN based. Due to the less training examples in our dataset, we have used data augmentation to improve the training of the model. This allows the network to learn invariances such as spatial, transformational, and rotational invariances. Dosovitskiy et al. [7] have shown in the unsupervised learning, how data augmentation helps in learning invariance.

Rest of the paper is organized as follows: Sect. 2 contains the related work of segmentation in medical images, Sect. 3 contains the methodology of the proposed architecture, Sect. 4 describes the results, and Sect. 5 concludes the research work with future directions.

2 Related Work

Many semi-automated and automated methods are used for the segmentation of medical images [8] such as thresholding, region growing, classifier, clustering, Markov random field model, Artificial Neural Network (ANN), deformable models, and metamorphs model [9]. These methods are popular and still in use for small-scale analysis. For the segmentation of 2D biomedical images, the fully convolutional network-based architecture called U-Net is used [10]. Another network that is used for the segmentation in the medical image analysis is the Deep Contour Aware Networks (DCAN) [11]. This architecture is FCN with multilevel contextual features used for the segmentation of glands. The architecture has won 2015 MICCAI gland segmentation competition. Besides 2D image segmentation, 3D image segmentation model was also introduced for the volumetric segmentation of MR images of the prostate. The architecture is called V-Net [12].

The cascade fully convolutional neural network was introduced where three networks hierarchically segment substructures of the brain which is a whole tumor (W-Net), tumor core (T-Net), and sequentially enhancing tumor core (E-Net) [13]. The deep learning framework for interactive image segmentation of brain tumor was proposed. The model was trained on binary segmentation [14].

Many architectures proposed for the segmentation of medical images is inspired by the U-Net architecture. Brahim et al. [15] have used U-Net architecture for the segmentation of lungs in CT scan images and achieved the Dice coefficient index of about 0.9502. The U-Net-based convolutional neural network was proposed for the segmentation of lungs using X-Rays with manually prepared lung masks [16].

The CNN-based method with 3D filters with some modifications over the existing U-Net architecture is used for the segmentation of brain tumor and bone in hands using MR images. Modifications made in the U-Net architecture were multiple segmented maps created at different scales and use of element-wise summation to forward feature maps [17].

3 Methodology

3.1 Preprocessing

The dataset used to train our model is from Kaggle Competition "Finding and Measuring lungs in CT data". It consists of 267 CT scan images and lung masks of size 512×512 which is the standard size of the dicom images. The range of pixel values is varying for each CT scan image and the pixel values of their corresponding masks are in a range [0, 255]. Because of the varying range of pixel values, the CT scan images are normalized with their corresponding mask as shown in Fig. 1. Normalization is done to scale down the pixel values $f(x,y)$, of an image f, to a fixed range. The CT scan images and their masks are normalized to a range [0, 1] using Eq. (1).

$$f(x, y) = \frac{f(x, y) - min(f)}{max(f) - min(f)} \tag{1}$$

The dataset contains less training examples for the FS-Net to train well, so the images are augmented by shifting in width and height by 0.1, rotated by 45° and zoomed by 0.1, the same is done for their corresponding masks as shown in Fig. 2. Image augmentation helps the model to get generalized and robust by learning invariance, which also helps in reducing the overfitting. At every epoch, the images are augmented and given to our model for comprehensive learning which makes the model universal and performs well on highly varying input data.

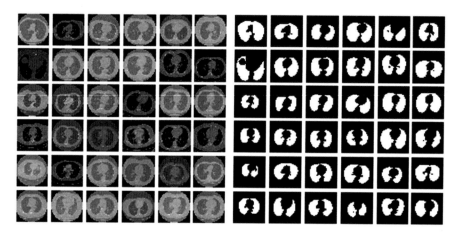

Fig. 1 Normalized CT scan slices and their corresponding masks

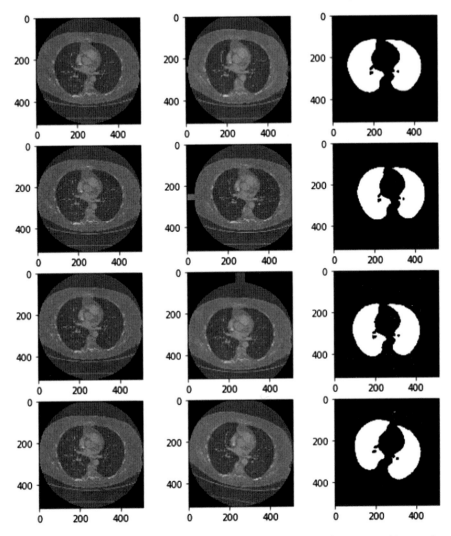

Fig. 2 Column 1 is the original CT scan images and Column 2 shows the augmented images that are zoomed, shifted in width, height by 0.1, and rotated by 45°. Column 3 shows their corresponding augmented masks

3.2 FS-Net Architecture

The architecture of our proposed model consists of the encoder and a corresponding decoder network. The visualization of our proposed model is shown in Fig. 3 and described in detail in Table 1.

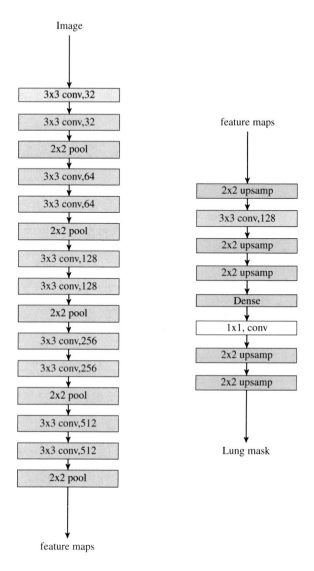

Fig. 3 Network architecture of FS-Net. **Left**: encoder network. **Right**: decoder network

The encoder network consists of 10 convolutional layers with 3×3 kernel size. The convolutional layer is the convolution of two signals, element-wise multiplication, and sum of the image $g(i, j)$ and filter $f(i, j)$.

$$f[i, j] * g[i, j] = \sum_{n_1=-\infty}^{\infty} \sum_{n_2=-\infty}^{\infty} f[n_1, n_2] \cdot g[i - n_1, j - n_2] \qquad (2)$$

Each convolutional layer in the encoder network produces a set of feature maps which are given as input to the following layers that are Rectified Linear Unit (ReLU)

Table 1 Summary of the FS-Net

Input	Output	Kernel	Activation	Parameters
Input	$1 \times 512 \times 512$	–	–	–
Conv1	$32 \times 512 \times 512$	3×3	ReLU	320
Conv2	$32 \times 512 \times 512$	3×3	ReLU	9248
Maxpool1	$32 \times 256 \times 256$	2×2	–	–
Conv3	$64 \times 256 \times 256$	3×3	ReLU	18496
Conv4	$64 \times 256 \times 256$	3×3	ReLU	36928
Maxpool2	$64 \times 128 \times 128$	2×2	–	–
Conv5	$128 \times 128 \times 128$	3×3	ReLU	73856
Conv6	$128 \times 128 \times 128$	3×3	ReLU	147584
Maxpool3	$128 \times 64 \times 64$	2×2	–	–
Conv7	$256 \times 64 \times 64$	3×3	ReLU	295168
Conv8	$256 \times 64 \times 64$	3×3	ReLU	590080
Maxpool4	$128 \times 32 \times 32$	2×2	–	–
Conv9	$512 \times 32 \times 32$	3×3	ReLU	1180160
Conv10	$512 \times 32 \times 32$	3×3	ReLU	2359808
Maxpool5	$512 \times 16 \times 16$	2×2	–	–
Upsample1	$512 \times 32 \times 32$	2×2	–	–
Conv11	$128 \times 32 \times 32$	3×3	ReLU	589952
Upsample2	$128 \times 64 \times 64$	2×2	–	–
Upsample3	$128 \times 128 \times 128$	2×2	–	–
Dense	$128 \times 128 \times 128$	–	–	16512
Conv12	$1 \times 128 \times 128$	1×1	Sigmoid	1153
Upsample4	$1 \times 256 \times 256$	2×2	–	–
Upsample5	$1 \times 512 \times 512$	2×2	–	–

and 2×2 max-pooling layer. ReLU is the nonlinearity activation function which is applied to each element of the feature maps and squashes the negative values to zero, i.e., $f(x) = max(0, x)$. It converges faster and is computationally efficient [18].

To make the representations manageable, we downsample the feature maps into pooling features/subsamples by a factor of 2 but doubles the feature channel at each downsampling step. Max-pooling operates over each activation map independently and takes those pixel values forward in the network which are prominent. Several layers of max-pooling make robust classification as it achieves more translational invariance and correspondingly there is the loss of spatial resolution of the feature maps. In the decoder network, the architecture consists of the upsampling layers followed by the convolutional layer. The decoder upsamples the features maps that are achieved by the encoder network and gives sparse feature maps as an output. The features maps are connected to each and every activation unit of dense layers which are the convolved by the filter of a convolutional layer to output dense feature maps.

The output of the last convolutional layer with kernel size 1×1 gives an output of $1 \times 128 \times 128$ which is further upsampled to get the desired 512×512 lung mask.[1]

The upsampling layers increase the resolution of images that are downsampled by the max-pooling layers. Therefore, a number of max-pooling layers and upsampling layers are kept equal. The features extracted in the encoder network are given to the upsampling layers, which process the given input and upscale its resolution by 2. The last convolutional layer of encoder network gives the output of shape $512 \times 32 \times 32$ which is downsampled by max-pooling to $512 \times 16 \times 16$. In the decoder network, the first layer is an upsampling layer which upscales the output of $512 \times 16 \times 16$ back to $512 \times 32 \times 32$ and does the same for the rest of the layers. We have used a total of 5 upsampling layers equal to the number of max-pooling layers, to receive the size of the output at the last layer of this network, same as that of the input image. Unlike U-Net, FS-Net does not transfer the entire feature maps to the decoder so there is less memory consumption and there is no concatenation of these feature maps to the decoder feature maps of upsampling layers which in turn reduces the overall training time of the model. In the last convolutional layer of the model, the activation function used is sigmoid activation that squashes the value to the range [0, 1] and gives a probability output. The segmentation is predicted on the basis of maximum probability at each pixel.

3.3 Training, Loss, and Dice Coefficient

The dataset is split into 70% training set and 30% test set. The model learns from the training set and provides an unbiased evaluation of the model on the test set. The desired model has performed well on both the sets. The hyperparameters such as learning rate, batch size, and epochs are set to 0.001, 8, and 170, respectively. We have used Adam as an optimization function that converges the loss function of the model fast and gives better results by rectifying the problems like vanishing gradient and high variance which are the causes of fluctuating loss function. The performance metrics taken for the proposed model is given by Eqs. (3) and (4).

$$Dice\ coefficient\ (Dice\ coeff) = 2 * \frac{|X \cap Y|}{|X \oplus Y|} \tag{3}$$

where X is the lung area obtained by the segmentation based on our network and Y is the ground truth obtained by manual segmentation.

It is not easy to combine all the properties of the model into a single real number evaluation metric, thus it is very useful to set up satisficing and optimizing metrics for the estimation of the performance of the model. Along with the optimizing metric, *Dice coeff*, we have also taken the satisficing metric, *training time* in consideration, which gives us the better representation to choose the best model. For the

---

[1] Detailed Computational Graph of FS-Net is available at http://github.com/abhigoogol/FS-Net.

Table 2 Comparison results of the FS-Net and U-Net

Architecture	Optimizer	Dice coeff	Loss	Epoch	Time	Parameters
FS-Net	Adam	0.9549	0.0498	170	51 min	7,846,081
U-Net	Adam	0.9629	0.0472	170	2 h 26 min	5,319,265

satisficing metric, we need a threshold value for the comparison and to get this value; we first train the U-Net and calculate its training time. The training time taken by U-Net is considered as the threshold value and we have to compare the time taken by FS-Net for training with this value. The main aim is to maximize the Dice coeff with the subject that the training time for FS-Net should be less than the U-Net. The loss function used in this model measure the performance of the model whose output is a probability value between 0 and 1. The desired output of the function should be minimum, i.e., smaller the loss more accurate is the prediction.

$$log\ loss = -\frac{1}{N}\sum_{i=0}^{\infty}[y_i log(\hat{y}_i) + (1 - y_i)log(1 - \hat{y}_i)] \tag{4}$$

where y_i is the actual output and $\hat{y}_i$ is the predicted output.

4 Results

We have computed the results of the proposed FS-Net architecture and compared it with the state-of-the-art U-Net. The time taken by U-Net for training is 2 h 26 min which is considered as the threshold value. The proposed model is considered better if it satisfies the following condition.

$$Model = Better\ \{if\ Dice\ coeff = max\ \&\ Time < threshold\}. \tag{5}$$

Table 2 shows that the FS-Net and U-Net have almost the same dice coefficient and loss but FS-Net takes significantly less time to train. FS-Net is $3X^2$ times faster to train and requires less memory as compared to U-Net. The key feature to be noted from the results is that the proposed model gives the promising dice coefficient score and is computationally efficient when compared to U-Net so FS-Net is preferred as the better model with high dice score and low resource consumption (Fig. 4).

Figure 5 shows the predictions of the lung mask $m(x, y)$ on test data and we performed pixel-wise multiplication of the predicted lungs masks with actual images $g(x, y)$ to segment the lung $s(x, y)$ in the CT data, $s(x, y) = m(x, y) * g(x, y)$.

---

[2] Actual training time of FS-Net is 2.863 times faster than U-Net. Model is trained on Nvidia K80.

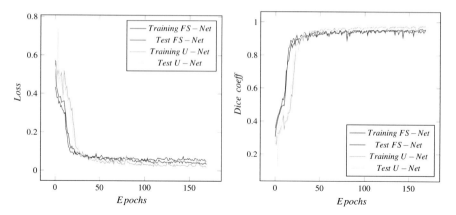

Fig. 4 FS-Net versus U-Net loss (**Left**) and Dice coeff (**Right**). The loss and dice score of the two models is almost same but time taken by FS-Net is less than U-Net

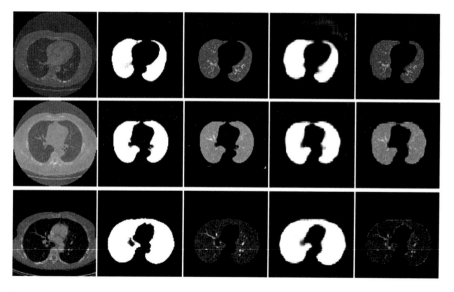

Fig. 5 Column 1 shows the original CT scans, Column 2 and 3 are the predicted lung masks and lung segments of U-Net. Column 4 and 5 show the predicted lung masks and lung segments of FS-Net

5 Conclusion

In our paper, we have presented a new architecture called FS-Net for the fast segmentation of lungs from CT scan images. It is a CNN-based architecture which consists of the encoder network for the extraction of prominent features for the better segmentation and decoder network to retain the resolution of the images to output lung mask. FS-Net has shown a significant decrease in training time without sacrificing the dice coefficient and loss as compared to U-Net. The proposed model performs as good as the U-Net in almost one-third training time. In the future, our work is to compare our model with many other segmentation models like Segnet, FCN, Deeplabs nets, GANs, and RefineNet. We will also implement ensemble learning for segmentation in the medical images where we can take different deep neural networks for segmentation and calculate their performance metrics to achieve better results.

References

1. Krizhevsky, A., Sutskever, I., Hinton, G.E.: Imagenet classification with deep convolutional neural networks. NIPS. pp. 1106–1114 (2012). https://doi.org/10.1145/3065386
2. Long, J., Shelhamer, E., Darrell, T.: Fully convolutional networks for semantic segmentation. In: Proceedings of the IEEE Conference on Computer Vision and Pattern Recognition, pp. 3431–3440 (2015). https://doi.org/10.1109/TPAMI.2016.2572683
3. Wang, N., Li, S., Gupta, A., Yeung, D.: Transferring rich feature hierarchies for robust visual tracking (2015). arXiv:1501.04587
4. Boykov, Y.Y., Jolly, M.P.: Interactive graph cuts for optimal boundary and region segmentation of objects in ND images. In: Proceedings Eighth IEEE International Conference on Computer Vision, ICCV 2001, vol. 1. IEEE (2001). https://doi.org/10.1109/ICCV.2001.937505
5. Ramesh, N., Yoo, J.-H., Sethi, I.K.: Thresholding based on histogram approximation. IEE Proc. Vis. Image Signal Process. **142**(5), 271–279 (1995). https://doi.org/10.1049/ip-vis:19952007
6. Noviko, A.A., Lenis, D., Major, D., Hladuvka, J., Wimmer, M., Buhler, K.: Fully convolutional architectures for multi-class segmentation in chest radiographs. IEEE Trans. Med. Imaging **37** (2018). https://doi.org/10.1109/TMI.2018.280608
7. Dosovitskiy, A., Springenberg, J.T., Riedmiller, M., Brox, T.: Discriminative unsupervised feature learning with convolutional neural networks. NIPS (2014)
8. Pham, D.L., Xu, C., Prince, J.L.: Current methods in medical image segmentation. Annu. Rev. Biomed. Eng. **2**, 315–337 (2000). https://doi.org/10.1146/annurev.bioeng.2.1.315
9. Huang, X., Tsechpenakis, G.: Medical image segmentation. Information Discovery on Electronic Health Records, Chapter 10 (2009)
10. Ronneberger, O., Fischer, P., Brox, T.: U-Net: Convolutional networks for biomedical image segmentation. In: MICCAI 2015, pp. 234–241 (2015). https://doi.org/10.1007/978-3-319-24574-4_28
11. Chen, H., Qi, X., Yu, l., Dou, Q., Qin, J., Heng, P.: Deep contour-aware networks for accurate gland segmentation. Med. Image Anal. **36**, 135–146 (2017). https://doi.org/10.1109/CVPR.2016.273
12. Milletari, F., Navab, N., Ahmadi, S.A.: V-Net: Fully convolutional networks for volumetric medical image segmentation. In: 2016 Fourth International Conference on 3D Vision, pp. 565–571 (2016). https://doi.org/10.1109/3DV.2016.79
13. Wang, G., Li, W., Ourselin, S., Vercauteren, T.: Automatic Brain Tumor Segmentation using Cascaded Anistropic Convolutional Neural Networks. BrainLes MICCAI (2017). https://doi.org/10.1007/978-3-319-75238-9_16
14. Wang, G., Li, W., Ourselin, S., Vercauteren, T., Zuluaga, M.A., Pratt, R., Patel, P.A., Aertsen, M., Doel, T., David, A.L., Deprest, J.: Interactive medical image segmentation using deep learning with image-specific fine-tuning. IEEE Trans. Med. Imaging(2018), arXiv:vol.abs/1710.04043, 2017. https://doi.org/10.1109/TMI.2018.2791721
15. Skourt, B.A., Hassani, A., Majda, A.: Lung ct image segmentation using deep neural networks. In: The First International Conference on Intelligent Computing in Data Sciences, pp. 109–113 (2018). https://doi.org/10.1016/j.procs.2018.01.104

16. Kalinovshky, A., Kovalev, V.: Lung image segmentation using deep learning methods and convolutional neural networks. In: 13th International Conference on Pattern Recognition and Information Processing, pp. 21–24 (2016)
17. Kayalibay, B., Jensen, G., Smagt, P.: CNN-based segmentation of medical imaging data. Computing Research Repository (CoRR) (2017) arXiv:vol.abs/1701.03056
18. Nair, V., Hinton, G.E.: Rectified linear units improve restricted boltzmann machines. In: ICML'10 Proceedings of the 27th International Conference on International Conference on Machine Learning, pp. 807–814

Improved Flower Pollination Algorithm for Linear Antenna Design Problems

Rohit Salgotra, Urvinder Singh, Sriparna Saha and Atulya K. Nagar

Abstract Flower pollination algorithm (FPA) is an evolutionary nature-inspired optimization technique, which mimics the pollinating behavior of flowers. FPA has a simple structure and has been applied to numerous problems in different fields of research. However, it has been found that it has poor exploration and exploitation capabilities. In this paper, to mitigate the problems of original FPA, a modified algorithm namely adaptive FPA (AFPA) has been proposed. In the modified algorithm, a four-fold population division has been followed for both global and local search phases. Moreover, to balance the local and global search, switching probability has been decreased exponentially with respect to iterations. For experimental testing, this algorithm has been further applied to antenna design problems. The aim is to optimize linear antenna array (LAA) in order to achieve minimum SLL in the radiation pattern to avoid antenna radiation in the undesired directions. The results of the proposed algorithm for same problems are compared with the results of popular algorithms such as particle swarm optimization (PSO), tabu search (TS), self-adaptive differential evolution (SADE), Taghchi's method (TM), cuckoo search (CS), and biogeography-based optimization (BBO). The simulation results clearly indicate the superior performance of AFPA in optimizing LAA.

Keywords Flower pollination algorithm · Adaptive parameters · Linear antenna design · Antenna arrays

R. Salgotra (✉) · U. Singh
Thapar Institute of Engineering & Technology, Punjab, India
e-mail: rohit.salgotra@thapar.edu

U. Singh
e-mail: urvinder@thapar.edu

S. Saha
Indian Institute of Technology Patna, Patna, India
e-mail: sriparna@iitp.ac.in

A. K. Nagar
Liverpool Hope University, Liverpool, UK
e-mail: atulya.nagar@hope.ac.uk

© Springer Nature Singapore Pte Ltd. 2020
K. N. Das et al. (eds.), *Soft Computing for Problem Solving*,
Advances in Intelligent Systems and Computing 1048,
https://doi.org/10.1007/978-981-15-0035-0_7

79

1 Introduction

Flower pollination algorithm (FPA) is a recently introduced algorithm in the field of nature-inspired algorithms. This algorithm takes inspiration from the flower species found in nature. Flowers are one among the most fascinating species prevalent on this earth and are found to be there from around 125 million years that is from the Cretaceous period [1]. These species have dominated the landscapes due to their continuous pollinating nature. They follow pollination either by biotic or abiotic means but mostly they are pollinated by a certain set of pollinator, which ultimately helps them to maintain their species count. The pollinators also get benefitted as they know which flower species is a continuous source of nectar and which is not. This process is also called as flower constancy. So here, pollination process, pollinator behavior, and flower constancy have been summed to formulate the FPA [1]. The algorithm since its inception has found application in large number of problems and has proved its worth in numerous fields of research. In FPA, the performance is based on the phenomena of exploration and exploitation. The exploration process is the global pollination phase which is carried out by using Lèvy fights-based component whereas exploitation is carried out by local pollination phase. On a whole, exploration is meant for exploring whole search space and exploitation is for exploring some specific areas of search space. One more parameter is the switch probability which determines the extent of exploitation and exploration in FPA. A higher value of probability can extend the exploration phase and lead to better searching capabilities whereas exploitation will be degraded at the same time and can diverge the solution from original one. On the other hand, a lower value of probability can prolong the exploitation operation and reduce the exploitation capabilities. Thus, a well balanced probability value is required for the better performance of an algorithm. Based on this, various research articles have been proposed in the literature to improve the performance of FPA. The major modification introduced in the recent past include [2–10].

One of the recently introduced FPA variants is the adaptive Lèvy FPA (ALFPA) [11]. This algorithm employs the concept of fast evolutionary mutation operators to find global optimal solution. More details about fast evolutionary mutation operators can be had from [12]. The algorithm consists of four search equations in the global and local search phase in order to find global optimal solution. The global pollination equations are inspired from the concepts of Lèvy, Cauchy, and Gaussian distribution functions whereas local pollination equations correspond to the simple random solutions. The algorithm also uses a linearly decreasing probability in order to add a balance between the global pollination and local pollination phase. Further, it has been found that the algorithm is considered as an extension of differential evolution (DE) and when compared with the likes of other recently introduced algorithms, it was found to provide highly competitive results. In present work, ALFPA is extended to the field of antenna array design. The algorithm though has been slightly changed with respect to the probability values in order to have a better balanced local and global search. The new search equation used for probability is the dynamically decreasing exponential distribution based random solution.

The antenna is the most fundamental component of wireless communication. It is imperative to have a proper antenna for good wireless link. Antenna design is a tedious and complicated job. The design of antenna is nonlinear and multimode problem and hence, number of optimization techniques have been used in the past to synthesize antennas [12–17]. In this paper, linear antenna array (LAA) is designed or synthesized using AFPA to have a radiation pattern with minimum or reduced sidelobe level (SLL). The organization of the paper includes an introduction in Sect. 1. Section 2 details about the recently proposed AFPA algorithm, Sect. 3 is concerned with the various design examples whereas, in the final Sect. 4, the paper is concluded.

2   Adaptive Flower Pollination Algorithm

FPA consists of two phases that are local pollination or local search phase and global search phase or the global pollination. The local search phase is dominated by the use of the exploitation process whereas global pollination or global search phase has the dominion of exploration phase. Both these phases are based on four rules given in [2] and these rules are derived from the behavior of flowers and pollinators found in nature. Now in order to control both these phases, the probability switch acts as a common factor. This factor mainly decides the extent of both global and local pollination. For most of the cases, it has been found that a higher value of probability provides much better results when compared to the lower or intermediate values. The FPA is a standard algorithm and in order to make it state-of-art algorithm, numerous modifications have been proposed. ALFPA is one such algorithm aimed at providing a better global and local pollination along with adaptive switch probability. The algorithm starts by initializing a random population, followed by the global and local pollination phase. The initialization step is same as that used in the standard FPA and has not been discussed in detail [18].

In global pollination phase, the concept of adaptive Lèvy inspired from [18], has been used in present work. This concept has been derived for finding the best fit solution from a pool of four search equations. For Lèvy distribution, the probability distribution is as given in Eq. (1)

$$f_{\text{Lèvy}(\lambda,\gamma)}(\lambda) = \frac{1}{\pi} \int_0^\infty e^{-\gamma q^\lambda} \cos(q\lambda) dq \tag{1}$$

here, $1 < \lambda < 2$, $\gamma > 0$ and is taken as 1. Also, four search equations based on λ, are used. The search equations reduces to Cauchy for $\lambda = 1$, Gaussian for $\lambda = 2$, and two random Lèvy values namely 1.3 and 1.5. These four values lead to four search equations as given by

$$x_{i,1}^{t+1} = x_i^t + G(\lambda)(T^* - x_i^t) \tag{2}$$

$$x_{i,2}^{t+1} = x_i^t + C(\lambda)(T^* - x_i^t) \tag{3}$$

$$x_{i,3}^{t+1} = x_i^t + L_1(\lambda)(T^* - x_i^t) \tag{4}$$

$$x_{i,4}^{t+1} = x_i^t + L_2(\lambda)(T^* - x_i^t) \tag{5}$$

where $G(\lambda), C(\lambda), L_1(\lambda)$, and $L_1(\lambda)$ are Gaussian distributed random number, Cauchy distributed random number, and adaptive Lèvy-based random number 1 and 2, respectively. Here, all the search equations are subjected to only 25% of the population and then, the solutions from each group are merged to evaluate the new solution.

In local search phase, two search equations have been used to find the new solution. This new solution is found by using equations as given below:

$$x_{i,5}^{t+1} = x_i^t + \epsilon\left(x_j^t - x_k^t\right) \tag{6}$$

$$x_{i,6}^{t+1} = x_i^t + \epsilon\left(x_l^t - x_m^t\right) \tag{7}$$

Here, x_j^t, x_k^t, x_l^t and x_m^t corresponds to four random solutions generated from the whole population with $j \neq k \neq l \neq m$ and ϵ is in the range of [0, 1].

The above given Eqs. (6) and (7) are subjected to each half of the population and is used to evaluate the new solution. The above-said solutions are then merged to define the full solution. Apart from this, the third parameter which is the controlling factor of exploitation and exploration is the probability switch. In present case, the value of switch probability is chosen based on exponentially decreasing function. This function is given by [19]

$$p(t) = p_{min} + (p_{max} - p_{min}) \cdot e^{-\left(\frac{t}{t_{max}/10}\right)} \tag{8}$$

where p_{min} and p_{max} are taken in the range of [0, 1], e is the exponential function, t is the current iteration, and t_{max} is the total number of iteration or generations. The reason for the use of exponentially decreasing function is that it helps the algorithm in converging slowly during the initial iterations and change faster toward the final stages. The pseudo-code for AFPA algorithm is given in Algorithm 1.

Begin:
Initialize population: *n/4*
Define probability
Define fitness of problem under consideration, *f(x)*
Find initial best flower: *T.*
do
Until i < total number of iterations
   for i=1:n
     if rand < p
       A d−dimensional step vector based on Gaussian based is drawn.
       global pollination using $\mathbf{x}_{i,1}^{t+1}, \mathbf{x}_{i,2}^{t+1}, \mathbf{x}_{i,3}^{t+1}$ *and* $\mathbf{x}_{i,4}^{t+1}$
        evaluate, find best and replace it by x_i^{t+1}
     else
       set ϵ **to uniformly distributed in the range of** [−1, 1]
       local pollination using $\mathbf{x}_{i,5}^{t+1}$ *and* $\mathbf{x}_{i,6}^{t+1}$
        evaluate, find best and replace it by x_i^{t+1}
     end if
       if x_i^{t+1} better than x_i^t, $x_i^t = x_i^{t+1}$
      end if
      Update switch probability using equation (8)
     end for
   Update best
   i=i+1;
End

Algorithm 1: Pseudo−code of proposed AFPA

---

3 Design Examples

This section details the implementation of AFPA to the LAA design. An LAA is an antenna array in which antenna elements are arranged along a single axis. A symmetrical LAA has 2N element placed symmetrically about the origin. The elements are arranged in a straight line and are divided into two groups of N elements each and that too on either side of the origin. The main aim of this kind of symmetry is to provide a highly dimensional radiation pattern in the desired direction and at the same time, helps in reducing the computational complexity of the antenna under consideration. The array factor (AF) decides the radiation pattern of an antenna and for LAA it is given as

$$AF(\phi) = 2 \sum_{n=1}^{N} I_n \cos[kx_n \cos(\phi) + \varphi_n] \tag{9}$$

where x_n, φ_n, I_n, are amplitude excitation, position, and phase of nth element, respectively, $k = 2\pi/\lambda$ is the wavenumber and AF is the array factor. For the case of equally

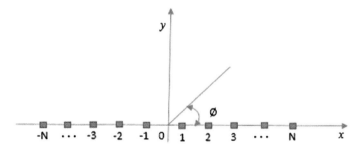

Fig. 1 A symmetric linear antenna array

spaced antenna arrays, each element is placed at a $\lambda/2$ distance from the other, that is, the position of each element is kept fixed whereas in case of unequally spaced antennas, the position of elements is different but the antennas have uniform excitation. A symmetrical LAA is shown in Fig. 1

In this work, equally spaced LAA having uniform element spacing are optimized. The amplitude excitations are varied for achieving desired radiation pattern with reduced SLL. The fitness function for achieving the desired objective using AFPA is given by

$$\text{minimize} \, fit = \max\{20\log|AF(\phi)|\} \tag{10}$$

subject to $\phi \in [\phi_1, 180°]]$ where $[\phi_1, 180°]$ is the region other than main lobe, i.e., sidelobe region.

In the first example, a 10-element equally spaced LAA is optimized for achieving minimum SLL. The amplitude excitations are varied between [0, 1] uniformly. The population size for AFPA is taken as 40 and the algorithm is run for 1000 iterations. The algorithm is run for 51 times and the statistical values for each case has been reported in Table 1. Since it is a symmetric array, only five amplitude excitations need to be optimized. The optimized excitations are tabulated in Table 2. The simulation results of AFPA are compared with that of other well-known algorithms and are

Table 1 Statistical performance of AFPA

	Best	Worst	Mean	Median	Std
10-element LAA	−25.3306	−25.3168	−25.3257	−25.3261	0.0030
16-element LAA	−33.4873	−33.1086	−33.3745	−33.3852	0.0790
24-element LAA	−37.1916	−36.0111	−36.7035	−36.6987	0.2949

Table 2 Optimized amplitude excitations for 10-element LAA

Element	1	2	3	4	5
Amplitude	1.0000	0.8979	0.7178	0.5002	0.3833

Table 3 Comparison results for 10-element LAA

Method	PSO [12]	TM [15]	BBO [14]	AFPA
SLL (dB)	−24.62	−24.88	−25.21	−25.33

presented in Table 3. The SLL achieved by AFPA is −25.33 dB which is least in the comparison Table 3. The convergence characteristics of AFPA are given in Fig. 2. The radiation pattern for AFPA optimized antenna is shown in Fig. 3.

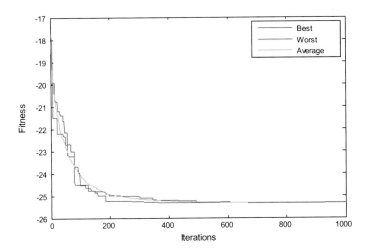

Fig. 2 Convergence characteristics of 10-element LAA

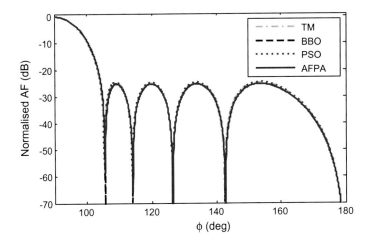

Fig. 3 Radiation pattern for 10-element LAA

In the next example, a 16-element LAA is optimized for the same objective optimization problem. The constraints are kept the same with respect to the previous example. The AFPA optimized amplitude excitations are presented in Table 4. The comparison between the performance of AFPA and PSO [12], TS [13], BBO [14], SADE [15], TM [15], ALO [16] is given in Table 5 which indicates the better performance of AFPA. The convergence profile for AFPA for optimizing 16-element LAA is shown in Fig. 4. Figure 5 gives the radiation pattern of the optimized antenna.

The last example presents the optimization of 24-element antenna for obtaining minimum SLL by optimizing antenna element excitations. The optimized excitations are given in Table 6. The performance of AFPA is compared in terms of SLL achieved in Table 7. The SLL achieved by AFPA is −37.19 dB whereas the SLL of PSO [12], TS [13], TM [15], CS [17], SADE [15] is higher than AFPA. The convergence graph of AFPA for this optimization is shown in Fig. 6. The radiation pattern of AFPA, PSO [14], and TS [15] are shown in Fig. 7.

Table 4 Optimized amplitude excitations for 16-element LAA

Element	1	2	3	4
Amplitude	1.0000	0.9467	0.8458	0.7115
Element	5	6	7	8
Amplitude	0.5575	0.4076	0.2666	0.1996

Table 5 Comparison of results for 16-element LAA

Method	TS [13]	TM [15]	ALO [16]	PSO [12]	BBO [14]	SADE [15]	AFPA
SLL (dB)	−26.20	−31.21	−30.85	−30.70	−33.06	−31.06	**−33.49**

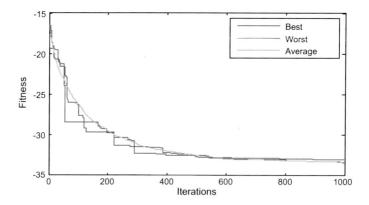

Fig. 4 Convergence characteristics of 16-element LAA

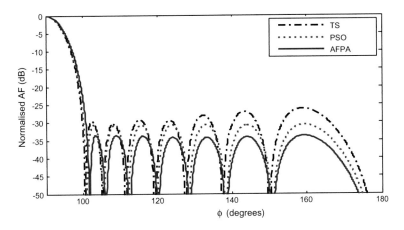

Fig. 5 Radiation pattern for 16-element LAA

Table 6 Optimized amplitude excitations for 24-element LAA

Element	1	2	3	4	5	6
Amplitude	1	0.9703	0.9131	0.8470	0.7591	0.6442
Element	**7**	**8**	**9**	**10**	**11**	**12**
Amplitude	0.5400	0.4324	0.3377	0.2348	0.1815	0.1274

Table 7 Comparison of results for 24-element LAA

Method	PSO [12]	CS [17]	SADE [15]	TS [13]	TM [15]	AFPA
SLL (dB)	−34.50	−34.50	−35.21	−27.54	−35.25	−37.19

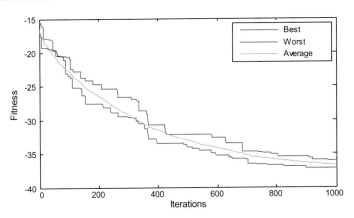

Fig. 6 Convergence characteristics of 24-element LAA

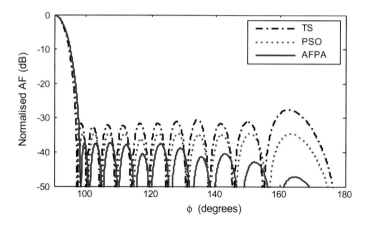

Fig. 7 Radiation pattern for 24-element LAA

4 Conclusion

This paper deals with the extension of a recently proposed ALFPA algorithm, with adaptive probability, to the field of LAA. The algorithm has been named as AFPA and uses linearly decreasing probability for adapting the switch probability. Three cases namely 10 element, 16 element, and 24 element of LAA have been exploited. The algorithm has been statistically tested for mean, best, worst, and standard deviation values for all these cases and it has been found that AFPA is found to provide good results consistently. Moreover, the results are also compared with other major algorithms such as TS, PSO, SADE, TM, BBO, and others and are found to be superior. Future work may include extension of the algorithm to various other fields such as pattern search, image segmentation, economic load dispatch, and others.

Acknowledgements This work is sponsored under INSPIRE Fellowship (IF160215) by Directorate of Science and Technology, Govt. of India.

References

1. Yang, X.S.: Flower pollination algorithm for global optimization. In: International Conference on Unconventional Computing and Natural Computation, pp. 240–249. Springer, Berlin, Heidelberg (2012)
2. Sakib, N., Kabir, M.W.U., Subbir, M., Alam, S.: A comparative study of flower pollination algorithm and bat algorithm on continuous optimization problems. Int. J. Soft Comput. Eng. **4**(2014), 13–19K (2014)
3. Salgotra, R., Singh, U.: A novel bat flower pollination algorithm for synthesis of linear antenna arrays. Neural Comput. Appl., 1–14 (2016)

4. Sharawi, M., Emary, E., Saroit, I.A., El-Mahdy, H.: Flower pollination optimization algorithm for wireless sensor network lifetime global optimization. Int. J. Soft Comput. Eng. **4**(3), 54–59 (2014)

5. Singh, U., Salgotra, R.: Synthesis of linear antenna array using flower pollination algorithm. Neural Comput. Appl., 1–11 (2016)

6. Wang, R., Zhou, Y.: Flower pollination algorithm with dimension by dimension improvement. Math. Probl. Eng. (2014)

7. Zhou, Y., Wang, R., Luo, Q.: Elite opposition-based flower pollination algorithm. Neurocomputing **188**, 294–310 (2016)

8. Valenzuela, L., Valdez, F., Melin, P.: Flower pollination algorithm with fuzzy approach for solving optimization problems. In: Nature-Inspired Design of Hybrid Intelligent Systems, pp. 357–369. Springer International Publishing, Cham (2017)

9. Ram, J.P., Rajasekar, N.: A novel flower pollination based global maximum power point method for solar maximum power point tracking. IEEE Trans. Power Electron. (2016)

10. Salgotra, R., Singh, U.: Application of mutation operators to flower pollination algorithm. Expert. Syst. Appl. (2017)

11. Khodier, M., Al-Aqeel, M.: Linear and circular array optimization: a study using particle swarm intelligence. PIER B **15**, 347–373 (2009)

12. Merad, L., Bendimerad, F., Meriah, S.: Design of linear antenna arrays for sidelobe reduction using the tabu search method. Int. Arab. J. Inf. Technol. **5**(3), 219–222 (2008)

13. Sharaqa, A., Dib, N.: Design of linear and elliptical antenna arrays using biogeography based optimization. Arab. J. Sci. Eng. **39**(4), 2929–2939 (2013)

14. Dib, N., Goudos, S., Muhsen, H.: Application of taguchi's optimization method and self-adaptive differential evolution to the synthesis of linear antenna arrays. PIER **102**, 159–180 (2010)

15. Saxena, P., Kothari, A.: Ant lion optimization algorithm to control side lobe level and null depths in linear antenna arrays. AEU-Int. J. Electron. Commun. **70**(9), 1339–1349 (2017)

16. Khodier, M.: Optimisation of antenna arrays using the cuckoo search algorithm. IET Microw. Antennas Propag. **7**(6), 458–464 (2013)

17. Yao, X., Liu, Y., Lin, G.: Evolutionary programming made faster. IEEE Trans. Evol. Comput. **3**(2), 82–102 (1999)

18. Chen, G., Huang, X., Jia, J., Min, Z.: Natural exponential inertia weight strategy in particle swarm optimization. In: The Sixth World Congress on Intelligent Control and Automation, 2006. WCICA 2006, vol. 1, pp. 3672–3675. IEEE (2006)

19. Singh, U., Salgotra, R.: Optimal synthesis of linear antenna arrays using modified spider monkey optimization. Arab. J. Sci. Eng. **41**(8), 2957–2973 (2016)

Texture-Based Fuzzy Connectedness Algorithm for Fetal Ultrasound Image Segmentation for Biometric Measurements

S. Jayanthi Sree◉ and C. Vasanthanayaki

Abstract Fuzzy connectedness segmentation approach guided by texture properties of the image is proposed for segmenting fetal organs such as femur, cranial bones, and abdomen from ultrasound images. This semiautomatic segmentation technique is proposed for fetal biometric measurements of biparietal diameter, head circumference, occipital diameter, femur length, and abdominal circumference. The texture information in the ultrasound images guides the fuzzy connectedness algorithm for efficient segmentation of fetal structures and thereby accurate biometric measurements. The proposed algorithm is compared with the manual segmentation of an expert and evaluation is performed with respect to region-based and distance-based metrics. The performance evaluation indicates that the proposed technique is comparable to manual segmentation results across all gestational ages.

1 Introduction

2D Ultrasound imaging is the primary modality for fetal health monitoring during entire span of pregnancy since, it is less expensive, safe, and free from radiations. Even with the vast development of medical imaging systems and diagnosis, fetal mortality and morbidity is still high. Fetal growth is monitored using parameters such as fetal weight, gestational age, and estimated delivery date. There are a number of techniques to arrive at these parameters. One of the techniques is the method using fetal biometric measurements such as femur length, biparietal diameter, head circumference, occipital diameter, crown-rump length, and abdominal circumference. These parameters and measurements are an indication of fetal growth and give a hint of any abnormality present. At present, these measurements are made manually by the experts, or at least the experts have to inscribe an ellipse or circle around cranial

S. Jayanthi Sree (✉)
Government College of Technology, Coimbatore 641013, India
e-mail: jayanthisrees@gmail.com

C. Vasanthanayaki
Government College of Engineering, Salem 636011, India

© Springer Nature Singapore Pte Ltd. 2020
K. N. Das et al. (eds.), *Soft Computing for Problem Solving*,
Advances in Intelligent Systems and Computing 1048,
https://doi.org/10.1007/978-981-15-0035-0_8

bones and abdomen or indicate the end points of femur. The manual measurement is a hindrance to the busy schedule of the clinicians. Further, the experts could suffer from repetitive stress injury, since they do the same thing for the entire day with each scan taking at least 30 min.

2D ultrasound images are fuzzy in nature, affected by noise such as shadows, mirror noise, and predominantly speckle noise. Further, the manual measurements also depend on the quality of the scan which in turn depends on the trimester and BMI of the mother. During later trimester, the fluid level of the uterus decreases resulting in low-quality scans. Since ultrasound is impeded by fatty tissues, ultrasound scans of larger BMI mothers are of low quality. Bony structures appear bright in ultrasound scan compared to fluids or tissues. Hence, segmentation of skull and femur bones is easier compared to abdomen segmentation which is a challenging task.

In this paper, a semiautomatic technique based on fuzzy connectedness algorithm guided by texture information has been proposed. The paper is organized as follows: Sect. 2 gives a survey of existing ultrasound fetal segmentation techniques. Section 3 explains the concepts of fuzzy connectedness and extraction of texture features. Section 4 describes the proposed methodology. Experimental results are discussed in Sect. 5 followed by the conclusion.

2 Literature Review

An automatic approach to locate femur based on the shape and size of the recognized objects was proposed by Thomas et al. [1]. Shan et al. proposed class separable fetal femur segmentation [2]. Gray scale opening with a priori knowledge-based structuring element was proposed by Shirmali et al. [3]. Wang et al. [4] proposed an entropy-based femur segmentation technique.

A semiautomatic technique based on active contour model was proposed by Chalana et al. [5] wherein the seed point and rough skull edge are given as input by the user. Hanna et al. [6] proposed a technique based on thresholding combined with Hough transform to obtain the head contour. Lu et al. [7] classified pixels using K means clustering, then morphological erosion and opening were performed for object recognition and Hough transform was used to fid head contour ellipse. Foi et al. [8] proposed an ellipse model for fitting the head contour by minimizing cost function using global multi-scale, multi-start, Nelder–Mead algorithm. Stebbing [9] proposed boundary fragment model to determine centroid and contour of the skull. Sun's [10] method includes finding circular shortest path, ellipse fitting for head contour detection. Namburete et al. [11] used a machine learning framework that employs feature set, inclusive of local statistics and shape information of pixel clusters (formed using simple linear iterative clustering) to segment fetal head. Mathews et al. [12] used shape-based thresholds combined with Chamfer matching and Hough transform for head segmentation. Anto et al. [13] proposed a Random Forest method with supervised learning classification to segment the head contour for low-cost settings.

Jardim et al. [14] proposed low-order parametric deformable models combined with maximum likelihood formulation to find rough borders of head and femur followed by statistical estimation framework wherein femur is taken as three-point spline and head is taken as eight-point spline. Ponomarev et al. [15] made use of the relative brightness of skull bones compared to background for segmentation of these objects using multilevel thresholding followed by recognition of segmented objects using two introduced shape-based descriptors and Linear Support Vector Machine. Zang et al. [16] presented a multi-scale and multi-oriented filter to extract the features corresponding to fetal body structures. Head and femur are detected from texton features and multi-scale local brightness. Least square ellipse fitting is used for head detection. Boundaries are connected for femur detection.

Most of these techniques rely on using morphological processing which uses low-level features of images such as edges and intensity levels. These techniques are effective in segmenting strong signal response producing structures such as femur and head. Some of these techniques such as Chalana et al. [5] and Jardim et al. [14] consider segmentation as optimization process which can get stuck at local minima. Here in this paper, fuzzy connectedness approach has been used. This region-based segmentation utilizes global relation of pixel hanging togetherness and local adjacency relation among pixels. These relationships are characterized using texture features for efficient segmentation of not only bony structures, such as head and femur, but also abdomen.

3 Preliminaries of Texture Features and Fuzzy Connectedness Segmentation

3.1 Texture Features

Texture is the spatial arrangement of the colors or intensities in an image and is an important feature based on which segmentation and classification of region of interest can be performed. Texture analysis can be performed by two methods: structural approach and statistical approach. In structural approach texture is described using a set of texels in some regular or repeated arrangement. In statistical approach, texture is described as a quantitative value defining the arrangements of intensities in a region. Identifying the texels in real images is very tedious compared to artificially generated images [17]. Statistical approaches describing texture include local binary pattern, gray-level co-ocurrance matrices, from which features such as energy, entropy, contrast, homogeneity, and correlation can be obtained. In this work, Law's texture energy measures [18] have been used to extract the texture information of the ultrasound image.

Law's approach uses local masks to identify various types of textures in an image. A set of nine 5×5 convolution masks is used to obtain the texture energy. These 5×5 masks are computed from combinations of four vectors given by Table 1.

Table 1 Vectors used to extract Law's texture energy measures

Vector name	Vector	Purpose
L5 (level)	[1 4 6 4 1]	Center weighted local average
E5 (edge)	[− 1 −2 0 2 1]	Detecting edges
S5 (spot)	[− 1 0 2 0 − 1]	Detecting spots
R5 (ripple)	[1 − 4 6 −4 1]	Detecting ripples

Table 2 Law's texture energy masks

Nine masks		
L5R5/R5L5	L5S5/S5L5	L5E5/E5L5
E5S5/S5E5	E5E5	
S5S5	E5R5/R5E5	
R5R5	S5R5/R5S5	

The outer product of these vectors gives the 2 D convolution. Among the 16 masks which can be formed from the four vectors, some of them are symmetric pairs which can be combined by averaging, thus obtaining nine convolution masks as given in Table 2.

The convolution of the nine masks with the image is performed using Eq. (1). Let the image be H of size M × N with h(i, j) representing individual pixels at coordinates (i, j) and let the mask be F of size 5 × 5 with f(i, j) indicating the element of the mask at position (i, j). Let the output of the convolution known as feature image be given by G.

$$g(i,j) = h(i,j) * f(i,j) = \sum_{m=-2}^{m=2} \sum_{n=-2}^{n=2} h(m, n)f(i + m, j + n) \qquad (1)$$

The texture energy images which are analogous to Fourier power spectrum [19] are obtained by calculating the standard deviation around each pixel according to Eq. (2).

$$s(i,j) = \frac{1}{(2n + 1)^2} \sum_{k=i-n}^{i+n} \sum_{l=j-n}^{j+n} |g(k, l) - mean| \qquad (2)$$

where n = 3 and mean is given by Eq. (3).

$$mean = \frac{1}{MN} \sum_{ij}^{MN} g(i, j) \qquad (3)$$

3.2 Fuzzy Connectedness Segmentation

The fuzzy connectedness [20] image segmentation is able to give a better segmentation result by tackling the fuzziness of the ultrasound images [21]. The main concept of this region-based approach is characterizing the degree of local hanging togetherness of pixels within an image with their spatial relationship and intensity similarities within a particular area under consideration.

There are two relations in fuzzy connectedness technique:

- Local hanging togetherness (affinity)
- Global hanging togetherness (connectedness).

The first step in fuzzy connectedness method is assigning the degree of local fuzzy relation called affinity which describes the local hanging togetherness between any two spatially close pixels a and b.

The hard adjacency, μ_α [22] of two pixels is given by Eq. (4).

$$\mu_\alpha(a, b) = \begin{cases} 1, & \text{if } a \text{ and } b \text{ are identical or differ} \\ & \text{in exactly one coordinate by } 1 \\ 0, & \text{otherwise} \end{cases} \tag{4}$$

The affinity of two adjacent pixels depends on the homogeneity of the region and closeness of the intensity values of the pixels to the mean intensity value of the area under consideration. The affinity between pair of pixels, a and b, is given by Eq. (5).

$$\mu_k(a, b) = \mu_\alpha(a, b)(w_1 G_1(f(a) - f(b)) + w_2 G_2(a - b)) \tag{5}$$

where $G_j(x)$ may be any function, for example, Gaussian. f(a) and f(b) are intensities of pixels a and b, respectively and a and b are the positions of the pixels.

The next step in the fuzzy connectedness approach is defining the global relation called fuzzy connectedness using local affinities. The degree of connectedness between two pixels, a and b, is calculated as the largest strength of all possible paths between the given pixels as given in Eq. (6).

$$\mu_c(a, b) = \max_{p_{ab} \in P_{ab}} \mu_k(p_{ab}) \tag{6}$$

where P_{ab} denotes set of all possible paths p_{ab} from a to b. Each path between two pixels, a and b, consists of orderly arrangement of nearby pixels starting from a and finishing in b. The strength of the path is the smallest affinity of any pair of the pixels in the sequence along the path given by Eq. (7).

$$\mu_N(p_{ab}) = \min_{j=1,2..m-1} \mu_k\left(a_j, a_{j+1}\right) \tag{7}$$

The result of applying this local and global hanging togetherness of the fuzzy connectedness approach is the connectivity map. The particular object of interest is obtained from the connectivity map by thresholding process.

4 Texture-Based Fuzzy Connectedness Segmentation of Fetal 2D Ultrasound Images

Fuzzy connectedness approach has difficulty in delineating regions with weak boundaries [23]. Hence, in this work, texture guided fuzzy connectedness approach has been proposed for accurate semiautomatic segmentation of fetal organs such as femur, head, and abdomen. The user is required to input a single pixel as seed point in case of femur, two or more seed points in case of head and abdomen. All points are required to lie at the center of boundary regions of the contours. The following are the steps involved in the proposed methodology:

Step 1: Extraction of Texture Energy Images
The nine texture energy images are extracted from the input fetal ultrasound image using Law's texture energy measures as discussed in Sect. 3.1.

Step 2: Filtering of Texture Energy Measures
The nine texture energy measures are Gaussian filtered. This is to ensure that there are no spurious or sharp changes in the feature images.

Step 3: Normalization of Texture Energy images
This is performed to ensure that all the texture features have the same weights. Normalization is done by subtracting the mean from all texture energy image pixels and dividing by the standard deviation.

Step 4: Apply Fuzzy Connectedness Segmentation using Texture Energy images
The ultrasound fetal images are applied with initial seed point as per the organ or region to be segmented. The fuzzy connectedness segmentation as discussed in Sect. 3.2 is applied by finding the fuzzy affinity between the pair of pixels, a and b, based on texture features calculated as given in Eq. (8) [instead of Eq. (5)].

$$\mu_k(a,b) = \mu_\alpha(a,b)(wG(f(a) - f(b)) + w_1 G_1(E_1(a-b)) + \cdots + w_9 G_9(E_9(a-b)))$$

$$(8)$$

where w, w_1, w_2 …w_9 are weights, G,G_1 …G_9 are Gaussian functions, E_1, E_2 …E_9 are the texture energy images obtained. The texture energy features of a corresponding pixel location are being compared with its corresponding pair to find the affinity between the pixel pairs which is then used to compute the connectedness. The flowchart depicting the flow of the proposed methodology is given in Fig. 1. The fuzzy connectedness approach implementation using Dijkstra's algorithm [24] is as follows.

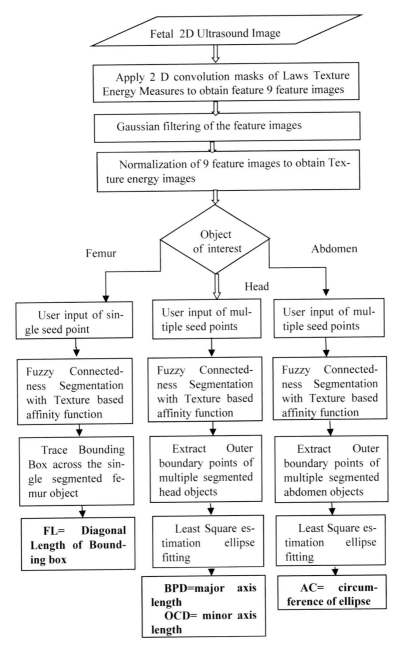

Fig. 1 Flowchart depicting the steps involved in fetal biometric measurements using texture-based fuzzy segmentation algorithm

---

Input: C(input image), o(seed point), μ_k(fuzzy affinity)

Output: K connectivity scene C_0=(C_0, f_0(connectedness value))

Auxillary data: A priority queue Q of pixels

Make the elements of C_0 to 0 except o(seed point) which is made 1

Push o(seed point) to Q

While Q is not empty do

Select and Remove a pixel c from Q for which f_0(c) is maximal

For every pixel e such that μ_k(c,e)>0 do

f_{val}= min(f_0(c), μ_k(c,e))

if f_{val}>f_0(e) then

f_0(e)=f_{val}

Update e in Q(or push if not yet in)

end if

end for

end while

4.1   Fetal Biometric Measurements

The user is required to input the initial seed pixel for texture-based fuzzy connectedness segmentation. For femur segmentation, a single seed pixel inside the region of interest will suffice. For head and abdomen segmentation, the outer boundary is not continuous; hence, more than one seed pixel at the disconnected boundaries is necessary. All these disconnected boundaries are segmented by the proposed segmentation approach as discussed in previous section. After the segmentation process, biometric measurements are to be made to extract the fetal parameters.

Femur Length

To extract the femur length, after femur segmentation, a bounding box is traced over the segmented femur object. This bounding box is rectangular in shape due to the shape and size of the femur. The diagonal length of the bounding box gives the femur length.

Biparietal Diameter, Head Circumference, Occipital Diameter, and Abdominal Circumference

The shape of head contour is elliptical and abdominal contour is circular. The outer boundary points of the segmented head and abdominal objects are extracted and fitted with an ellipse fitting algorithm which uses least squares estimation. For head contour, the major axis length of the fitted ellipse gives the Biparietal Diameter (BPD), the minor axis length of the fitted ellipse gives the Occipital Diameter (OCD), the circumference of the fitted ellipse gives the Head Circumference (HC). For abdomen contour, the circumference of the fitted ellipse gives the Abdominal Circumference (AC) of the fetus.

5   Experimental Results and Discussions

The proposed, texture-based fuzzy connectedness segmentation of fetal ultrasound image has been implemented in MATLAB 8.4 (R2015a). The technique has been applied on 50 ultrasound images of various gestational ages (20 fetal images, 20 cranial images, 10 abdominal images). The segmentation results are compared with a manual expert delineated segmentation. The fetal biometric measurements (FL, BPD, OFD, HC, and AC) are also compared with the manual expert measurements. The results and measurements from the proposed method are tabulated in Table 3.

The evaluation metrics which are used to compare the segmentation results with the expert manual delineation are region-based metrics to obtain precision and accuracy: specificity (True Positive) and sensitivity (True Negative) and distance-based metrics such as Maximum Symmetric Contour Distance (MSD) and Root Mean Square Symmetric Contour Distance (RMSD) to access local variability between proposed segmentation results and expert delineation [25]. The evaluation metrics are given in Table 4. It can be seen that the proposed textured-based FC method gives good results for head segmentation since head appears with high intensities. Fetal abdomen and femur segmentation using proposed method gives slightly less accurate results compared to manual segmentation due to varying appearance of these structures in ultrasound images.

The segmentation results of fetal head and abdomen are given in Fig. 2. The various intermediate outputs such as after applying proposed technique, extraction of outer boundary points, and ellipse fitting results are shown. The segmentation results of fetal femur are given in Fig. 3. The L5L5 and R5R5 texture energy image formation can be clearly seen with the intermediate results of feature images, filtering, and normalization. This texture image is utilized in the texture-based fuzzy connectedness approach for efficient segmentation of fetal objects.

Table 3 Results and measurements of the proposed technique

Object of interest	Segmentation object	Ellipses from segmentation	Biometric measurements
Head	–	Yes	BPD, HC, OFD
Femur	Yes	–	FL
Abdomen	–	Yes	AC

Table 4 Quantitative Evaluation of the proposed method for various objects of interest

Object of interest	Precision (%)	Sensitivity (%)	Specificity (%)	MSD (mm)	RMSD (mm)
Head	96	98	98	2.16	1.08
Femur	70	72	98	6.3	2.04
Abdomen	89	90	98	4.6	2.4

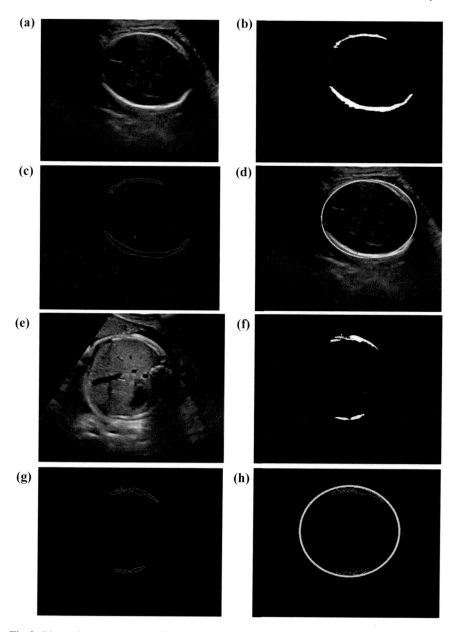

Fig. 2 Biometric measurements of fetal head and abdomen. **a** Input head ultrasound image. **b** Result of applying proposed method. **c** Extracting outer head boundary points. **d** Ellipse fitting of head contour (red) with manual contour (yellow). **e** Input abdomen ultrasound image. **f** Result of applying proposed method. **g** Extracting outer abdomen boundary points. **h** Ellipse fitting of abdomen contour (red color in comparison with manual delineation (yellow color)

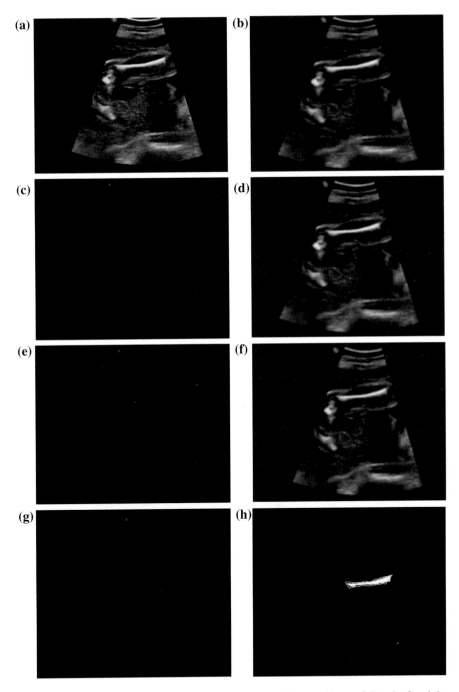

Fig. 3 Biometric measurements of fetal femur. **a** Input head ultrasound image. **b** Result of applying texture mask L5L5. **c** Result of applying texture mask R5R5. **d** and **e** Gaussian filtering of (**b**) and (**c**), respectively, (**f**) and (**g**) Normalization of (**d**) and (**e**), respectively,. **h** Fetal femur segmented output (white) in comparison with manual segmentation (in yellow)

6 Conclusion

In this paper, texture-based fuzzy connectedness approach has been proposed for accurate segmentation of fetal femur, head, and abdomen. This is a semiautomatic technique requiring users to input the initial seed pixels. The leaky nature of the fuzzy connectedness approach is eliminated by addition of texture information for the segmentation. The segmented objects are post-processed to derive the biometric measurements of fetus such as femur length, biparietal diameter, head circumference, occipital frontal diameter, and abdominal circumference. The segmentation results and measurements are evaluated against the manual delineation. The evaluation results show that the proposed technique is comparable to the manual expert segmentation with head segmentation showing higher performance. Extending the method to developing a fully automatic method for fetal biometric measurements will be beneficial to the fetal medicine community.

References

1. Thomas, J.G., Peters II, R.A., Jeanty, P.: Automatic segmentation of ultrasound images using morphological operators. IEEE Trans. Med. Imaging **10**(2) (1991)
2. Priestly Shan, B., Madheswaran, M.: Extraction of fetal biometrics using class separable shape sensitive approach for gestational age estimation. In: Proceedings of International Conference on Computer technology and development (2009)
3. Shrimali, V., Anand, R.S., Kumar, V.: Improved segmentation of ultrasound images for fetal biometry using morphological operators. In: Proceedings of International Conference of IEEE Engineering in Medicine and Biology (2009)
4. Wang, C.-W., Chen, H.-C., Peng, C.-W., Hung, C.-M.: Automatic femur segmentation and length measurement from fetal ultrasound images. In: Proceedings of Challenge US: Biometric Measurements from Fetal Ultrasound Images. ISBI (2012)
5. Pathak, S.D., Chalana, V., Kim, Y.: Interactive automatic fetal head measurements from ultrasound images using multimedia computer technology. Ultrasound Med. Biol. **23**(5), 665–673 (1997)
6. Hanna, C.W., Youssef, A.B.: Automatic measurements in Obstetrics ultrasound images. In: Proceedings of International Conference on Image Processing, vol. 3 (1997)
7. Lu, W., Tan, J., Floyd, R.C.: Fetal head detection and measurement in ultrasound images by an iterative randomized hough transform. Ultrasound Med. Biol. **31**(7), 929–936 (2005)
8. Foi, A., Maggioni, M., Pepe, A., Tohka, J.: Head contour extraction from fetal ultrasound images by difference of gaussians revolved along elliptical paths. In: Proceedings of Challenge US: Biometric Measurements from Fetal Ultrasound Images. ISBI 2012, pp. 1–3 (2012)
9. Stebbing, R.V., Manigle, J.E.: A boundary fragment model for head segmentation in fetal ultrasound. In: Proceedings of Challenge US: Biometric Measurements from Fetal Ultrasound Images, ISBI 2012, pp. 9–11 (2012)
10. Sun, C.: Automatic fetal head measurement from ultrasound images using circular shortest paths. In: Proceedings of Challenge US: Biometric Measurements from Fetal Ultrasound Images. ISBI 2012, pp. 13–15 (2012)
11. Ana, I., Namburete, L., Alison Nobel, J.: Fetal cranial segmentation in 2D ultrasound images using shape properties of pixel clusters. In: IEEE International Symposium on Biomedical Imaging (2013)

12. Mathews, M., Deepa, J., James, T., Thomas, S.: Segmentation of head from ultrasound fetal image using chamfer matching and hough transform based approaches. Int. J. Eng. Res. Technol. **3**(5) (2014)
13. Anto, E.A., Amoah, B., Crimi, A.: Segmentation of ultrasound images of fetal anatomic structures using Random Forest for low cost settings. In: Proceedings of Conference of IEEE Engineering in Medicine and Biology, vol. 6 (2015)
14. Jardim, S.V.B., Figueiredo, M.A.T.: Automatic contour estimation in fetal ultrasound images. ICIP **2**(2), 1065–1068 (2003)
15. Ponomarev, G.V., Gelfand, M.S., Kazanov, M.D.: A multilevel thresholding combined with edge detection and shape based recognition for segmentation of fetal ultrasound images. In: Proceedings of Challenge US: Biometric Measurements from Fetal Ultrasound Images, ISBI (2012)
16. Zang, L., Ye, X., Lambrou, T., Duan, W., Allinson, N., Dudley, N.J.: A supervised texton based approach for automatic segmentation and measurement of the fetal head and femur in 2 D ultrasound images. Phys. Med. Biol. **61**, 1095–1115 (2016)
17. Shapiro, L.G., Stockman, G.C.: Computer Vision. Prentice Hall (2001)
18. Laws, K.I.: Texture Image Segmentation. Ph. D. dissertation in Engineering, University of South California, Los Angeles (1980)
19. Richard, W.D., Keen, C.G.: Automated texture based segmentation of ultrasound images of the prostrate. Comput. Med. Imaging Graph. **20**(3), 131–140 (1996)
20. Udupa, J.K., Samarasekera, S.: Fuzzy connectedness and object definition: theory, algorithms, and applications in image segmentation. Graph. Model. Image Process. **58**(3), 246–261 (1996)
21. Rueda, S., Knight, C.L., Papageorghiou, A.T., Alison Noble, J.: Feature based fuzzy connectedness segmentation of ultrasound images with an object completion step. Med. Image Anal. **26**, 30–46 (2015)
22. Sonka, M., Hlavac, V., Boyle, R.: Image Processing Analysis and Machine Vision. Cengage Learning (2015)
23. Xian, M., Cheng, H.D., Zhang, Y.: A fully automatic breast ultrasound image segmentation approach based on neutroconnectedness. In: 22nd International Conference on Pattern Recognition (2014)
24. Nyul, L.G.: Fuzzy Techniques for Image Segmentation. Department of Image Processing and Computer Graphics, University of Szeged (2008)
25. Rueda, S., Fathima, S., Knight, C.L., et al.: Evaluation and comparison of current fetal ultrasound image segmentation methods for biometric measurements: a grand challenge. IEEE Trans. Med. Imaging **10**(10) (2013)

THD Analysis of Flying-Capacitor Multilevel Converter Using Fuzzy Logic Controller

M. Priya, Ponnambalam Pathipooranam and K. Muralikumar

Abstract The Flying Capacitor Multilevel Converter (FCMC) brings quite signifi-cant attention in the recent years. It is easier to extend the topology to a high number of levels when compared to other multilevel topology, namely the diode-clamped multilevel converter (DCMC). This is due to difficulties of DCMC for achieving capacitor voltage balance. On the other hand, the voltages in the capacitors of the FCMC can be con- trolled due to the availability of redundant switching states. The study of the FCMC with 5, 7, 9 and 11-levels are carried out. In order to get bet-ter performance of medium voltage (MV) and high power electric drive, the dv/dt and the Total Harmonic Distortions (THD) should be minimized by the introduction of more levels in the output voltage. The Phase-Shifted Carrier Pulse Width Mod-ulation Technique (PSC-PWM) are used to control the Root Mean Square (RMS) output voltage. For more effective, the output voltage is mainly controlled by Fuzzy Logic Controller (FLC). This FLC is intended for different voltage levels of FCMC to Control the RMS output voltages. The simulation results shows the performance of RMS voltage and significant reduction of THD is controlled by FLC.

Keywords Flying capacitor multilevel converter · PSC-PWM · THD · Fuzzy logic controller · RMS voltage

1 Introduction

Multilevel converter (MC) shows a significant role in electronics field and it is broadly used in renewable energy applications for converting DC to AC and used in industrial applications [1]. MCs not only allows the use of renewable energy sources but also

M. Priya (✉) · P. Pathipooranam · K. Muralikumar
Vellore Institute of Technology, Vellore 632014, Tamilnadu, India
e-mail: murthypriya.eee@gmail.com

P. Pathipooranam
e-mail: p.ponnambalam@gmail.com

K. Muralikumar
e-mail: kolamuralikumar@gmail.com

© Springer Nature Singapore Pte Ltd. 2020
K. N. Das et al. (eds.), *Soft Computing for Problem Solving*,
Advances in Intelligent Systems and Computing 1048,
https://doi.org/10.1007/978-981-15-0035-0_9

attains high power ratings. Some of the configurations available for developing the MCs like FCMC, which is a usage of extra capacitor clamped to power switches to provide DC voltage levels. To supply high capabilities from the inverters, especially during the power outages due to reducing the switching states by providing a clamped capacitor [2]. One of the major of advantages of using FCMC is its capacity to function at higher voltages, than the blocking capacity of each power cells and be made up of diodes and switching elements. In current coefficient of each limb is differing in polarity and equivalent. So, here not any net changes in the charge of capacitors [3]. The capacitor voltage difference and cells are maintained within a safe band and hence the capacitor voltages not having chance of unbalancing. The switching strategy of FCMC is to synthesize a sinusoidal waveform at the output and every voltage is applied at the output with the electrical angle [4, 5]. It requires amplitude at the output and application of angle gives low harmonic distortions [6]. It has switching strategies more than one, which are existing for a single voltage level. In FCMC, the balancing capacitor voltages are producing the same output voltage. Different combinations of capacitors may contain for permitting preferential charging and discharging of the capacitors [7]. It has a supple choice of capacitors, which makes simple to operate the capacitor voltages [8].

The open loop outputs are analyzed with 5, 7, 9, 11 levels of the FCMC and also the THDs are related with each other. By adopting a fuzzy controller a closed loop circuit is designed for transient response comparison of 5, 7, 9, 11 level FCMC and RMS value is performed by using a fuzzy toolbox in Simulink in builted MATLAB software.

2   Flying-Capacitor Multilevel Converter (FCMC)

2.1   5-Level FCMC

The 5-level FCMC topology is as shown in Fig. 1. A number of inverters meet the maximum voltages and control priorities with fewer harmonics at outputs. The FCMC topology has positive and negative groups and having four switching cells. For each cells, it has a connection of complementary switches that is (S_{1A}, S_{1B}),

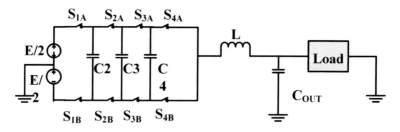

Fig. 1   5-level FCMC

Fig. 2 5-level FCMC output
voltage across the load

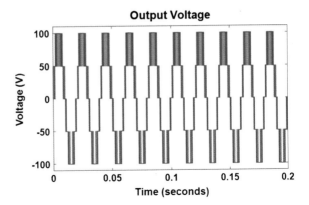

Fig. 3 FCMC 5-level FFT
analysis

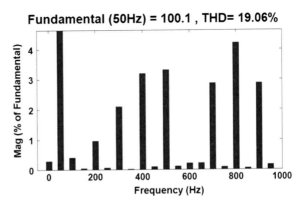

(S_{2A}, S_{2B}), (S_{3A}, S_{3B}), (S_{4A}, S_{4B}). C_2, C_3, C_4 are known as flying capacitors and it has VC_2, VC_3, VC_4 voltages respectively. Similarly, the above voltages are regulated with 3Vdc/4, Vdc/2, Vdc/4 respectively.

The output voltage of 5-level FCMC's presented in Fig. 2. For capturing the output voltage the phase disposition pulse width Modulation (PD-PWM) technique is used. This system gives appropriate voltages of the FCMC and gives suitable outputs when the system is used in the closed loop control. The equivalent method are used in the following levels of FCMCs.

The 5-level FCMC fast Fourier transform (FFT) analysis is shown in above Fig. 3. From this analysis, the THD of 5-level FCMC is 19.06% is settled. By using a appropriate filter circuit, the THD values are reduced. In order to conclude the harmonics order and harmonics magnitude on X-axis and Y-axis respectively, whereas the essential frequency is 50 Hz for the circuit in simulation.

2.2 7-Level FCMC

In 7-level FCMC circuit diagram is in Fig. 4. From the 5-level FCMC, it can extended to 7-level FCMC by increasing another two flying capacitors and switching cells

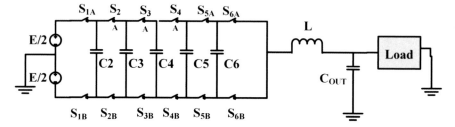

Fig. 4 7-level FCMC

Fig. 5 7-level FCMC output
voltage across the load

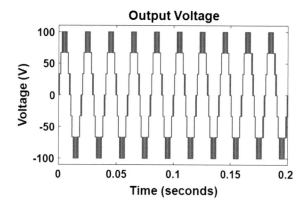

Fig. 6 FFT analysis for
7-level FCMC

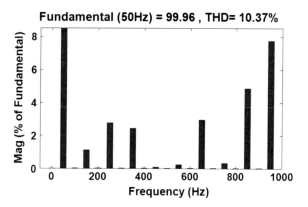

in the circuit with same value taken for the capacitors. The 7-level FCMC output
waveform is as presented in Fig. 5.

The 7-level FCMC FFT analysis is shown in Fig. 6. By this analysis, the THD
of 7-level FCMC is 10.37% is settled. By using a appropriate filter circuit, the THD
values are reduced. It can be concluded that the THD is 10.37% for 7-level FCMC
and for 5-level FCMC is 19.06% of THD. Then the decreased THD from 5-level to 7-
level FCMC is 8.69%, which is shown as harmonics order and harmonics magnitude

on X-axis and on Y-axis respectively, whereas the essential frequency is 50 Hz for the circuit in simulation.

2.3 9-Level FCMC

The 9-level FCMC circuit diagram is shown in Fig. 7. From the 7-level FCMC, it can extended to 9-level FCMC by increasing another two flying capacitors and switching cells in the circuit with same value taken for the capacitors. The 9-level FCMC output waveform is as presented in Fig. 8.

The 9-level FCMC FFT analysis is in Fig. 9. By this analysis, the THD of 9-level FCMC is 9.90% is settled. By using an appropriate filter circuit, the THD value is reduced. The THD for 5-level is 19.06% and the THD for 7-level is 10.37%. The THD is decreased after 5-level to 7-level FCMC is 8.69% and for 7-level FCMC to 9-level FCMC is decreased to 0.47%. From the circuit in order to conclude, harmonics order and harmonics magnitude on X-axis and on Y-axis respectively, whereas the essential frequency is 50 Hz for the circuit in simulation.

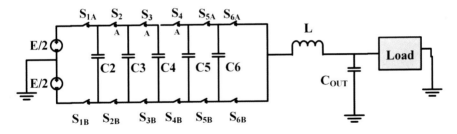

Fig. 7 9-level FCMC

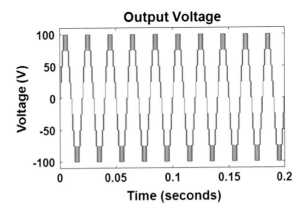

Fig. 8 9-level FCMC across the load output voltage

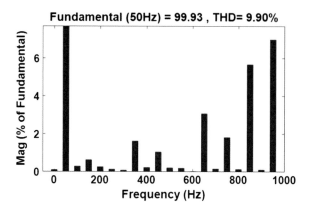

Fig. 9 FFT analysis for 9-level FCMC

2.4 11-Level FCMC

In 11-level FCMC circuit diagram is in Fig. 10. From 9-level FCMC, it can extended to 11-level FCMC by increasing another two flying capacitors and switching cells in the circuit with same value taken for the capacitors. The 11-level FCMC output waveform is as presented in Fig. 11.

The FFT analysis of 11-level FCMC is as shown in Fig. 12. From this analysis, the THD of 11-level FCMC is 7.45% is settled. By using a appropriate filter circuit, the THD value is reduced. The THD for 5-level is 19.06%, for 7-level is 10.37% and for 9-level is 9.90%. The THD is decreased from 5-level FCMC to 7-level FCMC is 8.69%, for 7-level FCMC to 9-level FCMC is de- creased to 0.47% and for 9-level FCMC to 11-level FCMC is decreases to 2.45%. From the circuit in order to conclude, harmonics order and harmonics magnitude on X-axis and on Y-axis respectively, whereas the essential frequency is 50 Hz for the circuit in simulation. Table 1 shows that the analysis of different levels of THD for different levels of FCMC. From the Table 1 to conclude with, the rate of THD is reduced when the higher level of FCMC configured.

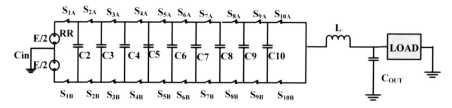

Fig. 10 11-level FCMC

Fig. 11 11-level FCMC output voltage across the load

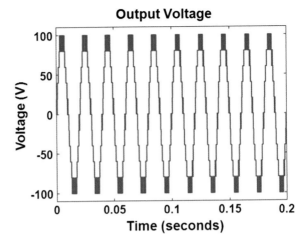

Fig. 12 FFT analysis for 11-level FCMC

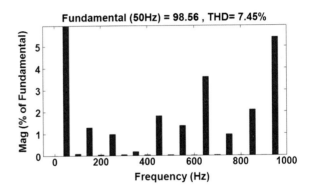

Table 1 THD analysis of different levels for FCMC

S. no	Number of levels	THD (%)
1	5	19.06
2	7	10.37
3	9	9.90
4	11	7.45

3 Fuzzy Controller for FCMC

The design of a fuzzy controller for FCMC is as presented in Fig. 13. The FLC, dissimilarly conventional controllers [9] does not involve a mathematical model of the structure has been controlled. However for understanding the systems, the control requirements are essential [10].

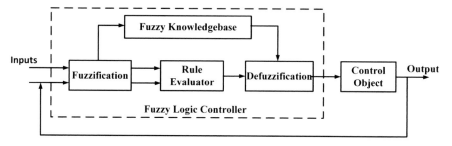

Fig. 13 Internal structure of FLC

The FLC designer needs to define what are the information's that flows into the structure, now the statistics is administered, and what information flows out of the structure [11, 12]. The FLC composed three blocks essentially. i.e. (i) Fuzzifier (ii) Inference Engine (iii) Defuzzifier.

3.1 5-Level FCMC of Fuzzy Controller

The fuzzy logic controller for the 5-level FCMC is designed. Which observes the range of the input membership function and it is different from fuzzy controller design for stack multilevel converter. The range of input membership functions is −70 to +70 for 5-level FCMC. The fuzzy controller design experiments with a simulated circuit for 5-level FCMC and the results are obtained as exposed in Fig. 14 and Fig. 15 respectively.

Figure 14 presents the corresponding output voltage for RMS value and base voltage of 5-level FCMC of Fuzzy Controller. From the output RMS value the reference voltage is noted. It is observed that the voltage at output has the settling time. Figure 15 illustrates the voltage at output across the load circuit for 5-level FCMC. The 5-level output waveform is observed to have pulse width of 0.2 s. The voltage

Fig. 14 5-level FCMC reference voltage and RMS output voltage for fuzzy controller

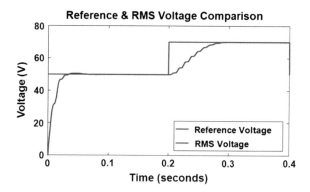

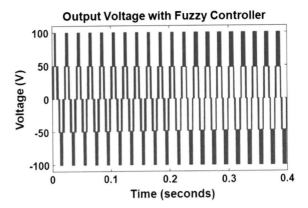

Fig. 15 5-level FCMC output voltage across the load for fuzzy controller

level is considerably low to produced 50 V of the output voltage and subsequently for 0.2 s the reference voltage is noted as 70 V. After some time the RMS output voltage settles at 70 V. The following transient response is designed for fuzzy controller experiments with 5-level FCMC is takes the reference voltage and is retained at 50 V.

3.2 7-Level FCMC of Fuzzy Controller

The fuzzy logic controller for the 7-level FCMC is designed. The limits of input member function is changed to −80 to +80. The fuzzy controller design experiments with a simulated circuit for 5-level FCMC and the results are obtained as presented in Fig. 16 and Fig. 17 respectively. Figure 16 shows the reference voltage and corresponding RMS output voltage for fuzzy controlled 7-level FCMC.

The output voltage can be observed the RMS output voltage for the base voltage but for the voltage at output has settling time. Figure 17 illustrates the output voltage across the load circuit for 7-level FCMC. The 7-level output waveform is observed to

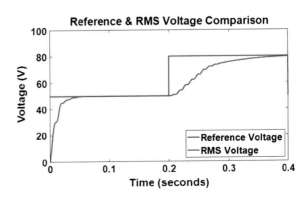

Fig. 16 7-level FCMC reference voltage and RMS output voltage for fuzzy controller

Fig. 17 7-level FCMC
output voltage across the
load for fuzzy controller

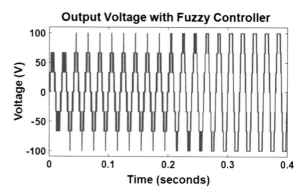

have pulse width of 0.2 s. The voltage level is considerably low to produced 50 V of
the output voltage and subsequently for 0.2 s the reference voltage is noted as 80 V.
After some time the RMS output voltage settles at 79.54 V. The following transient
response is designed for fuzzy controller experiments with 7-level FCMC is takes
the reference voltage and is retained at 50 V.

3.3 9-Level FCMC of Fuzzy Controller

The fuzzy logic controller for 9-level FCMC is designed. The limits of input member
function is changed to −90 to +90. The fuzzy controller design experiments with
a simulated circuit for 7-level FCMC and the outcomes are obtained in Fig. 18 and
Fig. 19 respectively.

Figure 18 illustrates the base voltage and corresponding RMS output voltage for
fuzzy controlled 9-level FCMC. For the reference voltage the output RMS voltage
is noted. It is observed that the voltage at output has the settling time. Figure 19

Fig. 18 9-level FCMC
reference voltage and RMS
output voltage for fuzzy
controller

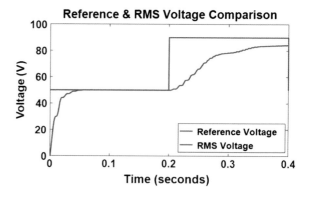

Fig. 19 9-level FCMC output voltage across the load for fuzzy controller

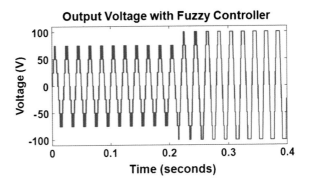

illustrates the voltage at output across the load circuit for 9-level FCMC. The 9-level output waveform is observed to have pulse width of 0.2 s. The voltage level is considerably low to produced 50 V of the output voltage and subsequently for 0.2 s the reference voltage is noted as 90 V. After some time the RMS output voltage settles at 84.4 V. The following transient response is designed for fuzzy controller experiments with 9-level FCMC is takes the reference voltage and is retained at 50 V.

3.4 11-Level FCMC of Fuzzy Controller

The fuzzy logic controller for the 11-level FCMC is designed. The limits of input member function is changed to −100 to +100. The fuzzy controller design experiments with a simulated circuit for 11-level FCMC and the outcomes are obtained as illustrated in Fig. 20 and Fig. 21 respectively.

Figure 20 illustrates the referred base voltage and corresponding output RMS voltage for fuzzy controlled 11-level FCMC. For the referred base voltage the output RMS voltage is noted. It is observed that the output voltage has the settling time.

Fig. 20 11-level FCMC reference voltage and RMS output voltage for fuzzy controller

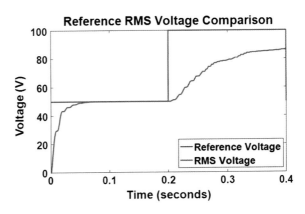

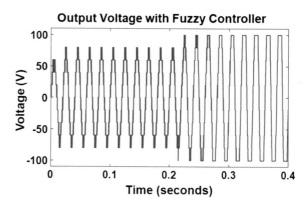

Fig. 21 11-level FCMC output voltage across the load for fuzzy controller

Figure 21 illustrates the output voltage across the load circuit for 11-level FCMC. The 11-level output waveform is observed to have pulse width of 0.2 s. The voltage level is considerably low to produced 50 V of the output voltage and subsequently for 0.2 s the reference voltage is noted as 100 V. After some time the RMS output voltage settles at 86.7 V. The following transient response is designed for fuzzy controller experiments with 11-level FCMC is takes the reference voltage and is retained at 50 V.

4 Conclusion

This Multi-Level converter is evaluated for various levels and their outputs are taken into consider with their THD. The PSC-PWM technique is used in the circuit to obtain improved voltage balancing with capacitors. The performance of 5-level, 7-level, 9-level and 11-level FCMC is studied. The THD is obtained for 5-level FCMC is 19.06%, for 7-level FCMC is 10.37%, for 9-level FCMC is 9.90% and for 11-level FCMC is 7.45% respectively. Suitable fuzzy controller is designed for all the FCMC's and the performance analysis are verified. Hence the FLC performance is reduced error to zero in the circuit by taking response from the output.

References

1. Lai, J.-S., Peng, F.Z.: Multilevel converters-a new breed of power converters. In: Industry Applications Conference, 1995. Thirtieth IAS Annual Meeting, IAS'95, Conference Record of the 1995 IEEE, vol. 3, pp. 2348–2356. IEEE (1995)
2. McGrath, B.P., Holmes, D.G.: Analytical modelling of voltage balance dynamics for a flying capacitor multilevel converter. In: Power Electronics Specialists Conference, 2007. PESC 2007, pp. 1810–1816. IEEE (2007)
3. Muralikumar, K.: Analysis of fuzzy controller for H-bridge flying capacitor multilevel converter. In: Proceedings of Sixth International Conference on Soft Computing for Problem Solving: SocProS 2016, vol. 1, p. 307. Springer (2017)

4. Reddy, V.P., Muralikumar, K., Ponnambalam, P., Mahapatra, A.: Asymmetric 15-level multi-level inverter with fuzzy controller using super imposed carrier PWM. In: Power and Advanced Computing Technologies (i-PACT), 2017 Innovations, pp. 1–6. IEEE (2017)
5. Viswanath, Y., Muralikumar, K., Ponnambalam, P., Kumar, M.P.: Symmetrical cascaded switched-diode multilevel inverter with fuzzy controller. In: Soft Computing for Problem Solving, pp. 121–137. Springer, Singapore (2019)
6. Cecati, C., Ciancetta, F., Siano, P.: A multilevel inverter for photovoltaic systems with fuzzy logic control. IEEE Trans. Industr. Electron. **57**(12), 4115–4125 (2010)
7. Shanthi, B., Natarajan, S.P.: FPGA based fuzzy logic control for single phase multilevel inverter. Int. J. Comput. Appl. **9**(3), 10–18 (2010)
8. Meynard, T.A., Fadel, M., Aouda, N.: Modeling of multilevel converters. IEEE Trans. Ind. Electron. **44**(3), 356–364 (1997)
9. Ruderman, A., Reznikov, B., Margaliot, M.: Analysis of a flying capacitor converter: a switched systems approach. In: 13 EPE-PECM Conference Proceedings (2008)
10. Self, K.: Designing with fuzzy logic. IEEE Spectr. **27**(11), 42–44 (1990)
11. Ponnambalam, P., Muralikumar, K., Vasundhara, P., Sreejith, S., Challa, B.: Fuzzy controlled switched capacitor boost inverter. Energy Procedia **1**(117), 909–916 (2017)
12. Rabee'H, A., Salih, S.M.: Control of cascade multilevel inverter using fuzzy logic technique. In: 2010 2nd IEEE International Symposium on Power Electronics for Distributed Generation Systems (PEDG), pp. 96–101. IEEE (2010)

Adaptive Neuro-Fuzzy Inference System for Predicting Strength of High-Performance Concrete

L. V. Prasad Meesaraganda, Nilarghya Sarkar and Nilanjan Tarafder

Abstract This study examines the performance of Adaptive Neuro-Fuzzy Inference System (ANFIS) for estimation of compressive strength of High-Performance Concrete (HPC) from given mix proportion. An ANFIS model merges advantages of both ANN and Fuzzy Logic. A total of 54 experimental datasets were used, where 36 datasets were used in training and 18 datasets were used for validating the model. Six input parameters include water binder ratio, age of testing, silica fumes, coarse and fine aggregate and superplasticizer, whereas compressive strength is the single output parameter. The experimental and obtained results were compared. The result illustrates that ANFIS model can be used as an alternative method to predict the compressive strength of high-performance concrete.

Keywords High-performance concrete · ANFIS model · Training and testing · Compressive strength

1 Introduction

Concrete is a composite material. It is most extensively used in the construction industry as construction material due to its good durability and strength criteria. Due to the rapid growth of construction industry in recent years, the massive volume of concrete is produced. Almost 10 billion tonnes (12 billion tonnes) of concrete is used in construction industry every year. Therefore, human safety is one of the important criteria which must be satisfied while producing a large volume of concrete. To fulfil this condition, maintenance of strength and durability criteria of the concrete is a major principle which must be satisfied. So effort was made to develop High-Performance Concrete (HPC) which mainly emphasis on strength and durability criteria. HPC focuses mainly on the strength and durability criteria. Certain characteristics of conventional concrete were developed to achieve HPC. Characteristics of HPC involve high strength and high durability (water absorption, permeability, acid

L. V. Prasad Meesaraganda (✉) · N. Sarkar · N. Tarafder
Civil Engineering Department, NIT Silchar, Silchar, Assam, India
e-mail: prasadsmlv@gmail.com

© Springer Nature Singapore Pte Ltd. 2020
K. N. Das et al. (eds.), *Soft Computing for Problem Solving*,
Advances in Intelligent Systems and Computing 1048,
https://doi.org/10.1007/978-981-15-0035-0_10

attack, etc.) as well as service life [1, 2]. HPC can be used to build more serviceable structure at comparatively lower cost.

Different soft computing techniques are very useful in solving complex problems. But the accuracy level can be increased if we combine two or more different soft computing techniques. In this study, the neural network is combined with fuzzy logic to form ANFIS model, which is a hybrid intelligent model and provides better accuracy. In this hybrid model, neural network learning and recognizing pattern combine with fuzzy logic's human-like reasoning technique [3].

2 Literature Review

Yeh [1] gave a technique of optimizing the mix proportions of HPC for specified workability and compressive strength, with the help of the artificial neural network. To determine the compressive strength, models were developed. Other parameters such as initial slump, slump after 45 min, initial slump flow and slump flow after 45 min were also predicted. Tesfamariam and Najjaran [4] suggested the use of ANFIS for estimation of concrete strength. The performance of ANFIS is checked with experimental concrete mix proportion datasets obtained from the works, and according to which range of value of r_2 range is determined which lies from 0.970 to 0.999. Parichatprecha et al. [5] aimed at determining the effect of the amount of water–binder ratio, water, cement as well as the use of mineral admixture (silica fume and fly ash) as a fractional replacement of cement on the durability criteria of HPC with the help of ANN. The model which is prepared has the capability to predict data within its range, but this range can be easily increased by retraining the proposed model with furthermore input data. Ozcan et al. [6] predicted compressive strength of concrete with silica fume by using Artificial Neural Network (ANN) and fuzzy logic. 3, 7, 28, 180 and 500 days compressive strength were foreseen using Fuzzy Logic and ANN model. The results concluded that models of ANN and fuzzy logic may be used to calculate the compressive strength for this type of concrete.

Sobhani et al. [7] developed artificial neural network and ANFIS model which are used for forecasting of the compressive strength of concrete at 28 days with no-slump. The result shows the proposed models of ANN and ANFIS are more feasible for forecasting the compressive strength. Muthupriya et al. [8] developed two types of HPC with fly ash containing silica fume and metakaolin. Input variable includes a specimen, cement silica fume, fly ash, water, sand, coarse aggregate and superplasticizer, whereas output variable contains compressive strength. Two neural network models are well prepared, namely, ANN-I and ANN-II. ANN-I contents one hidden layer and ANN-II contents to hidden layer. A comparative study was done between these two models. Chou and Pham [9] proposed an ensemble model to predict HPC's compressive strength. The performance of Support Vector Machine (SVM), ANN, classification and regression trees, etc., were applied for the development of this ensemble models. Analysed results illustrate that best expectation performance was obtained by establishing a combination of two or more models using the ensemble technique.

Gayathri et al. [10] developed a model using ANN for the prediction of Flexural strength, split tensile strength and compressive strength of HPC. Multilayer feedforward ANN model is used with 50,000 epochs. The result shows that ANN has a high possibility for prediction of split tensile test, compressive strength at 28 days and flexural strength of HPC. Reddy [11] developed three different models on ANFIS, ANN and Multiple Linear Regression (MLR) for the calculation of compressive strength at 28 days for recycled aggregate concrete. 14 different input parameters were taken. These parameters were used for the study of compressive strength at 28 days. The current study suggested that ANN and ANFIS model are more precise for calculation of compressive strength than multiple linear regressions.

From the extensive literature review it is identified that limited amount of research has been produced in the direction of high-performance concrete using analytical modelling. So this research work focuses on using ANFIS model to reduce the cost of construction, conserve energy and minimizing the wastage of material for achieving the required strength and by making number of trial mixes of concrete. The results reveal that experimental data can be estimated to a notably close extent Via ANFIS model.

3 Research Significance

The use of ANFIS model in predicting the material properties of high-performance concrete is a new approach. The neural network approach is somewhat accurate for a wide range of data. But it is not always possible to generate a huge amount of data in the laboratory due to various problems such as cost inefficiency, shortage of availability of raw materials, time-consuming, the requirement of large number of labours, etc. ANFIS model provides better accuracy where there is a shortage of data. ANFIS model involves self-adaptability for good accuracy due to the involvement of both the properties of neural network and fuzzy logic to the model. The primary objective of this work is to propose an ANFIS model with experimental data and to find the optimum amount of mineral admixture required. Also, the proposed model can be used for accurately predicting the data in future.

4 Methodology and Materials

4.1 Materials

In this work, cement which is used is 43Garde OPC conforming to IS 8112-1989 [12] and the utilization of silica fumes has reduced the consumption of cement. Silica fume having a bulk density of 750–850 kg/m3 and surface area 20,000 m2/kg. Fine aggregate of Zone-II was used in the experiment [13, 14] and specific gravity and water absorption were obtained as 2.52 and 1.15%, respectively. Coarse aggregate used has a nominal size of 20 mm [14] and its specific gravity along with water absorption was obtained as 2.61 and 0.46%. In this thesis, naphthalene-based water-reducing admixture (superplasticizer) as per IS 9103:1999 and ASTM C 494 was used [15].

5 Laboratory Test

The mix design was prepared using ACI 211.4R [16, 17]. High-performance concrete was developed for 50, 55 and 60 MPa strength and silica fumes were added with cement with 0, 2.5, 5, 7.5, 10 and 12.5% of initial cement content. For each trial mix, three cubes were casted. Compressive strength test is one of the fundamental tests of concrete. It provides an idea about overall characteristics of concrete. Different factors govern the compressive strength of concrete. They are quality of raw materials, water–binder ratio, controlled environmental conditions, use of mineral and chemical admixtures, etc. [18]. Compressive strength was tested in the laboratory for 28 days, 56 days and 90 days. A total of 54 experimental datasets were used, where 36 datasets were used in training and 18 datasets were used for validating the model. 24 h water absorption test was conducted for checking the durability criteria of concrete.

6 ANFIS Architecture

ANFIS has been primarily used for this project. Here fusion of fuzzy logic and ANN takes place. Artificial Neural Network (ANN) is a computing tool based on the process of genetic neural networks. The ANN techniques are applicable to the problems of civil engineering, because of their potentiality of learning straightly from examples. The correct response to deficient work, their ability to extract the results from minimal data and their generalized results production are the other important properties of ANN [19]. The capabilities mentioned above give rise to ANN, a very commanding mechanism to determine solution for several engineering problems, where data is insufficient [20]. The basic idea of ANN-based mathematical model for material performance is to educate an ANN system using that material for series of experiments including enough information in the results, about the material's behaviour and to succeed as a material model [21]. Such trained ANN system replicates the outcome of experiments and also able to estimate the outcome in other experiments through their simplification potential [22]. Fuzzy logic is an approach for computing based on the degrees of truth, instead using true or false logic. It is important to generate efficient fuzzy if-then rule in modelling because they directly affect the performance of the system [23]. This system includes a set of fuzzy linguistic rules which may be given by experts. These can also be taken out from mathematical data [24]. ANFIS mainly deals with Takagi–Sugeno fuzzy logic and this method comprises the hybrid system. ANFIS model is chosen as the backbone for validating experimental work. The detail description of ANFIS model is provided below (Fig. 1).

Figure 2 and Fig. 3 illustrates the architecture and structure of an ANFIS model having two input parameters and the mechanism of fuzzy-reasoning, respectively. For straightforwardness, let us adopt that the Fuzzy Inference System (FIS) has double input parameters u and v and single output parameter z. In the first-order fuzzy model of Takagi–Sugeno, a simple instruction set with dual IF-THEN rules is provided below:

$$\text{Rule I : If } u \text{ is } C_1 \text{ and } v \text{ is } D_1, \text{ then } g_1 = m_1 x + n_1 y + o_1 \qquad (1)$$

$$\text{Rule II : If } u \text{ is } C_2 \text{ and } v \text{ is } D_2, \text{ then } g_2 = m_2 x + n_2 y + o_2 \qquad (2)$$

Here u and v are inputs, C_1, D_1, C_2 and D_2 are fuzzy set and g_1 and g_2 are crisp function. Here g_1 and g_2 are the polynomial functions for the input variables of u

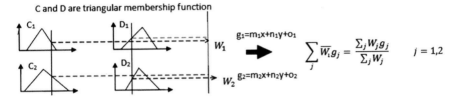

C and D are triangular membership function

Fig. 1 Functioning of ANFIS model

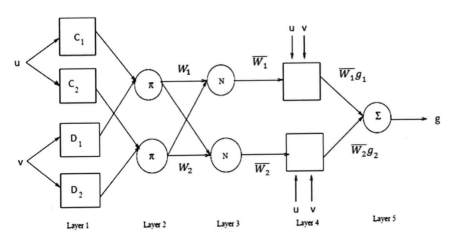

Fig. 2 Structure of ANFIS

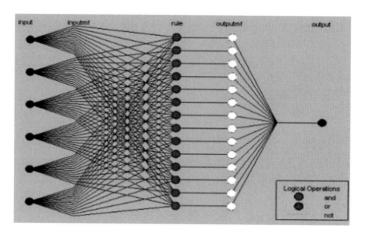

Fig. 3 Architecture of ANFIS model

and v. Here, m_1, m_2, n_1, n_2, o_1 and o_2 are design parameters which are determined at the time of training process. When $g(u, v)$ is a constant function, the Sugeno fuzzy model which is developed is of order zero. This can be treated as a distinctive case of Mamdani FIS. The above two rules state the first-order Sugeno FIS. In the figure below, square nodes are represented as an adaptive node and circular nodes are represented as a fixed node. The corresponding ANFIS structure shown as follows:

Layer 1: Each node 'j' of the layer is a square adaptive. The output of this layer is the fuzzy value of the original crisp value provided as input.

$$O_{(1,j)} = \mu C_j(u), \quad j = 1, 2 \tag{3}$$

$$O_{(1,j)} = \mu D_{j-2}(v), \quad j = 3, 4 \tag{4}$$

Here, u and v are input to node j and C_j and D_{i-2} are linguistic (fuzzy sets: small, medium, large, etc.). For Gaussian membership function

$$\mu C_j = \exp\left[-\left(\frac{x - e_j}{d_j}\right)^2\right]$$

where d_j and e_j are the parameters of membership function.

Layer 2: Each node of the layer is fixed circular and levelled as π. Here AND operator is used for the fuzzification of input. Output in this layer is presented by

$$O_{(2,j)} = W_i = \mu C_i(x) * \mu D_i(y), \quad i = 1, 2 \tag{5}$$

These are known as rule's firing weight.

Layer 3: All the nodes in the layer is a fixed circular, which is represented with N. In this layer, ratio of firing weight to all the firing weight is done.

$$O_{(3,j)} = \overline{W_i} = \frac{W_j}{W_1 + W_2}, \quad j = 1, 2 \tag{6}$$

The output of this layer is identified as normalized firing weight.

Layer 4: The nodes of this layer are square adaptive node. The output is the multiplication of first-order Sugeno model (polynomial function of the first order) and the normalized firing weight.

$$O_{(4,j)} = \overline{w_i} f_j = \overline{W_i}(m_1 u + n_1 v + o_1), \quad j = 1, 2. \tag{7}$$

where $\overline{w_i}$ is the Layer 3 output and p_1, q_1, r_1 are design parameters.

Layer 5: It is a single circular fixed node which is denoted by $\sum$. This node provides the output of ANFIS model.

$$O_{(5,j)} = \sum_j \overline{W_i} g_j = \frac{\sum_j W_j g_j}{\sum_j W_j} \tag{8}$$

In this Fuzzy Logic, the key learning rule is the backpropagation gradient descent and in which error signals are calculated recursively to the input layer from the output layer. It is same as the general feedforward neural network system. The recent development in the ANFIS is the hybrid learning method, which uses both gradient descent and least square methods to find practicable set of antecedent and consequent parameters and this latest method has been implemented in this study for the ANFIS model developed. Fuzzy Logic Toolbox with MATLAB version 7.1 was used in this method. The relationship of compressive strength as an output and six inputs that are water–powder ratio, fine aggregate, coarse aggregate, glass fibre and superplasticiser in ANFIS is shown in Fig. 4.

Similarly, Fig. 5 shows input membership function before training, Figs. 6, 7, 8, 9, 10 and 11 show membership function after learning for all six input parameters, respectively, and Fig. 12 shows the rule viewer showing first 30 rules of total 729 rules.

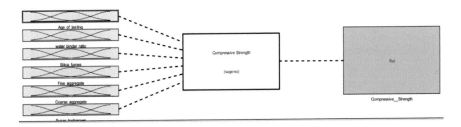

Fig. 4 Input–output data

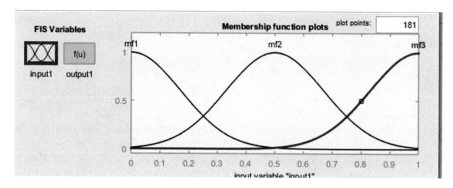

Fig. 5 Membership function before training

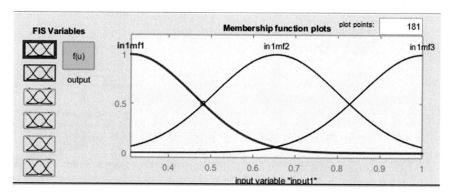

Fig. 6 Membership function for age of testing

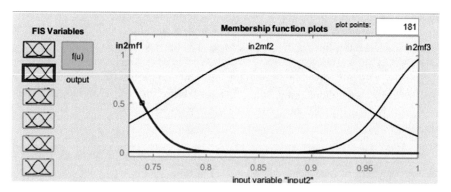

Fig. 7 Membership function for water–binder ratio

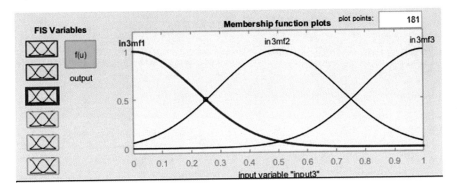

Fig. 8 Membership function for silica fumes

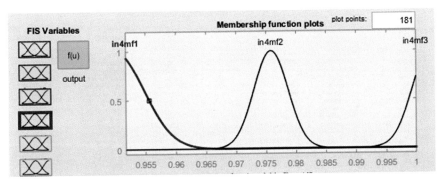

Fig. 9 Membership function for fine aggregate

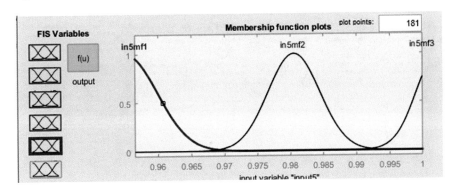

Fig. 10 Membership function for coarse aggregate

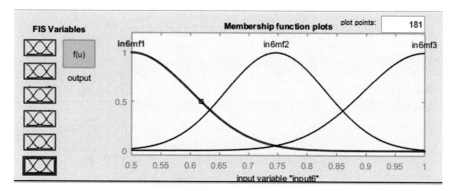

Fig. 11 Membership function for superplasticizers

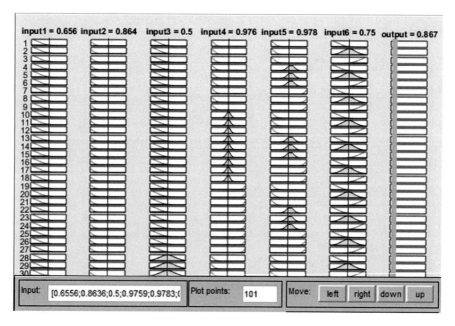

Fig. 12 Rule viewer

These rules were made during the training of ANFIS model. The calculation of the output (compressive strength) is done with the help of these rules. These rules can be made either by training with available input data or manually developing the rules. These rules can be edited, changed removed and added according to the requirement of the model with the help of rule editor.

7   Training and Validation of ANFIS Model

Training and testing data in ANFIS model set is used for the establishment of the relation between input and output. In this project, ANFIS model is developed to predict one output from a combination of six inputs. For this model, 36 experimental data are used for training ANFIS model, whereas 18 experimental data were used in testing. The experiment was conducted with 50, 55 and 60 MPa strength of concrete. Silica fumes were added with cement with 0, 2.5, 5, 7.5, 10 and 12.5% of initial cement content. Compressive strength was tested in the laboratory for 28 days, 56 days and 90 days. ANFIS model was prepared with six input parameters and one output parameter. Six input parameters as shown in Table 1 include (a) Age of testing; (b) Water–binder ratio; (c) Silica fumes; (d) Coarse aggregate; (e) Fine aggregate and (e) Superplasticizer. Compressive strength is the output parameter.

Training of ANFIS model was done with 0, 2.5, 7.5 and 12.5% content of silica fumes for 50 MPa concrete for 28, 56 and 90 days; 2.5, 5 10 and 12.5% content of silica fumes for 55 MPa concrete for 28, 56 and 90 days; 0, 5, 7.5 and 10% content of silica fumes for 60 MPa concrete for 28, 56 and 90 days; whereas experimentation was done with 5 and 10% content of silica fumes for 50 MPa concrete for 28, 56 and 90 days; 0 and 7.5% content of silica fumes for 55 MPa concrete for 28 days, 56 days and 90 days; 2.5 and 12.5% content of silica fumes for 60 MPa concrete for 28 days, 56 days and 90 days. During experimentation, 10% was obtained as the optimum quantity of silica fume.

The data provided for training the ANFIS model is arranged in ascending order with respect to the output parameter. Normalized format is applied to present the data used for training and testing. The values of parameter used in ANFIS model are shown in Table 2 and the normalized data used for testing are shown in Table 3. The predicted result for training data and testing data are shown in Figs. 13 and 14.

Table 1 Boundary range of experimental data

Component content	Minimum	Maximum	Range	Mean	Standard deviation
Water–binder ratio	0.265	0.364	0.099	0.312	0.028
Silica fumes (kg per cum)	0	71.25	71.25	32.81	22.79
Aggregates finer (kg per cum)	690	725	35	708.33	14.47
Aggregates coarser (kg per cum)	1100	1150	50	1126.7	20.74
SP (l L per cum)	4	8	4	5.922	1.296
Testing interval in days	28	90	62	58	25.589
Compressive strength (MPa)	56.12	82.3	26.18	68.84	7.18

Table 2 Values of parameter used in ANFIS model

Input layer neurons	6 (number)
Output layer neurons	1 (number)
Hidden layer	4 (number)
Membership functions (MFs) for each input	3 (number)
Input MF type	Gaussmf
Output MF type	Linear
Optimum method	Hybrid
Error tolerance	0.5
Epochs	2500
Number of rules	729

Table 3 Normalized data used for testing of ANFIS model

Sl. No.	Age of testing	Water–binder ratio	Silica fumes (kg/m^3)	Fine aggregates (kg/m^3)	Coarse aggregates (kg/m^3)	SP (l/m^3)
1	0.311111	0.952381	0.336842	0.951724	1	0.55
2	0.311111	0.903837	0	0.97931	0.982609	0.65
3	0.311111	0.909091	0.673684	0.951724	1	0.6
4	0.311111	0.840786	0.552561	0.97931	0.982609	0.725
5	0.622222	0.952381	0.336842	0.951724	1	0.55
6	1	0.952381	0.336842	0.951724	1	0.55
7	0.622222	0.903837	0	0.97931	0.982609	0.65
8	0.622222	0.909091	0.673684	0.951724	1	0.6
9	0.311111	0.798093	0.2	1	0.956522	0.9
10	1	0.909091	0.673684	0.951724	1	0.6
11	1	0.903837	0	0.97931	0.982609	0.65
12	0.311111	0.727151	1	1	0.956522	1
13	0.622222	0.840786	0.552561	0.97931	0.982609	0.725
14	1	0.840786	0.552561	0.97931	0.982609	0.725
15	0.622222	0.798093	0.2	1	0.956522	0.9
16	0.622222	0.727151	1	1	0.956522	1
17	1	0.798093	0.2	1	0.956522	0.9
18	1	0.727151	1	1	0.956522	1

Fig. 13 Predicted result for training data

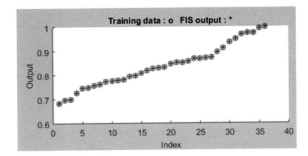

Fig. 14 Predicted result for testing data

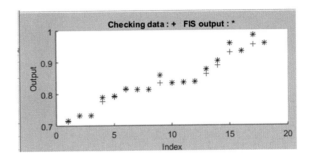

8 Result and Discussion

ANFIS model was trained with 36 experimental data and 18 experimental data were used for validation. 2500 epochs were used during the training process as shown in Figs. 15 and 16. The regression (R^2) value of training data was 99.99% and testing data was 98.87%. Below graph shows a correlation of both training and testing data. It is clear from the above table that data predicted by ANFIS model is very close to the experimental result. Statistical values provided in the table below shows the

Fig. 15 Experimental and predicted strength during the training phase

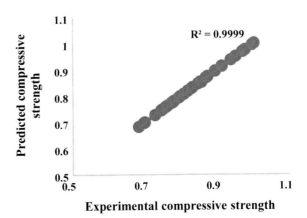

Fig. 16 Experimental and predicted strength during the testing phase

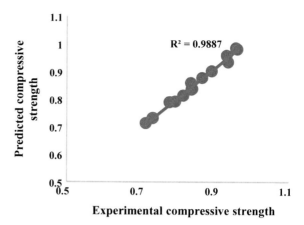

Table 4 Performance of ANFIS model

ANFIS database	R^2	MAD	MSE	RMSE	MAPE
Training	0.9956	0.020581	0.004621	0.067974	0.030226
Testing	0.9887	0.711677	1.391394	1.179574	0.967742

efficiency of ANFIS model. As the training input increases, the model will be able to perform more accurate prediction of the output parameter. During the selection of data properties of each node must not differ.

Table 4 shows R^2 (Regression), Mean Average Deviation (MAD), Mean Square Error (MSE), Root Mean Square Error (RMSE) and Mean Average Percentage Error (MAPE). The results in Table 4 show that these values are within the range. The overall performance of the proposed ANFIS model is accurate. This ANFIS model can be implemented for the prediction of HPC's compressive strength. The input values must be chosen within the range input provided during training. This range can be increased by training with additional new input data.

9 Conclusion

From the investigations, a model was developed to predict compressive strength (28 days, 56 days and 90 days) of HPC by using ANFIS and the experimental dataset is used for training and testing the ANFIS model. Training is done with 36 datasets and testing is done with 18 datasets. Input MF was taken as gaussmf and output membership function was taken as linear. A simple feedforward backpropagation technique was used to model problems involving non-linear variables.

The developed model R, MAD, MSE, RMSE and MAPE values are calculated for training and testing, found that all are within the permissible limits. The regression (R^2) value of training data was 99.56% and testing data was 98.87%. Statistical parameters for proposed ANFIS model includes 0.020581 (MAD), 0.004621 (MSE), 0.067974 (RMSE), 0.030226 (MAPE) for training phase and 0.711677 (MAD), 1.391394 (MSE), 1.179574 (RMSE), 0.967742 (MAPE) for testing phase.

The concrete industry can take advantage of the proposed models to obtain a reliable estimate of the elastic modulus from high-performance concrete. ANFIS model can predict compressive strength of concrete with satisfactory performance. This process reduces time, labour and wastage of material. The above-proposed model can be extended and improved by adding a number of input parameters as well as increasing their range of data if sufficient amount of data is available.

Acknowledgements I would like to express my deep sense of gratitude to Director, National Institute of Technology Silchar in anticipation of his kind support by providing Technical Support and I express my heartfelt gratitude to TEQIP III NIT Silchar for providing financial assistance and encouragement to do Innovative research.

References

1. Yeh, I.C.: Design of high performance concrete mixture using neural networks and nonlinear programming. Cem. Concr. Res. **28**(12), 1797–1808 (1998)
2. Smarzewski, P., Barnat-Hunek, D.: Property assessment of hybrid fiber-reinforced ultra-high-performance concrete. Int. J. Civ. Eng. Published online on 16th February 2017, https://doi.org/10.1007/s40999-017-0145-3
3. Sadrmomtazi, A., Sobhani, J. Mirgozar, M.A.: Modeling compressive strength of EPS lightweight concrete using regression, neural network and ANFIS. Constr. Build. Mater. **42**, 205–216 (2013)
4. Tesfamariam, S., Najjaran, H.: Adaptive network-fuzzy inferencing to estimate concrete strength using mix design. J. Mater. Civ. Eng. **19**(7), 550–560 (2007)
5. Parichatprecha, R., Nimityongskul, P.: Analysis of durability of high performance concrete using artificial neural networks. Constr. Build. Mater. **23**(2), 910–917 (2009)
6. Özcan, F., Atiş, C.D., Karahan, O., Uncuoğlu, E., Tanyildizi, H.: Comparison of artificial neural network and fuzzy logic models for prediction of long-term compressive strength of silica fume concrete. Adv. Eng. Softw. **40**(9), 856–863 (2009)
7. Sobhani, J., Najimi, M., Pourkhorshidi, A.R., Parhizkar, T.: Prediction of the compressive strength of no-slump concrete: a comparative study of regression, neural network and ANFIS models. Constr. Build. Mater. **24**(5), 709–718 (2010)
8. Muthupriya, P., Subramanian, K., Vishnuram, B.G.: Prediction of compressive strength and durability of high performance concrete by artificial neural networks. Int. J. Optim. Civil Eng. **1**, 189–209 (2011)
9. Chou, J.S., Pham, A.D.: Enhanced artificial intelligence for ensemble approach to predicting high performance concrete compressive strength. Constr. Build. Mater. **49**, 554–563 (2013)
10. Gayathri, U., Shanmugapriya, T., Suveka, V.: Predicting the strength of high performance concrete using artificial neural network. Int. J. Appl. Eng. Res. **11**(3), 3–8 (2016)
11. Reddy, T.C.S.: Predicting the strength properties of slurry infiltrated fibrous concrete using artificial neural network. Front. Struct. Civ. Eng.

12. Indian Standard Code IS: 8112-2013. Specifications for 43 grade ordinary Portland cement, Bureau of Indian Standards, New Delhi, India (2013)
13. Indian Standard Code IS: 2386-1997 Methods of test for aggregates for concrete, reprinted, Bureau of Indian Standards, New Delhi, India (2002)
14. Indian Standard Code IS: 383-1970. Specification for coarse and fine aggregates from natural sources for concrete, Bureau of Indian Standards, New Delhi, India (2002)
15. Prasad, M.L.V., Saha, P., Laskar, A.I.: Behaviour of self compacting reinforced concrete beams strengthened with hybrid fiber under static and cyclic loading. Int. J. Civ. Eng. 16(2), 168–179 (2018)
16. ACI Committee report ACI 211.4R-93, Guide for Selecting Proportions for High-Strength Concrete with Portland Cement and Fly Ash. Annual Book of ASTM Standards
17. Laskar, A.I., Talukdar, S.: A new mix design method for high performance concrete. Asian J. Civ. Eng. (Build. Hous.) 9(1), 15–23 (2008)
18. Reddy, T.C.S.: Predicting the strength properties of slurry infiltrated fibrous concrete using artificial neural network. Front. Struct. Civ. Eng. (2017). https://doi.org/10.1007/s11709-017-0445-3
19. Saridemir, M.: Prediction of compressive strength of concretes containing metakaolin and silica fume by artificial neural networks. Adv. Eng. Softw. 40, 350–355 (2009)
20. Kostic, S., Vasovic, D.: Prediction model for compressive strength of basic concrete mixture using artificial neural networks. Neural Comput. Appl. 26, 1005–1024 (2015)
21. Bilim, C., Atis, C.D., Tanyildizi, H., Karahan, O.: Predicting the compressive strength of ground granulated blast furnace slag concrete using artificial neural network. Adv. Eng. Softw. 40, 334–340 (2009)
22. Najigivi, A., Khaloo, A., Iraji Zad, A., Rashid, S.A.: An artificial neural networks model for predicting permeability properties of nano silica–rice husk ash ternary blended concrete. Int. J. Concr. Struct. Mater. 7(3), 225–238 (2013)
23. Allali, S.A., Abed, M., Mebarki, A.: Post earthquake assessment of buildings damage using fuzzy logic. Eng. Struct. 166, 117–127 (2018)
24. Dambrosio, L.: Data based fuzzy logic control technique applied to a wind system. Energy Procedia 126(9), 690–697 (2017)

Optimization of Target Oriented Network Intelligence Collection for the Social Web by Using k-Beam Search

Aditya Shaha and B. K. Tripathy

Abstract Target Oriented Network Intelligence Collection (TONIC) is a problem which deals with acquiring maximum number of profiles in the online social network so as to maximize the information about a given target through these profiles. The acquired profiles, also known as leads in this paper, are expected to contain information which is relevant to the target profile. TONIC problem has been solved by modeling it as search problem and using heuristics to direct the best-first search on the social graph. The problem with this approach is that in case of dense neighbors of the target profile the computation of the heuristic can be significantly expensive. In this paper, we have introduced a k-beam search Heuristic which significantly mitigates this overhead.

Keywords Social network intelligence · Heuristics search · Targeted crawling · Computation optimization · Artificial intelligence search

1 Introduction

The advent of Online Social Networks (OSN) has transformed the way humans used to communicate and have become an indispensable part of our lifestyle. These online social networks have not only transformed the way humans communicate but also revolutionized other facets of social life like target-based advertising, hiring, acquiring news, etc. With these boons online social networks are also adulterated by the perpetrators of fake news, hate speech, and unscrupulous contents. These people need to be identified and restrained from spreading such heinous content which is dangerous to the social decorum. The solution to the TONIC problem can be used as

A. Shaha
School of Computer Science and Engineering, VIT Vellore, Vellore, India
e-mail: adityapankaj.shaha2015@vit.ac.in

B. K. Tripathy (✉)
School of Information Technology and Engineering, VIT Vellore, Vellore, India
e-mail: tripathybk@vit.ac.in

© Springer Nature Singapore Pte Ltd. 2020
K. N. Das et al. (eds.), *Soft Computing for Problem Solving*,
Advances in Intelligent Systems and Computing 1048,
https://doi.org/10.1007/978-981-15-0035-0_11

an aid by the Government agencies to find these notorious profiles and report them to the respective online social networks for further actions.

With the progress of the online crawlers, it has now become possible to gain information about any user profile on the social network using an ensemble of crawling algorithms. But the increasing privacy and data breach issues have coerced the users of the online social networks to restrict the access to the limited number of profiles. Although this restriction acts as a protective wall from unexpected intrusions, the malefactor can also be protected by these provisions. TONIC problem deals with solving the case when the profile of the target has restricted its access to public and we need to gain information about this target without directly accessing its profile.

The social network phenomenon of Homophily shows that ties in a social network are formed between similar people [1]. So when the access to the target profile is not available it is safe to prognosticate that the information about the user profile can be extracted from his friends. The profiles that contain any information about the target profile are called *leads* in TONIC. The *List of Friends(LoF)* of these leads are considered as *potential leads(pl)*. The social networks enable people to share, tag and, like posts that are shared with them. Online social networks allow people to share information with their friends or circles. Although restricting access to one's profile is possible, the online social networks do not provide any mechanism to impede the access to one's friend's profile. These ideas form the basis of solution to TONIC Problem which tries to gain information of the restricted target node from all the friends of target that contain information pertaining to the target and are publicly accessible.

The TONIC problem has been dismantled into two orthogonal problems. The first problem deals with finding profiles that ostensibly contain information about the target. Such profiles are called leads. The second task is to analyze these leads and extract information about the target if such information exists. The second task, referred in this paper as profile acquisition, can be performed using standard IE and web scrapping techniques. The paper's central focus is on optimizing the task of choosing the leads to be explored next, thus by optimizing the first task of finding leads by making minimum number of API-calls, referred in this paper as budget.

The solution to the TONIC by Stern et al. [2] was given by modeling it as search problem which searched for leads in an unknown graph and whose search was guided by a set of novel heuristics. The unknown graph is discovered on the fly and the explored graph is referred to as Currently Known Graph (CKG). The problem with the above heuristics is when the list of friends is huge it is not feasible to check the efficacy of all the potential leads with every small topological change in the currently known graph. This paper tries to solve the problem when the LoF of leads are dense. Instead of acquiring a single profile at every iteration, the algorithm acquires k-top profiles using k-beam heuristic search. This paper shows that there exists a value of k for all subgraphs in an OSN for which the topological changes do not significantly affect the selection of the potential lead to be explored.

2 Related Work

Previous papers have solved TONIC with Artificial Intelligence Search Techniques, structuring the problems as a heuristic search problem performing targeted search using best-first algorithm. The heuristics use the CKG to pilot its search. The heuristics used to solve TONIC only consider the underlying topology of the CKG and hence can be dismantled into two orthogonal problems and solved independently.

In [2] BTF, Basic TONIC Framework (BTF) focuses the search on known leads and their neighbors are also called as list of friends (lead). The framework provided various heuristics like the random heuristic in which the algorithm chose the potential lead (pl) to be acquired randomly, FIFO Heuristic in which the (pl) was chosen based on first-in-first-out basis, clustering coefficient heuristic in which the (pl) was chosen based on highest known clustering coefficient, maximum degree heuristic in which the (pl) was chosen based on highest Known degree and Bayesian promising lead in which the promising factor of lead was calculated based on the probability of randomly selected a (pl) of lead being a lead. The Bayesian promising factor was proved to perform the best. In [3] the Extended TONIC Framework (ETF), ETF also allowed the search through the extended social circles of a target. This incorporated the search algorithm to perform some exploration rather than always enforcing the heuristics. Reference [4] introduced a combination of BTF and ETF. It proposed a new term ETF(n) which only explored the non-leads if they were n step away from a known lead. They empirically showed that ETF(n) for n > 1 does not fetch significant benefits as the increase in the percentage of leads degrades after n = 1. In [5] the flaws of exploration–exploitation dilemma in search were tackled with the concept of Volatile Multi-Armed bandit (VMAB) problem. Using VMAB, there was no need for an ad hoc heuristic function to guide the search in the social graph. The algorithm used past experience in the form of profile acquired previously during the search. In [6] the TONIC approach with VMAB was used to usher a social network crawler that aimed to solve exploration and exploitation dilemma. In [7] the author presented some methods of candidate selection of beam search approaches, which avoided searching all the nodes available thus optimizing the search. He successfully shows that beam search algorithm retains their advantage in efficiency while reducing the overhead of searching the entire search space. In [8] strategies of information retrieval in web crawlers were reviewed and on the basis of user customized parameters, the comparative analysis of different information retrieval strategies was performed. In [9] the efficiency of focused based crawling using best-first algorithm was analyzed. In [10] different search strategies to be used in networks which follow power law were discussed. In [11] the authors proposed a framework which helped in the management of the social network honeypots which aided the detection of Advanced Persistent Threat (APTs) at the reconnaissance phase. They showed that the attackers employ targeted crawling similar to reconnaissance to create genuine honeypots. In [9] the authors provided an analysis of the cost-effectiveness of strategies used to monitor organizational social networks and detect the socialbots that penetrate a target organization [12]. They showed that the attackers can acquire list of social network profiles either by the targeted crawling used in [2].

In this paper, we have introduced a k-beam search heuristic which can be used as an add-on to any aforementioned heuristic to reduce the computational overhead. The k-beam heuristic will return k potential leads instead of a single best lead which will be acquired first and then the topological changes will be made to the currently known graph. In case of dense neighbors, this heuristic works well because the best potential leads mostly belong to the same set of parent leads and performing repetitive computation of the heuristics is a mere computational overhead. This paper implements the k-beam search heuristic as a wrapper for Bayesian promising lead heuristic which performs best [2] among all the other heuristics.

3   Proposed k-Beam Heuristic Search Algorithm

The aim of the proposed algorithm is to minimize the number of computations required to decide the potential lead to be explored using a heuristic. The algorithm returns top k potential leads to be explored instead of the best potential lead to be explored. In case of dense graphs, where the *LoF* of the leads are outsized, the k-beam search makes it feasible to enforce the underlying heuristic on the social graph. Also, because of the density of the leads is high, the probability of the **current best lead** being the **next best lead** in successive searches is huge.

Algorithm 1 k-Beam Heuristic for Best-first Search

Input: *Target*, the target profile
Input: *Budget*, the number of allowed API-calls
Input: *InitialLeads*, the set of initial leads
Input: k, the number of potential leads acquired at a time
FoundLeads ← ϕ
  OPEN ← *InitialLeads*
  CLOSED ← ϕ
  while *(OPEN not empty AND Budget > 0)* **do**
    Kbest ← *ChooseKBest(OPEN,k)*
    Add *Kbest* to *CLOSED*
    Acquire *Kbest*
    for *best in KBest* **do**
      if *best is a lead* **then**
        Add *LOF(best)* \ *CLOSED* to *OPEN*
        Add *best* to *FoundLeads*
      end
      Decrease *Budget* by 1
    end
  end
end
return *FoundLeads*

---

where the function *ChooseKBest(OPEN,k)* can be any heuristic. In this paper, for the purposes of explanation *ChooseKBest(OPEN,k)* is *k-Beam Bayesian Promising Lead Heuristic*

3.1 Bayesian Promising Lead Heuristic

The Bayesian promising lead heuristic is an extension of the promising lead heuristic and aggregating promising lead heuristic proposed in [2]. Promising lead heuristic measures the promising factor of a lead. It is probability that a randomly drawn neighbor of *lead* is also a *lead*. The paper defines this measure as ratio of *potential leads(pl)* connected to *lead* that are *lead*.

Bayesian aggregation heuristic was introduced to overcome the shortcomings of *AvgP* heuristic which diminished the effect of a prominent lead and *MaxP* which only considered the most promising lead and ignored the rest.

$$BysP(pl) = 1 - \prod_{m \in L(pl)} (1 - pf(m))$$

where

$$pf(m) = \frac{L(m)}{NL(m) + L(m)}$$

where *L(m)* are lead neighbors of *lead m* and *NL(m)* are the non-lead neighbors of *lead m*

3.1.1 Example

In this example, a custom graph (V,E) is analyzed, having
10 *Vertices*, viz., {0, 1, 2, 3, 11, 12, 21, 22, 31, 32} and
14 *Edges*, viz., {(0, 1), (0, 2), (0, 3), (1, 2), (0, 11), (0, 12), (1, 11), (1, 12), (0, 21), (0, 22), (2, 21), (2, 22), (3, 31), (3, 32)}.
The *Target* Node is **0** and the *InitialLeads* are {1, 2, 3}. The given *budget* is 4 (Figs. 1 and 2).

3.2 k-Beam Bayesian Promising Lead Heuristic

The k-Beam Bayesian Promising Lead Heuristic (KBysP) is a wrapper on the original Bayesian promising lead heuristic proposed in [2] in which only one best potential lead to be explored was returned. But in case of KBysP top k-potential leads are

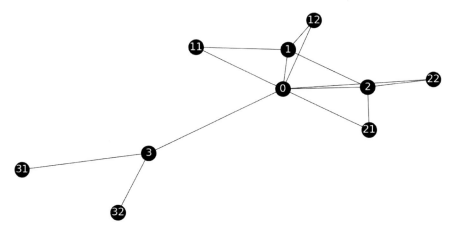

Fig. 1 The entire graph with target as 0

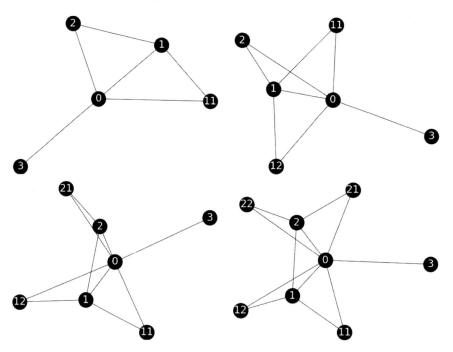

Fig. 2 Top-left graph is CKG after acquisition of node **11**, top-right graph is CKG after acquisition of node **12**, bottom-left graph is CKG after acquisition of node **21**, bottom-right graph is CKG after acquisition of node **22**

Table 1 Status of *FOUNDLEADS, CLOSED, OPEN* after every acquisition

Budget	FOUNDLEADS	CLOSED	OPEN
4	{1, 2, 3}	{1, 2, 3}	{11, 12, 21, 22, 31, 32}
3	{1, 2, 3, 11}	{1, 2, 3, 11}	{12, 21, 22, 31, 32}
2	{1, 2, 3, 11, 12}	{1, 2, 3, 11, 12}	{21, 22, 31, 32}
1	{1, 2, 3, 11, 12, 21}	{1, 2, 3, 11, 12, 21}	{22, 31, 32}
0	{1, 2, 3, 11, 12, 21, 22}	{1, 2, 3, 11, 12, 21, 22}	{31, 32}

Table 2 Status of *budget, BysP(pl), best* after every acquisition

Budget	BysP(pl)	Best
4	{32: 0.0, 11: 0.66, 12: 0.66, 21: 0.66, 22: 0.66, 31: 0.0}	11
3	{32: 0.0, 12: 0.87, 21: 0.66, 22: 0.66, 31: 0.0}	12
2	{32: 0.0, 21: 0.66, 22: 0.66, 31: 0.0}	21
1	{32: 0.0, 22: 0.80, 31: 0.0}	22
0	{32: 0.0, 31: 0.0}	ϕ

Table 3 Status of *FOUNDLEADS, CLOSED, OPEN* after every acquisition

Budget	FOUNDLEADS	CLOSED	OPEN
4	{1, 2, 3}	{1, 2, 3}	{11, 12, 21, 22, 31, 32}
3	{1, 2, 3, 11}	{1, 2, 3, 11}	{12, 21, 22, 31, 32}
2	{1, 2, 3, 11, 12}	{1, 2, 3, 11, 12}	{21, 22, 31, 32}
1	{1, 2, 3, 11, 12, 21}	{1, 2, 3, 11, 12, 21}	{22, 31, 32}
0	{1, 2, 3, 11, 12, 21, 22}	{1, 2, 3, 11, 12, 21, 22}	{31, 32}

acquired at once assuming that their acquisition does not significantly influence the acquisition of other potential leads (Tables 1 and 2).

It is experimentally shown in the paper that there is no significant difference between BysP and KBysP when the neighborhood of the target is dense. By acquiring k-leads at a time, we are optimizing the task of deciding the next potential lead which can be very expensive in case of very dense neighborhood (Tables 3 and 4).

KBysP is applied on example Sect. 3.1.1 with **k = 2** and the following results were observed (Fig. 3).

Observations

1. The results of normal BysP and KBysP are identical in case of leads found after each acquisition for the value of k = 2
2. In case of BysP the function *ChooseKBest(OPEN)* had to be called for all the acquisitions, *i.e.,* four times which in turn calculated *PromisingFactor (lead)* {3, 4, 5, 6, 7} times in case of all the acquisitions. Hence a total of 20 times. In case of KBysP the function *ChooseKBest(OPEN)* had to be called for all the

Table 4 Status of *budget*, *BysP(pl)*, *Kbest* after every acquisition

Budget	BysP(pl)	Kbest
4	{32: 0.0, 11: 0.66, 12: 0.66, 21: 0.66, 22: 0.66, 31: 0.0}	11, 12
2	{32: 0.0, 21: 0.66, 22: 0.66, 31: 0.0}	21, 22
0	{32: 0.0, 31: 0.0}	ϕ

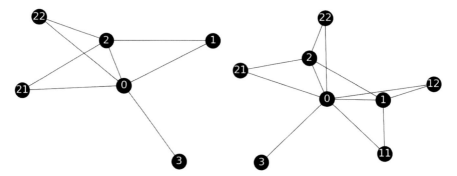

Fig. 3 Left-hand graph is the CKG after acquiring nodes {21, 22} and right-hand graph is the CKG after acquiring nodes {11, 12}

acquisitions, *i.e.,* four times which in turn calculated *PromisingFactor (lead)* {3, 5} times in case of all the acquisitions. Hence a total of eight times.

3. In case of BysP the function *CalculateBysP(pl)* was called {6, 5, 4, 3, } *i.e.,* 18 times for all the OPEN nodes. In case of KBysP *CalculateBysP(pl)* was called {6, 4}, *i.e.,* 10 times for all the OPEN nodes.

4 Evaluation

The Google+ dataset was used to analyze the verity of the algorithm. The Google+ dataset is collected by Fire et al. [11] and includes 211 K anonymized nodes with 1.5M links between them. The dataset was randomly sampled for 100 Nodes as target nodes in the experiment, having a minimum of 50 friends (leads). All the 100 target nodes have three randomly sampled leads initialized. The algorithm was executed five times with the budgets {5, 10, 15, 25, 50}.

It was observed that both BysP and KBysP had similar results in case of average number of leads found. But KBysP executed much faster and almost required **50%** less BysP function calls. The value for k in KBysP was taken to be 2.

From the Fig. 4 we can see that the performance of *KBysP* is much better than traditional *BySP*. Along with the performance Fig. 5 shows that the execution efficiency of the algorithm is also improved as the KBysP makes significantly less function calls on an average as compared to the traditional BysP.

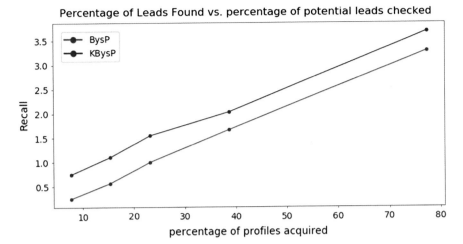

Fig. 4 Percentage of leads found versus percentage of potential leads checked

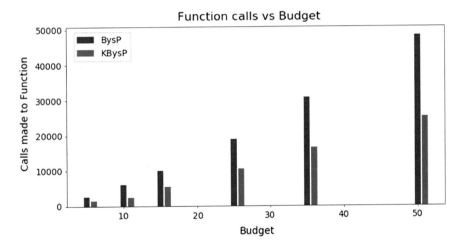

Fig. 5 Function calls versus budget

5 Conclusion

TONIC can be a very efficient tool when it comes to maximize the information about a target profile by acquiring the maximum number of leads (friends of targets) in an unknown social graph. The drawbacks of using TONIC directly with its heuristics is that the computations become extremely expensive when the density of the leads increase. We can use the k-beam search heuristic in this case, where the acquisitions are done k at a time thus improvising the speed of the algorithm. The parameter k needs to be empirically decided or can be prognosticated with some prior knowledge of the topology.

Further work would include learning k as a hyperparameter based on the reward the algorithm gets for a particular value of k. Also, k can be changed as the density decrease or increase. Learning when to change k and how much k should be changed at each epoch are some of the questions that can be tackled.

References

1. Bisgin, H., Agarwal, N., Xu, X.: World Wide Web **15**, 213 (2012). https://doi.org/10.1007/s11280-011-0143-3
2. Stern, R., Samama, L., Puzis, R., Beja, T., Bnaya, Z.: TONIC: Target oriented network intelligence collection for the social web. In: 27th AAAI Conference on Artificial Intelligence, pp. 1184–1190 (2013)
3. Samama-kachko, L., Stern, R., Felner, A.: Extended framework for target oriented network intelligence collection. In: SoCS, pp. 131–138 (2014)
4. Target Oriented Network Intelligence Collection (TONIC) By: Liron Samama-Kachko Supervised by: Dr. Rami Puzis, Dr. Roni Stern (2014)
5. Bnaya, Z., Puzis, R., Stern, R., Felner, A.: Volatile multi-armed bandits for guaranteed targeted social crawling. In: Late Breaking Papers at the Twenty-Seventh AAAI Conference on Artificial Intelligence, pp. 8–10 (2013)
6. Bnaya, Z., Puzis, R., Stern, R., Felner, A.: Bandit algorithms for social network queries. In: Proceedings-SocialCom/PASSAT/BigData/EconCom/BioMedCom 2013, pp. 148–153 (2013)
7. Xu, Z.W., Liu, F., Li, Y.X.: The research on accuracy optimization of beam search algorithm. In: 2006 7th International Conference on Computer-Aided Industrial Design and Conceptual Design, CAIDC. (2006). https://doi.org/10.1109/CAIDCD.2006.329467
8. Saini, C., Arora, V.: Information retrieval in web crawling: a survey. In: 2016 International Conference on Advances in Computing, Communications and Informatics, ICACCI 2016, pp. 2635–2643 (2016). https://doi.org/10.1109/ICACCI.2016.7732456
9. Rawat, S., Patil, D.R.: Efficient focused crawling based on best first search. In: Proceedings of the 2013 3rd IEEE International Advance Computing Conference, IACC 2013, pp. 908–911 (2013). https://doi.org/10.1109/IAdCC.2013.6514347
10. Adamic, L.A., Lukose, R.M., Puniyani, A.R., Huberman, B.A.: Search in power-law networks. Phys. Rev. E **64**, 046135 (2001)
11. Paradise, A., Shabtai, A., Puzis, R., Elyashar, A., Elovici, Y., Roshandel, M., Peylo, C.: Creation and management of social network honeypots for detecting targeted cyber attacks. IEEE Trans. Comput. Soc. Syst. **4**(3), 65–79 (2017)
12. Paradise, A., Shabtai, A., Puzis, R.: Detecting organization-targeted socialbots by monitoring social network profiles. Netw. Spat. Econ. 1–31 (2018)

Compressed Air Energy Storage Driven by Wind Power Plant for Water Desalination Through Reverse Osmosis Process

M. B. Hemanth Kumar and B. Saravanan

Abstract The need for an increase in the freshwater has led to water desalination technologies and its advancements. The expenses arise for the reverse osmosis (RO) desalination system for all the energy consumption parties has led to the implementation of RO membranes. The renewable energy sources provide a non-combustible source due to depleting conventional energy sources. In this work wind power integrated with RO systems for providing clean water by utilizing sustainable energy resource with a compressed air energy storage is proposed. This investigates the accomplishment of compressed air energy storage through an RO membrane connected to pressure vessel based on solution diffusion concept. Here the energy storage will act as a buffer due to variation in the wind velocity. The performance is validated by varying: wind velocity, storage tank volume, and RO elements, pressure inside the tank and pressure limits. The compressed air energy storage is found to provide better water production and water quality compared with conventional RO system connected with wind alone. By maintaining initial pressure in the tank and lower pressure limit the salt rejection was achieved at 98.5%. This shows the effectiveness of compressed air energy storage in combination with wind energy conversion which is better suited for water desalination process with RO.

Keywords Reverse osmosis · Compressed air energy storage · Desalination · RO membrane · Wind velocity

1 Introduction

Water desalination is an important factor for freshwater requirement by implementing RO membranes. A conventional RO consumes more energy and the expenses are also more for desalination process. The continuous usage of fossil fuels affects the

M. B. Hemanth Kumar (✉) · B. Saravanan
School of Electrical Engineering, VIT, Vellore, India
e-mail: hemanthkumar.b@vit.ac.in

B. Saravanan
e-mail: bsaravanan@vit.ac.in

© Springer Nature Singapore Pte Ltd. 2020
K. N. Das et al. (eds.), *Soft Computing for Problem Solving*,
Advances in Intelligent Systems and Computing 1048,
https://doi.org/10.1007/978-981-15-0035-0_12

climatic conditions [1] and also introduces effluents. To overcome these problems, it is necessary to utilize renewable energy source for meeting the energy requirement. Wind energy [2] coupled to compressed air energy system can be implemented for desalination process for obtaining clean water. A survey was done by the World Health Organization and it was estimated that about 1.1 billion peoples are unable to get proper drinking water and in future it may increase to 3 billion [3]. The demand for water doubles for every 20 years and the portable water production from the sea is becoming a sign of shortage in the world [4]. The salt concentration is typically around 35,000–45,000 mg/L and the actual amount for portable water must be 500 mg/L as per the Environmental protection [5]. Desalination is nothing but removing the salt particles from the available water to make it as fresh so that it can be consumed for usage. The desalination can be achieved in two ways 1. Heat exchange which consists of phase change for the solution and 2. Membrane method. With the increase in fuel consumption the heating process has been replaced by membrane method [6] in present practice. Nowadays commonly used membranes involve nanofiltration (NF), electrodialysis (ED), and reverse osmosis (RO) which accounts for about 69% being implemented in united states. Reverse osmosis involves the mixture to flow through the membrane made of porous material from one side and the other side delivers the separated product. This process is implemented since the membrane excludes the monovalent ions from the solution and allows the water to flow through it. To separate in a proper manner external pressure should be applied for overcoming osmotic pressure. If the RO membrane used is very efficient then it can able to remove majority of salt content in the water somewhere up to 99.8%. The RO performance is greatly affected by the feed water concentration and under what type of conditions it is being operated. The paper is structured as follows: in Sect. 2 importance of renewable energy is highlighted, in Sect. 3 wind energy for water desalination is focused and Sect. 4 deals with compressed air energy storage combined with wind is shown. Section 5 consist of results and followed by conclusion in Sect. 6.

2 Renewable Energy for Desalination

The continuous usage of fossil fuels for RO desalination has led to an increase in the pollutants in the atmosphere and further responsible for greenhouse effect. So, there is a need to integrate renewable energy sources to operate these systems. One of the challenges that need to be addressed is the coupling of renewable sources into desalination process in order to provide continuous power supply. The renewable energy is clean and inexhaustible and it can be used as energy source for RO process or desalination. Many researches have been done to improve the reliability [7] of renewable energy source such as wind, solar, geothermal, and tidal [8]. Due to different geographical location the availability of wind [9, 10] and solar may vary and it is used only when the renewable energy is sufficiently available till then fossil fuel conventional plants are used instead [11]. The solar PV system is used for RO application and successfully commissioned for seawater and brackish water in the countries such as Italy and Spain. Since the solar PV is having high capital cost as they require many PV arrays and require regulators. Compared to PV, Wind energy has large potential for small area and the energy production is proportional to hub height. The amount of energy drawn from grid is reduced by connecting wind when compared to solar PV system for desalination application. In [12] presented that the production cost of freshwater can be decreased up to 20% for a wind RO plant when the average wind speed is more than 5 m/s. The wind RO [13] can be installed at remote locations where there is sparsity of electric supply. But due to intermittence nature of wind it's a big challenge to provide continuous supply for desalination system. There are two basic types of energy storage mechanism, i.e., by using batteries, diesel engines, and a pressure stabilizer. So, we interconnect a compressed air energy storage to provide continuous supply when there is less wind speed and achieve continuous power generation for the RO plant to operate.

3 Wind Energy for Desalination

The RO operated from a wind energy system requires an optimum number of membranes and its operational characteristics. The modeling used here for the study is film theory modeling and it is operated by wind energy. The complete model consists of wind turbine, high-pressure pump, conductivity probes, and data acquisition system. The pressure is controlled in the membrane cell using automatic needle valve. The data is collected and named as RO control with 10 ms resolution. The wind pattern is measured and plotted in Fig. 1 and the average wind speed for this pattern is 6.75 m/s.

The feedwater flow at the membrane is calculated using the expression and the block diagram is presented in Fig. 2.

Fig. 1 Wind speed pattern for the desalination process

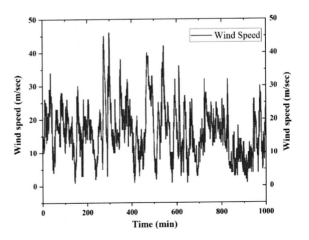

Fig. 2 Feed water flow into
the RO membrane

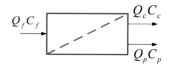

$$Q_f = \frac{R}{Q_p} \qquad (1)$$

The crossflow rate into the RO membrane is calculated as shown below

$$Q_f = Q_c + Q_p \qquad (2)$$

The total energy required to pump the feed water can be written as

$$W_{\text{pump}} = Q_f \cdot \Delta p \qquad (3)$$

Consider a wind turbine having an efficiency around 60% of power being transmitted then energy available at pump is

$$W_{\text{pump}} = 0.6 W_T \qquad (4)$$

The power produced by the wind turbine [14] at particular wind speed is given by

$$W_T = \frac{1}{2} \cdot \rho_{\text{air}} \cdot C_p \cdot A_{\text{rotor}} \cdot V_{\text{wind}}^3 \qquad (5)$$

where W_T gives the power and Rho represents the air density and it depends on surrounding conditions, A is the swept area in m² and V is the wind speed.

Due to variation in the wind speed the power generated also varies and the pump flow changes continuously. The performance of the pump depends on pump speed and the affinity law is used to examine the performance. When the impeller diameter is fixed then the relation between discharge and power of the pump is given as

$$\frac{Q_1}{Q_2} = \frac{N_1}{N_2} \text{ and } \frac{W_1}{W_2} = \frac{N_1^3}{N_2^3} \qquad (6)$$

where Q and N are the discharge rate and pump speed and it is reframed for determining the relation between discharge rate and pump power

$$\frac{Q_1}{Q_2} = \left(\frac{W_1}{W_2}\right)^{1/3} \qquad (7)$$

The presented block diagrams in Fig. 3 consists of a wind turbine connected to RO desalination system. The wind turbine generated sufficient power when the wind

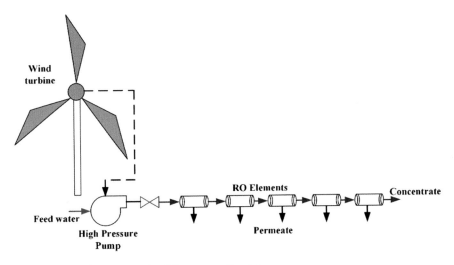

Fig. 3 Wind turbine connected to RO system without storage

speed is above rated speed to the desalination plant. The major drawback of this design is that the fluctuations in the discharge due to dynamic wind speed which is not satisfied and because of this the membranes are affected to changes in feed flow rates. As there is no storage system in the present model it cannot be in continuous operating mode when there is no wind speed and suppose operated from conventional power plant it increases the operating cost of desalination. So, in order to overcome this problem, energy storage is introduced which takes advantage of operating the RO desalination under low wind speed without depending on the energy from grid supply.

Without the energy storage, the water discharge from the pump enters the membrane which is connected in series and each membrane will get the flow from before membrane element. The salt that has been removed from the water is calculated using film theory and also permeate concentration were found. The flowchart shows the desalination process using wind energy.

Water permeability and salt permeability calculation: the salt properties of water are the characteristics for an RO membrane. The transport parameter is an important factor to study the operation of RO and the permeate flux [15] depends on permeability coefficient of water for the membrane and the salt reject ability depends on salt permeable coefficient. For determining the permeability coefficient of water, water flux experiment is conducted and during this a deionized water is pumped at a pressure of 1 psi into the membrane cell. The flowchart for the desalination process is shown in Fig. 4.

It is then operated for 2 h for obtaining membrane compaction and the permeability coefficient can be obtained based on the diffusion theory by using permeate flux expression [16] and it is given by

Fig. 4 Flow chart for
desalination process for the
wind data

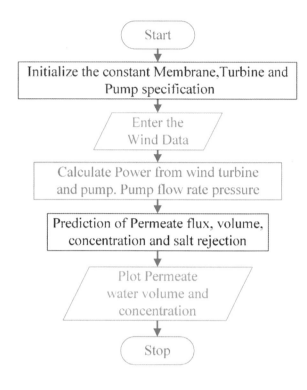

$$J_v = A \cdot (\Delta p - \Delta \pi_m) \tag{8}$$

where permeate water flux is represented as J_v, applied pressure as Δp and osmotic
pressure as $\Delta \pi_m$ in kPa.

By using solute flux equation the salt permeability for the membrane is determined. It's a function of solute concentration and salt permeability at the surface of
the membrane and permeate water and is given by

$$J_s = K_s \cdot (C_w - C_p) \tag{9}$$

where C_w and C_p represent membrane concentration and permeate salt concentration.

The mass transfer coefficient is an important parameter which decides the performance of RO membrane and is written as

$$k = 0.807 \left\{ \frac{\gamma D^2}{L} \right\}^{1/3} \tag{10}$$

where D represents the diffusion coefficient and k gives the mass transfer coefficient
in (m/s), and L is the channel length in m.

The Fig. 5 shows the RO system integrated with wind energy conversion system
and compressed energy storage system. The process is done in two stages, in stage one

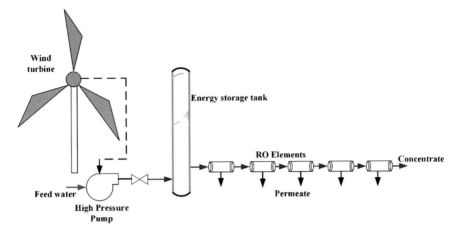

Fig. 5 Wind turbine connected with compressed storage for the desalination process

feed water is sent into the pressed vessel in order to make the gas gets compressed inside the pressure vessel and this increases the vessel pressure. In state two the pressurized feed water is allowed through the RO unit. So, this process is performed in order to avoid dynamic wind fluctuations which makes discontinuous operation of RO system. The advantage of operating in this fashion is to avoid the membrane damage due to fluctuating energy supply. As we are using a compressed storage tank it is easy to construct and the maintenance is also free when compared to other storage devices like batteries, fuel cells, etc.

4   Compressed Air Energy Storage with Wind Energy

The only disadvantage is the membrane downtime underfilling process when compared to continuous operating mode.

By varying the input parameters, the performance of the system is analyzed for its optimum operation. The storage tank volume decides the amount of wind energy for filling the tank and also the time required to fill the tank completely. The system performance by varying the tank volume is analyzed and presented.

5   Results and Discussion

The performance of conventional RO desalination is shown in Fig. 6 and 7. It shows that the permeate concentration and its flow for the different number of membrane elements. It's clear that the permeate flow has increased by increasing number of membrane elements for steady wind speed but it's not same for variable wind-driven

Fig. 6 Flow of permeate for steady and dynamic wind speed condition

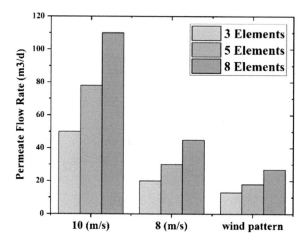

Fig. 7 Salt concentration of permeate for steady and dynamic wind speed condition

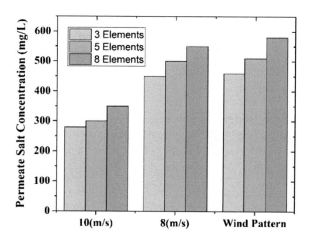

system. When the wind speed is constant at 10 m/s then the water production is high with a constant pressure of 985 psi. The same system is unable to develop the same pressure required for desalination when the wind speed has reduced to 6 m/s. so it requires high wind speed for the desalination process to function properly and it is not practically possible due to intermitent nature of wind. So as the average wind speed is high it can increase the water production compared with less average wind speed. The permeate volume is about 75% less than the steady wind speed condition so it is confined to operate the system in a continuous operation mode for higher water production. It is known that the salt concentration available in water is inversely proportional to water flux into the membranes. When the wind speed is less it induces low water flux thereby increasing permeate concentration. When the storage device is introduced into wind energy system the pressure variation can be seen in the storage tank as shown in Fig. 8. The pressure in the tank increases to

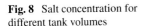
Fig. 8 Salt concentration for different tank volumes

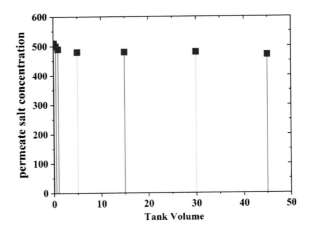

1000 psi during filling and pressure gradually decreases during desalination process. From Fig. 6 it can be observed that as the membrane element numbers are increased it substantially increases the membrane are and the permeate flow rate is also increased. When the flow rate is increased the water production in gradually increased.it can also be concluded that when the membrane elements are connected in series the permeate water increases due to the crossflow of water due to high salt concentration is tends to become feed flow for the next membrane. For maintaining better water quality, the low-pressure limit must be increased and membrane elements should be added in more numbers for more water production.

6 Conclusion

A desalination system supplied from the wind energy conversion system has been studied and found to be convincing compared to conventional desalination methods. A conventional RO system needs continuous supply of electricity which adds cost to the consumer whereas the proposed method reduces the cost by implementing wind energy whenever there is sufficient wind velocity. Under varying wind velocity and also under nonavailability of wind, compressed air energy storage serves to operate the RO continuously without depending on electric supply. In this manner fluctuation in wind speed can be overcome and desalination of water can be carried out without interruption. In future integration of energy recovery system can also be implemented and analysis can be made on the performance.

References

1. Kumar, M.B.H., Saravanan, B.: Impact of global warming and other climatic condition for generation of wind energy and assessing the wind potential for future trends. In: 2017 Innovations in Power and Advanced Computing Technologies (i-PACT), Vellore, pp. 1–5 (2017)
2. Hemanth Kumar, M.B., Saravanan, B.: Power quality improvement for wind energy conversion system using composite observer controller with fuzzy logic. Int. J. Intell. Syst. Appl. (IJISA), **10**(10), 72–80 (2018)
3. Greenlee, L.F., Lawler, D.F., Freeman, B.D., Marrot, B., Moulin, P.: Reverse osmosis desalination: water sources, technology, and today's challenges. Water Res. **43**, 2317–2348 (2009)
4. Distefano, T., Kelly, S.: Are we in deep water? water scarcity and its limits to economic growth. Ecol. Econ. **142**, 130–147 (2017). https://doi.org/10.1016/j.ecolecon.2017.06.019
5. Forstmeier, M., Mannerheim, F., D'Amato, F., Shah, M., Liu, Y., Baldea, M., Stella, A.: Feasibility study on wind-powered desalination. Desalination **203**, 463–470 (2007)
6. Charcosset, C.: A review of membrane processes and renewable energies for desalination. Desalination **245**, 214–231 (2009)
7. Gilau, A.M., Small, M.J.: Designing cost-effective seawater reverse osmosis system under optimal energy options. Renew. Energy **33**, 617–630 (2008)
8. Kershman, S.A., Rheinländer, J., Neumann, T., Goebel, O.: Hybrid wind/PV and conventional power for desalination in Libya—GECOL's facility for medium and small scale research at Ras Ejder. Desalination **183**, 1–12 (2005)
9. Hemanth Kumar, M.B.: An improved resonant fault current limiter for distribution system under transient conditions. Int. J. Renew. Energy Res. (IJRER) **7**(2), 547–555 (2017)
10. Kumar, M.B.H., et al.: Review on control techniques and methodologies for maximum power extraction from wind energy systems. IET Renew. Power Gener. **12**(14), 1609–1622 (2018)
11. Eltawil, M.A., Zhengming, Z., Yuan, L.: A review of renewable energy technologies integrated with desalination systems. Renew. Sustain. Energy Rev. **13**, 2245–2262 (2009)
12. El-Ghonemy, A.: Waste energy recovery in seawater reverse osmosis desalination plants. Part 2: case study. Renew. Sustain. Energy Rev. **16**, 4016–4028 (2012)
13. Madireddi, K., Babcock, R., Levine, B., Kim, J., Stenstrom, M.: An unsteady-state model to predict concentration polarization in commercial spiral wound membranes. J. Memb. Sci. **157**, 13–34 (1999)
14. Ganesh, C., Anupama, S., Hemanth Kumar, M.B.: Control of wind energy conversion system and power quality improvement in the sub rated region using extremum seeking. Indones. J. Electr. Eng. Inform. (IJEEI) **4**(1), 14–23 (2016)
15. Elimelech, M., Bhattacharjee, S.: A novel approach for modeling concentration polarization in crossflow membrane filtration based on the equiva lence of osmotic pressure model and filtration theory. J. Memb. Sci. **145**, 223–241 (1998)
16. Sassi, K.M., Mujtaba, I.M.: Simulation and optimization of full scale reverse osmosis desalination plant. Comput. Aided Chem. Eng. **28**, 895–900 (2010)

Fully Fuzzy Semi-linear Dynamical System Solved by Fuzzy Laplace Transform Under Modified Hukuhara Derivative

Purnima Pandit and Payal Singh

Abstract Semi-linear dynamical systems draw attention in many useful real world problems like population model, epidemic model, etc., they also occur in various applications involving parabolic equations. Now, when the modelling of such applications has inbuilt possibilistic uncertainty, it can be efficiently realized using fuzzy numbers. In this paper, we modify the existing Hukuhara derivative and give the pertaining results for it. We also redefine the Fuzzy Laplace Transform (FLT) and use it to solve such fully fuzzy semi-linear dynamical system.

Keywords Fuzzy semi-linear dynamical system · Fuzzy differential equation (FDE) · Fuzzy Laplace transform (FLT) · Modified Hukuhara derivative (mH-derivative) · Fuzzy convolution theorem

1 Introduction

Differential equations play a significant role to model the problem of science, physics, engineering, finance, economics, etc. While modelling, there may be possibilistic uncertainty in measuring or stating a parameter value that is involved. Such uncertainty can be depicted using fuzzy numbers, giving rise to fuzzy differential dynamical system.

The fuzzy theory was introduced by Zadeh in his explanatory paper in 1965 [1]. The fuzzy derivative was first given by Chang and Zadeh [2] and followed up by Dubois and Prade [3], they used extension principle to define fuzzy derivative.

P. Pandit
Department of Applied Mathematics, Faculty of Technology & Engineering, The Maharaja Sayajirao University of Baroda, Vadodara, Gujarat, India
e-mail: pkpandit@yahoo.com

P. Singh (✉)
Department of Applied Sciences, Faculty of Engineering and Technology,
Parul University, Vadodara, Gujarat, India
e-mail: singhpayalmath@gmail.com

© Springer Nature Singapore Pte Ltd. 2020
K. N. Das et al. (eds.), *Soft Computing for Problem Solving*,
Advances in Intelligent Systems and Computing 1048,
https://doi.org/10.1007/978-981-15-0035-0_13

The first systematic study of fuzzy dynamical system was given by Nazaroff [4]. An initial approach to solve fuzzy differential equation (FDE) was based on Hukuhara derivative, given by Puri and Ralescu [5] and studied in several papers. But fuzzy differential equations under Hukuhara derivative has a limitation that they may have solutions with increasing length of support or the solution is no fuzzier. Some authors used Zadeh's extension principle to solve fuzzy differential equations [6, 7]. The most popular approach to solve fuzzy differential equation is based on generalized Hukuhara derivative [8]. Uniqueness and existence of fuzzy differential equations are given in [9–13]. Analytic solution of Fuzzy differential equations based on another kind of derivative, numerical solution or by transform technique are as in [14–32]. Initially authors worked on FDE with only fuzzy initial condition but gradually the results were developed for systems with imprecise parameters, i.e. fully fuzzy systems as in [33–37].

The aim of this paper is to obtain fuzzy solution of fully fuzzy semi-linear dynamical system, given by as in (1), by redefining fuzzy Laplace Transform,

$$\dot{\tilde{Y}} = \tilde{A} \otimes \tilde{Y} \oplus \tilde{f}\left(t, \tilde{Y}\right); \tilde{Y}(0) = \tilde{Y}_0 \tag{1}$$

where $\tilde{Y} = \begin{bmatrix} \tilde{y}_1 \\ \vdots \\ \tilde{y}_n \end{bmatrix}$ is a $n \times 1$ state vector, $\tilde{A} = \begin{bmatrix} \tilde{a}_{11} & \cdots & \tilde{a}_{1n} \\ \vdots & \ddots & \vdots \\ \tilde{a}_{n1} & \cdots & \tilde{a}_{nn} \end{bmatrix}$ is $n \times n$ constant

fuzzy matrix, $\tilde{f} : I \times E^n \rightarrow E^n$ is continuous fuzzy valued mapping and the symbols $\otimes$ and $\oplus$ denotes the fuzzy product and fuzzy addition, respectively, as defined in the preliminaries.

The advantage of modified Hukuhara derivative (mH-derivative) is that it gives bounded and unique solution as time increases. We give characterization theorem for solving FDE using mH-derivative.

Fuzzy Laplace Transform (FLT), was first given by Allahviranloo [38] in 2010 and used by other authors [37, 39–41]. FLT is defined newly since the function is assumed to be differentiable under mH-derivative. We establish some crucial properties like fuzzy convolution theorem, Laplace transform of fuzzy derivative and inverse fuzzy Laplace transform using a new definition of fuzzy Laplace transform.

In this paper, we use three approaches to solve system (1) by FLT, in the first approach we neglect the nonlinear term and convert system (1) into homogeneous fuzzy dynamical system. In second approach, system (1) is linearized by the Taylor's expansion and it becomes non-homogeneous fuzzy dynamical system. The solution of such systems by FLT is given in [37]. In last approach, we convert system (1) into system of Volterra integral equations with the help of FLT [42] and after that we apply iterative scheme and solve an example that explains capability of the proposed method. This paper contains following sections, preliminaries, main result, application and conclusion.

2 Preliminaries

Let $(\mathbb{P}(R^n), d')$ be the nonempty compact, convex and complete metric space. Let $C, D \in R^n$. the distance between C and D is defined by Housdorff metric.

$$d'(C, D) = \max\left\{ \sup_{c \in C} \inf_{d \in D} \|c - d\|, \sup_{c \in C} \inf_{d \in D} \|c - d\| \right\}$$

Let $E^n = \{\tilde{u} : R^n \to [0\ 1]$ such that $\tilde{u}$ satisfies following properties$\}$

- $\tilde{u}$ is normal.
- $\tilde{u}$ is a fuzzy convex.
- $\tilde{u}$ is upper semicontinuous.
- $\overline{\text{supp}\,(\tilde{u})} = \left\{ x \in R^n / \tilde{u}(x) \geq 0 \right\}$ is compact.

For $0 < \alpha \leq 1$, denote $\tilde{u}^\alpha = \{x \in R^n / \tilde{u}(x) \geq \alpha\}$ then from the above properties follows, $\tilde{u}^\alpha \in \mathbb{P}(R^n) \forall \alpha \in [0\ 1]$.

2.1 Fuzzy Number in Parametric Form

A fuzzy number in parametric form is an ordered pair of the form $\tilde{u} = (\underline{u}, \bar{u})$ where $0 \leq \alpha \leq 1$ satisfying the following conditions:

- $\underline{u}$ is a bounded left continuous increasing function in the interval $[0, 1]$
- $\bar{u}$ is a bounded left continuous decreasing function in the interval $[0, 1]$

- $\underline{u} \leq \bar{u}$

For each α, $\underline{u} = \bar{u}$ then u is crisp number.

For triangular fuzzy number, u must be the form of $(l, c, r) \in R$ with $l \leq c \leq r$ then $\underline{u} = (l + (c - r)\alpha)$ and $\bar{u} = (r - (r - c)\alpha)$.

2.2 Fuzzy Multiplication, Addition and Difference

For $\tilde{u}, \tilde{v} \in E^n$ and $\lambda \in R$ the sum $\tilde{u} \oplus \tilde{v}$ and the product $k \otimes \tilde{u}, \tilde{u} \otimes \tilde{v}$ is defined as

$$\left[\tilde{u} \oplus \tilde{v}\right] = \left[\tilde{u}\right] \oplus [\tilde{v}] = [\underline{u}, \bar{u}] + [\underline{v}, \bar{v}] = [\underline{u} + \underline{v}, \bar{u} + \bar{v}],$$

$$\left[\tilde{u} \otimes \tilde{v}\right] = \left[\underline{u}, \bar{u}\right]\left[\underline{v}, \bar{v}\right] = [\min(\underline{u}\bar{v}, \underline{u}\,\underline{v}, \bar{u}\underline{v}, \bar{u}\bar{v}), \max(\underline{u}\bar{v}, \underline{u}\,\underline{v}, \bar{u}\underline{v}, \bar{u}\bar{v})]$$

$$\left[k \otimes \tilde{u}\right] = k[\underline{u}, \bar{u}] = [k \cdot \bar{u}, k \cdot \underline{u}] \text{ for all } \alpha \in [0\ 1]$$

$$\left[\tilde{u} - \tilde{v}\right] = [\min(\underline{u} - \underline{v})\max(\bar{u} - \bar{v})]$$

2.3 Fuzzy Continuity

We extend this definition as in [11], for system of equations, if $\tilde{f} : I \times E^n \to E^n$ then $\tilde{f}$ is fuzzy continuous at point (t_0, z_0), for any fixed number $\alpha \in [0\ 1]$ and any $\epsilon > 0, \exists \delta(\epsilon, \alpha) s.t \, d'\left(\tilde{f}(t, \tilde{z}), \tilde{f}(t_0, \tilde{z}_0)\right) < \epsilon$ whenever $|t - t_0| < \delta(\epsilon, \alpha)$ and $d'([\tilde{z}], [\tilde{z}_0]) < \delta(\epsilon, \alpha) \forall t \in I$ and $\tilde{z} \in E^n$.

2.4 Equicontinuity

As in [22], $\tilde{f}$ is equicontinuous function, i.e. $\left|\widetilde{f_{\pm}}(t, x, y) - \widetilde{f_{\pm}}(t, x_1, y_1)\right| < \epsilon, \forall \alpha \in [0\ 1]$ whenever $\|(t, x, y) - (t_1, x_1, y_1)\| < \delta$ and uniformly bounded on any bounded set.

2.5 Hukuhara Derivative

As in [5], consider, a fuzzy mapping $\tilde{f} : (a, b) \to E^n$ and $t_0 \in (a, b)$ then $\tilde{f}$ is said to be differentiable at $t_0 \in (a, b) \exists$ an element $\dot{\tilde{f}}(t_0) \in E^n$ such that for all $h > 0$ sufficiently small $\exists \tilde{f}(t_0 + h) \ominus \tilde{f}(t_0), \tilde{f}(t_0) \ominus \tilde{f}(t_0 - h)$ exist and the limits,

$$\lim_{h \to 0+} \frac{\tilde{f}(t_0 + h) \ominus \tilde{f}(t_0)}{h} = \lim_{h \to 0-} \frac{\tilde{f}(t_0) \ominus \tilde{f}(t_0 - h)}{h} = \dot{\tilde{f}}(t_0)$$

Let $\tilde{x}, \tilde{y} \in E^n$. if there exists $\tilde{z} \in E^n$ such that $\tilde{x} = \tilde{y} + \tilde{z}$, then $\tilde{z}$ is called the H-difference of $\tilde{x}$ and $\tilde{y}$ and it is denoted by $\tilde{x} \ominus \tilde{y}$. $\tilde{x} \ominus \tilde{y} \neq \tilde{x} + (-1)\tilde{y}$. In this paper "$\ominus$" stands for always H-difference.

2.6 Modified Hukuhara Derivative

The concept of Hukuhara differentiability is not applicable for function $\tilde{f}(t) = \tilde{c} \otimes g(t)$ when $\dot{g}(t) < 0$ where $\tilde{f} : (a, b) \to E^n$, $\tilde{c} \in E^n$ and $g(t) \in R^n$. We obtain the solution of such function under Generalized Hukuhara differentiability (gH-derivative). The only disadvantage of gH-derivative is, that it does not provide us with unique solution for FDE, we have to choose one solution from the set of solutions, that fits the problem more appropriately.

So, instead of using gH-derivative for solving of FDE, we modify the definition of Hukuhara differentiability so that we can estimate fuzzy derivative of the equation such as $\tilde{f}(t) = \tilde{c} \otimes g(t)$ where $\dot{g}(t) < 0$. Advantage of this derivative is it provides us with unique solution and the solution remains fuzzy as time increases.

Definition A function $\tilde{f} : (a, b) \to E^n$ is said to be modified Hukuhara differentiable at $t_0 \in (a, b)$ $\exists$ an element $\dot{\tilde{f}}(t_0) \in E^n$ such that for all $h > 0$ sufficiently small and the limits,

$$\lim_{h \to 0+} \frac{\tilde{f}(t_0 + h) \ominus \tilde{f}(t_0)}{h} = \lim_{h \to 0-} \frac{\tilde{f}(t_0) \ominus \tilde{f}(t_0 - h)}{h} = \dot{\tilde{f}}(t_0)$$

where,

$$\lim_{h \to 0+} \frac{\tilde{f}(t_0 + h) \ominus \tilde{f}(t_0)}{h} =$$

$$\left[\begin{array}{l} \min\left\{ \lim_{h \to 0} \frac{\left(\underline{f}(t_0+h) - \overline{f}(t_0)\right)}{h}, \lim_{h \to 0} \frac{\left(\overline{f}(t_0+h) - \tilde{f}(t_0)\right)}{h}, \lim_{h \to 0} \frac{\left(\tilde{f}(t_0+h) - \tilde{f}(t_0)\right)}{h}, \lim_{h \to 0} \frac{\left(\tilde{f}(t_0+h) - \underline{f}(t_0)\right)}{h} \right\}, \\ \max\left\{ \lim_{h \to 0} \frac{\left(\underline{f}(t_0) - \underline{f}(t_0-h)\right)}{h}, \lim_{h \to 0} \frac{\left(\underline{f}(t_0) - \tilde{f}(t_0-h)\right)}{h}, \lim_{h \to 0} \frac{\left(\tilde{f}(t_0) - \tilde{f}(t_0-h)\right)}{h}, \lim_{h \to 0} \frac{\left(\tilde{f}(t_0) - \underline{f}(t_0-h)\right)}{h} \right\} \end{array} \right]$$

$$\lim_{h \to 0-} \frac{\tilde{f}(t_0) \ominus \tilde{f}(t_0 - h)}{h} =$$

$$\left[\begin{array}{l} \min\left\{ \lim_{h \to 0} \frac{\left(\underline{f}(t_0) - \underline{f}(t_0-h)\right)}{h}, \lim_{h \to 0} \frac{\left(\underline{f}(t_0) - \tilde{f}(t_0-h)\right)}{h}, \lim_{h \to 0} \frac{\left(\tilde{f}(t_0) - \tilde{f}(t_0-h)\right)}{h}, \lim_{h \to 0} \frac{\left(\tilde{f}(t_0) - \underline{f}(t_0-h)\right)}{h} \right\}, \\ \max\left\{ \lim_{h \to 0} \frac{\left(\underline{f}(t_0) - \underline{f}(t_0-h)\right)}{h}, \lim_{h \to 0} \frac{\left(\underline{f}(t_0) - \tilde{f}(t_0-h)\right)}{h}, \lim_{h \to 0} \frac{\left(\tilde{f}(t_0) - \tilde{f}(t_0-h)\right)}{h}, \lim_{h \to 0} \frac{\left(\tilde{f}(t_0) - \underline{f}(t_0-h)\right)}{h} \right\} \end{array} \right]$$

Existence of Derivative

We now show that the existence of mH-derivative for the function $\tilde{f}(t) = \tilde{c} \otimes g(t)$ for any $g(t)$ was shown in [8] that the fuzzy derivative of $\tilde{f}(t) = \tilde{c} \otimes g(t)$ is, $\dot{\tilde{f}}(t) = \tilde{c} \otimes \dot{g}(t)$ if $\dot{g}(t) > 0$ but does not exist if $\dot{g}(t) < 0$. For $\dot{g}(t) < 0$ using Generalized Hukuhara differentiability, the solution was obtained but the limitation is that it is not unique in the same article [8].

Now by using the proposed mH-differentiability, we solve it as follows,

If, $\dot{g}(t) < 0$ and $\tilde{c} = \left[\underline{c}, \overline{c}\right]$, then,

$$\lim_{h \to 0+} \frac{\tilde{f}(t_0 + h) \ominus \tilde{f}(t_0)}{h} =$$

$$\left[\begin{array}{l} \min\left\{ \lim_{h\to 0} \frac{\left(\underline{cg}(t_0+h)-\underline{cg}(t_0)\right)}{h}, \lim_{h\to 0} \frac{\left(\underline{cg}(t_0+h)-\overline{cg}(t_0)\right)}{h}, \lim_{h\to 0} \frac{\left(\overline{cg}(t_0+h)-\underline{cg}(t_0)\right)}{h}, \lim_{h\to 0} \frac{\left(\overline{cg}(t_0+h)-\underline{cg}(t_0)\right)}{h} \right\} \\[2mm] \max\left\{ \lim_{h\to 0} \frac{\left(\underline{cg}(t_0+h)-\underline{cg}(t_0)\right)}{h}, \lim_{h\to 0} \frac{\left(\underline{cg}(t_0+h)-\overline{cg}(t_0)\right)}{h}, \lim_{h\to 0} \frac{\left(\overline{cg}(t_0+h)-\overline{cg}(t_0)\right)}{h}, \lim_{h\to 0} \frac{\left(\overline{cg}(t_0+h)-\underline{cg}(t_0)\right)}{h} \right\} \end{array} \right]$$

$\dot{g}(t) < 0 = g(t_0 + h) - g(t_0)$ is negative quantity so min and max value of right hand side limit is $-\overline{c}\,\dot{g}(t)$ and $-\underline{c}\,\dot{g}(t)$.

Similarly, for left hand side limit,

$$\lim_{h \to 0-} \frac{\tilde{f}(t_0) \ominus \tilde{f}(t_0 - h)}{h} =$$

$$\left\{ \begin{array}{l} \min\left\{ \lim_{h\to 0} \frac{\left(\underline{cg}(t_0)-\underline{cg}(t_0-h)\right)}{h}, \lim_{h\to 0} \frac{\left(\underline{cg}(t_0)-\overline{cg}(t_0-h)\right)}{h}, \lim_{h\to 0} \frac{\left(\overline{cg}(t_0)-\overline{cg}(t_0-h)\right)}{h}, \lim_{h\to 0} \frac{\left(\overline{cg}(t_0)-\underline{cg}(t_0-h)\right)}{h} \right\} \\[2mm] \max\left\{ \lim_{h\to 0} \frac{\left(\underline{cg}(t_0)-\underline{cg}(t_0-h)\right)}{h}, \lim_{h\to 0} \frac{\left(\underline{cg}(t_0)-\overline{cg}(t_0-h)\right)}{h}, \lim_{h\to 0} \frac{\left(\overline{cg}(t_0)-\overline{cg}(t_0-h)\right)}{h}, \lim_{h\to 0} \frac{\left(\overline{cg}(t_0)-\underline{cg}(t_0-h)\right)}{h} \right\} \end{array} \right\},$$

Now, $g(t_0) - g(t_0 - h) < 0$ is a negative quantity so again min and max value of right hand side limit is $-\overline{c}\dot{g}(t)$ and $-\underline{c}\,\dot{g}(t)$.

Hence, left hand and right hand limit exist and are equal.

So, $\dot{\tilde{f}}(t) = \tilde{c} \otimes \dot{g}(t)$ is modified Hukuhara differentiable function.

Example Consider the IVP with fuzzy initial condition,

$$\frac{dy}{dt} = -y;\ y(0) = (0.96, 1, 1.01) = (0.96 + 0.04\alpha,\ 1.01 - 0.01\alpha)$$

An attempt to solve such an example was done by [20], which was corrected by [22] using gH-derivative. In which the best solution was selected from the obtained ones. Using mH-derivative, it can be solved as,

$$\left[\dot{\underline{y}}, \dot{\overline{y}}\right] = -\left[\underline{y}, \overline{y}\right],\ \forall t > 0$$

Comparing both the sides, we get,

$$\underline{\dot{y}} = \min\left[-\bar{y}, -\underline{y}\right]$$

$$\overline{\dot{y}} = \max\left[-\bar{y}, -\underline{y}\right]$$

i.e. $\underline{\dot{y}} = -\bar{y}$ and $\overline{\dot{y}} = -\underline{y}$

Solving these equations, we get,

$$\underline{y}(t) = c_1 e^t + c_2 e^{-t}$$

$$\bar{y}(t) = -c_1 e^t + c_2 e^{-t}$$

After putting the initial condition, they become,

$$\underline{y}(t) = (0.025\alpha - 0.025)e^t + (0.985 + 0.015\alpha)e^{-t}$$

$$\bar{y}(t) = -(0.025\alpha - 0.025)e^t + (0.985 + 0.015\alpha)e^{-t}$$

Which is exactly the same for $\alpha = 0$ as in [22].

2.7 Fuzzy Riemann Integration

As in [43], Let $\tilde{f}(t)$ be *a* fuzzy valued function on $[a \, \infty)$ represented by $\left[\underline{f}(t), \bar{f}(t)\right]$ for any fixed value of $\alpha \in [0 \, 1]$, assume $\left[\underline{f}(t), \bar{f}(t)\right]$ both are Riemann integrable on $[a \, b]$ and assume there are two positive functions $\underline{M}$ and $\bar{M}$ such that $\int_a^b \left|\underline{f}(t)\right| \le \underline{M}$ and $\int_a^b \left|\bar{f}(t)\right| \le \bar{M}$ then $\tilde{f}(t)$ is improper fuzzy Riemann integrable on $[a \, \infty)$ and the improper fuzzy Riemann integral is fuzzy number.

$$\int_a^\infty \tilde{f}(t)dt = \left[\int_a^\infty \underline{f}(t)dt, \int_a^\infty \bar{f}(t)dt\right]$$

2.8 New Characterization Theorem for Solution of FDEs by Using ODEs

In [22, 44], authors solved FDEs by converting them into ODEs assuming the monotonicity of function. We redefine the characterization theorem in such a way so that this theorem is applicable for all kind of function.

Definition Let $\tilde{F} : I \rightarrow E^n$ be Hukuhara differentiable and $\tilde{F} = \left[\underline{F}, \bar{F}\right]$ then the boundary function $\underline{F}, \bar{F}$ are differentiable then $\dot{\tilde{F}} = \left[\underline{\dot{F}}, \bar{\dot{F}}\right]$, where $\underline{F} = \min F\left(t, \underline{x}, \bar{x}\right)$, $\bar{F} = \max F\left(t, \underline{x}, \bar{x}\right)$ and $\underline{\dot{F}} = \min \dot{F}\left(t, \underline{x}, \bar{x}\right)$, $\bar{\dot{F}} = \max \dot{F}\left(t, \underline{x}, \bar{x}\right)$.

Let us consider the fuzzy initial value problem (FIVP),

$$\dot{\tilde{x}} = \tilde{f}(t, x); \tilde{x}(0) = \tilde{x}_0, \tag{2}$$

where, $\tilde{f} : I \times E^n \rightarrow E^n$ and $\tilde{x}_0 \in E^n$.

Now from the above definition, we can convert FIVP into system of ODEs. Let $\tilde{x}(t) = \left[\underline{x}, \bar{x}\right]$ be Hukuhara differentiable then $\dot{\tilde{x}} = \left[\underline{\dot{x}}, \bar{\dot{x}}\right], \tilde{f} = \left[\underline{f}, \bar{f}\right]$ where $\underline{f}\left(t, \underline{x}, \bar{x}\right) = \min f\left(t, \underline{x}, \bar{x}\right)$, $\bar{f}\left(t, \underline{x}, \bar{x}\right) = \max f\left(t, \underline{x}, \bar{x}\right)$.

So, system (2) gets converted into the system (3).

$$\left. \begin{array}{l} \underline{\dot{x}}(t) = \underline{f}\left(t, \underline{x}, \bar{x}\right); \underline{x}_0 = \min\left[\underline{x}_0, \bar{x}_0\right] \\ \bar{\dot{x}}(t) = \bar{f}\left(t, \underline{x}, \bar{x}\right); \bar{x}_0 = \max\left[\underline{x}_0, \bar{x}_0\right] \end{array} \right\} \tag{3}$$

Theorem Suppose FIVP (2) where $\tilde{f} : I \times E^n \rightarrow E^n$ is such that,

(1) $\tilde{f}$ is equicontinuous function, i.e. $\left|\widetilde{f_+}(t, x, y) - \widetilde{f_+}(t, x_1, y_1)\right| < \epsilon, \forall \alpha \in [0 \ 1]$

Whenever $\|(t, x, y) - (t_1, x_1, y_1)\| < \delta$ and uniformly bounded on any bounded set.

(2)   There exist $L > 0$, such that $\left|\widetilde{f_\pm}(t, x, y) - \widetilde{f_\pm}(t, x_1, y_1)\right| < L \max\{|x - u|, |y - v|\}$

Then the FIVP (2) and system of ODEs (3) are equivalent.

Proof The equicontinuity of $\tilde{f}$ implies continuity of the function $\tilde{f}$ and second condition guarantees $\tilde{f}$ is Lipschitz as follows:

$$\sup \max\left\{\left|\underline{f}\left(t, \underline{x}, \bar{x}\right) - \underline{f}\left(t, \underline{y}, \bar{y}\right)\right|, \left|\bar{f}\left(t, \underline{x}, \bar{x}\right) - \bar{f}(t, \underline{y}, \bar{y})\right|\right\}$$

$$\leq L \sup \max\left\{\left|\underline{y} - \underline{x}\right|, |\bar{y} - \bar{x}|\right\}$$

i.e.,

$$D(\tilde{f}(t, \tilde{x}), \tilde{f}(t, \tilde{y}) \leq D(\tilde{x}, \tilde{y})$$

So, from the above condition, FIVP (2) has unique solution. The solution of the FIVP (2) is Hukuhara differentiable and so by definition, the functions $\underline{x}(t), \bar{x}(t)$ are the differentiable and solution of system (3).

Conversely, suppose $\underline{x}(t), \bar{x}(t)$ are the solution of system (3), since function is Lipschitz it guarantees uniqueness of solution, hence by decomposition theorem, we can construct fuzzy solution of $\underline{x}(t), \bar{x}(t)$ for each $\alpha \in [0, 1]$ refer Klir and Yuan [45]. So this is fuzzy solution of FIVP (2).

Hence, both system (2) and (3) are equivalent.

3 Fuzzy Laplace Transform Under mH-Derivative

3.1 Fuzzy Laplace Transform

Fuzzy Laplace Transform technique is very useful in solving FDEs, Fuzzy partial differential equation and their corresponding initial and boundary value problem. In this paper, we redefine the Fuzzy Laplace Transform (FLT) in a different manner and establish some other properties with the help of new definition.

Definition Fuzzy valued function $\tilde{f}(t) = \left[\underline{f}(t), \bar{f}(t)\right]$ in parametric form, is bounded and piecewise continuous on the interval $[0, \infty)$ and suppose that $\tilde{f}(t) \otimes e^{-st}$ is improper fuzzy Riemann integrable, then $\int_0^\infty e^{-st} \otimes \tilde{f}(t)dt$ is called fuzzy Laplace Transform and it is defined as,

$$\tilde{F}(s) = L\left(\tilde{f}(t)\right) = \int_0^\infty e^{-st} \otimes \tilde{f}(t)dt$$

$$\tilde{F}(s) = L\left(\tilde{f}(t)\right) = \lim_{t \to \infty} \int_0^t e^{-st} \otimes \tilde{f}(t)dt$$

Taking, alpha cut on both sides,

$$L\left(\tilde{f}(t)\right) = \lim_{t \to \infty} \int_0^t e^{-st} \otimes \left[\underline{f}(t), \bar{f}(t)\right]dt$$

$$L\left(\left[\underline{f}(t), \bar{f}(t)\right]\right) = \lim_{t \to \infty} \left[\int_0^t e^{-st}\underline{f}(t)dt, \int_0^t e^{-st}\bar{f}(t)dt\right]$$

$$\underline{F}(s) = L\left[\underline{f}(t)\right] = \min\left\{\lim_{t \to \infty} \left[\int_0^t e^{-st}\underline{f}(t)dt, \int_0^t e^{-st}\bar{f}(t)dt\right]\right\}$$

$$\bar{F}(s) = L[\bar{f}(t)] = \max\left\{\lim_{t\to\infty}\left[\int_0^t e^{-st}\underline{f}(t)dt, \int_0^t e^{-st}\bar{f}(t)dt\right]\right\}$$

Fuzzy inverse Laplace Transform is defined as,

$$L^{-1}[\underline{F}(s)] = \min\left[\underline{f}(t), \bar{f}(t)\right]$$

$$L^{-1}[\bar{F}(s)] = \max\left[\underline{f}(t), \bar{f}(t)\right]$$

3.2 Existence of Fuzzy Laplace Transform

The fuzzy Laplace integral $\int_0^\infty e^{-st} \otimes \tilde{f}(t)dt$ exist in the sense of improper fuzzy Riemann integral provided $\tilde{F}(t) = \left[\underline{f}(t), \bar{f}(t)\right]$ is of exponential order for this condition, $\lim_{t\to\infty}\frac{\underline{f}(t)}{e^{pt}} = 0$ and $\lim_{t\to\infty}\frac{\overline{f}(t)}{e^{pt}} = 0$ the relation is required to hold for some constant p, $\underline{M}$ and $\bar{M}$,

$$\left|\underline{f}(t)\right| \leq \underline{M}e^{pt}.\left|\bar{f}(t)\right| \leq \bar{M}e^{pt}$$

In addition, $\underline{f}(t)$ and $\bar{f}(t)$ both should be piecewise continuous on each finite interval of $[0, \infty)$.

Theorem 3.1 Let $\tilde{f}(t) = \left[\underline{f}(t), \bar{f}(t)\right]$ be piecewise continuous on every finite interval $t \geq 0$ and satisfy $\left|\underline{f}(t)\right| \leq \underline{M}e^{pt}$, $\left|\bar{f}(t)\right| \leq \bar{M} e^{pt}$ for some constant p, $\underline{M}$ and $\bar{M}$, then $L\left(\tilde{f}(t)\right) = \left(L\left[\underline{f}(t)\right], L[\bar{f}(t)]\right)$ exist for $s > p$, $\lim_{s\to\infty} L\left[\underline{f}(t)\right] = 0$ and $\lim_{s\to\infty} L[\bar{f}(t)] = 0$

Proof From this inequality,

$$\left|\underline{f}(t)\right| \leq \underline{M}e^{pt}$$

$$\int_0^\infty e^{-st}\left|\underline{f}(t)\right|dt \leq \int_0^\infty e^{-st}\underline{M}e^{pt}dt$$

$$L\left[\underline{f}(t)\right] \leq \frac{\underline{M}}{s-p}$$

As $s \to \infty$, $L\left[\underline{f}(t)\right] \to 0$

Similarly,

$$\left|\bar{f}(t)\right| \le \bar{M}e^{pt}$$

$$\int_0^\infty e^{-st}\left|\bar{f}(t)\right|dt \le \int_0^\infty e^{-st}\bar{M}e^{pt}dt$$

$$L\left[\bar{f}(t)\right] \le \frac{\bar{M}}{s-p}$$

As $s \to \infty$, $L\left[\bar{f}(t)\right] \to 0$

Hence, $L\left(\tilde{f}(t)\right)$ exist.

3.3 Fuzzy Laplace of Derivative

In the following, we give the result regarding the existence of FLT under mH-derivative.

Theorem 3.2 If $\tilde{f}(t) = \left[\underline{f}(t), \bar{f}(t)\right]$ be continuous fuzzy valued function, $\lim_{t\to\infty} e^{-st}\underline{f}(t) \to 0$ and $\lim_{t\to\infty} e^{-st}\bar{f}(t) \to 0$ for a large value of s and $\dot{\tilde{f}}(t)$ is piecewise continuous then $L\left(\dot{\tilde{f}}(t)\right)$ exist, and is given by,

$$L\left(\dot{\tilde{f}}(t)\right) = sL\left(\tilde{f}(t)\right)\ominus \tilde{f}_0$$

Proof In Theorem 3.1, we already proved $L\left(\tilde{f}(t)\right)$ exist because $\tilde{f}(t)$ is of exponential order and continuous and $\dot{\tilde{f}}(t)$ is piecewise continuous. So fuzzy Laplace derivative is given as,

$$L\left(\dot{\tilde{f}}(t)\right) = \int_0^\infty e^{-st}\dot{\tilde{f}}(t)dt$$

$$L\left(\dot{\tilde{f}}(t)\right) = \lim_{t\to\infty}\int_0^t e^{-st}\dot{\tilde{f}}(t)dt$$

Taking, alpha cut on both sides,

$$L\left(\left[\underline{\dot{f}}(t), \overline{\dot{f}}(t)\right]\right) = \lim_{t \to \infty} \int_0^t e^{-st}\left[\underline{\dot{f}}(t), \overline{\dot{f}}(t)\right]dt$$

$$L\left(\left[\underline{\dot{f}}(t), \overline{\dot{f}}(t)\right]\right) = \lim_{t \to \infty}\left[\int_0^t e^{-st}\underline{\dot{f}}(t)dt, \int_0^t e^{-st}\overline{\dot{f}}(t)dt\right]$$

Now integration by parts, we get,

$$L\left(\underline{\dot{f}}(t)\right) = \min\left\{sL\left(\underline{f}(t)\right) - \underline{f}(0), sL\left(\underline{f}(t)\right) - \bar{f}(0), sL\left(\bar{f}(t)\right) - \underline{f}(0), sL\left(\bar{f}(t)\right) - \bar{f}(0)\right\}$$

$$L\left(\overline{\dot{f}}(t)\right) = \max\left\{sL\left(\underline{f}(t)\right) - \underline{f}(0), sL\left(\underline{f}(t)\right) - \bar{f}(0), sL\left(\bar{f}(t)\right) - \underline{f}(0), sL\left(\bar{f}(t)\right) - \bar{f}(0)\right\}$$

Then, by first decomposition theorem as in Klir and Yuan [45].

$$L\left(\tilde{\dot{f}}(t)\right) = sL\left(\tilde{f}(t)\right) \ominus \tilde{f}_0$$

Example This example is taken from [38].

$$\dot{y} = -y, 0 \le t \le T; y(0) = \left(\underline{y}(0), \bar{y}(0)\right) = (-a(1 - \alpha), a(1 - \alpha))$$

The solution of the above problem is given by FLT under generalized Hukuhara differentiability in [38]. Two solutions were obtained and author selected the one solution from set of two solutions that fits the problem appropriately.

When we solve this problem under modified Hukuhara differentiability, then we obtain a unique solution with bounded support.

$$L[\dot{y}] = -L[y]$$

$$L\left(\underline{\dot{y}}(t)\right) = \min\left\{sL\left(\underline{y}(t)\right) - \underline{y}(0), sL\left(\underline{y}(t)\right) - \bar{y}(0), sL(\bar{y}(t)) - \underline{y}(0), sL(\bar{y}(t)) - \bar{y}(0)\right\} = -L(\bar{y}(t))$$

$$L(\bar{\dot{y}}(t)) = \max\left\{sL\left(\underline{y}(t)\right) - \underline{y}(0), sL\left(\underline{y}(t)\right) - \bar{y}(0), sL(\bar{y}(t)) - \underline{y}(0), sL(\bar{y}(t)) - \bar{y}(0)\right\} = -L\left(\underline{y}(t)\right)$$

Thus, fuzzy solution of the problem is,

$$\underline{y}(t) = -a(1 - \alpha)e^{-t}, \bar{y}(t) = a(1 - \alpha)e^{-t}$$

3.4 Fuzzy Convolution Theorem

The proof of the main theorem involves the FLT of product of two fuzzy functions for which theorem is proved below.

Theorem 3.3 Let $\tilde{f}(s)$ and $\tilde{g}(s)$ denote the fuzzy inverse Laplace transforms of $\tilde{f}(t)$ and $\tilde{g}(t)$, respectively., Then the product given by $\tilde{f}(s)\tilde{g}(s)$ is the fuzzy inverse Laplace transform of the convolution of $\tilde{f}$ and $\tilde{g}$, is given by,

$$L\left(\tilde{f}(t) * \tilde{g}(t)\right) = \tilde{f}(s) * \tilde{g}(s)$$

Proof

$$L\left(\tilde{f}(t) * \tilde{g}(t)\right) = \int_0^t e^{-st}\left[\tilde{f}(t).\tilde{g}(t)\right]dt$$

$$L\left(\left[\underline{f}(t), \bar{f}(t)\right] * \left[\underline{g}(t), \bar{g}(t)\right]\right) = \int_0^t \left[\underline{f}(t), \bar{f}(t)\right]\left[\underline{g}(t), \bar{g}(t)\right]dt$$

From fuzzy multiplication, we can write,

$$L\left[\underline{f}(t) * \underline{g}(t)\right] = \min \int_0^t e^{-st}\left[\underline{f}(\tau)\underline{g}(t-\tau), \underline{f}(\tau)\bar{g}(t-\tau), \bar{f}(\tau)\underline{g}(t-\tau), \bar{f}(\tau)\bar{g}(t-\tau)\right]dt d\tau$$

$$L[\bar{f}(t) * \bar{g}(t)] = \max \int_0^t e^{-st}\left[\underline{f}(\tau)\underline{g}(t-\tau), \underline{f}(\tau)\bar{g}(t-\tau), \bar{f}(\tau)\underline{g}(t-\tau), \bar{f}(\tau)\bar{g}(t-\tau)\right]dt d\tau$$

For solving above integration, we use substitution,

$$t - \tau = u, \tau = v$$

$$L\left[\underline{f}(t) * \underline{g}(t)\right] = \min \int_0^t e^{-st}\left[\underline{f}(v)\underline{g}(u), \underline{f}(v)\bar{g}(u), \bar{f}(v)\underline{g}(u), \bar{f}(v)\bar{g}(u)\right]du dv$$

$$L[\bar{f}(t) * \bar{g}(t)] = \max \int_0^t e^{-st}\left[\underline{f}(v)\underline{g}(u), \underline{f}(v)\bar{g}(u), \bar{f}(v)\underline{g}(u), \bar{f}(v)\bar{g}(u)\right]du dv$$

Then, by first decomposition theorem as in Klir and Yuan [45],

$$L\left(\tilde{f}(t) * \tilde{g}(t)\right) = \tilde{f}(s) * \tilde{g}(s)$$

4 Main Result

Now we give the main result pertaining to the solution of the semi-linear such as system (1) by using Fuzzy Laplace Transform under mH-derivative.

Main Theorem: If $\tilde{f} : I \times E^n \rightarrow E^n$ is fuzzy valued continuous function and Lipschitz then the solution of system (1) which is given below, exist and unique.

$$\dot{\tilde{Y}} = \tilde{A} \otimes \tilde{Y} \oplus \tilde{f}\left(t, \tilde{Y}\right); \tilde{Y}(0) = \tilde{Y}_0$$

where,

$$\tilde{A} = [\underline{A}, \bar{A}], \tilde{Y} = [\underline{Y}, \bar{Y}], \quad \tilde{Y}(0) = [\underline{Y}_0, \bar{Y}_0]$$

Proof Before proving above the theorem, for the most generalized form, we take up particular cases and prove them as lemmas.

Lemma 1 Suppose $\tilde{f}\left(t, \tilde{Y}\right) = 0$ in the system (1) then the solution of system (1) is given by

$$\underline{Y} = \underline{Y}_0 e^{\underline{B}t}$$

$$\bar{Y} = \bar{Y}_0 e^{\bar{B}t}$$

and $\underline{Y} \leq \bar{Y}$.

Proof If $\tilde{f}\left(t, \tilde{Y}\right) = 0$ then,

$$\dot{\tilde{Y}} = \tilde{A} \otimes \tilde{Y}; \tilde{Y}(0) = \tilde{Y}_0 \tag{4}$$

By taking, α—cut of Eq. (4),

$$\left[\dot{\underline{Y}}, \dot{\bar{Y}}\right] = [\underline{A}, \bar{A}][\underline{Y}, \bar{Y}]; [\underline{Y}(0), \bar{Y}(0)] = [\underline{Y}_0, \bar{Y}_0]$$

Now by fuzzy multiplication,

$$\dot{\underline{Y}} = \min(\underline{A}\,\underline{Y}, \underline{A}\bar{Y}, \bar{A}\underline{Y}, \bar{A}, \bar{Y}); \underline{Y}_0$$
$$\dot{\bar{Y}} = \max(\underline{A}\,\underline{Y}, \underline{A}\bar{Y}, \bar{A}\underline{Y}, \bar{A}, \bar{Y}); \bar{Y}_0$$

Now let us denote, $\underline{B} = \min(\underline{A}, \bar{A}), \bar{B} = \max(\underline{A}, \bar{A})$,

$$\dot{\underline{Y}} = \underline{B}\,\underline{Y}; \underline{Y}_0$$

$$\dot{\underline{Y}} = \bar{B}\bar{Y}; \bar{Y}_0$$

Now by fuzzy Laplace Transform,

$$sL[\underline{Y}] - \underline{Y}_0 = \underline{B}L[\underline{Y}]$$

$$sL[\bar{Y}] - \bar{Y}_0 = [\bar{B}]L[\bar{Y}]$$

Thus, applying inverse fuzzy Laplace Transform and we obtain the solution.

$$\underline{Y} = \underline{Y}_0 e^{\underline{B}t}$$

$$\bar{Y} = \bar{Y}_0 e^{\bar{B}t}$$

Lemma 2 $\tilde{f}\left(t, \tilde{Y}\right) \neq 0$ then, $\tilde{f}$ in the system (1) can be linearized around equilibrium point by Taylor's expansion.

Proof System (1) is given as,

$$\dot{\tilde{Y}} = \tilde{A} \otimes Y + \tilde{f}\left(t, \tilde{Y}\right)$$

with the initial condition, $\tilde{Y}(0) = \tilde{Y}_0$

First taking, $\alpha-$ cut of the above system,

$$\dot{\underline{Y}} = \min\left(\underline{A}\,\underline{Y}, \underline{A}\bar{Y}, \bar{A}\underline{Y}, \bar{A}\bar{Y}\right) + \underline{f}(t, \underline{Y}, \bar{Y})$$
$$\dot{\bar{Y}} = \max\left(\underline{A}\,\underline{Y}, \underline{A}\bar{Y}, \bar{A}\underline{Y}, \bar{A}\bar{Y}\right) + +\bar{f}(t, \underline{Y}, \bar{Y})$$

where, $\underline{f}(t, \underline{Y}, \bar{Y}) = \min f\left(t, \underline{Y}, \bar{Y}\right)$ and $\bar{f}(t, \underline{Y}, \bar{Y}) = \max f\left(t, \underline{Y}, \bar{Y}\right)$

Put $\dot{\underline{Y}} = 0, \dot{\bar{Y}} = 0$ and we get $(\underline{Y}e, \bar{Y}e)$ equilibrium point then applying Taylor's expansion and we get,

$$\dot{\underline{Y}} = \min(\underline{A}\,\underline{Y}, \underline{A}\bar{Y}, \bar{A}\underline{Y}, \bar{A}\,\bar{Y}) + \underline{f}\left(\underline{Y}e, \bar{Y}e\right) + \frac{\partial \underline{f}}{\partial \underline{Y}}(\underline{Y} - \underline{Y}e) + \frac{\partial \underline{f}}{\partial \bar{Y}}(\bar{Y} - \bar{Y}e)$$

$$\dot{\bar{Y}} = \max(\underline{A}\,\underline{Y}, \underline{A}\bar{Y}, \bar{A}\underline{Y}, \bar{A}\,\bar{Y}) + \bar{f}\left(\underline{Y}e, \bar{Y}e\right) + \frac{\partial \underline{f}}{\partial \underline{Y}}(\underline{Y} - \underline{Y}e) + \frac{\partial \bar{f}}{\partial \bar{Y}}(\bar{Y} - \bar{Y}e)$$

$$\dot{\underline{Y}} = \underline{B}\,\underline{Y} + \underline{f}\left(\underline{Y}e, \bar{Y}e\right) + \frac{\partial \underline{f}}{\partial \underline{Y}}(\underline{Y} - \underline{Y}e) + \frac{\partial \underline{f}}{\partial \bar{Y}}(\bar{Y} - \bar{Y}e)$$

$$\dot{\bar{Y}} = \bar{B}\,\bar{Y} + \bar{f}\left(\underline{Y}e, \bar{Y}e\right) + \frac{\partial \bar{f}}{\partial \underline{Y}}(\underline{Y} - \underline{Y}e) + \frac{\partial \bar{f}}{\partial \bar{Y}}(\bar{Y} - \bar{Y}e)$$

$$\dot{\underline{Y}} = \underline{C}\,\underline{Y} + \underline{D} \quad \& \quad \dot{\bar{Y}} = \bar{C}\,\bar{Y} + \bar{D}$$

where,

$$\underline{C}\,\underline{Y} = \underline{B}\,\underline{Y} + \frac{\partial \underline{f}}{\partial \underline{Y}}\underline{Y} + \frac{\partial \underline{f}}{\partial \bar{Y}}\bar{Y}, \ \underline{D} = \underline{f}(\underline{Y}e, \bar{Y}e) + \frac{\partial \underline{f}}{\partial \underline{Y}}\underline{Y}_e + \frac{\partial \underline{f}}{\partial \bar{Y}}\overline{Y}_e$$

$$\bar{C}\,\bar{Y} = \bar{B}\,\bar{Y} + \frac{\partial \bar{f}}{\partial \underline{Y}}\underline{Y} + \frac{\partial \bar{f}}{\partial \bar{Y}}\bar{Y}, \ \bar{D} = \bar{f}(\underline{Y}e, \bar{Y}e) + \frac{\partial \bar{f}}{\partial \underline{Y}}\underline{Y}_e + \frac{\partial \bar{f}}{\partial \bar{Y}}\overline{Y}_e$$

Now, by the first decomposition theorem as in Klir and Yuan [45],

$$\dot{\tilde{Y}} = \tilde{C}\tilde{Y} + \tilde{D} \tag{5}$$

Lemma 3 System (1) can be converted into Volterra integral equation as given below,

$$[\underline{Y}] = e^{\underline{B}(t)}\underline{Y}_0 + e^{\underline{B}(t)}\int_0^t e^{-\underline{B}(x)}\underline{f}(t, \underline{x}, \bar{x})dx$$

$$[\bar{Y}] = e^{\bar{B}(t)}\bar{Y}_0 + e^{\bar{B}(t)}\int_0^t e^{-\bar{B}(x)}\underline{f}(t, \underline{x}, \bar{x})dx$$

where, $\underline{B} = \min(\underline{A}, \overline{A})$, $\underline{B} = \max(\underline{A}, \overline{A})$

and,

$$\underline{Y}(t) = \lim_{i \to \infty}\underline{Y}_i$$

$$\bar{Y}(t) = \lim_{i \to \infty}\bar{Y}_i$$

where,

$$[\underline{Y}_{i+1}] = \underline{Y}(0)e^{\underline{B}t} + e^{-\underline{B}(t)}\int_0^t e^{-\underline{B}(x)}\underline{f}(t, \underline{x}_i, \overline{x^i})dx$$

$$[\overline{Y}_{i+1}] = e^{\bar{B}(t)}\underline{Y}(0) + e^{-\bar{B}(t)}\int_0^t e^{-\bar{B}(x)}\bar{f}(t, \underline{x}_i, \overline{x^i})dx$$

Then system (1) has fuzzy solution if its integral equations have fuzzy solution, i.e. $\underline{Y_{i+1}} \leq \overline{Y_{i+1}}$.

Proof Now for the system as given by (1), take α − cut and applying Fuzzy Laplace Transform

$$\dot{\underline{Y}} = \min(\underline{A}\,\underline{Y},\, \underline{A}\bar{Y},\, \bar{A}\underline{Y},\, \bar{A}\bar{Y}) + \underline{f}\left(t, \underline{Y}, \bar{Y}\right)$$

$$\dot{\bar{Y}} = \max = (\underline{A}\,\underline{Y},\, \underline{A}\bar{Y},\, \bar{A}\underline{Y},\, \bar{A}\bar{Y}) + + \bar{f}\left(t, \underline{Y}, \bar{Y}\right)$$

where,

$$\underline{f} = \min f\left(t, \underline{Y}, \bar{Y}\right),\ \bar{f} = \max f\left(t, \underline{Y}, \bar{Y}\right)$$

Now the system (1) becomes,

$$\dot{\underline{Y}} = \underline{B}\,\underline{Y} + \underline{f}\left(t, \underline{Y}, \bar{Y}\right)$$

$$\dot{\bar{Y}} = \bar{B}\,\bar{Y} + \bar{f}\left(t, \underline{Y}, \bar{Y}\right)$$

$$sL[\underline{Y}] = \underline{Y}_0 + \underline{B}L[\underline{Y}] + L\left[\underline{f}\left(t, \underline{Y}, \bar{Y}\right)\right]$$

$$sL[\underline{Y}] - \underline{B}L[\underline{Y}] = \underline{Y}_0 + L\left[\underline{f}\left(t, \underline{Y}, \bar{Y}\right)\right]$$

$$(s - I\underline{B})L[\underline{Y}] = \underline{Y}_0 + L\left[\underline{f}\left(t, \underline{Y}, \bar{Y}\right)\right]$$

$$L[\underline{U}] = \frac{\underline{Y}_0}{(s - I\underline{B})} + \frac{L\left[\underline{f}\left(t, \underline{Y}, \bar{Y}\right)\right]}{(s - I\underline{B})}$$

Now by fuzzy convolution theorem and inverse fuzzy Laplace Transform

$$[\underline{Y}] = L^{-1}\frac{\underline{Y}_0}{(s - I\underline{B})} + L^{-1}\frac{L\left[\underline{f}\left(t, \underline{Y}, \bar{Y}\right)\right]}{(s - I\underline{B})}$$

$$[\underline{Y}] = \underline{Y}_0 e^{\underline{B}t} + e^{\underline{B}(t)}\int_0^t e^{-\underline{B}(\tau)}\underline{f}\left(t, \underline{\tau}, \bar{\tau}\right)dx$$

Similarly,

$$[\bar{Y}] = e^{\bar{B}(t)}\bar{Y}_0 + e^{\bar{B}(t)}\int_0^t e^{-\bar{B}(\tau)}\bar{f}\left(t, \underline{\tau}, \bar{\tau}\right)dx$$

Now applying the iterative scheme, we get,

$$\left[\underline{Y_{i+1}}\right] = \underline{Y}(0)e^{\underline{B}t} + e^{-\underline{B}(t)} \int_0^t e^{-\underline{B}(\tau)} \underline{f}\left(t, \underline{\tau_i}, \overline{\tau^i}\right)dx$$

$$\left[\overline{Y_{i+1}}\right] = \underline{Y}(0)e^{\bar{B}(t)} + e^{-\bar{B}(t)} \int_0^t e^{-\bar{B}(\tau)} \bar{f}\left(t, \underline{\tau_i}, \overline{\tau^i}\right)dx$$

Now by successive approximate theorem [42], we can show the convergence of the above equations.

Proof of Main Theorem

If the system (1) is homogeneous then its solution can be given as in Lemma 1. If the system (1) has a nonlinear term in right hand side, then local solution suffices as in Lemma 2 or otherwise we get the solution shown in Lemma 3.

5 Application

Now we apply these three cases to solve fuzzy Prey–Predator model [46],

$$\dot{\tilde{x}} = \widetilde{0.1}x - \widetilde{0.005}xy; \ \dot{\tilde{y}} = -\widetilde{0.4}y + \widetilde{0.008}xy \tag{6}$$

with an initial condition, $\tilde{x}_0 = \widetilde{130}$ and $\tilde{y}_0 = \widetilde{40}$

$$\widetilde{0.1} = (0.05 + 0.05\alpha, 0.15 - 0.05\alpha),$$
$$\widetilde{0.005} = (0.004 + 0.001\alpha, 0.006 - 0.001\alpha),$$
$$\widetilde{0.4} = (0.3 + 0.1\alpha, 0.5 - 0.1\alpha),$$
$$\widetilde{0.008} = (0.007 + 0.001\alpha, 0.009 - 0.001\alpha),$$
$$\widetilde{130} = (120 + 10\alpha, 150 - 20\alpha) \text{ and } \widetilde{40} = (20 + 20\alpha, 50 - 10\alpha)$$

Solution

Case 1

Neglecting the nonlinear term in (6) that is considering the homogeneous system only.

$$\dot{\tilde{x}} = \widetilde{0.1}x; \ \dot{\tilde{y}} = -\widetilde{0.4}y \tag{7}$$

with an initial condition, $\tilde{x}_0 = \widetilde{130}$ and $\tilde{y}_0 = \widetilde{40}$

After applying the proposed scheme, the graph is obtained as follows (Fig. 1), Other iterations are given in Table 1.

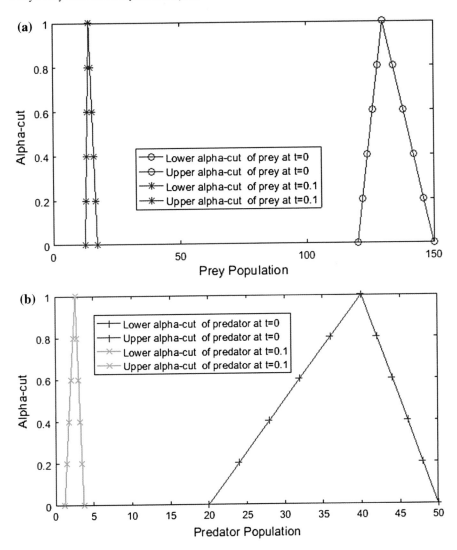

Fig. 1 **a** Fuzzy number representation of Prey population at $t = 0$ and 0.1. **b** Fuzzy number representation of Predator population at $t = 0$ and 0.1

Table 1 Number of Prey and Predator in the first approach

Time	Number of Prey	Number of Predator
0	(120, 130, 150)	(20, 40, 50)
0.1	(12.61, 14.36, 17.42)	(1.21, 2.68, 3.704)
0.2	(25.23, 28.73, 34.85)	(2.426, 5.36, 7.408)
0.3	(37.84, 43.10, 52.28)	(3.63, 8.04, 11.11)
0.4	(50.46, 57.46, 69.71)	(4.85, 10.72, 14.81)
0.5	(63.07, 71.83, 87.13)	(6.06, 13.4, 18.52)
1	(126.15, 143.67, 174.27)	(12.13, 26.81, 37.04)

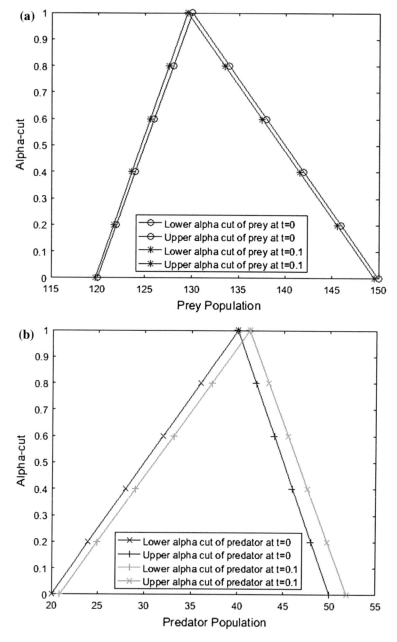

Fig. 2 **a** Fuzzy number representation of Prey population at $t = 0$ and 0.1. **b** Fuzzy Number representation of Predator population at $t = 0$ and 0.1

Table 2 Number of Prey and Predator in the second approach

Time	Number of Prey	Number of Predator
0	(120, 130, 150)	(20, 40, 50)
0.1	(118.66, 127.10, 147.54)	(24.45, 46.28, 59.152)
0.2	(119.54, 128.93, 149.12)	(21.79, 42.54, 53.68)
0.3	(119.27, 128.35, 148.63)	(22.68, 43.80, 55.51)
0.4	(118.98, 127.74, 148.10)	(23.57, 45.05, 57.33)
0.5	(118.66, 127.10, 147.54)	(24.45, 46.28, 59.56)
1.0	(116.67, 123.43, 144.177)	(28.78, 53.31, 68.05)

Case 2
First, linearize the Eq. (6) around equilibrium point by Taylor's expansion, we obtain the linearized form as below,

$$x = -\widetilde{0.25}\,\tilde{x} + \tilde{5}; \ y = \widetilde{0.16}\,\tilde{y} - \tilde{8} \tag{8}$$

$$\tilde{x}(0) = \widetilde{130}, \ \tilde{y}(0) = \widetilde{40}$$

Applying the proposed scheme, we get (Fig. 2),
Other iterations are given in Table 2,

Case 3
In this method, first, we take α—cut of Eq. (6) then convert these equations in (6) into Volterra integral equation by fuzzy Laplace Transform, then apply iterative scheme and we obtain fuzzy solution (Fig. 3).

Other iterations are given in Table 3 (Table 4),

Figure 4 represents the comparative study of Prey–Predator model in approach 3 and iterative scheme at core.

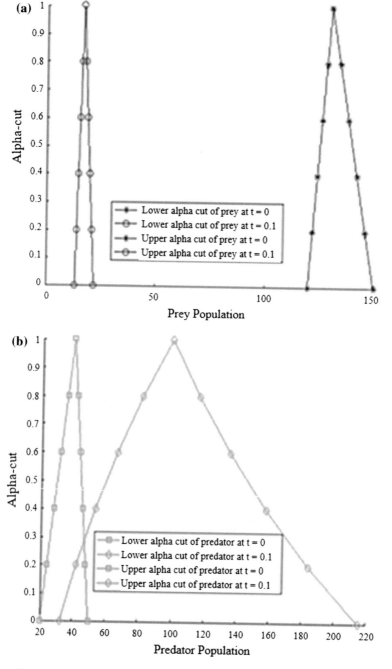

Fig. 3 **a** Fuzzy number representation of Prey population at $t = 0$ and 0.1. **b** Fuzzy number representation of Predator population at $t = 0$ and 0.1

Table 3 Number of Prey and Predator in the third approach

Time	Number of Prey	Number of Predator
0	130	40
0.1	118.49	60.56
0.2	114.774	66.2506
0.3	114.19	65.18
0.4	116.41	65.229
0.5	117.199	60.27

Table 4 Comparison of the third approach and iterative scheme

Time	Iterative scheme		Approach-3	
t	Prey	Predator	Prey	Predator
0	130	40	130	40
0.1	128	42.56	118.49	60.56
0.2	127.24	45.23	114.774	66.2506
0.3	125.64	48.03	114.19	65.18
0.4	123.88	50.94	116.41	65.229
0.5	121.89	53.95	117.199	60.27

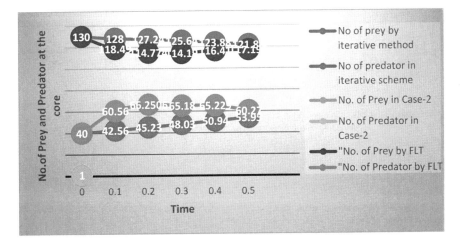

Fig. 4 Comparative graph

6   Conclusion

In this paper, we have proposed a new derivative Modified Hukuhara derivative (mH-derivative) for FDE. The proposed mH-derivative removes drawback of previously defined derivatives. We established Fuzzy Laplace Transform, it's derivative and convolution theorem under mH-derivative and gave characterization theorem for converting FDEs to ODEs. The result shows that the proposed method is capable of obtaining unique solution of dynamical system bounded support for all time under modified Hukuhara differentiability.

References

1. Zadeh L.A.: Fuzzy sets and systems. In: Proceedings of Syrup on Systems Theory. Polytechnic Institute Press, Brooklyn, NY (1965)
2. Chang, S.L., Zadeh, L.A.: On fuzzy mapping and control. IEEE Trans. Syst., Man Cybern. **2**, 30–34 (1972)
3. Dubois, D., Prade, H.: Towards fuzzy differential calculus, part 3, differentiation. Fuzzy Sets Syst. **8**, 225–233 (1982)
4. Nazaroff, G.J.: Fuzzy topological polysystems. J. Math. Anal. Appl. **91**, 478–485 (1973)
5. Puri M.L., Ralescu: Differential of fuzzy functions. J. Math. Anal. Appl. **91**, 321–325 (1983)
6. Buckley, J.J., Feuring, T.: Fuzzy differential equations. Fuzzy Sets Syst. **110**, 43–54 (2000)
7. Buckley, J.J., Jowers, L.: Simulating Continuous Fuzzy Systems. Springer, Berlin, Heidelberg (2006)
8. Bede, B., Gal, S.G.: Generalization of the differentiability of fuzzy number valued function with application to fuzzy differential equation. Fuzzy Sets Syst. **151**, 581–599 (2005)
9. Kaleva, O.: Fuzzy differential equations. Fuzzy Sets Syst. **24**, 301–317 (1987)
10. Seikala, S.: On the fuzzy initial value problem. Fuzzy Sets Syst. **24**(3), 319–330 (1987)
11. Song, S., Wu, C.: Existence and uniqueness of solutions to Cauchy problem of fuzzy differential equations. Fuzzy Sets Syst. **110**, 55–67 (2000)
12. Lupulescu, V.: Initial value problem for fuzzy differential equations under dissipative conditions. Inf. Sci. **178**, 4523–4533 (2008)
13. Nieto, J.J.: Rodríguez-Lopez, Euler polygonal method for metric dynamical systems. Inf. Sci. **177**, 4256–4270 (2007)
14. Hullermiere, E.: Numerical methods for fuzzy initial value problems. Int. J. Uncertain., Fuzziness Knowl.-Based Syst. **7**(5), 439–461 (1999)
15. Ma, M., Friedman, M., Kandel, A.: Numerical solutions of fuzzy differential equations. Fuzzy Sets Syst. **105**, 133–138 (1999)
16. Nieto, J.J.: The cauchy problem for continuous fuzzy differential equations. Fuzzy Sets Syst. **102**, 259–262 (1999)
17. Abbasbandy, S., Allahviranloo, T.: Numerical solutions of fuzzy differential equations by taylor method. Comput. Methods Appl. Math. **2**, 113–124 (2002)
18. Abbasbandy, S., Allahviranloo, T.: Numerical solutions of fuzzy differential equations by runge-kutta method of order 2. Nonlinear Stud. **11**(1), 117–129 (2004)
19. Abbasbandy, S., Allahviranloo, T., Darabi, P.: Numerical solutions of N-order fuzzy differential equations by runge-kutta method. Math. Comput. Appl. **16**, 935–946 (2011)
20. Allahviranloo, T., Ehmady, N., Ehmady E.: Numerical solutions of fuzzy differential equations by predictor-corrector method. Inf. Sci. **177**, 1633–1647 (2007)
21. Parandin, N.: Numerical solutions of fuzzy differential equations by runge-kutta method of 2nd order. J. Math. Ext. **7**, 47–62 (2013)

22. Bede, B.: Note on numerical solutions of fuzzy differential equations by predictor corrector method. Inf. Sci. **178**, 1917–1922 (2008)
23. Abbasbandy, S., Allahviranloo, T., Lopez, O., Nieto, J.J.: Numerical methods for fuzzy differential inclusions. Comput. Math Appl. **48**, 1633–1641 (2004)
24. Jayakumar, T., Kanagarajan, K., Indrakumar, S.: Numerical solution of nth-order fuzzy differential equation by runge-kutta method of order five. Int. J. Math. Anal. **6**, 2885–2896 (2012)
25. Ghazanfari, B., Shakerami, A.: Numerical solution of fuzzy differential equations by extended Runge-Kutta like formula of order 4. Fuzzy sets and syst. **189**, 74–91 (2012)
26. Bede, B., Rudas, I.J., Bencsik: First order linear fuzzy differential equation under generalized differentiability. Inform. Sci. **177**, 1648–1662 (2007)
27. Georgion, D.N., Nieto, J.J., Rodrigue-Lopez, R.: Initial value problem for higher order fuzzy differential equations. Nonlinear Anal. **63**(4), 587–600 (2005)
28. Buckley, J.J., Feuring, T.: Fuzzy initial value problem for nth-order linear differential equations. Fuzzy Sets Syst. **121**(2), 247–255 (2001)
29. Mosleh, M.: Fuzzy neural network for solving a system of fuzzy differential equations. Appl. Soft Comput. **13**, 3597–3607 (2013)
30. Mosleh, M., Otadi, M.: Simulation and evaluation of fuzzy differential equations by fuzzy neural network. Appl. Soft Comput. **12**, 2817–2827 (2012)
31. Purnima, P., Payal, S.: Prey-Predator model and fuzzy initial condition. Int. J. Eng. Innov. Technol. (IJEIT) **3**(12) (2014)
32. Buckley, J.J., Feuring, T., Hayashi, Y.: Linear System of first order ordinary differential equations: fuzzy initial conditions. Soft. Comput. **6**, 415–421 (2002)
33. Oberguggenberger, M., Pittschmann: Differential equations with fuzzy parameters. Math. Comput. Model. Dyn. Syst. **5**(3), 181–202 (1999)
34. Purnima, P., Payal, S.: Numerical technique to solve dynamical system involving fuzzy parameters. Int. J. Emerg. Trends Technol. Comput. Sci. (IJETTCS) **6**(4), 051–057 (2017). ISSN 2278-6856
35. Xu, J., Zhigao, L., Neito, J.J.: A class of linear differential dynamical system with fuzzy matrices. J. Math. Anal. Appl. **368**, 54–68 (2010)
36. Ghazanfari, B., Niazi, S., Ghazanfari, A.G.: Linear matrix differential dynamical system with fuzzy matrices. Appl. Math. Model. **36**, 348–356 (2012)
37. Pandit, P., Payal, S.: Fuzzy Laplace transform technique to solve linear dynamical system with fuzzy parameters. In: Proceeding International Conference on "Research and Innovations in Science, Engineering and Technology" ICRISET-2017 (2017)
38. Allahviranloo, T., Ahmadi: Fuzzy Laplace transforms. Soft Comput. **14**, 235–243 (2010)
39. Salahshour, S., Allahviranloo, T.: Applications of fuzzy Laplace transforms. Soft. Comput. **17**, 145–158 (2013)
40. Eljaoui, E., Mellani, S., Saadia Chadli, L.: Solving second order fuzzy differential equation by the fuzzy Laplace transform method. Adv. Differ. Equ. (2015). https://doi.org/10.1186/s13662-015-0414-x
41. Hayder, A.K., Ali, H.F.M.: Fuzzy Laplace transforms for derivatives of higher orders. Math. Theory Model. **4**(2) (2014)
42. Sita, C.: On the solutions of first and second order nonlinear initial value problems. In: Proceedings of the World Congress on Engineering 2013, vol. I, WCE 2013, July 3–5, London, U.K. (2013)
43. Wu, H.C.: The improper fuzzy Riemann integral and its numerical integration. Infom. Sci. **111**, 109–137 (1999)
44. Kaleva, A note on fuzzy differential equations. Nonlinear Anal. **64**, 895–900 (2006)
45. Klir, G.J., Yuan, B.: Fuzzy Sets and fuzzy Logic: Theory and Applications. Prentice Hall, Englewood Cliffs, NJ (1995)
46. Akin, O., Oruc, O.: A Prey Predator model with fuzzy initial values. Hacet. J. Math. Stat. **41**(3), 387–395 (2012)

Comparison of Performance of Four-Element Microstrip Array Antenna Using Electromagnetic Bandgap Structures

K. Prahlada Rao, R. M. Vani and P. V. Hunagund

Abstract This paper discusses the comparison of performance of conventional microstrip array antenna using two different types of two-dimensional electromagnetic bandgap structures. The electromagnetic bandgap structures employed are fork type and plus shape. Design of miniaturized and low-cost four-element microstrip array antenna with reduced mutual coupling is the main objective of this paper. The former and latter electromagnetic bandGap structures produced miniaturization of 28.20 and 22.42%, respectively. Both the electromagnetic bandgap structures have accounted for considerable decrease in mutual coupling of conventional microstrip array antenna. The radiation characteristics of the modified array antenna depict a decrease in back lobe power. The array antennas are designed using Mentor Graphics IE3D simulation software. The experimental results are taken using vector network analyzer.

1 Introduction

Wireless communication is the transfer of information between two points in space which are not electrically connected. The antenna designers are focusing on fabricating antennas with low profile, less weight, and low cost. This assures the reliability and effectiveness of the antenna under test. Microstrip antennas and arrays are possible solutions toward achieving these objectives. These antennas are also easily compatible with other electrical devices and systems. Materials with dielectric constant ranging from 2.2 to 14 are used to fabricate the microstrip antennas. A large

K. Prahlada Rao (✉) · P. V. Hunagund
Department of PG Studies and Research in Applied Electronics, Gulbarga University, Gulbarga 585106, Karnataka, India
e-mail: pra_kaluri@rediffmail.com

R. M. Vani
University Science Instrumentation Center, Gulbarga University, Gulbarga, Karnataka, India

© Springer Nature Singapore Pte Ltd. 2020
K. N. Das et al. (eds.), *Soft Computing for Problem Solving*,
Advances in Intelligent Systems and Computing 1048,
https://doi.org/10.1007/978-981-15-0035-0_14

181

number of dielectric materials like Rogers RT-Duroid, alumina, glass epoxy, PTFE-woven glass fiber, etc. Dielectric substrate plays a very crucial role in miniaturization and impedance matching [1–7].

Microstrip antenna arrays are superior to the single element antennas in terms of bandwidth, gain, directivity, and radiation characteristics. However, these antenna arrays suffer from a major limitation of severe mutual coupling values between the array elements because of surface waves that are generated due to the TE and TM modes resonating at very high frequencies. These waves travel inside and are confined to the substrate. When the array elements are closely spaced these surface waves can travel to the adjacent antenna elements and result in undesirable coupling. This negative effect on the microstrip array antennas can be minimized using different techniques, namely, using metamaterials, defective ground structures, electromagnetic bandgap (EBG) structures, etc. EBG structures are usually periodic arrangement of unit cells. They are an effective tool to suppress surface wave propagation, thereby enhancing the efficiency and radiation properties of the array antenna. They can be one, two, or three dimensional in nature [8–13].

2 Design of Conventional Microstrip Array Antenna

The conventional microstrip array antenna (CMAA) consists of four identical radiating patches designed at a frequency of 6 GHz. The dielectric used to fabricate the array antenna is FR-4 glass epoxy with a dielectric constant of 4.2 and loss tangent of 0.0245. The dimensions of each radiating patch are L_p and W_p equal to 15.73 and 11.76 mm, respectively. Quarter wave transformer is having dimensions of (L_t and W_t) 6.47 mm × 0.47 mm. The dimensions of the feed are L_f and W_f equal to 6.52 mm × 3.05 mm. Using this design, array antenna parameters return loss, resonant frequency, bandwidth, and virtual size reduction can be measured (Fig. 1).

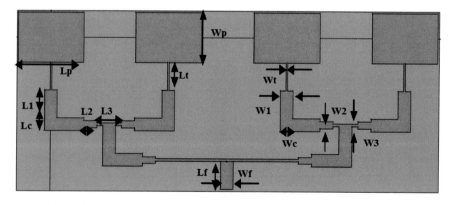

Fig. 1 Schematic of CMAA

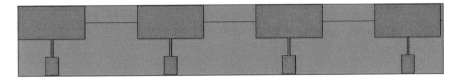

Fig. 2 Schematic of CMAA for the measurement of mutual coupling

For closely spaced array elements, interference is very high (> -20 dB). Interference between the array elements is measured in terms of mutual coupling. To measure mutual coupling the four equally spaced array antennas are fed separately as shown in Fig. 2. The distance between the antennas is equal to quarter of a wavelength. In Fig. 1 also the distance between the array elements is maintained the same.

The proposed technique involves the loading of EBG structures in the finite ground plane of CMMA. The EBG structures loaded are of slot type.

3 Configurations of EBG Shapes

To examine the performance of CMAA with EBG structures, two novel EBG slot structures are used. The first type of EBG structure employed is the fork shape slot EBG and the second one is the plus shape slot EBG (Fig. 3).

In the fork shape slot unit cell, $A = 10$ mm, $B = 2.5$ mm, $C = 2.5$ mm, and $D = 5$ mm. In the plus shape slot unit cell, $P = 9$ mm and $Q = 2$ mm.

The former EBG structure is designed by arranging the fork shape slot unit cell in an array of 4 rows and 14 columns. The periodicity is $S = 3$ mm (Fig. 4).

The latter EBG structure is designed by arranging the plus shape slot unit cell in an array of 4 rows and 9 columns. The unit cells are repeated after every $S_1 = 5$ mm (Fig. 5).

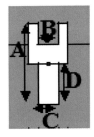
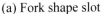
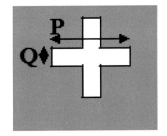

(a) Fork shape slot   (b) Plus shape slot

Fig. 3 Unit cells of EBG structures. **a** Fork shape slot. **b** Plus shape slot

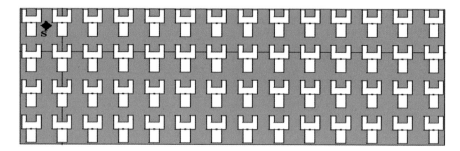

Fig. 4 Fork shape slot EBG structure

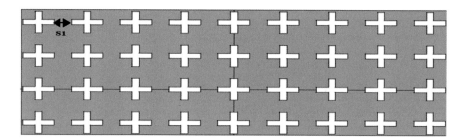

Fig. 5 Plus shape slot EBG structure

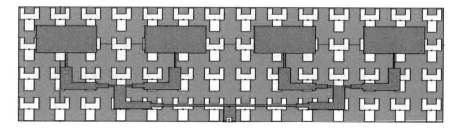

Fig. 6 CMAA with fork shape EBG

4 Array Antenna Design Using EBG Structures

The EBG structures discussed in Sect. 3 are integrated into the ground plane of CMAA. The relevant schematics are depicted in Figs. 6, 7, 8 and 9, respectively.

5 Photographs of the Fabricated Array Antennas

See Figs. 10, 11, 12, 13, 14 and 15.

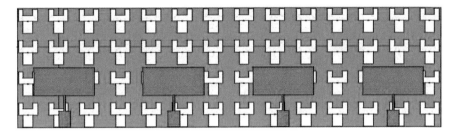

Fig. 7 CMAA with fork shape EBG for mutual coupling measurement

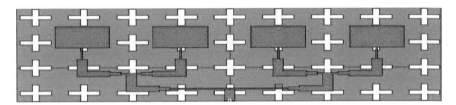

Fig. 8 CMAA with plus shape EBG

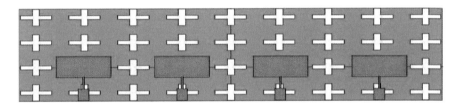

Fig. 9 CMAA with plus shape EBG for mutual coupling measurement

(a) Front view   (b) Back view

Fig. 10 CMAA

(a) Front view   (b) Back view

Fig. 11 CMAA for mutual coupling measurement

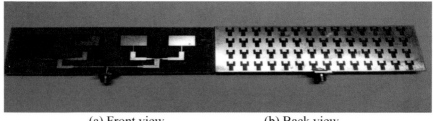

(a) Front view   (b) Back view

Fig. 12 CMAA with fork shape EBG structure

6 Results and Discussion

Connectors are used to connect the vector network analyzer and the array antennas. In this paper, female SMA (SubMiniature version A) type connectors are employed. The performance of CMAA with both the EBG structures loaded is compared to judge the better candidate. Various performance parameters are taken into consideration. Figures 16, 17 and 18 show the graphs of return loss and mutual coupling versus frequency of CMAA while Figs. 19, 20, 21, 22, 23, and 24 depict the graphs of CMAA with EBG structures.

6.1 Resonant Frequency

Return loss is represented by the S-parameter S_{11}. From Figs. 16, 17, and 18 the fundamental resonant frequency of CMAA is 5.53 GHz. A look at Figs. 19, 20, and 21 depict that the fundamental resonant frequency of array antenna with fork shape slot is 3.97 GHz.

(a) Front view   (b) Back view

Fig. 13 CMAA with fork shape EBG structure for mutual coupling measurement

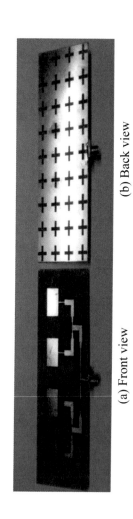

(a) Front view   (b) Back view

Fig. 14 CMAA with plus shape EBG structure

(a) Front view   (b) Back view

Fig. 15 CMAA with fork shape EBG structure for mutual coupling measurement

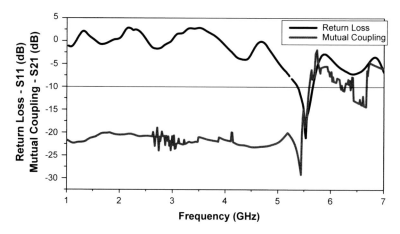

Fig. 16 Plot of return loss and mutual coupling—S_{21} versus frequency of CMAA

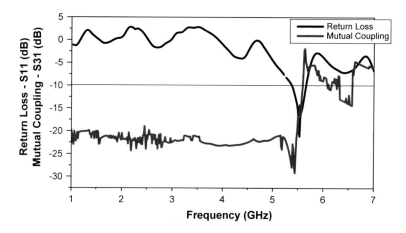

Fig. 17 Plot of return loss and mutual coupling—S_{31} versus frequency of CMAA

However, with plus shape slot EBG structures the array antenna is producing a fundamental resonant frequency of 4.29 GHz as depicted in Figs. 22, 23, and 24.

The fundamental resonant frequencies of 3.97 and 4.29 GHz obtained by using fork shape and plus shape EBG structures correspond to a virtual size reduction of 29.01 and 22.42%. The fork shape slot EBG structure is resulting in a better virtual size reduction.

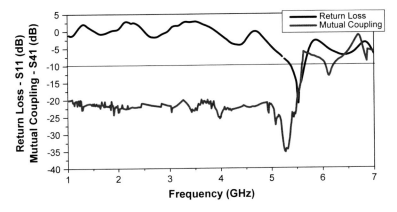

Fig. 18 Plot of return loss and mutual coupling—S_{41} versus frequency of CMAA

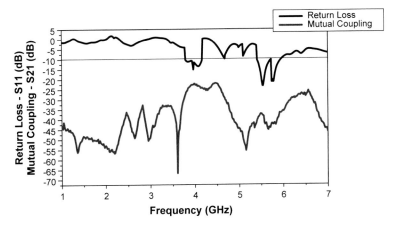

Fig. 19 Return loss and mutual coupling—S_{21} against frequency of CMAA with fork shape. EBG structure

6.2 Bandwidth

The parameter bandwidth is calculated by subtracting the lower frequency point from the upper frequency point where the return loss is -10 dB value. The bandwidth of CMAA is equal to 273 MHz as depicted in Figs. 16, 17, and 18, respectively. The overall bandwidth (%) is determined by using the formula

$$\text{Bandwidth}(\%) = \frac{\text{Bandwidth}}{\text{Resonant Frequency}} \times 100 \qquad (1)$$

This results in bandwidth (%) of 4.89%.

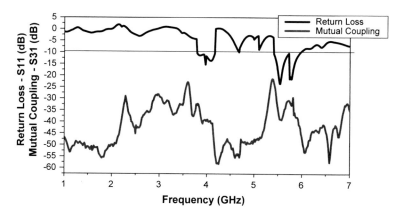

Fig. 20 Return loss and mutual coupling—S_{31} versus frequency of CMAA with fork shape. EBG structure

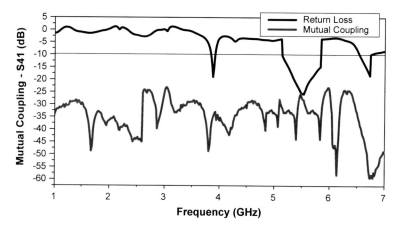

Fig. 21 Return loss and mutual coupling—S_{41} versus frequency of CMAA with fork shape. EBG structure

From Figs. 19, 20, and 21 the fork shape slot EBG structure produces bandwidths of 410 and 600 MHz, respectively. Hence the overall bandwidth (%) is equal to 19.08%. On the other hand, the plus shape slot EBG structure produces bandwidths of 110, 200, and 170 MHz, respectively. In this case the overall bandwidth (%) is calculated as 9.53%.

In terms of overall bandwidth (%), the fork shape slot EBG structure is producing superior performance compared to its counterpart.

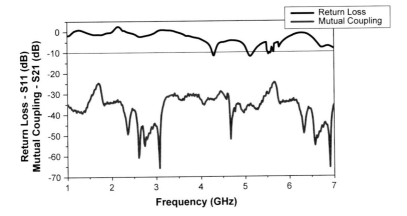

Fig. 22 Return loss and mutual coupling—S_{21} against frequency of CMAA with plus shape. EBG structure

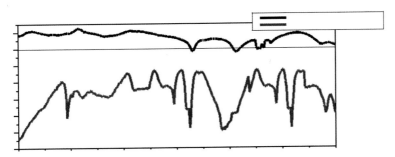

Fig. 23 Return loss and mutual coupling—S_{31} against frequency of CMAA with plus shape. EBG structure

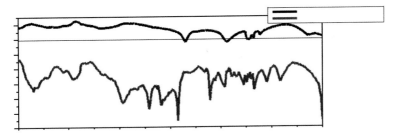

Fig. 24 Return loss and mutual coupling—S_{41} against frequency of CMAA with plus shape. EBG structure

6.3 Mutual Coupling

From Figs. 16, 17, and 18, at the fundamental frequency of CMAA, i.e., 5.53 GHz, the return loss and the mutual coupling plots are overlapping and the values of the mutual coupling coefficients (S_{21}, $S_{31,}$ and S_{41} are greater than -20 dB) (S_{21} = -16.95 dB, S_{31} = -14.22 dB, and S_{41} = -17.30 dB, respectively). This implies that there is severe interference between the antenna 1 and the antennas 2, 3, and 4. Whenever the mutual coupling value is more than -20 dB, it implies the level of interference is high. On the other hand when it is less than -20 dB, it implies the level of interference is low.

With the incorporation of fork shape slot and plus shape slot EBG structures, the return loss and mutual coupling plots are no more overlapping at the frequency of 5.53 GHz. This is a sign of reduction of interference between the transmitting and receiving antennas. The mutual coupling coefficients are reduced to S_{21} = -38.04 dB, S_{31} = -38.64 dB, S_{41} = -38.4 dB using fork shape slot EBG structure and S_{21} = -30.42 dB, S_{31} = -23.70 dB, S_{41} = -35.8 dB using plus shape slot EBG structure. These S-parameter values indicate that mutual coupling values are reduced to a lesser value using fork shape slot EBG than plus shape slot EBG. Lesser the values of mutual coupling coefficients, better is the performance of the array antenna.

With the above discussion, the CMAA is performing better using fork shape slot EBG structure than using plus shape slot EBG structure in terms of reduction of mutual coupling.

6.4 Radiation Pattern

At the angle of $270°$, the amount of backward power radiated is measured. Using fork shape slot EBG structure, the backward power radiated is equal to -9 dB as compared to -5.5 dB using plus shape slot EBG. Without EBG, the array antenna is radiating a backward power of -4.5 dB. Therefore, the decrease in the amount of back lobe radiation is lesser using fork shape slot EBG than plus shape slot EBG (Fig. 25).

Front to back ratio (FBR) is calculated by deducting the back power (dB) from the front power (dB). The front power radiated by the array antenna without EBG is -2 dB. With the introduction of fork shape and plus shape EBG structures, the front power is equal to -2.5 and 0.5 dB, respectively. The front to back ratio of the array antenna without EBG is equal to 2.5 dB. The values calculated using fork shape and plus shape EBG are 6.5 and 6 dB, respectively. Among the two EBG structures employed, fork shape EBG structure produces better FBR value compared to its opponent. The significance of FBR is it specifies how effectively the antenna is radiating power in the desired or wanted or forward direction and in the undesired

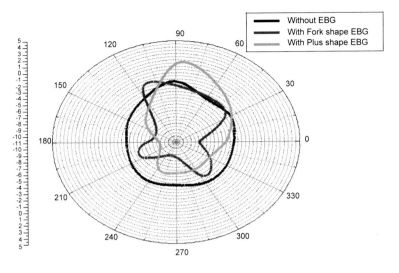

Fig. 25 Radiation patterns

or unwanted or backward direction. Hence, higher the FBR of an antenna it leads to the better antenna.

7 Conclusion

In this paper, the study of performance of CMAA with the novel 2D-EBG structures has been proposed and examined experimentally. The fork shape slot and plus shape slot EBG structures have been designed. The former EBG structure has overcome the limitations of CMAA better than the latter EBG structure. The CMAA has performed better with fork shape EBG structure than with plus shape EBG structure.

References

1. Balanis, C.A.: Antenna Theory, Analysis and Design, 2nd edn. Wiley (1997)
2. Bahl, I.J., Bhartia, P.: Microstrip Antennas. Artech House (1980)
3. Scott, J.: Lecture notes of EEET1071/1127 Microwave and Wireless Passive Circuit Design
4. Yang, F., Rahmat-Samii, Y.: Electromagnetic Band Gap Structures in Antenna Engineering. Cambridge University Press (2009)
5. Elsheakh, D.N., Abdallah, E.A., Iskander, M.F., Elsadek, H.A.: Microstrip antenna array with new 2D-electromagnetic band gap structure shapes to reduce harmonics and mutual coupling. Prog. Electromagn. Res. C **12**, 203–213 (2010)
6. Benikhelf, F., Boukli-Hacene, N.: Mutual coupling reduction in microstrip antenna arrays using EBG structures. Int. J. Comput. Sci. Issues, **9**(4, 3), 265–269 (2012)

7. Alsulami, R., Song, H.: Double-sided microstrip circular antenna array for WLAN/WiMAX applications. J. Electromagn. Anal. Appl. **5**, 182–188 (2013)
8. Saxena, D., Agarwal, S., Srivastava, S.: Low cost E-shaped microstrip patch antenna array for WLAN. Int. J. Adv. Res. Electr., Electron. Instrum. Eng. **3**(4), 8831–8838 (2014)
9. Bait-Suwailam, M.M., Siddiqui, O.F., Ramahi, O.M.: Mutual coupling reduction between microstrip patch antennas using slotted-complementary split-ring resonators. IEEE Antennas Wirel. Propag. Lett. **9**, 876–878 (2010)
10. Chauhan, S., Singhal, P.K.: Enhancement of bandwidth of rectangular patch antenna using multiple slots in the ground plane. Int. J. Res. Electron. Commun. Technol. **1**(2), 30–33 (2014)
11. Zainud-Deen, S.H., Badr, M.E., El-Deen, E., Awadalla, K.H., Sharshar, H.A.: Microstrip antenna with defected ground plane structure as a sensor for landmines detection. Prog. Electromagn. Res. B **4**, 27–39 (2008)
12. Verma, A.: EBG structures and its recent advances in microwave antenna. Int. J. Sci. Res. Eng. Technol. **1**(5), 84–90 (2012)
13. Rana, R., Vyas, N., Verma, R., Kaushik, V., Arya, A.K.: Dual stacked wideband microstrip antenna array for Ku-band applications. **4**(6),132–135 (2014)

Model Development for Strength Properties of Laterized Concrete Using Artificial Neural Network Principles

P. O. Awoyera⬤, J. O. Akinmusuru, A. Shiva Krishna, R. Gobinath, B. Arunkumar and G. Sangeetha

Abstract This study develops predictive models for determination of strength parameters of laterized concrete made with ceramic aggregates, based on the principle of Artificial Neural Networks (ANN). The model development follows the results of the experimental phase (covering compressive and split-tensile strengths), where numerous materials were used in varying proportions: ceramics (fine and coarse fractions), river sand, and granite were substituted between 0 and 100%, laterite between 0 and 30%, and curing ages between 3 and 91 days. The cement proportion was maintained at 100%, and the water–cement ratio was 0.6. The model development was performed in MATLAB based on the Levenberg–Marquardt (LM) principles, where input data were separated in ratio 70%:15%:15% for learning, testing, and validation phases, respectively. After several trials, the selected model architecture, based on satisfactory performance in terms of means square error, contains eight-input layer, ten-hidden layer, and two-output layer neurons.

Keywords Ceramics · Laterized concrete · Levenberg–Marquardt · MATLAB · Modeling · Strength properties

1 General Introduction

Recent advances in research have revealed a number of materials which can be used as partial or complete replacement of cement, aggregates, and reinforcement bar in concrete. The alternative materials, which are mostly sourced from industrial and construction wastes, have a major benefit of ensuring sustainability of the built

P. O. Awoyera (✉) · J. O. Akinmusuru
Department of Civil Engineering, Covenant University, Ota, Nigeria
e-mail: paul.awoyera@covenantuniversity.edu.ng

A. Shiva Krishna · R. Gobinath
SR Engineering College, Warangal, Telangana, India

B. Arunkumar · G. Sangeetha
Center for Artificial Intelligence and Deep Learning, SR Engineering College, Warangal, Telangana, India

© Springer Nature Singapore Pte Ltd. 2020
K. N. Das et al. (eds.), *Soft Computing for Problem Solving*,
Advances in Intelligent Systems and Computing 1048,
https://doi.org/10.1007/978-981-15-0035-0_15

environment [1, 2]. With several investigations continually conducted to examine the possibility of reusing industrial and construction wastes, ceramics obtained from construction and demolition activities is one of those materials found to possess engineering properties that are somewhat similar to the natural aggregates.

A novel mixture, incorporating ceramic aggregate and laterite (a natural soil material in Sub-Saharan African countries), has been investigated by the author in parallel studies [3, 4]. Those studies revealed the potential of the materials as sustainable alternatives to the conventional aggregates. It is to be noted that the concrete containing portion of laterite as natural sand is mostly called laterized concrete [5]. Laterite has been used with other materials, such as laterized concrete produced using blended fly ash, where Ogunbode et al. [6] reported that flexural strength of the concrete increased with increasing curing age, but decreased with increasing laterite and fly ash content. Thus, the completed studies on the potential use of ceramics and laterite as ingredients in concrete have reported strength and microscale properties of the concrete. However, to the best of our knowledge, there is no available predictive model that can forecast the performance of the concrete. Therefore, this study develops predictive models that optimize the compressive and split-tensile strengths of laterized concrete covering the various materials substitution levels.

1.1 Artificial Neural Networks: General Concept

The modeling approach based on Artificial Neural Network (ANN) principles is an advancement to the known regression statistics approach. In ANN, completed experimental data are used to develop a model in a more comprehensive manner than the regression methods. The way ANN works is an information processing paradigm [7], which functions as the human nervous system and the brain process information [8, 9]. One major novelty in ANN is the way it processes system of information.

The ANN architecture comprises several highly dependent processing elements, which are also known as neurons. The neurons work together to solve problems. ANN is similar to human in its operation, it learns by the provided examples through its ability in pattern recognition or data classification [10]. The application of ANN also can be found in areas such as categorization, prediction and forecasting, and optimization [11].

ANN is mostly utilized for developing models for forecasting the behavior of nonlinear systems [12]. The general expression on which the ANN principle is based is

$$\text{output} = \sum_{n=0}^{n} X_n W_n - b \tag{1}$$

where W_n is weight, X_n is input, and b is a bias.

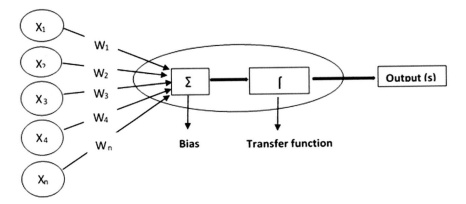

Fig. 1 Architecture of an ANN model

A study by Pala et al. [13] has established that ANN could predict the response of a system no matter the accuracy or completeness of the data input. There are three major phases of executing ANN analysis; these include training, testing, and model validation. In the training phase, both the weights and biases are adjusted in order to obtain the desired output. The testing phase covers the series of reply by the network to input without altering the model [10].

A sample architecture of an ANN model is presented in Fig. 1.

ANN data analysis involves trial and error at every phase, in order to establish a suitable model for the data. However, the model that can be selected must have minimum mean square error (MSE) based on comparison of predicted outputs and actual outputs.

The suitability of ANN for civil engineering application has been reported in a number of experiments, where it has been utilized for developing predictive models for construction application [9, 14–17], concrete shrinkage, and corrosion characteristics [18].

The application of ANNs, a proven modeling tool, for predicting properties of concrete elements has been widely investigated. From the published data in literature, it was reported that ANN results are more reliable than regression models. However, the use of ANNs for predicting the performance of modified concrete elements is not overly explored. Therefore, the application of ANNs to ceramic-laterized mortar and concrete, in which many material mixtures are varied at a time, can be an interesting development. In essence, this will facilitate the use of the materials being tested for construction applications.

The modeling of strength characteristics of concrete was done using regression model in last few decades. However, the use of ANN was introduced recently, which takes care of the limitation of regression models. Due to the nonlinear behavior of concrete, so model development using linear regression is deficient for predicting strength of concrete. In this regard, the use of ANN for predicting strength of laterized concrete is needed.

2   Materials and Methods

Figure 2 shows the raw materials that were used in the experimental phase of this study. The aggregates were preliminarily treated, by air-drying in the laboratory for 7 days, prior to use. This was done in order to dry off the in situ moisture from the materials before they are utilized. The properties of the raw materials were already determined in parallel studies [19–21], the same has been summarized in Table 1. At the experimental phase, the mix proportion adopted for the binder, fine aggregate, and coarse aggregate is 1:1.5:3, with 0.6 water/cement ratio. The mix is suitable for concrete containing laterite as aggregate, in that, it takes care of the marginality of any of the aggregates.

In developing a predictive model for the strength properties of the laterized concrete, the strength results obtained from experimental analysis of concrete made from different mix proportions of the materials were utilized.

That is, the strength results obtained, at every substitution level of the raw materials, or variation of factors, are as follows:

Fig. 2   Aggregates, **a** river sand, **b** ceramic fine, **c** laterite, **d** ceramic coarse, **e** granite

Table 1   Materials' characteristic features

Properties	River sand	Laterite	Fine ceramic	Coarse ceramics	Granite
Specific gravity	2.61	2.13	2.26	2.31	2.87
Water absorption (%)	2.24	4.70	2.52	0.55	0.23
Fineness modulus (%)	2.24	1.80	2.20	6.88	6.95

Table 2 Inputs and outputs data for modeling

Input data	Minimum	Maximum
Cement (%)	100	100
Sand (%)	0	100
Ceramic fine fraction (%)	0	100
Laterite (%)	0	30
Granite (%)	0	100
Ceramic coarse fraction	0	100
w/c (%)	0.6	0.6
Age (days)	3	91
Outputs		
Compressive strength (kN/m^2)		
Split-tensile strength (kN/m^2)		

i. Ceramics fine was used as a replacement for sand at 0, 25, 50, 75, and 100%,
ii. Laterite was used as a replacement for fine aggregate (sand), at 0, 10, 20, and 30%, and
iii. Ceramics coarse was used as a replacement for gravel at 0, 25, 50, 75, and 100%.
iv. Curing ages at 3, 7, 14, 28, and 91 days.

The upper and lower limits of the materials were selected based on the results of trial mixes. Thus, for the modeling, the input and output parameters which were used during training and testing phases are contained in Table 2.

The ANN model development was performed using MATLAB tools, where the error backpropagation and recall algorithm, as described by Zurada [22] was adopted. The feedforward backproportion model was utilized. Figure 3 shows the flow chart of ANN development. Sampling comprises 125 data samples obtained from completed parallel studies for ANN development.

The input data were separated in ratio 70%:15%:15% for learning, testing, and validation phases, respectively. A number of trials were made before finalizing the model for validation. This is necessary in order to ensure accuracy of the model.

3 Results and Discussion

Using the ANN principle, a suitable model that can predict both compressive strength and the split-tensile strength of ceramic-laterized concrete has been developed. Since the precision of ANN prediction largely depends on the selected network architecture [24], a number of trials were made before the final model was selected. The input data for the ANN model were cement, river sand, laterite, ceramic fine aggregate, granite, ceramic coarse aggregate, water/cement ratio and curing age, these factors normally have significant effect on the properties of the hardened ceramic-laterized concrete. While the output parameters are compressive and split-tensile strengths

Fig. 3 Flowchart for
modeling using ANN in
MATLAB. *Source* Awoyera
[23]

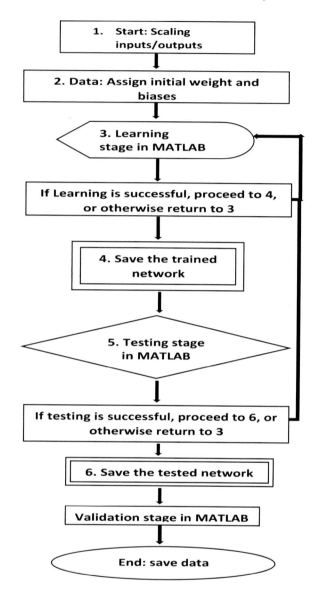

that were obtained after each curing regimes. The best ANN architecture selected
for the model is presented in Fig. 4. Based on the principles of ANN [21], the best
model was selected using minimum MSE criteria; in that, out of all the trials that
were performed, the model yielded lowest MS. In other words, better performance
of ANN model can only be ensured if R^2 is higher and MSE is lower in a model [17],
and is more suitable for prediction.

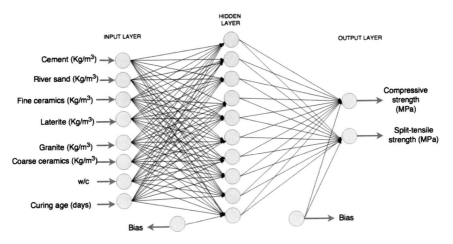

Fig. 4 Developed ANN model

The selected model architecture, 8-10-2, has eight-input neurons (input layer), ten-hidden neurons (hidden layer), and two predefined output neurons (output layer). Table 3 presents the evaluation of the performance of the best ANN architecture. Both MSE and R^2 were determined using Eqs. 2 and 3 [7, 15, 16]:

$$MSE = \frac{\sum_{i=1}^{n}(o_i - t_i)^2}{n} \tag{2}$$

$$R^2 = \frac{\sum(o - t)^2}{\sum(o - o_{mean})^2} \tag{3}$$

where n represents total data sets, o is network output, t is target output, and o_{mean} = network output average.

Figure 5a, b shows the distribution of the predicted data values, the actual data values, and their corresponding error values, for compressive strength and split-tensile strength of the data sets studied, respectively. A strong statistical correlation exists between the predicted data and experimental data (actual), because of negligible variation from both data. Figure 6 presents the matching of the predicted and original data for the train, test, validation and all data sets. This result shows that the predicted data was in agreement with the experimental data. The R^2 values in all the four cases

Table 3 ANN architecture performance

Final model	Training		Test		Validation	
	MSE	R^2	MSE	R^2	MSE	R^2
8-10-2	0.00603	0.99984	0.00566	0.9983	0.00564	0.99846

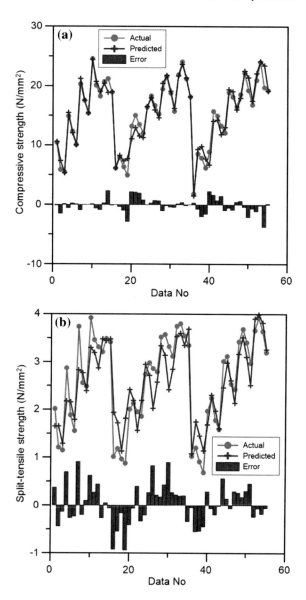

Fig. 5 Predicted and actual values **a** compressive strength, **b** split-tensile strength for all data

were very near to 1. Thus, this is a huge indication that the input and output values are within proximity.

Furthermore, the predicted strength has minimal percent error, which adjudges that the model is statistically reliable for prediction.

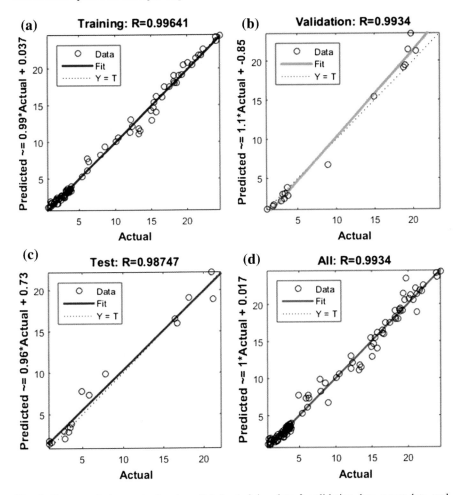

Fig. 6 Data correlations: actual and predicted **a** training data, **b** validation data, **c** test data, and **d** all data

4 Conclusion

This research develops predictive models for determination of strength properties of laterized concrete using artificial neural network principles. The conclusions drawn from the study are listed as follows:

i.   In line with the need for sustainability in the construction industries, this study has demonstrated the significance of combining ceramics, laterite and other conventional aggregates for concrete production. Also, based on the principles of ANN, a best model from the analysis has been selected based on its minimum MSE value compared to every other trials.

ii. The selected ANN architecture from this study was 8-10-2, which has 8 input neurons, 10 hidden neurons, and 2 predefined output neurons. The model developed has the capacity to predict with accuracy. The predicted data values were close to the experimental data values. In addition, the R^2 values in all the four cases (tests data, trained data, validated data, and all data) were near to 1, which thus signifies that there is a strong statistical correlation between the input and output values of the selected model.

References

1. Selvarajkumar, P., Murthi, P., Gobinath, R., Awoyera, P.O.: Eco-friendly high strength concrete production using silica mineral waste as fine aggregate—an ecological approach. Ecol. Environ. Conservat. **24**(2), 909–915 (2018)
2. Banu, T., Chitra, G., Gobinath, R., Awoyera, P.O., Ashokumar, E.: Sustainable structural retrofitting of corroded concrete using basalt fiber composite. Ecol. Environ. Conservat. **24**(3), 353–357 (2018)
3. Awoyera, P.O., Akinmusuru, J.O., Dawson, A.R., Ndambuki, J.M., Thom, N.H.: Microstructural characteristics, porosity and strength development in ceramic-laterized concrete. Cement Concrete Compos. **86** (2018)
4. Awoyera, P.O., Dawson, A.R., Thom, N.H., Akinmusuru, J.O.: Suitability of mortars produced using laterite and ceramic wastes: mechanical and microscale analysis. Constr. Build. Mater. **148** (2017)
5. Olusola, K.: Some Factors Affecting Compressive Strength and Elastic Properties of Laterite Concrete, Unpublished Ph.D Thesis. Department of Building, Obafemi Awolowo University, Nigeria (2005)
6. Ogunbode, E., Ibrahim, S., Kure, M., Saka, R.: flexural performance of laterized concrete made with blended flyash cement (Fa-Latcon). Greener J. Sci. Eng. Technol. Res. **3**(4), 102–109 (2013)
7. Tanyildizi, H., Özcan, F., Atis, C.D., Karahan, O., Uncuog, E.: Comparison of artificial neural network and fuzzy logic models for prediction of long-term compressive strength of silica fume concrete. Adv. Eng. Softw. **40**, 856–863 (2009)
8. Zhang, G., Patuwo, B.E., Hu, M.Y.: Forecasting with artificial neural networks: the state of the art **14**, 35–62 (1998)
9. Sobhani, J., Najimi, M., Pourkhorshidi, A.R., Parhizkar, T.: Prediction of the compressive strength of no-slump concrete: a comparative study of regression, neural network and ANFIS models. Constr. Build. Mater. **24**(5), 709–718 (2010). Accessed from http://dx.doi.org/10.1016/j.conbuildmat.2009.10.037
10. Shafabakhsh, G., Jafari Ani, O., Talebsafa, M.: Artificial neural network modeling (ANN) for predicting rutting performance of nano- modified hot-mix asphalt mixtures containing steel slag aggregates. Constr. Build. Mater. J. **85**, 136–143 (2015)
11. Parichatprecha, R., Nimityongskul, P. (2009). Analysis of durability of high performance concrete using artificial neural networks. Constr. Build. Mater. **23**(2), 910–917. Accessed from http://dx.doi.org/10.1016/j.conbuildmat.2008.04.015
12. Alshihri, M.M., Azmy, A.M., El-bisy, M.S.: Neural networks for predicting compressive strength of structural light weight concrete. Constr. Build. Mater. **23**(6), 2214–2219. Accessed from http://dx.doi.org/10.1016/j.conbuildmat.2008.12.003
13. Pala, M., Özbay, E., Öztas, A., Yüce, M.: Appraisal of long-term effects of fly ash and silica fume on compressive strength of concrete by neural networks. Constr. Build. Mater. **21**(2), 384–394 (2007)

14. Bhatti, M.A.: Predicting the compressive strength and slump of high strength concrete using neural network **20**, 769–775 (2006)
15. Chen, L.: Grey and neural network prediction of concrete compressive strength using physical properties of electric arc furnace oxidizing slag. J. Environ. Eng. Manag. **20**(3), 189–194 (2010)
16. Garzón-roca, J., Marco, C.O., Adam, J.M.: Compressive strength of masonry made of clay bricks and cement mortar: estimation based on neural networks and Fuzzy Logic. Eng. Struct. **48**, 21–27 (2013)
17. Sadrmomtazia, A., Sobhanib, J., Mirgozar, M.: Modeling compressive strength of EPS lightweight concrete using regression, neural network and ANFIS. Constr. Build. Mater. **42**, 205–216 (2013)
18. Hodhod, O., Ahmed, H.I., Hodhod, O.A., Ahmed, H.I.: Modeling the corrosion initiation time of slag concrete using the artificial neural network modeling the corrosion initiation time of slag concrete using the artificial neural network, November 2016, pp. 8–12
19. Awoyera, P.O., Akinmusuru, J.O., Ndambuki, J.M.: Green concrete production with ceramic wastes and laterite. Constr. Build. Mater. **117**, 29–36 (2016)
20. Anandaraj, S., Rooby, J., Awoyera, P.O., Gobinath, R.: Structural distress in glass fibre-reinforced concrete under loading and exposure to aggressive environments. Constr. Build. Mater. (2018)
21. Murthi P., Awoyera, P.O., Palanisamy, S., Dharsana, D., Gobinath, R.: Using Silica mineral waste as aggregate in a green high strength concrete: workability, strength, failure mode and morphology assessment. Australian J. Civil Eng. (2018)
22. Zurada, J. (1992). Introduction to artificial neural systems. Info Access Distribution Ltd.
23. Awoyera, P. (2017). Predictive models for determination of compressive and split-tensile strengths of steel slag aggregate concrete. Mater. Res. Innovat. **22**(5), 287–293. Accessed from http://dx.doi.org/10.1080/14328917.2017.1317394
24. Ni, H.-G., Wang, J.-Z.: Prediction of compressive strength of concrete by neural networks. Cement Concr. Res. **30**(8), 1245–1250 (2000). Accessed from http://www.sciencedirect.com/science/article/pii/S0008884600003458

Wind Power Forecasting Using Parallel Random Forest Algorithm

V. Anantha Natarajan and N. Sandhya Kumari

Abstract Wind power keeps on developing and it is broadly observed as the sustainable power source best capable to compete with fossil fuel electricity generation. In the near past wind power forecasting has improved the situation for the estimation of power production in wind farms. In general wind power forecasts are generated in two different ways one namely using Numerical Weather Prediction (NWP) and the other one using physical forecasting methods. Physical forecasting is deeply dependent on meteorological facts and the data from the NWP. The approach of physical method is vulnerable by the way that wind speeds are estimated a few feet over the ground can fluctuate. Statistical scheme encompasses models likewise ANN, SVM, and etc. less reliant on accuracy of Numerical Weather Predictions (NWP), yet relies more extremely on historical information of wind speed at respective areas. To compute large accurate wind energy forecast a good amount of real-time observations of historical observations from the wind farms becomes essential. Wind power forecasts using Support Vector Machines (SVM) and Artificial Neural Networks (ANN) suffers from slow training speed, and poor generalization ability. This paper aims at conducting experiments to assess the performance and test the suitability of the Random Forest Algorithm for wind power forecasting. The prediction results are seeming to be close with the actual wind power generated at the wind farms and it is more accurate when compared to the results of the ANN.

Keywords Wind power forecasting · Numerical weather predictions · Random forest algorithm · Support vector machines · Artificial neural networks

V. A. Natarajan (✉) · N. S. Kumari
Department of Computer Science and Engineering, Sree Vidyanikethan Engineering College, Tirupati, India
e-mail: vananthanatarajan@vidyanikethan.edu

N. S. Kumari
e-mail: sandhyanihitha@gmail.com

© Springer Nature Singapore Pte Ltd. 2020
K. N. Das et al. (eds.), *Soft Computing for Problem Solving*,
Advances in Intelligent Systems and Computing 1048,
https://doi.org/10.1007/978-981-15-0035-0_16

209

1 Introduction

Wind power has normally turned out to be fastest developing energy asset around the globe [1]. The Global Wind Energy Council (GWEC) has predicted that by 2021 the capacity of global wind will reach 800 GW and also it was seen in the year 2017 alone wind farms with a capacity of 60 GW production has been installed. The global wind power production is increasing cumulatively by 12.6% and at present it has reached a total of 486.8 GW [2]. Globally around 90 countries are meeting their power requirements using wind energy and 9 countries among them have a production capacity above 10k MW.

For planning the power distribution from clean renewable power production, forecasting of wind power production becomes essential. The forecasting of wind power refers to the estimation of the expected production in future from a group of wind turbines. At any moment adjust must be kept up between electricity generation and consumption in the electricity grid, otherwise disorders in power supply or quality may happen. Accurate wind power forecasts are critical in diminishing the event or length of reductions which mean to cost funds, enhanced worker security and relieving the physical effects of outrageous climate on wind power systems.

The short-term prediction using advanced approaches for wind power require predictions of meteorological factors. The predictions of meteorological factors greatly rely upon wind direction and wind speed, and also on humidity, pressure and temperature for wind power forecast. The estimation of power few days ahead of production involves prediction of hourly power generation at different wind farms based on historical measurements such as wind speed, direction and power generated. The ability to forecast wind power helps to achieve full compliance of customer energy demands and would help in proper design of the power network. Freshly, quite a lot of techniques [3] have been established for wind power forecasting. Current methods can be classified as time series, statistical and physical modeling techniques based on the forecasting models. Recently, investigates utilize a combination of statistical and physical strategies to get optimal forecast frameworks.

Some algorithms adopted in literature for short-term forecasting of wind power are Support Vector Machine (SVM), Artificial Neural Networks(ANN), Complex Value Recurrent Neural Networks, Gaussian Process, Support Vector Regression with Cuckoo Optimization Algorithm and Genetic Algorithm, Long Short-Term Memory Recurrent Neural Networks, Polynomial model tree learning algorithm, Adaptive Neuro Fuzzy Inference system model (ANFIS), Local Neuro Fuzzy Model (LNF) and Back Propagation Supervised Training Algorithm. When compared with statistical method these algorithms have higher prediction accurateness, but they need a mass of raw data and exist problems, such as the slow training speed, poor generalization ability. The fore mentioned algorithms are compared and reviewed for short-term forecasting of wind power in Table 1.

Table 1 Comparison of short-term forecasting of wind power approaches

Input and data source	Remarks
Wind direction, temperature, wind speed, air pressure and power output PIS (Plant information system), SCADA (supervisory control and data acquisition system) [4]	Method: CRNN (Complex valued recurrent neural network model) Advantage: The output signal of this method does not simply depend on the present input signals of the network. However, in its training process has an internal memory Disadvantage: The Recurrent Neural Network training time is more than that of the Static Neural Network
Location, wind direction, temperature, wind speed, Humidity and air pressure SCADA system with NWP [1]	Method: Gaussian process Advantage: This process is able to find ultimate prediction performance when near limited amount of training of the data model. It is utilised to fabricate the connection between corrected wind power and wind speed Disadvantage: If it will utilize the vast amount of information and effect an issue of slow training process
Actual Wind Speed Collected historical Data from Shandong wind farm in china [5]	Method: PSO-SVR, COA-SVR and GA-SVR These three algorithms are utilized to improve the performance of the SVR technique and gain the optimal generalization ability
Wind Direction (sine and cosine), air pressure, Wind speed and wind power Wind power and NWP information from yilan wind farm [6]	Method: Generalized regression neural network The proposed cluster method is estimated successfully in prediction of wind power This method training samples improves accurateness of short term forecasting of wind power
Wind measures of speed and power noted with an interval of 15 min on top of land at 70 meters North to east China wind farm from year 2014 [7]	LSTM Recurrent Neural Networks
Wind Direction and Wind Speed Sotavento wind farm in Spain four different months of 2010 [8]	Method: Local Neuro fuzzy approach trained by the Polynomial model tree learning algorithm (POLYMOT) This model allows to model highly nonlinear wind power time series in an efficient and accurate manner

(continued)

Table 1 (continued)

Input and data source	Remarks
Wind Speed, wind Direction and history wind power Zhejiang Provincial Electric Power Test Research Institute [3]	Method: Empirical mode decomposition (EMD) and radial basis function neural networks model (RBFNN) RBFNN output is linear weighted sum of hidden unit output and its learning rate faster compared with other neural networks This model use RBF as activation function and neurons in the input space area are very small, so it needs more radial basis neurons [9]
Wind Speed SCADA of Gold wind micro grid wind farm Beijing, China and NWP models [10]	Method: Back propagation supervised training Algorithm Neural network is utilised the back propagation supervised training Algorithm and capture the difficult input/output dealings This algorithm to assess a lesser amount of accurateness of the wind power prediction [11]

Irrespective of the research work analyzed in the literature survey, new techniques and approaches of input and output data manipulations are still in demand with a specific goal to improve accurateness of the prediction and reduce the vulnerability in forecasting of wind power while keeping particularly acceptable computation time [12]. In the proposed work for wind power prediction Artificial Neural Networks and Parallel Random Forest algorithm are used.

The rest of the paper is structured as follows. In Sect. 2 the description about the dataset utilized for testing and training the mathematical model is given. A short brief about the algorithms used for wind power forecasting is presented in Sect. 3. Section 4 defines about the experiments conducted and the results obtained and the Sect. 5 finally concludes the paper.

2   Data Description

The data for wind power forecasting was collected for about three years from seven different wind farms. The data consist of historical power measurements are hourly temporal resolution (wind speed, direction, zonal (u) and meridional (v)), as well as meteorological forecasts of the wind components with a high level of availability over that period in each of the wind farm. Meteorological measurements were gathered at 10 m above ground level for the zonal (u) and meridional (v) components of surface wind. The data has 104,832 samples collected from the year 2009 to 2012. These samples are divided into training and testing set using cross validation.

3 Wind Power Forecasting Methods

There are two wind power forecasting methods implemented in this paper to evaluate the accuracy of forecasting of wind power. For comparability, these methods have been trained with the same input variables of training data, tested with same test dataset and the horizon of 48 h ahead for wind power forecasting. The problem of wind power forecasting is formulated as a regression problem as follows [13].

$$y_p = \beta_0 + \sum_{i=1}^{m} \beta_i x_i + \varepsilon \tag{1}$$

where m is the no of variables in the dataset, ε is the random error, β_0 is the regression constant, y_p represents predict wind power, x_i represents number of features and β_i correlation coefficient of x_i.

A. *Wind Power Forecasting Using Neural Network*

Artificial Neural Network (ANN) is inspired by the characteristic of particular cells that have a couple of wellsprings of information sources that can be activated by certain external method [14]. The neuron or node in an ANN gets an arrangement of weighted data sources and procedures their aggregate with its activation capacity and passes the outcome of the activation to neuron in the down next layer. The input to activation is represented as a dot product of input as follows:

$$\emptyset\left(\sum_i w_i a_i\right) = \emptyset\left(w^T a\right) \tag{2}$$

The sigmoidal activation function is used in the ANN model and is denoted as

$$\emptyset\left(w^T a\right) = \frac{1}{1 + \exp\left(-w^T a\right)} \tag{3}$$

The objective is to train the multi-layer network using training data i.e. labelled data such that it produces the appropriate outputs for un-labelled data if given a set of test inputs. For the duration of training time the model learns the accurate weights of connections to supply the target output for a given unknown test input. The trained weights and its network form a function that works on input info [15].

The weights on the input connections are introduced to some random weights because the training of the network relies upon the initial weights, it can be significance to train the system a few times utilizing different starting points. Figure 1 presents the schematic view of the training and testing process of ANN.

Training: learning weights in supervised mode data(x_i,y_i)

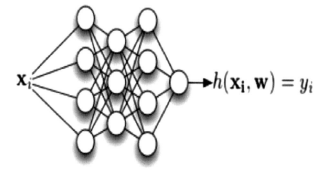

Testing: predict y_{test} for given x_{test}

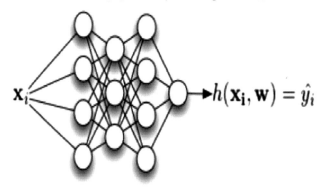

Fig. 1 Multi-layer neural network

Training: learning weights in supervised mode data (x_i, y_i)

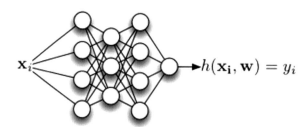

Testing: predict y_{test} for given x_{test}.

For finding the optimal set of weights to fit the training data the best way to assess the fitness is to compute the least squares loss (or error) over our dataset as follows

$$L(W) = \sum_i (h(x_i, W) - y_i)^2 \qquad (4)$$

For finding optimal set of weights $L(W)$ must be minimized. It is a closed-form resolution for conventional least squares, but generally it is possible to minimize $L(W)$ using gradient descent.

B. *Wind power forecasting Using Parallel Random Forest Algorithm*

The conventional Random Forest Regression algorithm has some inherent parallel characteristics and it can be implemented in parallel form easily [16]. The Random forest is an ensemble of several independent CART decision trees and the efficiency of the parallel implementation will be better when compared to its sequential implementation. As it is an ensemble model the trees of the Random forest model can be built in parallel. When the amount of data to be processed by the model is high then this parallel building of trees is not alone enough since the large volume of amount of data must be processed by a single processor. The simple and effective solution to this problem is to use Map Reduce programming framework to process the data in a distributed and parallel environment. The enormous capacity of facts is fragmented into pieces and each piece of fact is controlled by an altered processor core or an altered machine. The data is processed in two steps i. Map—produces a set of key-value pairs in which the key is an index for the data value and ii. Reduce—process the key-value pairs generated by each Map function and output the required results.

The random splitting of data into chunks makes it biased and it is not possible to control the degree of bias. If trees are built on these biased chunks of data it leads to problem of over fitting of the final random forest. The vertical partitioning of data is used since the RF algorithm is independent of feature variables [17]. The training data is split into several feature subsets. Consider if the size of the training data set S is $N \times M$ then there are $x_0, x_1, \ldots, x_{m-2}$ independent variables and x_{m-1} is a dependent variable present in the data set. Now the data is split into $(M - 1)$ feature subset and each subset is loaded into a separate data node in a Hadoop multi cluster setup. Every single subgroup of facts is handled by a distinct Map task and the trees are constructed in parallel.

4 Experiments and Results

In general, any wind power forecasting mechanism typically has errors in the range of 10–15% MAE-Mean Absolute Error for a wind farm [18]. Even with this erroneous forecasting mechanism, large scale wind management studies have exhibited that utilizing short term wind power predictions will help to improve the overall power generation and transmission system operation by reducing the un-served renewable energy and in parallel it helps to maintain the level of system stability. The experimental setup helps to generate a short–term wind power forecast. Three years of historical data is considered for training and testing the model. For evaluating the performance, the models are trained with one year, two years, and three years of data respectively.

A. Artificial Neural Network

The feed-forward network is modelled with input layer having 04 neurons, hidden layers with 15 neurons and the output layer with one neuron. The Artificial Neural Network was trained with Levenberg–Marquardt algorithm also known as the damped least-squares algorithm. The algorithm does not compute the Hessian matrix directly but instead computes it using the Jacobian Matrix and gradient vector.

The loss function used in our experiments can be expressed as

$$f = \sum e_i^2, \; i = 0, 1, 2, 3 \ldots .m \tag{5}$$

where 'm' is the number of sample in the dataset. The Jacobian matrix of the loss function can be expressed as in (5).

$$J_{i,j} f(w) = \frac{de_i}{dw_j} (i = 1, 2, ..m \; \& \; j = 1, 2, \ldots ., n) \tag{6}$$

where 'm' is the number of sample in the dataset and 'n' is the number of inputs in the datasets. The gradient vector of the error function can be estimated using

$$f = 2J^T \cdot e \tag{7}$$

where 'e' represents the vector of all error terms. Using the above computed Jacobian matrix and gradient vector using Eqs. (6)and (7) respectively the Hessian Matrix H can be computed as given below.

$$HF = 2J^T \cdot J + I \tag{8}$$

After iteration the weights are updated as shown below

$$w_{i+1} = w_i - (J_i^T \cdot J_i + J_i I)^{-1} \cdot \left(2J_i^T \cdot e_i\right)$$
$$i = 0, 1, \ldots , \tag{9}$$

The experiments aim at finding the smallest network that has the best ability of generalization to handle the unknown test data. The cross-validation pruning is used to estimate the optimal number of hidden units and the feature subset. The cross-validation pruning follows a *leave-k-out* methodology by which the model is iteratively re-estimated leaving a subset of data each time. Finally, the model with the lowest error is chosen.

B. *Parallel Random Forest Algorithm*

The training of decision trees is usually done with a single training set which generates strongly correlated trees. There is a possibility of generating duplicate trees if the training algorithm is deterministic. From this time bootstrap selection is utilised for de-correlating the trees by training them with altered datasets. The bootstrapping procedure increases the performance of the model by reducing its variance at the same time lacking aggregate the bias. The guesses made on a single tree are high sensitive to noise in the training set but the usual of de-correlated tree is not.

For training the random forest model a random subset of the features is used [19]. This reduces the correlation between the trees when one or more strong predictors of the response variable are present in the training set for many of the trees. The trees in the random forest are trained with $p/3$ (rounded up) features where 'p' is the number of features present in the training data set.

In typical random forest training based on the nature and size of the training dataset few hundreds to thousands of trees are used. An optimal value for the number of trees can be determined using cross-validation approach. But for analysing the performance of the random forest for different tree size the experiments are conducted with three different trees sizes.

The forecast outcomes 48 h ahead is presented in Figs. 2 and 3 which exhibits the predicted and actual wind power in seven different wind farms. The error in the RF predictions are due to the chaotic nature of wind. More accurate forecasts are generated by Random Forest algorithm except some spurious differences of the predicted and actual wind power.

C. *Simulation Results*

In this section some benchmarks and the outcomes of power prediction frameworks are analysed, the neural network model and parallel random forest model simply uses the historical values of wind direction and speed. These historical data are used to train the models for short-term forecasting of wind power and the parallel random forest model when compared with neural network model has less error at all 7 wind farms (Fig. 4).

This figure above demonstrates that it does not take a large number of trees to stabilize the forest prediction error estimates. Increasing the number of trees beyond this level does not reduce the error in the predictions. The evaluation of root mean squared error has been tabulated in Tables 2 and 4.

D. *Forecasting Accuracy Evaluation*

The deterministic WPFs are compared and evaluated on accurateness of the prediction. In case of forecasting the prediction error 'e_{t+k}' estimated for a prediction at a time '$t + k$' can be represented as the difference between the actual wind power measured '$y_{Pactual}$' on that time and the predicted wind power 'y_p'. The error criteria received in the performance assessment of the deterministic WPF is Root Mean Square Error (RMSE). It gives an additional illustrative evaluation of the forecasting error for the total approval time frame. The RMSE keeps record of the deviation of the errors more absolutely.

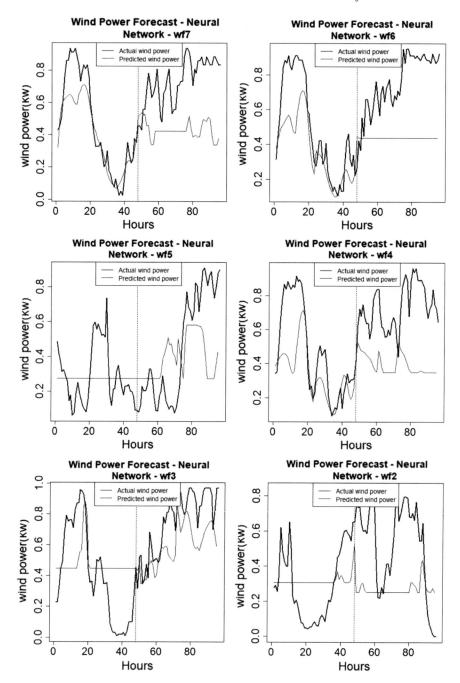

Fig. 2 Results of predictions with neural network (3 years of data)

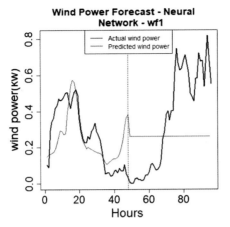

Fig. 2 (continued)

$$Error = y_p - y_{pactual} \tag{10}$$

y_p refers to predict wind power and $y_{pactual}$ refers actual wind power and the difference between actual and predicted wind power is the error.

$$RMSE = \sqrt{\frac{1}{n} \sum_{j=1}^{n} \left(y_{pj} - y_{pactual\,j} \right)^2} \tag{11}$$

The above equations '*m*' is the no of variables in the dataset, '*n*' is the number of samples in the dataset. Table 2 indicates that performance of the neural network model measured by the RMSE and Table 3 indicates that performance of the parallel random forest model measured by the RMSE.

The parallel random forest algorithm figures shown for short-term wind power prediction is feasible and it achieves a better performance in forecasting accuracy than neural network model. Especially presents an outstanding performance in 1–48 h forecast horizon. In Terms of RMSE, the neural network model has 20–30% error at 7 wind farms with improved accuracy in parallel random forest algorithm which has only 3–7% error.

Table 4 indicates that performance of the neural network model measured by the accuracy using RMSE and Table 5 indicates that performance of the parallel random forest model measured by the accuracy using RMSE.

The parallel random forest algorithm achieves a better performance in forecasting accuracy than neural network model. Especially presents an outstanding performance in 1–48 h forecast horizon.

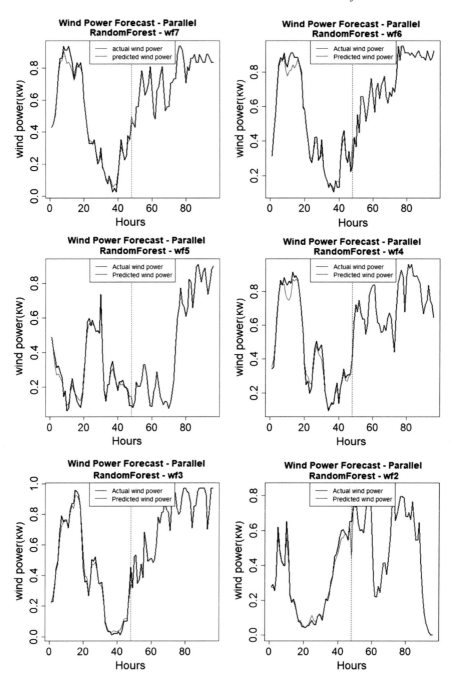

Fig. 3 Results of predictions with parallel random forest (3 years of data)

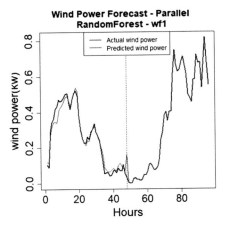

Fig. 3 (continued)

Fig. 4 Random forest
generalization error. (Error
convergence along the
number of trees in the forest)

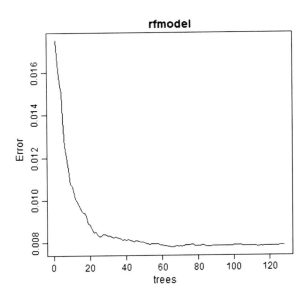

5 Conclusion

Short-term forecasting of power is powerfully effects on the protection and finances
of the electricity grid. The experiments conducted in this research is based on statistical models for forecasting several hours ahead using only historical data including
wind direction, date, wind speed, u and v collected at 7 wind farms. The conventional random forest algorithm is modified and implemented as Parallel Random
Forest Algorithm using Hadoop MapReduce framework for accurateness short-term
forecasting of wind power. Based on the experimental results it is clear that the

Table 2 Root Mean Square error for neural network

RMSE			
Wind farm	Number of layers		
	2	4	6
Wf1	0.23	0.22	0.19
Wf2	0.22	0.23	0.27
Wf3	0.27	0.15	0.17
Wf4	0.31	0.26	0.22
Wf5	0.27	0.25	0.30
Wf6	0.28	0.24	0.26
Wf7	0.24	0.25	0.21

Table 3 Root Mean Square error for parallel random forest

RMSE			
Wind farm	Number of trees		
	32	64	128
Wf1	0.0505	0.0498	0.0496
Wf2	0.0398	0.0394	0.0391
Wf3	0.0798	0.0796	0.0796
Wf4	0.0407	0.0402	0.0398
Wf5	0.0459	0.0463	0.0460
Wf6	0.0665	0.0659	0.0660
Wf7	0.0463	0.0467	0.0467

Table 4 Accuracy for Neural Network

Accuracy (%)			
Wind farm	Number of layers		
	2	4	6
Wf1	73	77	81
Wf2	77	76	72
Wf3	74	84	83
Wf4	69	73	78
Wf5	74	74	70
Wf6	73	76	73
Wf7	76	75	79

Table 5 Accuracy for parallel random forest

Accuracy (%)			
Wind farm	Number of trees		
	32	64	128
Wf1	94.95	95.02	95.04
Wf2	96.02	96.06	96.09
Wf3	92.02	92	92.02
Wf4	95.93	95.98	96.02
Wf5	95.41	95.37	95.4
Wf6	93.35	93.41	93.4
Wf7	95.37	95.33	95.33

implementation of the proposed approach when compared with existing method of neural networks for forecasting of wind power produces more accurate results. The prediction results are closer to the actual power produced and it is recommended for the wind power forecasting. The RMSE has very less value, outperforming other prediction methods available in the literature. The principle disadvantage of the proposed approach is lack of ability of forecasting wind gust, since wind is measured stable for the duration of the entire time frames. However, sudden wind deviations might be injurious for wind turbines. Additionally, Accurateness diminishes as the time frames growths, building the model suitable simply for the short term.

References

1. Chen, N., Qian, Z., Nabney, I.T. and Meng, X.: Wind power forecasts using gaussian processes and numerical weather prediction. IEEE Trans. Power Syst. (2014)
2. GWEC: Global Statics. http://gwec.net/publications/global-wind-report-2/global-wind-report-2016
3. Zhenga, Z.W., Chena, Y.Y., Zhoua, X.W., Huoa, M.M., Zhaoc, B., Guod, M.Y.: Short-term wind power forecasting using empirical mode decomposition and RBFNN. Int. J. Smart Grid Clean Energy (2012)
4. Liu, Z., Gao, W., Wan, Y.H., Muljadi, E.: Wind Power Plant Prediction by Using Neural Networks
5. Wang, J., Zhou, Q., Jiang, H., Hou, R.: Short-term wind speed forecasting using support vector regression optimized by Cuckoo Optimization Algorithm. Hindawi Publishing Corporation Mathematical Problems in Engineering, vol. 2015, Article ID 619178, p. 13 (2015)
6. Dong, L., Wang, L., Khahro, S.F., Gao, S., Liao, X.: Wind power day-ahead prediction with cluster analysis of NWP. Renew. Sustain. Energy Rev.
7. Wu, W., Chen, K., Qiao, Y., Lu, Z.: Probabilistic Short-term Wind Power Forecasting Based on Deep Neural Networks (2017)
8. Faghihnia, E., Salahshour, S., Ahmadian, A., Senu, N.: Developing a local neurofuzzy model for short-term wind power forecasting. Hindawi Publishing Corporation Advances in Mathematical Physics (2014)

9. JIANG, Yu., Xingying, C.H.E.N., Kun, Y.U., Yingchen, L.I.A.O.: Short-term wind power forecasting using hybrid method based on enhanced boosting algorithm. J. Mod. Power Syst. Clean Energy **5**(1), 126–133 (2017)
10. Tesfaye, A., Zhang, J.H., Zheng, D.H., Shiferaw, D.: Short-term wind power forecasting using artificial neural networks for resource scheduling in microgrids. Int. J. Sci. Eng. Appl. (2016)
11. Men, Z., Yee, E., Lien, F.S., Wen, D., Chen, Y.: Short-Term Wind Speed and Power Forecasting using an Ensemble of Mixture Density Neural Networks. Defence R&D Canada External Literature
12. Zhang, Y., Kim, S.J., Giannakis, G.B.: Short-Term Wind Power Forecasting Using Nonnegative Sparse Coding. Renewable Energy & the Environment (IREE) Grant RL-0010-13, University of Minnesota, and NSF Grants. CCF-1423316 and CCF-1442686
13. Zheng, D., Eseye, A.T., Zhang, J., Li, H.: Short-term wind power forecasting using a double-stage hierarchical ANFIS approach for energy management in microgrids. Protection and Control of Modern Power Systems (2017)
14. Li, S., Wang, P., Goel, L.: Wind power forecasting using neural network ensembles with feature selection. IEEE Trans. Sustain. Energy (2015)
15. Shekhawat, A.S.: Wind power forecasting using artificial neural networks. Int. J. Eng. Res. Technol. (IJERT)
16. Zhou, Z., Xiaohui, L., Wu, H.: Wind Power Prediction based on Random Forests Advances in Computer Science Research
17. Chen, J., Li, K., Tang, Z., Bilal, K., Yu, S., Weng, C., Li, K.: A parallel random forest algorithm for big data in a spark cloud computing environment. IEEE Trans. Parallel Distrib. Syst. **28**(4) (2017)
18. Wang, X., Guo, P., Huang, X.: A Review of Wind Power forecasting Models. ICSGCE 2011: 27–30 September 2011, Chengdu, China (2011)
19. Chang, W.Y.. Short-term wind power forecasting using the enhanced particle swarm optimization based hybrid method. Energies **6**, 4879–4896 (2013)

Leukemia Cell Segmentation from Microscopic Blood Smear Image Using C-Mode

Neha Singh and B. K. Tripathy

Abstract Blood-related diseases such as leukemia are very dreadful diseases and detection of such diseases must be carried out at very early stage. In manual method of leukemia detection, experts check the microscopic images. This is time-consuming process which depends on the person's skill and does not have standard accuracy. The automated leukemia detection system analyzes the microscopic blood smear image and overcomes these drawbacks of manual detection. Many literature surveys are done, and it was found that average accuracy up to 84–87% is achieved to date. In this technique for automating leukemia, we are applying a new algorithm called C-mode as a method of segmentation which is required in the process of detecting early symptoms of diseases via medical imaging analysis. Improving the present method by application of soft computing to provide accuracy is the focus of the paper. Next, we can achieve by applying C-mode on RGB image, this removes number of steps which was earlier needed to process. Third, the execution speed increases as C-mode reduces the number of iteration. The result obtained provides more accuracy and is used for feature extraction.

Keywords Blood cell · White blood cells · Leukemia · Segmentation · K-means · C-mode · Blast cell

N. Singh
School of Computing Science and Engineering, VIT University Vellore, Vellore, Tamil Nadu, India
e-mail: 04neha.singh@gmail.com

B. K. Tripathy (✉)
School of Information Technology and Engineering, VIT University Vellore, Vellore, Tamil Nadu, India
e-mail: tripathybk@vit.ac.in

© Springer Nature Singapore Pte Ltd. 2020
K. N. Das et al. (eds.), *Soft Computing for Problem Solving*,
Advances in Intelligent Systems and Computing 1048,
https://doi.org/10.1007/978-981-15-0035-0_17

1 Introduction

Medical imaging has turned out to be a standout among the most imperative representation and understanding techniques in science and pharmaceutical over the previous decade. This time has seen a large advancement of new, effective instruments for recognizing, putting away, transmitting, breaking down, and showing Medical image is making advancement in digital image processing field [1]. The most testing part of medical imaging lies in the advancement of incorporated frameworks for the utilization of the clinical part. The odds of getting influenced by this disease are quickened because of progress in propensities in the general population, for example, use of tobacco, weakening of dietary propensities, the absence of exercises, and numerous more. Planning, execution, and approval of complex medical frameworks require a tight interdisciplinary joint effort between doctors and engineers. A standout among the most dreaded of human diseases is cancer. Leukemic is a sort of blood cancer, and on the off chance that it is distinguished late, it will come about in death. Leukemia happens when a considerable measure of anomalous white platelets is delivered by the bone marrow. Leukemic begins in blood-shaping tissues, often the bone marrow. It prompts the overproduction of irregular white platelets, the part of the immune system which safeguards the body against disease. These WBC if developed immaturely and are greater in number than the RBC and the balance between both is disturbed this leads to the development of cancer [2–4]. At present, the detection of the disease is carried out by taking a blood sample and testing it in labs by a pathologist. The method includes techniques like CBC or bone marrow aspiration.

In this paper, for automating leukemia we are applying a new algorithm called C-mode as a method of segmentation which is required in the process of detecting early symptoms of diseases via medical imaging analysis. The rest of this paper is composed as pursues. In Sect. 2, literature survey is explained. In Sect. 3, process and proposed division calculation are clarified. In Sect. 4, results and discussion are presented, and finally, Sect. 5 comprises conclusions of work.

2 Literature Survey

Microscopic image is used for detecting blood disorders via visual inspection. From that detection, certain diseases related to blood disorder are classified. There are many drawbacks of visual leukemia detection such as difficulty in getting consistent results from visual inspection, the perception of the observer, experience in performing the task, level of alertness, and degradation of speed [5]. Only qualitative outputs can be achieved through visual inspection for further research. So, the advantage of automation of visual sample inspection is to aid pathologists to recognize anomalies in blood samples competently, precisely, and quicker with the assistance of most recent innovations. Blood images are used for diagnosis as they are cheap and do not require expensive testing and lab equipment.

Automatic image process system is desperately required and might overcome connected constraints in a visual examination, diseases can be detected and diagnosed at an earlier stage. Out of all the main processes of image processing [6, 7], segmentation is considered the most vital and critical step in the automatic system because all later steps are highly reliant on the image segmentation step. The system will focus on white blood cells disease, leukemic, focus is to improve preprocessing, and segmentation technique.

The target of the paper is to improve the segmentation technique that can thus improve accuracy and reduces the time for processing as compared to another method. It is the most important part of image processing. Segmentation is a technique of decomposing an image into a sub-image, whose union is from the original image. It is particularly performed to extract the region of interest that is meaningful with respect to the application. There are various methods for segmentation and are classified as follows:

- Region-Based
- Edge-Based
- Threshold
- Clustering
- Model-Based

Several techniques developed till date for image segmentation proposed a module-based system; single-cell selector is his first module, WBC identification is carried in his second module, and the third module is used to identify blast/normal cell [8]. A general survey of WBC segmentation techniques was presented by Adollah et al. [9]. By using contrast enhancement and thresholding [10]. Mohamed et al. presented a technique for WBC nucleus segmentation. Two-level segmentation for segmenting nucleus and cytoplasm was applied in [11]. Here, in this they used morphological analysis for separating the nucleus, whereas separating the cytoplasm is based on gray level thresholding. Disadvantage of this method is that it is applied on sub-images to ease the implementation. Otsu's segmentation was used by Khobragade et al. [12]. First, they used a linear color enhancement technique and then image filtering and edge detection.

2.1 Clustering-Based Segmentation

Clustering is a trendy method being adopted for many purposes in days. Clustering is a troublesome issue which has been tended to by different analysts in expanded regions, for example, pattern recognition, data mining, biology, marketing, image processing, and so forth. The name clustering itself means grouping, a grouping of similar items or elements. Clustering(grouping) is done to separate similar data from dissimilar one. Clustering on dataset may lead to the creation of one or more groups depending on the property of similarity. The number of the groups can be certain depending on user need or uncertain means unsupervised depending upon data properties, the dynamic behavior of forming a group without mentioning the actual number of groups.

Based on this method of grouping there are two categories of clustering, one is called supervised and the other is called unsupervised. Supervised learning is like first giving training and then giving actual test data to test and for the group, from clustering point, we need to give an initial number of clusters. Unsupervised learning is something dynamic, and there is no need to mention an initial number of clusters, it will form automatically. It is like when a fish is born it automatically learns to swim no one teach them.

The first clustering algorithm was proposed by Lloyd in 1957 as a technique for pulse-code modulation and used by MacQueen in 1967, since then the development of clustering algorithm was tremendous and utilized as a part of numerous spaces. There are two broad classes of clustering. In hard clustering, each element belongs to exactly one cluster. That is one element can belong to the cluster (true) or does not belong to the cluster. This is more common, and K-means is an excellent example of such type. Second, in soft clustering, each element may belong to more than one cluster. Means there is a situation that an element may have say 60% similarity in one cluster and 40% similarity in another cluster. Fuzzy clustering is an example of soft computing.

Based on this there has been various algorithm developed such as K-means, fuzzy clustering, and applications of rough set in clustering called as rough set clustering (RCM) and hybrid models such as rough fuzzy C-means, fuzzy rough C-means, possibilistic C-means, and fuzzy possibilistic C-means. Some of the applications of such algorithm particularly in segmentation are given below: The input image is converted into HSV color model and segmentation is carried out by K-means clustering whose centroid and variance is thus used for the Gaussian parameter [13]. Automated blood count was carried out using the two-stage process, in the first stage input image is converted into HSV and in second stage K-means is followed by EM [14]. In [15] established an automatic segmentation technique for microscopic bone marrow WBC images. They use the FCM algorithm and mathematical morphology. A color-based image segmentation was proposed [16], they focused on nucleus as the region of interest and segmentation is performed using K-means clustering followed by nearest neighbor classification in L × a×b specifically for feature extraction shape features that are a Hausdorff dimension and contour signature is implemented for classifying a lymphocytic cell nucleus. In [17], the authors improved the segmentation method by using the fuzzy C-means algorithm, then used shape and texture feature for feature extraction and then applied SVM for classification. Rough fuzzy C-means was developed by Mitra et al. [18]. They have used the algorithm on both the dataset numeric and image for segmentation.

3 Process

The reason for the lymphocyte (ordinary and ALL) cell segmentation is to remove the lymphocyte nucleus and cytoplasm from other diverse parts in a minute blood spread image. Blood spread pictures comprise leukocyte cells "nucleus and cytoplasm", red platelets, platelets, and background. It is exceptionally hard to early identify ALL infection. This is because of the extensive variety of size and morphology of the lymphocyte cells, and the way that ALL cell morphology and size are so near typical lymphocytes in the early phases of the ailment. The lymphocyte cell's core seems darker than the background, and red platelets. In addition, expansive contrasts exist in the morphology and size of the lymphocyte core and cytoplasm.

3.1 K-Means

This algorithm minimizes the total distance of data points to the cluster center of the cluster they are assigned to. The algorithm starts with an initial dataset and with initial cluster centers. Distance from each data point is calculated to each center and to which the data point is near it is assigned to that cluster. Again, the mean is calculated for new centroid from the grouped data and the iteration proceeds until the new centroids match with the old one. The disadvantage with this algorithm is different initial partitions can result in different final clusters and it does not work well with clusters (in the original data) of different sizes and different densities. The centroid is calculated by

$$v_i = \sum_{j=1}^{k} \sum_{i=1}^{n} \left\| x_i^j - c_j \right\|^2 \tag{1}$$

3.2 C-Mode

C-mode is a modified algorithm. It has been modified from K-means. It provides many advantages like the speed of execution is increased as the number of iteration is less. Instead of means value, the mode value is selected for choosing the centroid. So, within the cluster instead of choosing the mean of data points, the most frequent value is taken as the new centroid. The number of iterations that was increasing due to few single or double precision leading to new iteration is now reduced as a direct data point is picked as the centroid. Both algorithms rely heavily on picking the right value for K the decision made by K-means is not strong because the centroid is chosen is not among the data point, so the iteration increases but as in comparison,

C-mode forces the centroids to make this decision which results into much defined clustered [19].

Algorithm

  i.   Start with choosing random centroid.
  ii.   Calculate distance matrix using the Euclidean distance from the centroid to each data point.
  iii. Each data points are grouped into a cluster if that distance to the data point and the centroid is min compared to all centroid.
  iv.   On grouping, the new centroid is selected based on mode value within the group.
  v.   Repeat Step iii until centroids stabilize or several iterations reached its highest value.
  vi.   Stop.

The following are the flowchart of the proposed module: (Fig. 1).

A. *Image acquisition*

The process starts with acquiring the image. The image is a microscopic image, so before applying another method, preprocessing of the image is carried out. We have followed the image processing technique and the steps are divided into four major parts. 1. Acquisition, 2. Preprocessing, 3. Segmentation, and 4. Feature extraction [20]. The image is acquired through online pathologist site (Fig. 2).

B. *Image preprocessing*

The obtained image is preprocessed and enhanced. The preprocessing steps involve clearing the noise and smoothing. Special filtering is used to remove noise and smoothing the image. We have divided the process into two levels after preprocessing that is cell extraction and nucleus extraction which is carried using two-level segmentation.

C. *Cell Segmentation*

The image acquired is an RGB image and to extract the nucleus we convert it into L × a×b color space. So, we convert the RGB color module into two color spaces. The advantage of converting into Lab is that it is device-independent color space and it describes mathematically all perceivable colors that aspire to perceptual uniformity. So, all pixel containing RGB is converted into green–red and blue–yellow and L is for luminance. Then we apply the simple K-means algorithm on "an" and "b" component to get the blast. As K-means is an iterative process and group data based on mean value, we divide the set of pixels into two clusters so that we obtain a group of the cluster for background and another for the blast. Here we obtained cell blast using this process. Then some morphological operations are performed. Morphological operations are used as image enhancement technique. This is also useful in feature extractions like texture and shape feature. The morphological operations such as dilation and hole filling were done this gives more clear and prominent image.

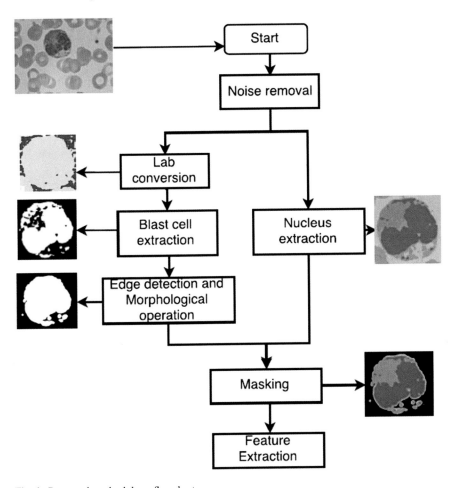

Fig. 1 Proposed methodology flowchart

Fig. 2 Microscopic blood
smear image

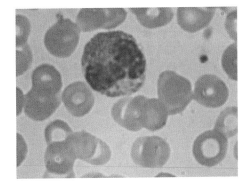

Fig. 3 Output

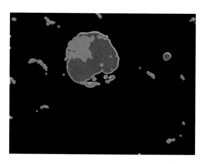

D. *Nucleus Segmentation*

The nucleus segmentation is carried using a modified algorithm called C-mode. The preprocessed RGB image is taken as input to the algorithm and output obtained is the cell nucleus. This does not require several processes to get the output. C-mode is a modified algorithm of K-means.

E. *Resultant output*

Then masking of the image is done. The cell nucleus is masked on blast cell. This gives the desired output result (Fig. 3).

4 Results and Discussion

We took 20–30 microscopic images and processes it using our methodology. It was found that the execution speed and accuracy both had increased by using this methodology. The following is the result of execution in which we input both cancerous and normal microscopic blood smear images and obtain the result. The output of the result is shown in Fig. 4. We used a C-mode algorithm for extracting the nucleus. It has not only reduced the number of preprocessed steps required before clustering but also increased the execution time. Here, we took two types of blood cells, normal lymphocyte and reactive lymphocyte.

The C-mode algorithm was generally used for textual data and categorical data but it has not been applied for clustering on the image. For nucleus extraction direct image was passed to K-means and C-mode and the result obtained gives the following advantages: 1. It preserves the shape and size of the cell nucleus and it is not needed to perform the morphological operation. 2. It gives noise-free and clear image. 3. The execution spends gets reduced due to less number of iterations. Figure 4 shows the difference when K-means and C-mode are applied to extract nucleus.

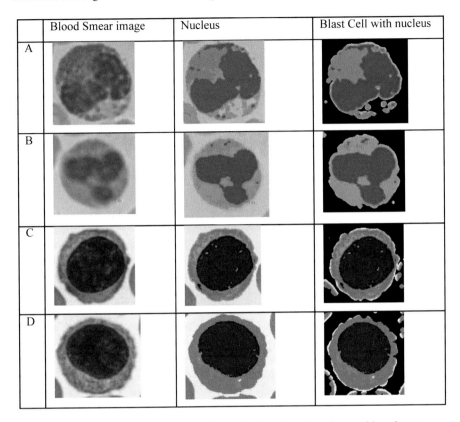

	Blood Smear image	Nucleus	Blast Cell with nucleus
A			
B			
C			
D			

Fig. 4 Microscopic blood smear images, **a**, **b** reactive lymphocyte, **c**, **d** normal lymphocyte

4.1 Comparision Between C-Mode and K-Means

A. **Visual Image comparisons**
See Fig. 5.

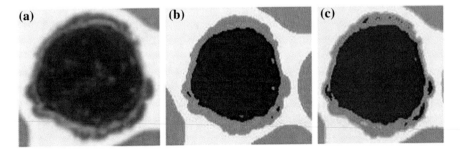

Fig. 5 **a** Actual cell **b** C-mode processed image **c** K-means processed image

B. *Execution time comparisons*

The processing speed of both algorithms was tested and it was found that on an average C-mode is faster and takes approximate 22 s execution times with is 48% faster than K-means.

4.2 Feature Extraction

Once the output is ready it can be sent to feature extractions module. There are many features which are necessary [21–24]. There feature helps in identifying the cancerous cell depending on its shape, size, morphological structure, color, texture, intensity, Statistic, etc. These features are fetched to the classifier so that it will classify cells into infected to normal. A cancerous cell is identified by its irregular shape, size, and other property. Generally, normal resting lymphocytes area unit little cells with condensed chromatin granule and a tiny low quantity of pale stainability living substance. The nucleus of a resting leucocyte is simply slightly larger than a red blood corpuscle and reactive lymphocytes show a spread of morphologic options. Reactive lymphocytes with immunoblast-like morphology square measure massive cells with high nuclear-cytoplasmic ratios, condensed body substance, and deeply stainability protoplasm. Another style of reactive WBC has less condensed body substance and luxuriant, pale blue protoplasm which will seem to hug adjacent red blood cells. These cells also are known as Downey kind II cells. They will be seen during a sort of conditions; however, the square measure typically exaggerated in mono owing to herpes infection. We have used Hausdorff dimension, textural basted, and statistical-based feature; texture feature is calculated using GLCM (Table 2). The result obtained is shown in Table 1, and Hausdorff dimension for full image and crop image are calculated and shown in Table 3.

Hausdorff dimension is used to measure the perimeter roughness. Here, we have used the box- counting method. It is observed that the HD of the noncancerous cell is less than that cancerous cell.

Table 1 Execution speed of algorithm	S. no.	K-mean(s)	C-mode(s)
	1	28.5339	15.3610
	2	21.3220	19.8949
	3	17.9190	15.4839
	4	66.9939	26.0290
	5	26.8309	31.8389
	Average	32.3199	21.7215

Table 2 Statistical textual-based feature

	Features	Cancerous cell	Normal cell
	Images		
1	Contrast	0.2613	0.3430
2	Correlation	0.1090	0.1034
3	Energy	0.7543	0.6754
4	Homogeneity	0.9344	0.9076
5	Mean	0.0040	0.0029
6	Standard deviation	0.0927	0.1042
7	Entropy	0.6173	0.7948
8	RMS	0.0928	0.1042
9	Variance	1.2997	0.0108
10	Smoothness	0.9247	0.8485
11	Kurtosis	6.6290	6.8846
12	Skewness	0.5598	0.4064
13	IDM	1.2079	1.7404
14	Cluster prominence	1.4812	2.188

5 Conclusion

In this paper, the focus was to achieve image segmentation by using soft computing technique such as C-mode and to find the outcome of the algorithm. It was found that the algorithm works correct and as compared to K-means which gives better visual result and is better in execution speed. C-mode is faster and in this experiment it took approximate 22 s execution times which is 48% faster than K-means. The system was built by incorporating both clustering technique C-mode and k-means. The accuracy and morphological structure of the nucleus were preserved and thus the output was helpful in feature extraction. "Blood image dataset" was taken from [25]. Future work will be carried out in improving the system to make the system more efficient and to classify any type of cancerous cell.

Table 3 Hausdorff Dimension (HD)

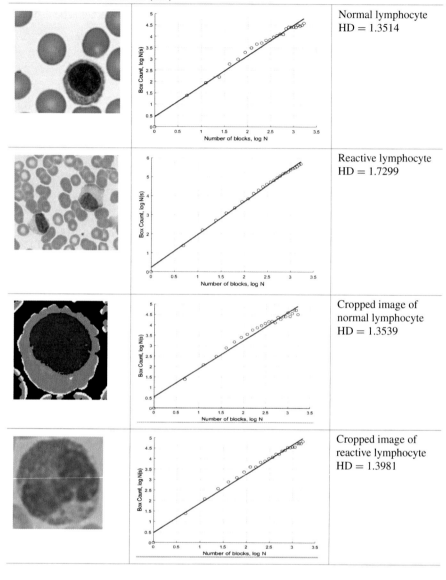

References

1. Valêncio, C.R., Tronco, M.N., Bonini-Domingos, A.C., Bonini-Domingos, C.R., Traina, C., Traina, A.J.: Knowledge extraction using visualization of hemoglobin parameters to identify thalassemia. In: Proceedings of 17th IEEE Symposium in Computer-Based Medical Systems, pp. 523–528 (2004)
2. Sabino, D.M.U., da Fontoura Costa, L., Rizzatti, E.G., Zago, M.A.: A texture approach to leukocyte recognition. Real-Time Imaging **10**(4), 205–216 (2004)
3. Colunga, M.C., Siordia, O.S., Maybank.: Leukocyte recognition using EM-algorithm. In: International Conference on Artificial Intelligence, Mexico, pp. 545–555 (2009)
4. Kasmin, F., Prabuwono, A.S., Abdullah, A.: Detection of leukemia in human blood sample based on microscopic images: a study. J. Theor. Appl. Informat. Technol. **46**(2) (2012)
5. Di Ruberto, C., Dempster, A., Khan, S., Jarra, B.: Analysis of infected blood cell images using morphological operators. Image Vis. Comput. **20**(2), 133–146
6. Gonzalez, R.C.: Digital Image Processing. Pearson Education India (2009)
7. Adollah, R., Mashor, M.Y., Nasir, N.M., Rosline, H., Mahsin, H., Adilah, H.: Blood cell image segmentation: a review. In: 4th Kuala Lumpur International Conference on Biomedical Engineering, pp. 141–144 (2008)
8. Mohamed, M., Far, B., Guaily, A.: An efficient technique for white blood cells nuclei automatic segmentation. In: IEEE International Conference on Systems, Man, and Cybernetics, pp. 220–225 (2012)
9. Sadeghian, F., Seman, Z., Ramli, A.R., Kahar, B.H.A., Saripan, M.I.: A framework for white blood cell segmentation in microscopic blood images using digital image processing. Biol. Procedures Online **11**, 196 (2009)
10. Khobragade, S., Mor, D.D., Patil, C.Y.: Detection of leukemia in microscopic white blood cell images. In: International Conference in Information Processing, pp. 435–440 (2015)
11. Ramakrishnan, N., Sinha, A.G.: Blood cell segmentation using EM algorithm. In: Proceedings of 3rd Indian Conference Computer Vision Graph, pp. 445–450 (2002)
12. Sinha, N., Ramakrishnan, A.G.: Automation of differential blood count. In: Conference on Convergent Technologies for the Asia-Pacific Region, vol. 2 (2003)
13. Theere-Umpon, N.: White Blood Cell Segmentation and Classification in Microscopic Bone Marrow Images. Lecture Notes in Computer Science—FSKD, pp. 787–796 (2005)
14. Mohapatra, S., Patra, D.: Automated cell nucleus segmentation and acute leukemia detection in blood microscopic images. In: International Conference on Systems in Medicine and Biology, pp. 49–54 (2010)
15. Mohapatra, S., Samanta, S.S., Patra, D., Satpathi, S.: Fuzzy based blood image segmentation for automated leukemia detection. In: International Conference on Fuzzy based blood in Devices and Communications, pp. 1–5 (2011)
16. Sobti, S., Shah, V., Tripathy, B.K.: A refined rough fuzzy clustering algorithm. In: IEEE International Conference in Computational Intelligence and Computing Research, pp. 1–4 (2014)
17. Huang, Z.: A Fast clustering algorithm to cluster very large categorical data sets in data mining. In: DMKD (1997)
18. Mitra, S., Banka, H., Pedrycz, W.: Rough–fuzzy collaborative clustering. IEEE Trans. Syst. Man Cybern. Part B Cybern. **36**(4), 795–805 (2006)
19. Osowski, S., Markiewicz, T., Marianska, B., Moszczyński, L.: Feature generation for the cell image recognition of myelogenous leukemia. In: 12th European Conference in Signal Processing, pp. 753–756 (2004)
20. Theerapattanakul, J., Plodpai, J., Pintavirooj, C.: An efficient method for segmentation step of automated white blood cell classifications. IEEE TENCON Conference, pp. 191–194 (2004)
21. Mohapatra, S., Patra, D., Satpathy, S.: Unsupervised blood microscopic image segmentation and leukemia detection using color based clustering. Int. J. Comput. Informat. Syst. Indust. Manag. Appl. 477–485 (2012)

22. Tripathy, B.K., Tripathy, A., Rajulu, K.G.: Possibilistic rough fuzzy C-means algorithm in data clustering and image segmentation. In: IEEE International Conference on Computational Intelligence and Computing Research, pp. 1–6 (2014)
23. Tripathy, B.K., Tripathy, A., Govindarajulu, K.: On PRIFCM algorithm for data clustering, image segmentation and comparative analysis. In: IEEE International on Advance Computing Conference, pp. 333–336 (2015)
24. Bhargava, R., Tripathy, B.K., Tripathy, A., Dhull, R., Verma, E., and Swarnalatha, P.: Rough intuitionistic fuzzy c-means algorithm and a comparative analysis. In: Proceedings of the 6th ACM India Computing Convention, pp. 23 (2013)
25. Image Dataset (n.d.). Accessed from Mathworks File Exchange from http://www.mathworks. com/matlabcentral/fileexchange/36634

Implementation of Exploration in TONIC Using Non-stationary Volatile Multi-arm Bandits

Aditya Shaha, Dhruv Arya and B. K. Tripathy

Abstract Target Oriented Network Intelligence Collection (TONIC) is a problem which deals with acquiring maximum number of profiles in the online social network so as to maximize the information about a given target through these profiles. The acquired profiles, also known as leads in this paper, are expected to contain information which is relevant to the target profile.In the past, TONIC problem has been solved by modelling it as a Volatile Multi-arm bandit problem with stationary reward distribution. The problem with this approach is that the underlying reward distribution in case of TONIC changes with each exploration which needs to be incorporated for making future acquisitions. The paper shows a successful solution to the TONIC problem by modelling it as Volatile Bandit problem with non-stationary reward distributions. It illustrates a new approach and compares it's performance with other algorithms.

Keywords Volatile multi-arm bandit problems · Exploration-exploitation dilemma · Discounted upper confidence bound · Computation optimization · Reinforcement learning

1 Introduction

Target Oriented Network Intelligence Collection (TONIC) is a problem formulated by Stren et al. [1] deals with the OSN (Online Social Networks) e.g. Facebook, Google+ etc. A TONIC problem defines a target profile. In the Social Network

A. Shaha · D. Arya
School of Computer Science and Engineering, VIT Vellore, Vellore, India
e-mail: aditya.shaha.p@gmail.com

D. Arya
e-mail: aryadhruv@gmail.com

B. K. Tripathy (✉)
School of Information Technology and Engineering, VIT Vellore, Vellore, India
e-mail: tripathybk@vit.ac.in

© Springer Nature Singapore Pte Ltd. 2020
K. N. Das et al. (eds.), *Soft Computing for Problem Solving*,
Advances in Intelligent Systems and Computing 1048,
https://doi.org/10.1007/978-981-15-0035-0_18

the profiles are connected by friendship relationship. The target profile has blocked public access and hence directly retrieving information from the target profile is not possible. Every social network restricts the API calls that one can make to the service. This restriction is defined as budget in TONIC. So the task is given a target profile, and some initial friends of the target profile (also called as leads), the agent has to acquire maximum number of friends of the target profile within the given budget.

TONIC problem has been solved by modeling it as a search problem in an unknown graph. The acquiring of a profile is synonymous to scraping data from that profile and the goal is to maximize the number of profile which has the information pertinent to target. The solution to the TONIC problem is based on the social network concept of homophily which states that people with similar interest tend to be friends with each other. With this underlying assumption, Stren et al. [1] came up with 6 Heuristics to guide the Best-first search in an unknown graph. Of these 6 Heuristics the Bayesian Promise was observed to perform the best. However, these search solutions relied heavily upon the heuristics and failed in case when the underlying subgraph did not follow the heuristic completely. In such cases, it was expected for the algorithm to explore for new options.

Multi-arm bandit problem is a reinforcement learning problem which tackles the exploration-exploitation dilemma in reinforcement learning. In the problem a bandit machine with K independent arms is given as an input. At each time step, the agent can play only one arm and he will be rewarded for his action which will be sampled from a stationary probability distribution function of the arm. The agent is unaware of the exact probability distribution function that governs the rewards for each arm. The task of the agent is to learn the arm that has maximum expectation for its probability distribution function, with minimum time steps. The agent's task is to minimize the regret which is defined as the difference between the rewards obtained by playing the optimal arm and rewards obtained by following the current algorithm. The minimization is achieved by balancing the exploitation, the learned information, and exploration acquiring new information. If the agent only exploits his current knowledge he might not be able to achieve the maximum rewards and if the agent is only exploring, he won't be able to the exploit the acquired information which can maximize his rewards.

In 2013, Banya et al. [2] defined volatile multi-arm bandits (VUCB1) as an extension of multi-arm bandits where there was a possibility of new arm appearing and disappearing at each time-step. As a result at each time-step the optimal arm changes. They modeled the TONIC problem as a volatile bandit problem to incorporate exploration. The arms in this case were equivalence classes, such that the profiles connected to the same leads were a part of same equivalence class. These equivalence classes can be removed or added as new leads are found. The VUCB1 Problem by Volatile Upper Confidence Bound Algorithm (VUCB). The problem with this approach was that the algorithm assumes the underlying rewards distribution for each arm to be stationary. But at each time step as the some of the potential leads are being pruned and some are being added, the rewards that any arm can deliver keeps on changing for each time step. Here the rewards is defined as the probability of randomly sampling a potential lead from an arms that has maximum value for VUCB algorithm, to be a lead.

In this paper a variant of VUCB called Volatile Discounted Upper Confidence Bound (VUCB-D) has been used to take into account the changing probability distributions of the rewards. Section 2 explores the related work associated with solving TONIC Problem. Section 3 explains in detail how the previous approaches used Multi-arm bandits to solve the TONIC problem. In Sect. 4 the approach has been delineated and a new algorithm has been proposed. Section 5 evaluates the results of the new algorithm and compares it with the results of the existing solutions. The work has been concluded in Sect. 6.

2 Related Work

Previous papers have solved TONIC with Artificial Intelligence Search Techniques, structuring the problems as a heuristic search problem performing targeted search using best-first algorithm. The Heuristic based Best-First Search approach only exploits the heuristic and does not use exploration which limits the performance of the algorithm. By modeling the problem as VUCB1, the algorithm encourages exploration but it failed to incorporate for the changing underlying reward distributions of the arms.

In [1] BTF, Basic TONIC Framework (BTF) uses the known leads to prognosticate the next acquisition such that the possibility of a potential lead (pl), which is a node connected to the lead, being the lead is very high. This is accomplished through various heuristics like the Random Heuristic in which the algorithm chose the potential lead (pl) to be acquired randomly, FIFO Heuristic in which the (pl) was chosen based on First-in-First-out basis, Clustering Coefficient Heuristic in which the (pl) was chosen based on Highest Known Clustering Coefficient, Maximum Degree Heuristic in which the (pl) was chosen based on Highest Known Degree and Bayesian Promising lead in which the promising factor of lead was calculated based on the probability of randomly selected a (pl) of lead being a lead. The Bayesian Promising factor was proved to perform the best. In [3] the Extended TONIC Framework (ETF), ETF also allowed the search through the extended social circles of a target. This incorporated the search algorithm to perform some exploration rather than always enforcing the heuristics. Reference [2] Introduced a combination of BTF and ETF. It proposed a new term ETF(n) which only explored the non-leads if they were n step away from a known-lead. They empirically showed that ETF(n) for n > 1 does not fetch significant benefits as the increase in the percentage of leads degrades after n = 1. In [4] the TONIC approach with VMAB was used to direct a social network crawler that aimed to solve exploration and exploitation dilemma. In [5] the flaws of exploration-exploitation dilemma in search were tackled with the concept of Volatile Multi-Armed bandit (VMAB) problem using a novel algorithm called Volatile Upper Confidence Bound 1 (VUCB1). Using VUCB1, there was no need of an ad hoc heuristic function to guide the search in the social graph. The algorithm used past experience in the form of profile acquired previously during the search. The concept of Mortal Multi-Armed Bandits designed for online advertising was introduced in [6]. It assumed that arms can expire or become available. However, the number of arms is fixed, i.e., whenever one arm disappears, a new arm appears. In [7] used the Upper Confidence Bound to solve multi arm bandits which tackled the exploration-exploitation dilemma. In this paper, it was shown that UCB can be applied when an algorithm has to make exploitation-versus-exploration decisions based on uncertain information provided by a random process. In [8] Auer et al. showed that the optimal logarithmic regret was also achievable uniformly over time, with simple and efficient policies, and for all reward distributions with bounded support. They provided a Finite Time Analysis for UCB-1, UCB-2, and UCB-1 Normal in this paper. In [9] Gerrivier et al. introduced another variant of UCB-1 for non-stationary reward distribution that considers a sliding window of previous rewards for computing UCB. In [10] Kocsis et al. introduced a variant of UCB for non-stationary reward distribution called UCB-D (Discounted UCB).

In this paper TONIC Problem has been modeled as a non-stationary Volatile Multi arm bandit problem where the reward distribution of each arm changes with every acquisition and also new arms appear and disappear at every time-step. The arms have been initiliazed using the promising factor instead of by pulling them i.e. making acquisitions of the potential leads from that equivalence class because that it decreases the number of API calls and also gives us a rough estimate about the reward expectation from that arm. It has been observed that the performance of this algorithm is better than the originally proposed VUCB1.

3 TONIC as a Multi-arm Bandit Problem

It must be noted that TONIC can be treated as a multi-arm bandit (MAB) problem. MAB problem is a classic reinforcement learning problem where we have a set of arms which generate rewards on being pulled. In the case of TONIC, the set of known leads can be treated as arms. Finding a lead from the neighborhood of an arm generates a reward. Every known lead has some probability of generating a reward. This probability depends on the number of leads and non-leads connected to it. Once an arm has been picked, one of it's neighbors can be randomly sampled to choose the next node to explore. Thus, TONIC can be treated as the problem of picking the arm which will generate the best reward. Some algorithms for solving TONIC using the MAB approach as covered in the next subsections.

3.1 Upper Confidence Bound

Upper Confidence Bound Method is used for solving the exploration-exploitation dilemma in Reinforcement learning. These methods are deterministic policies which augment the algorithm proposed by Lai and Robbins (1985) to a non-parametric context. These methods were introduced and analyzed by Agarwal (1995). The task of the agent is to play an arm in the tth that maximizes the upper bound of the confidence interval from the expected stationary reward $\mu(i)$. That is, instead of randomly selecting arms, that arm is to be selected (based on previous experience) which is expected to give us the best expected reward. The Expected Reward is calculated from the rewards observed till time $t - 1$. The most popular of the UCB algorithm is UCB-1, in which the arm with the maximum value for $\bar{X}_t(i) + c_t(i)$ is selected to be pulled next. In this equation $\bar{X}_t(i) = (N_t(i))^{-1} \sum_{s=1}^{t} X_{(s)}(i) \mathbb{1}_{\{I_s=t\}}$ also called the empirical mean and $c_t(i) = B\sqrt{\varepsilon log(t)/N_t(i)}$ which is called the padding function. In the padding function B is the upper bound on the reward and $\varepsilon > 0$ is some constant.

In the above equation $N_t(i) = \sum_{s=1}^{t} \mathbb{1}_{I_{\{s=t\}}}$ which denotes the number of times arm i has been chosen.

Algorithm 1 Upper Confidence Bound Algorithm

Input: T, Time Horizon
Input: K, Total Number of Arms
Input: $\{a_1, a_2, a_3...a_k\}$, set of K Arms
while $(t <= K)$ **do**
| Play arm $I_t = a_t$
end
while $(t > K \text{ and } t < T)$ **do**
| $I_t = a_{\underset{1 \leq i \leq K}{argmax} \{\bar{X}_t(i)+c_t(i)\}}$
end
return $a_{\underset{1 \leq i \leq K}{argmax} \{\bar{X}_t(i)+c_t(i)\}}$

3.2 Stationary Volatile Upper Confidence Bound (VUCB)

Volatile Upper Confidence Bound is an augmentation of the original Upper Confidence Bound in which the arms can appear and disappear at each time-step. [2] VUCB1 uses same policy as UCB1 but only on the available arms. If no arm appears in the time horizon T, VUCB1 behaves exactly as UCB1.

When a new arm appears the algorithm follows the following 3 steps

1. Pull the new arm immediately.
2. Reset the $T_i(n)$ (the number of times arm i was pulled) of all other existing arms a_i to 1 but keeping their reward averages.
3. The total number of pulls (labeled by the turn variable n) is set to the current number of available arms.

4 TONIC as a Non-stationary Volatile Multi-arm Bandit Problem

The Upper Confidence Bound algorithm assumes that the underlying probability distribution of rewards does not change with time. However, this condition does not hold for TONIC. Consider the graph in Fig. 1. Let 0 be the target node and let 14 and 3 be the given leads. The probability of picking a lead in arm 14 is $2/8 = 0.25$ and in arm 3 is $2/5 = 0.40$. Suppose, the algorithm picks 3 as the best arm and 11 as the potential lead. Since 11 is a lead, it is added to the collection of leads and removed from the list of potential leads of both the arms. The probability of picking a lead in arm 14 and arm 3 is now $1/7 = 0.14$ and $1/4 = .25$ respectively. The probability distribution has changed abruptly. Therefore, the underlying assumption of UCB does not hold. While UCB makes a good balance between exploitation and exploration, it does not take abrupt probability distribution changes into account.

4.1 Proposed Volatile Upper Confidence Bound-Discounted Search Algorithm

As illustrated above, the underlying probability distribution of rewards changes every time a potential lead is picked. If the potential lead that has been picked is a lead, the probability of finding a lead in that arm again goes down. Since the number of non-leads outnumber the leads in the neighbourhood of any arm in general, the effect of removal of non-lead can be assumed to be negligible. The proposed algorithm ensures that the Upper Confidence Bound (UCB) for every arm is updated whenever an arm is picked. This is done using a discount factor which reduces the estimate of the Upper Confidence Bound. In every iteration, the Discounted Upper Confidence Bound of every arm is calculated using the equation (reference). The arm with the maximum UCB is picked, then a neighbor of this arm is randomly sampled and it is checked whether it is a lead. Note that for this algorithm to work, an estimate of the UCB of arm is needed initially. This is usually done by picking every arm for the first

Fig. 1 Number of leads found on an average versus budget

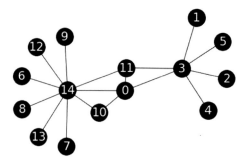

few iterations. However, due to the volatility of arms, this could potentially result in a large number of API calls. Instead the Bayesian Promise Heuristic has been used here to initialize the UCB.

Algorithm 2 Discounted Volatile Upper Confidence Bound for TONIC

Input: *Target*, the target profile
Input: *Budget*, the number of allowed API-calls
Input: *InitialLeads*, the set of initial leads
Input: ξ, the hyper parameter that controls exploration
Input: γ, the discounting factor
FoundLeads $\leftarrow$ *InitialLeads*
  CLOSED $\leftarrow \phi$
  $X \leftarrow PromisingFactorInit(InitialLeads)$
  $N[i] \leftarrow 1$ for every initial lead i
  while *(Budget > 0)* **do**
  | $bestArm_t \leftarrow \underset{1 \leq i \leq K}{argmax} \ \overline{X}_t(\gamma, i) + c_t(\gamma, \xi, i)$
  | $N[bestArm_t] \leftarrow N[bestArm_t] + 1$
  | $best \leftarrow$ randomly sampled neighbour of $bestArm_t$
  | Add *best* to *CLOSED*
  | Acquire *best*
  | **if** *best is a lead* **then**
  | | $X[bestArm_t] \leftarrow X[bestArm_t] + 1$
  | | Add *best* to *FoundLeads*
  | | $X[best] \leftarrow \underset{1 \leq i \leq K}{max} \ X[i]$
  | | $N[best] \leftarrow 1$
  | **end**
  | Decrease *Budget* by 1
  end
return *FoundLeads*

---

1. In the above algorithm *PromisingFactorInit(InitialLeads)* is used to calculate the promising factor of the *initial leads* defined as

$$pf(m) = \frac{L(m)}{NL(m) + L(m)} \tag{1}$$

   where *L(m)* are lead neighbors of *lead m* and *NL(m)* are the non-lead neighbors of *lead m*
2. $\bar{X}_t(\gamma, i)$, $N_t(\gamma, i)$, $c_t(\gamma, \xi, i)$ and $n_t(\gamma)$ are defined in Sect. 4.2.

4.2 Discounted Upper Confidence Bound

The Discounted Upper Confidence Bound algorithm modifies UCB-1 to account for abrupt changes in reward probability distribution. One additional hyper-parameter—

the discount factor $\gamma \in (0, 1)$ is added. The gist of the algorithm is that the previous rewards become less important with time since the probability distribution has changed. Therefore the previous rewards of each arm are discounted when calculating the current UCB.

The equation for the discounted empirical average is given by

$$\overline{X}_t(\gamma, i) = \frac{1}{N_t(\gamma, i)} \sum_{s=1}^{t} \gamma^{t-s} X_s(i) \mathbb{1}_{\{I_s=i\}} \tag{2}$$

where

$$N_t(\gamma, i) = \sum_{s=1}^{t} \gamma^{t-s} \mathbb{1}_{\{I_s=i\}} \tag{3}$$

and the discounted padding function is

$$c_t(\gamma, \xi, i) = 2B \sqrt{\frac{\xi \log n_t(\gamma)}{N_t(\gamma, i)}} \tag{4}$$

where

$$n_t(\gamma) = \sum_{i=1}^{K} N_t(\gamma, i) \tag{5}$$

where $\mathbb{1}_{\{I_s=i\}}$ is 1 when arm i was selected at time s.

4.3 Application on Synthetic Example

In this section, the execution of the three algorithms on the graph Fig. 2 has been illustrated. The *TargetNode* in the given graph is **0**. At the start, the set *InitialLeads* is {15 and 9}. The budget has been fixed to 6. Table 1 shows execution of BysP.

Fig. 2 The entire graph

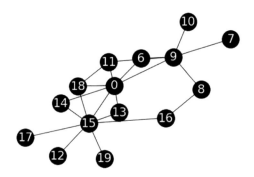

Table 1　Status of *FOUNDLEADS, CLOSED* after every acquisition for BysP

Budget	FOUNDLEADS	CLOSED
6	{15, 9}	{9, 12, 15}
5	{15, 9}	{9, 12, 13, 15}
4	{15, 9, 13}	{9, 12, 13, 14, 15}
3	{15, 9, 13, 14}	{9, 12, 13, 14, 15, 16}
2	{15, 9, 13, 14}	{9, 12, 13, 14, 15, 16, 17}
1	{15, 9, 13, 14}	{9, 12, 13, 14, 15, 16, 17, 18}
0	{15, 9, 13, 14, 18}	{9, 12, 13, 14, 15, 16, 17, 18}

Table 2　Status of *FOUNDLEADS, CLOSED* after every acquisition for VUCB1

Budget	FOUNDLEADS	CLOSED
6	{15, 9}	{9, 15}
5	{15, 9}	{9, 12, 15}
4	{15, 9}	{9, 10, 12, 15}
3	{15, 9, 18}	{9, 10, 12, 15, 18}
2	{15, 9, 18, 11}	{9, 10, 11, 12, 15, 18}
1	{15, 9, 18, 11, 13}	{9, 10, 11, 12, 13, 15, 18}
0	{15, 9, 18, 11, 13}	{9, 10, 11, 12, 13, 15, 17, 18}

Table 3　Status of *FOUNDLEADS, CLOSED* after every acquisition for VUCB-D

Budget	FOUNDLEADS	CLOSED
6	{9, 15}	{9, 15}
5	{9, 15, 6}	{9, 6, 15}
4	{9, 15, 6, 11}	{9, 11, 6, 15}
3	{9, 15, 6, 11, 18}	{6, 9, 11, 15, 18}
2	{9, 15, 6, 11, 18}	{6, 9, 10, 11, 15, 18}
1	{9, 15, 6, 11, 18, 14}	{6, 9, 10, 11, 14, 15, 18}
0	{9, 15, 6, 11, 18, 14, 13}	{ 6, 9, 10, 11, 13, 14, 15, 18}

The column *Budget* indicates the remaining budget and the column *FOUNDLEADS* shows the corresponding labels of the leads found till that instance. All the nodes that have been explored till that instance are shown in the *CLOSED* column. Both VUCB1 (Table 2) and BysP (Table 1) end up with 5 leads after 6 API calls. In comparison, VUCB-D (Table 3) ends up with 7 leads for this particular graph (Fig. 3).

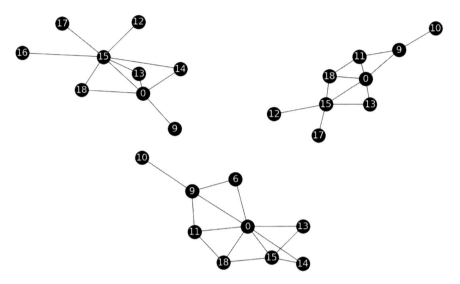

Fig. 3 Final Known states of graphs for each algorithm on given example, Top-Left graph is for *BysP*, Top-Right graph is for *VUCB1*, Bottom graph is for *VUCB-D*

Observations

1. VUCB-D outperforms BysP and VUCB1 when the number of leads are outnumbered by non-leads in the neighborhoods of leads
2. The proposed algorithm also also manages to strike a balance between exploitation of known good leads and exploration of other leads.

5 Evaluation

The proposed algorithm was evaluated on the Google+ dataset [10]. The dataset is a snapshot of the Google+ network and contains 211 k anonymized nodes with 1.5M links between them. The dataset was randomly sampled for 100 target nodes. The only those target nodes were selected that had more than 30 and less than 70 friends. For every target node, 3 initial leads were randomly picked. The algorithm was run with Budgets {50, 65, 75, 90, 100}. All algorithms were executed with the same set of initial leads and target.

It was observed that VUCB-D outperforms VUCB1 in most cases. The performance of VUCB-D and BySP is very similar but VUCB-D outperforms BySP in terms of execution time.

In Fig. 4 it can be seen that for a low budget of 50, BysP outperforms both VUCB-D and VUCB1. However, as the budget increases, the number of leads found by both BysP and VUCB-D are similar. In Fig. 5 it can bee seen the execution time for all

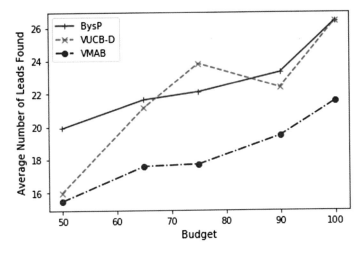

Fig. 4 Number of leads found on an average versus budget

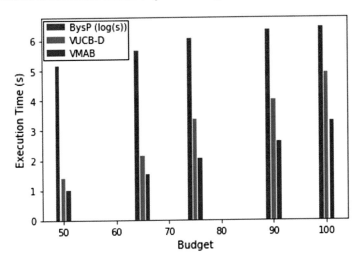

Fig. 5 Execution time versus budget

three algorithms for the given budgets. Note that the log of the execution time of BysP has been shown to fit it into the given space. It can be seen that both VUCB-D and VUCB1 are much faster than BysP.

6 Conclusion

Discounted Upper Confidence Bound works effectively when the underlying probability distribution function of the arms generating rewards is changing. It is computationally cheap as compared to other heuristics and maintains exploration exploitation balance when searching in a unknown search space. VUCB-D was successful in accounting for the changing underlying reward distribution in TONIC and was able to explore new states after exploiting the heuristically best state.

The future work would include learning the hyper-parameter γ and ξ which guide the exploration. Also we wish to prove that there exists a probabilistic bound for our algorithm and that the system always performs better when exploration is incorporated which has been successfully shown empirically in this paper.

References

1. Stern, R., Samama, L., Puzis, R., Beja, T., Bnaya, Z.: TONIC: Target oriented network intelligence collection for the social web. In: 27th AAAI Conference on Artificial Intelligence, pp. 1184–1190 (2013)
2. Samama-Kachko, L.: Target Oriented network intelligence collection (TONIC) (2014)
3. Samama-kachko, L., Stern, R., Felner, A.: Extended Framework for Target Oriented Network Intelligence Collection. In: (SoCS), pp. 131–138 (2014)
4. Bnaya, Z., Puzis, R., Stern, R., Felner, A.: Bandit algorithms for social network queries. In: Proceedings-SocialCom/PASSAT/BigData/EconCom/BioMedCom 2013, 148153 (2013)
5. Chakrabarti, D., Kumar, R., Radlinski, F., Upfal, E.: Mortal multi-armed bandits. In: Neural Information Processing Systems, pp. 273–280 (2008)
6. Bnaya, Z., Puzis, R., Stern, R., Felner, A.: Volatile multi-armed bandits for guaranteed targeted social crawling. In: Late Breaking Papers at the Twenty-Seventh AAAI Conference on Artificial Intelligence, pp. 8–10 (2013)
7. Auer, P.: Using confidence bounds for Exploration Exploitation trade-offs. JMLR **3**, 397–422 (2002)
8. Auer, P., Cesa-Bianchi, N., Fischer, P.: Mach. Learn. **47**, 235 (2002). https://doi.org/10.1023/A:1013689704352
9. Garivier, A., Moulines, E.: On upper-confidence bound policies for non-stationary bandit problems (2008). arXiv:0805.3415
10. Kocsis, L., Szepesvri, C.: Discounted UCB. In: 2nd PASCAL Challenges Workshop, pp. 784–791 (2006)

Fuzzy Logic-Based Model for Predicting Surface Roughness of Friction Drilled Holes

N. Narayana Moorthy and T. C. Kanish

Abstract The nontraditional hole-making process Friction Drilling (FD) receives major attention nowadays because of its operational efficiency in terms of unpolluted, chipless hole making and in fact, the holes are drilled in single step. It is a cumbersome and challenging task to predict surface finish of the work material in the final stages of operation. This difficulty arises because of nonlinear interactions between the process parameters and nonuniform nature of the heat caused by friction which occurred between the conical drill bit rotating at high speed and the workpiece. Since this process is having ambiguities and uncertainties, a model based on fuzzy logic has been developed for the prediction of surface roughness of drilled holes in the FD process. Operating parameters such as rotational speed of the spindle, feed rate, and workpiece temperature are the three membership functions chosen to propose this fuzzy model. These functions are assigned for each input of the model. This fuzzy logic model is verified by two firsthand set of parameter values. The results opine that the established fuzzy model is well in agreement with the investigational data with the maximum deviation of 3.81%. Furthermore, three-dimensional surface plots are developed using this fuzzy model to reveal the influence of individual process parameters on the surface ambiguities. The outcomes of the study attest that the three-dimensional surface plots are much useful for selecting input parameter combinations to achieve the required surface roughness.

Keywords Friction drilling · Chipless hole making · Surface roughness · Fuzzy logic · Surface plots

N. Narayana Moorthy (✉)
School of Mechanical Engineering, Vellore Institute of Technology (VIT), Vellore 632014, Tamil Nadu, India
e-mail: n2moorthy@gmail.com

T. C. Kanish
Centre for Innovative Manufacturing Research, Vellore Institute of Technology (VIT), Vellore 632014, Tamil Nadu, India
e-mail: tckanish@vit.ac.in

© Springer Nature Singapore Pte Ltd. 2020
K. N. Das et al. (eds.), *Soft Computing for Problem Solving*,
Advances in Intelligent Systems and Computing 1048,
https://doi.org/10.1007/978-981-15-0035-0_19

251

1 Introduction

In recent years, automobile industries, creating holes of thin-walled metal components for various structural elements are gaining more attention. Normally, the methodologies adopted for drill-hole-making process is focused more on twisted drills based traditional methods. However, the lifetime of the twist drills is much shortened as a result of its rapid tool wear. Further this also leads to flow chip wall, which, in turn, causes poor surface of the drill hole [1]. In order to overcome these drawbacks, researchers have developed a new nontraditional chipless hole-making process called "Friction Drilling". This process acquires less tool wear and thus provides better tool life. Also, this process aids in producing good surface reasonably lesser wastage of materials compared to the traditional methods. Instead of material being cut, the process forms collar and bush which leaves enough surface area for threading. During the friction drilling process, prediction of hole surface roughness generated is a vital role.

Hitherto, many research works have been focused on developing analytical [2], empirical, and optimization techniques [3] to identify the optimal process parameters of friction drilling. It is evident from the literature that the research works have been confined only on the features of thermomechanical vicissitudes by means of finite element approaches [4], experimental investigations [5], mechanics of drill-hole formation [6], studies on machining characteristics [7], tool wear monitoring [8, 9], varying the tool materials, and coated tools [1, 10] to improve the wear resistance and analysis of surface roughness using microscopy investigations [11] obtained during friction drilling process.

As stated earlier, there exist nonlinear interactions between the process parameters and nonuniform nature of the frictional heat developed between the friction drilling tool and material. This creates obscurities and uncertainties in this process. These issues are resolved by obtaining optimization through soft computing techniques which are gaining more attraction among the researchers [12]. In the current study, an attempt is made to develop a fuzzy model to understand and predict the appropriate process parameters on the surface roughness during drill hole through friction drilling process. Three-dimensional surface plots developed via. fuzzy model to establish the interface between each input process parameters on surface roughness are evaluated and employed in this study.

2 Experimental Details

In this study, the friction drilling trials are performed using a precision CNC vertical milling machine. The schematic diagram of the experimental setup of friction drilling is shown in Fig. 1. A nonferrous alloy, beryllium copper is chosen for the study due to its excellent mechanical property and corrosion resistance. The tungsten carbide

Fig. 1 Schematic diagram of friction drilling experimental setup

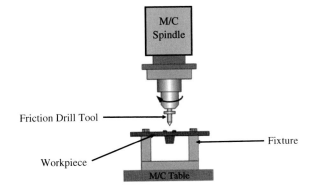

Table 1 Process parameters chosen for the study and the echelons

Code	Experimental parameters	Unit	Echelons		
			1	2	3
A	Workpiece temperature	°C	30	50	70
B	Spindle speed	RPM	2500	3000	3500
C	Feed rate	mm/min	40	50	60

(WC) tool material is employed so that the cutting-edge hardness is retained at high machining temperatures during experimental study.

Based on the earlier findings and pilot studies, the influencing parameters such as workpiece temperature, rotation speed of conical tool spindle, and feed rate are chosen in this current investigation. The experiments are executed based on Taguchi L9 orthogonal array deploying three echelons of experimental variables as shown in Table 1. In this study, the initial temperature of the workpiece is kept as one of the process parameters. As the temperature is also chosen as one of the input variables, the samples are also preheated to various temperatures in the calibrated induction furnace, as mentioned in Table 1. The other parameters such as the thickness of the workpiece (3 mm), tool diameter (10 mm), and workpiece dimension are set constant in all the trials.

3 Results and Discussion

The experimental results that include the impact of friction drilling on the surface roughness, fuzzy logic model, and three-dimensional surface plots for selecting input parameter combinations to achieve the required surface roughness are discussed in this section.

Table 2 Response table

S. no	A (°C)	B (rpm)	C (mm/min)	Ra (μm)
1	30	2500	40	0.1857
2	30	3000	50	0.2579
3	30	3500	60	0.6111
4	50	2500	50	0.2579
5	50	3000	60	0.4948
6	50	3500	40	0.4603
7	70	2500	60	0.4247
8	70	3000	40	0.2589
9	70	3500	50	0.5153

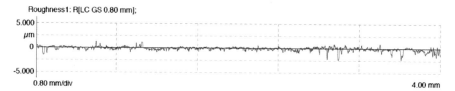

Fig. 2 Profile shows the surface roughness of the friction drilled hole (Ra = 0.2579 μm) at cutting condition A = 50 °C; B = 2500 rpm; C = 50 mm/min

3.1 Surface Roughness Measurement

In the current investigation, the surface roughness for the friction drilled holes is considered as response variable during the process. At the end of each experiment, the surface roughness (Ra) of individual drill holes are measured using a stylus type Profilometer (Mahr Marsurf GD120). The measured values are depicted in Table 2. The surface roughness obtained at the cutting condition of a particular case A = 50 °C; B = 2500 rpm; C = 50 mm/min is illustrated in Fig. 2.

3.2 Development of Fuzzy Logic Model For Predicting Surface Roughness

For the past few decades, the fuzzy logic theory fascinated many researchers in almost all fields of engineering [13]. The relation between the three input process parameters, namely, workpiece temperature (A), spindle speed (B), and feed rate (C) with the output parameter surface finish (Ra) are employed to build the fuzzy inference system and is depicted in Fig. 3.

The steps involved in the fuzzy model consist of elements such as fuzzification (defines membership functions for framing vague rules), fuzzy rule (a vague rule of

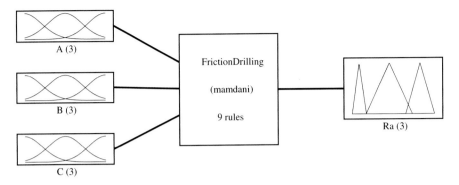

Fig. 3 Fuzzy inference system

Table 3 Membership function and linguistic variables adopted in the fuzzy model

Inputs				
Notation	Membership function	Units	Linguistic variables	Choice
A	Workpiece temperature	°C	Low (L), Medium (M), High (H)	30–70
B	Spindle speed	rpm		2500–3500
C	Feed	mm/min		40–60
Output				
Ra	Surface finish roughness	μm	Low, Medium, High	0.1857–0.6111

choice for processing), defuzzification (implements the functionality of the rules and facts to make a reasonable decision), and validation of fuzzy model.

3.2.1 Fuzzification

In this process, membership functions are designated to each process parameter for the numerical measurement analysis and to classify the fuzzy sets. The fuzzy set is defined using triangular membership function by assigning the linguistic variables such as Low (L), Medium (M), and High (H) for each input parameter. Three linguistic variables, namely Low, Medium, and High are assigned for the output parameter (Ra). The linguistic variables assigned for both input and output limits are presented in Table 3.

3.2.2 Membership Function

It is denoted as a curve, which points each input space that is mapped to a membership value between 0 and 1 [14]. The membership functions are formed for the process parameters and the output variables are represented in Table 2. There are several types of membership function such as triangular, trapezoidal, sigmoid, and Gaussian. In this study, the Gaussian shape of membership function was selected for the input process parameters to accomplish flatness in the input variables. The triangular membership function is selected for the output variable because of its progressively growing and declining characteristics. The input selected and output membership function curves are illustrated in Fig. 4.

3.2.3 Fuzzy Rules

Based on the results obtained from the experimental studies, the rules are framed. The rules contain a set of "if-then" statements with the input parameter as initial temperature, spindle speed and feed, and output as surface roughness (Ra). The input parameters are assigned by three fuzzy subdivisions; whereas the output is assigned by three indices. The rules are given below

Rule No. 1: IF [A = Low], [B = Low], and [C = Low] THEN (Ra) will be Low.
Rule No. 2: IF [A = Low], [B = Medium], and [C = Medium] THEN (Ra) will be Low.
Rule No. 3: IF [A = Low], [B = High], and [C = High] THEN (Ra) will be High.
Rule No. 4: IF [A = Medium], [B = Low], and [C = Medium] THEN (Ra) will be Low.
Rule No. 5: IF [A = Medium], [B = Medium], and [C = High] THEN (Ra) will be High.
Rule No. 6: IF [A = Medium], [B = High], and [C = Low] THEN (Ra) will be Medium.
Rule No. 7: IF [A = High], [B = Low], and [C = High] THEN (Ra) will be Medium.
Rule No. 8: IF [A = High], [B = Medium], and [C = Low] THEN (Ra) will be Medium.
Rule No. 9: IF [A = High], [B = High], and [C = Medium] THEN (Ra) will be High.

Based on the above rules, an inference engine is developed. Two processes, namely, the assumption (uses **IF** part) and aggregation (uses **THEN** part) are involved in processing the inference engine and then defuzzification takes place.

3.2.4 Defuzzification

Defuzzification is the process of converting fuzzy values into crisp value. In this study, the defuzzification process arises based on the Centroid of Area (COA) method which interprets a membership function of a fuzzy set and provides a specific crisp value. These fuzzy values provide better Ra value which, in turn, leads to better prediction.

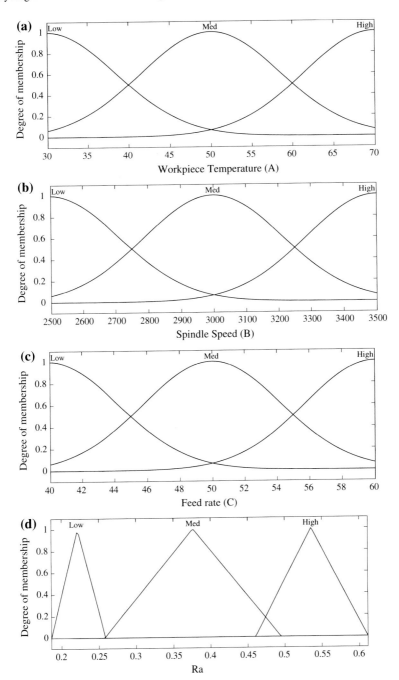

Fig. 4 Membership function plots of input parameters and output variable **a** Workpiece temperature (A), **b** Spindle Speed (B), **c** Feed Rate (C), **d** Output Variable Surface Roughness (Ra)

Table 4 Validation of developed fuzzy model with experimental results

Exp. no.	A (°C)	B (rpm)	C (mm/min)	%ΔRa		% Error
				Expt.	Fuzzy model	
1	40	3250	55	0.4319	0.4373	1.23%
2	60	2850	45	0.2652	0.2551	3.81%

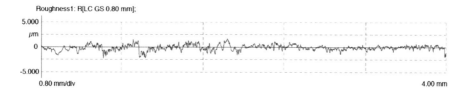

Fig. 5 Surface roughness of friction drilled hole (Ra = 0.4319 μm) for the validation experiment at cutting condition, A = 60 °C; B = 2850 rpm; C = 45 mm/min

3.2.5 Authentication of Fuzzy Model

The fuzzy model developed for friction drilling is tested using two new data of process specifications which have not been employed in the current experimental data. The results of Ra retrieved from both the experimental trials and the predicted fuzzy model value are shown in Table 4. The surface roughness profile of the friction drilled hole is illustrated in Fig. 5 for the cutting condition of A = 60 °C; B = 2850 rpm; and C = 45 mm/min. The results disclose that this fuzzy model provides the results which are in agreement with the experimental values. Further, the results show that the fuzzy model developed provides better prediction of surface roughness (Ra) during the friction drilling process.

3.3 Interaction Between the Process Parameters

To obtain an in-depth understanding of the influence of each variable on the surface roughness in the friction drilling process, a parametric study is also carried out with the proposed fuzzy model. To conduct parametric studies, three-dimensional surface plots (Fig. 6) are established based on the fuzzy model. From the plots, it is observed that there are some nonlinear interactions between the process parameters on surface roughness. Figure 6 shows the relationship between different workpiece temperatures, various feed rates, and spindle speed on surface roughness. The trends show that the nonlinear interaction between the process parameters, when feed rate or spindle speed increases, the surface roughness values get slightly increased. It is understood that this trend is because of the nonuniform nature of the heat developed

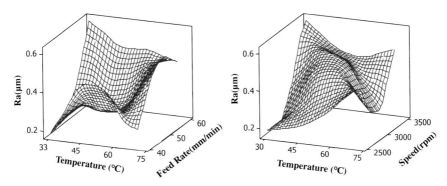

Fig. 6 Three-dimensional surface plots for Ra

due to friction between the rotating conical tool bit and the workpiece. The developed three-dimensional surface plots are very useful for selecting input parameter combinations to achieve the required surface roughness.

4 Conclusions

In this present study, a fuzzy model is developed and proposed to envisage the surface roughness of the friction drilled hole which is one of the vital, functional needs of the industrial components as it greatly affects the resistance to corrosion and wear. The three input process specifications, namely, speed of the tool rotation, feed rate, and workpiece temperature are utilized for proposing this fuzzy model. The following are the conclusions drawn from the current investigation.

- A fuzzy model is developed and proposed to determine the surface roughness of the friction drilled hole with the use of experimental data. The proposed fuzzy logic model is validated with two diverse sets of experimental specifications other than the actual parameters experimental data set. The comparison of experiment value and fuzzy model's value shows that the developed model values are close to the experimental value with maximum deviation of 3.81%.
- Based on the proposed fuzzy model, 3D surface plots are devised to depict the influence of individual process specifications on surface roughness. The developed three-dimensional surface plots are very useful in choosing the input parameter combinations to achieve the required surface roughness.
- It is established from the three-dimensional surface plots, that there are some nonlinear interactions available between the process parameters, when feed rate or spindle speed increases, the surface roughness values slightly increased. This is owing to the reason that a nonuniform nature of the heat is developed because of friction between the rotating conical tool bit and thin workpiece.

- The proposed fuzzy logic model will be beneficial in depicting the role and influence of variables on the Ra values during the friction drilling process. The proposed fuzzy logic-based model is recommended to envisage and to investigate the process parameters involved in the friction drilling process for attaining the anticipated surface finish.

References

1. Lee, S.M., et al.: Friction drilling of austenitic stainless steel by uncoated and PVD AlCrN-and TiAlN-coated tungsten carbide tools. Int. J. Mach. Tools Manufact. **49**.1, 81–88 (2009)
2. Qu, Jun, Blau, Peter J.: A new model to calculate friction coefficients and shear stresses in thermal drilling. J. Manuf. Sci. Eng. **130**(1), 014502 (2008)
3. Ku, W.-L., et al.: Optimization in thermal friction drilling for SUS 304 stainless steel. Int. J. Advanc. Manufact. Technol. **53**.9–12, 935–944 (2011)
4. Miller, Scott F., Shih, Albert J.: Thermo-mechanical finite element modeling of the friction drilling process. J. Manuf. Sci. Eng. **129**(3), 531–538 (2007)
5. Miller, S.F., et al.: Experimental and numerical analysis of the friction drilling process. J. Manufact. Sci. Eng. **128**.3, 802–810
6. Krasauskas, P., et al.: Experimental analysis and numerical simulation of the stainless AISI 304 steel friction drilling process. Mechanics **20**(6), 590–595 (2014)
7. Chow, H.M., Lee, S.M., Yang, L.D.: Machining characteristic study of friction drilling on AISI 304 stainless steel. J. Mater. Process. Technol. **207**.1–3, 180–186 (2008)
8. Miller, Scott F., Blau, Peter J., Shih, Albert J.: Tool wear in friction drilling. Int. J. Mach. Tools Manuf **47**(10), 1636–1645 (2007)
9. Mutalib, M.Z.A. et al.: Characterization of tool wear in friction drilling. J. Tribol. **17**:93–103 (2018)
10. Kerkhofs, M., et al.: The performance of (Ti, Al) N-coated flowdrills. Surf. Coat. Technol. **68**, 741–746 (1994)
11. Miller, Scott F., Shih, Albert J., Blau, Peter J.: Microstructural alterations associated with friction drilling of steel, aluminum, and titanium. J. Mater. Eng. Perform. **14**(5), 647–653 (2005)
12. Chandrasekaran, M., et al.: Application of soft computing techniques in machining performance prediction and optimization: a literature review. Int. J. Advanc. Manufact. Technol. **46**(5-8), 445–464 (2010)
13. D'Errico, G.E.: Fuzzy control systems with application to machining processes. J. Mater. Process. Technol. 109.1–2, 38–43 (2001)
14. Jang, J.SR.: ANFIS: adaptive-network-based fuzzy inference system. IEEE Trans. Syst. Man Cybernet **23.3**, 665–685 (1993)

Face Recognition and Classification Using GoogleNET Architecture

R. Anand, T. Shanthi, M. S. Nithish and S. Lakshman

Abstract Face recognition is the most important tool in computer vision and an inevitable technology finding applications in robotics, security, and mobile devices. Though it is a technology of the past, state-of-the-art machine learning (ML) techniques have made this technology game-changing and even surpass human counterparts in terms of accuracy. This paper focuses on applying one of the advanced machine learning tools in face recognition to achieve higher accuracy. We created our own dataset and trained it on the GoogleNet (inception) deep learning model using the Caffe and Nvidia DIGITS framework. We achieved an overall accuracy of 91.43% which was fairly high enough to recognize the faces better than the conventional ML techniques. The scope of the application of deep learning is enormous and by training a huge volume of data with massive computational power, accuracy greater than 99% can be achieved. This paper will give a glimpse of deep learning, from creation of dataset to training and deploying the models, and the method can be applied for dataset corresponding to any field, be it medicine, agriculture or manufacturing, reducing the human effort and thus triggering the revolution of automation.

1 Introduction

Face recognition has been, for the past few years, the frontrunner in many technological fields like robotics, biometrics and security. Though face recognition dates back to past few decades, it has gained more interest recent years due to the state-of-the-

R. Anand · T. Shanthi · M. S. Nithish · S. Lakshman (✉)
Department of Electronics and Communication Engineering, Sona Signal
and Image Processing Research Center, Sona College of Technology, Salem, India
e-mail: lakshman2497@gmail.com

R. Anand
e-mail: anand.r@sonatech.ac.in

T. Shanthi
e-mail: shanthi@sonatech.ac.in

© Springer Nature Singapore Pte Ltd. 2020
K. N. Das et al. (eds.), *Soft Computing for Problem Solving*,
Advances in Intelligent Systems and Computing 1048,
https://doi.org/10.1007/978-981-15-0035-0_20

261

art accuracy achieved. The accuracy is fairly high enough to compete with human counterparts and even surpass them in some cases. Such high accuracies became possible only after using Convolutional Neural Nets (CNN) in face recognition [1]. The CNN then evolved to deep learning which is simply a more sophisticated neural net. Therefore, the seriousness with which face recognition algorithms are being developed has gone over the roof and more sophisticated models are developed and deployed by the researchers both in big corporate tech giants and in academia. Each year, one model created by some organization/team replaces another of the previous year in terms of accuracy and complexity. Models have achieved accuracies of a whopping 99.63% (FaceNet [2] by Google) which is more accurate than a human. In order to trigger an AI revolution and disrupt a wide range of tech sectors the tools for developing and deploying deep learning models have been simplified by tech titans like google, amazon, and facebook. With the huge computational power available they develop many deep learning models and give us the flexibility to customize it to our applications. In this paper, we take one of the advanced and a more versatile model called the GoogleNet or the Inception model. We created our own dataset by scrapping images from the internet and organizing it to a dataset. We will focus on changing the input layer of the inception model, customizing it to our dataset. The model has been trained using nvidia DIGITS and has achieved the desired accuracies.

2 Literature Review

Eigenfaces [3, 4] and Fisherfaces [5] were some of the first methods to propose an idea for recognizing faces. These two methods worked with gray scale images which are nothing more than a series of numbers, each number corresponding to some intensity level. So these methods treated images as vector. By treating the images as samples of data, we can perform a principal components analysis [6] and obtain the eigenvectors which make up the basis of the vector space. Each face has unique features and these features are what eigenvectors represent. The eigenvectors actually represent the strongest characteristics of the faces in the dataset. When a dataset is provided for training, the eigenvectors are extracted from the set as a whole. Whenever it is provided with a new image, it detects a face, extracts the eigenvectors and matches it with existing eigenvectors and thus determines to which person the face belongs. The change in illumination in faces within the same dataset was one of the major problems that these methods faced. When one face's illumination differed from the others', it made it hard to extract the eigenvectors. This problem was overcome by the Local Binary Pattern Histogram (LBPH) method. The LBPH [7, 8] method takes a different approach than the eigenfaces method. In LBPH each image is analyzed independently. The LBPH method is somewhat simpler, in the sense that we characterize each image in the dataset locally; and when a new unknown image is provided, we perform the same analysis on it and compare the result to each of the images in the dataset. In LBPH, an image is taken and each pixel in it is taken and assigned as the central value (or threshold value) and the pixels surrounding it are given a binary value based on the threshold value. When all pixels are converted to binary values, a histogram [9] is formed. But the problem with LBPH is that it does not read each and every feature of the face, as CNN does, which results in relatively lower accuracy.

Neural network is a technique that mimics the neurons in the human brain, which is the most advanced and powerful system ever known to man. They started to gain popularity when the closeness with which they resembled the neurons came to light and the accuracy of their output surpassed that of its predecessors. Le et.al proposed a model in which convolution operations are performed prior to the neural nets, which was named LeNet. Convolution operations are performed to extract certain features from the data provided, therefore assigning weights to the neural nets. This operation reduced the time as well as the complexity of the following layers. This model by Le was in the end of '90s when the computing power of systems was not so high and the dataset for training was not properly confined, due to which the model was not widely used. But in the last decade, with computational powers of computers being so high, the model was again brought to light and modified according to the needs. One such model which surfaced during the last decade was AlexNet [10] which contained seven layers and provided a decent accuracy. After this, there were many attempts to develop the perfect model. One such model worth mentioning is ZF Net. Then the following year, there was ZF NET [11] (or Zeiler Fergus Net) which was a slight modification to AlexNet. Instead of using 11×11 sized filters in the first layer (which is what AlexNet implemented), ZF Net used filters of size 7 $\times$ 7 and a decreased the stride value. The reason behind this modification is that a smaller filter size in the first convolution layer helps to retain a lot of original pixel information in the input volume. A filtering of size 11×11 proved to be skipping a lot of relevant information, especially as this is the first convLayer. But not until 2015, when GoogleNet [12] was created and was high accuracy achieved. GoogleNet used 17 layers in its model of which 10 were inception layers. An inception layer is one in which operations like convolution, concatenation, and maxpooling are performed in parallel. This approach of using inception layers helped GoogleNet achieve the highest-achieved accuracy of 99.63%.

3 Dataset Creation

Dataset creation is very important in deep learning as the output of the training model, viz., the accuracy of the prediction/classification depends hugely on the amount of data fed for training. So, the more the data provided for training, the more accurate the prediction. In case of face recognition, in order to train a deep learning model the following has to be done.

- Provide as many unique faces as possible
- Provide the maximum no. of faces for each individual.

We created our own dataset taking the faces of celebrities concerning sportsper-sons and actors. The work involved was the collection of seven individuals' faces with each photo being almost close to the dimension 256 × 256. The dimension should not be too small because the bottleneck structure of GoogleNet would reduce the dimension further and the model will not provide the desired accuracy. After the collection was done, we put together all of the photos and made sure that no two faces of the same individual were identical and nor were below the required dimensions.

4 Proposed Work

The inception model provides state-of-the-art accuracy but the model is flexible, as we can change the hyperparameters in any layer of the model customizing it to our needs. In this paper, we attempt to modify the architecture more suitable to the dataset created. The size of the input images is 256 × 256. In the first layer of the architecture, convolution operation is performed with a filter size of 34 × 34 and a stride of value 2. This reduces the dimension of the image to 112 × 112. The size of the output due to convolution operation can be obtained as follows.

$$N_{out} = \frac{N_{in} + 2p - k}{s} + 1 \tag{1}$$

wherem N_{in} = input size, N_{out} = output size, p = convolution padding size, k = convolution kernel size and s = convolution stride size. Further maxpooling with a patch size of 3 × 3 and stride of value 1.5 reduces the dimension to 56 × 56. This is succeeded by another convolution and pooling operation followed by inception layers. The structures of the inception layers were not changed and it was the same as in GoogLeNet. This is similar to the bottleneck approach wherein each layer reduces the dimensions. The last few layers of the inception consist of fully connected layers made of neural nets. The linear and softmax functions are applied to each perceptron to include nonlinearities. The last layer of the neural net consists of seven neurons, which outputs different probabilities to different persons depending on the input image. The neuron with highest probability represents the person which model has correctly classified. The architecture flow of our proposed method which is shown in Table 1.

4.1 Caffe and DIGITS

The dataset prepared above was trained on the GoogLeNet deep learning model or Inception v1. The inception model is a very robust state-of-the-art model with very high accuracy. Hence, a number of deep learning platforms give support for this model. One such platform is the Caffe [13] and we used it because of the simplicity

Table 1 Architecture flow of our proposed method

Type	Patch size/stride	Output size
Convolution	34 × 34/2	112 × 112 × 64
Max pool	3 × 3/2	56 × 56 × 64
Convolution	3 × 3/1	56 × 56 × 192
Max pool	3 × 3/2	28 × 28 × 192
Inception (3a)		28 × 28 × 256
Inception (3b)		28 × 28 × 480
Max pool	3 × 3/2	14 × 14 × 480
Inception (4a)		14 × 14 × 512
Inception (4b)		14 × 14 × 512
Inception (4c)		14 × 14 × 512
Inception (4d)		14 × 14 × 528
Inception (4e)		14 × 14 × 832
Max pool	3 × 3/2	7 × 7 × 832
Inception (5a)		7 × 7 × 832
Inception (5b)		7 × 7 × 1024
Avg pool	7 × 7/1	1 × 1 × 1024
Drop out (40%)		1 × 1 × 1024
Linear		1 × 1 × 7
Softmax		1 × 1 × 7

Fig. 1 Hierarchy of deep learning platforms and hardware

and ease of use. No surprise that it was developed for the above-stated purpose in UC Berkely. Caffe is a deep learning framework originally developed for developing deep learning applications using C++. With the support of Pycaffe the caffe framework supports python. The Caffe framework was cloned from github and installed on the system. DIGITS is another deep learning platform that is unparalleled in the ease of deployment and rapid prototyping of any deep learning model. The DIGITS platform was developed by GPU manufacturer Nvidia. The platform works best on the systems equipped with Nvidia GPU's but also supports CPU based systems. Unlike Caffe, the DIGITS is not a standalone platform. The DIGITS platform depends on the Caffe which in turn operates over python. The flow of Hierarchy of deep learning platforms and hardware is shown in Fig. 1.

Like Caffe, the DIGITS was also installed and the correct path to the Caffe for the operation of DIGITS was specified. When run, the DIGITS run as a local server in any browser of that system relying on the underlying CPU/GPU and memory. After running DIGITS the browser displays a user interface where the dataset can be loaded and the required deep learning model can be selected. All the parameters of the model during training can be visualized in the form of graphs where all metrics are available. After the required level of training at any point in time the resulting model can be downloaded.

4.2   Training Process

Training a deep learning model for a given dataset falls under supervised learning. For more sophisticated models like inception the number of hyperparameters are more and training such models require huge computational power and thousands of hours of training. To reduce this complexity, training method called transfer learning is adopted. In transfer learning, the inception model is pretrained by Google with accuracy close enough. Then the weight of the fully connected layers in the inception model is adjusted for the given dataset to achieve the required accuracy. Our hardware was a traditional laptop PC with Intel core i3 CPU and 4GB RAM running Ubuntu v16. The DIGITS was launched and the path to the dataset consisting of RGB images of different labels was specified. A total of 350 images were used for training, out of which 25% of the images were reserved for validation. The dataset is now prepared for training over a deep learning model. Then a classification model was created with the required dataset and the numbers of iterations were specified around 100. The classification model was selected as GoogLeNet with 256×256 architecture. The training was initiated and the model was trained for the given dataset. The PC was kept in a well-ventilated and cool environment with proper power backup and monitored time to time for accuracy and loss. The training was carried over 50 hours straight. All the parameters available as graph were visualized and the accuracy began to saturate after 100 iterations. After 113 iterations the training was stopped and the resultant model was downloaded. Now the downloaded model can be integrated with any other python program and the required operations can be performed.

For every iteration, the model understands the dataset better and the results can be inferred from the graph. The losses drastically reduce after a few iterations and then the slope of the loss decreases. The losses become low after a hundred iterations. Similarly, accuracy is low for the early iterations and the accuracy begins to saturate after hundred iterations. Table 2 shows the parameters that have evolved throughout the training process.

Table 2 Evaluation of parameters with different Epochs

Parameters	Epoch #25	Epoch #50	Epoch #75	Epoch #100	Epoch #113
Accuracy (val)	36.25	37.5	46.25	42.5	50
Accuracy top-5 (val)	86.25	90	86.25	88.75	92.5
Loss (train)	1.91165	1.4028	0.7599	0.88255	0.5612
Loss1/loss (train)	1.57042	1.1918	0.64438	1.0050	0.76085
Loss2/loss (train)	1.56094	1.216523	0.65539	1.03687	0.5499

5 Results and Discussion

Fig. 2 which is fed as input (Testing Image) to the trained model of our method. The model correctly predicts the person as Messi (person in the image is the football legend Lionel Messi). The Table 3 shows the prediction accuracy of test images.

The image passes through various layers of the inception model and the visualization of the image in each of the layers can be viewed below. The bottleneck approach of inception model is evident from the data obtained. In each of the layers, the dimension of the image is reduced. The convolution operation looks for features such as shades for eyes, forehead, and cheeks. The first layer uses kernel size of 34 × 34 with a stride of 2. Pooling and normalization further reduce the dimension of the image. The second convolution operation uses a kernel size of 3 × 3 with a stride of

Fig. 2 Testing image

Table 3 Prediction rate

Testing image name	Prediction accuracy (%)
Messi	86.52
Federer	13.33
Vidyut	0.13
Tyson	0.01
Ajith	0.01

Table 4 Confusion matrix

Predicted class/Actual class	Ajith	Federer	Messi	Surya	Tyson	Vidyut	robertjr	Per-class accuracy (%)
Ajith	7	0	1	1	0	0	1	70
Federer	0	6	1	1	0	1	1	60
Messi	1	1	7	1	0	0	0	70
Surya	0	1	0	8	0	1	0	80
Tyson	0	1	0	0	9	0	0	90
Vidyut	1	3	0	0	0	6	0	60
Robertjr	0	0	0	0	0	1	9	90

2. After these operations a number of inception layers start training on the images and also reduce the dimensionality. In the final layer, classifier with a softmax activation classifies the individuals.

5.1 Confusion Matrix

Once the testing is over the confusion matrix can be created for the obtained model. In confusion matrix a separate set of images, not the ones used for training, belonging to the identities or persons for which the model has been trained is used to evaluate the model. All the images are grouped into folders, each folder consisting of images of respective persons. The same is created as a dataset in DIGITS. Now, the dataset is converted into a text file that consists of the image paths of all the images. Now the trained model is selected and the text file is given as input to classify all the images at once. Now the model classifies the images and creates a table as shown in Table 4.

The first column lists the individuals who have been used to evaluate the model. Per-class accuracy gives how accurate the model was correctly recognizing the person. The overall accuracy obtained by the model was 91.43% for Top-five predictions.

6 Conclusion

Deep learning is more than yet another machine learning technique. It has a huge scope and the potential to disrupt every sector we know ranging from medicine, defense, manufacturing, mobile services. Face recognition is one such application of deep learning and has already disrupted the field of computer vision. Since face recognition is the simplest application, i.e., to classify the images, it acts as a benchmark to test and experiment all the new machine learning models being developed.

By doing the above work, we explored the different frameworks and platforms in deep learning and knew how to customize any deep learning model depending on our applications. This paperwork gives a clear view of deep learning from the creation of dataset to training using a model and deploying it in real time. These steps can be applied to datasets obtained from any field and corresponding actions like classification, decision-making, and control can be made. Deep Learning is the future and it will disrupt all the fields with automation.

References

1. Lawrence, S., et al.: Face recognition: A convolutional neural-network approach. IEEE Trans. Neural Netw. **8**(1) 98–113 (1997)
2. Schroff, F., Kalenichenko, D., Philbin, J.: Facenet: a unified embedding for face recognition and clustering. In: Proceedings of the IEEE Conference on Computer Vision and Pattern Recognition (2015)
3. Zhang, Jun, Yan, Yong, Lades, Martin: Face recognition: eigenface, elastic matching, and neural nets. Proc. IEEE **85**(9), 1423–1435 (1997)
4. Belhumeur, P.N., Hespanha, J.P., Kriegman, D.J.: Eigenfaces versus fisherfaces: recognition using class specific linear projection. Yale University New Haven United States (1997)
5. Turk, M.A., Pentland, A.P.: Face recognition using eigenfaces. In: Proceedings CVPR'91, IEEE Computer Society Conference on Computer Vision and Pattern Recognition, IEEE (1991)
6. O'Rourke, N., Psych, R., Hatcher, L.: A step-by-step approach to using SAS for factor analysis and structural equation modeling. Sas Institute (2013)
7. Turk, Matthew, Pentland, Alex: Eigenfaces for recognition. J. Cogn. Neurosci. **3**(1), 71–86 (1991)
8. Yang, Bo, Chen, Songcan: A comparative study on local binary pattern (LBP) based face recognition: LBP histogram versus LBP image. Neurocomputing **120**, 365–379 (2013)
9. Dalal, N., Triggs, B.: Histograms of oriented gradients for human detection. In: IEEE Computer Society Conference on Computer Vision and Pattern Recognition, 2005 CVPR 2005, vol. 1. IEEE (2005)
10. Krizhevsky, A., Sutskever, I., Hinton, G.E.: Imagenet classification with deep convolutional neural networks. In: Advances in Neural Information Processing Systems (2012)
11. Zeiler, M.D., Fergus, R., Visualizing and understanding convolutional networks. In: European Conference on Computer Vision. Springer, Cham (2014)
12. Szegedy, C., et al.: Going deeper with convolutions. In: Proceedings of the IEEE Conference on Computer Vision and Pattern Recognition (2015)
13. Jia, Y., et al.: Caffe: convolutional architecture for fast feature embedding. In: Proceedings of the 22nd ACM International Conference on Multimedia. ACM (2014)

Enhancing Public Health Surveillance by Measurement of Similarity Using Rough Sets and GIS

Priyansh Jain, Harshal Varday, K. Sharmila Banu and B. K. Tripathy

Abstract Public healthcare plans are essential to regions present around the world and deriving optimal strategies to combat the negative effects is one of the tasks involved in the process. Measuring similarity between two regions based on a common attribute is a popular problem, and this paper aims to find a method for the same, the common attribute being a key statistic taken from public health surveillance records. The method uses the application of rough measurement concept to determine the similarity between different regions having disease-linked death rates, which in turn can effectively be used to derive contingency plans in the case of a relevant event or before the next one takes place. We also take advantage of GIS tools to help in the processing and visualization of spatial data, and this paper discusses the role of GIS in public health surveillance as well.

Keywords Public health surveillance · Geographic information system (GIS) · Similarity · Rough sets · Rough membership · Geoprocessing

1 Introduction

This paper mainly discusses the use of rough measurement concept in finding similarity measure between two different geographic regions based on a common attribute value. The research aims to first spatially represent disease severity of a chosen disease dataset with georeference, using GIS tools which are efficient in working

P. Jain (✉) · H. Varday · K. Sharmila Banu · B. K. Tripathy
Vellore Institute of Technology, Vellore 632014 Tamil Nadu,, India
e-mail: priyansh.jain2016@vitstudent.ac.in

H. Varday
e-mail: harshalvarday1@gmail.com

K. Sharmila Banu
e-mail: sharmilabanu.k@vit.ac.in

B. K. Tripathy
e-mail: tripathibk@vit.ac.in

© Springer Nature Singapore Pte Ltd. 2020
K. N. Das et al. (eds.), *Soft Computing for Problem Solving*,
Advances in Intelligent Systems and Computing 1048,
https://doi.org/10.1007/978-981-15-0035-0_21

271

with geospatial data and associated geoprocessing. Following which, the methodology is discussed wherein we use membership calculation to determine the similarity between two regions selected.

The paper is separated into the following sections: Sect. 1—gives basic information about the concepts and technologies used. Section 2—contains background about the chosen dataset and GIS tools. Section 3—contains the methodology used in the research. Section 4—contains the results. Section 5—contains the analysis. Section 6—contains the conclusion.

1.1 Introduction to Rough Sets

The concept of rough set was first described by Pawlak [1], which happens to be an approximation of a conventional/crisp set. Formally, the approximation is present in terms of set pairs that can provide both the lower and upper estimation of the original set. The standard rough set theory was introduced by Pawlak [1] in 1982, the approximations themselves are crisp sets. However, the variations describe the approximating sets to take the form of fuzzy sets as well. This sets the foundation for a new mathematical approach to the concept of vagueness or imprecision. This vagueness is expressed by a boundary region of a set. This concept can be defined by means of topological operations, interior and closure, called approximations.

Rough sets have been proposed for a very wide variety of applications. In particular, the rough set approach seems to be important for artificial intelligence and cognitive sciences, especially in machine learning, knowledge discovery, data mining, expert systems, approximate reasoning, and pattern recognition.

1.2 GIS Tools

Geographical Information Systems (GIS) are tools that allow to map, store, present, analyze, and manipulate spatial data. These tools also allow for interactive python console and scripts, so that repetitive and complicated tasks can be automated. GIS tools are mainly used in urban planning—transport, environmental, public health, and disaster management. One of the major use of GIS is in public health surveillance, which is collecting, analyzing, monitoring, and interpreting public health data for planning healthcare policies and strategies to act upon even before disaster impacts the regions, and try to predict outbreak of diseases.

1.3 Public Health Surveillance and GIS

Public Health Surveillance is used in planning, most urban planning requires the use of GIS, namely—transport, environmental, public health, and disaster management. One of the major use of GIS tools is in public health surveillance, which is collecting, analyzing, monitoring, and interpreting public health data for planning healthcare policies and strategies to act as a backup plan before the diseases actually impacts, and try predicting the outbreak of diseases, and accordingly plan health care for the probable outbreak and reduce the outbreak from spreading and impacting other individuals and regions. Also, it allows to set goals for public health care, in terms of policies and strategies. Apart from these applications, these tools are also used for crime mapping, planning defense strategy, planning sustainable development, real estate, community planning, transport, and logistics planning. One of the main requirements for GIS tools is geospatial data, for which there needs to be accessed to Spatial Data Infrastructures (SDIs) whose role is to acquire and maintain geospatial data in a universally agreed format so that it is easier for wide range of research purposes. Many large institutes have developed systems to collect data about different diseases, syndromes, and outbreaks, so as to keep a record for future study and analyze pattern to prevent further transmission. Nowadays most regions have their local databases for diseases, which are generally not in sync with the central system for disease management. SDIs also play a significant role in training and awareness about geospatial data, its use and importance of GIS tools. Even though the tools are feature-rich, the lack of awareness and training renders the features unused by end users. This comes in handy when there are emergencies like forest fires, weather risk prevention, and tracking water in case of storms/heavy rains for planning and tracking the paths for the rains or fire and map out evacuation route [2].

2 Background

2.1 Problems Associated with Finding Similarity Between Regions

The problems associated with finding similarity between two different spatial regions are not uncommon. There are methods like Moran's I and Geary's C which find the spatial autocorrelation by comparing neighboring points and giving out an aggregate degree of correlation. These methods are not applicable in cases where we might need to compare datasets with discrete values, or cases where we need to compare nonadjacent regions for similarity [3, 4]. Then there is semivariogram method which finds the variance between the points as a function of spatial difference between them. For the case of finding the similarity/correlation between two different non-neighboring regions this method cannot be used [4].

2.2 GIS Tools—Survey

GIS is widely used in Community Health Assessment (CHA) which helps in determining health issues and thereby helping to decide allocation of government resources and initiatives. CHA is broken down into the following:

• Geographic community identification.
• Identification of health factors within the community.
• Identification of bordering communities of interest.
• Identification of health factors within the bordering community.
• Comparing the factors between the community and bordering community.
• Identification of aggregate community.
• Identification of health factors in aggregate community.
• Comparing factors within community and aggregate community.

The tools used were ArcGIS, Epi-Map, Forestry GIS, and conventional map along with statistical software like Excel, SPSS, SAS, Stata, etc. Most GIS tools despite having advanced features for data manipulation and analysis are being used only for simple functions of data representation only. The more complex functions of computations and manipulation are carried out in external tools separated from GIS tools. This is mainly due to lack of training and awareness of GIS tools [5].

2.3 Necessary Infrastructure for Providing Resources to GIS Tools

Use of GIS is broadly divided into two categories, namely health geography which includes health outcomes and epidemiology applications. Spatial Data Infrastructures (SDI) include people and institute who are focused toward gathering data, standardizing it, storing, and maintaining the data. Data with spatial reference transforms into spatial data, and it makes it easier for further geoanalysis. SDI are expected to provide standardized service for the delivery of spatial data to researchers and academicians through a streamlined request process. The availability of data is taken for granted by users, be it free or for a nominal fee. SDI acts like a central authority which maintains spatial data and connects with local authorities to ensure the data remains in sync at the center. Although SDI require a large initial investment, it gives long-term returns, comparable to objectives of geolibraries. The basic requirements for SDI are to develop geospatial culture, spreading awareness among people and organizations, and provide training in the use of geospatial data and further processing. It also has to keep the resources available for the general public, be it tools, licenses, material, etc. SDI has to make sure that there are no confidentiality issues with respect to the data that they maintain and distribute, which might lead to trouble [2]. One of the important things is to adhere to a nationally accepted standard for maintaining spatial data.

Table 1 Sample data from countywise fatality rate in US per 100,000 due to heart diseases and stroke

State name	County name	Rate
Iowa	Mills	279.53
Illinois	Marshall	370.34
Utah	Salt Lake	296.66

There are other essential problems associated with spatial data apart from the need for a standard, organized infrastructure to capture spatial data. These problems include privacy of spatial data, given spatial data is highly sensitive. Securing the data can be done by using encryption or without it. Another significant problem can be validation of the analysis result by healthcare professionals. There might be significant changes even with minuscule change in the spatial data [6].

2.4 About the Dataset

The dataset chosen provides the countywise average fatality rate due to heart diseases and stroke in the county per 100,000 deaths. Table 1 shows sample data provided in the dataset. The data contains the state name, county name, Federal Information Processing Standard (FIPS) county codes, and finally the average fatality rate due to heart diseases and stroke in the county per 100,000 deaths. The data is obtained from Center for Disease Control and Prevention, a US federal agency under health department. It aims at protecting public health by controlling and preventing diseases. It also provides additional information about the diseases and provides access to related information to general public.

3 Methodology

3.1 Rough Sets

Rough set is defined by topological operations called approximations, thus this definition also requires advanced mathematical concepts [1]. Rough sets are formal approximation of crisp sets. It gives lower and upper approximation of the original set. This concept deals with rough membership of individual elements of the universal set with respect to different sets.

Rough set theory is the portrayal of the different limits present in the sets, by specifying their vagueness. This uses the boundary regions—that is, the boundary regions of a set decide whether the set is crisp or rough. The emptiness of a boundary region shows that the set is crisp. It can be noted that the vagueness that is specified in the rough set theory is different from uncertainty [1].

There are many interesting applications to rough sets, and they have a fundamental role in AI systems and their subconcepts like Machine Learning and Pattern Recognition. The most interesting advantage is that there are absolutely no prerequisites required for the concept to work its charms.

The rough membership of any element $x \in U$ (Universal set) with respect to any set $A \subseteq U$ denotes the degree with which x belongs to the set A. Mathematically

$$\mu_A(x) = \frac{|A \cap [x]_R|}{|[x]_R|} \tag{1}$$

Here, $[x]_R$ denotes equivalence class of the element x with respect to U (set of all similar elements).

When we consider membership in our case, it is basically the rough membership that is calculated on the basis of set intersections in our case. If we were to use a simple 10×10 grid system that has colors belonging to a finite set of three colors, red, brown and green, and these exist for each unit that we call a polygon, the membership of a region could be calculated based on the total number of polygons that have a similar color in the subregion as compared to the entire region of our polygons. The above formula when applied gives us our result. It is interesting to note that the results come out to be pretty accurate, and what we build upon on this paper is the main application that helps us to calculate the same for maps. The similarity calculated can hence be used to derive a contingency plan for the similar regions as well.

3.2 Rough Measurement

The concept of rough measurement finds the similarity between two different sets based on the rough membership values of all the elements with respect to the two sets. Let us assume set $A \subseteq U$ and set $B \subseteq U$, where U is universal set of elements. Then the similarity between the sets A and B is given by these formulae.

Case 1:

$$Sim D_R(A, B) = \begin{cases} 1, & A = B = \phi \\ \frac{2\sum_{i=1}^{n} \min\{a_i, b_i\}}{\sum_{i=1}^{n}(a_i + b_i)}, & else \end{cases} \tag{2}$$

Case 2:

$$Sim D_R(A, B) = \begin{cases} 1, & A = B = \phi \\ \frac{\sum_{i=1}^{n} \min\{a_i, b_i\}}{\sum_{i=1}^{n} \max\{a_i, b_i\}}, & else \end{cases} \tag{3}$$

We will explore our results with both the cases.

Fig. 1 High level overview
of the process flow

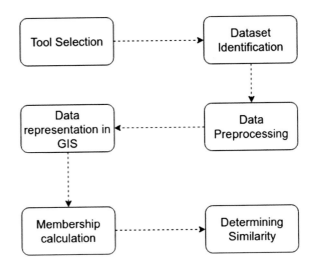

3.3 Process Flow

The algorithm will use the above-specified methods to calculate the similarity between two or more regions, taken two at a time. An overview of the steps taken is as in Fig. 1.

A diagrammatic representation of the method employed is given below, and this caters to the Case 1 in the previous section. Case 2 is not described, as the changes are rather obvious (Fig. 2).

4 Results

For sample test, we take multiple test cases to test the rough membership function. For the first case, we determine the similarity of two states that have comparable death rates among counties. The second case, we find the similarity of two states having counties with completely different fatality rates. The test cases include pair of states that are neighboring, non-neighboring, approximately similar fatality rate, and different fatality rates. We take into consideration different pairs of test cases to prove that rough measurement concept works in all the test cases irrespective of proximity of the pair of regions, or the difference in fatality rates of regions.

We apply the rough measurement concept to find similarity between different states based on the fatality rate. We first represent the data on the map using QGIS [7] and categorize the fatality rate into five categories. This is done so as to have lesser data values to work on. Then we proceed by finding the memberships of different counties, and later we find similarity value between different states. Below are the

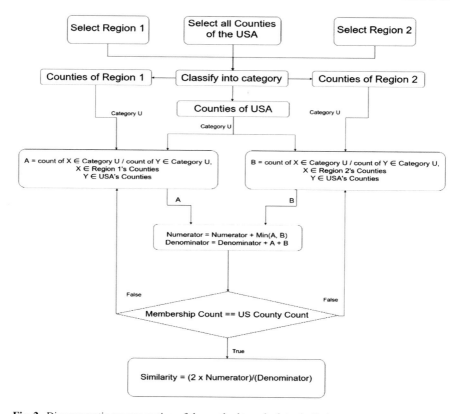

Fig. 2 Diagrammatic representation of the method to calculate similarity

screenshots from QGIS showing different test cases and corresponding fatality rates. The test cases include the similarity measure for the following:

1. Oklahoma and Arkansas—High degree of similarity
2. Mississippi and Alabama—High degree of similarity
3. Colorado and Oklahoma—Low degree of similarity
4. North Dakota and South Dakota—High degree of similarity (Figs. 3, 4, 5, 6, and 7; Table 2).

5 Analysis and Discussions

First, we represent the data about fatality rate due to heart and stroke in the US obtained from the CDC [8]. The data is represented in QGIS [7], by mapping the county boundary layers with the dataset using foreign key concept. This maps the

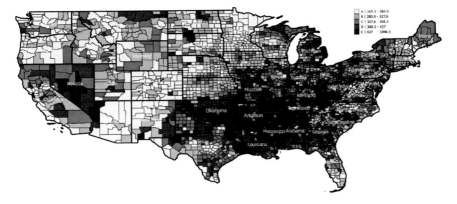

Fig. 3 It shows the representation of the data in QGIS on a map layout, with fatality rate categorized into five categories

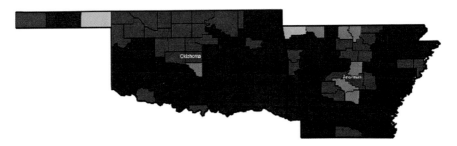

Fig. 4 Regions of Oklahoma and Arkansas

Fig. 5 Regions of
Mississippi and Alabama

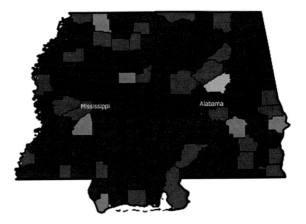

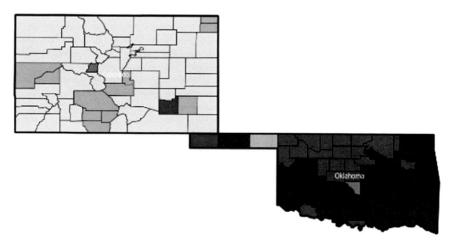

Fig. 6 Regions of Colorado and Oklahoma

Fig. 7 Regions of North
Dakota and South Dakota

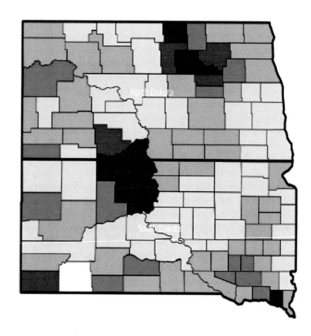

Table 2 Similarity values
between different states based
on average countywise
fatality rates

Region 1	Region 2	Similarity (%)
Oklahoma	Arkansas	85.36
Mississippi	Alabama	79.51
Colorado	Oklahoma	02.17
North Dakota	South Dakota	73.52

Fig. 8 The state of Louisiana

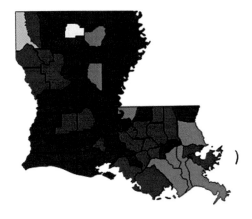

fatality rate of a county with the county map boundary. This gives a visual representation of the fatality rate. The fatality rate is then categorized into five categories, to give a better idea.

Second, we find the similarity measure between different regions. Since there are no other favorable methods to find similarity between two states, we use rough membership function. We find out memberships of each county with respect to different regions (states) by applying membership function to all the counties based on the category of the average fatality rate. This gives us membership values of each county, which then is used to calculate the similarity between different regions (states). Then using the rough membership concept, we find the similarity between the different states of the USA (Fig. 8).

Figure 9 represents the similarity calculated through both algorithms, Wang [3] and Shi [3], and another similarity index based on set theory, the Braun-Blanquet Index [9, 10].

It is observed that Shi's method tends to give out a higher value of the similarity, for the same values pertaining to corresponding states compared. This is because, in the above formulae, we have the denominator as the summation of maximum of two elements in Wang's method, and the summation of mean of two elements in Shi's method, and thus the latter is always greater than the former, since maximum of a set of numbers is always greater than or equal to average.

One could infer from the findings that if a threshold value is used in the measurement, Shi's method would yield a set with more elements as compared to the other methods discussed.

6 Conclusion and Future Work

The use of rough membership calculation and the generation of similarity for different subregions was shown in the paper. QGIS was used to plot the subregions for given datasets, and hence fatality rates across the regions were visualized. We calculated

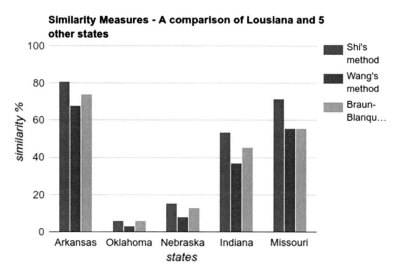

Fig. 9 Comparison of similarity of Louisiana's heart disease death rates with five other states

the rough membership with respect to different subzones, and approximate the lower and upper values, followed by the calculation of similarity. This calculation does not have to apply for just two states, we can theoretically perform this on a subregion of any size. Also the subregions can be neighbors or non-neighbors. The similarity thus obtained proves to be a very useful result, for it can provide a solid ground for deriving contingency plans. This helps save time, money, and effort required for making contingency plans for the regions with high degree of similarity. This method can be applied to any dataset with some amount of data cleaning and normalization. In the instance of a change of datasets, we can use the abovementioned implementation in healthcare studies, economic studies regarding not just the fatality rates caused by heart diseases, but also other diseases or general causes. This study can be extended to cluster regions based on similarity. Also, the method can be expanded to incorporate multiple features like population density among other features. A more accurate similarity model can hence be generated.

References

1. Pawlak, Z.: Rough sets. Int. J. Comput. Inform. Sci. **11**(5), 341–356 (1982)
2. Boulos, M.N.K.: Towards evidence-based, GIS-driven national spatial health information infrastructure and surveillance services in the United Kingdom. Int. J. Health Geogr. **3**(1) (2004)
3. Liao, W.: The rough method for spatial data subzone similarity measurement. J. Geogr. Inf. Syst. **4**(01) (2012)
4. Gunaratna, N., Liu, Y., Park, J.: Spatial autocorrelation. J. Recuperado el **2** (2013)

5. Scotch, M., et al.: Exploring the role of GIS during community health assessment problem solving: experiences of public health professionals. Int. J. Health Geogr. **5**(1) (2006)
6. Sharmila Banu, K., Tripathy, B.K.: Data analytics in spatial epidemiology: a survey. J. Technol. **78**(10), 159–165 (2016)
7. Quantum Geographic Information System (software). https://www.qgis.org/
8. Centers for Disease Control and Prevention. https://www.cdc.gov/
9. Lenarčič, A.: Rough sets, similarity and optimal approximations. Ph.D. thesis, McMaster University, June (2017)
10. Braun-Blanquet, J: Pflanzensoziologie. Springer (1928)

Data Analytics Implemented over E-commerce Data to Evaluate Performance of Supervised Learning Approaches in Relation to Customer Behavior

Kailash Hambarde, Gökhan Silahtaroğlu, Santosh Khamitkar, Parag Bhalchandra, Husen Shaikh, Govind Kulkarni, Pritam Tamsekar and Pranita Samale

Abstract Online purchase portals have a spectacular opportunity for business expansion. E-commerce portals have data repositories pertaining to online transactions that could be analyzed through data analytics to find valuable insight for further expansion of business as well as targeted marketing. This study has made an attempt for the implementation of data analytics over the shared data set of Turkey-based e-commerce company. Precisely, a comparative analysis of supervised machine learning algorithms has been worked out for predicting customer behavior and products being brought. Their efficiency has been found out and they have been ranked purpose wise. The implementations of algorithms are carried out in Python.

K. Hambarde (✉) · S. Khamitkar · P. Bhalchandra · H. Shaikh · G. Kulkarni · P. Tamsekar · P. Samale
School of Computational Sciences, SRTM University, Nanded, Nanded, India
e-mail: kailas.srt@gmail.com

S. Khamitkar
e-mail: s_khamitkar@yahoo.com

P. Bhalchandra
e-mail: srtmun.parag@gmail.com

H. Shaikh
e-mail: husen09@gmail.com

G. Kulkarni
e-mail: govindcoolkarni@gmail.com

P. Tamsekar
e-mail: pritamtamsekar@gmail.com

P. Samale
e-mail: pranitapatil217@gmail.com

G. Silahtaroğlu
School of Business and Management Science, I.M. University, Istanbul, Turkey
e-mail: gsilahtaroglu@medipol.edu.tr

© Springer Nature Singapore Pte Ltd. 2020
K. N. Das et al. (eds.), *Soft Computing for Problem Solving*,
Advances in Intelligent Systems and Computing 1048,
https://doi.org/10.1007/978-981-15-0035-0_22

Keywords Customer behavior analytics · Classification · Random forest

1 Introduction

Due to the advent of the Internet, the Internet service backbone has evolved as infrastructural mainstay, thereby allowing people to get connected with each other as well as with markets for their daily needs. This has led to online shopping syndrome where people browse e-commerce companies' websites and purchase products as per their convenience. This resulted in a variety of products being available before users which could be bargained and purchased on discounts. That is why, today everyone prefers online shopping. This is possible thanks to payment availability through banks in 24 * 7 * 365 style. All online shopping portals have excellent and simple interfaces that allow users to quickly search their products and purchase them in simple steps. All such portals maintain their customer data for future use in terms of mobile numbers, e-mail addresses, physical addresses, credit and debit card details, IP addresses of customer's mobile phones/computers, etc. This data help them to maintain and establish new channels for direct marketing using their previous purchasing behavior. But analyzing customer's behavior is very complex as it involves many variables to take into account. Recently researchers have come up with new approaches to analyze such complex logic by using machine learning techniques [1]. The machine learning techniques are a subset of Artificial Intelligence which empowers e-commerce companies to better predict purchasing behavior of customers. The machine learning algorithms are classified as supervised and unsupervised. This underlined work is dealt with supervised machine learning algorithms. The major objective of this investigation is to apply different classification algorithms on the obtained data set to identify the hidden pattern of consumer behavior and determine, out of all standard techniques, the best classification techniques [2]. We have implemented eight mostly used classification algorithms like Logistic Regression, Random Forest, SVM, Gradient Boosting, XGB Classifier, Decision Tree, Naïve Bayesian, and K-Neighbors, over the collected data set. The results can enable us to understand the choice of algorithm for a particular problem. The present investigation is useful for marketing tactics to promote products. Further, this study is also useful for the prediction of customers who often react to the product offers based on the previous historical data.

2 Related Works

In [1], Gökhan Sılahtaroğlu and Hale Dönertaşli have presented an article on the investigation and forecast of e-customers' behavior by using machine learning. Their objectives were to predict customer behavior pattern that will or will not buy a product. For this objective, they shared similar data and applied two supervised machine learning algorithms, namely Decision Tree and ANN. They used confusion matrix for evaluating classifiers performance. The author found that ANN performed well on that data set. In [3], T. K. Das has come up with a new customer classification predictive model based on machine learning techniques. Objectives of this paper were to identify the customers who more likely respond to the offer of a company based on customers' past purchasing behavior. For this, they collected bank details of customers from publicly available UCI machine learning repository. For experimental purpose, the author used supervised machine learning algorithms such as Naïve Bayes, KNN, and SVM for classifying bank customers. For evaluation purpose, the author used confusion matrix, and found that Naïve Bayes performed well on the data set. In another similar work [4], the author analyzed Watson's Turkey and Ukraine-based e-commerce site data set for classifying customer's behavior pattern. For this project, the author used Weka software. One more similar work [5] presents a comparative analysis on supervised machine learning algorithm. For that, the author came up with a case study such as credit-card fraudulent-transaction detection. Algorithms such as Random Forest, Stacking Classifier, XGB Classifier, Gradient Boosting, Logistic Regression, MLP Classifier, SVM, Decision Tree, KNN and Naïve Bayes have been used in this study. Among all the algorithms, Random Forest has given the highest accuracy at 95% and the lowest accuracy has been given by Naïve Bayes at 91%. Several classification algorithms such as Support Vector Machine, Random Forest, Logistic Regression, Decision Tree, and NN were used for shopping list prediction [6]. The truthful prediction of customer's shopping provides significant influence on effective customized services and one-to-one marketing actions in retail industry. This could lead to exploiting customer's satisfaction and company profitability. Click stream database model consists of three attributes which probably include (1) unique session identifier, (2) date and time of user action, and (3) URL of the node visited [2]. Parameters such as customer age and gender play a vital role in retailing, marketing, and purchasing habits. By analyzing mouse movements of his/hers, we can predict the gender of an online customer [7].

3 Understanding Data Set

The data used for this investigation work is used from Turkey-based e-commerce company. This was shared by author number 1 for academic and further research purpose. The explanation of data variables is as per citation [1]. We have followed the standard data mining procedure during experimentation setup. The steps are explained in detail in citation [4] and mainly include date collection (Turkey-based ecommerce company data set was considered for the experimental purpose [1]), data preprocessing, classification, data analysis, and evaluation.

The overall methodology is illustrated below in Fig. 1.

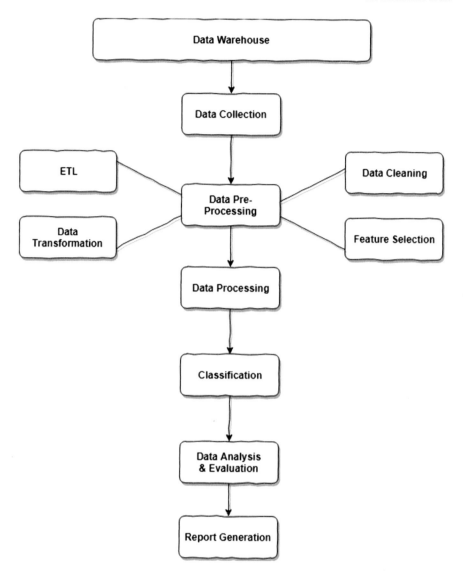

Fig. 1 Proposed research methodology

4 Results and Discussions

The performance of the model basically is based on the normalcy, accuracy, specificity, and recall which act as a decision support system in the marketing domain. With the help of confusion matrix, the performance measures of the proposed model

Table 1 Confusion matrix

		Predicted	
		Class 1 (negative)	Class 1 (positive)
True	Class 1 (negative)	TN	FP
	Class 1 (positive)	FN	TP

are discussed. Confusion matrix reveals the nature of the actual and predicted classifications information obtained from the several algorithms executions on the data set. Table 1 reveals the nature of the confusion matrix for a binary classifier.

1. Accuracy Overall: how often is the classifier correct? [3]: Table 3 shows the comparative analysis of used algorithms; among all algorithms, Random Forest gives the highest accuracy of 94.81 and Naïve Bayes gives, 73.86, the lowest accuracy on this data set.

$$\text{Accuracy} = \frac{TP + TN}{TP + TN + FP + FN} \tag{1}$$

2. Misclassification or error rate: Overall, how often is it wrong? Table 3 shows the comparative analysis of used algorithms; among all algorithms, Random Forest gives the lowest error rate of 0.04 and Naïve Bayes gives 0.24, the highest error rate on this data set.

$$\text{Error Rate} = \frac{FP + FN}{TP + TN + FP + FN} \tag{2}$$

3. Recall: It is the probability of correctly predicting positive examples. Table 3 shows the comparative analysis of used algorithms; among all algorithms, Naive gives the highest recall 1.0, and K-Neighbors give 94.66, the lowest recall on this data set.

$$\text{Recall} = \frac{TP}{TP + FN} \tag{3}$$

4. Specificity: It defines the probability of correctly predicting negative cases. In Table 3, as compared to all algorithms, Random Forest gives the highest specificity at 90% and Naïve Bayes gives 32.95.

$$\text{Specificity} = \frac{TN}{TN + FP} \tag{4}$$

5. True Positive Rate (TPR): It is the fraction of positive cases that were correctly identified. In Table 2, the comparative analysis reveals that the Random Forest algorithm classified 2962 instances as true positive. Further, the lowest instances

Table 2 Comparison of true positive, false positive, true negative, and false negative of classifiers

Classifier techniques	True positive	False positive	True negative	False negative
Logistic Regression	2872	119	331	110
Random Forest	2962	35	318	117
Support Vector Machine	2877	119	331	110
Gradient Boosting	2920	71	323	118
XGB Classifier	2951	40	302	139
K-Neighbors	2877	114	279	162
Decision Tree	2877	113	319	122
Naïve Bayes	2094	897	441	0

were classified by the Naïve Bayes algorithm with 2094 instances.

$$TPR = \frac{TP}{TP + FN} \tag{5}$$

6. False Positive Rate (FPR): It represents the proportion of negative cases that were incorrectly classified as positive. In Table 2, among all the classifiers, Random Forest classifies 35 instances as false positive and Naïve Bayes 897 instances.

$$FPR = \frac{FP}{FP + TN} \tag{6}$$

7. True Negative Rate (TNR): It is the proportion of negative cases that were correctly classified. Table 2 shows the scenario where Naïve Bayes classified 441 instances as TNR, and K-Neighbors classified 279 instances.

$$TNR = \frac{TN}{TN + FP} \tag{7}$$

8. False Negative Rate (FNR): It is the proportion of positive cases that were incorrectly classified as negative seen in Table 2. In this scenario, Naive Bayes classified 0 instances as FNR and K-Neighbor classified 162 instances as FNR (Table 3).

$$FNR = \frac{TN}{TN + FP} \tag{8}$$

Below, Figs. 2, 3, and 4 further confirm the above results and discussions (Fig. 5).

Table 3 Comparison of accuracy, error rate, specificity, and recall of classification techniques

Classifier techniques	Accuracy (%)	Error rate (%)	Specificity (%)	Recall (%)
Logistic Regression	93.32	0.06	96.02	96.31
Random Forest	94.81	0.04	90.00	98.83
Support Vector Machine	93.32	0.06	73.55	96.31
Gradient Boosting	94.66	0.05	81.87	96.11
XGB Classifier	94.78	0.05	88.3	95.5
K-Neighbors	91.00	0.08	70.99	94.66
Decision Tree	93.15	0.06	73.84	95.96
Naïve Bayes	73.86	0.26	32.95	1.0

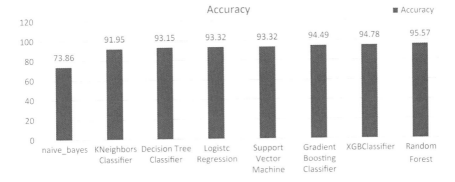

Fig. 2 Accuracy comparison

Fig. 3 Error rate comparison

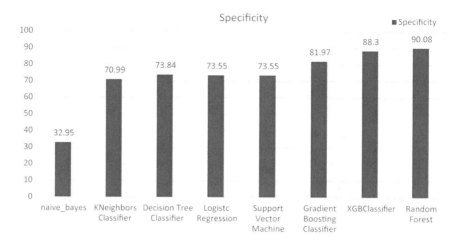

Fig. 4 Specificity comparison

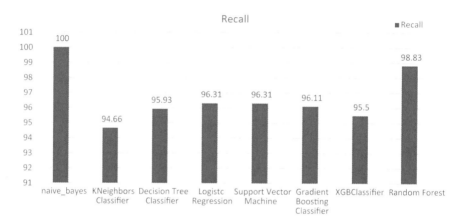

Fig. 5 Recall comparison

5 Conclusion

The presented work dealt with the comparative analysis of major supervised machine learning algorithms applied over the shared data set obtained from an e-commerce company of Turkey. Further, this study aimed to target customer's behavior patterns related to product purchasing. The comparative analysis of supervised algorithms reveal that the Random Forest classifiers produced the most significant results with maximum accuracy and specificity. The lowest accuracy was witnessed in Naïve Bayes with lowest sensitivity, accuracy, and specificity. Further, Naïve Bayes also achieved the highest error rate with a large number of false positive and false negative cases when compared to their peers. The overall result obtained through the analysis reveals that the effective marketing of the product can be achieved with the help of the Random forest classifier by targeting the customer behavior pattern. The outcome of the presented work can be useful for campaigning by marketing departments to expand business promotional offers and to avoid the wasteful expenditure of sending offers and promotions to nonrespondents. Detailed information about the algorithm and python libraries which are used in study can be found in [8] and [9].

References

1. Sılahtaroğlu, G., Dönertaşli, H.: Analysis and prediction of E-customers' behavior by mining clickstream data. In: 2015 IEEE International Conference on Big Data (Big Data), pp. 1466–1472. IEEE
2. Sılahtaroğlu, G.: Predicting gender of online customer using artificial neural networks. Int. J. Adv. Electron. Comput. Sci. 2(10) (2015). ISSN: 2393-2835
3. Das, T.K.: A customer classification prediction model based on machine learning techniques. In: 2015 International Conference on Applied and Theoretical Computing and Communication Technology (iCATccT), pp. 321–326. IEEE
4. Altunan, B., Arslan, E.D., Seyis, M., Birer, M., Üney-Yüksektepe, F.: A data mining approach to predict E-commerce customer behaviour. In: The International Symposium for Production Research, pp. 29–43. Springer, Cham
5. Peker, S., Kocyigit, A., Eren, P.E.: An empirical comparison of customer behavior modeling approaches for shopping list prediction. In: 2018 41st International Convention on Information and Communication Technology, Electronics and Microelectronics (MIPRO). IEEE
6. Senecal, S., Kalczynski, P.J., Nantel, J.: Consumers' decision-making process and their online shopping behavior: a clickstream analysis. J. Bus. Res. 58(11), 1599–1608
7. Vicari, D., Alfó, M.: Model based clustering of customer choice data. Comput. Stat. Data Anal. 71, 3–13
8. Han, J., Pei, J., Kamber, M.: Data Mining: Concepts and Techniques. Elsevier
9. Géron, A.: Hands-on Machine Learning with Scikit-Learn and Tensor Flow: Concepts, Tools, and Techniques to Build Intelligent Systems. O'Reilly Media, Inc

Optimal Renewable Energy Resource Based Distributed Generation Allocation in a Radial Distribution System

Kola Sampangi Sambaiah and T. Jayabarathi

Abstract Distributed generation (DG) allocation is the most promising source for reducing network loss and enhancing bus voltage stability in a distribution system. Because of the vast availability and nonpolluting character of renewable energy resource, it is gaining more attention nowadays. The most widely used renewable-based DG (RDG) is wind turbine (WT) and solar photovoltaic (SPV). Power generation patterns of the WT and SPV modules are random and nonlinear because the power output of WT and SPV modules are dependent on wind speed and solar irradiation. These require a probabilistic model to represent the actual power generation. The present paper reflects the potency of WT and SPV modules for reducing system losses and enhancing voltage stability. A new hybrid gray wolf optimizer (HGWO) is proposed to solve the DG allocation problem. The proposed optimization method is tested on IEEE 12- and 15-bus radial distribution system (RDS) and it is found that the proposed HGWO has more potency in terms of loss reduction and voltage stability enhancement compared to the existing techniques.

Keywords Renewable energy source · Optimal DG allocation · Meta-heuristic optimization · Hybrid gray wolf optimizer · Wind and solar DG · Renewable distributed generation

1 Introduction

Demand for electrical power is rapidly increasing day by day due to population explosion and urbanization. In the present scenario, the existing capacity of transmission lines is not suitable for increasing power demand. Since network upgradation involves huge investment, an alternative option for network upgradation deferment

K. S. Sambaiah · T. Jayabarathi (✉)
School of Electrical Engineering, Vellore Institute of Technology, Vellore, India
e-mail: tjayabarathi@vit.ac.in

K. S. Sambaiah
e-mail: sambaiahks@gmail.com

© Springer Nature Singapore Pte Ltd. 2020
K. N. Das et al. (eds.), *Soft Computing for Problem Solving*,
Advances in Intelligent Systems and Computing 1048,
https://doi.org/10.1007/978-981-15-0035-0_23

is by using distributed generation which supplies customer demand locally. Distributed generation is a localized power generation source directly connected to the distribution system [1].

Conventional power generation by DG utilizes a reciprocal engine, diesel generator, gas turbine, and micro-turbine. Due to increased concern toward environmental changes and global warming, RDG has been opted as an alternative for conventional DG. Primarily, energizing of RDG units are carried out by solar, wind, and fuel cell. In the market, currently, several DG technologies are available; among these, solar and wind-based DG technologies are vastly used.

In general, the structure of the electrical distribution system is radial. Radial distribution system (RDS) consists of the main feeder which begins from the substation and connects customers at different load points. In RDS, the ratio of resistance to reactance (R/X) causes more power losses. Optimal allocation of DG will improve system overall efficiency by grid reinforcement, reducing operating cost and power losses, the deferment in system upgrades, and enhancing reliability, voltage stability, and integrity of the system. Aforementioned characteristics are deteriorating when DG is allocated at a nonoptimal location. It will be more crucial for sustaining the stability and reliability of the system while dealing with multiple DG allocation problems [2]. Many analytical and heuristic or meta-heuristic optimization techniques are used to solve multiple DG allocation problems. Analytical techniques used for optimal DG allocation are continuation power flow (CPF) and Kalman filter algorithm [3–5]. Meta-heuristic optimization techniques such as evolutionary programming (EP), genetic algorithm (GA), differential evolution (DE), particle swarm optimization (PSO), ant colony optimization (ACO), cuckoo search algorithm (CSA), bacterial foraging optimization (BFO), intelligent water drop (IWD), invasive weed optimization (IWO), flower pollination algorithm (FPA), and ant lion optimization (ALO) are performing comparatively better [6].

In [3], the impact of DG technologies on bus voltage stability is investigated and a maximum loading or voltage collapse point is found by using a power flow continuation. However, the DG size is not considered. Lee et al. [4] used power loss index for DG location and Kalman filter algorithm for finding the DG size which reduces computational time by minimum samples. In [5], an analytical technique is used for DG location; this technique is a non-iterative technique which is free from convergence problem.

In the present paper, power loss minimization is the main objective and evaluates corresponding bus voltage stability. Optimal DG allocation is carried out by an HGWO. The proposed HGWO performance is compared with different optimization techniques. However, in the present paper, stochastic nature of the WT and SPV module outputs are modeled using a probabilistic approach [7–9]. The proposed HGWO is tested for IEEE 12- and 15-bus RDS. The proposed HGWO performance is compared with the existing techniques on the basis of power loss reduction and voltage stability enhancement.

2 Power Generation Modeling of Wind and Solar Power

Power generation output of WT and SPV modules purely depend on geographical location, wind speed, ambient temperature, and solar irradiance. In the present section for better understanding, wind and solar power generation are modeled.

2.1 Modeling of WT

Classification of WTs based on wind speed are of two categories, namely fixed and variable speed WT. In general, variable speed WTs are used as DG units and real power output produced depends on wind speed [9]. The electrical power output generated by a typical WT is given by

$$P_{WT} = \begin{cases} 0 & v_{aw} > v_{cout} \text{ or } v_{aw} < v_{cin} \\ P_{rated} * \left(\frac{v_{aw} - v_{cin}}{v_N - v_{cin}} \right) & v_{cin} \leq v_{aw} \leq v_N \\ P_{rated} & v_N \leq v_{aw} \leq v_{cout} \end{cases} \tag{1}$$

where v_{cin}, v_N, and v_{cout} are cut-in, nominal, and cut-out speed of WT, respectively; the average wind speed is given by v_{aw}; rated power output P_{rated} of WT is given by

$$P_{rated} = 0.5 * \left(\rho A v_{aw}^3 C_p \right) \tag{2}$$

where A is the rotor swept area; v_{aw} is the wind speed; ρ is the air density; and C_p is the WT power coefficient.

2.2 Modeling of SPV Module

Modeling of the SPV module mainly depends on solar irradiance and ambient temperature. It is known that a single SPV module is unable to generate a large amount of electrical power. Hence, the series and parallel connections of these modules form an array. These series and parallel connections increase generated electrical power by increasing current and voltage level. The maximum power output of an SPV array having $N_s \times N_p$ SPV modules is given by

$$P_{SPV} = N_s N_p P_{mmd} \tag{3}$$

where N_s and N_p are the number of series and parallel SPV modules; P_{mmd} is the generated maximum electrical power which is represented as

$$P_{mmd} = V_{oc} * FF * I_{sc} \tag{4}$$

where I_{sc} is the short circuit current, FF is the fill factor, and V_{oc} is the open-circuit voltage of the SPV module. I_{sc}, FF, and V_{oc} are the functions of the SPV module temperature and solar irradiance; and these are given by

$$V_{oc} = \frac{V_{noc}}{1 + k_2 * \ln\left(\frac{G_n}{G_a}\right)} \left(\frac{T_n}{T_a}\right)^{k_1} \tag{5}$$

$$I_{sc} = I_{nsc}\left(\frac{G_a}{G_n}\right)^{k_3} \tag{6}$$

$$FF = \left(1 - \frac{R_{ss}}{\frac{V_{oc}}{I_{sc}}}\right) \frac{\frac{V_{oc}}{\frac{nKT}{q}} - \ln\left(\frac{V_{oc}}{\frac{nKT}{q}} + 0.72\right)}{1 + \frac{V_{oc}}{\frac{nKT}{q}}} \tag{7}$$

where G_a and G_n are module actual and nominal solar irradiance; T_a and T_n are module actual and nominal temperature, respectively; V_{noc} and I_{nsc} are SPV module nominal short-circuit voltage and open-circuit current; R_{ss} is the SPV module series resistance; k_1, k_2, and k_3 are different constants; these constants show solar irradiance, photo-current, and cell temperature nonlinear relationships; n is the density factor; T is the SPV module temperature; K is the Boltzmann constant; and q is the electron charge [9].

3   Problem Formulation

After the modeling of WT and SPV module power output generated, the problem involves recognizing the location, the size and the type of RDG to be integrated. Allocation of RDG at various locations (buses) of RDS and checking whether the operational constraints are within the limits or not is done by maintaining system integrity. The major objective of allocating RDG to an RDS is to reduce power loss and enhance voltage stability.

3.1   Problem Objective

The major objectives of the present work are to reduce network power loss and enhance voltage stability of an RDS.

3.1.1 Network Power Loss

Total network active power loss of an RDS having N number of lines is given by

$$P_N^{\text{Tloss}} = \sum_{b=0}^{N-1} \left(\frac{P_{b,b+1}^2 + Q_{b,b+1}^2}{|V_b|^2} \right) * R_{b,b+1} \tag{8}$$

where $P_{b,b+1}$ and $Q_{b,b+1}$ are the active and reactive power flows between the buses b and $b + 1$, respectively in kW and kVAR; V_b is the voltage at the bus b; $R_{b,b+1}$ is the line resistance between the buses b and $b + 1$.

3.1.2 Voltage Stability

Since voltage stability indices are the exact measure at off-line studies, the voltage stability function (VSF) is given by

$$\text{VSF} = \sum_{b=1}^{N-1} (2 * V_{b+1} - V_b) \tag{9}$$

where $V_1 = 1$ p.u. is the substation voltage magnitude. The value of VSF makes the system more stable when it is higher.

3.2 Objective Function

The major objective function of the present problem is to reduce power loss and enhance voltage stability.

$$F = f \left\{ \sum_{b=2}^{N} \sum_{j \in \text{type}} P_{\text{DG},bj} * n_b \right\} \tag{10}$$

where j is the DG type and n_b is the DG size at the bus b.

3.3 Constraints

The equality and inequality constraints for the present problem are given by

$$P_{ss} + \sum_{b=2}^{N} \sum_{j \in type} P_{DG,bj} * n_b - \sum_{b=2}^{N} P_{D,b} - P_{loss} = 0 \tag{11}$$

$$Q_{ss} - \sum_{b=2}^{N} Q_{D,b} - Q_{loss} = 0 \tag{12}$$

$$P_{DG,bj} * n_{min,b} \leq P_{DG,bj} * n_b \leq P_{DG,bj} * n_{max,b} \tag{13}$$

$$V_{min} \leq V_b \leq V_{max} \tag{14}$$

where P_{ss} and Q_{ss} are substation active and reactive; P_{loss} and Q_{loss} are the active and reactive power losses; $n_{min,b}$ and $n_{max,b}$ are the number of minimum and maximum DG units able to connect at the bus b. $V_{min} = 0.95$ p.u. and $V_{max} = 1.05$ p.u.

4 Solution Technique

Optimal allocation of renewable energy source based DGs to an RDS is a constrained nonlinear problem. Several analytical, heuristic or meta-heuristic techniques are performing better for these type of problems [6]. In the present paper, a new HGWO is used for solving optimal allocation problem which is a combination of gray wolf optimizer (GWO) and DE techniques.

4.1 The Gray Wolf Optimizer

GWO was proposed by Mirjalili et al. [10] in 2014. It is based on swarm intelligence. The two main operators employed in this technique are encircling and hunting.

4.1.1 Encircling Prey

In the group of gray wolves, the distance between any wolf and the prey is given by

$$\vec{E} = \left| \vec{C} \cdot \vec{X}^{prey}(t) - \vec{X}^{wolf}(t) \right| \tag{15}$$

$$\vec{C} = 2 * \vec{p}_1 \tag{16}$$

where $\vec{X}^{\text{prey}}$ and $\vec{X}^{\text{wolf}}$ are the position vectors of the prey and the gray wolf; t represents iteration number; $\vec{p}_1$ is the random vector with range $[0, 1]$ having the same dimensions as that of position vectors.

4.1.2　Hunting

Encircling operators (15) and (16) give information about the prey and how close the wolves are from the prey. This is given by

$$\vec{X}^{\text{wolf}}(t + 1) = \vec{X}^{\text{prey}}(t) - \vec{A} \cdot \vec{E} \tag{17}$$

$$\vec{A} = a * (2 * \vec{p}_2 - 1) \tag{18}$$

where a is the linearly reducing variable over the sequence of iterations from 2 to 0; $\vec{p}_2$ is the random vector with range $[0, 1]$ having the same dimensions as that of vectors $\vec{X}^{\text{wolf}}$, $\vec{X}^{\text{prey}}$, and $\vec{E}$.

The prey position $\vec{X}^{\text{prey}}$ or the searching optimizer is unknown in the solution landscape; it is considered that wolves α, β, and γ have the best knowledge about the prey position. These three wolves' positions are used to update the remaining (Omega) wolves' position. The distance between the best wolves and to any wolf $\vec{X}^{\text{wolf}}$ is given by

$$\vec{E}_\alpha = \left| \vec{C}_1 \cdot \vec{X}_\alpha - \vec{X}^{\text{wolf}} \right|, \ \vec{E}_\beta = \left| \vec{C}_2 \cdot \vec{X}_\beta - \vec{X}^{\text{wolf}} \right|, \ \text{and} \ \vec{E}_\gamma = \left| \vec{C}_2 \cdot \vec{X}_\gamma - \vec{X}^{\text{wolf}} \right| \tag{19}$$

For the above distances, the new position of the wolf can be obtained as follows:

$$\vec{X}_1^{\text{wolf}} = \vec{X}_\alpha - \vec{A}_1 \cdot \vec{E}_\alpha, \ \vec{X}_2^{\text{wolf}} = \vec{X}_\beta - \vec{A}_2 \cdot \vec{E}_\beta \ \text{and} \ \vec{X}_3^{\text{wolf}} = \vec{X}_\gamma - \vec{A}_3 \cdot \vec{E}_\gamma \tag{20}$$

$$\vec{X}^{\text{wolf}}(t + 1) = \left(\frac{\vec{X}_1^{\text{wolf}} + \vec{X}_2^{\text{wolf}} + \vec{X}_3^{\text{wolf}}}{3} \right) \tag{21}$$

Repeatedly applying these two operators gives the prey location or the best solution.

4.2　The Hybrid Gray Wolf Optimizer

The performance of the GWO for standard benchmark functions is reported in [10]. However, the present problem has discrete characteristics because of discretely varying location and size. The search capability of the GWO is enhanced by using DE operators namely crossover and mutation. The procedure of proposed hybrid algorithm for the present DG allocation problem is given as.

4.2.1 Crossover

The crossover operator employed for the present problem is taken from [11]. The crossover operator applied to the ith wolf and obtained jth component is given by

$$X_j^i = \begin{cases} X_j^{\text{prob}} & \text{if } \text{rand}_j^i < C_{\text{prob}} \\ X_j^i & \text{else} \end{cases} \quad j = 1, 2, \ldots, E \, i = 1, 2, \ldots, N^{\text{wolf}} \quad (22)$$

where $\text{prob} \in [1, 2, \ldots, N^{\text{wolf}}]$, $\text{prob} \neq i$, and $\text{rand}_j^i \in [0, 1]$ is generated randomly $\forall j, i$. E is each the vector solution dimensionality. N^{wolf} is the total number of wolves or solutions.

However, for the present problem, the crossover rate C_{prob} is changing dynamically. This is given by

$$C_{\text{prob}} = 0.2 * F^{i,\text{best}} \quad (23)$$

$$F^{i,\text{best}} = \frac{F^i - F^{\text{best}}}{F^{\text{worst}} - F^{\text{best}}} \quad i = 1, 2, \ldots, N^{\text{wolf}} \quad (24)$$

where the ith wolf fitness value; F^{best} and F^{worst} are the best and the worst fitness values for the current iteration in the pack of wolves. From Eqs. (23) and (24), the C_{prob} value is 0 for the best wolf and 0.2 for the worst wolf. Since the proportionality relation between crossover and vector, the vector solution of the best wolf is unaffected.

4.2.2 Mutation

The mutation operator applied to the ith wolf and obtained jth component is given by

$$y_j^i = \begin{cases} y^{\text{gbest},j} + r_d\left(y_j^p - y_j^q\right) & \text{if } \text{rand}_j^i < \mu_{\text{prob}} \\ y_j^i & \text{else} \end{cases} \quad (25)$$

$$\mu_{\text{prob}} = 0.05 * F^{i,\text{best}} \quad (26)$$

where $p, q \in [1, 2, \ldots, N^{\text{wolf}}]$, $p \neq q \neq i$, r_d and $\text{rand}_j^i \in [0, 1]$ are generated randomly $\forall j, i$. μ_{prob} is the mutation probability or mutation rate; $y^{\text{gbest},j}$ is the global best wolf in the entire iterative process for jth component. In the iteration process, the comparison between the best wolf and global best wolf is carried out. However, if the present best wolf or current solution is better than the already obtained global best wolf or global best solution, it is replaced by the present best wolf or current solution. From Eqs. (25) and (26), the μ_{prob} value is 0 for the best wolf and 0.05 for the worst wolf.

4.3 Optimal DG Allocation Through HGWO Algorithm

Appropriate allocation of DG in a distribution system reduces power loss and enhances voltage stability. The major control variables in the present problem are location, size, and type. The fitness value is evaluated by using Eq. (10) and the load flow is carried out by a direct approach [12]. The flowchart of the proposed algorithm is shown in Fig. 1.

For the proposed HGWO algorithm for the DG allocation problem, the stepwise procedure is as follows:

Step-0: In the present initialization step, choose the number of iterations, wolves' population N^{wolf}, DG locations, and size. Generate initial wolves' population N^{wolf} that satisfies all the constraints.

Step-1: Run the load flow and identify the power loss. By using the objective function (10), evaluate the fitness value. Recognize the α, β and γ wolves and also the best global solution P^{gbest}. However, $P^{\text{gbest}} = P^{\alpha}$ in the very first iteration.

Step-2: Use encircle operator (15) and hunting operator (17) for this problem to compute

$$\vec{E}_{\alpha} = \left| \vec{C}_1 \cdot \vec{X}_{\alpha} - \vec{X}^i \right|, \ \vec{E}_{\beta} = \left| \vec{C}_2 \cdot \vec{X}_{\beta} - \vec{X}^i \right| \text{ and } \vec{E}_{\gamma} = \left| \vec{C}_2 \cdot \vec{X}_{\gamma} - \vec{X}^i \right| \quad (27)$$

$$\vec{P}_1^{\text{wolf}} = \vec{P}_{\alpha} - \vec{A}_1 \cdot \vec{E}_{\alpha}, \ \vec{P}_2^{\text{wolf}} = \vec{P}_{\beta} - \vec{A}_2 \cdot \vec{E}_{\beta} \text{ and } \vec{P}_3^{\text{wolf}} = \vec{P}_{\gamma} - \vec{A}_3 \cdot \vec{E}_{\gamma} \quad (28)$$

Evaluate the population of the next generation $P(t + 1)$:

$$\vec{P}^i(t + 1) = \left(\frac{\vec{P}_1^{\text{wolf}} + \vec{P}_2^{\text{wolf}} + \vec{P}_3^{\text{wolf}}}{3} \right) \quad (29)$$

Step-3: By using the crossover (22) and mutation (25) operators, replace the X_j^i with P_j^i.

Step-4: Check for DG constraints violation. If violated, fix them in the limits.

Step-5: Check whether the stopping criterion is satisfied or not. If yes, stop and display the results; otherwise, repeat Steps 1–5.

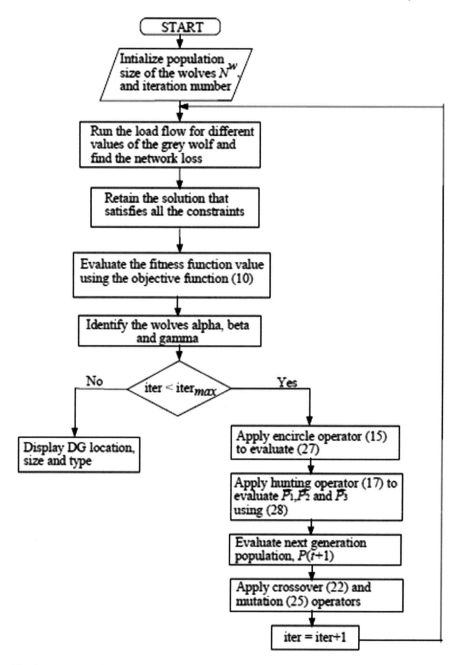

Fig. 1 Flowchart of the proposed algorithm

5 Results and Discussion

The present problem of optimal DG allocation using HGWO algorithm is tested on IEEE 12- and 15-bus RDS. The scope of the present paper is restricted to the optimal allocation of WT and SPV modules in an RDS. The proposed HGWO and load flow are implemented in MATLAB®. The results are presented and discussed for the test case next.

5.1 *Test Systems*

IEEE 12-bus RDS has 12 buses and 11 distribution lines. The system total active and reactive loads are 435 kW and 405 kVA, respectively. The base values are 11 kV and 10 MVA. Figure 2 shows the single line diagram (SLD) of the IEEE 12-bus RDS. The base case active and reactive power losses are 20.7135 kW and 8.041039 kVAR. IEEE 15-bus RDS has 15 buses and 14 distribution lines. The system total active and reactive loads are 1,226 kW and 1251 kVA, respectively. Figure 3 shows the SLD of

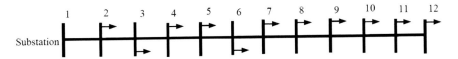

Fig. 2 SLD of a 12-bus RDS

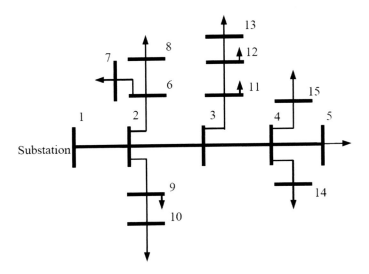

Fig. 3 SLD of a 15-bus RDS

IEEE 15-bus RDS. The base case active and reactive power losses are 61.7656 kW and 57.2686 kVAR [8].

5.2 Renewable DG Allocation

The generated output power of the WT and SPV modules is modeled by using probabilistic approach. Required parameters and equivalent value of WT and SPV modules are obtained from [8]. Renewable energy sources considered in the present paper are suitable to obtain the optimal value of DG size. The proposed HGWO is compared with the PSO algorithm in terms of location, size, and type of DG which is presented in Table 1. The proposed HGWO is compared with the PSO algorithm in terms of power loss and voltage stability for 12- and 13-bus test systems which are presented in Tables 2 and 3.

Table 1 Comparison of the proposed HGWO with PSO and GWO in terms of DG location, size, and type

Test system	PSO [8]			GWO			HGWO		
	DG location	DG size (kW)	DG type	DG location	DG size (kW)	DG type	DG location	DG size (kW)	DG type
12-bus	Bus-5	211.42	Solar	Bus-6	182.1	Solar	Bus-6	183.3	Solar
	Bus-10	158.58	Solar	Bus-9	138.9	Solar	Bus-9	138	Solar
15-bus	Bus-3	840	Wind	Bus-3	870.7	Wind	Bus-3	674.7	Wind
	Bus-8	336	Wind	Bus-5	496.1	Wind	Bus-5	566.9	Wind
				Bus-11	255.7	Wind	Bus-10	418.5	Wind

Table 2 Comparative study of the proposed HGWO with PSO and GWO algorithms for IEEE 12-bus

	Without DG	PSO [8]	GWO	HGWO
Total power loss (kVA)	22.22	12.5	**10.27**	**10.04**
VSF	10.7113	10.8698	**10.8994**	**10.9086**

Table 3 Comparative study of the proposed HGWO with PSO and GWO algorithms for IEEE 15-bus

	Without DG	PSO [8]	GWO	HGWO
Total power loss (kVA)	84.23	54.8	**40.1**	**39.24**
VSF	13.5720	13.5337	**13.8039**	**13.8199**

When the SPV modules are allocated at optimal locations with appropriate sizes in a 12-bus RDS, the power loss reduction and voltage stability enhancement are achieved. It is found that the proposed HGWO gives a total power loss of 10.04 kVA and voltage stability of 10.9086 p.u. When compared to PSO, it has achieved 19.68% reduction in total power loss and 0.36% increment in voltage stability for 12-bus RDS. The voltage stability enhancement achieved by SPV modules allocation in a 12-bus RDS using the proposed HGWO is shown in Fig. 4. When the WTs are allocated at optimal locations with appropriate sizes in a 15-bus RDS, the power loss reduction and voltage stability enhancement are achieved. It is found that the proposed HGWO gives a total power loss of 39.24 kVA and voltage stability of 13.8199 p.u. When compared to PSO, it has achieved 28.39% reduction in the total power loss and 2.11% increment in voltage stability for 15-bus RDS. The voltage stability enhancement achieved by WTs allocation in a 15-bus RDS is shown in Fig. 5.

The GWO and HGWO convergence characteristics in terms of power loss for a 12-bus RDS is shown in Fig. 6. The GWO and HGWO convergence characteristics in terms of power loss for a 15-bus RDS is shown in Fig. 7; here, the proposed HGWO has fast convergence characteristics compared to GWO. However, the proposed HGWO and GWO have almost similar values of total power loss and voltage stability.

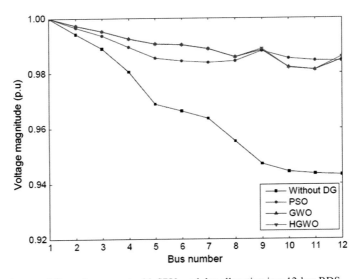

Fig. 4 Voltage stability enhancement with SPV modules allocation in a 12-bus RDS

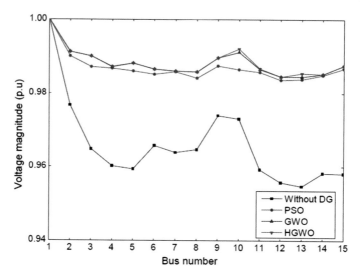

Fig. 5 Voltage stability enhancement with WT allocation in a 15-bus RDS

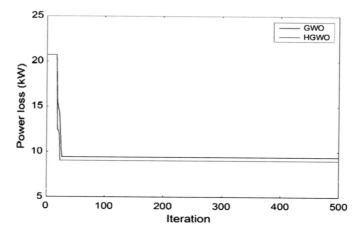

Fig. 6 Illustration of the GWO and HGWO convergence characteristics for a 12-bus RDS

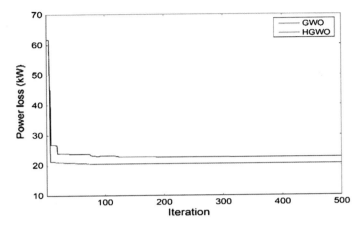

Fig. 7 Illustration of the GWO and HGWO convergence characteristics for a 15-bus RDS

6 Conclusion

Optimal allocation of renewable energy resource-based DG in a radial distribution system is a complex real-world optimization problem. Several researchers have developed many optimization algorithms to solve this type of optimization problem in a fast and efficient way. In this paper, a new HGWO is proposed to solve the optimal allocation problem. The major objective of the present problem is power loss reduction and evaluation of the corresponding voltage stability. The proposed HGWO simulated test results are compared with the existing optimization algorithms and found to be better. It can be concluded from the results that the proposed HGWO is highly suitable for optimal allocation of renewable energy resource-based DG in an RDS. Optimal mix of renewable energy resources is the future research on this topic.

References

1. Ackermann, T., Ran Andersson, G., Soder, L.: Distributed generation: a definition. Electr. Power Syst. Res. **57**, 195–204 (2001)
2. Kashem, M.A., Ledwich, G.: Multiple distributed generators for distribution feeder voltage support. IEEE Trans. Energy Convers. **20**(3), 676–684 (2005)
3. Hedayati, H., Nabaviniaki, S.A., Akbarimajd, A.: A method for placement of DG units in distribution networks. IEEE Trans. Power Deliv. **23**(3), 1620–1628 (2008)
4. Lee, S., Park, J.: Selection of optimal location and size of multiple distributed generations by using Kalman filter algorithm. IEEE Trans. Power Syst. **24**(3), 1393–1400 (2009)
5. Wang, C., Nehrir, M.H.: Analytical approaches for optimal placement of distributed generation sources in power systems. IEEE Trans. Power Syst. **19**(4), 2068–2076 (2004)
6. Sambaiah, K.S.: A review on optimal allocation and sizing techniques for DG in distribution systems. Int. J. Renew. Energy Res. (IJRER) **8**(3), 1236–1256 (2018)

7. Kayal, P., Chanda, C.K.: Optimal mix of solar and wind distributed generations considering performance improvement of electrical distribution network. Renew. Energy **75**, 173–186 (2015)
8. Kayal, P., Chanda, C.K.: Placement of wind and solar based DGs in distribution system for power loss minimization and voltage stability improvement. Int. J. Electr. Power Energy Syst. **53**, 795–809 (2013)
9. Atwa, Y.M., El-Saadany, E.F., Salama, M.M.A., Seethapathy, R.: Optimal renewable resources mix for distribution system energy loss minimization. IEEE Trans. Power Syst. **25**(1), 360–370 (2010)
10. Mirjalili, S., Mohammad, S., Lewis, A.: Grey wolf optimizer. Adv. Eng. Softw. **69**, 46–61 (2014)
11. Sanjay, R., Jayabarathi, T., Raghunathan, T., Ramesh, V., Mithulananthan, N.: Optimal allocation of distributed generation using hybrid grey wolf optimizer. IEEE Access **5**, 14807–14818 (2017)
12. Teng, J.: A direct approach for distribution system load flow solutions. IEEE Trans. Power Deliv. **18**(3), 882–887 (2003)

PV Module Temperature Estimation by Using ANFIS

Challa Babu and Ponnambalam Pathipooranam

Abstract The recent advantages in the thermal extraction schemes of Photovoltaic (PV) systems became more feasible for domestic applications. The amount of thermal energy available at the PV backside is the source for all thermal extraction/utilizing schemes. But, till now, there is no accurate theoretical method to find the module temperature by any mathematical equation to be adaptable for all the conditions. In this work, we introduced the soft computing based estimation using Adaptive Neural Fuzzy Inference Systems (ANFIS) tool in MATLAB to estimate the module temperature of PV with respect to change of irradiance, ambient temperature, and wind velocity.

Keywords PV/thermal systems · Temperature estimation · ANFIS · Error reduction

1 Introduction

The incoming energy from the sun is utilized for many applications either using light or heat energy. But, from the past decade, the research is moving toward the hybrid systems, which combine both heat and light energies. The hybrid PV/Thermal (PV/T) systems are majorly classified as the concentrated and flat plate systems. The concentrated ones majorly accumulate the large heat energy and part of light energy is converted as electrical energy by semiconducting materials. The flat plate systems concentrate majorly on the light energy collection and converting it into electrical energy. Both the systems have their own merits and demerits. The selection of the type of collector is based on the application. In this work, we used the flat plate system to estimate the module temperature [1].

C. Babu · P. Pathipooranam (✉)
School of Electrical Engineering, VIT, Vellore, India
e-mail: P.Ponnambalam@gmail.com

C. Babu
e-mail: babu2342@gmail.com

© Springer Nature Singapore Pte Ltd. 2020
K. N. Das et al. (eds.), *Soft Computing for Problem Solving*,
Advances in Intelligent Systems and Computing 1048,
https://doi.org/10.1007/978-981-15-0035-0_24

There are many correlations present in the literature to estimate the module temperature [2–5]. But all the mathematical equations having the empirical values or estimated values with a minimum error of 2–5%. To reduce the amount of error for the estimation, soft computing based estimation is a good choice. The research in the ANFIS tool is producing the results with the accuracy of 99% which depends on the training data [6].

The efficiency of the PV system depends on the amount of energy generated per given solar inputs such as irradiance and temperature. The PV system utilizes the light energy to convert it as useful electrical energy but the heat energy is passed through each layer and it is dissipated to the atmosphere, i.e., the energy is wasted. In the recent advancements in hybrid PV configurations, PV/Thermal systems become more efficient systems to utilize the thermal energy of the PV modules and it can be used directly or indirectly as per the application demand. Among the PV/Thermal systems, PV with Thermoelectric Generator (TEG) configuration is more popular [7]. For these thermal systems, the thermal energy present at the backside of the PV panel is the major input source to make the effective design configuration to extract heat energy.

In this work, the ANFIS tool is used to predict the backside of the PV panel temperature. The neural network trained with the data (irradiance, ambient temperature, wind velocity, and module temperature) is collected from the literature. In that, irradiance, ambient temperature, and wind velocity are considered as the inputs and module temperature is considered as the output.

This paper is organized as follows: Sect. 2 has the PV module temperature dependent parameters and its influence, Sect. 3 has the ANFIS introduction, training, and rules of the system, Sect. 4 has the results and discussion of module temperature with respect to different estimating methods.

2   PV Module Temperature

The amount of energy generated per given solar input depends majorly on irradiance and temperature [8]. All the PV manufacturers are providing the temperature coefficients of voltage/current/power at the Standard Test Condition (STC) condition, i.e., 1000 W/m^2, 25 °C. If the coefficient value is negative, it indicates decrement in that value. The heat transfer of the PV module is shown in Fig. 1.

The PV module temperature depends on many atmospheric and physical parameters and are follows:

a. Irradiance,
b. Ambient temperature,
c. Wind velocity,
d. Humidity, and
e. Module mounting.

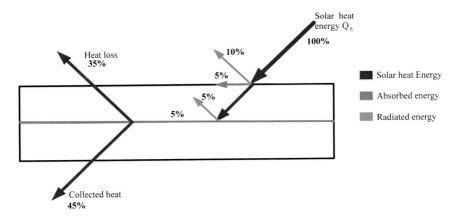

Fig. 1 Heat flow of the PV module

Among the parameters, in this work, we consider the major parameters: irradiance, ambient temperature, and wind velocity. The humidity and module mounting doesn't impact the module temperature by <1% [9] such that, these two parameters are neglected for the analysis.

3 Prediction Using ANFIS Tool

The latest advance in soft computing techniques made is the hybrid combination of fuzzy and neural network. The ANFIS tool tunes the fuzzy rules as per the training data provided to the neural network [10]. The ANFIS structure designed with five layers is shown in Fig. 2.

In that, the first layer is to take the inputs. The second layer is the input membership function to map the input data with output data by using different membership functions. The third layer generates the rules as per the tuning of the membership function. The fourth layer is the output membership function used to map the input data as per the fuzzy rules to reduce the estimation error. The final layer is the output layer which is the estimated value for the given inputs.

The following steps are executed for the estimation of PV module temperature by using ANFIS tool in MATLAB.

Step 1:

Load the solar irradiance, ambient temperature, wind velocity, and the module temperature to the network. The ambient temperature is varied from 5 to 60 °C with a step size of 1 °C; solar irradiance is varied from 100 to 1000 W/m² with a step size of 10 W/m²; and wind velocity is varied from 1 m/s to 15 m/s with a step size of 0.25 m/s. A total of 204 input data sets is used for the neural net training.

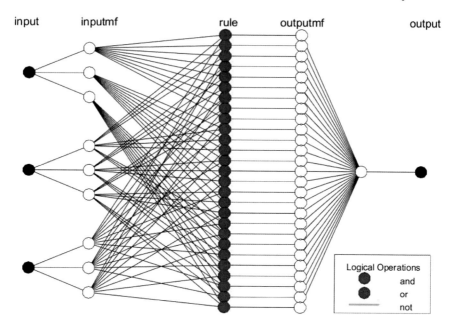

Fig. 2 ANFIS structure

Step 2:

In the input membership function taken by selecting the grid participation to enable the selection of number of membership function and its type. In this case, we used three [3 3 3] triangular membership functions to map the input data. The fuzzy rules are auto-tuned to produce the corresponding output for the given input variables. The fuzzy rules are shown in Fig. 3.

Step 3:

For training the network, Back Propagation Algorithm (BPA) is used with zero tolerance. The training is done with 2000 epochs. The training error is shown in Fig. 4.

Step 4:

For testing of the network, we used the input values of irradiance, ambient temperature, and wind velocity. The prediction accuracy after testing for the training data is shown in Fig. 5.

Performance of ANFIS in the estimation:

The performance of ANFIS for the estimation of the PV module temperature is analyzed with the surface figures. The Surface images are the 3-D images drawn

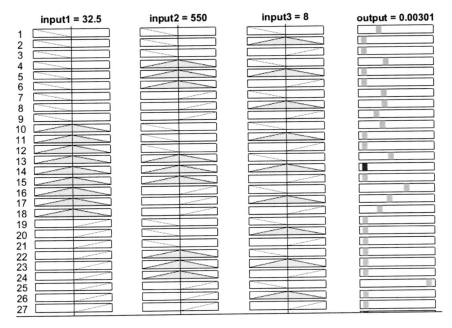

Fig. 3 Fuzzy rules for ambient temperature, irradiance, and wind velocity

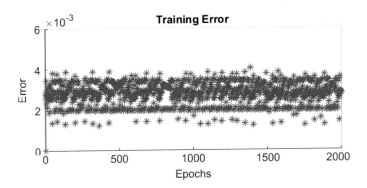

Fig. 4 Training error of ANFIS

between any of the two different inputs of Input1 (ambient temperature), Input2 (irradiance), and Input3 (wind velocity).

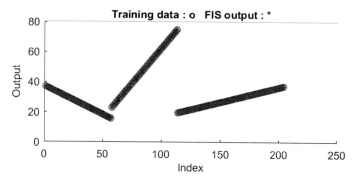

Fig. 5 Testing of ANFIS with trained data

4 Results and Discussion

With the change in the ambient temperature, the PV module temperature also changes linearly. The amount of increment in the ambient temperature approximately increases the module temperature by 35% depending on the time of concentration. The input change in the wind velocity has the inverse relation with the module temperature. This is because the wind velocity decreases the amount of heat accumulated on the module surface. The amount of increment in wind velocity causes the maximum change of 10% decrement in the module temperature. The input change in the irradiance causes the linear change in the module temperature. The amount of increment in irradiance approximately increases the module temperature by 20%.

The performance of ANFIS is analyzed with the surface images of different inputs versus the PV module temperature. The effectiveness of the estimation is described by the amount of error. In this estimation, the amount of error for the module temperature estimation is too small. The overall output of the PV module temperature with respect to ambient temperature and irradiance clearly conveys that the module temperature increases with increase in both the inputs, i.e., Input1 and Input2 as shown in Fig. 6a. The module temperature with respect to irradiance and wind velocity clearly conveys that the module temperature increases by a small range of change in both the inputs, i.e., Input2 and Input3 as shown in Fig. 6b. The module temperature with respect to the ambient temperature and wind velocity clearly conveys that the module temperature decreases drastically with change in both the inputs, i.e., Input1 and Input3 as shown in Fig. 6c.

The rules of fuzzy system are tuned to get the accurate estimation by the neural network. In this case, we used the test data for checking the estimation accuracy of ANFIS. The test data has 100 random values of different input configurations. The ANFIS model predicts the PV module temperature with an error less than 1%. The performance the system increases with the increase in the training data sets accuracy. The estimated values are tested with the existing data present in the literature. The estimation of the PV module temperature is analyzed in a theoretical manner to calculate the error of the system.

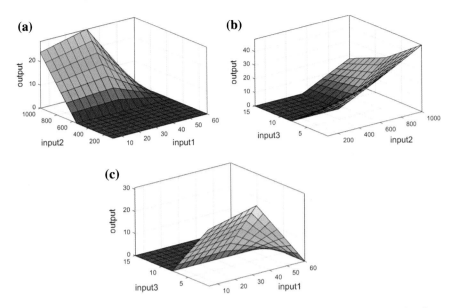

Fig. 6 Surface images of PV module temperature versus **a** ambient temperature and irradiance **b** irradiance and wind velocity **c** ambient temperature and wind velocity

5 Conclusion

The PV module temperature is the source for the all the thermal extraction systems like PV/Thermal-based PV-TEG configurations. In this work, the ANFIS tool in MATLAB is used for the estimation of the PV module temperature. The results convey that the accuracy of this estimation is around 99% for the trained data sets. The module temperature relation with different inputs conveys that the module temperature is always greater than the ambient temperature by around 20 °C. The error of prediction is below 1% for the random input data sets of ambient temperature, irradiance, and wind velocity.

References

1. Babu, C., Ponnambalam, P.: The role of thermoelectric generators in the hybrid PV/T systems: a review. Energy Convers. Manag. **151**, 368–385 (2017)
2. Skoplaki, E.P., Palyvos, J.A.: Operating temperature of photovoltaic modules: a survey of pertinent correlations. Renew. Energy **34**, 23–29 (2009)
3. Skoplaki, E., Palyvos, J.A.: On the temperature dependence of photovoltaic module electrical performance: a review of efficiency/power correlations. Sol. Energy **83**, 614–624 (2009)
4. Skoplaki, E., Boudouvis, A.G., Palyvos, J.A.: A simple correlation for the operating temperature of photovoltaic modules of arbitrary mounting. Sol. Energy Mater. Sol. Cells **92**, 1393–1402 (2008)

5. Muzathik, A.M.: Photovoltaic modules operating temperature estimation using a simple correlation. Int. J. Energy Eng. **4**, 151 (2014)
6. Kharb, R.K., Shimi, S.L., Chatterji, S., Ansari, M.F.: Modeling of solar PV module and maximum power point tracking using ANFIS. Renew. Sustain. Energy Rev. **33**, 602–612 (2014)
7. Babu, C., Ponnambalam, P.: The theoretical performance evaluation of hybrid PV-TEG system. Energy Convers. Manag. **173**, 450–460 (2018)
8. Bayrak, F., Abu-Hamdeh, N., Alnefaie, K.A., Öztop, H.F.: A review on exergy analysis of solar electricity production. Renew. Sustain. Energy Rev. **74**, 755–770 (2017)
9. Fudholi, A., Sopian, K., Yazdi, M.H., Ruslan, M.H., Ibrahim, A., Kazem, H.A.: Performance analysis of photovoltaic thermal (PVT) water collectors. Energy Convers. Manag. **78**, 641–651 (2014)
10. Tektaş, M.: Weather forecasting using ANFIS and ARIMA models. Environ. Res. Eng. Manag. **51**, 5–10 (2010)

Modified Artificial Potential Field Approaches for Mobile Robot Navigation in Unknown Environments

Ngangbam Herojit Singh, Salam Shuleenda Devi and Khelchandra Thongam

Abstract Navigation is the important task of any mobile robot. In this paper, a modified potential field method for mobile robot navigation has been proposed for unknown environments. Here, the shortest distance between the goal position and robot has been considered to establish a new repulsive potential functions. The developed repulsive potential functions assure that the goal is the global minimum of the total potential field. To demonstrate the efficiency of the proposed method, computer simulations have been carried out through MATLAB software.

Keywords Navigation · Potential field method · Mobile robot · Global minimum

1 Introduction

Mobile robot navigation is an emerging researches topic in the field of robotics. For effective environment robot navigation, an efficient navigation method is required. Many researchers have developed various navigation method based on fuzzy logic, neural networks, particle swarm optimization, hybrids methods, genetic algorithm, etc. To improve the artificial potential field method in mobile robot path planning, a traditional artificial potential field method with chaos optimization has also been used [1]. The potential field function is utilized as target function of chaos optimization and two-stage chaos is used. To prevent from deadlock, mobile robot navigation

N. H. Singh (✉) · S. S. Devi · K. Thongam
National Institute of Technology Mizoram, Aizawl, Mizoram, India
e-mail: herojitng@gmail.com

S. S. Devi
e-mail: shuleenda26@gmail.com

K. Thongam
National Institute of Technology Manipur, Langol, Manipur, India
e-mail: thongam@gmail.com

© Springer Nature Singapore Pte Ltd. 2020
K. N. Das et al. (eds.), *Soft Computing for Problem Solving*,
Advances in Intelligent Systems and Computing 1048,
https://doi.org/10.1007/978-981-15-0035-0_25

is done using an artificial potential field [2]. This method is capable of escaping from different problems such as non-reachability and deadlock while navigating. The modification of traditional artificial potential method is made on the effective front-face obstacle information related with the velocity direction. Using additional control force into the artificial potential field method, an Unmanned Aerial Vehicle (UAV) path planning problems in 2D environment is resolved [3]. Here, the path planning problem is changed into an optimization problem. Further, the original constrained optimization problem is transformed into an unconstrained optimization problem. An Adaptive Neuro-Fuzzy Inference System (ANFIS) controller [4] has been presented for mobile robot navigation in unknown static environments. Here, obstacle distances given by the sensor are taken as an input of the ANFIS and the robot steering angle is the output. Nie et al. [5] used APF for formation control and the safe distance along with repulsive force is used for UAV formation control confronting collision avoidance. For path planning of mobile robot [6], a modified APF has been applied. To overcome the issues of traditional APF, new point-APF is proposed. It generates the new point of attractive force when the obstacles blocked a path which makes the straight line to the goal position. Moreover, a modified APF method [7] toward online path planning is used which introduce dynamically repulsive APF to the standard APF, at each time potential gradient descent algorithm (PGDA) gets trapped in a local minimum. It removes any blocking configuration in the environment. Farid et al. [8] proposed a mobile robot navigation using harmonic potential field method. Ngangbam et al. [9] proposed a navigation method for mobile robot in a static environment.

Based on various intelligent navigation techniques [10], a comparative study has been performed which is able of navigating in dynamic and static environments. Soft computing techniques have been used for mobile robot navigation and obstacle avoidance in static environments only. From the various literature, we have motivated to propose the novel modified APF method for mobile robot navigation in unknown environments.

In this paper, the safe distance from the obstacles and the target attractive force for the target position has been set. The simulations results show the effectiveness of the proposed method. The remainder of the paper is organized as follows. Section 2 represents the kinematics and the definition of the formation. Section 3 provides the modified APF method. In Sect. 4, the simulation scenarios are presented. Conclusion is made in the last section.

2 Kinematic Model of Wheeled Mobile Robot

The geometric description of mobile robot in 2D environment is shown in Fig. 1. Here, the mobile robot wheels are assumed to be nondeforming and moves without slip on the surface. To achieve the motion and orientation, the two wheels are independently driven by two motors. The kinematic model of the wheeled mobile robot is given as

Fig. 1 Kinematic model of
the differential drive
two-wheeled mobile robot

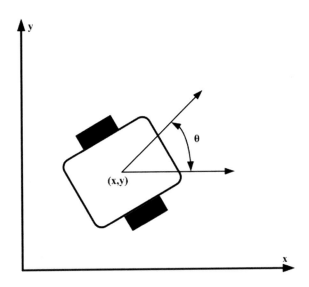

$$\begin{pmatrix} \overline{x} \\ \overline{y} \\ \overline{\theta} \end{pmatrix} = \begin{pmatrix} V\cos\theta \\ V\sin\theta \\ \omega \end{pmatrix} \tag{1}$$

where V represents the linear velocity and the angular velocity of the mobile robot
is ω.

3 Modified Artificial Potential Field (APF)

The proposed method has been used to navigate a mobile robot from source to
destination position. For simplification of the problem analysis, two assumptions are
formed

i. First, Robot is of mass point.
ii. Second, robot moves in 2D workspace. Here, the position in the workspace is
presented by $c = [x, y]$ that is known.

3.1 Attractive Potential Function

The attractive potential function is given below:

$$U_{att}(c) = \frac{1}{2}\delta\rho^n(c, c_{target}) \tag{2}$$

where δ is a positive scaling factor, $\rho''(c, c_{target})$ is the distance between the robot c and the target c_{target} and n = 2.

The attractive potential force $(F_{att}(x))$ is given by

$$F_{att}(x) = \nabla U_{att}(x) = \delta(x_{target} - x) \tag{3}$$

where $\nabla = \frac{\partial}{\partial x}\vec{i} + \frac{\partial}{\partial y}\vec{j} + \frac{\partial}{\partial z}\vec{k}$

3.2 Repulsive Potential Function

The repulsive potential function is given below:

$$U_{rep}(obs_i) = \begin{cases} \frac{1}{2}\alpha_i(\frac{1}{\rho(c,c_{obs_i})} - \frac{1}{\rho_o})^2 & if \quad \rho(c, c_{obs_i}) \leq \rho_o \\ 0 & if \quad \rho(c, c_{obs_i}) > \rho_o \end{cases} \tag{4}$$

$F_{rep}(obs_i) = -\nabla U_{rep}(obs_i)$

where $i = 1 \ldots m$, m is number of obstacles, α_i is the positive scaling factor, ρ_o = positive constant denoting the obstacle on the robot and $\rho(c, c_{target})$ is the minimum distance from the robot 'c' to the obstacle.

Total potential influences on the robot (U_{total}) is given by

$$U_{total} = U_{att} + \sum_{i=1}^{m} U_{rep}(obs_i) \tag{5}$$

Similarly, the total force applied on the robot is given by

$$F_{total} = F_{att} + \sum_{i=1}^{m} F_{rep}(obs_i) \tag{6}$$

The problem of APF in robot navigation is the local minimum problem, i.e., the target is not the minimum of the total potential function. To overcome this problem, we developed a new repulsive potential function.

3.3 New Repulsive Potential Function

As from the traditional APF, the global minimum of the total potential function is not at the target position. The repulsive potential force increases due to presence of obstacle near the target. To maintain the global minimum at the target in environment, we developed a new repulsive potential function given by

$$U_{rep}(obs_i) = \begin{cases} \frac{1}{2}\alpha_i(\frac{1}{\rho(c,c_{obs_i})} - \frac{1}{\rho_o})^2 \rho^i(c, c_{target}) & if \quad \rho(c, c_{obs_i}) \le \rho_o \\ 0 \quad if \quad \rho(c, c_{obs_i}) > \rho_o \end{cases} \tag{7}$$

where $\rho^i(c, c_{target})$ are the minimum distance from the robot 'c' and obstacles i. $\rho(c, c_{target})$ is the distance between the robot and the target.

The total potential U_{total} can be obtained from Eq. (5). For $m = 2$ and $\delta = \alpha_i = 1$, there is only one minimum exist which is at the target.

4 Simulation Results

The simulation results help to demonstrate the effectiveness of modified APF for mobile robot navigation. The results are achieved with Intel(R)Core(TM) i5 CPU with 3.20 GHz 3.33 GHz, 4 GB RAM, running MATLAB R2015b under Windows 7 Ultimate, 64-bit Operating System. In this application, the 4×3 square unit (pixel) is taken as the robot size. I is the initial position of the robot and G is the goal position. The x- and y-axis of the robot workspace ranges from 0 to 500, where the horizontal axis is the x-axis and the vertical one is the y-axis. The black rectangular and oval objects are the obstacles. Figures 2, 3, 4, and 5 show robot motion in different complex environments using proposed method. Figures 6, 7, and 8 show robot motion in different complex environments using firefly algorithm in [12].

Fig. 2 Robot navigation in first environment

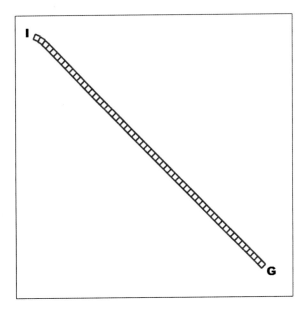

Fig. 3 Robot navigation in
second environment

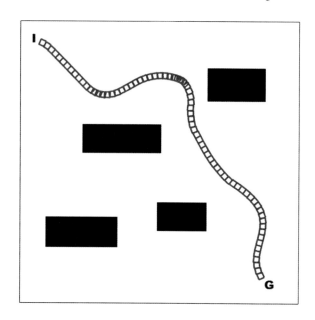

Fig. 4 Robot navigation in
third environment

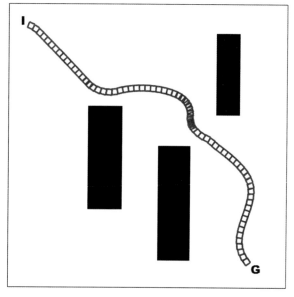

Fig. 5 Robot navigation in fourth environment

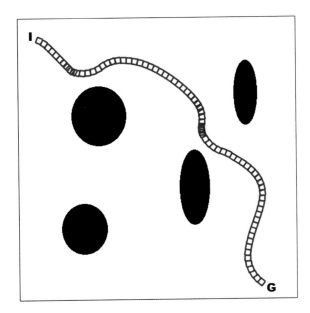

Fig. 6 Robot navigation for firefly algorithm [12] in second environment

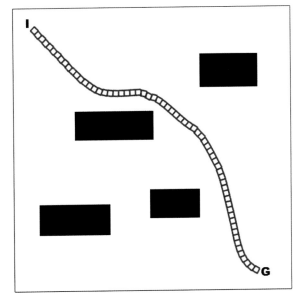

Fig. 7 Robot navigation for firefly algorithm [12] in third environment

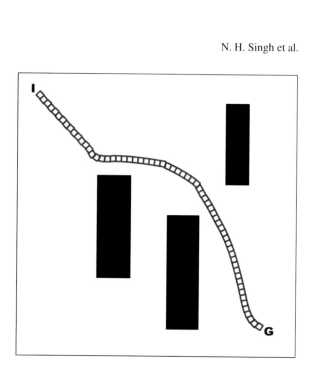

Fig. 8 Robot navigation for firefly algorithm [12] in fourth environment

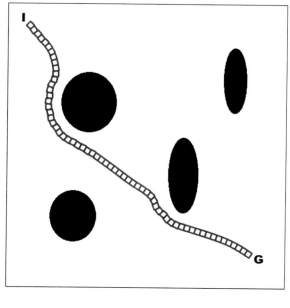

5 Evaluation and Discussion

In Figs. 2, 3, 4, 5, 6, 7, and 8, the initial position I of the robot is $x = 40$, $y = 40$ and the goal position G of the robot is $x = 450$, $y = 450$.

The maximum root speed is 10 units and the degree of calculating degree is 3. The attractive and repulsive potential scaling are 300,000 each. The minimum attractive potential at any point is 0.5.

We have considered four different environments to test our proposed method. The first environment is of obstacles free environment and robot travels in straight line, which is shown in Fig. 2. The second environment is having four different sizes of rectangular shape obstacles and robot avoids the obstacles to reach the goal position, which is shown in Fig. 3. The third environment is having three different sizes of rectangular shape obstacles and robot avoids the obstacles to reach the goal position, which is shown in Fig. 4. And the last environment is having four different sizes of oval shape obstacles and robot avoids the obstacles to reach the goal position, which is shown in Fig. 5.

6 Comparison

Table 1 shows the comparative analysis of various navigation methods for different environments based on two parameters (Path Length Computational Time). The diagrammatic path length of our method in three different environments is shown in Figs. 3, 4, and 5. And another method, namely, firefly Algorithm, in [12] of same three environments is shown in Figs. 6, 7, and 8, respectively.

7 Conclusion

The modified potential field method for navigation of mobile robot has been presented for unknown environments. Considering the shortest distance between the goal position and the robot, new repulsive potential functions have been established.

Table 1 Comparative analysis of various methods for different environments

Environment	Firefly algorithm in [12]		Proposed method	
	Path length (pixels)	Computational time (second)	Path length (pixels)	Computational time (second)
Second	820.485	1.622	802.141	1.610
Third	802.775	1.6543	800.995	1.599
Fourth	825.396	1.7515	819.386	1.581

Simulations are performed in four different environments for demonstrating the effectiveness and efficiency of proposed method. Comparative analysis is done with the other renown existing method. In future, the proposed method can be applied in moving goal position.

References

1. Chen, W., Wu, X., Lu, Y., et al.: An improved path planning method based on artificial potential field for a mobile robot. Cybern. Inf. Technol. **15**, 181–191 (2015)
2. Weerakoon, T., Ishii, K., Nassiraei, A.A., et al.: An artificial potential field based mobile robot navigation method to prevent from deadlock. J. Artif. Intell. Soft Comput. Res. **5**, 189–203 (2015)
3. Chen, Y.B., Luo, G.C., Mei, Y.S., Yu, J.Q., Su, X.L., et al.: UAV path planning using artificial potential field method updated by optimal control theory. Int. J. Syst. Sci. **47**, 1407–1420 (2016)
4. Pandey, A., Kumar, S., Pandey, K.K., Parhi, D.R., et al.: Mobile robot navigation in unknown static environments using ANFIS controller. Perspect. Sci. **8**, 421–423 (2016)
5. Zunli, N., Xuejun, Z., Xiangmin, G., et al.: UAV formation flight based on artificial potential force in 3D environment. In: 29th Chinese Control and Decision Conference (CCDC) 28–30 May 2017. Chongqing, China (2017)
6. Lee, D., Jeong, J., Kim, Y.H., Park, J.B., et al.: An improved artificial potential field method with a new point of attractive force for a mobile robot. In: 2nd International Conference on Robotics and Automation Engineering 29–31 Dec 2017. Shanghai, China (2017)
7. Bounini, F., Gingras, D., Pollart, H., et al.: Modified artificial potential field method for online path planning applications. In: IEEE Intelligent Vehicles Symposium (IV) 11–14 June 2017. Los Angeles, CA, USA (2017)
8. Panati, S., Baasandorj, B., Chong, K.T.: Autonomous Mobile Robot Navigation Using Harmonic Potential Field. In: IOP Conference Series: Materials Science and Engineering, vol. 83, pp. 1–7 (2015)
9. Singh, N.H., Thongam, K.: Mobile robot navigation using fuzzy logic in static environments. Procedia Comput. Sci. **125**, 11–17 (2018)
10. Pandey, A., Pandey, S., Parhi, D.R.: Mobile robot navigation and obstacle avoidance techniques: a review. Int. Robot. Autom. J. **2**, 1–12 (2017)
11. Pradhan, S.K., Parhi, D.R., Panda, A.K., Behera, R.K.: Potential field method to navigate several mobile robots. Appl. Intell. **25**, 321–333 (2006)
12. Patle, B.K., Pandey, A., Jagadeesh, A., Parhi, D.R.: Path planning in uncertain environment by using firefly algorithm. Def. Technol. (2018). https://doi.org/10.1016/j.dt.2018.06.004

Analysis of BASNs Battery Performance at Different Temperature Conditions Using Artificial Neural Networks (ANN)

B. Banuselvasaraswathy, R. Vimalathithan and T. Chinnadurai

Abstract In Body Area Sensor Network (BASN), battery power management is an important issue to be addressed to extend the lifetime of the sensor with increased performance. The lifetime of the battery relies on several factors like charging and discharging cycles, Voltage rating, current ratings, and temperature. The temperature variation in the battery leads to increased performance but shortens its lifetime due to the internal chemical reaction that occurs inside the battery. Therefore, it is essential to analyze the temperature variations to enhance the battery lifetime as well as to improve the lifetime of entire sensor network. In this paper, BASNs battery efficiency is analyzed at different temperature profile, charging, and discharging cycles. The voltage is measured from the obtained results. As rise in temperature influence the battery discharging capacity. Thus, maintaining optimal temperature is very essential in BASNs battery to increase the lifetime of battery. Further, Artificial Neural Network (ANN) is developed to examine the experimental results to obtain the optimum battery operating temperature.

Keywords Body area sensor networks · Battery performance · Temperature variations · Artificial neural networks

1 Introduction

In recent days, the impact of advanced technology toward modern life is progressing and implementation of Wireless Sensor Networks (WSN) is growing enormously. WSN is extensively used in many places like hospitals, schools, agricultural monitoring, industrial monitoring, healthcare monitoring, etc. WSN consists of several

B. Banuselvasaraswathy (✉)
Department of ECE, Sri Krishna College of Technology, Coimbatore, India
e-mail: banu.saraswathy74@gmail.com

R. Vimalathithan
Department of ECE, Karpagam College of Engineering, Coimbatore, India

T. Chinnadurai
Department of ICE, Sri Krishna College of Technology, Coimbatore, India

© Springer Nature Singapore Pte Ltd. 2020
K. N. Das et al. (eds.), *Soft Computing for Problem Solving*,
Advances in Intelligent Systems and Computing 1048,
https://doi.org/10.1007/978-981-15-0035-0_26

hundreds to thousands of minute sensor to monitor several physical parameters from the examining environment. Meanwhile, different wearable and implantable devices are designed to monitor the daily activities of a person suffering from chronic diseases in order to make revolution in the field of healthcare applications. This leads to the evolution of BANs. In general, there exist two types of sensors, namely, implantable and wearable sensors. The implantable sensors are embedded into the human body for measuring various physiological signals. Whereas, the wearable devices are placed on the surface of the human body to measure the signals based on user's need. The wearable devices are blooming in today's modern world due to its small size, advanced technology, and efficient functionality. The weight of wearable devices should be light and comfortable for the patients to wear for long duration making ease for physicians to monitor and record patient's physiological signals continuously.

One of the most crucial challenges in employing WSN is its energy limitations. The lifetime and performance of the entire network relies on energy management of resources. Hence, it is significant to balance energy consumption within the entire network in order to enhance the network lifetime. Henceforth, the approach with the efficacy of balancing energy utilization of each sensor is highly preferred. Besides, self-organization of the networks is also essential to make decision by interacting with the neighboring nodes. Each sensor nodes are embedded with a power source operating with the help of battery. A sensor node does not remain functional if the energy in the battery has drained out as most of the nodes will be disconnected from the sink affecting the function of entire WSN resulting in loss of data transmission to the destination. In addition, the reduced battery storage capacity in sensor nodes have shorter network lifetime and needs frequent battery replacement. Hence, conserving energy and extending the lifetime of battery in sensor nodes are additional crucial issue that has to be considered to ensure that the data are received continuously by the sink for a prolonged period of time.

In this contest, the temperature variation and optimum temperature finding are essential in the WSNs to maximize the battery lifetime and performance. The Artificial Neural Network (ANN) is the best optimization method utilized to find the optimum temperature of the battery. ANNs are a mathematical simulation tool, comprising of neurons arranged in layers. ANNs are used to predict the battery charging time, cycle time, and temperature.

In this paper, battery temperature was considered at different cycle times and different charging periods. The resultant values are optimized using AI techniques of artificial neural networks (ANN) to predict the operating temperature of WSNs battery module. A linear model of ANN is used to model the correlation between input and output parameters. Moreover, the nonlinear operating points modeled by ANN are incorporated to improve the performance of the battery process.

2 Related Works

Energy conservation remains as a state of art of research in design of WSN. Hence, it becomes challenge for several researchers and developers to prolong the lifetime of battery without affecting the performance of system. Several researchers have discussed various energy-efficient methods to improvise the battery lifetime. They are as follows: Hooshmand et al. [1] introduced the lossy signal compression technique to minimize the data size of collected bio-signals to increase the monitoring time as well as to maximize the battery lifetime. Wood et al. [2] introduced AlarmNet based on WSNs-AAH. This technique enables the nodes during data transmission and disables the sensor nodes during sleep and idle state to minimize energy requirement. Akyildiz et al. [3] popularized the approach utilizing spatiotemporal correlation to enhance the efficiency of sensor node. The entire sensor nodes are partitioned into cluster and single sensor nodes are assigned to each cluster for surveillance time series and other nodes are turned off to save energy. Galzarano et al. [4] adopted QL-MAC protocol. This protocol is efficient under heavy traffic conditions since the duty cycle can be adjusted through Q learning based on neighbors traffic to reduce the energy consumption.

Carrano et al. [5] introduced a hardware model to turn on and off the motes to conserve energy, but the main limitation of the designed model is that the energy of battery gets depleted in limited days if it is used for continuous activity monitoring. Henceforth, Lin et al. [6] implemented a Logical Correlation-Based Sleep Scheduling Mechanism (LCSSM) in Ambient Assisted Homes (AAH) batteries. In this methodology, the sensors are deactivated if it does not sense any event in order to save energy. Rong and Pedram [7] designed an analytical model to estimate the capacity of Lithium–ion battery. With Dualfoil simulator [8] as reference assessment model the demonstration was carried out with minimum error rate of 5%. Hausmann and Depcik [9] introduced an approach by expanding Peukert's equation to determine the battery capacity. The variable current and temperature are also taken into consideration while performing experimental analysis. ANNs help in simulating biological neural networks by utilizing mathematical systems and the entire process is carried out by the neurons organized as layers. ANN is highly preferred for predicting power and capacity [10], gust effects are considered to determine the battery failure, battery life cycle, and storage [11]. The degradation behavior of battery at a different calendar aging was analyzed, and it is inferred that the function of battery is easily influenced by internal resistance. The resistance rises gradually with the temperature thereby accelerating the battery performance in turn reducing the lifetime of the battery [12]. The efficiency and performance of battery depends on parameter such as operating conditions, combination of electrochemical, and mechanical process of battery [13]. Waag et al. [14] discussed the reason behind the progression of internal resistance in the nickel–magnesium and cobalt-based Lithium–ion battery at different current, temperature, and SOC levels. Matsushima et al. [15] observed the evolution of the internal resistance in battery during trickle charge at high temperature and voltage level such as 45 °C and 4 V.

3   Motivation of Research

WSN is widely used in many applications and nodes are deployed in external environments as well as at the location subjected to more variations in temperature. Therefore, it is necessary to consider the influence of thermal effect and voltage behavior on the battery lifetime. At higher temperature, thermal effect stimulates chemical reaction within the battery resulting in effectiveness of the capacity of battery at increased temperature. Accordingly, operating battery at higher temperature weakens its charging and discharging life cycle. Hence, it is complex task to determine the battery behavior at various operating temperatures resulting in difficult scenario to estimate the lifetime of battery at several operating conditions. Conventionally, WSN researchers use different software to study the behavior of node and estimate the lifetime of battery approximately before deploying into real-time application. But unfortunately, the existing battery models failed to analyze the impact of several operating temperatures on battery's lifetime, hence, resulting in accurate results. This motivated the researchers to march toward designing BANs nodes with increased battery lifetime.

4   Proposed Methodology

4.1   Operation States in WSN

WSN comprises of four operation modes, namely, transmitting, receiving, listening/idle, and sleep states. During transmission mode, the transmitting part remains active and antenna radiates energy for data transmission consuming maximum power. In receiving mode, the receiving part remains active and the antenna starts to receive the data. However, the power consumed during this state is not high as much as for the transmission part. At idle state, the transceiver is ready to accept the data but does not receive any data currently. During this process, only the receiver circuitry remains active and other parts can be turned off to minimize energy consumption compared to other two states. In sleep state, all the circuits remain in off state but it requires certain amount of energy to bring the circuitry back into active modes.

Figure 1 shows the different operation states of WSN. It is clear that at least a minimum amount of energy is required to carry out all function resulting in deterioration of network lifetime. Hence, extending the lifetime of battery has to be addressed seriously in turn to enhance the network lifetime.

Fig. 1 Operation modes of
WSN

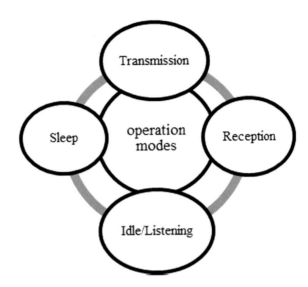

4.2 Lithium–Ion Battery Characteristics

Szente-Varga et al. [16] analyzed the performance of Ni-MH battery at different operating scenarios based on mathematical operations. The obtained results through the experiment carried out clearly show that this model can predict the behavior of battery only from the known discharge curve at constant temperature. Hence, Li-ion (Lithium–Ion) batteries are widely used due to its less self-discharge and longer cycle time. Moreover, the discharge effect does not affect the battery cycle time [17]. Erdinc et al. [18] demonstrated a dynamic model for Li-ion batteries. In this approach, both capacity and temperature effects are considered. Ye et al. [19] designed a mathematical model for Li-ion batteries. In this model electrochemical mechanism, mass transfer, electronic conduction, and effect of temperature on performance of battery are also taken into account (Table 1).

4.3 Influence of Temperature in Lithium–Ion Battery

As discussed by McGowan and Manwell [20], most chemical processes get accumulated at higher temperatures. These values are taken as higher k value in Kinetic Battery Model (KiBaM). According to Svante August Arrhenius (1859–1927) different temperature values will affect the battery lifetime and its voltage behavior for a long time. Temperature variation triggers many chemical reactions inside the battery, resulting in a reduction of efficiency. These chemical variations over temperature can be calculated using Arrhenius Eq. (1):

Table 1 Specification of lithium–ion battery

S. no	Parameters	Specification (Li-ion battery)
1	V max	4.2 V
2	V min	2.75 V
3	Capacitance/capacity	1600 mAh
4	ESR	140 mΩ
5	Weight	123 g
6	Charge operating temperature	0÷45 °C
7	Discharge operating temperature	−10÷50 °C
8	Size	68 × 48 × 5.7 mm

$$R_k = F_b . e^{-\frac{E_a}{k.T}} \tag{1}$$

where R_k is the constant value of regression, F_b pre-factor (S^{-1}), E_a is the activation energy in KJ/mol, k is the gas constant (8.314×10^{-3} kJ/mol K), and T is the temperature of the battery. This equation produces a constant reaction rate over the temperature. Batteries result in increased effective capacities at higher temperatures and decreased effective capacities operating at lower temperatures.

5 Results and Discussion

The effects of voltage at various time interval and cycles are analyzed. In addition, the discharge capacities of battery at different temperatures are also observed and plotted. The voltage consumed for data transmission in WSN has been monitored at different time interval (hours) and depicted as shown in Fig. 2a. In Fig. 2b, corresponding temperature versus battery discharging capacity has been analyzed at different time period (hours). From Fig. 2a and b it is inferred that, as the temperature rises, the amount of discharging capacity also considerably increases. The reason for variation in temperature is due to environmental condition as well as different body temperature from person to person wearing wearable device. The sensor senses the function of the human body and transmits the signal to the main server.

The packet size of the transmitting signal is large; hence, more amount of power is consumed for data transmission increasing the temperature of the battery and other components thereby reducing the lifetime of battery. Likewise, the power consumption for receiving the data is less compared to transmitting the data. As a result, the battery temperature will also influence the sensor performance, charging, and discharging cycles. The cycle time of the battery charging and discharging (chemical reaction of the battery) is driven by either voltage or temperature. The rise in temperature increases the chemical reaction of the battery at faster rate leading to

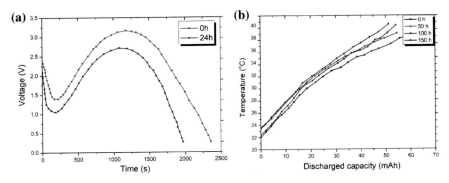

Fig. 2 Analysis of **a** voltage with respect to time, **b** discharge capacity of battery with temperature (°C)

increased battery performance. Further, the unwanted chemical reaction caused due to temperature variation reduces the lifetime of the battery. The voltage consumed for data transmission in WSN has been analyzed for different cycles and is illustrated in Fig. 3a. Different cycle times (discharging and charging) with the interval of 0, 15, 30, and 45 are also analyzed.

The different temperature values $-5, 0, 5,$ and $10\,°C$ and its corresponding voltage values are shown in Fig. 4. The output voltage produced from the battery decreases with increased temperature due to maximum effect on band gap and resistance than the rate of chemical reaction enhancing the lifetime of battery in warmer conditions. The capacity of battery is also influenced easily by the temperature since increased temperature stimulates the excess chemical reaction to progress thereby degrading the battery lifetime. In subzero temperature range ($-5\,°C$), the voltage of the battery gets reduced for a shorter period of time than compared with the remaining temperature ($0, 5,$ and $10\,°C$).

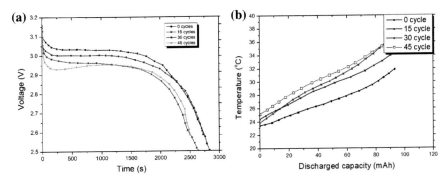

Fig. 3 Analysis of **a** voltage with respect to time, **b** discharge capacity of battery with temperature (°C)

Fig. 4 Effect of voltage with respect to time at different temperature (°C)

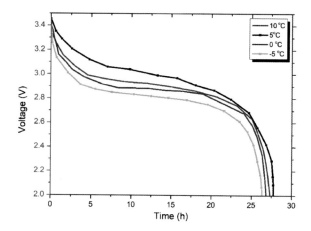

6   ANN Model

The ANN model was designed to find the optimum temperature in which the battery has to be operated to enhance the network lifetime. The back-propagation neural network (BPNN) technique is used to predict the temperature range. A typical feed-forward system consists of three layers: the input, the hidden, and the output layers. The activation of neuron depends on the weighted input signal. The input layer consists of two input neurons, namely, charging cycle and charging hours. It transmits the net input and output data to the next layer as follows Eqs. 2 and 3.

$$a_i^{(1)}(x) = net_i^{(1)} \tag{2}$$

$$b_i^{(1)}(x) = f_i^{(1)}[net_i^{(1)}(x)] = net_i^{(1)}(x), \quad i = 1, 2, 3 \tag{3}$$

where

$a_i^{(1)}$ input layer with node i,
$net_i^{(1)}$ sum of input layers and
$b_i^{(1)}$ hidden layer with node i.

The hidden layer neurons are then linked to the output neuron with a purelin transfer function. The multilayer perception network used Levenberg–Marquardt algorithm as training function; the sum of square errors was considered as a performance during the training of the network. The single output neuron with node, Nk, generates the best temperature value with a linear activation function as shown in Fig. 5. The output neuron was calculated by making a net sum of entering signals.

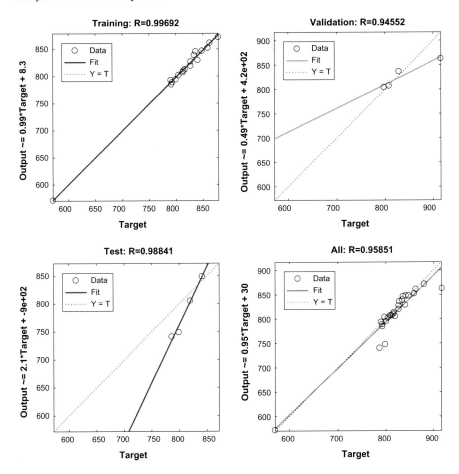

Fig. 5 Regression analysis

6.1 ANN Models: Evaluation and Error Analysis

In this analysis, a multilayer perception-based ANN model is designed by using MATLAB to predict the correlation between input parameters and temperature. The line of best fit was calculated using the regression coefficients and is shown Fig. 5.

The correlation coefficient value is found to be nearer to zero when compared with all models. It can be viewed in both the training and testing data results which were close to the distribution plot range. The R-value of "1" means a close relationship, and "0" indicates a random relationship. It was observed that a regression coefficient of R = 0.9692 was obtained for training data, R = 0.94552 for testing data, R = 0.98841 for validation data, and R = 0.95851 for all the data. Hence, accurate prediction of temperature is possible using this model.

The best validation performance obtained is 38.37 as shown in Fig. 6. The regression analysis was carried out to find the correlation coefficient. The correlation coefficient was used to measure the relationship between the measured and predicted values.

7 Conclusion

The energy consumption of a network has a direct impact on the network lifetime. The lifetime of entire network in turn depends on the battery power. But determining the lifetime of battery is a tedious process. There are multiple factors influencing the functional behavior of battery and degrading the network lifetime. In this work, the effect of temperature variation on the battery's model was taken into account and the battery performance was analyzed at different cycle times and time periods. The experimental results show that the temperature variations weakens the battery performance and lifetime. Moreover, maintaining a desired temperature is very essential in battery module. The best optimum battery operating temperature is obtained by developing an Artificial Neural Network (ANN). The predicted temperature value from ANN is 38.37 °C. This temperature will ensure the best operating temperature of the battery.

Future extension of this research is planned to implement different battery modules along with sensor nodes, loading and unloading condition of the battery, data transmission, and temperature module. This will provide a better understanding of the temperature influence on the battery performance.

Fig. 6 Performance plat for best validation

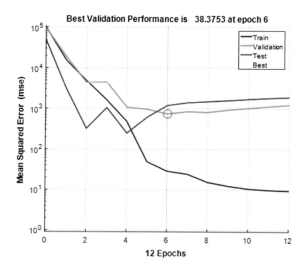

References

1. Hooshmand, M., Zordan, D.Del, Testa, D., Grisan, E., Rossi, M.: Boosting the battery life of wearables for health monitoring through the compression of biosignals. IEEE Internet Things J. **4**(5), 1647–1662 (2017)
2. Wood, A.D., Stankovic, J.A., Virone, G., Selavo, L., He, Z., Cao, Q., Stoleru, R.: Context-aware wireless sensor networks for assisted living and residential monitoring. IEEE Netw. **22**(4), 26–33 (2008)
3. Akyildiz, I.F., Vuran, M.C., Akan, O.B.: On exploiting spatial and temporal correlation in wireless sensor networks. Proc. WiOpt **4**, 71–80 (2004)
4. Galzarano, S., Liotta, A., Fortino, G. (2013) QL-MAC: A Q-learning based MAC for wireless sensor networks. In: International Conference on Algorithms and Architectures for Parallel Processing, pp. 267–275. Springer, Cham
5. Carrano, R.C., Passos, D.G., Magalhães, L.C.S., Célio Vinicius, N. (2014) Survey and taxonomy of duty cycling mechanisms in wireless sensor networks. IEEE Commun. Surv. Tutor. **16**(1), 181–194
6. Liu, C., Wu, K., Pei, J.: An energy-efficient data collection framework for wireless sensor networks by exploiting spatiotemporal correlation. IEEE Trans. Parallel Distrib. Syst. **18**(7), 1010–1023 (2007)
7. Rong, P., Pedram, M. (2006) An analytical model for predicting the remaining battery capacity of lithium-ion batteries. IEEE Trans. Very Large Scale Integration (VLSI) Syst. **14**(5), 441–451
8. Newman, J., FORTRAN Programs for the Simulation of Electrochemical Systems. http://www.cchem.berkeley.edu/jsngrp/fortran.html. Accessed 12 Dec 2016
9. Hausmann, A., Depcik, C.: Expanding the Peukert equation for battery capacity modeling through inclusion of a temperature dependency. J. Power Sour. **235**, 148–158 (2013)
10. Urbina, A., Paez, T.L., Jungst, R.G., Liaw, B.Y.: Inductive modeling of lithium-ion cells. J. Power Sour. **110**(2), 430–436 (2002)
11. Karami, H., Mousavi, M.F., Shamsipur, M., Riahi, S.: New dry and wet Zn-polyaniline bipolar batteries and prediction of voltage and capacity by ANN. J. Power Sour. **154**(1), 298–307 (2006)
12. Stroe, D.I., Swierczynski, M., Kær, S.K., Teodorescu, R.: Degradation behavior of lithium-ion batteries during calendar ageing—the case of the internal resistance increase. IEEE Trans. Ind. Appl. **54**(1), 517–525 (2018)
13. Vetter, J., Novák, P., Wagner, M.R., Veit, C., Möller, K.C., Besenhard, J.O., Hammouche, A.: Ageing mechanisms in lithium-ion batteries. J. Power Sour. **147**(1–2), 269–281 (2005)
14. Waag, W., Käbitz, S., Sauer, D.U.: Experimental investigation of the lithium-ion battery impedance characteristic at various conditions and aging states and its influence on the application. Appl. Energy **102**, 885–897 (2013)
15. Matsushima, T., Takagi, S., Muroyama, S., Horie, T. (2005) Lifetime and residual capacity estimate for Lithium-ion secondary cells for stationary use in telecommunications systems. In: Telecommunications Conference, INTELEC'05. Twenty-Seventh International, pp. 199–204. IEEE
16. Szente-Varga, D., Horvath, G., Rencz, M. (2006) Creating temperature dependent Ni-MH battery models for low power mobile devices. In: Proceedings of the 12th International Workshop on Thermal investigations of ICs (THERMINIC), pp. 27–29. Nice, France
17. Valle, O.T., Milack, A., Montez, C., Portugal, P., Vasques, F. (2013) Polynomial approximation of the battery discharge function in IEEE 802.15. 4 nodes: case study of MicaZ. In: Advances in Information Systems and Technologies, pp. 901–910. Springer, Berlin, Heidelberg

18. Erdinc, O., Vural, B., Uzunoglu, M. (2009) A dynamic lithium-ion battery model considering the effects of temperature and capacity fading. In: Proceedings of the 2009 International Conference on Clean Electrical Power (ICCEP), pp. 9–11. Capri, Italy
19. Ye, Y., Shi, Y., Cai, N., Lee, J., He, X.: Electro-thermal modeling and experimental validation for lithium ion battery. J. Power Sour. **199**, 227–238 (2012)
20. Manwell, J.F., McGowan, J.G.: Lead acid battery storage model for hybrid energy systems. Sol. Energy **50**(5), 399–405 (1993)

ASIC Implementation of Fixed-Point Iterative, Parallel, and Pipeline CORDIC Algorithm

Grande Naga Jyothi, Kundu Debanjan and Gorantla Anusha

Abstract In this paper, we proposed a Coordinate Rotation Digital Computer (CORDIC) algorithm for efficient hardware implementation of mathematical functions which can be carried out in a wide variety of ways for many digital signal processing applications. The CORDIC is a single unified algorithm for calculating many elementary functions such as trigonometric, hyperbolic, logarithmic function, exponential functions, multiplication, division, and so on. In this paper, a novel low power, low area, and high throughput fixed-point CORDIC algorithms are proposed. The standard CORDIC is also implemented for comparing the synthesis results. The proposed architecture scaling has been done using low area and low-power Scale Factor Correction Unit (SFCU). A low ADP SQRT-CSLA based ADD/SUB unit is proposed to overcomed the disadvantages of the basic ADD/SUB unit used in the standard CORDIC. The ROM lookup table size is also reduced to half. Extensive simulations are performed to verify the functionality. The standard and proposed CORDIC architectures are simulated in cadence NC launch and synthesized in cadence RC tool using TSMC GPDK 45 nm technology and area, power, and delay are calculated. The area and power consumption of the proposed CORDIC architecture are less when compared with standard CORDIC design.

Keywords CORDIC algorithm · Pipeline · Parallel · Digital signal processing

G. Naga Jyothi (✉) · K. Debanjan
VIT University, Vellore, India
e-mail: grande.nagajyothi@vit.ac.in

K. Debanjan
e-mail: debanjankundu@ymail.com

G. Anusha
GCT Coimbatore, Coimbatore, TamilNadu, India
e-mail: anushagorantla3@gmail.com

© Springer Nature Singapore Pte Ltd. 2020
K. N. Das et al. (eds.), *Soft Computing for Problem Solving*,
Advances in Intelligent Systems and Computing 1048,
https://doi.org/10.1007/978-981-15-0035-0_27

1 Introduction

Digital Signal Processing (DSP) plays preeminent role than the microprocessors due to several advantages like multiple instructions in multiply accumulate can be performed in single cycle and a large number of addressing modes is not required. The microprocessors are more flexible and available at less cost, but the speed is very slow when compared with the online DSP applications. To get rid of this, reconfigurable logic computer architectures are used to increase the speed and reduce the hardware architectures at low costs which are more competitive for the traditional software approach. For microprocessor-based systems many algorithms are optimized and are not well mapped for hardware systems. Hardware system solutions are not more preferable than the software system solution. Software system solution is more dominated and are well suited. Among these, hardware and structurally efficient algorithms are a class of iterative solutions for mathematical functions and trigonometric functions. In these they use block of shifts and adders to perform the operation.

The trigonometric functions are performed by using the vector rotation. CORDIC will perform trigonometric operation. The other function is square root function, which is executed by using the incremental expression of desired function. The square root functions can be performed easily with an extension of the hardware architecture, which is not possible in CORDIC. The trigonometric and square root functions both have some similarities in their function but differ in their architecture and accuracy. The CORDIC function will have an additional bit of accuracy for each iteration. CORDIC was an attractive choice for computing elementary functions instead of using various polynomial methods.

CORDIC made its first hardware appearance in HP's Pocket calculator HP 35, the arithmetic Co-processor of Intel 8087, RADAR signal processors, and Robotics. In 1959, Volder [1] introduced the Coordinate Rotation Digital Computer (CORDIC) algorithm first time then CORDIC algorithm was generalized and unified by Walther [2] and later more research has been performed on CORDIC. The unified CORDIC algorithm utilizes the coordinate rotations to evaluate a lot of computational complex mathematical functions such as:

1. Trigonometric function
2. Hyperbolic function
3. Exponential function
4. Logarithmic functions
5. Multiplication
6. Division and
7. Square root.

CORDIC is the powerful algorithm used to perform fundamental operations which are in the form of $(a \pm b \gg 2)$ and it has simple hardware architecture to perform mathematical operation due to its unique property in nature. At the time of invention of the CORDIC, the multipliers and divider were very expensive in hardware.

CORDIC algorithm can be implemented with less hardware architecture when compared with the conventional method and it is more suitable for high-speed applications with low cost. CORDIC scheme was mainly used in digital applications of real-time navigation which is more flexible.

The eminent of the paper is organized as follows. Section 2 describes the mathematical background of CORDIC and describes proposed Iterative CORDIC, Pipelined CORDIC, and Parallel CORDIC architectures. Section 3 presents the simulation and synthesis results of proposed CORDIC. Finally, Sect. 4 concludes the paper.

2 Background

CORDIC was designed by Volder. He suggested CORDIC for determining the trigonometric relation present in the plane coordinate rotation to the special purpose and also solves the conversion of rectangular to polar coordinate. The operation present in CORDIC is in digital form which is equivalent of analog resolver [3, 4]. The drawbacks present in analog devices will be overcome by the CORDIC. A prototype digital computer, CORDIC-I, was built at Convair, a division of General Dynamics Corporation Texas Fort Worth. It consists of a special serial arithmetic block having three adders, three shift registers, and special interconnections. Originally, CORDIC was programmed to solve the following equations:

$$X' = l(x \cos \theta - y \sin \theta) \tag{1}$$

$$Y' = l(x \cos \theta + x \sin \theta) \tag{2}$$

or

$$R = l\sqrt{x^2 + y^2} \tag{3}$$

$$\theta = \tan inv \frac{y}{x} \tag{4}$$

where l is constant.

The first set of equations is solved in a CORDIC operation called Rotation operation Mode (RM) and the second set is solved in Vectoring operation Mode (VM) [5–7]. In vectoring operation, the coordinate components of a vector axes are chosen and the magnitude and the angle of the original vector with the x-axis are calculated [8–10]. In rotation operation, the two-dimensional vector and a rotation angle are chosen, and after rotation the coordinate components of the vector are calculated.

Essentially, the fundamental computing method used in both vector and rotation operations is a step-by-step order of micro-rotations which gives a complete rotation by a known angle as in the rotation operation. Else the result in zeroing the final angle

Fig. 1 Coordinate systems of a unified CORDIC

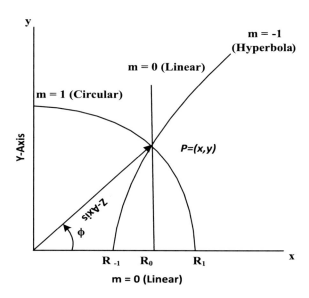

of the vector as in the vectoring operation. These micro rotation angles are taken for implementing adder and shift operations. The micro-rotations are implemented in iterations. The number of iterations needed is equal to the number of significant bits that presents the rotated components. On behalf of rotating a vector along a circular curve, the vector can be rotated along a hyperbola or a line, as shown in Fig. 1, where Ø is the "angle" and R_0, R_1 and R_{-1} are the "radii" of the vector P = (x, y) in hyperbolic, circular, and linear coordinate systems, respectively. The normalized radius can be geometrically given as the distance from the origin to the intersection point of the coordinate curve with the x-axis. The generalized angle is equal to twice the area enclosed by the vector x-axis, and coordinate curve, divided by the radius squared, as shown below:

$$\phi = \frac{(2 * area \ under \ curve)}{R^2} \tag{5}$$

A set of unified equations that explains the coordinate components of the rotated vector was described by Walther. These equations are characterized in terms of the coordinate system [11–13]. Hence many fundamental functions can be calculated using CORDIC, such as multiplication, divisions, hyperbolic, and square root functions.

$$R = \sqrt{x^2 + my^2} \tag{6}$$

$$\phi = \frac{1}{\sqrt{m}} * \tan inv \frac{\sqrt{my}}{x} \tag{7}$$

where

Fig. 2 Flow diagram for
CORDIC classification

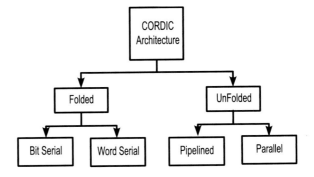

The iterative equations of the CORDIC can be arranged in various ways. The user can choose and design based upon the number of iterations and the application requirement. The CORDIC can be classified [7] as:

1. Sequential/Iterative Architecture
2. Parallel Architecture
3. Pipelined Architecture

The general classification of the CORDIC architectures is presented in the Fig. 2. The folded architectures are basically area efficient and very simple but they take a large number of clock cycles to give the final output for a given input. Hence, they are slow. There are unfolded architectures in which the iterative architecture is unrolled for the n number of iterations. Therefore, they consume large hardware resources as compared to the folded architectures. The computation speed for these architectures is very high as compared to the previous architectures.

The unfolded architecture is further subdivided in to the parallel and pipelined versions. The structure of iterative architecture is shown in Fig. 3 and the working operation of CSLA and XOR is shown in Fig. 4. In the parallel architectures shown in Fig. 5, the structure for iterative CORDIC is instantiated multiple times simultaneously without any registers at the successive output stages of the iterations. Therefore, the frequency of operation of the parallel architectures is less. The pipelined architectures of CORDIC is similar to the parallel architecture. The only difference is that the pipelined architectures places registers at the output stages of every iteration. Due to which the maximum delay path gets reduced, hence it operates at larger frequency than the iterative and the parallel architectures. The parallel architecture of CORDIC is shown in Fig. 5. The pipelined architectures [4–6] has disadvantage of an initial latency equivalent to the number of iterations for the first output. The pipelined architecture for CORDIC is shown in Fig. 6.

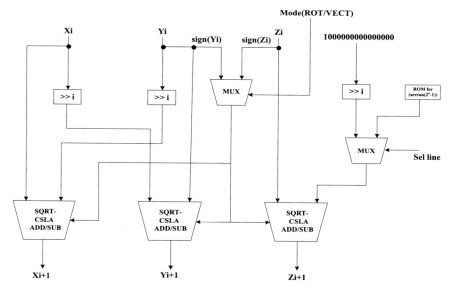

Proposed CORDIC Core for Single Iteration

Fig. 3 Iterative CORDIC architecture

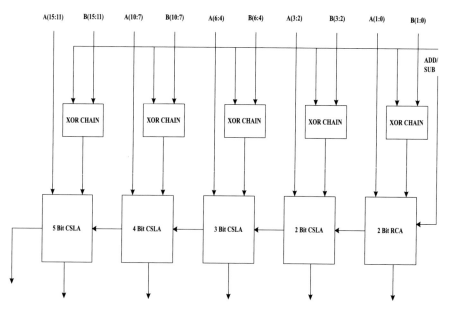

Fig. 4 CSLA and XOR gate

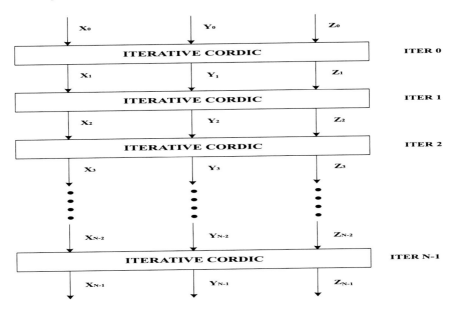

Fig. 5 Parallel CORDIC architecture

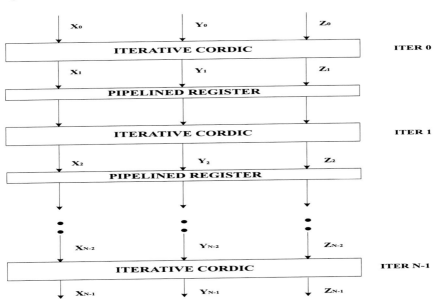

Fig. 6 Pipelined CORDIC architecture

3   Simulation and ASIC Synthesis Results:

This section consists of the simulation waveforms and synthesis results for the various design of CORDIC blocks. The simulation waveforms have been obtained by performing the simulation using Cadence NC Launch tool and synthesis results have been obtained by synthesizing the design blocks through Cadence RC tool using TSMC gpdk 45 nm technology. Simulation results for the proposed 16-bit fixed-point iterative, parallel, and pipelined CORDIC are shown in Figs. 7, 8 and 9. The synthesis results for parallel and pipelined CORDIC are shown in Figs. 10 and 11. The area, power, and delay of the standard CORDIC (designed by using basic add/sub

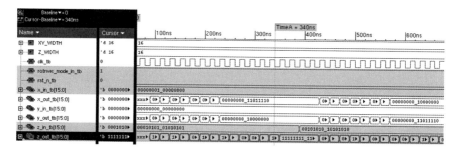

Fig. 7 Iterative CORDIC simulation waveform

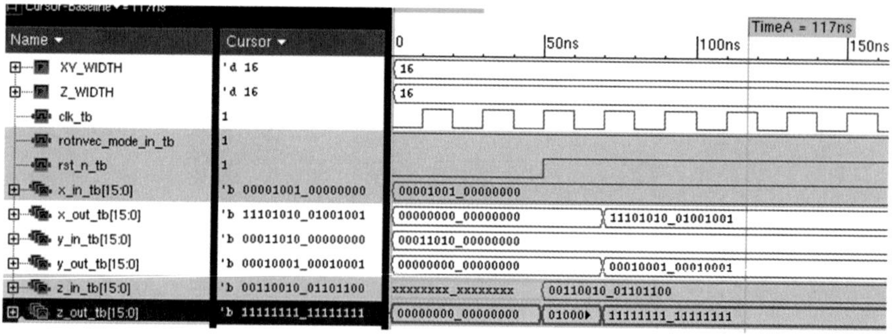

Fig. 8 Parallel CORDIC simulation waveform

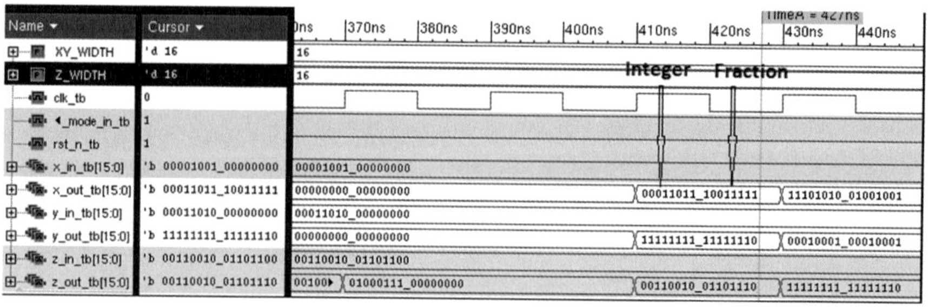

Fig. 9 Pipelined CORDIC simulation waveform

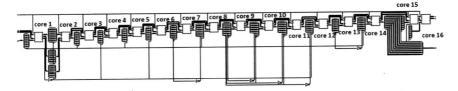

Fig. 10 Synthesized netlist of Cascaded 16 stages of the Core for Parallel CORDIC

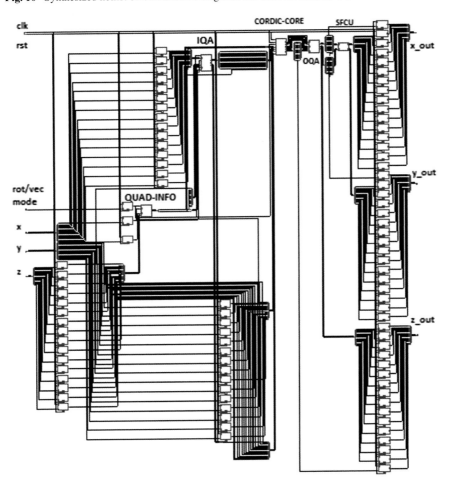

Fig. 11 Pipelined CORDIC synthesized netlist schematic

Table 1 Synthesis results comparison of proposed iterative, parallel, and pipelined CORDIC

Design	Area (μm^2)	Power (mW)	Delay (ns)
Standard iterative CORDIC	4108	0.27	7.43
Proposed iterative CORDIC	2712	0.16	5.48
Standard parallel CORDIC	9833	1.45	52
Proposed parallel CORDIC	7660	1.013	40
Standard pipeline CORDIC	16798	13.3	3.09
Proposed pipelined CORDIC	12640	9.1	2.887

Fig. 12 Area report

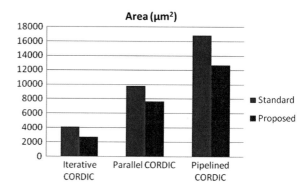

Fig. 13 Power report

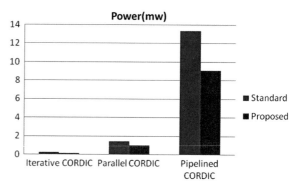

unit, multiplier for scale factor correction and ROM for all iterations) and the proposed CORDIC are compared and tabulated in Table 1. The simulation waveform for iterative CORDIC gives output after 17 clock cycles for 16-bit CORDIC. Simulation waveform for parallel CORDIC gives output every clock cycle for 16-bit CORDIC. The proposed pipelined CORDIC simulation waveform gives output for every clock cycle after passing latency of 18 clock cycles for first input for 16-bit CORDIC. Figures 12 and 13 explains the area report and power report for the iterative, pipelined and parallel CORDIC algorithm.

4 Conclusion

We have designed 16-bit fixed-point iterative, parallel, and pipelined CORDIC architectures. This CORDIC can be used for singular value decomposition. It is favorable for DSP applications as it is designed for fixed-point arithmetic. This architecture consists of the SQRT-CSLA-based ADD/SUB unit, which is area, power, and delay efficient in comparison to the standard one. As we know, in case of pipelined and parallel architectures the core stages are unrolled. Each iteration step uses three ADD/SUB units. So, the number of ADD/SUB will be multiplied by the number of stages. Hence, it can lead to a good amount of reduction in overall area. The architecture replaced the traditional multiplier units with the proposed Scale Factor Correction Unit (SFCU). The SFCU consumes very less area and power in comparison to the multiplier. Synthesis results of the proposed CORDIC architecture shows the significant reduction in area and power compared to the standard CORDIC. Proposed parallel CORDIC architecture reduced 50% of area and 40% reduction in power when compared with standard iterative CORDIC, the proposed pipelined CORDIC design in comparison to the standard design, the reduction in area and power is 24.7% and 31.1%, respectively.

References

1. Volder, J.: The CORDIC trigonometric computing technique. IRE Trans. Electron. Comput. **EC-8**(3), 330–334 (1959)
2. Walther, J.S.: A unified algorithm for elementary functions. In: Proceedings of the, 1971, Spring Joint Computer Conference, pp. 379–385. Atlantic city, N.J. USA (1971)
3. Andraka, R.J.: Building a high Performance bit-serial processor in an FPGA. In: Proceedings of Design Super Con '96, pp. 5.1–5.21 (1996)
4. Deprettere, E., Dewilde, P., Udo, R.: Pipelined CORDIC architecture for fast VLSI filtering and array processing. In: Processing ICASSP'84, pp. 41.A.6.1–41.A.6.4. (1984)
5. Hu Y.H., Naganathan, S.: An angle recording method for CORDIC algorithm implementation. IEEE Trans. Comput. **42**, 99–102 (1993)
6. Sibul, L.H., Fogelsanger, A.L.: Application of cordinate rotation algorithm to singular value decomposition. In: IEEE International Symposium on Circuits and Systems, pp. 821–824 (1984)
7. Bhakthavatchalu, R., Nair, P.: Low power design techniques applied to pipelined parallel and iterative CORDIC design. IEEE 1978-1-4244-8679-3/11
8. George Lee, C.S., Chang, P.R.: A maximum pipelined CORDIC architecture for inverse kinematic position computation. IEEE J. Robot. Autom. **Ra-3**(5) (1987)
9. Timmermann, D., Hahn, H., Hosticka, B.: Modified CORDIC algorithm with reduced iterations. Electron. Lett. **25**(15), 950–951 (1989)
10. Lo, H., Lin, H., Yang, K.: A new method of implementation of VLSI CORDIC for sine and cosine computation. In: Proceedings of the 1995 IEEE International Symposium on Circuits and Systems. Seattle, WA, USA (1997)
11. El-Guibaly, F.: Analysis of the CORDIC algorithm. In: Cyprus International Conference Computer Application to Engineering Systems, pp. 12–17. Nicosia, Cyprus (1991)
12. Chen, C.Y., Liu, W.C.: Architecture for CORDIC algorithm realization without ROM lookup tables. Department of Electronic Engineering, I-Shou University Kaohsiung, Taiwan
13. NagaJyothi, G., SriDevi, S.: Distributed arithmetic architectures for FIR filters-A comparative review. In: 2017 International Conference on Wireless Communications, Signal Processing and Networking (WiSPNET). IEEE (2017)

Elephant Herding Optimization Based Neural Network to Predict Elastic Modulus of Concrete

B. S. Adarsha, Narayana Harish, Prashanth Janardhan
and Sukomal Mandal

Abstract The concrete requirement is proliferating, especially in developing countries. The elastic modulus of concrete plays a vital role as a design parameter in construction applications. The experimental procedures for finding the elastic modulus of concrete are quite complicated and expensive. Thus, researchers have always been looking for better and efficient methods to replace the traditional methods. In the present study, the issue is addressed by hybridized soft computing technique called Elephant Herding Optimization Based Artificial Neural Network (EHO-ANN). The developed model is then validated with linear regression and standard empirical formulas used for estimation of elastic modulus of concrete. The performances are evaluated by statistic measures like Correlation Coefficient (CC), Root Mean Square Error (RMSE) and Mean Absolute Percent Error (MAPE). The developed EHO-ANN model (Train CC 0.9102 and Test CC 0.9095) has outperformed linear regression (Train CC 0.8287 and Test CC 0.7575) and empirical formula (Train CC 0.8170 and Test CC 0.7645) in predicting the elastic modulus of concrete. Hence, the EHO-ANN model can be used as an alternative method to predict the elastic modulus of concrete.

Keywords Artificial neural network · Elephant herding optimization · Elastic modulus of concrete · Prediction

1 Introduction

When the economy starts to grow, the construction industry usually replicates it. Almost every civil structure surrounding us utilizes concrete. Thus, it plays a huge role in the construction industry. This is due to its vast versatility, comparatively

B. S. Adarsha · N. Harish (✉)
Department of Civil Engineering, M S Ramaiah Institute of Technology, Bengaluru, India
e-mail: harishnnitk@gmail.com

P. Janardhan
Department of Civil Engineering, National Institute of Technology Silchar, Assam, India

S. Mandal
Department of Civil Engineering, Presidency University, Bengaluru, India

© Springer Nature Singapore Pte Ltd. 2020
K. N. Das et al. (eds.), *Soft Computing for Problem Solving*,
Advances in Intelligent Systems and Computing 1048,
https://doi.org/10.1007/978-981-15-0035-0_28

353

inexpensiveness, and its ability to withstand most of the loading conditions. The elastic modulus of concrete has a significant influence on strains under load and in turn affects the displacement of the structure. High elastic modulus is essential when deflection is undesirable and low elastic modulus is preferred when flexibility is needed. Hence, it is essential to know the elastic modulus to know the behavior of concrete.

In the past several decades, many researchers have carried out experiments to find modulus of elasticity of different types of concrete [1–3]. But, compared to compressive strength tests, the experimental procedure [ASTM C469/C469M] to measure the elastic modulus is quite complicated and expensive. To make things much more straightforward, researchers have found ways to estimate elastic modulus of concrete through an empirical approach. This has led to the formulation of equations to define the relationship between the compressive strength and elastic modulus [4–9], thus making compressive strength an indication of the elastic modulus of concrete [10, 11].

But the empirical formulas are influenced by materials in the concrete and national preference. Due to these various complications, it is necessary to find alternate efficient methods that are capable of replacing existing traditional methods for predicting the elastic modulus of concrete. In the past two decades, soft computing is being used for solving some of the hardest problems which are being confronted today that can tolerate the uncertainty, imprecision, approximation, and partial truth of the solution [12]. A wide range of soft computing methods has been used already, as numerous databases are available in the field of civil and particularly, in the field of structural concrete. Demir and Korkmaz [13] have applied theory of fuzzy logic for the prediction of the modulus of elasticity of high-strength concrete. The automatic regression methods such as Genetic Programming (GP), Artificial Bee Colony Programming (ABCP), and Biogeographically Based Programming (BBP) have been successfully utilized for predicting the elastic modulus of Recycled Aggregate Concrete (RAC) [14].

To reap benefits of different techniques, Rinchon et al. [15] have adopted a hybrid model of Artificial Neural Network (ANN) and Genetic Algorithm (GA) for prediction and optimization of the ultimate bond strength of reinforced concrete. Ahmadi-Nedushan [16] has used a hybrid of Adaptive Neuro-Fuzzy Inference System (ANFIS) to predict the elastic modulus of normal and high-strength concrete. Inspired by herding behavior of elephant groups, Wang et al. [17] proposed Elephant Herding Optimization (EHO) technique and benchmarked its performance with Biogeography Based Optimization (BBO), Differential Evolution (DE), and GA. EHO has also been used for training ANN, and it provided better classification accuracy compared to crow search algorithm and Levenberg–Marquardt BP (LM-BP) [18]. Recently, Meena et al. [19] observed some limitations in the existing EHO technique and suggested some modifications, which further improved the ability of EHO as an optimization technique.

In the present study, the elasticity modulus of concrete is predicted using modified Elephant Herding Optimization Based Artificial Neural Network (EHO-ANN). Further, this study is compared with traditional methods that are used in the field of civil engineering for predicting elastic modulus of concrete like linear regression and standard empirical formulas in terms of Correlation Coefficient (CC), Root Mean Square Error (RMSE), and Mean Absolute Percent Error (MAPE).

2 Data Collection

Elastic modulus is expressed as a function of compressive strength as per codal provisions [4–9], hence the compressive strength of concrete is used to predict the elastic modulus of concrete in this study. 80 datasets were collected from previous literature [1, 14, 20]. These datasets have been randomly divided into two groups of 75 and 25%, which are used for training and testing the model, respectively.

2.1 Back Propagation Neural Network

The working procedure of ANN is explained in [21]. In the present study, initially, Backpropagation Neural Network (BPNN) with LM algorithm [22] is developed to predict the elastic modulus of concrete. LM algorithm has been chosen because it is the fastest and the first choice BP algorithm, even though it uses more memory compared to other algorithms. Further, it has been proved to be a reliable ANN training algorithm based on previous studies [24, 25]. The transfer function used in both hidden and output layer is a hyperbolic tangent sigmoid function (Tansig).

2.2 Elephant Herding Optimization

Elephant Herding Optimization (EHO) is a nature-inspired optimization technique, which has been inspired by the herding behavior of elephants. The very complex behavior of elephants has been simplified for the purpose of optimization technique. Elephants are known to be social animals. Their population is composed of several numbers of clans. In each clan, elephants only move under the influence of matriarch, which is often the oldest elephant. For the purpose of optimization technique, a number of elephants in each clan are assumed to be the same. The matriarch is assumed to hold the best solution in the herd of elephants, while the male elephant's position is decoded to be the worst solution. The behavior of elephants was mathematically modeled by Wang et al. [17]. This mathematical model was further improved by Meena et al. [19], which has been adopted in this study. Steps of the optimization cycle are as follows:

Step-1: Clan updating operator

Suppose there is a total of C number of clans and N number of elephants in each clan, then the position of ith elephant, $i = 1, 2, 3, \ldots, N$ in the c_jth clan, c_jth $= 1, 2, 3, \ldots, C$ is given as $Z_{cj,i}$.

The position of each elephant other than matriarch and matured male elephant (which hold the best and worst solution in each clan, respectively) is updated.

$$Z_{new,cj,i} = Z_{cj,i} + \alpha(Z_{best,cj,i} - Z_{cj,i}) \times r \tag{1}$$

where $Z_{new,cj,i}$ and $Z_{cj,i}$ are the new and present position of ith elephant in c_jth clan. α is a scale factor between 0 and 1. $Z_{best,cj,i}$ is the best position in c_jth clan (matriarch) and r is a random number between 0 and 1.

The position of matriarch elephant is updated around the current best position using Eq. (2).

$$Z_{new,cj,i} = Z_{best,cj,i} + \beta \times Z_{center,cj} \tag{2}$$

$$Z_{center,cj} = \sum_{i=1}^{n} Z_{cj,i}/n_z \tag{3}$$

where $Z_{new,cj,i}$ is the new matriarch position and $Z_{best,cj,i}$ is the best position obtained by the matriarch of all clan till now. β is a scale factor between 0 and 1 and n_z gives the total number of elephants in each clan.

Here, the best performing elephant (matriarch) of a clan's position is updated to be in between the best position obtained by the clan so far, and the center of all elephants in the clan using Eq. (2).

Step-2: Clan separating operator

The newly generated elephant would occupy the position near to matriarch of the respective clan.

$$Z_{worst,cj,i} = Z_{fitness_{cj}}$$
$$Z_{fitness_{cj}} = \mu \times Z_{local,cj} \tag{4}$$

where, $Z_{worst,cj,i}$ is the position of worst or matured ith male elephant in c_jth clan. $Z_{local,cj}$ is the local best position of the elephant in c_jth clan and μ is random proximity factor between 0.9 and 1.1.

In this step, the worst performing elephant of a clan is dropped and replaced by a "calf" within 10% of matriarch's position using Eq. (4).

Step-3: The above steps are repeated until the convergence criteria are attained.

2.3 Elephant Herding Optimization Artificial Neural Network

The hybrid of global and local optimization techniques to train ANN has proved to be very effective [23, 26]. BPNN makes small jumps in the search environment. As a result, it has a great ability to locate nearby local optima. But EHO has the ability to find more neighborhoods than BPNN. Thus, a hybrid of ANN and EHO utilizes the strong global searching potential of EHO and also strong local searching potential of BPNN. The flow chart of EHO-ANN is shown in Fig. 1.

Initially, ANN is trained by EHO based on the defined number of populations, clans, and iterations. The network will be trained using Eqs. (1)–(4) appropriately. EHO will run until a number of generations wears out. Once the EHO loop is completed, the obtained weights and biases are carried forward to BPNN as default weights and biases. If the obtained results do not satisfy the required criteria, again the whole process is repeated for different population size, a number of clans, and number of generations until the required criteria is achieved.

3 Results and Discussions

In the present study, BPNN and EHO-ANN models are developed to predict the elastic modulus of concrete. The predicted results are compared with linear regression and standard empirical formulas by means of statistical measures namely CC, RMSE, and MAPE, which are defined as follows:

$$CC = \frac{\sum\limits_{i=1}^{n} (E_m - \overline{E_{mm}})(E_p - \overline{E_{pm}})}{\sqrt{\sum\limits_{i=1}^{n} (E_m - \overline{E_{mm}})^2} \times \sqrt{\sum\limits_{i=1}^{N} (E_p - \overline{E_{pm}})^2}} \tag{5}$$

$$RMSE = \sqrt{\frac{1}{n} \sum\limits_{i=1}^{n} (E_m - E_p)^2} \tag{6}$$

$$MAPE = \frac{\sum\limits_{t=1}^{n} \left| \frac{E_m - E_p}{E_m} \right|}{n} \times 100 \tag{7}$$

here, E_m and E_p represents the measured and predicted elastic modulus of concrete, respectively, $\overline{E_{mm}}$ and $\overline{E_{pm}}$ are the mean value of measured and predicted observations, n is the number of observations. The collected dataset is randomly divided into training by considering about 75% dataset and the remaining dataset for testing. All soft computing models are run using MATLAB® 2016a installed in HP Pavilion

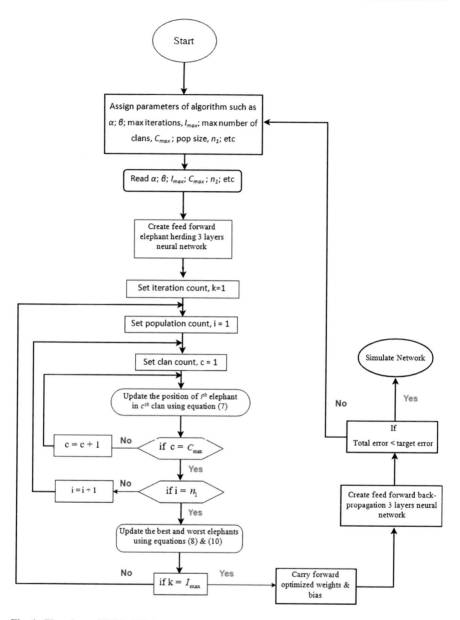

Fig. 1 Flowchart of EHO-ANN

Table 1 Statistical performance of BPNN model

Network		CC	RMSE	MAPE
1-4-1	Train	0.8837	2.957	8.50
	Test	0.8429	3.184	7.50

Intel® Core ™ i5 CPU @ 2.30 GHz and 8 GB RAM and 64-bit windows 10 operating system.

3.1 Performance of Backpropagation Neural Network

The statistical results obtained for BPNN is shown in Table 1. After various trials, four number of hidden nodes yielded a better result and proved to be optimum for this particular dataset.

Based on various trails, it was also observed that BPNN results fluctuated randomly on each trail. This is mainly due to ANN getting stuck in local optima [23]. Since the elastic modulus is a critical parameter, adoption of BPNN in such sensitive cases can be risky. To reduce this fluctuation and to further improvise the result, EHO can be adopted. Also hybridizing different soft computing techniques has shown to outperform stand-alone models due to its complementary nature [23, 26]. In this study, EHO has been hybridized with BPNN, with the aim to nullify the drawbacks of each other and develop a reliable prediction model.

3.2 Performance of Elephant Herding Optimization Artificial Neural Network

In order to compare with BPNN, the same number of hidden nodes has been used. Initially, ANN is trained by EHO, the obtained weights and biases are then carried forward as default parameters for BPNN. When EHO trains ANN, the weights are considered to be state vectors and biases as a cost function. EHO generates weights and biases of ANN and keeps adjusting them until it runs out of generations which area fixed value.

Based on various trails, it is observed that a number of generations, population, and clan size are essential parameters for training ANN by EHO. As the number of generations runs out, the obtained weights and biases will be carried forward to BPNN. Thus, the entire process would complete one EHO-ANN loop. The statistical results obtained for EHO-ANN are shown in Table 2.

Since BPNN is trained by default weights which are obtained by EHO, the results on each trail are almost constant and further improvised.

Table 2 Statistical performance of EHO-ANN model

Generations	Population	Clan		CC	RMSE	MAPE
30	100	10	Train	0.9102	2.559	7.02
			Test	0.9095	2.23	5.81

3.3 Performance Evaluation of Linear Regression, Empirical Formulas and Soft Computing Models

In order to compare soft computing models with traditional prediction methods, the same training and testing data are separately considered as two different data for linear regression and empirical formulas. Initially, linear regression is carried out to find the relationship between compression strength and elastic modulus of concrete in MS Excel 2010® and the Eq. (8) was obtained for training dataset and the same equation was utilized for the testing dataset.

$$y = 0.189x + 19.48 \tag{8}$$

The statistical performances of all models are shown in Table 3. Considering empirical formulas, it can be seen that CC, RMSE, and MAPE are way down the line compared to soft computing models. Out of all the empirical formulas, CSA [11] performed better with a CC 0.7645, RMSE 5.494, and MAPE of 16.17 for testing data. But even this is quite unreliable compared to soft computing model with CC more than 0.8429, RMSE and MAPE less than 3.1843 and 7.5098, respectively, for testing data. Also, it is observed from Table 3 that, soft computing models are better than linear regression.

In terms of soft computing model comparison, the EHO-ANN model performed better than BPNN. This is because the hybrid technique makes use of both EHO

Table 3 Statistical results for empirical, linear regression, and soft computing models

Prediction model	Training data			Testing data		
	CC	RMSE	MAPE	CC	RMSE	MAPE
Linear regression	0.8287	3.4588	9.9192	0.7575	4.0762	10.8499
ACI [6]	0.8170	7.0648	20.8819	0.7645	6.0184	16.7673
CSA [7]	0.8170	5.9230	17.6000	0.7645	5.4949	16.1721
CEB [8]	0.8138	9.3349	29.5021	0.7627	7.4593	21.4709
IS [9]	0.8170	9.0266	26.2731	0.7645	7.3033	18.7605
TS [10]	0.8170	8.4660	26.0511	0.7645	6.6225	18.1339
EURO [11]	0.8138	7.5197	23.1359	0.7627	5.8099	16.3140
BPNN	0.8818	2.9576	8.5010	0.8429	3.1843	7.5098
EHO-ANN	0.9102	2.5592	7.0243	0.9095	2.2306	5.8159

and BP strong global and local searching potential, respectively. Further, the hybrid model is quite reliable in sensitive cases. Figures 2, 3, 4 shows the CC between observed and predicted elastic modulus for linear regression, CSA, and EHO-ANN for training and testing, respectively.

The performance of soft computing model, empirical approach (CSA), and linear regression for predicted elastic modulus of concrete are compared with test experimental datasets as shown in Fig. 5. From Fig. 5, it is observed that the EHO-ANN has a similar trend with experimental data; hence, EHO-ANN can be used for predicting the elastic modulus of concrete.

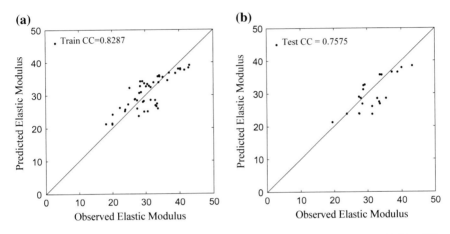

Fig. 2 Observed and predicted elastic modulus for linear regression: **a** Train CC and **b** Test CC

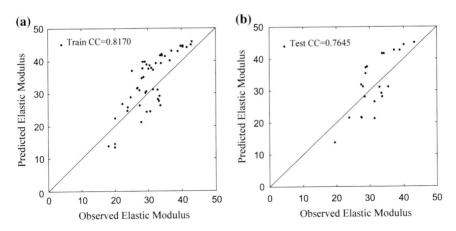

Fig. 3 Observed and predicted elastic modulus for CSA: **a** Train CC and **b** Test CC

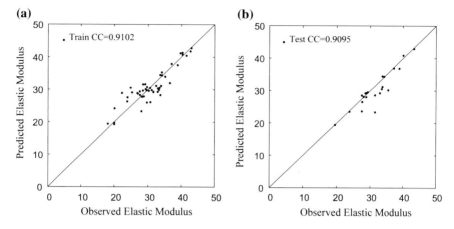

Fig. 4 Observed and predicted elastic modulus for EHO-ANN: **a** Train CC and **b** Test CC

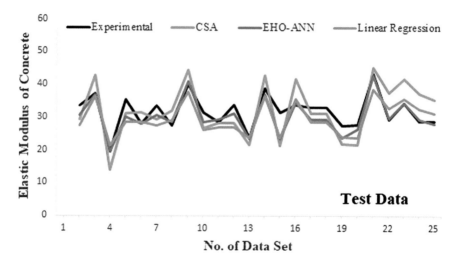

Fig. 5 Comparison of linear regression, CSA and EHO-ANN model with experimental data

4 Conclusions

- BPNN provided good results and proven to be a reasonable prediction technique in the past. But the performance fluctuates over a wide range on each trail and gives a poor performance at times, due to getting stuck in local optima. So, it is quite risky to be used in sensitive situations.
- Hybrid model of EHO-ANN performed better than a stand-alone model of BPNN. This appears to be highly influenced by the strong local searching potential of BP and strong global searching potential of EHO.

- EHO-ANN did not show many fluctuations in its performance over a large number of trails since BPNN was provided with predefined weights and biases obtained from EHO.
- Validating performance of various standard empirical formulas for predicting elastic modulus of concrete against soft computing models, it clearly shows the superiority of soft computing tools over traditional methods.
- The hybrid model of EHO-ANN can be utilized to obtain a reliable solution in the prediction of elastic modulus thereby making EHO-ANN as an alternative approach for prediction of the modulus of elasticity of concrete.
- Hence, soft computing techniques are more reliable to predict the elastic modulus of concrete compared to traditional methods used in the field of civil engineering.

References

1. Sarıdemir, M.: Effect of silica fume and ground pumice on compressive strength and modulus of elasticity of high strength concrete. Constr. Build. Mater. **49**, 484–489 (2013)
2. Mesbah, H.A., Lachemi, M., Aitcin, P.C.: Determination of elastic properties of high-performance concrete at early ages. Mater. J. **99**(1), 37–41 (2002)
3. Kocab, D., Kucharczykova, B., PetrMisak, P., Kralikova, M.: Development of the elastic modulus of concrete under different curing conditions. Proc. Eng. **195**, 96–101 (2017)
4. American Concrete Institute: Building code requirements for structural concrete (ACI 318M-95) and commentary (ACI318RM-95) (1995)
5. Standard, C.S.A.: A23. 3-04. Canadian Standard Association, 232 (2004)
6. Code, C.F.M.: Comite euro-international du beton. Bull. d'information **213**, 214 (1993)
7. Indian Standard, I.S.: 456: 2000. Plain and reinforced concrete code of practice (2000)
8. Turkish Standards Institute: Requirements for design and construction of reinforced concrete structures. TS500-2000 (2000)
9. De Normalisation, C.E.: Eurocode 2: design of concrete structures—part 1-1: general rules and rules for buildings. Belgium, Brussels (2004)
10. Demir, F.: Prediction of elastic modulus of normal and high strength concrete by artificial neural networks. Constr. Build. Mater. **22**(7), 1428–1435 (2008)
11. Topçu, İ.B., Bilir, T., Boğa, A.R.: Estimation of the modulus of elasticity of slag concrete by using composite material models. Constr. Build. Mater. **24**(5), 741–748 (2010)
12. Ibrahim, D.: An overview of soft computing. Proc. Comput. Sci. **102**, 34–38 (2016)
13. Demir, F., Korkmaz, K.A.: Prediction of lower and upper bounds of elastic modulus of high strength concrete. Constr. Build. Mater. **22**(7), 1385–1393 (2008)
14. Golafshani, E.M., Behnood, A.: Automatic regression methods for formulation of elastic modulus of recycled aggregate concrete. Appl. Soft Comput. **64**, 377–400 (2018)
15. Rinchon, J.P.M., Concha, N.C., Calilung, M.G.V.: Reinforced concrete ultimate bond strength model using hybrid neural network-genetic algorithm. In: IEEE 9th international conference on humanoid, nanotechnology, information technology, communication and control, environment and management (HNICEM), pp. 1–6 (2017)
16. Ahmadi-Nedushan, B.: Prediction of elastic modulus of normal and high strength concrete using ANFIS and optimal nonlinear regression models. Constr. Build. Mater. **36**, 665–673 (2012)
17. Wang, G.G., Deb, S., Coelho, L.D.S.: Elephant herding optimization. In: 3rd international symposium on december computational and business intelligence (ISCBI), pp. 1–5. IEEE (2015)

18. Sahlol, A.T., Ismail, F.H., Abdeldaim, A., Hassanien, A.E.: Elephant herd optimization with neural networks: a case study on acute lymphoblastic leukemia diagnosis. In: IEEE 12th international conference on computer engineering and systems (ICCES), pp. 657–662 (2017)
19. Meena, N.K., Parashar, S., Swarnkar, A., Gupta, N., Niazi, K.R.: Improved elephant herding optimization for multiobjective DER accommodation in distribution systems. IEEE Trans. Industr. Inf. **14**(3), 1029–1039 (2018)
20. Moretti, J.F., Minussi, C.R., Akasaki, J.L., Fioriti, C.F., Pinheiro Melges, J.L., Mitsuuchi Tashima, M.: Prediction of modulus of elasticity and compressive strength of concrete specimens by means of artificial neural networks. Acta Sci. Technol. **38**(1) (2016)
21. Mandal, S., Rao, S., Harish, N.: Damage level prediction of non-reshaped berm breakwater using ANN, SVM and ANFIS models. Int. J. Naval Arch. Ocean Eng. **4**(2), 112–122 (2012)
22. Levenberg, K.: A method for the solution of certain non-linear problems in least squares. Q. Appl. Math. **2**(2), 164–168 (1944)
23. Mavrovouniotis, M., Yang, S.: Training neural networks with ant colony optimization algorithms for pattern classification. Soft. Comput. **19**(6), 1511–1522 (2015)
24. Bal, L., Buyle-Bodin, F.: Artificial neural network for predicting drying shrinkage of concrete. Constr. Build. Mater. **38**, 248–254 (2013)
25. Dantas, A.T.A., Leite, M.B., de Jesus Nagahama, K.: Prediction of compressive strength of concrete containing construction and demolition waste using artificial neural networks. Constr. Build. Mater. **38**, 717–722 (2013)
26. Zhang, J.R., Zhang, J., Lok, T.M., Lyu, M.R.: A hybrid particle swarm optimization–backpropagation algorithm for feedforward neural network training. Appl. Math. Comput. **185**(2), 1026–1037 (2007)

Adaptive Sensor Ranking Based on Utility Using Logistic Regression

S. Sundar, Cyril Joe Baby, Anirudh Itagi and Siddharth Soni

Abstract Wireless Sensor Networks (WSN) consists of several tens to hundreds of nodes, interacting with each other. Thus, they have multiple communications between them, transferring and receiving several packets of data to each other. In order to reduce the overall traffic in the network and lessen the presence of redundant node data, this paper proposes an adaptive sensor ranking method by evaluating the task necessity, utility, and region coverage of a particular node in a given WSN. Logistic regression has been used to adaptively train the WSN to assign a status to node as on or off, thereby, decreasing the overall data transmission into the network, while still accounting for the entire range of the WSN.

Keywords Wireless sensor networks · Logistic regression · Adaptive sensor ranking · Coverage configuration protocol

1 Introduction

WSN refers to a large collection of cheap, low-power, and intelligent spatially placed or dispersed nodes (containing sensors) which are used for purposes like recording various physical parameters like temperature, humidity, sound, wind, etc. [1]. They rely on spontaneous signal formation and wireless connectivity and bear resemblance to Ad Hoc networks. These nodes often have a scarce and nonrenewable energy source to power themselves.

For applications like tracking and target detection, a Region of Interest (RoI) has to be defined for the coverage of an area. In such applications, a sufficient number of nodes of the network have to be activated for ensuring that that every point of the RoI is within a given maximal distance from at least one of the nodes that are active. Also, it is assumed that the number of nodes deployed is more than the number of

S. Sundar (✉) · C. J. Baby · A. Itagi · S. Soni
School of Electronics Engineering, Vellore Institute of Technology, Vellore, India
e-mail: Sundar.s@vit.ac.in

C. J. Baby
e-mail: cyrilbabyjoe@gmail.com

© Springer Nature Singapore Pte Ltd. 2020
K. N. Das et al. (eds.), *Soft Computing for Problem Solving*,
Advances in Intelligent Systems and Computing 1048,
https://doi.org/10.1007/978-981-15-0035-0_29

nodes that lie strictly in the RoI. These algorithms help the nodes conserve energy significantly and increase the lifetime of the network. Overhead communication for control and routing has to be minimal to reduce energy consumption and decrease packet traffic. Often the coverage configuration is restructured in certain intervals to rationalize the energy of the nodes and also account for any possible node failure.

The WSNs that are in use today have multiple communications or hops between them, transferring data among themselves and the receiving data from the gateway. These methods involve the usage of Coverage Configuration Protocol (CCP), with an adaptive sensor ranking. The CCP is the protocol that has to be used for communication between the different nodes of a WSN with one of the major requirements being that the sensing coverage quality is maintained even when the redundant sensors are inactive or turned off [2]. This means that the active nodes need to keep the degree of coverage without any kind of voids in sensing with a number of on-duty sensor that is minimal in number. Also, in CCPs, the judgment that whether the region that the node is sensing is fully covered by its neighbors is a key and is known as off-duty eligibility rule.

Adaptive sensor ranking refers to the process of allotting a rank to a sensor node accordingly depending upon the RoI and the portion of the RoI that the sensor has covered in order to turn off the redundant sensors and thus, helping in optimizing the energy usage. This concept of adaptive sensor ranking can be highly improvised by applying machine learning to it. Machine learning algorithms give the networks, the ability to learn by the means of large amount of relevant database gathered from the results obtained after the execution of the machine learning algorithm(s) several times and hence the results, after feature extraction, help the network to learn and perform a particular allotted task with high precision and optimization. In this paper, a logistic regression model is able to evaluate the necessity of a sensor node and lets the node decide whether it should join the network or withdraw from it depending on the results obtained. This would result in a drastic amount of optimization of energy consumption in the network as a whole and will increase the lifetime of the network to a whole new extent.

2 Literature Review

The area of coverage of a WSN is a critical factor that determines the usability of the WSN and the effectiveness of the WSN. Several applications demand a specific form of coverage of RoI. In order to ascertain such coverage, the nodes must organize their schedules, sensor activation intervals, data, and overhead signals. CCP may be distinct to a network, in accordance to its application, that is set up to maximize RoI coverage while minimizing overall node energy utilization by means of optimizing the sensor nodes necessity, task requirement, and usage of the sensor node in the WSN.

Nodes running CCP are assumed to wake up periodically and declare their availability using hello packets [3]. After waking up, a node enters a listen phase for a specified period, during which it obtains the status of the state of its neighbors, their presence, and position. The status of the neighbor nodes is obtained because of all the nodes in active state declare their availability by transmitting their hello packets. At the end of the listen phase, each node concludes whether its sensing range is covered by neighboring active nodes or not. Based on that, it evaluates itself to sleep mode if it is not required, by doing so, it enters withdraw state. Otherwise, it enters a join phase in which it persists to be a part of the network until a timer expires. Previous works have shown good promise with adaptive sensor ranking techniques wherein, the probability of a sensor nodes status as active was related to joining timer of the node [4]. The global rank of a node in the network was computed by considering the distance between the particular node and its neighboring nodes such that the presence of nearby nodes will make the rank decrease and absence of nearby nodes would give a high probability to make the node active, and effectively increase its rank. Inverse distance weights was the basis for this concept [5].

In recent works, minimum k-coverage algorithms have shown viable results in constraint-based coverage of RoIs by WSNs in certain applications. This is where each point in an RoI is covered by at least k number of sensor nodes. A few variants of the k-Coverage protocols are the Self-Scheduling driven k-Coverage (SSCk) protocol and the Triggered-Scheduling driven k-Coverage (TSCk) protocol [6]. The former protocol suggested that the activation of each sensor be calculated based on the local information it gathers on its sensing neighbors in order to k-Cover the RoI. The latter protocol allowed a node to trigger the necessary number of its neighboring nodes to achieve k-coverage.

To achieve the two-objective coverage problem of WSNs, differential evolution methods have been used [7]. They are powerful real parameter optimizers, which when utilized correctly can provide the optimal solution for most geometrically realizable coverage scenarios. Mobility Aware Reputation Node Ranking (MARNR) technique can be used to identify sensor node mobility rate and node reputation after clustering them effectively [8]. The ranking is assigned based on node reputation and its mobility rate.

A probabilistic approach for a CCP model has proven to be quite reliable and robust [9]. This protocol is distributed and local, allowing rapid reconfiguring of the network to meet its Quality of Service (QoS) requirements. It works on the foundation of the assumption that the detection ability of a sensor node is the decay function of the sensing range projected on the Neyman–Pearson probabilistic detection model. However, their model also worked on the assumption that the nodes are static and may not reconfigure correctly if any or all the nodes are mobile.

3 Proposed Approach

In this paper, the sensor ranking of a node is calculated to assign its status to join the WSN and contribute its sensor information to it or to withdraw from the network and sleep. The region where the sensor nodes information is required is called the RoI. The RoI is a subsector or a part of the WSN coverage area. The sensor ranking is predominantly determined by the following parameters:

1. Region of Interest Coverage: Ratio of the area of interest covered by the sensor node to the area of the coverage of the sensor node. It factors in the geographical area to be covered, dimensionally.
2. Sensor Utility: This factor indicates the importance of a sensor node and its requirement, if any, in the network. It defines how much area is left uncovered when the particular sensor node is removed from the network. It is the ratio of area left uncovered to the total area of the network.
3. Task Necessity: This factor is considered in a WSN with more than one layer. Each layer has different sensors; that is, each layer is required to perform a different sensing function. The task necessity factor is a binary variable that defines the necessity of a sensor node for a given sensing task in a given region of interest. That is, either the particular sensor functioning is required or not.

These parameters expresse the weights (importance) of the sensor, with these parameters obtained from each node, a machine learning algorithm known as logistic regression is used to train and predict if a particular node is going to join or withdraw from the network, while still not compromising the overall coverage of the WSN.

4 Sensor Ranking

The workspace is represented by a 100 by 100 graph along the horizontal (x) axis and vertical (y) axis. The sensor nodes are randomly dispersed by giving them arbitrary x and y coordinates. The range of a sensor node as a circle by defining its radius and the RoI is defined as any geometric shape by giving its vertices. Figure 1 illustrates a model workspace, with the green stars as the WSN nodes and the rectangular area as the RoI.

5 Logistic Regression

Logistic regression is one of the most popular and computationally efficient machine learning algorithms [9–11]. It works on the basis of hypothesis function optimization as shown in Eq. (1)

Fig. 1 Sample workspace
consisting of 12 nodes

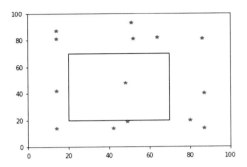

$$h = \text{sigmoid}(w_0 + w_1 x_1 + \ldots + w_n x_n) \tag{1}$$

where w_i are the weight attached to the features x_i ($i = 1, 2, 3, \ldots, n$) and where i is the number of features. The sigmoid function is the activation function defined by Eq. (2)

$$\text{sigmoid}(y) = \frac{1}{1 + e^{-y}} \tag{2}$$

The hypothesis function is optimized by minimization of the cost function as shown in Eq. (3)

$$\text{Cost} = \frac{-1}{n} \sum_{i=1}^{n} y_i \log(h(x_i)) + (1 - y_i)\log(1 - h(x_i)) \tag{3}$$

In this paper, the cost function has been focused by adding a regularization parameter λ as shown in Eq. (4).

$$C = \frac{-1}{n} \sum_{i=1}^{n} (y_i \log(h(x_i)) + (1 - y_i)\log(1 - h(x_i))) + \frac{\lambda}{2n} \sum_{i=1}^{n} (w_i)^2 \tag{4}$$

where y_i is the output of the hypothesis function when xi is the input. The weights get updated iteratively. The updation is as shown in Eq. (5) after differentiating the cost function updating the weights by a learning rate (α).

$$w_i := w_i - \alpha \left(\frac{\partial C}{\partial w_i} \right) \tag{5}$$

This weight updation repeats till C has reached its local minimum or preferably its global minimum. It is at this point that the hypothesis function is optimized and is said to give the most accurate prediction. In this paper, the training was for 1000 iterations with a learning rate of 0.01. The results of which are discussed in the following section. After the logistic regression is executed, the algorithm returns the sensor ranking of each sensor node as a probabilistic value of the nodes' requirement. Higher the value, more likely it is that that particular node is required.

6   Results and Discussions

The algorithm was implemented in software simulation and in a hardware implementation. In this paper, the workspace setup had the following software simulations:

1. Simulation 1, as seen in Fig. 2, had the following parameters:

   a. Area of the WSN: 100 m × 100 m
   b. Range of a sensor node: 25 m
   c. Types of sensors: 1
   d. Number of sensor nodes: 12
   e. RoI area: 2500 m^2

2. Simulation 2, as seen in Fig. 4, had the following parameters:

   a. Area of the WSN: 200 m × 100 m
   b. Range of a sensor node: 35 m
   c. Types of sensors: 1
   d. Number of sensor nodes: 25
   e. RoI area: 5625 m^2

3. Simulation 3, as seen in Fig. 6, had the following parameters:

   a. Area of the WSN: 200 m × 100 m
   b. Range of a sensor node: 35 m
   c. Types of sensors: 1
   d. Number of sensor nodes: 25
   e. RoI area: There is no specific RoI

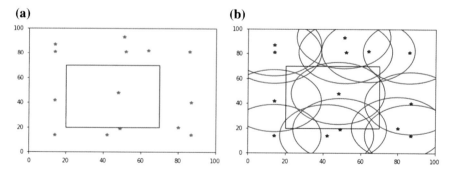

Fig. 2 a Workspace consisting of 12 sensor nodes, **b** same workspace displaying the nodes' coverage

4. Simulation 4, as seen in Fig. 6, had the following parameters:

   a. Area of the WSN: 200 m × 100 m
   b. Range of a sensor node: 35 m
   c. Types of sensors: 2
   d. Number of sensor nodes: 25
   e. RoI area: 5625 m^2

The nodes were randomly dispersed and their parameters were assigned in accordance to their position with respect to the RoI and the sensor's purpose in that RoI. Figures 2, 4, 6, and 8 illustrate the workspace defined and used in this paper for Simulations 1, 2, 3, and 4. The overall visible area is the entire coverage area of the WSN, the green stars are the sensor nodes with type-1 sensor function and the green circle-shaped nodes are having type-2 sensing function (in Fig. 6). The red circles illustrate the coverage range of each node and the rectangle is the RoI.

After logistic regression algorithm was executed, results of Simulation 1 are seen in Fig. 3, results of Simulation 2 are seen in Fig. 5, results of Simulation 3 are seen in Fig. 7, and results of Simulation 4 are seen in Fig. 9. These values are tabulated and compared in Table 1. In all the cases, the red-colored nodes withdraw from the network and enter sleep mode and the blue-colored nodes join the network (Figs. 4, 6, 8).

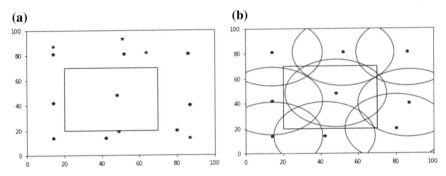

Fig. 3 Simulation 1 after running the algorithm: **a** blue-colored sensors join the network and red-colored sensors withdraw and sleep; **b** The workspace without the sensors that have withdrawn from the network

Table 1 Results of the simulations

N	R	A	T	RC	NA	E
12	25	100 × 100	1	Yes	9	25
25	35	200 × 200	1	Yes	17	32
25	35	200 × 200	1	No	16	36
25	35	200 × 200	2	No	18	25

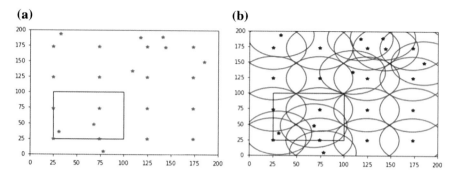

Fig. 4 Simulation 3 with (**a**) 25 nodes and (**b**) along with their coverage areas

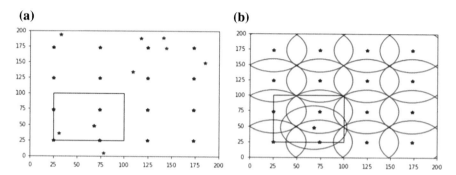

Fig. 5 Simulation 2 after running the algorithm: **a** blue-colored sensors join the network and red-colored sensors withdraw and sleep; **b** The workspace without the sensors that have withdrawn from the network, along with their coverage areas

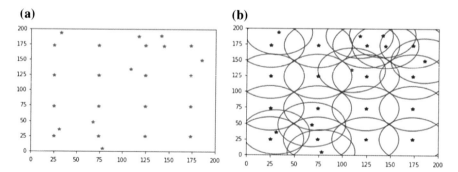

Fig. 6 Simulation 3 before running the algorithm: **a** with 25 node and no RoI, **b** along with their coverage ranges in red color

In Fig. 7 (Simulation 3), 9 out of 25 withdraw from the network and still manage to cover the entire coverage area with significant reduction in redundant data. In these three cases, the sensor utility of each node is 1, meaning that all the sensors are performing the same sensing function as desired within the RoI. In the fourth case, the sensor utilities of some nodes are 1 and others are 0. That is, some of the sensors are not required, irrespective of their location. And as seen in Fig. 9 (Simulation 4), 9 out of 25 nodes are made to withdraw by factoring in their utility as well.

Table 1 contains the details of all the software simulations, the number of nodes in each case, the parameters of the nodes as well. In Table 1, N = total number of nodes, R = radius of coverage of each node (in m), A = WSN coverage dimension (in m × m), T = types of sensors in the WSN, RC = ROI consideration status, NA = number of active nodes, E = energy saved (in %).

For the hardware implementation, the 4 sensor nodes were communicating using User Datagram Protocol (UDP) communication. UDP is a multicast supporting, transport layer protocol [12]. Although UDP has no congestion control methods, it is best suited for low-bandwidth networks such as the one established in this paper.

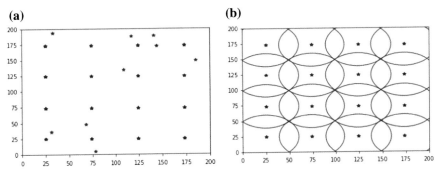

Fig. 7 Simulation 3 after running the algorithm: **a** blue-colored sensors join the network and red-colored sensors withdraw and sleep, **b** with only the nodes that have joined the network

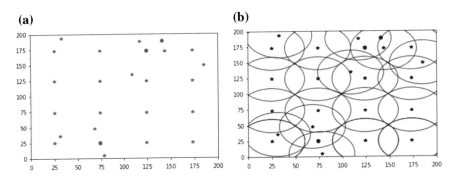

Fig. 8 Here, the workspace consists of (**a**) 25 nodes and there are two different types of sensor nodes, (**b**) each with their own coverage radius

The hardware implementation had the following configuration:

a. Area of the WSN: 100 m × 50 m
b. Range of a sensor node: 35 m
c. Number of sensor nodes: 4
d. RoI area: 400 m².

Figure 10a displays the setup in configuration 1 with the RoI as shown. In this configuration, after the algorithm was executed, the WSN decided to make node 3 and

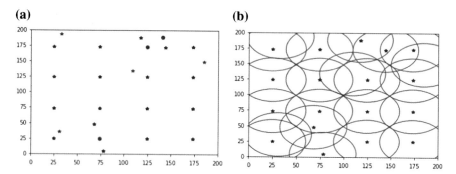

Fig. 9 Simulation 4 after running the algorithm: **a** blue-colored sensors join the network and red-colored sensors withdraw and sleep; **b** The workspace without the sensors that have withdrawn from the network, along with their coverage areas

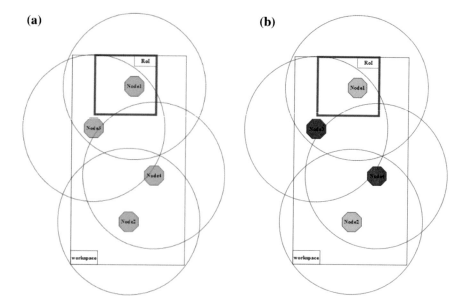

Fig. 10 a The workspace for the hardware implementation with first RoI, **b** after the algorithm is executed, the red-colored nodes withdraw from the network and the blue-colored sensors join

(a) **(b)**

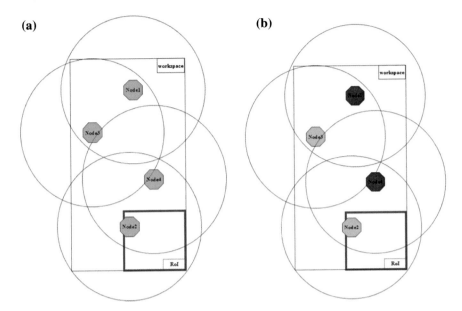

Fig. 11 **a** The workspace for the hardware implementation with first RoI, **b** after the algorithm is executed, the red color nodes withdraw from the network and the blue color sensors join

node 4 (the red-colored nodes) to withdraw from the network and make them sleep, as seen in Fig. 10b. By doing so, 50% of the energy consumption of the network goes down and so does the traffic overhead. Figures 10a and 11a show the configurations that were set up. The circles in red show the radius of coverage of the sensor nodes, the green hexagons are the sensor nodes themselves, the black rectangular box is the entire workspace, and overall WSN coverage area and the blue box illustrate the RoI.

Figure 11a displays the setup in configuration 2 with a second RoI as shown. In this configuration, after the algorithm was executed, the WSN decided to make node 1 and node 4 (the red-colored nodes) to withdraw from the network and make them sleep, as seen in Fig. 11b. By doing so, 50% of the energy consumption of the network goes down and so does the traffic overhead. In both the cases, the RoIs were adequately covered and overage coverage was maintained.

7   Conclusion

The proposed work is a way to optimize the sensor usage in a WSN. The proposed algorithm uses machine learning to determine the necessity of a sensor node in a WSN based on various parameters and region of interest . It can be seen that the algorithm optimizes the algorithm by determining the least number of sensor nodes to be active for 1-Coverage of given region of interest. The selection of sensors that has to be active is sensitive to changes in RoI.

The algorithm optimizes the WSN and reduces the overall data transmission and hence makes the network power efficient. This proposed algorithm can be used in low-power wireless systems where power optimization of the network is a primary concern. This algorithm can also be used in WSNs with high expected lifetime. It can also be used in wireless networks to reduce the transmission traffic in the network by enabling only the necessary transmissions.

References

1. Prathap, U., Shenoy, P.D., Venugopal, K.R., Patnaik, L.M.: Wireless sensor networks applications and routing protocols: survey and research challenges. In: 2012 International symposium on cloud and services computing (ISCOS), pp. 49–56. IEEE (2012)
2. Ammari, Habib M., Das, S.K.: Centralized and clustered k-coverage protocols for wireless sensor networks. IEEE Trans. Comput. 61(1), 118–133 (2012)
3. Rajaboina, R., Reddy, P.C., Kumar, R.A., Venkatramana, N.: Performance comparison of TCP, UDP and TFRC in static wireless environment. Int. J. Inf. Comput. Secur. 8(2), 158–180 (2016)
4. Santini, S.: Using adaptive sensor ranking to reduce the overhead of the coverage configuration protocol. In: 2010 Sixth international conference on intelligent sensors, sensor networks and information processing (ISSNIP), pp. 109–114. IEEE (2010)
5. Shepard, D.: A two-dimensional interpolation function for irregularly-spaced data. In: Proceedings of the 23rd ACM national conference/annual meeting, August 1968, pp. 517–524 (1968)
6. Ammari, H.M., Das, S.K.: On the design of k-covered wireless sensor networks: self-versus triggered sensor scheduling. In: IEEE International Symposium on World of Wireless, Mobile and Multimedia Networks and Workshops, 2009. WoWMoM 2009, pp. 1–9 (2009)
7. Xu, Y., Xiaohui, W., Han, Z.: Improved differential evolution to solve the two-objective coverage problem of wireless sensor networks. In: 2016 Chinese on control and decision conference (CCDC), pp. 2379–2384. IEEE (2016)
8. Jai, B.P., Anandamurugan, S.: Mobility aware reputation node ranking (MARNR) for efficient clustering at hot spots in wireless sensor networks (WSN). In: 2013 international conference on information communication and embedded systems (ICICES), pp. 312–317. IEEE (2013)
9. Ying, T., Yang, O.: Probabilistic based k-coverage configuration in large-scale sensor networks. In: 2010 2nd international conference on information science and engineering (ICISE), pp. 2257–2260. IEEE (2010)
10. Anjuman, P., Khullar, V. Sentiment classification on Big Data using Naïve Bayes and logistic regression. In: International conference on computer communication and informatics (ICCCI-2017), pp. 1–5. IEEE (2017)
11. Hu, Q., Shi, Z., Zhang, X., Gao, Z.: Fast HEVC intra mode decision based on logistic regression classification. In: 2016 IEEE international symposium on broadband multimedia systems and broadcasting (BMSB), pp. 1–4. IEEE (2016)
12. Cheng, W., Hüllermeier, Eyke: Combining instance-based learning and logistic regression for multilabel classification. Mach. Learn. 76(2–3), 211–225 (2009)

Detection of Dementia from Brain Tissues Variation in MR Images Using Minimum Cross-Entropy Based Crow Search Algorithm and Structure Tensor Features

N. Ahana Priyanka and G. Kavitha

Abstract Dementia causes cognitive dysfunction and deterioration of brain. Alzheimer Disease (AD) and Mild Cognitive Impairment (MCI) are most common forms of dementia. Globally, it is estimated that about 47 million people are affected by dementia. Various researches suggest that AD and MCI share a number of equally severe cognitive deficits, but the pathophysiology has not yet been addressed in a comprehensive way. An attempt is made to observe the prognosis difference in these disorders and to analyze the tissue variation in T1-weighted MR brain images. Samples used in this analysis are obtained from IXI, MIRIAD, and ADNI 2 database. Initially, skull stripping is carried out using Robust Brain Extraction Tool (ROBEX), Brain Extraction Tool (BET), and Brain Surface Extractor (BSE). Further, segmentation of brain tissues is performed using multilevel minimum cross-entropy based Bacteria Foraging Algorithm (BFO) and Crow Search Algorithm (CSA). Various geometric features and Structure Tensor (ST) features are extracted from White Matter (WM), Gray Matter (GM), and Cerebrospinal Fluid (CSF) for normal, MCI, and AD to observe the structural changes. The result shows that ROBEX performs better delineation of brain. Minimum cross-entropy based CSA achieves better segmentation than BFO based on similarity measures and computation time. Further, ST features extracted from the brain tissues are able to show anatomical variation effectively than geometric features. It is identified from the ANOVA test that structure tensor features of GM shows better variation to discriminate normal, MCI, and AD images. Hence, this framework could be used to differentiate normal, MCI, and AD images such as cognitive disorders effectively.

Keywords Dementia · Bacteria foraging algorithm · Crow search algorithm · Structure tensor

N. Ahana Priyanka (✉) · G. Kavitha
Department of Electronics Engineering, Madras Institute of Technology, Chromepet, Chennai 600044, India
e-mail: ahanachellian@gmail.com

G. Kavitha
e-mail: kavithag_mit@annauniv.edu

© Springer Nature Singapore Pte Ltd. 2020
K. N. Das et al. (eds.), *Soft Computing for Problem Solving*,
Advances in Intelligent Systems and Computing 1048,
https://doi.org/10.1007/978-981-15-0035-0_30

377

1 Introduction

Dementia is a progressive chronic impairment of cognitive ability. The progression of dementia varies from person to person. The current diagnosis of dementia is a combination of clinical criteria which includes neurological examination, mental status test, and image-based analysis. The structural changes in brain due to dementia are difficult to identify by external assessment. Magnetic Resonance (MR) images provide information about brain substructures and its surrounding tissues [1]. The main technical challenge in neuroimaging-based analysis is to differentiate demented subject from normal based on atrophy interventions. The biomarker of dementia generally includes cerebral atrophy and ventricular enlargement. Majority of studies show that changes in brain tissues lead to variation in medial temporal lobe and parietal areas. Hence, there is a need to observe the microstructural changes in demented subject to identify its prognosis.

Brain tissue segmentation generates valid information about brain tissues. The brain tissues are bounded together and are mostly complex to analyze [2]. Various automated and semiautomatic segmentation methods such as thresholding, clustering, texture, and region-based methods [3] have been used in the literature for brain tissue segmentation. Among them, thresholding is commonly used. Thresholding emphasizes the key feature from the image which gives possible threshold level to perform segmentation. The entropy can provide optimized threshold information about the distribution of gray levels and appropriate partition of the target image. However, computational complexity increases for multilevel thresholds. Metaheuristic algorithms could be used to overcome these difficulties. These algorithms are based on behavioral nature of an organism to find an optimum threshold.

In this analysis, an attempt has been made to study the brain tissue (GM, WM, and CSF) changes in normal, MCI, and AD using metaheuristic algorithms such as Bacterial Foraging Optimization (BFO) and Crow Search Algorithm (CSA). BFO and CSA are inspired from food foraging behavior of animals. Bacteria are able to predict the nutrient from dynamical behavior of movement. However, in CSA, behavior is based on the food source and memory of theft position. It avoids premature convergence and critical flaws which results in a better performance. The effectiveness of these social algorithms relies on search space, diversification, and intensification. These social algorithms are used to obtain multilevel thresholds for segmentation. Here, minimum cross-entropy is used as a fitness function to provide near-optimal threshold to segment the brain tissues. The segmentation results are validated with similarity measures. The segmented images are used to obtain geometric and structure tensor feature that can differentiate normal, MCI, and AD. This framework helps to diagnose the progression of dementia disorders.

2 Methodology

This section briefs about the database, skull stripping process, CSA, and BFO methods. The complete flow of the proposed work is shown in Fig. 1.

Initially, MR brain images are subjected to preprocessing. In preprocessing, ROBEX tool is used for skull stripping. After delineation of skull, multilevel threshold-based segmentation is carried out. The multilevel threshold is obtained by minimum cross-entropy based BFO and CSA methods. The brain tissues are segmented into WM, GM, and CSF using the optimum procedures. The segmentation results based on the optimization algorithms are compared using similarity measures. The features such as geometric and structure tensor are extracted from the segmented regions. These features are used to analyze the changes in brain tissue for normal, MCI, and AD.

2.1 Database

The T1-weighted axial view Magnetic Resonance (MR) brain images used in this work includes normal, MCI, and AD subjects. The normal images are obtained from Information Extraction from Images (IXI) database, Minimal Interval Resonance Imaging (MIRIAD) database, and Alzheimer's Disease Neuroimaging Initiative (ADNI 2) database are used to extract AD and MCI subjects, respectively. This work involves both male and female subjects.

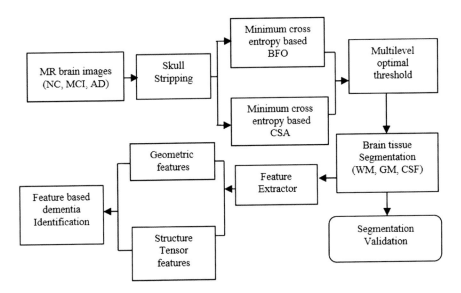

Fig. 1 Flow diagram of the proposed work

2.2 Skull Stripping

The images are preprocessed using skull stripping to remove the non-brain tissues. The extraction of brain tissue requires brain extraction algorithms due to the closeness of pixel intensities with the adjacent regions. Robust Learning-Based Brain Extraction (ROBEX) is a tool for brain extraction [4]. This tool is based on the combination of discriminative and generative model. It uses random forest method to detect voxels that are located on the brain boundary. Random walker and adaptive thresholding algorithm are used to prevent leakage and detect the brain boundary at resection cavities. The skull stripped image is subjected to optimized threshold-based segmentation of brain tissues.

2.3 Segmentation Using Multilevel Optimization Algorithm

The minimum cross-entropy thresholding algorithm selects the threshold by minimizing the cross-entropy between the original image and its thresholded version [5]. The optimized multilevel threshold is obtained with the help of modeling behavior of crow and bacteria. In this context, minimum cross-entropy is used as objective function to search an optimal threshold to segment the brain tissues.

The Bacterial Foraging Optimization (BFO) [6] relies on the selection mechanism of nutrient in varying population. The BFO is a non-gradient optimization method that follows the process such as chemotaxis, reproduction, and elimination. The bacterial population is randomly initialized to different positions to search a threshold. The bacterial movement of swarming and tumble take place in chemotaxis. Each bacterium in the population is assigned a fitness value by chemotaxis. Each time, fitness solutions are updated to get a better threshold than the previous. According to the reproduction step, only the bacteria with higher fitness value can survive and population is reinitialized over the entire search space. Elimination and dispersal are responsible for discard and generation of bacteria.

Crow search algorithm is very popular due to its search of food and memory behavior [7]. The parameters considered for CSA are total population, Awareness Probability (AP), and Flight Length (Fl). The position and memory of the crow are initialized to get a threshold. The total number of crows is represented as N which can be considered randomly. In this work, N is considered as (20). The search for the food takes place in n dimension. The fitness function is evaluated by the determination of quality of position. The feasibility of position is verified by means of AP (0.1) which ranges from 0 to 1. If the value of AP increases, the algorithm fails to catch the global optima and gets trapped in local optima. If the fitness of the position is not feasible, the new position is found. If the new position is feasible, the memory is updated else the same position is remembered. Fl (2) of the crow is used to obtain a global optimum solution. The termination criterion is based on the best position of the memory in terms of solution to the optimization problem.

2.4 Validation

In this work, seven similarity measures are used to quantify the similarity between segmentation and ground truth which are represented in Table 1. The ground truth images are obtained from experts. The similarity measures [8] such as Dice Coefficient (DC), Jaccard Coefficient (JC), Hawkins and Dotson (HD), Sokal and Sneath (SS), Hamann(HN), Baroni-Urbani and Buser (BB), and Rogers and Tanimoto (RT) are used. These measures are used to validate the degree of correspondence between ground truth images ($P_{i,j}$) and segmented images ($Q_{i,j}$). The contingency values used for validate the segmentation are a = number of times $P_{i,j} = 1$ and $Q_{i,j} = 1$, b = number of times $P_{i,j} = 0$ and $Q_{i,j} = 1$, c = number of times $P_{i,j} = 1$ and $Q_{i,j} = 0$, and d = number of times $P_{i,j} = 0$ and $Q_{i,j} = 0$. Here, 1 and 0 represent the presence and absence of pixels.

2.5 Feature Extraction

The geometric measures are used to quantify anatomical variation within a class of images to understand the disease progression [9]. These geometric features include area, perimeter, convex area, solidity, eccentricity, equidiameter, and form factor. Area indicates the actual number of pixel in the segmented region. Perimeter calculates the distance around the boundary of the segmented region. The convex area gives the number of pixels in the convex image. Solidity can be calculated as the ratio of area to convex area. Eccentricity is the measure of aspect ratio. The ratio between the major axis length and minor axis length represents the eccentricity. Equidiameter specifies the diameter of a circle with the same area as the region. Form factor gives the ratio of the area of an image to the area of a circle with the same perimeter as the image

Structure tensor features are used to detect the variation [10] in cortical and subcortical regions of normal, MCI, and AD. The Jacobian matrix formed by the gradient of brain images is used to calculate the structure tensor. It contains the partial derivatives of the images to observe the variation in an image. Tensor help to determine the structures in the regions by identifying its edges, lines, and flow. Orientation (θ) corresponds to the direction of largest Eigenvector of the tensor matrix. Coherence (Ch) measure is used to relate the length of orientation vector to the length of gradient vector. The coherence value varies from 0 to 1. Anisotropy index measure gives a relation between the lengths of the orientation vector and the length of the gradient vector. The values of anisotropy measure vary from 0 to 1 where 1 indicates highly oriented structures.

Table 1 Similarity measures formulas

Dice coefficient	Jaccard coefficient	Hawkins and Dotson	Sokal and Sneath	Hamann measure	Baroni-Urbani and Buser	Rogers and Tanimoto
$\frac{a}{a+b+c}$	$\frac{2*a}{(2*a+b+c)}$	$\frac{1}{2}\left(\frac{a}{a+b+c}+\frac{d}{b+c+d}\right)$	$\frac{a}{(a+2*(b+c))}$	$\frac{(a+d)-(b+c)}{(a+b+c+d)}$	$\frac{\sqrt{ad}+a}{\sqrt{ad}+a+b+c}$	$\frac{a+d}{a+d+2(b+c)}$

2.6 Statistical Analysis

Tissue difference between normal, MCI, and AD are examined by Analysis of Variance (ANOVA) test for geometric features and structure tensor features. Tests are performed with significant level of $p < 0.05$. This model [11] helps to identify the significant interaction between brain region and disease status.

3 Results

In this work, T1-weighted images in an axial sliced section are considered. Totally 120 images are considered in this analysis that includes normal (40), MCI (40), and AD (40) subjects.

Skull stripping eliminates the non-brain tissues from MR brain images [12]. Here, the brain tissues are delineated using BET, BSE, and ROBEX Tools are represented in Fig. 2. The degree of similarity between the ground truth and the segmented image is quantitatively evaluated using various similarity measures. The skull striping validation for BSE, BET, and ROBEX are given in Table 2.

Dice coefficient is higher for ROBEX in all the considered images compared to BET and BSE. ROBEX results in higher sensitivity for normal and MCI than other tools. The specificity and precision are better for ROBEX in all the considered subjects. From the above table, it is observed that ROBEX produces higher accuracy in AD (0.833), MCI (0.867), and normal (0.887) than other two methods. Thus, a higher value of similarity measures for ROBEX indicates better delineation of the brain tissues. The segmentation result of BET and BSE are not used for further processing.

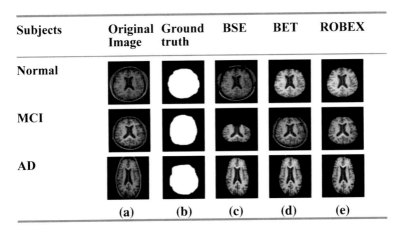

Subjects	Original Image	Ground truth	BSE	BET	ROBEX
Normal					
MCI					
AD					
	(a)	(b)	(c)	(d)	(e)

Fig. 2 Skull stripping results: **a** T1-weighted image, **b** Ground truth, **c** BSE, **d** BET, and **e** ROBEX

Table 2 Normalized similarity and performance measures for skull stripping

	Method	Dice	Sensitivity	Specificity	Precision	Accuracy
Normal	BET	0.854	0.731	0.9918	0.9835	0.885
	BSE	0.593	0.396	0.9552	0.8571	0.727
	ROBEX	0.859	0.745	0.9869	0.9742	0.887
MCI	BET	0.333	0.244	0.9991	0.9983	0.766
	BSE	0.731	0.551	0.9953	0.9830	0.852
	ROBEX	0.766	0.591	0.9993	0.9972	0.867
AD	BET	0.745	0.566	0.998	0.996	0.832
	BSE	0.745	0.566	0.998	0.996	0.832
	ROBEX	0.746	0.567	0.999	0.997	0.833

In addition, the quality of segmented region is evaluated with their size and shape. The geometric measures are calculated for the extracted whole brain using ROBEX is given in Table 3. Various studies suggest that there exist a prominent change in MCI due to structural nonuniform variation in brain that increases its area and perimeter. Form factor helps to identify the irregularity of shape in an image. It is evident from the above table that AD and MCI subjects have a minimum form factor when compared to normal. The form factor variation is high for normal (0.526) when compared to MCI (0.425) and AD (0.447). Solidity is measured as the proportion of area under the convex hull to area of the object. AD shows higher solidity due to its shape variation. From the above results, it is evident that ROBEX provides better delineation of brain for normal, MCI, and AD.

In this work, minimum cross-entropy based BFO and CSA algorithm are adopted to segment the tissue with optimized threshold values. These population-based algorithms search the optimal thresholds in its search space to segment the brain tissues. In Fig. 3, the results of tissue segmentation are represented. It is visualized that minimum cross-entropy based CSA outperforms in brain tissue segmentation than BFO. It is evident that BFO was not able to segment white matter clearly.

Figure 4 shows the segmentation accuracy for minimum cross-entropy based BFO and CSA. It is observed that minimum cross-entropy based BFO gives average accuracy of 88.6% for WM, 85% for GM, and 88.5% for CSF. Similarly, minimum cross-entropy based CSA resulted with accuracy of 90% for WM, 86% for GM, and 88.7% for CSF images. Here, accuracy is used to indicate the exact number of pixels that are correctly identified with respect to ground truth. The segmentation

Table 3 Normalized geometric measures for whole brain using ROBEX

Subjects	Area	Perimeter	Form factor	Solidity
Normal	0.626	0.441	0.526	0.313
MCI	0.624	0.572	0.425	0.277
AD	0.615	0.428	0.447	0.393

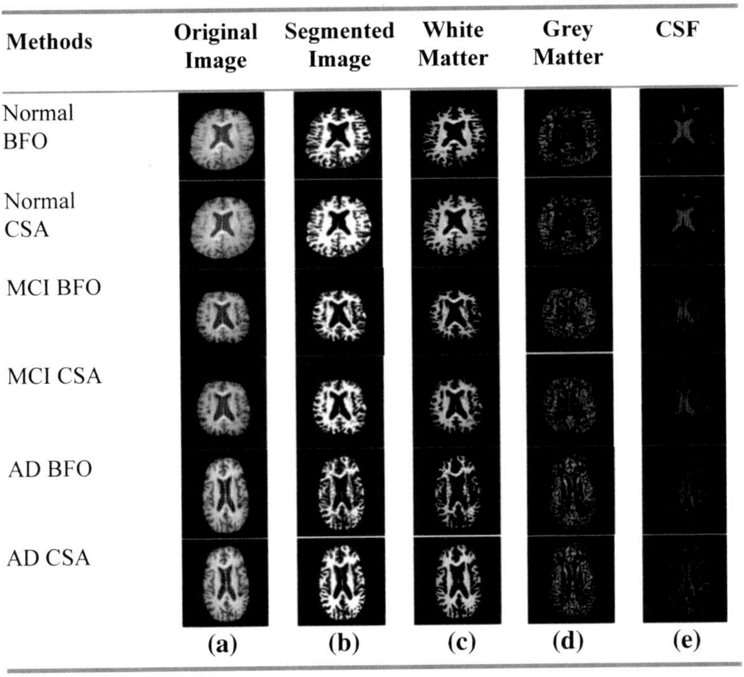

Methods	Original Image	Segmented Image	White Matter	Grey Matter	CSF
Normal BFO					
Normal CSA					
MCI BFO					
MCI CSA					
AD BFO					
AD CSA					
	(a)	(b)	(c)	(d)	(e)

Fig. 3 Segmentation result of minimum cross-entropy based BFO and CSA for Normal, MCI, and AD images: **a** original images, **b** segmented image, **c** white matter, **d** gray matter, and **e** CSF

Fig. 4 Accuracy of minimum cross-entropy based BFO and CSA

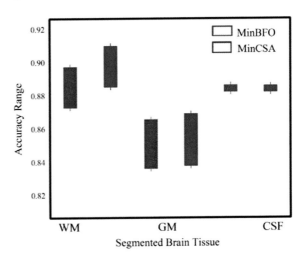

result shows that minimum cross-entropy based CSA gives better performance than BFO. Algorithms based on population method are complex with several elements and stochastic operations. The efficiency of these algorithms is observed by different parameters such as computation time and accuracy. The computation time is evaluated for each algorithm for normal, MCI, and AD images. The computation time variation for minimum cross-entropy based BFO and CSA is represented in Fig. 5a–c.

The graph indicates that the computation performance for minimum cross-entropy based BFO and CSA with respect to number of images. It is found that CSA requires less computation time than BFO. This is due to slow convergence rate in BFO. The computation time is higher for CSA for normal, MCI, and AD images when compared to BFO.

Further, the segmentation performance of entropy-based multilevel thresholding is analyzed using different similarity measures. Figure 6a–c shows the average values of various similarity measures obtained for normal, AD, and MCI subjects. The figure shows average similarity measures for WM, GM, and CSF. The JC is higher in AD (0.59) images compared to normal and MCI images. The DC is similar for normal and MCI images and slightly higher in AD (0.74) images. The HD is comparatively uniform for normal, MCI, and AD. A higher SS is obtained for MCI

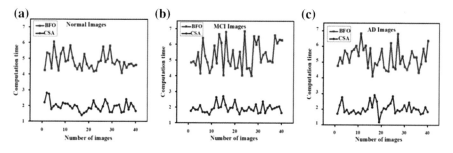

Fig. 5 Computation time for minimum cross-entropy based BFO and CSA: **a** Normal, **b** MCI, and **c** AD images

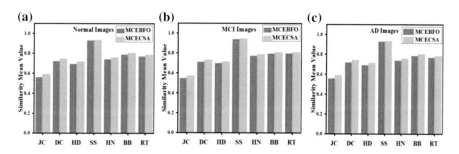

Fig. 6 Similarity measures for minimum cross-entropy based BFO and CSA: **a** Normal, **b** MCI, and **c** AD images

Table 4 Geometric features for various brain tissues

Tissues	Normalized value						
	Subjects	Area	Perimeter	Convex area	Solidity	Eccentricity	Equidiameter
WM	Normal	0.739	0.735	0.793	0.458	0.378	0.829
	MCI	0.463	0.504	0.535	0.366	0.447	0.616
	AD	0.619	0.637	0.682	0.502	0.809	0.723
GM	Normal	0.351	0.350	0.486	0.504	0.625	0.476
	MCI	0.279	0.397	0.482	0.317	0.624	0.457
	AD	0.138	0.185	0.168	0.574	0.873	0.254
CSF	Normal	0.072	0.250	0.129	0.379	0.821	0.121
	MCI	0.052	0.283	0.122	0.348	0.796	0.113
	AD	0.064	0.219	0.115	0.354	0.784	0.119

images. Similarly, the HN (0.79), BB (0.86), and RT (0.81) for MCI images. The values for all the similarity measures are lower in minimum cross-entropy based BFO than CSA. Overall better segmentation is observed in WM and GM compared to CSF. Therefore, minimum cross-entropy based CSA provides better segmentation and more accurate representation of brain tissues. It can be observed that there exists a significant improvement in segmentation result and computation time for minimum cross-entropy CSA based methods when compared to BFO. In order to further analyze the structural and texture variations in normal, MCI, and AD subjects, the geometric and structure tensor features are extracted from CSA segmented outcomes. Geometric measures could recognize the shape variation in normal, MCI, and AD images.

Anatomic variations are observed by various geometric features as shown in Table 4. Area, perimeter, convex area, and equivalent diameter of all brain tissues are lower for WM and CSF in AD and MCI. GM reduction is observed in AD images. This is due to the accumulation of extracellular amyloid plaque and intracellular neurofibrillary tangle in GM. WM damage is recognized as second causative factor for development of AD. It is observed that MCI subjects have lower WM area when compared to normal and AD. This is due to degeneration of WM in cortex region. The obtained results show that the solidity and eccentricity of WM and GM in AD subjects are higher than MCI and normal. Higher solidity indicates more irregular shape. The progressive decline of tissues shows higher eccentricity. From various studies, it has been observed that the AD pathology is highly correlated with WM and GM. In addition, CSF variation is also observed between normal, MCI, and AD subjects. From the studies, it is evident that WM and GM variations highly exist in MCI and AD [13]. This variation is due to the faster development of brain atrophy. These changes are highly influential in AD and MCI subjects for the development of dementia. Thus, it is observed that the average values of all these features better discriminate normal, MCI, and AD. The variation in geometric features is very less for normal and MCI. In order to better differentiate normal and MCI subjects, the texture pattern of brain tissue is further analyzed using structure tensor features.

Structure tensor is used to identify the texture variations between normal, MCI, and AD. The structure tensor feature includes coherence, orientation, and anisotropy index. The features are extracted and the average values are represented in Table 5.

It is observed that coherence, orientation, and anisotropy index of WM and GM are significantly low for MCI than normal. The orientation field gives information about changes in direction within a spatial region. This is due to the structural change during disease progression. Comparatively, these features are higher for WM and GM in AD than MCI. The coherence and orientation of CSF are higher in MCI than normal and AD. This is due to increase in level of tau in MCI subjects [14]. However, these features are slightly low in AD than MCI. Anisotropy index of AD is higher than normal and MCI. The texture pattern changes in WM and GM are prominent indicators of variation between MCI and AD. An ANOVA result was carried out for geometric and structure tensor features for WM, GM, and CSF. These results are represented in Tables 6 and 7.

The result shows that there exist a prominent variation in geometric and structure tensor features extracted from normal, MCI, and AD. The p-value extracted from the geometric measures indicates that there exists no prominent significance to observe the variation between the controls. The structure tensor feature of GM ($p < 0.00001$) is able to show the highest significant difference between normal, MCI, and AD subjects when compared to WM and CSF. Although structure tensor features show

Table 5 Structure tensor features form segmented brain tissues

Tissues	Normalized value			
	Subjects	Coherence	Orientation	Anisotropy Index
WM	Normal	0.414	0.145	0.385
	MCI	0.328	0.114	0.301
	AD	0.399	0.137	0.368
GM	Normal	0.441	0.147	0.405
	MCI	0.359	0.121	0.331
	AD	0.429	0.142	0.397
CSF	Normal	0.861	0.470	0.322
	MCI	0.937	0.540	0.317
	AD	0.935	0.511	0.373

Table 6 ANOVA test result for WM, GM, and CSF geometric measures

Tissue	Area	Perimeter	Convex area	Solidity	Eccentricity	Equidiameter
WM	0.0753	0.0273	0.0171	0.3015	0.0001	0.0420
GM	0.0003	0.0914	0.0021	0.2153	0.0002	0.0035
CSF	0.0002	0.0322	0.0406	0.7736	0.1021	0.0017

Table 7 ANOVA test result for WM, GM, and CSF structure tensor features

Tissue	Coherence	Orientation	Anisotropy index
WM	0.0016	0.0021	0.0043
GM*	**0.0001**	**0.0001**	**0.0001**
CSF	0.0001	0.0032	0.0001

*Indicate the significant variation

a variation in WM and CSF, the difference was not significant. The result indicates that structure tensor features provide higher significant value for brain tissues when compared to geometric features.

4 Conclusion

In this work, minimum cross-entropy based BFO and CSA based optimized multilevel segmentation is used to segment the brain tissue for normal, MCI, and AD images to analyze its variation. ROBEX method is able to skull strip the MR brain image better than BET and BSE based on similarity and performance measures. It is evident from the obtained similarity measure and computation time that CSA outperforms BFO. It is observed that CSA can efficiently describe the topological variations in brain tissue such as WM, GM, and CSF for normal, MCI, and AD images. The geometric features derived from the skull stripped whole-brain images shows a variation between the normal, MCI, and AD images. The geometric and structure tensor features extracted from the segmented tissue are able to quantify the individual tissue changes in normal, MCI, and AD subjects. Structure tensor features well discriminate normal, MCI, and AD subjects. It is observed that WM and GM variation is highly correlated with AD subjects. CSF variation is shown for MCI. Further, the decline of WM and GM in MCI indicates the higher progression of severity when compared to AD images. It is able to observe that there exists a significant GM variation for AD subjects. This analysis indicates better variation which is observed between normal vs MCI and normal vs AD. Hence, this framework could be used by the physicians for better preliminary diagnosis of severity in demented subjects.

References

1. Valkanova, V., Ebmeier, K.P.: Review on neuroimaging in dementia. Maturitas **79**, 202–208 (2014)
2. Guo, P.: Brain tissue classification method for clinical decision-support systems. Eng. Appl. Artif. Intell. **64**, 232–241 (2017)
3. Valverdea, S., Oliver, A., Rouraa, E., González-Villàa, S., Pareto, D., Vilanovac, J.C., Ramió-Torrentà, L., Rovira, À., Lladóa, X.: Automated tissue segmentation of MR brain images in the presence of white matter lesions. Medical Image Anal. **35**, 446–457 (2017)

4. Iglesias, J.E., Liu, C.-Y., Thompson, P.M., Zhuowen, T.: Robust brain extraction across datasets and comparison with publicly available methods. IEEE Trans. Med. Imaging **30**(9), 1617–1634 (2011)
5. Tang, K., Yuan, X., Sun, T., Yang, J., Gao, S.: An improved scheme for minimum cross entropy threshold selection based on genetic algorithm. Knowledge-Based Syst. **24**, 1131–1138 (2011)
6. Kora, P., Kalva, S.R.K.: Detection of bundle branch block using adaptive bacterial foraging optimization and neural network. Egypt. Inform. J. **18**, 67–74 (2017)
7. Oliva, D., Hinojosa, S., Cuevas, E., Pajares, G., Avalos, O., Gálvez, J.: Cross entropy based thresholding for magnetic resonance brain images using crow search algorithm. Expert Syst. Appl. **79**, 174–180 (2017)
8. Rácz, A., Andrić, F., BajuszKároly Héberger, D.: Binary similarity measures for fingerprint analysis of qualitative metabolomic profiles. Metabolomics **14**(29) (2018)
9. Anandh, K.R., Sujatha, C.M., Ramakrishnan, S.: A method to differentiate mild cognitive impairment and Alzheimer in MR images using eigen value descriptors. J. Med. Syst. **40**(1), 1–25 (2016)
10. Archana, M., Ramakrishnan, S.: Detection of Alzheimer disease in MR images using structure tensor. In: Annual international conference of the IEEE engineering in medicine and biology society, pp. 1043–1046 (2014)
11. Jensena, H.L.B., Lillenes, M.S., Rabano, A., Günther, C.-C., Riaz, T., Kalayou, S.T., Ulstein, I.D., Bøhmer, T., Tønjum, T.: Expression of nucleotide excision repair in Alzheimer's disease is higher in brain tissue than in blood. Neurosci. Lett. **672**, 53–58 (2018)
12. Kalavathi, P., Surya Prasath, V.B.: Methods on skull stripping of MRI head scan images a review. J. Digit. Imaging **29**(3), 365–379 (2016)
13. Dicks, E., Tijms, B.M., ten Kate, M., Gouw, A.A., Benedictus, M.R., Teunissen, C.E., Barkhof, F., Scheltens, P., van der Flier, W.M.: Gray matter network measures are associated with cognitive decline in mild cognitive impairment. Neurobiol. Aging **61**, 198–206 (2018)
14. Amlien, I.K., Fjell, A.M., Kristine, B., et al.: Mild cognitive Impairment: cerebrospinal fluid tau biomarker pathologic levels and longitudinal changes in white matter integrity. Radiology **266**(1), 295–303 (2013)

A Hybrid Approach for Intrusion Detection System

Neelam Hariyale, Manjari Singh Rathore, Ritu Prasad and Praneet Saurabh

Abstract Intrusion detection concept plays a vital role in personal computer (PC) security design. Intrusion detection system (IDS) essentially fits in as a unit which monitors the user activities, incoming traffic, and then distinguishes or classifies which one is intrusion and which one is normal or legitimate. Fundamentally IDS recognizes any misuse or unauthorized access which is essentially an attack to the crucial assets of the system or network. It detects the malicious traffic data on a network or a host. The task of desired classification often gets affected by the presence of noisy, redundant, and irrelevant data input. Occurrence of noise in dataset leads to poor classification as it increases the possibility of wrong detection of a class. High misclassification rate and low detection rate by the classifier in IDS enlarges feature space. To overcome this limitation, hybrid intrusion system (H-IDS) is proposed in this paper. H-IDS uses a hybrid strategy with support vector machine (SVM) and intelligent water drops (IWD) to execute the feature selection and classification techniques for IDS. Experimentations reveal that proposed H-IDS helps to accomplish the goal by attaining high classification, detection, and precision as compared to current state of the art.

Keywords IDS · Intrusion · IWD · Personal computer · SVM

N. Hariyale (✉) · M. S. Rathore · R. Prasad
Technocrats Institute of Technology (Advance), Bhopal 462021, MP, India
e-mail: neelamhariyale080@gmail.com

M. S. Rathore
e-mail: manjarisinghrathore@gmail.com

R. Prasad
e-mail: rit7ndm@gmail.com

P. Saurabh
Technocrats Institute of Technology, Bhopal 462021, MP, India
e-mail: praneetsaurabh@gmail.com

© Springer Nature Singapore Pte Ltd. 2020
K. N. Das et al. (eds.), *Soft Computing for Problem Solving*,
Advances in Intelligent Systems and Computing 1048,
https://doi.org/10.1007/978-981-15-0035-0_31

391

1   Introduction

Aim of intrusion detection system (IDS) is to superintend network and computer assets to scan the undesired activities and its misuse [1]. Security in this regard refers to the processes, practices, and methods of protecting various information systems to ensure safety of their hardware, software, and services provided [2]. Security for PC is deployed in various ways, e.g., user account access controls, biometric authentication, firewalls, etc. IDS is one of the major tools used to impart security [3]. IDS works as a security guard for any environment, which keeps an eye on all the activities happening in that environment and on finding any illegitimate or unexpected event informs the authorized host [4, 5].

A lot of research has been done in this domain but often it suffers from complexities and limitations of feature selection and classification processes in IDS [6]. Therefore, this work introduces hybrid intrusion detection system (H-IDS) that aims at improving feature selection techniques to enhance the execution of the classifier and achieve high-detection rate with greater precision. This research paper is organized in below-mentioned manner, Sect. 2 puts forward the related survey, Sect. 3 presents the proposed methodology, Sect. 4 presents results and analysis, and Sect. 5 includes the conclusion.

2   Related Work

IDS collects the network traffic from network and performs the preprocessing of collected data followed by feature selection process algorithm which ultimately selects the feature of preprocessed data. This is well illustrated in below Fig. 1.

Attacks are of two types active and passive. Active attacks directly confront the system while passive attacks only monitor the information on the system [7]. These attacks are identified and then classified through different methods. Various researchers have worked in this domain and suggested different techniques for better feature selection and attack detection. Support vector machine is one of the prominent machine learning techniques that aims proper utilization of classification and regression [8]. It is grounded based on the ideology of a hyperplane classifier or linear separability. Let us assume that it has N data samples for training $\{(x_1, y_1), (x_2, y_2), \ldots, (x_n, y_n)\}$. Here $x_i \in Rd$ and $y_i \in \{+1, -1\}$. Hyperplane is represented by (w, b), which implies weight vector and bias, respectively. With the help of the following equation we can classify x as a new object.

$$f(x) = \sin(w.x + b) = \sin\left(\sum_{i=1}^{N} \alpha_i y_i (x_i, x) + b\right) \tag{1}$$

Usually, the raw facts and figures are not linearly separable, but nonlinear facts and figures can be mapped with the help of a kernel function K to a greater dimensional

Fig. 1 IDS working model

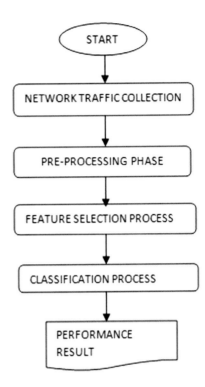

space so that they come out to be linearly separable in particularly that feature space. Following is the nonlinear decision function of SVM.

$$f(x) = \sin\left(\sum_{i=1}^{N} \alpha_i . y_i K(x_i, x) + b\right) \qquad (2)$$

Here, $K(x_i, x)$ denotes the kernel function. In this work, optimal subset of features is selected by the hybrid model of feature for the classification between normal and intrusion classes.

Heba et al. [9] proposed an IDS model that used principal component analysis (PCA) for features reduction and SVM for classification. PCA generated eigenvectors by applying projections to the input dataset. In another work, Ambusaidi et al. [10] implemented IDS with least square support vector machine (LSSVM). In this work linear and nonlinear dependent features can be managed through a feature selection algorithm based on mutual information. Thereafter, Chen et al. [11] presented a model for intrusion detection that combined rough set theory (RST) and SVM. They established that the method is useful in reducing the space density of the data. Preprocessing of the data and the reduction of the dataset is done by the RST method. The RST builds an information table or more formally a decision table which describes the features of processes. SVM has been used for the purpose of classification. Various other explores in this domain have presented issues of noisy, redundant, and irrelevant set of input datasets which hampers the performance of the classifiers. Several efforts have been taken to address this issue of managing errant dataset. But such efforts often lead to either loss of more sensitive features, relevant features or increase the computational time of the model, when one factor of relevance is preferred over the other. Some explore involving bioinspired techniques [12, 13] can be looked upon to find more realistic solution in this domain [14, 15]. Thus, there is a need for an IDS that can select the most optimal set of features for input dataset which further enhances the accuracy and efficiency of the classifier unit. The next section introduces the proposed hybrid intrusion detection system (H-IDS) that should improve feature selection techniques to enhance the execution of the classifier and achieve high detection rate with greater precision.

3   Proposed Methodology

This section presents hybrid intrusion detection system (H-IDS) to overcome the challenges of huge feature space comprising of noisy, reiterating, and inappropriate features. H-IDS optimizes the procedure of feature selection such that it augments overall performance of the classifier.

3.1   Working

H-IDS incorporates intelligent water drops (IWD) along with support vector machine (SVM) and achieves an optimal subset of attributes that evolve a reliable IDS to get increased detection rate, accuracy, and reduced false alarm rate. Figure 2 below shows the working of the proposed H-IDS. The first step of H-IDS starts functioning with KDD Cup 99 dataset. This dataset includes the network traffic data packets with both normal and attack instances. The next step is of preprocessing step that takes care of cleaning step that makes sure that input dataset is modified in a way to make its processing less complicated. The subsequent step is the hybrid method of feature selection, wherein the challenge of feature selection and classification are ensemble collectively to produce optimized result. This approach produces the desired result.

3.2   Intelligent Water Drop (IWD)

Intelligent water drop (IWD) proposed by Hosseini [16] is influenced by the flow of water drops in a river. IWD is a heuristic algorithm used for optimization and finding

Fig. 2 H-IDS working

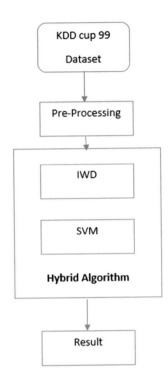

maximum and minimum values. The next subsection introduces the algorithm of the proposed H-IDS with IWD method.

3.3 H-IDS Algorithm

IDS problem is depicted in the form of a graph (N, E) which comprises N nodes. In this approach, the graph is generated by taking all the 41 features of KDD Cup 99 dataset as nodes. Each node is connected to each other node, i.e., there is a path connecting all the features with each other.

Step 1:

1.1 Initiation of the static and dynamic values

The optimal best solution TTB is initialized with worst value as $q(\text{TTB}) = -\infty$. The term Max Iteris is the highest possible iterations which are user specific. The iteration count, itercount keeps the count of the number of iterations. The value of this count is initialized to zero. The term for total water drops initialized as positive integer value which is NIWD. This term is basically a set equal to the nodes in the graph, parameters updating velocity are av, bv, and cv. Here, av $= \text{cv} = 1$ and bv $= 0.01$, parameters updating soil are, as, bs, and cs Here, as $= \text{cs} = 1$ and bs $= 0.01$, the local soil updating parameter is $\rho n = 0.9$. The global soil upgrading parameter is ρIWD $= 0.9$. The soil on every edge is represented by the term initSoil. The soil among any nodes i and j is depicted by soil (i, j) $=$ initSoil. The value of initSoil is initialized as 1000. The initial velocity of each IWD is represented by the term $\lceil$initVel$\rfloor$ ˆ($\lceil$IWD$\rfloor$ _k). It is generally initialized to value $= 4$.

1.2 Initialization of dynamic parameters

Each IWD maintains a list of traversed node list denoted by $V_C^{\text{IWD}_K} = \{\}$ which is vacant at the start. Each IWD has its own initial velocity denoted by vel$^{\text{IWD}_k}$. Every IWD has its own measure of soil which is initially set to zero.

Step 2: Distribute IWDs on the nodes of the graph in any order of choice as their first visited nodes.

Step 3: Revise the traversed node list of every IWD to accommodate the nodes currently traversed.

Step 4: Steps from 5.1 to 5.4 are repeated for the IWDs with partially available solutions.

Step 5: Phase to select the next node and apply local updates to values of soil and velocity.

Step 5.1: For IWD present on node i, select the next node j. This is done by choosing the node which fulfills the parameters of the problem area and does not belong to the traversed list of nodes V_Cˆ($\lceil$IWD$\rfloor$ _K) of the IWD, making use of the below

$$p_i^{\text{IWD}_k}(j) = \frac{f(\text{soil}(i, j))}{\sum_{l \neq V_C^{\text{IWD}_K}} f(\text{soil}(i, l))} \tag{3}$$

where $f(\text{soil}(i, j)$ the inverse value of the soil is computed among nodes i and j soil(i, j) represents the amount of soil on the local path among nodes i and j represented below

$$f(\text{soil}(i, j)) = 1/(\varepsilon_s + g(\text{soil}(i, j)))$$

$$g(\text{soil}(i, j)) = \begin{cases} \text{soil}(i, j) \text{ if } \min(\text{soil}(i, l)) \geq 0 \text{ where } l \neq V_c(\text{IWD}) \\ \text{soil}(i, j) - \min(\text{soil}(i, l)) \text{ where } l \neq V_c(\text{IWD}) \text{ else} \end{cases} \tag{4}$$

Then, newly traversed node j is appended to the node list destination series number

Step 5.2: Velocity of every IWD$_k$ at a particular time $(t + 1)$ traversing node i to node j, the velocity upgrades as described below. End if

$$\text{vel}^{\text{IWD}_k}(t + 1) = \text{vel}^{\text{IWD}_k}(t) + \frac{a_v}{b_v + c_v \text{soil}(i, j)} \tag{5}$$

Step 5.3: Soil carried by every IWD while traversing the path from node i to j:

$$\Delta \text{soil}(i, j) = \frac{a_s}{b_s + c_s.\text{time}(i, j; \text{vel}^{\text{IWD}_k}(t + 1))} \tag{6}$$

$$\text{time}\left(i, j; \text{vel}^{\text{IWD}_k}(t+1)\right) = \frac{\text{HUD}(i, j)}{\text{vel}^{\text{IWD}_k}(t+1)} \tag{7}$$

Step 5.4: Update values of the soil of the path directing node i to j which covered IWD along with soil carried through respective IWD:

$$\text{soil}(i, j) = (1 - \rho_n) \cdot \text{soil}(i, j) - \rho_n \cdot \Delta\text{soil}(i, j) \tag{8}$$

$$\text{soil}^{\text{IWD}} = \text{soil}^{\text{IWD}} + \Delta\text{soil}(i, j) \tag{9}$$

Step 5.5: The iteration's best solution TIB calculated out of total solutions obtained from every IWD is as follows:

$$T^{\text{IB}} = \arg(\max/\min) q(T^{\text{IWD}}) \forall T^{\text{IWD}} \tag{10}$$

Step 5.6: q represents the solution's quality function. Quality function is elucidated as a parameter bringing together the validity of the classifier and the content of the feature subset represented as

$$q(T^{\text{IWD}}) = \text{SL}(T^{\text{IWD}})/\text{DR}(T^{\text{IWD}}) \text{ and,}$$
$$T^{\text{IB}} = \arg(\min) q(T^{\text{IWD}}) \forall T^{\text{IWD}} \tag{11}$$

Step 5.7: Path's soil forming the current iteration's optimal solution given as

$$\text{soil}(i, j) = (1 + \rho_{\text{IWD}}) \cdot \text{soil}(i, j) - \rho_{\text{IWD}} \frac{1}{q(T^{\text{IWD}})} \tag{12}$$

where $q(\text{TIB})$ is the function defined in Eq. 9 and ρ_{IWD} is a +ve constant.
Step 5.8: Best solution given as

$$T^{\text{TB}} = \begin{cases} T^{\text{TB}} & \text{if } q(T^{\text{TB}}) \geq q(T^{\text{IB}}) \\ T^{\text{IB}}, & \text{otherwise} \end{cases} \tag{13}$$

Step 5.9: Number of iterations incremented by

$$\text{Iter}_{\text{count}} = \text{Iter}_{\text{count}}$$
$$+ \text{repeat steps from step 2 if,}$$
$$\text{Iter}_{\text{count}} < \text{MaxIter}$$

Step 5.10: Algorithm stops with the best solution $T^{\text{^TB}}$.

4 Result and Analysis

4.1 Dataset

The KDD Cup 99 dataset was collected as a zipped file package from the UCI repository. The dataset is a standard dataset originally generated during the years (1998–1999) by collecting TCP dump data traffic for 7 weeks. This dataset has been made freely available online for the purpose of training and experimentation. KDD Cup 99 dataset comprises various attacks are categorized in different subclasses of attacks.

4.1.1 User 2 Root (U2R)

It is a type of attack in which the attacker obtains access as an ordinary client on the framework either by hacking passwords or gain admittance to the base of the system framework. Buffer overflow, Http runnel, and Load module are prominent user 2 root attacks.

4.1.2 Denial of Service (DOS)

It is an attack which would not allow any authorized access to a resource by overburdening it and keeping it busy. Thus, it prevents legitimate access to resources. Mail bomb, Back Land, and Netpune are some of the examples of these attacks.

4.1.3 Remote 2 Local (R2L)

Attacker remotely situated gets some vulnerability of that system to get access as its local user. Imap, guess password, FTP-write, and PHF remain major remote 2 local attacks.

4.1.4 Probe

Effort by malicious user to collect information about the system or network to bypass its security controls. lpsweep, mscan, namp, and port sweep are well-known examples.

4.2 Performance Measures

A number of ratios are used to compute the throughput of the detector and permit comparability.

(i)   True positive (TP): The detector finds normal data when normal data is available.
(ii) True negative (TN): The detector finds abnormal data when abnormal data is available.
(iii) False positive (FP): The detector finds abnormal data at the time when normal sample is present.
(iv) False negative (FN): The detector finds normal data at the time when abnormal data are present.

On the basis of these parameters algorithm evaluation parameters are defined as follows:

$$\text{(i) Accuracy} : \frac{\mathbf{TP + TN}}{\mathbf{TP + TN + FP + FN}} \tag{14}$$

$$\text{(ii) Detection rate} : \frac{\mathbf{TP}}{\mathbf{TP + FN}} \tag{15}$$

$$\text{(iii) Precision} : \frac{\mathbf{TP}}{\mathbf{TP + FP}} \tag{16}$$

4.3 Classification Accuracy Comparison

It is a measure of correctly classifying the instances according to its class. Experiments are done to assess proposed H-IDS and then it is compared to the current state of the arts (SVM and K means). KDD Cup 99 dataset is used for all the experimentations for H-IDS and state of the arts (SVM and K means).

The above Table 1 and Fig. 3 represent classification of test dataset for both current state of the art and H-IDS with variation in KDD Cup 99 test dataset instances from

Table 1 Classification accuracy	Dataset size (no of instances)	SVM (%)	K-means (%)	H-IDS (%)
	1000	75.1	67.2	90.2
	2000	78.1	69.5	91.5
	3000	80.2	70.1	93.6
	4000	82.3	73.2	94.1
	5000	85.1	79.3	95.2

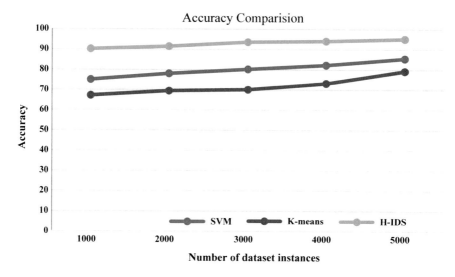

Fig. 3 Classification accuracy comparison

1000 to 5000. These results very clearly state that the proposed H-IDS outperforms the existing methods of classification like SVM and K-means. In all experiments for all instances proposed H-IDS reported higher classification rate as compared to SVM and K-means. These results also show that the new integrations of IWD and SVM in H-IDS are working proficiently and achieving higher classification rate.

4.4 Detection Rate Comparison

It tells correct detections among attack instances. Experiments are performed to measure the performance and subsequent impact of new integrations of IWD and SVM in H-IDS. KDD Cup 99 dataset is used for all the experimentations for H-IDS and state of the arts (SVM and K means).

Table 2 and Fig. 4 illustrate detection rate for current state of art and H-IDS with variation in KDD Cup 99 test dataset instances from 1000 to 5000. Experimental results very firmly show that proposed H-IDS gives higher detection rate for

Table 2 Detection rate comparison

Dataset size (no of instances)	SVM (%)	K-means (%)	H-IDS (%)
1000	75.2	69.2	89.2
2000	76.1	70.5	91.3
3000	79.2	71.1	92.6
4000	80.3	72.2	94.5
5000	82.1	78.1	95.1

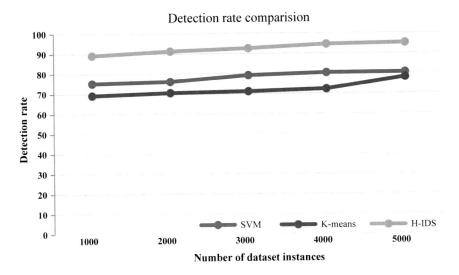

Fig. 4 Detection rate comparison

experiments. H-IDS reports a detection rate of 89.2% for 1000 test instances which increases to 95.1% for 5000 test instances that remain higher as compared to SVM and *K*-means. High detection rates vindicate that new incorporations like IWD and SVM in H-IDS help in achieving higher detection rate.

4.5 Precision Comparison

Precision can be defined as the fraction of relevant retrieved data points, it gives the measure of relevant classification. Experiments are done to assess the performance of proposed H-IDS and then it is compared to the current state of the arts (SVM and *K* means). KDD Cup 99 dataset is used for all the experimentations for H-IDS and state of the arts (SVM and *K* means).

The above Table 3 and Fig. 5 demonstrate precision rate of current state of arts and H-IDS with variation in KDD Cup 99 test dataset instances from 1000 to 5000. These

Table 3 Data Size Classification

Dataset size (no of instances)	SVM (%)	*K*-means (%)	H-IDS (%)
1000	72.2	66.5	88.6
2000	74.3	69.1	92.3
3000	74.9	71.3	94.5
4000	75.8	72.5	94.9
5000	78.1	79.3	95.3

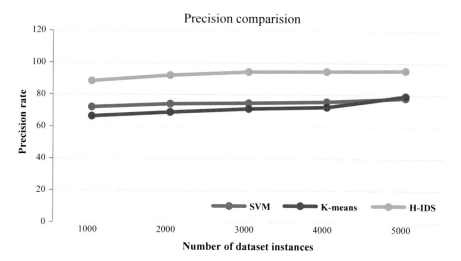

Fig. 5 Precision comparison

results very clearly state that the proposed H-IDS performs better than the existing methods and reports better and higher precision for all the variation in dataset.

5   Conclusion

Over the years attacks have evolved and so are the security tools. Intrusion detection system is a vital tool in security and mitigating threat perception. IDS recognize any misuse/unauthorized access or attack as it detects malicious traffic on a network or a host. Detection suffers from the presence of noisy, redundant, and irrelevant data inputs. Consequently, it results in high misclassification rate and low detection rate. This paper proposed a hybrid intrusion system (H-IDS) that used a hybrid strategy with support vector machine (SVM) and intelligent water drops (IWD) to execute the feature selection and classification techniques for IDS. Experimental results illustrate better classification rate with higher detection and precision of H-IDS compared to current state of art.

References

1. Aslahi-Shahri, B.M., Rahmani, R., Chizari, M., Maralani, A., Eslami, M., Golkar, M.J., Ebrahim, A.: A hybrid method consisting of GA and SVM for intrusion detection system. Neural Comput. Appl. **27**(6), 1669–1676 (2016)
2. Isaiah, O.A., Johnson, O.V., Mutiu, G.: Denial of service (DoS) attacks using PART Rule and decision table rule. J. Electr. Electron. Syst. **6**(2), 1–4 (2017)

3. Pozi, M.S.M., Sulaiman, N.M., Mustapha, N., Perumal, T.: Improving anomalous rare attack detection rate for intrusion detection system using support vector machine and genetic programming. Neural Process. Lett. **44**(2), 279–290 (2015)
4. Mabu, S., Chen, C., Lu, N., Shimada, K., Hirasawa, K.: An intrusion-detection model based on fuzzy class-association-rule mining using genetic network programming. IEEE Trans. Syst. Man Cybern. Part-C: Appl. Rev. **41**(1), 130–139 (2011)
5. Aghdam, H.M., Kabiri, P.: Feature selection for intrusion detection system using ant colony optimization. Int. J. Netw. Secur. **18**(3), 420–432 (2016)
6. Tsai, C.F., Hsu, Y.F., Lin, C.Y., Lin, W.Y.: Intrusion detection by machine learning: a review. Expert. Syst. Appl. Elseveir, 11994–12000 (2009)
7. Ravale, U., Marathe, N., Padiya, P.: Feature selection based hybrid anomaly intrusion detection system using K means and RBF kernel function. In: International Conference on Advanced Computing Technologies and Applications, Elsevier, pp. 428–435 (2015)
8. Farhat, V., McCarthy, B., Raysman, R., Holland & Knight L.L.P.: Cyber Attacks: Prevention and Proactive Responses, Practical Law Company on Its Intellectual Property & Technology web services at http://us.practicallaw.com/3-511-5848 (2016)
9. Heba, F.E., Darwish, A., Hassanien, E.A., Abraham, A.: Principle components analysis and support vector machine based intrusion detection system. In: IEEE International Conference on Intelligent Systems Design and Applications, pp. 363–367 (2010)
10. Ambusaidi, A.M., He, X., Nanda, P., Tan, Z.: Building an intrusion detection system using a filter-based feature selection algorithm. IEEE Trans. Comput. 2986–2998 (2016)
11. Chen, R., Cheng, F., Hsieh, C.: Using rough set and support vector machine for network intrusion detection. Int. J. Netw. Secur. Appl. (IJNSA) **1**(1), 1–12 (2009)
12. Saurabh, P., Verma, B., Sharma, S.: Biologically inspired computer security system: the way ahead, recent trends in computer networks and distributed systems security, CCIS, vol. **335**, pp. 474-484. Springer (2011)
13. Saurabh, P., Verma, B.: An efficient proactive artificial immune system based anomaly detection and prevention system. Expert. Syst. Appl., Elsevier **60**, 311–320 (2016)
14. Saurabh, P., Verma, B.: Immunity inspired cooperative agent based security system. Int. Arab. J. Inf. Technol. **15**(2), 289–295 (2018)
15. Saurabh, P., Verma, B., Sharma, S.: An Immunity inspired anomaly detection system: a general framework a general framework. In: 7th International Conference on Bio-Inspired Computing: Theories and Applications (BIC-TA 2012), vol. 202, AISC, Springer, pp. 417–428 (2012)
16. Shah-Hosseini, H.: The intelligent water drops algorithm: a nature-inspired swarm-based optimization algorithm. Int. J. Bio-Inspired Comput. **1**(1), 71–79 (2009)

Inspection of Crop-Weed Image Database Using Kapur's Entropy and Spider Monkey Optimization

V. Rajinikanth, Nilanjan Dey, Suresh Chandra Satapathy and K. Kamalanand

Abstract Image assessment measures are commonly employed in different domains to extract the helpful information to take essential decisions. This paper implements a soft-computing approach to examine the Benchmark Crop-Weed (BCW) images of Computer Vision Problems in Plant Phenotyping (CVPPP2014) challenge database. The proposed work executes a hybrid procedure based on Spider Monkey Optimization (SMO) algorithm and Kapur's multi-thresholding and the Watershed Segmentation (WS) based extraction. After extracting the Crop-Weed regions of BCW pictures, the superiority of the proposed tool is then assessed by implementing a relative study among extracted segment and its related ground-truth. Additionally, the prominence of SMO is validated against the Bat-Algorithm (BA) and Firefly-Algorithm (FA). The outcome of this study authenticates that SMO-based technique is competent in examining the BCW pictures with significant accuracy and precision.

Keywords Crop-Weed image · Condition monitoring · Spider Monkey Optimization · Image assessment · Validation

V. Rajinikanth (✉)
St. Joseph's AI Group, St. Joseph's College of Engineering, Chennai 600119, Tamilnadu, India
e-mail: rajinikanthv@stjosephs.ac.in

N. Dey
Department of Information Technology, Techno India College of Technology, Kolkata 700156, India

S. C. Satapathy
School of Computer Engineering, Kalinga Institute of Industrial Technology (Deemed to Be University), Bhubaneswar 751024, Odisha, India

K. Kamalanand
Department of Instrumentation Engineering, M.I.T Campus, Anna University, Chennai 600044, Tamilnadu, India

© Springer Nature Singapore Pte Ltd. 2020
K. N. Das et al. (eds.), *Soft Computing for Problem Solving*,
Advances in Intelligent Systems and Computing 1048,
https://doi.org/10.1007/978-981-15-0035-0_32

405

1 Introduction

In recent years, considerable measures are taken to increase the yield in agriculture. Smart cultivation is one of the procedure in which the plants are continuously supervised using dedicated monitoring procedures. Image assisted procedures are normally employed in the agriculture to monitor the weed, disease in crops, and damage in crops due to insects. A programmed drone with a high-resolution camera or autonomous robots can be employed to supervise the farming field [1, 2]. Usually, the drone/robot is engaged to collect various images related to the crops and then these images are then processed in a centralized monitoring station in order to take needed decision to preserve the healthy crops [3].

In literature, there exist a variety of procedures to monitor the condition and classification of crop based on the image assisted procedures. Burgos-Artizzu et al. (2011) implemented Otsu's thresholding-based procedure for the real-time examination of cropweed in maize fields [4]. Tellaeche et al. (2011) proposed SVM-based computer vision approach to identify the weed section based on RGB segmentation process [5]. Montalvo et al. (2013) proposed Otsu's-based weeds/crops identification procedure for images of maize fields [6]. Aitkenhead et al. (2003) implemented neural network-based examination procedure [7]. Herrera et al. (2014) discussed the shape descriptor and fuzzy-based weed classification procedure [8].

The work of this paper considers the Benchmark Crop-Weed (BCW) dataset offered by Haug and Ostermann (CVPPP 2014) for the examination [9]. The data collection procedure and the previous experiment implemented using this dataset can be found in [1, 3, 9]. This paper considers a hybrid procedure based on a multilevel thresholding scheme to enhance the picture and a mining procedure to extract the Region of Interest (ROI). In order to have a better outcome of the proposed procedure, a recently developed heuristic approach called the Spider Monkey Optimization (SMO) algorithm is considered to monitor the multi-thresholding task. After mining the ROI from the test image, a qualified assessment with the Ground-Truth (GT) is executed and the necessary image similarity values are recorded. Further, the performance of SMO is validated against the well-known heuristic procedures, such as Bat-Algorithm (BA) and the Firefly-Algorithm (FA). The investigational result of this study authenticates that SMO approach offers improved result for the considered dataset.

The remaining division of this study is ordered as follows: Sect. 2 presents the summary of the SMO and its implementation, Sect. 3 depicts the results achieved with the CVPPP2014 dataset, and the conclusion of the current work is described in Sect. 4.

2 Methodology

This division of paper presents various procedures implemented in this paper to segment the BCW images. The BCW images of CVPPP2014 challenge database include 60 RGB scale images of dimension 1296 × 966 pixels [9]. Every image is available with a binary mask and RGB scale Ground-Truth (GT). Figure 1 represents the methodology applied in this study. Initially, the test picture is preprocessed using the Spider Monkey Optimization guided Kapur's multi-thresholding and post-processing based on watershed algorithm. The thresholding procedure enhances the cropweed section and possibly eliminates the background region according to the chosen threshold. The enhanced segment is then extracted with the Watershed Segmentation (WS) approach. Later, the segmented ROI is then compared against the GT image and the image likeliness events are then recorded. Finally, the superiority of SMO approach is validated by replicating the work with well-known heuristic algorithms, like BA and FA existing in the literature.

2.1 Spider Monkey Optimization (SMO)

SMO is a heuristic approach invented by Bansal et al. (2014) based on the mimic of the group foraging activity followed by Spider Monkeys [10]. In recent years, the SMO is applied to find the best answers for a selection of optimization assignments [11–13]. In this work, the Lévy flight SMO (LSMO) discussed by Sharma et al. (2017) is considered to preprocess the BCW images [14]. Lévy flight (LF) is a proven search procedure commonly approved by the researchers to improve the presentation of heuristic search [15–17]. In LSMO, the LF strategy is considered to advance the exploitation facility of traditional SMO. The implementation procedure and the necessary pseudocode for the LSMO are clearly depicted in [14].

The LSMO consists of the following stages: (i) Leader stage (local & global), (ii) Learning stage (local & global), and (iii) decision making (local & global).

Let, N represents the population size, D denotes the search dimension, R represents a random number of size [0,1] and S_i denotes the agents of size ($i = 1, 2, 3, …, N$), $S_{\min j}$ and $S_{\max j}$ specifies the lower and upper values of S_i in jth direction.

The mathematical expression of the LSMO is as follows:

Initialization is presented in Eq. (1)

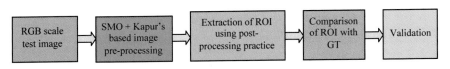

Fig. 1 Implementation of cropweed assessment procedure

$$S_{ij} = S_{\min j} + R(0, 1) * (S_{\max j} - S_{\min j}) \tag{1}$$

Leader stage (local)

$$S_{\text{new}ij} = S_{ij} + R(0, 1) * (\text{Local}_{kj} - S_{ij}) + R(-1, 1) * (S_{rj} - S_{ij}) \tag{2}$$

Learning (local)

A greedy search is implemented to revise the location of S based on the finest location.

Decision making (local)

$$S_{\text{new}ij} = S_{ij} + R(0, 1) * (\text{Global}_j - S_{ij}) + R(0, 1) * (S_{ij} - \text{Local}_{kj}) \tag{3}$$

Equations (2) and (3) denotes the leader and learning stages for the local phase and similar procedure exist for the global leader phase, learning, and decision making. Complete implementation guidance for LSMO is available in [14].

In this work, LSMO is considered to monitor the preprocessing procedure.

2.2 Kapur's Entropy

Recently, Kapur's multi-thresholding is widely implemented by the researchers to enhance the ROI from test pictures [18]. In the literature, KE assisted multilevel thresholding is implemented to threshold the gray as well as RGB scale pictures. In this work, the preprocessing task is executed with different thresholds (Th = 2, 3, 4, 5) and the matching outcome is considered to extract the ROI. In this task, the threshold values of the test picture are arbitrarily adjusted by the LSMO till the KE value reaches maximum value [19, 20]. The arbitrary threshold value, which offers the maximum KE is then recorded as the optimal threshold based on the assigned Th value. The mathematical relation for the KE can be expressed as in Eq. (4) and the essential parameter specifications are available in [18].

$$F_{\text{KE}\max}(\text{Th}) = \sum_{J=1}^{K} H_J^C \tag{4}$$

2.3 Watershed Segmentation

The ROI from BCW depiction can be mined with semi- and computerized actions [21]. Semiautomated approaches may require the assistance of worker to begin the mining task. Therefore, computerized procedures are generally chosen throughout picture examination. This study approves the Marker-Controlled Watershed Segmentation (MCWS) to extract ROI from BCW pictures. MCWS consists of edge detection, creation of WS, alteration of the indicator tempo to mark the pixel cluster for ROI, expansion of familiar pixels, and mining of acknowledged section.

2.4 Performance Evaluation

A comparative study among the extracted ROI and GT is forever suggested to critic the concert of the created image appraisal system alongside the result by the specialists. The prior studies also confirm that assessment with GT is required to calculate the Picture Likeliness Values (PLV) discussed in [22–26].

2.5 Validation

The consequence and the throughput of LSMO technique are validated with the mean of PLV. Normally, greater PLV will choose the proximity of excavated ROI alongside the GT. Hence, if PLV is fine, then the implemented portrait evaluation system may be used to inspect the other Crop-Weed pictures collected from the cultivated fields. Further, in this paper, the performance of LSMO is validated against the well-known LF guided BA and FA existing in [16, 17, 20, 22].

3 Result and Discussion

This division of the paper represents the results acquired. All the investigational work of this paper is realized using the Matlab software. This study considers 1296 × 966 sized RGB BCW database of CVPPP2014 challenge [9]. Initially, the LSMO constraints are allocated as follows: agents size = 40, exploration dimension = Th, stopping criteria for LSMO = F_{KEmax}, and total number of iterations = 2000. Similar algorithm parameters are applied for the BA and FA.

Figure 2 presents the example images adopted for the discussion. Figure 2a represents the pseudo name, Fig. 2b, c depicts the test image and its corresponding GT. Initially, the LSMO + KE-based multi-thresholding is implemented on the BCW images with Th = 2, 3, 4, 5 and the enhancement of Crop-Weed section and elimination of the background is visually examined. Initially, the image with pseudo number 09 is considered for the discussion and its results are shown in Fig. 3. Figure 3a–d shows the outcome of thresholding for Th = 2–4, respectively. Figure 3a, b provides better enhancement compared to Fig. 3c, d. After executing the multi-thresholding task, the WS approach is then implemented to extort the ROI (Crop-Weed segment) from preprocessed illustration. Figure 3e–h presents the outcome of WS, in which

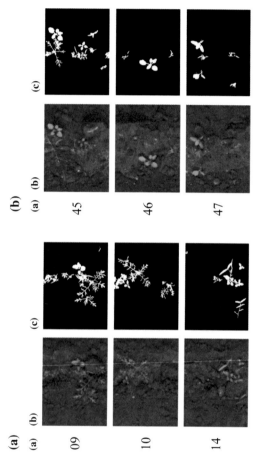

Fig. 2 Example images. **a** Artificial surname, **b** plant-weed image, **c** GT

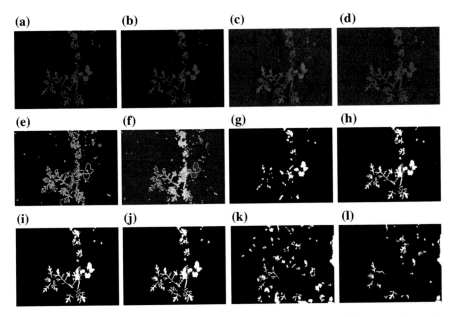

Fig. 3 Results obtained with the implemented SMO-assisted technique. **a–d** Preprocessing results obtained for the threshold 2–5, respectively, **e–h** results attained with the watershed procedure. **i–l** Outcome of the proposed tool for various threshold values

Fig. 3e represents the detected edge, Fig. 3f shows the watershed and Fig. 3g and h shows the marker rate and extracted ROI, respectively. Figure 3i–l shows the ROI extracted from Fig. 3a–d using the WS technique.

When comparing the extracted ROI with the GT, it is found that extracted image in Fig. 3j shows the maximal similarity with the GT image. Hence, LSMO + KE thresholding with Th = 3 is fixed for other test images available for the BCW dataset considered in this paper. Figure 4 depicts the ROI extracted using the proposed technique. After extracting the ROI, a comparative check between ROI and GT is then performed and the essential PLV are mined to verify the merit of proposed work. The comparable practice is then performed with the LF guided BA and FA and the results are then considered to validate the proposed procedure.

Table 1 presents the PLV obtained for the ROIs of Fig. 4 and its mean value is presented as Average 1. Related practice is performed for the other images of BCW dataset and its mean value is depicted as Average 2. The earlier work of Haug and Ostermann (2014) achieved an average accuracy of 85.9% and precision of 79.6% using Otsu's approach [9]. In this paper, hybrid image processing scheme based on the LSMO offers an average accuracy of 99.75% and precision of 80.27%. As depicted in Fig. 5, these values obtained with the BA and FA is also closer to the results of the LSMO. The product of this study authenticates that the proposed technique is very significant in the examination of RGB scale Crop-Weed images.

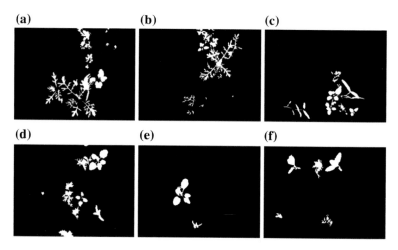

Fig. 4 Results attained with the chosen test images. **a–f** The extracted plant/weed section from the test image with pseudo name 09, 10, 14, 45, 46, and 47

Table 1 Performance measures obtained with the SMO algorithm

Image	Jaccard	Dice	Sensitivity	Specificity	Accuracy	Precision
09	0.5802	0.7343	0.9913	0.9977	0.9977	0.5831
10	0.6559	0.8076	0.9562	0.9926	0.9906	0.6758
14	0.7852	0.8125	0.9852	0.9875	0.9974	0.8073
45	0.7106	0.8116	0.9608	0.9885	0.9906	0.7225
46	0.7099	0.8304	0.9859	0.9982	0.9981	0.7172
47	0.6009	0.7507	1.0000	0.9988	0.9988	0.6009
Average 1	0.6738	0.7912	0.9799	0.9939	0.9955	0.6845
Average 2	0.7926	0.8327	0.9846	0.9961	0.9973	0.8027

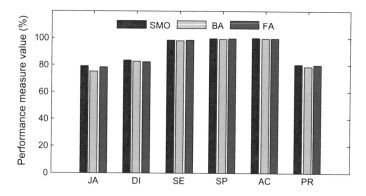

Fig. 5 Performance comparison of SMO, BA, and FA

4 Conclusion

This paper employs a hybrid image examination technique based on Kapur's multithresholding and watershed segmentation. In this work, benchmark CVPPP2014 challenge dataset containing 60 RGB scale imagery of size 1296 × 966 pixels are chosen for the evaluation. The chosen test pictures are mainly investigated using Th = 3 with LF guided SMO, BA, and FA. The results of this study also confirm that the WS approach is very efficient in extracting the ROI from the preprocessed picture. The outcome of this paper authenticates that the PLV attained with SMO is better than the BA and FA. Further, the hybrid image processing approach offers superior values of average accuracy and precision compared with Otsu's-based approach. In future, the proposed method can be used to inspect other Crop-Weed images recorder using the digital camera.

References

1. Haug, S., Michaels, A., Biber, P., Ostermann, J.: Plant classification system for crop/weed discrimination without segmentation. In: IEEE Winter Conference on Applications of Computer Vision (WACV), IEEE, pp. 1142–1149 (2014). https://doi.org/10.1109/wacv.2014.6835733
2. Bakhshipour, A., Jafari, A., Nassiri, S.M., Zare, D.: Weed segmentation using texture features extracted from wavelet sub-images. Biosys. Eng. **157**, 1–12 (2017). https://doi.org/10.1016/j.biosystemseng.2017.02.002
3. Haug, S., Ostermann, J.: Plant classification for field robots: a machine vision approach. In: Computer Vision and Pattern Recognition in Environmental Informatics, pp. 248 (2016). https://doi.org/10.4018/978-1-4666-9435-4.ch012
4. Burgos-Artizzu, X.P., Ribeiro, A., Guijarro, M., Pajares, G.: Real-time image processing for crop/weed discrimination in maize fields. Comput. Electron. Agric. **75**(2), 337–346 (2011). https://doi.org/10.1016/j.compag.2010.12.011
5. Tellaeche, A., Pajares, G., Burgos-Artizzu, X.P., Ribeiro, A.: A computer vision approach for weeds identification through support vector machines. Appl. Soft Comput. **11**, 908–915 (2011). https://doi.org/10.1016/j.asoc.2010.01.011
6. Montalvo, M., Guerrero, J.M., Romeo, J., Emmi, L., Guijarro, M., Pajares, G.: Automatic expert system for weeds/crops identification in images from maize fields. Expert Syst. Appl. **40**(1), 75–82 (2013). https://doi.org/10.1016/j.eswa.2012.07.034
7. Aitkenhead, M.J., Dalgetty, I.A., Mullins, C.E., McDonald, A.J.S., Strachan, N.J.C.: Weed and crop discrimination using image analysis and artificial intelligence methods. Comput. Electron. Agric. **39**(3), 157–171 (2003). https://doi.org/10.1016/s0168-1699(03)00076-0
8. Herrera, P.J., Dorado, J., Ribeiro, A.: A novel approach for weed type classification based on shape descriptors and a fuzzy decision-making method. Sensors **14**(8), 15304–15324 (2014). https://doi.org/10.3390/s140815304
9. Haug, S., Ostermann, J.: A crop/weed field image dataset for the evaluation of computer vision based precision agriculture tasks. Lect. Notes Comput. Sci. **8928**, 105–116 (2014). https://doi.org/10.1007/978-3-319-16220-1_8
10. Bansal, J.C., Sharma, H., Jadon, S.S., Clerc, M.: Spider monkey optimization algorithm for numerical optimization. Memetic Comput. **6**(1), 31–47 (2014)
11. Sharma, H., Hazrati, G., Bansal, J.C.: Spider monkey optimization algorithm. In: Studies in Computational Intelligence, vol. **779**, pp. 43–59 (2019). https://doi.org/10.1007/978-3-319-91341-4_4

12. Gupta, K., Deep, K., Bansal, J.C.: Spider monkey optimization algorithm for constrained optimization problems. Soft. Comput. **21**(23), 6933–6962 (2017)
13. Gupta, K., Deep, K., Bansal, J.C.: Improving the local search ability of spider monkey optimization algorithm using quadratic approximation for unconstrained optimization. Comput. Intell. **33**(2), 210–240 (2017)
14. Sharma, A., Sharma, H., Bhargava, A., Sharma, N., Bansal, J.C.: Optimal power flow analysis using Lévy flight spider monkey optimisation algorithm. Int. J. Artif. Intell. Soft Comput. **5**(2), 320–352 (2016)
15. Dey, et al.: Firefly algorithm for optimization of scaling factors during embedding of manifold medical information: an application in ophthalmology imaging. J. Med. Imaging Health Inform. **4**(3), 384–394 (2014). https://doi.org/10.1166/jmihi.2014.1265
16. Satapathy, S.C., Raja, N.S.M., Rajinikanth, V., Ashour, A.S., Dey, N.: Multi-level image thresholding using Otsu and chaotic bat algorithm. Neural Comput. Appl. **29**(12), 1285–1307 (2018). https://doi.org/10.1007/s00521-016-2645-5
17. Yang, X.S.: Nature-Inspired Metaheuristic Algorithms, 2nd edn. Luniver Press, Frome, UK (2011)
18. Kapur, J.N., Sahoo, P.K., Wong, A.K.C.: A new method for gray-level picture thresholding using the entropy of the histogram. Comput. Vision. Graph. Image Process. **29**, 273–285 (1985)
19. Dey, N., Rajinikanth, V., Ashour, A.S., Tavares, J.M.R.S.: Social group optimization supported segmentation and evaluation of skin melanoma images. Symmetry **10**(2), 51 (2018). https://doi.org/10.3390/sym10020051
20. Rajinikanth, V., Satapathy, S.C., Dey, N., Lin, H.: Evaluation of ischemic stroke region from CT/MR images using hybrid image processing techniques. In: Intelligent Multidimensional Data and Image Processing, pp. 194–219 (2018). https://doi.org/10.4018/978-1-5225-5246-8.ch007
21. Deng, G., Li, Z.: An improved marker-controlled watershed crown segmentation algorithm based on high spatial resolution remote sensing imagery. Lect. Notes Electr. Eng. **128**, 567–572 (2012)
22. Rajinikanth, V., Dey, N., Satapathy, S.C., Ashour, A.S.: An approach to examine magnetic resonance angiography based on Tsallis entropy and deformable snake model. Futur. Gener. Comput. Syst. **85**, 160–172 (2018). https://doi.org/10.1016/j.future.2018.03.025
23. Raja, N.S.M., Fernandes, S.L. Dey, N., Satapathy, S.C., Rajinikanth, V.: Contrast enhanced medical MRI evaluation using Tsallis entropy and region growing segmentation. J. Ambient. Intell. Humaniz. Comput. 1–12 (2018). https://doi.org/10.1007/s12652-018-0854-8
24. Rajinikanth, V., Satapathy, S.C., Fernandes, S.L., Nachiappan, S.: Entropy based segmentation of tumor from brain MR images–a study with teaching learning based optimization. Pattern Recogn. Lett. **94**, 87–95 (2017). https://doi.org/10.1016/j.patrec.2017.05.028
25. Rajinikanth, V., Raja, N.S.M., Satapathy, S.C., Dey, N., Devadhas, G.G.: Thermogram assisted detection and analysis of ductal carcinoma in situ (DCIS). In: International Conference on Intelligent Computing, Instrumentation and Control Technologies (ICICICT), IEEE, pp. 1641–1646 (2018). https://doi.org/10.1109/icicict1.2017.8342817
26. AlShahrani, A.M., Al-Abad, M.A., Al-Malki, A.S., Ashour, A.S., Dey, N.: Automated system for crops recognition and classification. In: Computer Vision: Concepts, Methodologies, Tools, and Applications, pp. 1208–1223. IGI Global (2018). https://doi.org/10.4018/978-1-5225-5204-8.ch050

Implementation of Fuzzy-Based Multicarrier and Phase Shifted PWM Symmetrical Cascaded H-Bridge Multilevel Inverter

K. Muralikumar, Ponnambalam Pathipooranam and M. Priya

Abstract Multilevel inverter technology garnered its attention in the electrical power industry for producing various levels of the output voltage and it is considered as a potential alternative for the high power and medium voltage energy control applications. The cascaded MLI is being used for producing less stress on the switching devices, less distorted output voltage, and for generating a lesser common-mode voltage. The phase shifted PWM is considered as the best solution for H-bridge cascaded inverters. This paper analyzes a simple and flexible phase shifted PWM technique for common-mode voltage reduction of a seven-level H-bridge cascade inverter topology. This cascaded H-bridge MLI system distributes power consistently between the modules and ensures the same power exploitation of inverter switches within a module. The functional verification of the seven-level inverter is done using MATLAB/Simulink software and also presented the fuzzy control system for controlling the system. The experimental work for open loop is done using D-Space.

Keywords Multilevel inverter · Cascaded H-Bridge · Multicarrier PWM · Phase shifted PWM · Fuzzy logic controller

1 Introduction

The multilevel inverters for Pulse Width Modulations (PWM) are increased significantly in high-performance power electronic applications and thereby reducing the utilization of high-power rating devices such as active power filter, drives, and static var compensators [1]. The present scenario introduces many political challenges, technical, and economics because of the way changing the designing and

K. Muralikumar (✉) · P. Pathipooranam · M. Priya
School of Electrical Engineering, VIT, Vellore 632014, Tamilnadu, India
e-mail: kolamuralikumar@gmail.com

P. Pathipooranam
e-mail: p.ponnambalam@gmail.com

M. Priya
e-mail: murthypriya.eee@gmail.com

© Springer Nature Singapore Pte Ltd. 2020
K. N. Das et al. (eds.), *Soft Computing for Problem Solving*,
Advances in Intelligent Systems and Computing 1048,
https://doi.org/10.1007/978-981-15-0035-0_33

415

managing of Electrical Energy Resources (EES) [2], i.e., generators, transmission, and distribution networks. From this technical point of view, the usage of Electrical Power Converters (EPCs) presents the challenging problems, additional power losses, include increasing topological complexity and Electromagnetic Interference (EMI) [3, 4]. Hence it is dropping the inclusive efficiency, quality of the package, and network stability. The main advantages of employing MLI on high-power rated devices are stated as follows: It can work on when DC bus voltage is high, the cascaded connection of switching devices, switching devices between multiple voltage levels and reduces output voltage harmonics [5, 6]. An MLI splits the single dc power either in a direct way or indirectly, and the outcome of each leg is greater than two distinct levels of voltage. In the multilevel inverter, three types of topologies are categorized—diode-clamped multilevel inverter, flying capacitor multilevel inverter, and cascaded multilevel inverter. The cascaded MLI is commonly used among the abovementioned configurations. It is due to the absence of the issues related to the DC link capacitor voltage [7], even then it requires various dc sources for motor drive-based applications.

The aim of the modified existing configuration is to enhance the quality at the inverter terminal of available energy. Then the Pulse Width Modulation (PWM) is an improvement of both reputation and applications of Multilevel Inverters (MLI), which becomes effectively different from the current inverter topologies [8]. In a recent period, they distribute applied voltages away from the number of cascaded power devices and were employed that mostly in elevation voltage and high power in industrial and traction applications [4]. Hence to overcome their voltage restrictions and permitting the eliminations in output transformers in Medium-High Voltage System (MHV). Hence, there is a modulated staircase of output voltage waveform without using a dissipative passive filter and bulky expensive [9]. The switching strategy in switching frequency of MLI is classified into two categories: fundamental switching frequency and the second is higher switching frequency. The carrier PWM techniques based on the sinusoidal waveform works with higher switching frequency and Phase shifted PWM (PS-PWM) techniques based on the sinusoidal waveform which works with low switching frequency. Generating fundamental frequency is basically equals to generating staircase form. By using the multicarrier PWM and phase shifted PWM, we can lessen the percentage of THD of the seven-level cascaded H-bridge inverter without changing the levels.

2 H-Bridge Cascaded Multilevel Inverter

Those from the operation of multilevel topologies, the CH-bridge multilevel inverter are cascaded for H-bridge, or to setup for H-bridges on a sequence arrangement. These cascaded H-bridge, the bridges are cascaded with each phase [2]. The total output voltage for the cascaded inverter is generally equivalent to the sum of the separate bridge output voltages and it can be controlled such that a staircase waveform can be generated. Sinusoidal is more tends to output voltage waveform by increasing

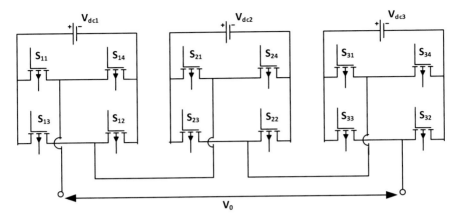

Fig. 1 Single-phase CH-Bridge inverter with seven levels

the H-bridges in a phase. Several H-bridges are cascading through the stepped output voltages and obtained output current by cascading in the cascaded MLI.

The purpose of single H-bridge of three-level output V_{dc}, 0, and $-V_{dc}$ is to add the extra H-bridges for the existing H-bridges, by increasing two number of levels, since three H-bridges inverters for seven-level output, which is to be cascaded and considered for 12 switches in the overall count of the cascaded inverters. The topology of single-phase CH-MLI is shown in Fig. 1 with seven levels and in every phase it consists of three H-bridges identically. For n-level $(n-1)/2$ is the identical H-bridges for cascaded H-bridge topology [10], which is used in each phase. H-bridge needs a DC bus for separate DC sources, so this is beneficial for accumulating energy for that topology from RES, like fuel cells and solar panels.

3 Modulation Control Schemes for Multilevel Inverter

The availability of multilevel inverter modulation control techniques can be isolated in two ways as shown in Fig. 2. One of the schemes is PWM with a low switching frequency that is fundamental switching frequency PWM and the other scheme is Pulse Width Modulation (PWM) operating at the high switching frequency, such as PWM-based multicarrier, Selective Harmonic Elimination (SHE) and space vector in PWM. In multilevel topology sinusoidal PWM need multiple carriers PWM [6, 7]. Precisely, for all individual DC sources require their own carrier wave. Most of the multilevels have been developed using carrier schemes for reducing the distortion in MLIs, considering on the triangular carrier for the conventional SPWM. One of the methods is using carrier disposition form and the other is phase shift multiple carrier signals. Generally, $(n-1)$ carriers need for n-level multilevel topology [9].

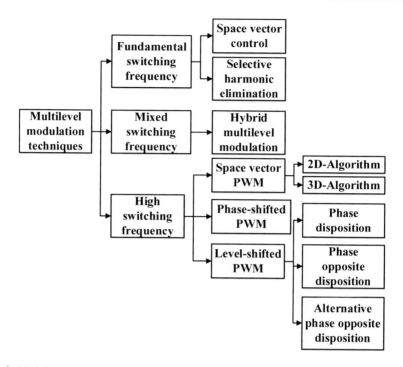

Fig. 2 Modulation control schemes for multilevel inverter

3.1 Multicarrier Modulation Technique

A standout among those types of the modulation signal (duty cycle) is describing the voltage sources with a triangle waveform of modulations is to intersect the illustration of signals. Alternative PWM approaches are differing with three types of phase relationships: Alternate Phase Opposition Disposition (APOD), Phase Disposition (PD), and Opposition Disposition (POD). For APOD, all carrier waveform is in out of phase with 180° of its nearby carrier. For POD, the carrier waveforms for all above the zero references are in phase which is in 180° out of phase and having below zero. And finally, for PD, the carrier wave is in phase.

3.1.1 APOD Alternate Phase Opposition Disposition PWM

The module Alternate Phase Opposition Disposition (APOD) PWM is shown in Fig. 3. 180° mode is for every carrier waveform with a neighbor carrier wave having out of phase. The method of APOD has some rules: if the number of level n = 7 are in out of phase with 180° of neighbor carriers then n−1 = 6, the carrier wave is arranged. It has three positive and three negative carriers, whose magnitude is +ve and −ve. If the reference is higher when compared to the first positive of the

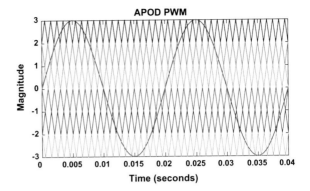

Fig. 3 Carrier arrangements—APOD PWM method

converter is switched to $+V_{dc}$. If the base voltage is higher when compared to the second positive carrier converter is switched to $2V_{dc}$. If the referred base voltage is higher when compared to the third positive carrier wave is uppermost of the converter is switched to $3V_{dc}$. Uncertainty for less reference, all positive carrier wave of the converter is switched to "0", as well as the negative carrier is of lesser waveforms. If the reference is less, then the first negative carrier wave of the converter is switched to $-V_{dc}$. If the referred base voltage is smaller, then second $-ve$ carrier wave of the inverter is switched to $-2V_{dc}$. If the referred voltage is less, the third $-ve$ carrier is the lowermost wave of the converter is switched to $-3V_{dc}$.

3.1.2 Phase Opposition Disposition (POD) PWM

POD PWM is shown in Fig. 4. The above all carrier waves have zero references which are with zero phase difference and if the carrier wave is below zero references then there exists a phase difference of 180°. For a number of levels n = 7, when the carriers are arranged as for n−1 = 6 and all the carrier waveforms beyond zero are in phase which has 180° out of phase, are in lower than zero. The three carrier waves with zero lines which are above the reference and three carrier waves with

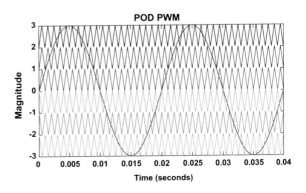

Fig. 4 Carrier arrangements—POD PWM method

Fig. 5 Carrier arrangements
for PD PWM technique

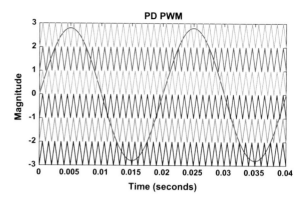

180° phase shift which is zero reference line and which comes under the rules of the Phase Opposition Disposition (POD). If the reference is higher, then the first + ve carrier wave of the inverter is transferred to $+V_{dc}$. If the reference is higher, then second +ve carrier wave of the inverter is transferred to $+2V_{dc}$. If the reference is higher, then third positive uppermost carrier wave of the converter is transferred to $+3V_{dc}$.

3.1.3 PD—Phase Disposition PWM

Phase Disposition(PD) PWM is shown in Fig. 5. All high-frequency waveforms are in zero phase difference with above and below zero references. Phase Disposition (PD) method follows the rules for a number of levels having n = 7, and which the carrier wave is arranged at n−1 = 6. So that the above zero and below zero of all carrier waveforms are in phase. It has the above three carrier waveforms of the zero line reference.

If the reference is larger, then the first +ve carrier wave of the converter is tuned to $+V_{dc}$. If the reference is larger, then second +ve carrier wave of the converter is transferred to $+2V_{dc}$. If the reference is larger, then the third +ve carrier waveform is uppermost of the converter is switched to $+3V_{dc}$. If the reference waveform is smaller, then all the +ve carrier wave as well as −ve carrier wave is less to the converter and is switched "0". If the base voltage is smaller, then first −ve carrier wave of the converter is switched to $-1V_{dc}$. If the reference is less, then the second −ve carrier waveform of the inverter is switched to $-2V_{dc}$. If the reference is lesser, then the third −ve is lowermost carrier wave of the converter is switched to $-3V_{dc}$.

3.1.4 Carrier-Based Phase Shifted PWM

The operating rules for PSCPWM for seven levels are when the carrier waveforms are arranged $(n-1) = 6$ as the number of levels n = 7. The carriers among Full Bridge

Fig. 6 Carrier arrangements for phase shifted PWM technique

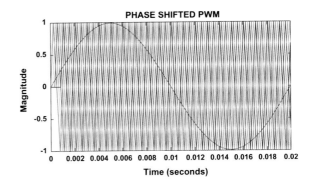

(FB) inverter are shifted by a phase angle of 90°. If the reference is higher, then carrier waveforms of the converter are transferred to the $+V_{dc}$ voltage level. If the reference base voltage is low, then uppermost carrier waveform of the converter is transferred to $V_{dc}/2$ and also greater than the all another carrier. If the reference base voltage is more, then the lowermost carrier form of the converter is switched to $V_{dc}/3$ and all other carriers are greater. If the reference is low, then the two upmost high-frequency carrier waveform for the converter is switched to "0" and two lowermost carriers are greater. If the reference is less, then the converter is lowermost carrier waveform of the converter is transferred to $-V_{dc}/2$ and all other carriers are low. Uncertainty for low reference, all carrier waveform of the converter is switched to $-V_{dc}$. The carrier-based phase shifted PWM is shown in Fig. 6, the same frequency of all the triangular carriers is surrounded with a similar amplitude of peak to peak, even though it presents some carrier wave between phase shift and two adjacent waves. Essentially the carrier signals $(n-1)$ are required for voltage levels "n" which are having the angle of $\theta = 360°/(m-1)$ phase shifted. The appropriate gate signals are modulating the signals with the relationship of carrier waveform that are generated. In broad, "n" voltage level of multilevel inverter with triangular carrier requires $(n-1)$ the modulation scheme of phase shifted multicarrier due to all of the triangular carriers which contains amplitude of similar frequency and similar peak to peak, but the two adjacent carriers and their phase shift impression is given by $Ph_{CR} = 360°/(n-1)$. The phase shift modulation got accepted by evaluating six-carrier signals with a single reference sine signal. The six-carrier signals are produced with a phase difference of 60°. The carrier signals are measured up to with sinusoidal and balancing sine, as a result, six-PWM signals are achieved.

4 Fuzzy Controller

The structure of the fuzzy controller organization is illustrated in Fig. 7, and the diagram provides information about the values of frequency and switching angles. Using the values by switching the angles and frequency values, the harmonic contents

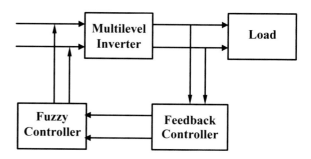

Fig. 7 Block diagram of a fuzzy logic controller

present in the system are to be eliminated. In this system, the feedback controller is used to manipulate the error voltage values. These error voltage values from the fuzzy control system are taken as the input. The voltage values are the feedback controller which is manipulated at different times from the waveform of voltage values. At different time values, the reference waveform of voltage values are taken to reference voltage values. The harmonics waveform is given for different time values for voltage values which are compared with the reference waveform. The equation given below is manipulated by voltage error values [11–14].

The fuzzy controller's information is the error e, i.e., $e = V_{ref} - V_0$ and the modification in error CE, i.e., $CE = e_n - e_{n-1}$, where V_0 is the cascaded Multilevel Inverter (MLI) output voltage, V_{ref} is the converted output voltage, and the subscript "N" incomes tentative cases. δ_{MN} is the variation of stability value brought by the fuzzy logic controller at the Nth challenging moment. To provide certain PWM signs MN, the measured flag MS is attained and continued to SPWM by operating δ_{MN}.

4.1 Inputs for Fuzzification

Certain principles of E and CE inputs are to be fuzzified. In this work, the seven triangular membership functions are chosen.

4.2 Rule Table of Membership Function

The E and CE membership functions and modulation index membership functions are shown in Figs. 8 and 9. The rule table of corresponding membership functions as shown in Table 1.

Fig. 8 E and CE member function

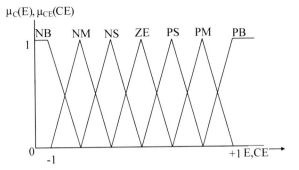

Fig. 9 Modulation index changing membership functions

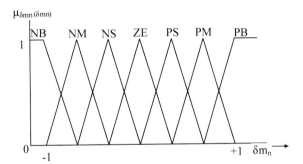

Table 1 Rule table of membership functions

CE	E						
	NB	NM	NS	ZE	PS	PM	PB
NB	NB	NB	NB	NB	NM	NS	ZE
NM	NB	NB	NB	NM	NS	ZE	PS
NS	NB	NB	NM	NS	ZE	PS	PM
ZE	NB	NM	NS	ZE	PS	PM	PB
PS	NM	NS	ZE	PS	PM	PB	PB
PM	NS	ZE	PS	PM	PB	PB	PB
PB	ZE	PS	PM	PB	PB	PB	PB

5 Results

5.1 Simulation Results

The single-phase seven-level CH-MLI for open loop is modeled in Simulink, utilizing the power system square set. The exchanging pulses for every H-bridge are produced

Fig. 10 The output voltage
of APOD PWM technique

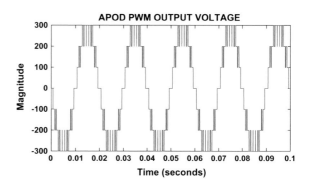

Fig. 11 FFT analysis of
APOD PWM technique

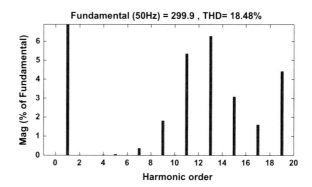

from various multicarrier sine PWM procedures. The output voltage waveforms for
multicarrier and phase shifted PWM techniques are shown in the above figures. The
simulation results for the output voltage of Cascaded seven-level H-bridge inverter
for APOD, POD, PD, and phase shifted pulse width modulation as shown in Figs. 10,
12, 14, and 16. From the figures, it formed seven levels and the output voltage on
y-axis and time-period on the x-axis. By using Fast Fourier Transform (FFT) analysis
for the seven-level cascaded H-bridge inverter as shown in Figs. 11, 13, 15, and 17.
From the FFT investigating the %THD esteem can be figured and the values are
tabulated in Table 2. From Table 2, it is noticed that the %THD is high (18.48%) for
APODPWM technique and low (8.11%) for phase shifted PWM technique (Table 2,
Figs. 18 and 19).

5.2 Hardware Results

In hardware, for seven levels, CH-MLI for open loop have done. The MOSFET
utilized IRF840 switches and TLP350 driver boards. The N-channel IRF840 switches
are used and its rating are 8A, 500 V, and 0.850 Ω. It is found that the MOSFET has

Fig. 12 Output voltage for
POD PWM technique

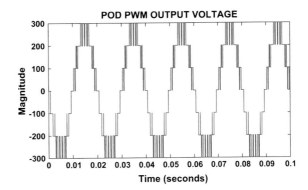

Fig. 13 FFT analysis of
POD PWM technique

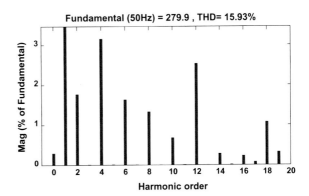

Fig. 14 Output voltage for
PD PWM technique

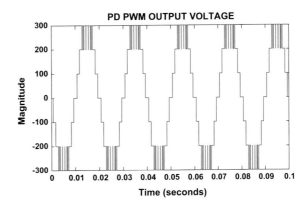

Fig. 15 FFT analysis of PD
PWM technique

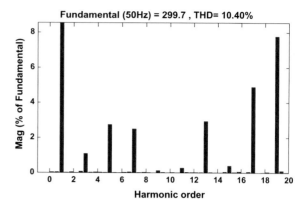

Fig. 16 The output voltage
for phase shifted PWM
technique

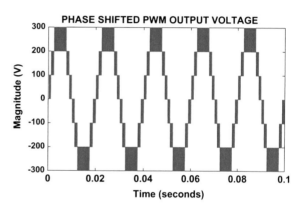

Fig. 17 FFT analysis of
phase shifted PWM
technique

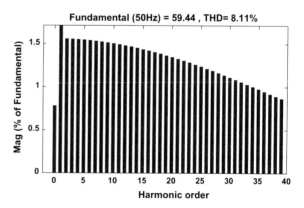

Table 2 THD comparison for multicarrier and phase shifted PWM techniques

Name of the PWM technique	Total Harmonic Distortion (%THD)
APOD	18.48
POD	15.93
PD	10.40
Phase Shifted	8.11

Fig. 18 The output voltage for phase shifted PWM with fuzzy controller

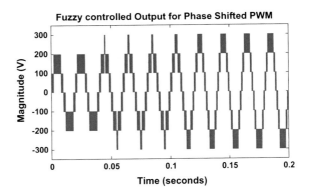

Fig. 19 RMS output voltage of the phase shifted PWM with fuzzy controller

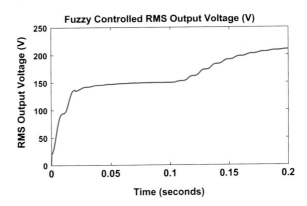

great breakdown voltage withstanding capability and it is used for switch controllers, exchanging converters, and relay drivers. These sorts can work genuinely with the foundation developed in circuits to 500 V; it can set the voltage of MOSFET IRF840. Specifications and requirements for the project are as follows: MOSFETs IRF840, driver boards TLP350, D-space kit, transformer 1:8 tapings with rating 230 V input and 0–15 V output, and DC sources are used to the requirement to our accountability process.

Fig. 20 Full setup of
seven-level CHBMLI in
D-space

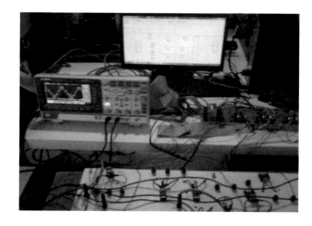

Fig. 21 Output voltage and
current waveform of
CHBMLI in D-space

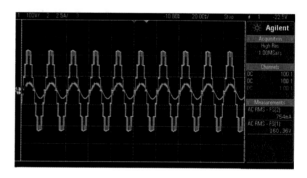

5.2.1 Single-Phase Cascaded H-Bridge Inverter with Seven Levels Using D-Space

Figure 20 shows the full setup of single-phase seven-level CHB MLI circuit kit in D-space. Three dc sources, MOSFET (IRF840), and driver (TLP350) are utilized and the output voltage is obtained using D-space processor it can withstand about 230 V in that programmable dc sources. Figure 21 demonstrates the yield RMS voltage (160.32 V) and current (754 mA) waveforms for seven-level CH-bridge multilevel inverter in D-Space.

6 Conclusion

In this paper, multicarrier and phase shifted sine PWM methods for seven-level inverter have been introduced. Execution factors corresponding %THD have been estimated, displayed, and broken down using both simulation and hardware implementations. It is initiated that the multicarrier APODPWM, multicarrier PODPWM, multicarrier PDPWM, and phase shifted PWM strategies, bring down %THD and also better performance is obtained in hardware. Contingent upon the execution measure fundamental, in particular, the use of favored MLI in light of the yield quality, fitting PWM must be utilized.

References

1. Franquelo, L.G., et al.: The age of multilevel converters arrives. IEEE Ind. Electron. Mag. **2**(2) (2008)
2. Rodriguez, J., Lai J.S., Peng, F.Z.: Multilevel inverters: a survey of topologies, controls, and applications. IEEE Trans. Ind. Electron. **49**(4), 724–738 (2002)
3. Tolbert, L.M., Peng, F.Z., Habetler, T.G.: Multilevel converters for large electric drives. IEEE Trans. Ind. Appl. **35**(1), 36–44 (1999)
4. Muralikumar, K.: Analysis of fuzzy controller for H-bridge flying capacitor multilevel converter. In: Proceedings of Sixth International Conference on Soft Computing for Problem Solving: SocProS 2016, vol. 1, p. 307. Springer (2017)
5. Calais, M., Agelidis, V.G.: Multilevel converters for single-phase grid connected photovoltaic systems-an overview. In: IEEE International Symposium on Industrial Electronics, 1998. Proceedings. ISIE'98. IEEE, vol. 1 (1998)
6. Ertl, H., Kolar, J.W., Zach, F.C.: A novel multicell DC-AC converter for applications in renewable energy systems. IEEE Trans. Ind. Electron. **49**(5), 1048–1057 (2002)
7. Kang, F.S., et al.: Multilevel PWM inverters suitable for the use of stand-alone photovoltaic power systems. IEEE Trans. Energy Convers. **20**(4), 906–915 (2005)
8. Reddy, V.P., Muralikumar, K., Ponnambalam, P., Mahapatra, A.: Asymmetric 15-level multilevel inverter with fuzzy controller using super imposed carrier PWM. In: 2017 Innovations in Power and Advanced Computing Technologies (i-PACT). IEEE, pp. 1–6 (2017)
9. Viswanath, Y., Muralikumar, K., Ponnambalam, P., Kumar, M.P.: Symmetrical cascaded switched-diode multilevel inverter with fuzzy controller. In: Soft Computing for Problem Solving, pp. 121–137. Springer, Singapore (2019)
10. Ponnambalam, P., Muralikumar, K., Vasundhara, P., Sreejith, S., Challa, B.: Fuzzy controlled switched capacitor boost inverter. Energy Procedia. **1**(117), 909–916 (2017)
11. Baker, R.H.: Electric power converter. US Patent No. 3,867,643 (1975)
12. Cecati, C., Ciancetta, F., Siano, P.: A multilevel inverter for photovoltaic systems with fuzzy logic control. IEEE Trans. Industr. Electron. **57**(12), 4115–4125 (2010)
13. Kevin, S.: Designing with fuzzy logic. IEEE Spect. **105**, 42–44 (1990)
14. Raviraj, V.S.C., Sen, P.C.: Comparative study of proportional–integral, sliding mode and fuzzy logic controllers for power converters. IEEE Trans. Ind. Appl. **33**(2), 518–525 (1997)

Derived Shape Features for Brain Hemorrhage Classification

Soumi Ray⓪ and **Vinod Kumar**

Abstract This paper presents the potential of derived shape features in the classification of the brain hemorrhage. Derived shape features are secondary features which are calculated from commonly used popular primary shape features. These features contain more relevant information having higher dependency on the shape of the target hemorrhage. Selection of high potential features is done to reduce the dimension of the input feature set to optimize classifier accuracy. The potential of these derived features is demonstrated and discussed with respect to the primary features.

Keywords Classification · Secondary features · Hemorrhage · Brain image · Shape features

1 Introduction

Classification of diseases has a significant contribution in diagnosis and treatment. It guides the practitioners in the right direction. Classification can be done in more than one way. Each way has its own angle to look at the disease. In case of brain hemorrhage, topographic information provides the required classification. Brain hemorrhages are classified into two major classes—extra-axial and intra-axial. Each of the classes is further divided into subclasses [1] as shown in Fig. 1.

Hemorrhage is the accumulation of blood in the local vicinity of rupture region. Depending on the tissue layers adherence, blood spreads in different patterns creating different shapes. Because of similar intensity and pattern of all classes of hemorrhages, they are not easily distinguishable by intensity and texture related features. Shape features have better potential in brain hemorrhage classification because of significant difference in nature of spread between different types of hemorrhage.

Several researchers have reported success high accuracy in the classification of brain hemorrhage using shape features [2–5]. Shape features describe the shape of

S. Ray (✉) · V. Kumar
Electrical Department, Indian Institute of Technology, Roorkee, Roorkee 247667, Uttarakhand, India
e-mail: Soumiray.eng@gmail.com

© Springer Nature Singapore Pte Ltd. 2020
K. N. Das et al. (eds.), *Soft Computing for Problem Solving*,
Advances in Intelligent Systems and Computing 1048,
https://doi.org/10.1007/978-981-15-0035-0_34

Fig. 1 Hemorrhage classes

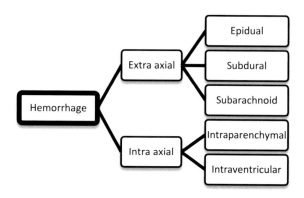

the ROI. Shape features are normally extracted from part of image to understand the local area information. Hemorrhage area, perimeter, longest axis, eccentricity, circularity, rectangularity, convexity, and much more information are extracted from the shape of ROI. Popularly used hemorrhage classifiers are decision tree [2, 6], Support Vector Machine (SVM) [7, 8], Multilayer Perception (MLP) [9], K-Nearest Neighbor (KNN) [10], K-means, and Expectation Maximization (EM) [5] methods.

For efficient classification, appropriate feature selection is most important. Though each feature contains some information about the target object, classification accuracy does increase proportionally with increase in number of input features. For continuous increase in number of features, a gradual deterioration is observed in classification accuracy after reaching the peak as shown in Fig. 2. The inclusion of features which are noncontributing or redundant must be removed to obtain maximum accuracy. This method is called feature optimization. Optimization is performed by features extraction and feature selection.

Feature extraction is a process of projecting the set of input features into some other plane of low dimensions. A new set of features are computed to be used as input of the classifier. Feature selection is relatively a simple method. Only the potential features are selected from the entire list of available input features. In this paper, we have combined both the techniques to gain higher performance of classifier with lower dimension input. Initially, we projected the available input features to another plane to create more relevant features, but put no control over dimension. The derived features are then optimized for classifier input.

Fig. 2 Performance of classifier

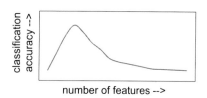

Table 1 Primary shape features

	Shape feature	Description
1	Area (A)	Count of total number of pixels in the hemorrhage patch is the measure of hemorrhage area
2	Perimeter (P)	Total length of boundary of the hemorrhage body is the perimeter of that hemorrhage
3	Major axis (M)	The longest available span in the hemorrhage body is considered as major axis of that hemorrhage
4	Minor axis (m)	Longest available span in the perpendicular direction of major axis is the minor axis of that hemorrhage
5	Equivalent circle (Ec)	The circle area of which is equal to the area of hemorrhage is denoted as equivalent circle
6	Best fit circle (Cb)	The smallest circle which can envelope all pixels in the hemorrhage
8	Best fit ellipse (E)	The smallest ellipse which can warp the entire hemorrhage body
9	Boundary (B)	The count of pixels form the bounding line around the shape

2 Primary Shape Features

Different types of hemorrhage have different shapes. EDH is biconvex and remains limited within one hemisphere. SDH is crescent in shape and can cross the suture line. The shape of SAH is irregular. IVH follows the shape of the ventricle with increase in its size. IPH can be of different shapes but not convex or crescent. As already discussed, the significant difference in shape between classes empowers classifier's accuracy.

The commonly used primary shape features are listed and described in Table 1. These features are the primary shape descriptors. They are evaluated directly from the Region Of Interest (ROI) of image.

3 Derived Shape Features

Each primary descriptor describes information independently. The problem arises with this independence when the same measured value for different shapes is offered. A similar case is discussed for elaboration. Four different types of hemorrhage are taken. In one case they have almost same hemorrhage area with different ROI area; in other cases they have different hemorrhage areas with almost same ROI area as shown in Fig. 3a, b. With this information it is difficult to identify the class of a hemorrhage. Using these two features, we have computed a higher order feature Information Packing Factor (IPF) which is more relevant. The derived feature depicts more class specific relevance as collated in Table 2 and Fig. 3c, d. To use this power, a set of derived features are evaluated from the primary features.

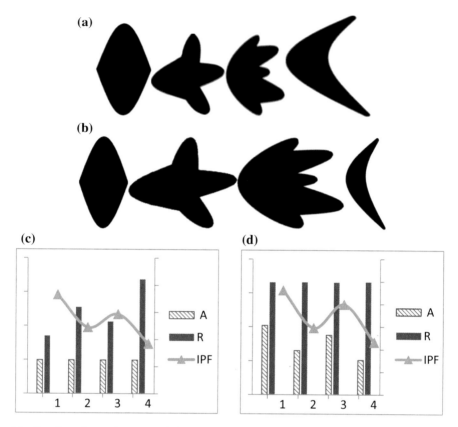

Fig. 3 **a** Same hemorrhage area, **b** Same ROI area, **c** IPF variation with constant ROI area, and **d** IPF variation with constant hemorrhage area

Table 2 Similarity in primary features' value

Image identifier	Image type	Hemorrhage area A	ROI size R	Ratio A/R
Figure 3a	Same ROI size	30539	**49715**	0.6143
		19536	**49776**	0.3925
		26339	**49646**	0.5305
		15345	**49830**	0.3079
Figure 3b	Same hemorrhage size	**24862**	42606	0.5835
		24890	63754	0.3904
		24825	53040	0.4680
		24825	84328	0.2944

3.1 Area to Perimeter Ratio (Ap)

This feature is derived from the area and perimeter of the hemorrhage body. For a straight line it is 1. For same area if the shape of hemorrhage is circular the ratio value will be maximum. Ap decreases with increase in non-circularity in shape. EDH offers maximum Ap due to its elliptical nature.

$$Ap = A/P$$

where A is the hemorrhage area and P is its perimeter.

3.2 Height (H) and Width (W)

Major and minor axes are rotation dependent. With rotation of hemorrhage body, these features also rotate in proportionally. To avoid angle correction, two simple features are used replacing these two axes. Irrespective of angle, the largest length across x and y axis are taken as height and width of the hemorrhage. Height (H) and Width (W) are basically the cosine component of Major (M) and Minor (m) axis, respectively. The concept is presented and elaborated in Fig. 4.

$$H = M \cos\theta \text{ and } W = m \cos\theta$$

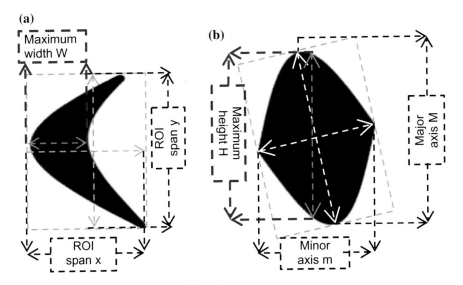

Fig. 4 a–b Major axis, minor axis, H, W, and hemorrhage spans

3.3 Span Ratio (Sr)

It is the ratio of height and width which are used in place of major and minor axis. This feature is actually the ratio of major and minor axes as the cosine part is equal for both.

$$\text{Sr} = H/W = (H^* \cos\theta)/(W^* \cos\theta) = (\text{Major axis})/(\text{min or axis})$$

3.4 Maximum Distribution Ratio (·X and ·Y)

The ratio between hemorrhage maximum span to ROI maximum span in each axis is the distribution ratio across the respective axis. When H is the hemorrhage span along y axis and W is along x axis then

$$\cdot x = W/x \quad \text{and} \quad \cdot y = H/y$$

where x and y are ROI span across x and y axis, respectively.

3.5 Circular Area Ratio (Cr) and Diameter Ratio (∂)

Diameter ratio is the ratio between the diameter of equivalent circle (d_1) and best fit circle (d_2) of a hemorrhage.

Circularity of hemorrhage is the measure of the shape's tendency to fit into a circle. The ratio of hemorrhage area and best fit circle becomes 1 when the shape is completely circular.

$$\text{Cr} = A/C = \text{Ec}/C = \pi r_1^2/\pi r_2^2 = r_1^2/r_2^2$$
$$\partial = d_1/d_2 = r_1/r_2 = \sqrt{A/C}$$

3.6 Elliptical Area Ratio (Er)

It is the ratio between hemorrhage area and best fit ellipse area. A convex shape is better fitted inside an ellipse increasing the value of Er. Generally, EDH offers higher Er than other hemorrhage patches.

Fig. 5 Skull vicinity

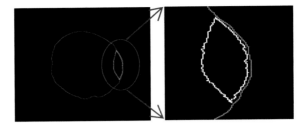

3.7   Skull Vicinity (V)

It is measured by taking the distance between boundary elements of hemorrhage and boundary of corresponding brain image. When hemorrhage boundary candidates remain within five-pixel distance of brain boundary, the logical value is recorded as positive [2]. The mathematical calculation is formulated as follows:

$$\%V = \left[\left(\sum \Delta d\{(1, 0) : |B_i - H_i| \le 5, 0\}\right)/ \sum H_i\right] * 100$$

B_i and H_i represent ith element of brain and hemorrhage boundary, respectively. For less than and equal to five pixels distance, Δd is considered as 1. Otherwise Δd is considered as 0. It presents how tightly or loosely the hemorrhage is attached to the skull as shown in Fig. 5.

3.8   Local_Information Packing Factor (Local_IPF) [11]

It is the information density of hemorrhage within ROI. Instead of evaluating IPF with respect to entire image size, for efficient calculation, IPF within ROI is evaluated.

3.9   Compactness (C) [11]

This feature is very close to IPF but not all background information within ROI is considered for compactness calculation. It only considers background pixels which are longitudinally located in-between hemorrhage pixels.

4   Optimization of Features

To study the potential of the features a combined set of primary and derived features is used to achieve the best performance of the classifier. The potential of a feature depends on its power of separating hemorrhages between classes. This potential is measured by comparing local and global variance.

Local variance is

$$\sigma_l^2 = \sum_i \frac{\sum_n (x_n - \mu_i)^2}{n}$$

Here, μ_i presents the mean of feature of ith class and n is the number of candidate images of that class. Similarly, global variance is

$$\sigma_o^2 = \sum_i \frac{\sum_n (x_n - \mu_o)^2}{n}$$

where μ_0 is the overall mean.

The separability S is

$$S = \frac{\sigma_o^2}{\sigma_l^2}$$

5   Result

5.1   Feature Selection

In this work, we have used 50 EDH, 50 SDH, 31 SAH, and 31 intra-axial hemorrhages. The hemorrhages are primarily categorized as major class and other class. In major class EDH and SDH are included. All other hemorrhages are clubbed into other class. The separability of different features are collated in Table 3 and graphically presented in Fig. 6.

The power is examined into two steps. First, the features having great potential to separate major class from rest are examined. In the second step, the features with higher potential to classify EDH from SDH are selected.

Table 3 Separability of different features

Primary features	Separability		Derived features	Separability	
	Target versus other	EDH versus SDH		Target versus other	EDH versus SDH
Perimeter	1.10	2.01	Skull vicinity V	3.19	0.99
Area	1.07	1.00	Span ratio Sr	1.28	1.03
W	1.01	1.04	Local_IPF	1.01	6.09
H	1.46	1.03	Circular area ratio Cr	1.45	1.39
Best fit circle	1.15	1.06	Elliptical area ratio Er	1.06	4.79
Best fit ellipse	1.08	1.05	Maximum distribution ratio $\cdot x$	1.03	4.61
Best fit rectangle	1.08	1.05	Maximum distribution ratio $\cdot y$	1.02	2.07
Radius of equivalent circle	1.12	1.02	Area to perimeter ratio Ap	1.04	1.22
			Compactness	1.03	3.40

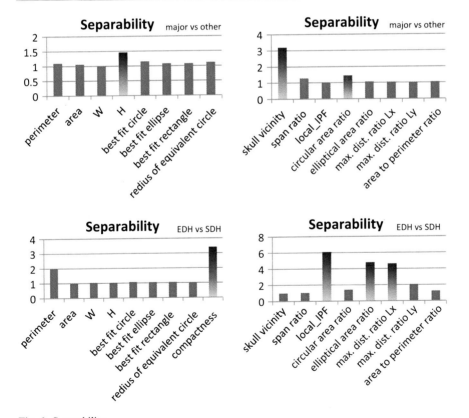

Fig. 6 Separability

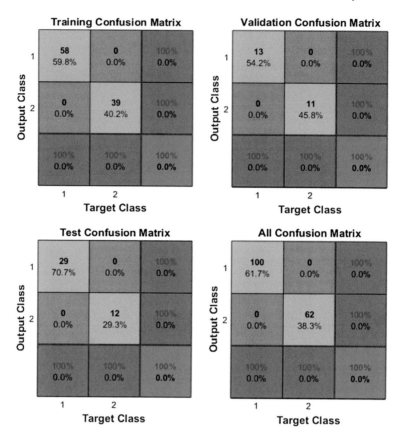

Fig. 7 Major versus other class

5.2 Classification

Skull vicinity followed by height and circular area ratio has demonstrated maximum power in separating major class from others. The major class is classified from others with 100% accuracy by skull vicinity as shown in Fig. 7. The major class is then classified into two classes EDH and SDH using local_IPF, elliptical area ratio, ·x, and compactness. They accurately classified the dataset as shown in Fig. 8.

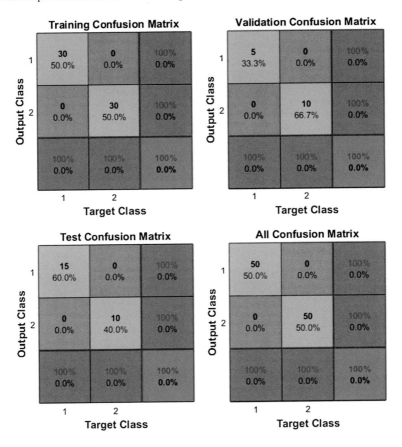

Fig. 8 EDH versus SDH

6 Conclusion

The selected features accurately classified the hemorrhages in respective groups. The proposed methodology has efficiently reduced the dimension of classifier input for hemorrhage classification system with the help of derived features.

Using only potential primary features the same classification process is repeated to verify derived features importance. The results are shown in Fig. 9a, b. To classify major against other three primary features are considered—height, best fit circle, and radius of equivalent circle. For EDH and SDH classification perimeter, best fit circle, best fit ellipse, and best fit rectangle are used because of their higher separability power. It is proved that the derived features are practically strong in case of classification of brain hemorrhage.

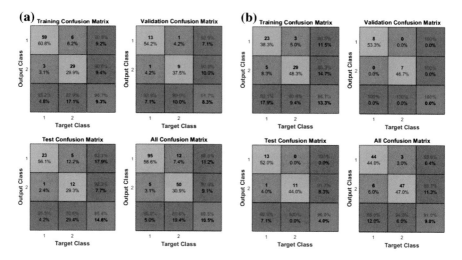

Fig. 9 **a** Major versus other classification using potential primary features **b** EDH versus SDH using potential primary features

References

1. Saatman, K.E., Duhaime, A.-C., Bullock, R., Maas, A.I., Valadka, A., Manley, G.T.: Classification of traumatic brain injury for targeted therapies. J. Neurotrauma **25**(7), 719–738 (2008)
2. Liao, C.C., Xiao, F., Wong, J.M., Chiang, I.J.: A knowledge discovery approach to diagnosing intracranial hematomas on brain CT: recognition, measurement and classification. In: International Conference on Medical Biometrics, pp. 73–82. Springer (2008)
3. Shahangian, B., Pourghassem, H.: Automatic brain hemorrhage segmentation and classification algorithm based on weighted grayscale histogram feature in a hierarchical classification structure. Biocybern. Biomed. Eng. **36**(1), 217–232 (2016)
4. Gong, T., Liu, R., Tan, C.L., Farzad, N., Lee, C.K., Pang, B.C., Tian, Q., Tang, S., Zhang, Z.: Classification of CT brain images of head trauma. In: IAPR International Workshop on Pattern Recognition in Bioinformatics, pp. 401–408. Springer (2007)
5. Lee, T.H., Fauzi, M.F.A., Komiya, R.: Segmentation of CT brain images using K-means and EM clustering. In: Fifth International Conference on Computer Graphics, Imaging and Visualisation, 2008. CGIV'08. IEEE, pp. 339–344 (2008)
6. Maduskar, P., Acharyya, M.: Automatic identification of intracranial hemorrhage in noncontrast CT with large slice thickness for trauma cases. In: SPIE Medical Imaging, International Society for Optics and Photonics, pp. 726011–726018 (2009)
7. Hsu, C.-W., Lin, C.-J.: A comparison of methods for multiclass support vector machines. IEEE Trans. Neural Netw. **13**(2), 415–425 (2002)
8. Srivastava, D.K., Sharma, B., Singh, A.: Classification of hematomas in brain CT images using support vector machine. In: Information and Communication Technology for Sustainable Development, pp. 375–385. Springer (2018)
9. Chowdhury, D.R., Chatterjee, M., Samanta, R.: An artificial neural network model for neonatal disease diagnosis. Int. J. Artif. Intell. Expert. Syst. (IJAE) **2**(3), 96–106 (2011)
10. Ramteke, R., Monali, Y.K.: Automatic medical image classification and abnormality detection using K-Nearest neighbour. Int. J. Adv. Comput. Res. **2**(4), 190–196 (2012)
11. Ray, S., Kumar, V.: Binary Image Features Proposed to Empower Computer Vision (2018). arXiv preprint arXiv:180808275

Prediction of Crime Rate Using Data Clustering Technique

A. Anitha

Abstract One of the most alarming and predominant aspects in our society is crime. Deterrence against crime is vital for the safety of each citizen of a nation. Analysis of crime in a systematic pattern can enable us to follow the trend of crime occurring and help us to prevent adversity. The main focus of this paper is to analyze the patterns of the data collected over a period of time about the crime against women by applying various clustering algorithms such as K-Means Custering, Agglomerative Custering, and Density-Based Spatial Clustering with Noise (DBSCAN). Clustering techniques are used massively along with the analysis, investigation, and uncovering the patterns for occurrence of different data. The comparative analysis of K-Means, Agglomerative Hierarchical Custering (AHC), and Density-Based Spatial Clustering (DBSCAN) were classified by training and test the real-time crime data against women were collected from the West Bengal, one of the famous state in India. The comparative analysis shows that the accuracy has been achieved as 96.1% of accuracy using DBSCAN clustering technique. This can help the police force to predict crimes which can occur in the future and take steps for accurate prevention.

Keywords Clustering · Crime rate · K-Means · DBSCAN · Agglomerative hierarchical clustering · Prediction

1 Introduction

Crime against women is a punishable offense that violates humanity and should be charged immediately by the law. Criminology is defined as a thorough study of crime and criminal motives. It is a multidisciplinary science that collects data on crime performances and perform investigations on them. The assaults and harassment upon women are increasing day by day and the police department should take proper

A. Anitha (✉)
School of Information Technology and Engineering,
VIT University, Vellore, India
e-mail: aanitha@vit.ac.in

© Springer Nature Singapore Pte Ltd. 2020
K. N. Das et al. (eds.), *Soft Computing for Problem Solving*,
Advances in Intelligent Systems and Computing 1048,
https://doi.org/10.1007/978-981-15-0035-0_35

measures to control and reduce these type of activities immediately. Identification and prevention of crimes are the chief challenges for the police force in the society as there exist a huge amount of data of criminal activities. For this reason, we need to provide methodologies and procedures to predict crimes that can happen in the future and take preventive measures accordingly. Data Mining is a field of study that defines various classification and clustering techniques that is useful for this purpose. Generally, clustering helps to group a set of abstract objects into classes of similar objects or patterns. In this paper, we are going to analyze three clustering techniques such as k-means cluster, aglomerative herarchical clustering, and dnsity-bsed custering which helps in identifying the crime pattern.

1.1 Basics of K-Means Clustering Algorithm

In general, K-means clustering is a partitioning technique for data analysis and to take care of observation based on the location and the distance between each data points. The commonly used k-means clustering is based on the Euclidean distance. This is considered as a special case of gradient descent method, where the cluster is adjusted at each learning step. The learning process involves considering the set of clusters each with h-dimensional centers, obtained by the number of input nodes in the input layer. The group center now becomes the center of the hidden neurons. The procedure for the k-means clustering is given as

1. Randomly select cluster "c".
2. Calculate the distance between the input vector and cluster center using Euclidean distance.
3. Assign the input value to the cluster center whose distance is minimum of all the cluster centers.
4. Recalculate the distance again until the same input value is reassigned.
5. Stop the process.

Partitioning the objects into totally unrelated groups (K) is obtained by a pattern that every object inside every cluster stay as close as remain to each other. K-means is one of the simplest methods for clustering. It classifies the given objects into the k clusters where k is the number of desired clusters. It first selects the initial center randomly for each cluster and put each data object to its nearest cluster centroid. After putting all data objects, centroid is recalculated with the cluster and the entire process is repeated. In each iteration, centroid change their location. This process is repeated until no centroid move in each iteration. As the result k clusters are found.

The sum-of-square criterion for the given data points in the data space is given as

$$\text{sum-of-square} = \sum_{i=1}^{K} \sum_{n \in S_i} |x_n - d_i|^2 \tag{1}$$

where

- K: Number of cluster
- S : Data space
- x_n: Data points
- d_i: Geometric centroid of the data points

1.2 Basics of Agglomerative Hierarchical Clustering

Agglomerative Hierarchical Clustering (AHC) is a lower to upper clustering method where clusters have subclusters, which in turn have subclusters, etc. Clusters generated in the initial partition is nested into the clusters generated in the later partitions. The subclusters give more knowledge during the data analysis, which mining the hidden data leads to knowledge discovery. The process of Agglomerative Hierarchical Clustering (AHC) is as follows:

1. Allocate each data objects in to separate clusters.
2. Distance matrix is calculated using Euclidean distance to evaluate pair-wise distances.
3. Build the distance matrix using the distance value from step 2.
4. Find out the pair of clusters with minimum distances.
5. Eliminate the negligible distance clusters and merge them.
6. Again evaluate the distances from the new clusters to all other clusters, and revise the distance matrix.
7. Repeat the process until the data objects reduce in to a single element.

1.3 Density-Based Clustering Technique

Density-based clustering technique was proposed by Martin Ester in 1996. It is the most useful clustering which deals with spaces. DBSCAN requires two parameters such as Epsilon and minpoints. The dense point in the cluster is called the starting point or the epsilon. The minimum number of points required to form dense region is the minpoints. If any data lies outside away from the dense area is called noise. The procedure to implement the DBSCAN is as follows:

1. Find the point in the epsilon range and find the minimum distance using Euclidean distance to find the neighborhood data.
2. Set the minpoints as minimum and create the clusters.
3. Identify the core points with more than minpoints neighbor data. Let C be the core point then the data points follow for the cluster is given, where x_i data points fall in the core points.

4. Find the connectivity of the core points using neighborhood data, and ignore the noncore points (noisy data).
5. Allocate each noisy data identified by the nearby cluster to discover more knowledge else eliminate it as noise and remove the data.

The paper has been planned as follows: Sect. 1 contains the introduction, whereas Sect. 2 discuss about the literature review. Section 3 explains the methodologies applied followed by Sect. 4, which provides the experimental analysis. The paper concludes with conclusion as Sect. 5.

2   Literature Review

Anitha and Acharjya have discussed the prediction of missing associations by hybridizing rough set with neural network [1]. Kurt Hornik et al. used R-extension package that uses a working space that uses fixed point and genetic algorithm (FPGA) solvers and implementing interfaces like CLUTO and Gmeans as two external solvers for the clustering feature extraction solvers used by Spherical k-means clustering technique [2]. Michael I. Jordan and Brian Kulis studied the Dirichlet Process Mixture Model (DPMM) and also the hard or exclusive clustering algorithms by the K-means algorithms which are derived from Bayesian nonparametric viewpoint [3]. They projected a process called hierarchical Dirichlet that provides high accuracy result, and it also reduces time complexity. Anitha and Acharjya has discussed the crop suitability prediction using rough set on fuzzy approximation space and neural network [4]. Navjot Kaur et al. made a comparative analysis using k-means clustering using ranking techniques and traditional k-means clustering with fixed number of clusters. They concluded that ranking based K-means efficient and effect with respect to time and space complexity over predicting the new objects [5]. Martti Juhola and Xingan Li utilized unsupervised self-organizing map to predict the crime by collecting data of 56 countries over 28 attributes [6]. The authors concluded that self-organizing maps are going to be a new crime mapping tool that could process huge amount of data. Anitha and Acharjya have predicted the customer choice using fuzzy rough set on two universal sets (FRSTUS) with radial basis function neural network (RBNN) [7].

Santhosh Baboo and Malathi used assorted data mining techniques like DBSCAN and K-means clustering and concluded that the tool can be used by the law enforcers to detect and avert crime more faster by identifying common crime patterns [8]. Malathi et al., used clustering and classification using Data Mining (DM) techniques to examine the crime data of police division to identify the crime trend and also suggest methods to reduce and prevent crimes. Francis O. Enem and Raphael Obi Okonkwo used various data mining techniques to aid law agencies to find out and stop terrorism [9]. They also deliberated the limitations of data mining in preventing offence in Nigeria and gave a conclusion that data mining can be only used to analyze crime in this province. Acharjya and Anitha have made a correlative study of statistical and rough computing models in predictive data analysis [10]. Mande et al. proposed a new procedure for crime analysis [11]. By using Generalized Gaussian Mixture Model they mapped the crime activities based on eye-witnessed specified feature. The model gave a unique and appropriate result.

3 Dataset Description and Proposed Research Model

This paper focus on using three clustering techniques such as K-means clustering, Agglomerative Hierarchal Clustering (AHC), and Density-Based Spatial Clustering with Noise (DBSCAN). For implementation of these clustering techniques, we collected crime dataset from 2001 to 2013 of West Bengal, India. The data are preprocessed and sorted according to our related crime records that are suitable for clustering were extracted and the consolidated database is formed. The above said three clustering techniques were implemented on the consolidated dataset, an experiment and comparative analysis have been discussed.

3.1 Dataset

The data was collected from State Crime Records Bureau West Bengal. The dataset consists of information about crimes against women of different districts of West Bengal during the year 2001 to 2013. The crime dataset consists of approximately 326 instances and 10 attributes a crime records with several attributes like seizure and abduction, extortion, endowment deaths, attempt to rape, gang rape, assaults on female, abuse to diffidence of women, unkindness by partner, importation of girls, etc. The total number of crimes against women in different districts of West Bengal during 2001–2013 were collected and presented in Tables 1 and 2. Table 1 represents the notations of various attributes whereas Table 4 depicts the consolidated sample data pertaining to our study.

3.2 Implementation of Proposed Model

The steps involved in implemeting the clustering technique of the empirical study is discussed in this section. Predicting the districts where the crime rate against the women features on real datasets is considered as the main objective of this article. We used K-Means Clustering (K-means), Density-based spatial clustering of applications with noise (DBSCAN), and the Agglomerative Hierarichal Clustering taken into consideration. The method is based on minimizing the Mean Square Error (MSE). The three clustering techniques as discussed in Sect. 1 are used to train the

Table 1 Notation representation table

Attributes	Abbreviation	Notation
District	DIS	a_1
Year	Year	a_2
Rape	Rape	a_3
Seizure and abduction	SAN	a_4
Enowment deaths	ED	a_5
Attack on female to outrage her diffidence	AFO	a_6
Abuse of diffidence	AD	a_7
Cruelty by partner/relatives	Cru	a_8
Importation of girls	IMG	a_9

Table 2 Sample dataset

Obj	a_1	a_2	a_3	a_4	a_5	a_6	a_7	a_8	a_9
x_1	Purab Midnapur	2009	41	74	40	33	0	475	0
x_2	Purulia	2003	53	0	9	30	0	170	0
x_3	Sealdah G.R.P	2002	1	6	3	7	0	55	0
x_4	Uttar Dinajpur	2001	0	1	0	3	0	225	0
x_5	24 Parganas North	2009	80	85	18	75	0	2378	0
x_6	24 Parganas North	2010	173	292	71	174	8	2879	5
x_7	Asansol	2010	258	215	48	113	0	310	0
x_8	Bankura	2010	18	13	19	13	0	279	3
x_9	Birbhum	2010	36	45	18	59	2	301	0
x_{10}	Burdwan	2001	44	29	25	48	0	279	7
x_{11}	Coochbehar	2012	72	69	28	93	0	301	0
x_{12}	Dakshin Dinajpur	2006	143	170	12	62	0	535	0
x_{13}	DARJEELING	2004	56	87	19	77	7	158	8
x_{14}	Hoogly	2010	36	75	2	57	0	92	0
x_{15}	Malda	2011	40	23	5	48	0	372	2

dataset. Based on the input attribute values, sum-of-squares are computed using weka tool as discussed in Eq. (1). The workflow of the proposed model is depicted in Fig. 1.

Implementation Using K-Means Clustering K-means clustering is an approach where interior patterns of crime dataset and associations among them are emphasized. This method is generally applied when there is a need to sort out a large crime database, make an outline of the required dataset. This approach helps to handle the large dataset by searching the right information and retrieve the preferred data that helps for prediction of crimes. This algorithm aims to generate groups in the dataset, where a variable K represents how many groups are there. This algorithm assigns each of the data points to any of the K number of groups and it is done by analyzing

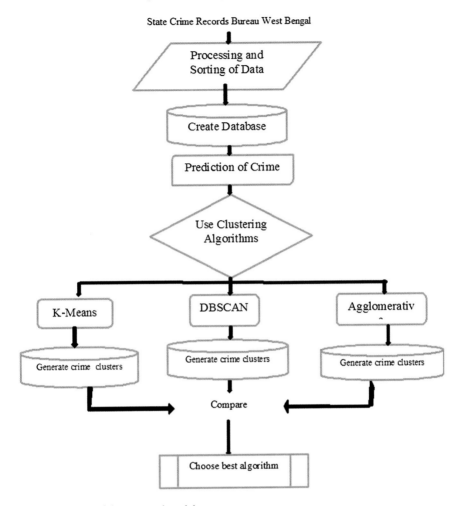

Fig. 1 Workflow of the proposed model

the provided features. Finally, the clustering is accomplished based on the similarity of the features. The results are represented in Figs. 2 and 3. Figure 2 represents the district attribute versus other attributes. Figure 3 depicts about the accuracy prediction using K-means clustering

The result of this method is as follows:

1. The centroid are generated for K clusters, which can be useful for labeling new data.
2. Every single cluster contains particular data points.

Experimental Analysis Using DBSCAN In the year of 1996, Hans-Peter Krieger, Xiaowei Xu, Martin Ester, and Jrg Sander came with the idea of DBSCAN algorithm.

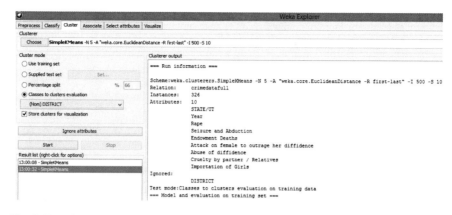

Fig. 2 Experimental analysis using K-Means Clustering

kMeans
===

Number of iterations: 7
Within cluster sum of squared errors: 26.04621203374859
Missing values globally replaced with mean/mode

Cluster centroids:

Attribute	Full Data	Cluster#				
		0	1	2	3	4
	(326)	(30)	(45)	(108)	(79)	(60)
STATE/UT	WEST BENGAL	WEST BENGAL	WEST BENGAL	WEST BENGAL	WEST BENGAL	WEST BENGAL
Year	2007.1564	2009.7	2011.4898	2002.8704	2006.5656	2010.8333
Rape	68.9325	238.1333	86.8367	32.1755	79.1013	22.4833
Seizure and Abduction	78.5368	296.1333	151.551	26.7037	57.6076	30.9667
Endowment Deaths	17.3405	54.3	18.4654	9.6389	20.8608	7.1667
Attack on female to outrage her diffidence	85.227	236.3333	155.5918	40.3056	83.5949	35.2167
Abuse of diffidence	6.5368	4	29	2.2963	3.2911	1.3667
Cruelty by partner / Relatives	456.3926	1981.1667	682.1837	157.6481	370.1266	160.5333
Importation of Girls	0.3773	0.4	0.5102	0.0648	1	0

Fig. 3 Accuracy prediction using K-Means Clustering

It is a density-based spatial clustering algorithm. This approach produces certain areas which have high concentration and forming clusters. Then it chooses arbitrary shaped clusters in spatial database with noise. A set of maximum densely connected points are defined as cluster. Figure 4 discusses about the clusters formed with respect to district (DIS) as its attribute. Figure 5 provides the accuracy prediction with respect to the cluster instance of the attribute district.

Agglomerative Hierarchical Clustering (ACH) In the field of statistics and data mining hierarchical clustering algorithm is such an approach which produces a hierarchy of clusters. Agglomerative hierarchical clustering algorithm is often called as a bottom-up strategy where each object starts by being assigned in their own cluster. Then these individual atomic clusters are integrated to form bigger cluster. This process continues until all the clusters are merged to make a sole parent cluster or until some crucial end ctiterias are achieved. This method never repeats any step, more

(Density-based Clustring Algorithm)
Input: Number of objects that need to be clustered, Eps and MinPts.
Output: Clusters of connected points with high density.

1. Select any arbitrary point, say x.
2. With respect to the Eps and MinPts, access all the points which can be density reachable from that point X.
3. If the point X is observed to be a core point, a cluster is generated.
4. Else If the point X is observed being a border point then no points can be density reachable from the point X and so it searches for the next data point of the data space and traverses it in the same manner
5. The process continues until the algorithm finishes visiting all the data in the data space.

Fig. 4 DBSCAN clustering with respect to district attribute

specifically if an object is already attached then it cannot be separated further. The accuracy calculated is represented in Fig. 6.

(Agglomerative Hierarchical Clustering Algorithm)
Input: Number of objects that need to be clustered.
Output: Clustered Instances.

1. Each object or data point is assigned to an individual and atomic cluster.
2. Distances between all the pairs of clusters are evaluated.
3. On the basis of distance values a distance matrix is formed.
4. Pair of clusters having the shortest distance is searched.
5. If found, then the pair is removed from the distance matrix and merged together.
6. Now the distances of all other clusters from the newly formed cluster are measured and the matrix is also updated accordingly.
7. Repeat the steps until there remains a single element in the matrix.

13 | BANKURA
13 | BIRBHUM
13 | BURDWAN
13 | COOCHBEHAR
13 | DAKSHIN DINAJPUR
13 | DARJEELING
13 | HOOGHLY
13 | HOWRAH
 8 | HOWRAH CITY
13 | HOWRAH G.R.P.
13 | JALPAIGURI
13 | KOLKATA
13 | MALDA
 1 | MIDNAPUR
13 | MURSHIDABAD
13 | NADIA
13 | PURULIA
13 | SEALDAH G.R.P.
13 | SILIGURI G.R.P.
13 | UTTAR DINAJPUR
12 | KHARAGPUR G.R.P.
12 | PASCHIM MIDNAPUR
12 | PURAB MIDNAPUR
 2 | BDN CP
 2 | BKP CP
 2 | JHARGRAM
 2 | SILIGURI_PC

Cluster 0 <-- 24 PARGANAS NORTH

clustered instances :   313.0   96.0123 %

Fig. 5 Accuracy prediction using DBSCAN

4 Comparative Analysis

Experimental analysis has been carried out to get efficiency of the proposed model. The experiments were conducted with a computer having Intel Pentium Processor, 8GB RAM, Windows 10 operating system, and MATLAB R2008a. For analysis purpose, data are collected from Crime bureau of West Bengal, India. We have chosen three clustering techniques such as Simple K-means, aggloromotive hierarichal, and density-based clustering. The number of objects is randomly taken by the weka tool. The clusters are formed with respect to the district attribute so that to predict the clustering accuracy for the rest of the attributes. The results of the clustering accuracy

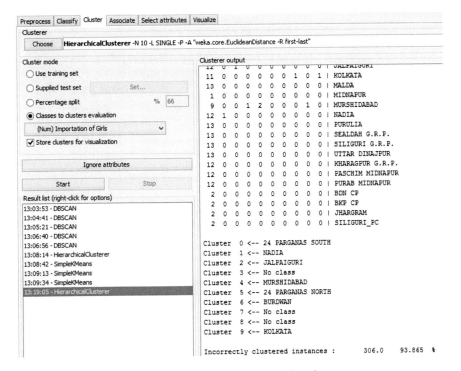

Fig. 6 Accuracy prediction using agglomerative hierarchial clustering

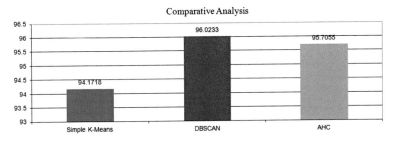

Fig. 7 Experimental comparative analysis

predicted are depicted in Figs. 3, 5, and 6. From the comparative analysis, it is clearly seen that the crime rate toward women with respective to various district and other parameters are clustered for a huge radius of geographical area. To consolidate, Fig. 7 gives the visual representation of the accuracy analysis. From Fig. 7 it is seen that the DBSCAN provides much clustering accuracy than the other algorithms

5 Conclusion

The poposed paper have fed the crime dataset in the three separate clustering techniques using Weka tool. These three clustering techniques are analyzed and compared to calculate the best accurate clustering approach to predict crime for the specified dataset. From the graphs provided, it is seen that K-means algorithm gives an accuracy of 94.172%, DBSCAN algorithm gives an accuracy of 96.0233%, and Agglomerative algorithm gives an accuracy of 95.7055%. Comparing these values, it is observed that DBSCAN algorithm gives the best accuracy of the considered dataset. The proposed method helps the agencies of law enforcement system to analyze the crime in an improved and accurate manner.

References

1. Anitha, A., Acharjya, D.P.: Neural network and rough set hybrid scheme for prediction of missing associations. Int. J. Bioinform. Res. Appl. **11**(6), 503–524 (2015)
2. Buchta, C., Kober, M., Feinerer, I., Hornik, K.: Spherical k-means clustering. J. Stat. Softw. **50**(10), 1–22 (2012)
3. Kulis, B., Jordan, M.I.: Revisiting k-means: new algorithms via Bayesian nonparametrics (2011). arXiv:1111.0352
4. Anitha, A., Acharjya, D.P.: Crop suitability prediction in Vellore District using rough set on fuzzy approximation space and neural network. Neural Comput. Appl. 1–18 (2017)
5. Kaur, N., Sahiwal, J.K., Kaur, N.: Efficient k-means clustering algorithm using ranking method in data mining. Int. J. Adv. Res. Comput. Eng. Technol. (IJARCET) **1**(3), 85 (2012)
6. Li, X., Juhola, M.: Country crime analysis using the self-organizing map, with special regard to demographic factors. AI Soc. **29**(1), 53–68 (2014)
7. Anitha, A., Acharjya, D.P.: Customer choice of super markets using fuzzy rough set on two universal sets and radial basis function neural network. Int. J. Intell. Inf. Technol. (IJIIT) **12**(3), 20–37 (2016)
8. Malathi, A., Baboo, S.S.: An enhanced algorithm to predict a future crime using data mining (2011)
9. Okonkwo, R.O., Enem, F.O.: Combating crime and terrorism using data mining techniques. In: 10th International conference IT people centred development. Nigeria Computer Society, Nigeria (2011)
10. Acharjya, D., Anitha, A.: A comparative study of statistical and rough computing models in predictive data analysis. Int. J. Ambient. Comput. Intell. (IJACI) **8**(2), 32–51 (2017)
11. Mande, U., Srinivas, Y., Murthy, J.V.R., Kakinada, V.V.: Feature specific criminal mapping using data mining techniques and generalized gaussian mixture model. Int. J. Comput. Sci. Commun. Netw. **2**(3), 375–379 (2012)

Identification of Astrocytoma Grade Using Intensity, Texture, and Shape Based Features

Arkajyoti Mitra, Prasun Chandra Tripathi and Soumen Bag

Abstract Astrocytoma is a common type of brain tumor that develops in the glial cells in cerebrum or astrocytes. In a malignant form, it is associated with high mortality. Identifying its grade helps the physicians to think about effective treatment. However, the irregular structure of this tumor type creates difficulty in the identification of its grade. Due to this, medical practitioners suggest additional examinations such as Magnetic Resonance Spectroscopy (MRS) and biopsy for accurate grade identification. In this work, we propose a method to identify astrocytoma grade from brain Magnetic Resonance Imaging (MRI). The proposed method can classify the tumor into low grade and high grade. The segmentation of the brain MRI is performed using spatial fuzzy clustering. We have used intensity, texture, and shape based features for classification. Five classifiers are used for the classification purpose. Our experiment results show that we can achieve an accuracy rate of 92.3% by integrating all three types of features together and applying a suitable classifier.

Keywords Astrocytoma · Brain MRI · Classification · Feature extraction · Segmentation · Tumor

1 Introduction

Magnetic Resonance Imaging (MRI) [1] is a painless and noninvasive procedure that provides extensive information about the inner organs (soft tissues) such as the brain, skeletal system, and reproductive system. An MRI encapsulates valuable information

A. Mitra (✉) · P. C. Tripathi · S. Bag
Department of Computer Science and Engineering, Indian Institute of Technology
(Indian School of Mines), Dhanbad, Dhanbad 826004, India
e-mail: arkaminator@gmail.com

P. C. Tripathi
e-mail: prasunchandratripathi@gmail.com

S. Bag
e-mail: bagsoumen@gmail.com

© Springer Nature Singapore Pte Ltd. 2020
K. N. Das et al. (eds.), *Soft Computing for Problem Solving*,
Advances in Intelligent Systems and Computing 1048,
https://doi.org/10.1007/978-981-15-0035-0_36

based upon numerous tissue parameters such as T1 (spin-lattice), T2 (spin-spin), and PD (proton density) giving us accurate brain tissue characterization. Brain tumors [2] are made up of cells that exhibit unstrained growth in the brain. The rate of survival can be enhanced if this gets detected in its embryonic stage.

Recent works have shown that image processing and pattern recognition methodologies are useful for the detection and characterization of brain tumors. In [3], researchers have proposed a method to characterize a brain tumor into two types: Gliomas and Meningioma using Probabilistic Neural Network. In [4], Computer-Aided Diagnosis (CAD) for brain tumors is presented. The method can classify a brain tumor into six types. The method utilizes texture- and histogram-based features. In [5], authors have used rough set firefly-based feature selection for tumor grade identification. The method has the ability to classify brain tumor into two types. In [6], authors have used rough granular computing and random forest for the classification of brain tumors. Researchers [7] have used bag-of-words method for the classification of tumors. In [8], authors have developed an expert system to classify astrocytoma. They have used Shuffling Frog Leaping algorithm for feature selection. The classification is carried out using three classifiers that are Support Vector Machines (SVM), Linear Vector Quantization (LVQ), and Naive Bayes. In [9], authors have used SVM to classify the astrocytoma grades from a brain MRI. We have observed that a very less number of works on grade identification of astrocytoma have been reported in the literature till now. These methods do not provide good classification accuracy for this tumor type due to its irregular appearance. In this paper, we have proposed a brain tumor classification method to detect astrocytoma. The proposed method works for two-class grade identification. The method performs the segmentation of brain MRI through spatial fuzzy clustering. Integration of three types of feature extraction methods is used for obtaining meaningful features from segmented MRIs, and finally five different classifiers are used for the classification.

The remaining article is categorized as follows: Sect. 2 describes the proposed methodology, experimental results and its analysis are discussed in Sect. 3, and finally, Sect. 4 concludes the article.

2   Proposed Methodology

The proposed method consists of three main phases such as preprocessing, feature extraction, and classification. Sections 2.1 to 2.3 describe these phases sequentially. The proposed method can be visualized as shown in Fig. 1.

2.1   Preprocessing

In this work, we have used skull stripping and brain MRI segmentation as a preprocessing step.

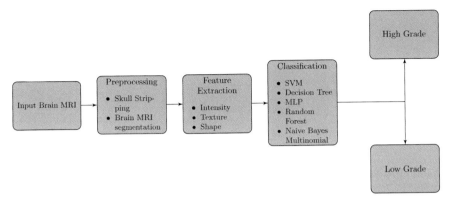

Fig. 1 Flow diagram of the proposed method

Skull Stripping A brain MRI scan also includes non-cerebral tissues such as skull, dura, meninges, and scalp. These tissues typically deteriorate the segmentation performance. Thus, the non-cerebral tissues in a brain MRI are removed before any further processing [10]. In this work, we have used intensity thresholding and morphological operations to extract the brain region from a brain MRI. The steps are as follows:

1. Input image (Fig. 2a) is binarized with a global threshold T as shown in Fig. 2b. The threshold T is taken in the range [0.12, 0.16].
2. Morphological erosion is applied on the binary image to separate the brain with skull part (Fig. 2c) and the largest region is selected as shown in Fig. 2d.
3. Morphological dilation is applied on the output of Step 2 as shown in Fig. 2e and region filling algorithm is executed to obtain brain mask (Fig. 2f). The brain region is extracted using the mask as shown in Fig. 2g.

Brain MRI Segmentation The segmentation of brain MRI is to partition the brain image into prominent tissues and abnormalities (if any). The important regions of brain MRI include Gray Matter (GM), White Matter (WM), and Cerebrospinal fluid (CSF). The segmentation of brain MRI is not a trivial task due to the presence of intensity inhomogeneity artifact and complex anatomy of the human brain. We have used spatial fuzzy C-means (sFCM) [11] for the segmentation of brain MRI. This algorithm includes the local information in the clustering process. Due to that, it provides robustness in the segmentation in the presence of intensity inhomogeneity artifact.

Suppose $Y = \{y_1, y_2, \ldots, y_i, \ldots, y_N\}$ is a brain MRI of size N and y_i denotes gray level of ith pixel. The sFCM algorithm partitions the image Y into C tissue classes $(1 < C < N)$ using the following iterative equations [11]:

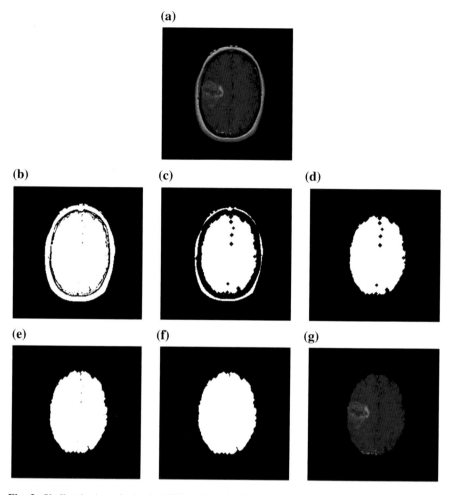

Fig. 2 Skull stripping of a brain MRI: **a** Input brain MRI, **b** Binarized MRI, **c** Eroded image, **d** Selected largest connected region, **e** Dilated image, **f** Region filled image, and **g** Skull stripped image

$$\mu_{ik} = \frac{1}{\sum_{j=1}^{C} \left(\frac{||y_i - c_k||^2 + \frac{\alpha}{N_R} \sum_{r \in N_i} ||y_r - c_k||^2}{||y_i - c_j||^2 + \frac{\alpha}{N_R} \sum_{r \in N_i} ||y_r - c_j||^2} \right)^{\frac{1}{m-1}}} \tag{1}$$

$$c_k = \frac{\sum_{i=1}^{N} \mu_{ik}^m \left(y_i + \frac{\alpha}{N_R} \sum_{r \in N_i} y_r \right)}{(1 + \alpha) \sum_{i=1}^{N} \mu_{ik}^m} \tag{2}$$

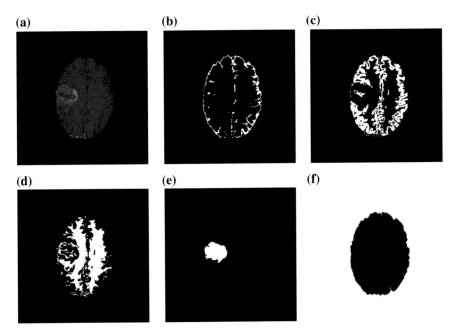

Fig. 3 Segmentation of a brain MRI: **a** Input image, **b** CSF, **c** GM, **d** WM, **e** Tumor, and **f** Background

where c_k represents cluster prototype of kth cluster, μ_{ik} represents fuzzy membership in kth cluster of pixel i ($\mu_{ik} \in [0, 1]$), m is the fuzzy weighting exponent ($m > 1$), N_i is a set of neighbors of pixel i, α is the trade-off parameter, N_R is the size neighborhood, and $||\bullet||$ denotes L2-norm. In this work, we have taken $C = 5$ (number of tissue classes) for CSF, GM, WM, tumor, and background as shown in Fig. 3.

2.2 Feature Extraction

Representing an image requires a large amount of data which in turn occupies large memory space and processing such data for classification purpose would take high computational time and resources. The most effective approach is to extract features from an image which gets rid of all unnecessary data and retains only the relevant information that are useful for classification. In our study, we have extracted five intensity-based features that are mainly based on the histogram of the grayscale image. We have also extracted 64 texture-based features and five shape-based features for improving the accuracy rate.

Intensity-Based Features In this study, we have considered five statistical features which are purely based on the pixel intensities of the grayscale image. They are Mean, Variance, Skewness, Kurtosis, and Energy. Mean gives the average of intensities present at each pixel. Variance gives the deviation of each pixel from the mean. It is calculated by squaring the difference between each intensity value present in the image with its mean and dividing the sum of squares by the total number of values present in the image. The normal distribution is having a skewness of zero and for any symmetric data, skewness should be near zero. It can be described as a measure of symmetry. Skewed right is indicated by the positive value of skewness. We can determine whether the data is peaked or flat relative to a normal distribution by the help of kurtosis.

Texture-Based Features Gray-Level Co-occurrence Matrix (GLCM) [12] is based on the number of occurrences of a pair of gray-scale values of pixels in the whole image. The gray levels can be set according to our requirement which we can normalize from the whole spectrum of gray values present in the image. In our study, we have found that 64 gray levels give the optimal performance result. Increasing to more than 64 levels does not contribute much in the classification process. We have observed that using GLCM elements directly as a feature gives significantly higher performance result than using second-order statistical features such as Correlation, Contrast, Homogeneity, and Entropy. We have evaluated the average of four GLCMs obtained in four directions ($0°$, $45°$, $90°$, and $135°$). The co-occurrence distance we have considered is four pixels from the center pixel across all directions. This means that when we consider the gray value of a pixel, we pair it up with the fourth pixel along the four different directions and count the occurrences of the pair in the entire image.

Shape-Based Features We have used four basic shape-based properties as a feature for our study. They are Area, Perimeter, Minor Axis, Major Axis, and Circularity. The major and minor axis give an idea about the orientation of the irregular-shaped spread-out tumors. The circularity feature also helps in focusing the outer bound region of the tumors.

2.3   Classification

After extracting features from the segmented images, we ran several classifiers upon these features. The mode of learning used in our experiments is supervised learning. In this learning method, the models or classifiers learn from a labeled dataset known as training data and try to predict the class of an unknown image from the testing data. Since our experiments are based on limited samples of labeled data, we have applied a special technique called k-fold cross-validation to check the accuracy of our models on the dataset. In cross-validation, we randomly shuffle the dataset into k folds or groups and then take a single group as our testing data and remaining groups as our training dataset. This helps in estimating the performance of the model on unknown data and also produces a less-biased result. For any dataset, the value of k should be chosen carefully, or else it might lead to overfitting which simply means that it will give high accuracy in the known dataset but will perform poorly on the unknown data. In our experiments, we have chosen $k = 5$, i.e., 5-fold cross-validation. The classifiers that we have tried to focus are SVM, Multilayer Perceptron (MLP), Random Forest, Decision Tree, and Naive Bayes Multinomial classifier.

3 Experimental Results and Analysis

3.1 Experimental Setup

In the experimental setup for tumor segmentation, the neighborhood window is taken as a square window of size 3×3. We have taken the fuzzy weighting exponent $m = 2$, number of cluster $C = 5$, and parameter $\alpha = 1$.

For each of the classifiers, the batch size is set to 100. In SVM, the tolerance parameter has been set to 0.001 and the kernel function chosen is Normalized Polynomial Kernel. In Decision Tree, we have split the attributes based on the Best First Search which is evaluated using RMSE (Root Mean Squared Error). In Random Forest, the number of iterations has been set to 100 and the execution slot is set to 1 with Random Tree as a base classifier. In MLP, we have set 38 nodes in the hidden layer in order to train using each feature. We have set the learning rate to be 0.3 with a momentum of 0.2. Here, the number of epochs has been set to 500 and a validation threshold has been set to 20. For Naive Bayes Multinomial function, we have set the number of decimal places rounding up to 2. We have also tested classification using different kernels for SVM as it was giving optimal results among all classifiers. We have used kernels such as RBF (Radial Basis Function) kernel, polynomial kernel, Puk (Pearson VII Universal Kernel) along with Normalized Polynomial Kernel. We have set values for different parameters as per our experimented analysis.

The proposed method was developed and successfully run in MATLAB R2017a and we have used Weka machine learning toolbox for running several classification algorithms. All of these were executed under Windows 8, 64-bit operating system running 2.5 GHz CPU, Intel Core i5-3210M processor, and 4GB RAM.

3.2 Dataset

The dataset has been acquired from Cancer Imaging Archive.[1] To encourage the development and evaluation of new algorithms in medical imaging, the archive provides various benchmark datasets. We have taken 168 brain images of astrocytoma.

---

[1] http://www.cancerimagingarchive.net/.

There are 94 cases for low-grade astrocytoma and 74 cases representing high-grade astrocytoma. The images are in T1-w contrast-enhanced form.

3.3 Result Analysis

The results of the preprocessing step are illustrated in Fig. 4. Different regions in the segmented image are shown using five colors: Red color shows the tumor region, Green color shows the gray matter, White color shows the white matter, Blue color shows the CSF, and Black color shows the background.

For any brain tumor classification, texture-based features are considered to be the major contributory element among all features. In our studies, we have considered the most common texture-based feature which is GLCM and on the sole basis of GLCM as a feature, we ran five different machine learning classifiers (SVM, Random Forest, MLP, Decision Tree, and Naive Bayes Multinomial). The maximum accuracy is achieved using the SVM classifier as depicted in Table 1. Next, we have integrated three types of features (intensity, texture, and shape based features) in our feature set. The classification results for the integrated feature set are presented in Table 2. It is noticed that SVM provides better performance compared to other classifiers.

We have also tested the classification accuracy using different kernels for SVM as depicted in Table 3. It is noticed that Normalized Polynomial Kernel gave better result compared to three other kernels. In general, the polynomial kernel is a kernel function commonly used with SVM that represents the similarity of vectors (training samples) in a feature space over polynomials of the original variables. It not only observes any feature and determines their similarity but also considers combinations of these features.

We have compared our performance in the proposed methodology with two existing methods in Table 4. It is noticed that our proposed method performs better than the other two existing methods (Zhao et al. [9] and Subashini et al. [8]). In earlier studies, only a few parameters were being used as features such as Magnetic Resonance (MR) and other clinical parameters. Earlier, GLCM was considered only for computing second-order statistical features such as Contrast, Correlation, Homogeneity, and Entropy. In our proposed method, we have taken into account all the well-known parameters such as intensity, texture, and shape as our features for better classification. We have also used GLCM as a feature vector directly into our feature set. The improvement in our performance is due to the integration of all these feature extraction methods together.

Fig. 4 Preprocessing of brain MRI. First column: input brain images. Second column: skull-stripped brain images. Third column: segmented brain images

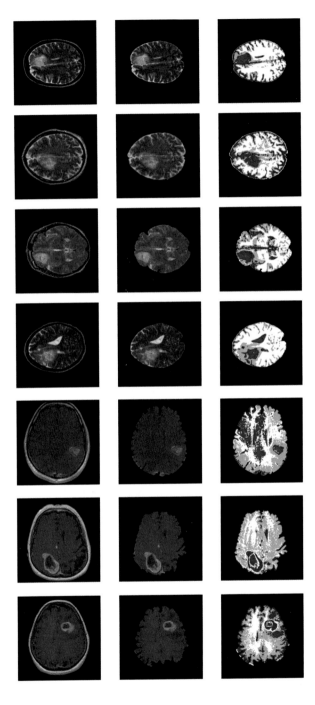

Table 1 Accuracy using GLCM features

Classifier	Accuracy (in %)		
	Low-Grade	High-Grade	Total
SVM	**94**	**83**	**89.9**
Decision tree	96	74	86.9
Random forest	93	85	89.8
Multilayer perceptron	89	81	85.7
Naive bayes multinomial	90	86	88.7

Table 2 Accuracy using all features

Classifier	Accuracy (in %)		
	Low-Grade	High-Grade	Total
SVM	**97**	**85**	**92.3**
Decision tree	96	77	87.5
Random forest	93	85	89.3
Multilayer perceptron	93	85	89.9
Naive bayes multinomial	90	86	88.7

Table 3 Accuracy using different SVM kernels

SVM Kernels	Accuracy (in %)		
	Low-Grade	High-Grade	Total
Normalized polynomial kernel	**97**	**85**	**92.3**
Polynomial kernel	96	81	89.9
Puk	92	87	90.5
RBF kernel	100	45	76.2

Table 4 Comparison with existing methods

Method	Number of images used	Accuracy (in %)
Proposed method	168	**92.3**
Zhao et al. [9]	106	82.1
Subashini et al. [8]	200	91.0

4 Conclusion

Astrocytoma is the deadliest type of brain tumor in young people. Identifying its grade has importance in effective diagnosis and prognosis. Radiologists face difficulty in its grade identification due to its uncertain appearance. In this work, we proposed a method to identify astrocytoma grade using prominent image features extracted from brain MRI. We extracted 74 intensity, texture, and shape based features for classification. Five classifiers have been used for classification. To evaluate the effectiveness of the proposed method, we have compared it with two existing methods. The accuracy rate can be further improved by introducing more number of samples in the dataset.

References

1. Magnetic resonance imaging. https://www.radiologyinfo.org/
2. Brain tumor research. https://www.braintumourresearch.org/
3. Georgiadis, P., Cavourous, D., Kalatzis, I., Daskalakis, A., Kagadis, G.C., Sifaki, K., Malamas, M., Nikiforidis, G., Solomou, E.: Improving brain tumor characterization on MRI by probabilistic neural networks and non-linear transformation of textural features. Comput. Methods Programs Biomed. **89**(1), 24–32 (2008)
4. Sachdeva, J., Kumar, V., Gupta, I., Khandelwal, N., Ahuja, C.K.: Segmentation, feature extraction, and multiclass brain tumor classification. J. Digit. Imaging **26**(6), 1141–1150 (2013)
5. Jyothi, G., Inbrani, H.: Hybrid tolerance rough set-firefly based supervised feature selection for MRI brain tumor image classification. Appl. Soft Comput. **46**, 639–651 (2016)
6. Koley, S., Sadhu, A.K., Mitra, P., Chakraborty, B., Chakraborty, C.: Delineation and diagnosis of brain tumors from post contrast T1-weighted MR images using rough granular computing and random forest. Appl. Soft Comput. **41**, 453–465 (2016)
7. Chung, J., Huang, W., Cao, S., Yang, R., Yang, W., Yun, Z., Wang, Z., Feng, Q.: Enhanced performance of brain tumor classification via tumor region augmentation and partition. PloS one **10**(10), e0140381 (2015)
8. Subashini, M.M., Sahoo, S.K., Sunil, V., Easwaran, E.: A non-invasive methodology for the grade identification of astrocytoma using image processing and artificial intelligence techniques. Expert. Syst. Appl. **43**, 186–196 (2016)
9. Zhao, Z.-X., Lan, K., Xiao, J.-H., Zhang, Y., Xu, P., Jia, L., He, M.: A new method to classify pathologic grades of astrocytomas based on magnetic resonance imaging appearances. Neurol. India **58**(5), 685 (2010)
10. Laha, M., Tripathi, P.C., Bag, S.: A skull stripping from brain MRI using adaptive iterative thresholding and mathematical morphology, In: Proceedings of International Conference on Recent Advances in Information Technology, pp. 1–6 (2018)
11. Ahmed, M.N., Yamany, S.M., Mohmed, N., Farag, A.A., Moriarty, T.: A modified fuzzy c-means algorithm for bias field estimation and segmentation of MRI data. IEEE Trans. Med. Imaging **21**(3), 193–199 (2002)
12. Torheim, T., Malinen, E., Kvaal, K., Lyng, H., Indahl, U.G., Anderson, E.K.F., Futsæther, C.M.: Classification of dynamic contrast enhanced MR images of cervical cancers using texture analysis and support vector machines. IEEE Trans. Med. Imaging, **33**(8), 1648–1656 (2014)

Early Prenatal Diagnosis of Down's Syndrome-A Machine Learning Approach

Esther Hannah, Lilly Raamesh and Sumathi

Abstract A chromosomal disorder called Down's syndrome is a disorder where the disability is seen at the intellectual level. It further shows up a prominent change in the appearance of the face, and often accompanied by an unhealthy muscle tone during infancy. Trisomy-21 is the cause of such conditions in many cases. This research article focuses to improve the quality of health care by using smart technologies. A smart healthcare system that is based on the use of machine learning methods in the detection of presence of trisomy-21 disorder in a fetus is implemented. The system is trained using medical data consisting of a well-defined set of features. The feature set consists of features representing both maternal and fetal data. The proposed Down Syndrome Detection (DSD) system produces better accuracy in terms of precision, recall, and F-measure in classifying an unknown test sample.

Keywords Supervised training · Classification · Decision trees · Feature extraction · Down's syndrome · Trisomy-21

1 Introduction

There is a rapid increase in the need for services like education and health care and resources like water, energy, etc. This growing need accounts for the increase in urbanization in developing countries. Hence, there is an urge to use resources in a smart manner which in turn leads to smarter cities. Using smart healthcare technologies improves the well-being of the residents in the city where it is implemented. The benefits of smart healthcare are increasing manifold. This research article focusses on a global birth defect called Down's syndrome. Considering human race, birth defects are said to affect almost 1 in every 33 infants born, which can amount to 3.2 million children every year. In developing countries, birth defects are estimated to be the primary cause of deaths among children below five years of age. Birth defects due to chromosomal disorders are reported to be the cause of about 9.5% of

E. Hannah (✉) · L. Raamesh · Sumathi
St. Joseph's College of Engineering, Chennai, India
e-mail: hanmoses@gmail.com

© Springer Nature Singapore Pte Ltd. 2020
K. N. Das et al. (eds.), *Soft Computing for Problem Solving*,
Advances in Intelligent Systems and Computing 1048,
https://doi.org/10.1007/978-981-15-0035-0_37

perinatal deaths and about 9.9% of stillbirths in India [1]. A fetus which is affected by trisomy-21 suffers from a chromosomal disorder which can affect the normal and social life of the unborn child [2–4].

Down syndrome happens when a part of chromosome 21 gets attached with other chromosomes very early in fetal development [5]. Fetuses that are affected by trisomy-21 disorder have three genes of chromosome 21 [6]. It is believed that possessing an extra copy of chromosome 21 hinders the normal and timely development of the fetus [7]. It is estimated that about 15 percent of people affected by trisomy-21 have a thyroid gland with reduced functioning and increased risk of hearing and vision problems [8, 9, 6]. In addition, it must be noted that a small percentage of children with trisomy-21 also develop leukemia. With an effort to focus on this global disorder, we propose a machine learning-based approach toward the detection of Down's syndrome. This smart system will definitely enable early detection of the Down syndrome in fetuses, thus working toward the betterment of the society and in turn to smarter health care.

Prenatal screenings are screening tests that help to detect the possibility of a fetus being affected by Down's syndrome [10, 11]. These calculate quantities of different substances present in the blood sample taken from the pregnant mother. Along with the maternal age, the blood values are used to predict the chance of carrying a child with trisomy-21 disorder. These screening tests are generally done during the first trimester and sometimes in the second trimesters. These screening tests can predict about 75–80% of fetuses with trisomy-21 [12–14].

2 Smart Healthcare Using Machine Learning

Smart healthcare management is one that converts health-related data into clinical sights that can consist of diagnoses, treatment, and maintenance of digital health details remotely [15, 16]. Smart health systems often refer to home health services, patient monitoring systems, etc. The word "Smart Healthcare" refers to the usage of intelligent and networking technologies in health care which can help to maintain and also improve the healthy living conditions of citizens. It enables a change in the perspective of highlighting the importance that "prevention is better than cure" [15].

Down Syndrome Detection (DSD) system will enable the delivery of a framework that provides digital base with artificial intelligence grip, can reform the existing health system, thus linking it tightly with a smart city. Nowadays, machine learning algorithms are widely used in a number of medical applications where intelligence is given to machines or systems in the form of supervised learning. Classification is perhaps one of the most widely applied supervised learning techniques where a classifier is trained using a training dataset, and a testing set validates the classifier. Classification techniques are found to be successful in cases where tasks easily done by humans need to be automated and the classifier uses the knowledge gained during the training process.

In this present work, a method of detecting the presence of trisomy-21 disorder in a fetus is modeled as a classification model where maternal and fetal data is classified into one of the two classes as "presence of DS" or "absence of DS" using the machine learning-based Down Syndrome Detection (DSD) mode. The present focus is on a robust technique for early prenatal diagnosis of trisomy-21 that exploits the advantages of feature extraction. A well-defined complete set of features that characterize the fetus and the mother is taken as training dataset. Decision tree-based classification is utilized in the detection of Down syndrome. Thus, the proposed work highlights the need for classifying the trisomy-21 disorder much earlier than the conventional prenatal screening. The system uses efficient preprocessing of data in order to produce clean, error-free data. Good preprocessing assures the system the delivery of high quality and accurate results which is the need of the hour.

The proposed smart system that is capable of detecting the chance of the birth of a Down's syndrome affected child can be installed in cities. Deployment of the smart system should be on the computers of hospitals, scan centers, and healthcare centers of the target city. The proposed work aims to serve the community at large by early detection of a developmental problem, right diagnosis, medical help, and awareness to families to resort to appropriate action and plan subsequent pregnancies. The system can be used for both inpatient or outpatient environments and ensures the availability of appropriate healthcare advice and consecutive decision-making at the right time. The proposed smart healthcare system can be used in both developed and developing nations, which upholds a promising scheme in the change toward smarter healthcare.

The remainder of this paper is given as follows: Sect. 3 brings out the current work done related to Down's syndrome. Section 4 proceeds to elaborate the proposed machine learning-based framework, then Sect. 5 depicts the experimental setup and discusses the evaluated results, and Sect. 6 gives the article a concluding note with scope for enhancements.

3   Related Work

One of the major challenges to a healthy population is chromosomal disorders. One such genetic disorder is Down syndrome. Not only does it affect the growth of the child, but poses to generate various medical issues. It is identified with modern technological advances in health care Down syndrome should be detected before birth or at least soon after birth [14]. Though Down's syndrome cannot be cured completely, adequate treatment and help can be provided to improve the quality of life. Since there is a rise in probability of giving birth to a child with Down syndrome, the importance of prenatal screening during the early weeks of pregnancy is evident [11, 5].

4 Proposed Architecture

Worldwide, the number of children affected by Down's syndrome is on the rise [17, 18], proving the fact that it is fatal in many developing countries due to lack of technological and medical facilities, negligence, and mainly due to little or no awareness toward the grave condition [19, 20]. This condition many a time come with heart diseases and other associated chronic conditions which can be treated when detected early. Surgery for treating such conditions in children of body weight less than ten kilograms often gives poor chance of survival or leading to complications like pulmonary hypertension. It is found that the survival rate of children with Down's syndrome goes down to 44% when the child has congenital heart disease [20].

The proposed smart Down Syndrome Detection (DSD) system can provide a definitive diagnosis by making use of artificial intelligence methods. The decision support system makes use of classification and is trained to classify a new fetal test data as being affected by trisomy-21 disorder or not, thus avoiding futile expectations and anxiety about the unborn baby. The inputs to the system would be prenatal screening test reports, maternal age, and the first-trimester ultrasound images. Once parents gain knowledge about the chromosomal disorder, they are left open to make decisions even as early as 11 weeks of gestational age. The following subsection discusses classification approach and its role in detecting the Down syndrome affected fetus.

4.1 Classification

Many classification algorithms exist in literature and decision trees are one of the popular methods due to the ease of its implementation [21–23]. It is a set of supervised learning algorithms that classifies data objects into their predefined class labels. Decision trees are one class of classification algorithms, whose most useful characteristics is their comprehensibility. The main benefits of decision trees and its role in classification is discussed in the following subsection.

4.2 Decision Trees

The DSD model is trained and tested with decision trees classification, which can be a form of multiple variable analysis. Decision trees have the capability to increase the performance of traditional forms of analysis, multidimensional forms, and a number of tools and techniques. The approach starts with extracting features which are preprocessed to remove noise and a training data set with patient data and class attribute is used for classifying each instance using the tree induction algorithm. Decision trees are successful in the classification of sentences in the field of text summarization [24]. The proposed work also uses decision trees for detecting Down's syndrome. The following subsection discusses the DSD model.

4.3 Down Syndrome Detection (DSD) Model

Researchers have found that about 75% of trisomy-21 fetuses have increased Nuchal Translucency (NT) thickness and 60–70% have absent nasal bone. The above two biomarkers can be detected using image processing techniques on the first-trimester ultrasound images. The smart Down Syndrome Detection (DSD) system uses supervised learning using decision trees in order to classify test data as being affected by Down's syndrome or not. Since prenatal screening tests and ultrasounds are now routinely offered to women of all ages, the proposed system can detect the Down's syndrome quite early than all existing prenatal diagnostic tests. Figure 1 shows the overall architecture of the proposed DSD approach.

The proposed machine learning-based DSD Model shown in Fig. 1 consists of the following two major phases namely: (i) Model Construction (ii) Model Usage. These phases are discussed in the following subsections:

4.3.1 Model Construction

In the model construction phase, both fetal and maternal data is collected and pre-processed. Preprocessing is an important and crucial step for cleaning any training dataset. The training dataset is created using the features extracted using the feature extraction process [25, 26]. The main role of the feature selection and extraction is to select and extract important features from blood report of the pregnant mother and the fetal ultrasound images. We have identified six important features for each patient, viz., maternal age, neural tube thickness of fetus, presence or absence of nasal bone, whether previously had a child with a genetic disorder, have a family history of Down's syndrome, and cell-free DNA blood test data.

Training the DSD Model

The DSD model is trained with a unique combination of maternal and fetal data. The class attribute "status" shows whether the selected patient data indicates the presence or absence of Down syndrome for each training tuple. The "status" is denoted as DS for presence of DS and NOT-DS for absence of DS.

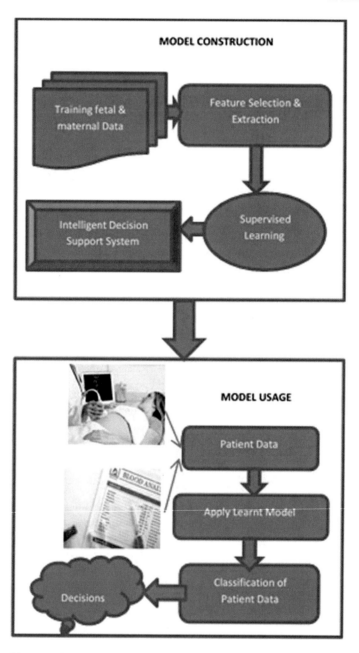

Fig. 1 Architecture of DSD

Tree-Based Rule Extraction

In the decision tree induction process, the six features extracted for each pregnant mother are preprocessed and separated to form the set of nodes for the tree. The information gain is calculated for every feature as shown in Eq. (1). This induction iterates until one of the termination conditions is reached. The class attribute "status" is either DS and NOT_DS for the set of training examples. The elements of class DS and NOT_DS are p and n, respectively. The quantity of information needed for a tuple in the training set, belonging to the classes DS and NOT_DS is shown in Eq. (1)

$$I(p, n) = -\frac{p}{p+n} \log_2 \frac{p}{p+n} - \frac{n}{p+n} \log_2 \frac{n}{p+n} \qquad (1)$$

Rules generated from the formed decision tree can then be used to classify unknown test data into one of the classes as DS or NOT_DS. The test data classified into DS class can represent fetal affected by Down's syndrome and hence, can be proceeded with further investigation and clinically co-related.

4.4 Model Usage

Once the DSD model is constructed using decision trees algorithm, the same model is used to classify a given test data into one of the classes: DS or NOT-DS. The DSD model was implemented using three flavors of decision trees such as J48, LMT, and random trees. J48 is the implementation of the Iterative Dichotomiser 3—ID3 algorithm. Logistic Model Tree (LMT) combines logistic regression along with supervised learning. A random forest is a group of decision trees where the final output is aggregated from the outputs of a number of decision trees. Regression trees have the ability to limit overfitting.

5 Experimental Setup

For training the above described DSD Model, we have used data collected from online medical websites over a period of 6 months. We have used 60% of the data for training the DSD system, and the remaining 40% for testing. The WEKA environment was used in the evaluation of the DSD system. The results produced by the DSD model for J48, LMT, and random trees are evaluated in terms of the accuracy measures namely, precision, recall, and F-score as given in Eqs. (2), (3), and (4). Recall indicates the percentage of the content in the system-classified data that is recalled correctly. The percentage of the content in the system-classified data which is relevant is given by precision [27]. Both recall and precision measures can be combined to show the

Table 1 Classifier Accuracies

Classifier	Training set method		10-fold cross-validation	
Method	% of correctly classified instance	% of incorrectly classified instance	% of correctly classified instance	% of incorrectly classified instance
J48	88.46	11.538	80.762	19.238
LMT	83.33	16.66	53.84	46.153
Random trees	80.77	19.23	73.076	26.923

performance of the proposed DSD system in terms of F-measure. For experiment purpose, the value of α is kept 1, showing F1 measure and their experimental setup is discussed in the following subsection.

$$Recall = True\,Positives/True\,Positives + False\ Negatives \qquad (2)$$

$$Precision = True\,Positives/(True\,Positives + False\,Positives) \qquad (3)$$

$$F - measure = \frac{(\alpha + 1)^{*}Recall^{*}Precision}{Recall + (\alpha^{*}Precision)} \qquad (4)$$

5.1   Results and Evaluation

The investigation of tree-based classifiers namely J48, LMT, and random trees on DSD was tested and evaluated using two methods: (i) training set test method (ii) 10-fold cross method. The results based on the accuracy metric is given below in Table 1 and the percentage of correctly classified instances along with the percentage of incorrectly classified instances is shown as a graphical representation in Fig. 2.

Table 1 shows the percentage of correctly classified instances of maternal and fetal data for the detection of the presence of Down's syndrome in the fetus. It is shown that J48 is classifying better than the other classifiers investigated.

In addition to accuracy, we have evaluated the DSD model using recall, precision, and F-measure as shown in Table 2. Evaluating the performance of the DSD system is done on three types of machine learning-based classifiers as shown in Table 2, with its graphical representation in Fig. 3.

It is evident that detection of down syndrome is able to detect with an F1 value of 0.884, 0.829, and 0.807 for the classifiers J48, LMT, and random trees, respectively. It is noted that J48 using decision trees outperforms the other two.

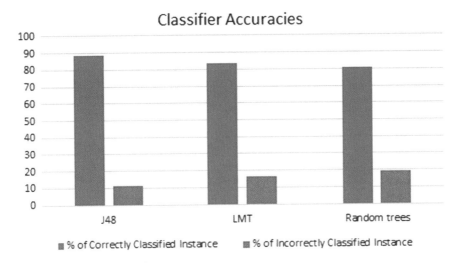

Fig. 2 Classifier accuracies

Table 2 Precision, recall, and F-measure

Training set method				10-fold cross-validation		
Method	Precision	Recall	F1	Precision	Recall	F1
J48	0.887	0.885	0.884	0.81	0.808	0.807
LMT	0.875	0.833	0.829	0.549	0.538	0.513
Random trees	0.81	0.808	0.807	0.732	0.731	0.73

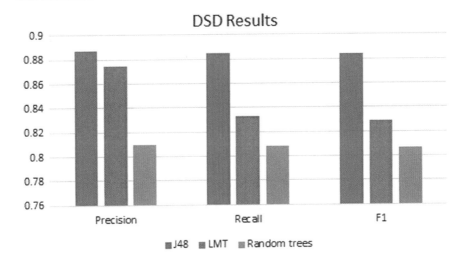

Fig. 3 Precision, recall, and F-measure of DSD

6 Conclusion

The proposed smart system allows pregnant women and their families to get the information they need to make an informed decision in a much shorter time, minimizing anxiety and stress. The DSD is definitely a means of changing the present healthcare technologies into smarter healthcare technology. The DSD system uses machine learning algorithm namely decision trees and implements three flavors of decision trees. Using supervised learning, the DSD system is able to detect the presence of Down's syndrome in a fetus. Once Down syndrome is detected by the system, the following options are available to parents: (i) AWARENESS: Some parents want to be informed about the status of their fetus ahead, so that they can be prepared for the arrival of the child as well as keep their extended families informed about their expected child. (ii) ADOPTION: There are some parents who would want to give the Down syndrome affected child for adoption and want to make arrangements for adoption earlier instead of opting for medical termination. (iii) TERMINATION: Some parents may want perform medical termination of the Down syndrome affected fetus, and thus discontinue the pregnancy. In future, the DSD system can make use of image processing techniques to analyze fetal ultrasound images, in order to improve precise estimation of neural tube thickness. Other classification methods can be explored to improve system accuracy.

References

1. Ramakrishnan, U., Grant, F., Goldenberg, T., Zongrone, A., Martorell, R.: Effect of women's nutrition before and during early pregnancy on maternal and infant outcomes: a systematic review. Paediatr. Perinat. Epidemiol. **26**(Suppl. 1), 285–301 (2012)
2. Christianson, A., Howson, C.P., Modell, B.: The hidden toll of dying and disabled children, March of Dimes, Global report on birth defects (2006)
3. Steingass, K.J., Chicoine, B., McGuire, D., Roizen, N.J.: Developmental disabilities grown up: down syndrome. J. Dev. Behav. Pediatr. **32**(7), 548–558 (2011)
4. Gupta, S., Arora, S., Trivedi, S.S., Singh, R.: Dyslipidemia in pregnancy may contribute to increased risk of neural tube defects—a pilot study in north Indian population. Indian J. Chem. Biochem. **24**(2), 150–154 (2009)
5. de Graaf, G., Buckley, F., Skotko, B.G.: Estimates of the live births, natural losses, and elective terminations with down syndrome in the United States. Am J Med Genet A. **167A**(4), 756–767 (2015)
6. Parker, S.E., Mai, C.T., Canfield, M.A., Rickard, R.: Updated national birth prevalence estimates for selected birth defects in the United States. Birth Defects Res. J. **88**(12), 1008–1016 (2010)
7. Schieve, L.A., Boulet, S.L., Kogan, M.D., Van Naarden-Braun, K., Boyle, C.A.: A population-based assessment of the health, functional status, and consequent family impact among children with Down syndrome. Disabily Health J. **4**(2), 68–77 (2011)
8. Prevention and control of birth defects in South-East Asia region. Strategic Framework. World Health Organization. 2013–2017
9. Sood, M., Agarwal, N., Verma, S., Bhargava, S.K.: Neural tubal defects in an east Delhi hospital. Indian J. Pediatr. **58**, 363–365 (1991)
10. Birth defects in South-East Asia, a public health challenge, situational analysis, World Health Organization (2013)

11. Grover, N.: Congenital malformations in Shimla. Indian J. Pediatr. **67**(4), 249–251 (2000)
12. Bhide, P., Sagoo, G.S., Moorthie, S., Burton, H., Kar, A.: Systematic review for birth prevalence of neural tube defects in India. Birth Defects Res. (Part A) **97**, 437–443 (2013)
13. Sharma, J.B., Gula, N.: Potential relationship between dengue fever and neural tube defects in a northern district of India. Int. J. Gynecol. Obstetrics. **39**, 291–295 (1992)
14. Cherian, A., Seena, S., Bullock, R.K., Antony, A.C.: Incidence of neural tube defects in the least developed area of India: a population based study. Lancet **366**, 930–931 (2005)
15. UNCTAD secretariat.: Issues Paper On Smart Cities and Infrastructure United Nations Commission on Science and Technology for Development, Inter-sessional Panel (2016)
16. Verma, M., Chhatwal, J., Singh, D.: Congenital malformations- a retrospective study of 10,000 cases. Indian J. Pediatr. **58**, 245–252 (1991)
17. Higuera, C., Gardiner, K.J., Cios, K.J.: Self-Organizing feature maps Identify proteins critical to learning in a mouse model of down syndrome. PLOS ONE (2015)
18. Chavez-Alvarez, R, Chavoya, A., Mendez-Vazquez, A.: Discovery of possible gene relationships through the application of self-organizing maps to DNA microarray databases. PLoS One **9**(4) (2014)
19. Sharma, R., Birth defects in India: hidden truth, need for urgent attention. Indian J. Hum. Genet. **19**(2), 125–129 (2013)
20. Boehringer, S., Guenther, M., Sinigerova, S., Wurtz, R.P., Horsthemke, B., Wieczorek, D.: Automated syndrome detection in a set of clinical facial photographs. Am. J. Med. Genet. (2011)
21. El-Qawasmeh, E.: Categorizing received email to improve delivery. Int. J. Comput. Syst. Sci. Eng. **26**(2) (2011)
22. Phyu, T.N.: Survey of classification techniques in data mining. In: Proceedings of the International Multi-Conference of Engineers and Computer Scientists International Association of Engineers, pp. 727–731 (2009)
23. Anyanwu M.N., Shiva, S.G.: Comparative analysis of serial decision tree classification algorithms. Int. J. Comput. Sci. Secur. (IJCSS) **3**(3) 231
24. Hannah, E., Mukherjee, S.: A classification based summarization (CBS) model for summarizing text documents. Int. J. Inf. Commun. Technol. **6**(3/4) (2014)
25. Çelik, E., İlhan, H.O., Ibir, A.: Detection and estimation of down syndrome genes by machine learning techniques. In: Signal Processing and Communications, Applications Conference (SIU) (2017)
26. Yong, S.P., Abidin, A.I.Z., Chen, Y.Y.: A Neural Based Text Summarization System, In: Proceedings of the 6th International Conference of DATA MINING (2005)
27. Chen, J., Chen, D., Lemon, O., A feature-based detection and tracking system for gaze and smiling behaviors. Int. J. Comput. Syst. Sci. & Eng. **26**(3) (2011)

Recent Research Advances in Black and White Visual Cryptography Schemes

T. E. Jisha and Thomas Monoth

Abstract Visual Cryptography (VC) is a type of image secret sharing scheme which decrypts an original secret image with Human Visual System (HVS). In this, the original image can be alienated into n shadows or shares and allocated to n participants; stacking any k shares reveals the secret image which ensures the security measures. In this paper, we examined the recent research advances in black and white VCSs. We reviewed the existing techniques and a comparative study of VC for binary images is presented. The study is performed with respect to different parameters and draws the current barriers related to the visual cryptography schemes.

Keywords Visual cryptography scheme · Pixel expansion · General access structure · Basis matrices

1 Introduction

Naor and Shamir [1] implemented a new type of (k, n) image secret sharing scheme, termed as Visual Cryptography Scheme (VCS). In this, an original secret image can be encrypted into n shares and overlapping of k shares out of n shares decrypts the original image. The scheme ensures the security measures and any $k-1$ shares cannot reveal the original image. The beauty of visual cryptography scheme is the decryption performed with no computation and it is done by HVS. The main key measures of VC are its contrast, pixel expansion, and security.

VCS is widely applied for transmitting the secret, complicated passwords, and images in armed forces, intelligence, electronic billing and tax payments, secret video conferencing, medical field, and e-banking. The problems related to the VC are image

T. E. Jisha (✉)
Department of Information Technology, Kannur University, Kannur 670 567, Kerala, India
e-mail: jishatevinoy@gmail.com

T. Monoth
Department of Computer Science, Mary Matha Arts & Science College, Kannur University, Mananthavady, 670 645 Wayanad, Kerala, India
e-mail: tmonoth@yahoo.com

© Springer Nature Singapore Pte Ltd. 2020
K. N. Das et al. (eds.), *Soft Computing for Problem Solving*,
Advances in Intelligent Systems and Computing 1048,
https://doi.org/10.1007/978-981-15-0035-0_38

479

contrast, pixel expansion, explicit codebook generation, participant number restriction, share alignment problem, secret recovery loss, etc. The factors mentioned above appeared in the VCS entail additional consideration. Here comes the significance of VC in research [2, 3]. Remaining part of the paper is structured as follows. In Sect. 2, we have described a study and definition of black and white visual cryptography. Section 3 describes the recent research advances in visual cryptography for binary images, Sect. 4 highlights the performance analysis of different black and white VCSs, and Sect. 5 draws the conclusions and the future works road map.

2 Visual Cryptography Scheme

In VCS, the original image is a group of binary (black or white) pixels. It is encrypted into n shares and each pixel is modified into m pixels. Any VCS can be explained with two $n \times m$ binary basis matrices, B^w and B^b.

The formation of B^w and B^b for k out of n ($k \leq n$) VCS is based on the parameter $1 \leq d \leq m$ and $\alpha > 0$, where d is a threshold value between 1 and m, α denotes the relative difference and m denotes the pixel expansion.

$$\alpha = \left(\omega_H\left(B^b\right) - \omega_H\left(B^w\right)\right)/m$$

where $\omega_H(B^b)$ and $\omega_H(B^w)$ are the hamming weight (number of ones) of the basis matrices B^b and B^w, respectively. The basis matrices B^b and B^w must satisfy the following three conditions:

The ORed m-vector V of any k of the n rows in

1. B^w satisfies $\omega_H(V) \leq d - \alpha .m$, where $\alpha .m$ denotes the contrast of the image.
2. B^b satisfies $\omega_H(V) \geq d$.
3. For any set $\{r_1, r_2,..., r_t\} \subset \{1, 2,...,n\}$ with $t < k$, the $t \times m$ matrices obtained by restricting B^w and B^b to rows $r_1, r_2,..., r_t$ are not distinguishable.

The clauses 1 and 2 are associated with contrast of the decrypted secret image and clause 3 is associated with security [3, 4].

3 Recent Research Advances in Visual Cryptography Schemes

Many published studies of VCS have concentrated on discussing binary images. Here, we have presented an assessment of the existing visual cryptography schemes for binary images after an extensive review process.

The (k, n) black and white visual secret sharing scheme acknowledged as VC was invented by Naor and Shamir in 1994, which encrypts a binary image into n

meaningless shares. The n shares can be distributed to n participants, one for each participant. The secret image can be decrypted by overlapping any k shares jointly using the human visual system. The secret image cannot be revealed by any k-1 or fewer transparencies [1]. Ito et al. invented a (k, n) image size invariant VCS with $m = 1$ to encrypt a binary image into the size invariant shares and decode into the same sized reconstructed image as the secret image [5].

Tzeng and Hu [6] have presented a new description for visual cryptography, in which the revealed images may be lighter or darker than backgrounds. They had run experiments on random access structures and their scheme indeed has better pixel expansion (contrast). In this paper, they had restricted their consideration to *2out of n*-threshold VCSs. A different approach, the probabilistic method, used to construct the VSS scheme was developed by Yang [7]. The major distinction between their scheme and conventional scheme is that, while the conventional scheme uses sub-pixels, their scheme uses pixels. The scheme is non-expansible as the secret image size and shares size are equal. Also, they used the occurrence of white pixels to illustrate the decrypted image's contrast.

Blundo et al. [8] proved that the VCSs can attain a lower bound in pixel expansion. Their schemes improved the ones presented in [6] in case of pixel expansion. Klein and Wessler [9] accomplished the reduced pixel expansion and contrast efficiency for generalized VCS with less than $2^n - 1$ subsets. Hajiabolhassan and Cheraghi [10] presented the finest pixel expansion with another lower bound and also, they obtained the smaller pixel expansion with an upper bound with regard to strong chromatic index.

Liu et al. [11] analyzed the demerits of existing definitions of visual cryptography scheme's contrast and presented a new description of contrast depending on their studies which enhances the clarity of the decrypted image. A VCS with efficient capacity and high contrast for the encoding of multiple secret images is proposed by Lee and Chiu [12]. They introduced a new encryption method to the $(2, 2) - 2$ Visual Secret Sharing Scheme for multiple images (VSSM) which encrypts concurrently two secret images into two transparencies. They took a hybrid encryption procedure for developing a deterministic scheme using codebook and randomly generated pixels. Liu et al. [13] proposed two Multi-pixel Encryption Size Invariant VCSs (ME-SIVCSs) that improved the clarity of the reconstructed image by decreasing the difference of the darkness levels and also avoided some known thin line problems.

Petrauskiene et al. [14] proposed dynamic visual cryptography based optical experimental technique for vibration generation equipment's control. In this scheme, the matching algorithms and initial stochastic phase deflection are used to embed an image into a stochastic background. It is examined by HVS at the vibration of structure in the predetermined values. The secret image can be reconstructed by the moiré fringes creation. Yan et al. [15] constructed three general threshold progressive VCS from three cases ($(2$ out of $2)$, $(2$ out of $n_x)$, and $(n_x$ out of $n_x)$) with no pixel expansion. They conducted analysis and experiments to assess the security conditions and efficacy of the developed schemes. In wider application, the scheme is loss tolerant and access control.

Arco et al. [16] followed a measure-independent approach and concentrated on the deterministic scheme. The characterization of measure-independent contrast optimal VCS is provided based on the structural properties of the schemes. Chiu and Lee [17] invented a *(k out of n)*—user-friendly threshold VCS (FVCS) for binary secret images according to the probabilistic *(k out of n)*—VCS with complementary cover images. They stated that this method is more efficient than the existing methods with respect to the decrypted secret image's contrast and of the meaningful shares. They also addressed the issues related to the VC like pixel expansion, image contrast, residual traces problem, interference problem in inter-images, and inadequate systematic encryption methods. The scheme is pixel-unexpanded.

Lee et al. [18] introduced a new scheme to encrypt two secret images into two meaningful shares with no pixel expansion by using a predefined codebook simultaneously. Secret images can be decrypted via turning and stacking. In this scheme, the value of black pixel in original image can be fully extorted. Ou et al. [19] introduced an XOR-based non-expansible VCS with directly generated meaningful transparencies. They implemented an algorithm by a simple $2^n \times n$ matrix for developing an *(n out of n)* VC (XOR-based). Using this algorithm, they examined a *(n, n)* XOR-based VCS with meaningful transparencies. They validated the method theoretically with sufficient proofs. They also solved the problems of OR-based VCS in case of pixel alignment and contrast. Palevicius and Ragulskis [20] introduced a visual image encryption scheme using holography generated by computer, moiré patterns, and dynamic VC with Gerchberg–Saxton algorithm. The decryption is done by HVS based on the harmonic oscillations' amplitude matching to a predetermined value. In this scheme, image cannot be divided into shares and it is embedded into a single random cover image.

A *(2, 2)* VCS used for transferring the confidential data through image based on fuzzy logic was proposed by Mudia and Chavan [21]. The encryption and decryption is done by rule-based share generation fuzzy logic. The safe transfer of confidential data and prevention of attack by unauthenticated persons is possible since the fuzzy logic does not hold an exact value. This scheme is mainly used for authentication. Hodeish et al. [22] proposed a new non-expansible *(k, n)* VCS on the basis of generation of codebook, matrix transpose, Boolean n-Vector, and XOR. The method provides perfect security, computational difficulty, storage problems, quick transfer, and undistorted reconstructed image. The share images can be generated by the Boolean n-vector based on the codebook.

Lakshmanan and Arumugam [23] developed basis matrices for a *(k, n)*- VCS with $2 \leq k \leq n$, where k and n are integers. As each participant in this scheme has same contrast with equal cardinality, they also attained a new rule for relative contrast and pixel expansion. They stated that their scheme provides most favorable contrast and slightest pixel expansion in case of $k = n$ and $n-1$. Singh et al. [24] proposed a XOR-based Visual Cryptography (VC) and eliminates the barriers in the existing schemes including image contrast, pixel expansion, explicit codebook generation, participant number restriction, share alignment problem, and secret recovery loss. It guarantees the perfect decryption of the original image with perfect visual quality.

They introduced three novel algorithms for construction of basis matrices, random transparencies, and transform random transparencies to meaningful transparencies.

Yang et al. [25] developed a general (k, n) Reversible Absolute Moment Block Truncation Coding(AMBTC) based VCS (RAVCS) for encrypting a black and white secret image into multiple AMBTC shadows. They constructed a *(k, k)* RAVCS scheme to encrypt a secret image by considering AMBTC image. The scheme was extended to split the same original image into multiple sets of shares by considering multiple images. The shadow distribution algorithms are used to construct a matrix which is used to distribute the multiple groups of shadows to members. By sharing the shadows of k or more participants, the secret image is decrypted by XOR or OR.

Guo and Zhou [26] constructed three visual cryptography schemes by hyper graph decomposition: a hyper star access structure, a normal form hyper star access structure, and a general access structure from a hyper star with center access structure, full threshold VCS, and its several decomposed normal forms, respectively. They presented the theoretical analysis of the schemes. Hua et al. [27] presented a VC-based novel multilevel security scheme and retains the beauty of the VC in the case of its decryption. They proposed a Region Incrementing (area expanded) VC Scheme (RIVCS) which encodes a secret image with respect to the network security. Each region of a secret image is owed with a certain confidentiality level. These levels are decrypted incrementally according to the participants of different combinations. They first developed the General AS (GAS) based region incrementing visual cryptography scheme. Then, they developed the algorithm for assigning the levels. Finally, for sharing the secret pixels, they implemented the encoding matrices. Experimentally, they proved that the scheme is appropriate for network security.

Jia et al. [28] introduced the transformation of Collaborative VCS (CVCS) into the multiple secret images VCS using a general access structure. The basis matrix formation in CVCS among two VCSs is invented into an integer linear programming problem that reduces the pixel expansion problem under the related constraints security and image quality. They constructed the association between additional VCSs and proved the efficacy of CVC scheme with experimental results. Shivani and Agarwal [29] introduced a novel Verifiable Progressive Visual Cryptography (VPVC) used for authentication, copyright protection along with confidential payload. They assured that their scheme removes the barriers in VCS like pixel expansion, noisy transparencies, explicit generation of codebook, and participants number limit. They also proved that the visual quality percentage of the reconstructed image remains as half without decreasing the secrecy level and localization of the share's forged region. Their scheme is a blending of Progressive VC (PVC) and Verifiable VC (VVC) with cheating prevented shares.

4 Comparative Study of Visual Cryptography Schemes

We compared the various black and white visual cryptography schemes with different parameters; the scheme used, pixel expansion, methods used, whether the decryption

Table 1 Comparison of different black and white visual cryptography schemes

Sl. no.	Author & Year	Scheme	Pixel expansion	Methods used	Decryption	Size of the reconstructed image	Merits/Demerits
1	Naor & Shamir (1994)	(2, 2) & (k, n)	$m = 4, 2^k, 2^{k-1}$	Boolean matrices	Without computation	Varied	Introduced new paradigm of cryptographic schemes called visual cryptography
2	Ito et al. (1999)	(k, n)	$m = 1$	Based on the conventional scheme	Without computation	Not varied	Image size invariant visual secret sharing scheme
3	Tzeng and Hu (2002)	(2, n)	$m = 4$	Random access structures	Without computation	Varied	Improved definition for visual cryptography better pixel expansion (contrast)
4	Yang (2004)	(2, 2), (2, 3) and (3, 3)	$m = 1$	Probabilistic method	Without computation	Not varied	Non-expansible
5	Blundo et al. (2006)	(2, n)	$m \geq |T_0|/2$	Nonredundant basis matrices	Without computation	Varied	Improved the ones presented in [Tzeng and Hu, 2002]

(continued)

Table 1 (continued)

Sl. no.	Author & Year	Scheme	Pixel expansion	Methods used	Decryption	Size of the reconstructed image	Merits/Demerits
6	Klein and Wessler (2007)	(k, n)	less than $2^n - 1$	General access structure	Without computation	Varied	Extended visual cryptography scheme
7	Hajiabolhassan and Cheraghi (2010)	(2, n)	New lower bound and upper bound	Using qualified sets' hyper graph	Without computation	Varied	Pixel expansion using upper bound with regard to strong chromatic index
8	Lee and Chiu (2011)	(2, 2)	$m = 1$	Hybrid encryption procedure using codebook	Without computation	Not Varied	2 VCSs for multiple images (VSSM) which encodes concurrently two secret images into two transparencies
9	Liu et al. (2012)	(k, n)	$m = 1$	General access structure	Without computation	Not Varied	Decreased the darkness levels and also avoided some known thin line problems

(continued)

Table 1 (continued)

Sl. no.	Author & Year	Scheme	Pixel expansion	Methods used	Decryption	Size of the reconstructed image	Merits/Demerits
10	Yan et al. (2014)	$(2, 2)$, $(2, n_x)$ (n_x, n_x)	$m = 1$	General threshold progressive visual cryptography scheme	Without computation	Not Varied	In wider application, the scheme is loss tolerant and access control
11	Arco et al. (2014)	$(2, n)$ & (n, n)	$m = 6$	Measure-independent approach	Without computation	Varied	Contrast optimal visual cryptography schemes
12	Chiu and Lee (2015)	(k, n)	$m = 1$	Based on the probabilistic method	Without computation	Not Varied	FVCS to solve the problems like pixel expansion, residual traces, inter-images interference, contrast, and security
13	Lee et al. (2015)	$(2, 2)$	$m = 1$	Using a predefined codebook	Decrypted via turning and stacking	Not Varied	Encrypt two secret images into two meaningful shares with no pixel expansion
14	Ou et al. (2015)	(n, n)	$m = 1$	Implemented basic algorithm by a simple $2^n \times n$ matrix	XOR	Not Varied	Non-expansible XOR-based VCS with directly generated meaningful shares

(continued)

Table 1 (continued)

Sl. no.	Author & Year	Scheme	Pixel expansion	Methods used	Decryption	Size of the reconstructed image	Merits/Demerits
15	Palevicius and Ragulskis (2015)	New scheme	$m = 1$	Based on holography, moiré methods, and dynamic VC	Using the computationally reconstructed field of amplitudes	Not Varied	Integration of dynamic visual cryptography with Gerchberg–Saxton algorithm
16	Mudia and Chavan (2016)	(2, 2)	$m = 1$	Based on rule-based share generation fuzzy logic	Decryption is done by rule-based generation	Not Varied	Mainly used for authentication
17	Hodeish et al. (2016)	(k, n)	$m = 1$	Constructed using codebook, matrix transpose, Boolean n-Vector, and XOR	XOR	Not Varied	Provides perfect security, eliminates the computational difficulty and storage problems, fast transmission, and undistorted decrypted image
18	Lakshmanan and Arumugam (2017)	(k, n)	$m = 2^r(n-1)(n-3)\cdots(n-2r+1)/r!$ if k is odd, $m = 2^r - 1n(n-2)\cdots(n-2r+2)/r!$ if k is even	Construction of basis matrices	Without computation	Varied	Provides optimal image quality and least pixel expansion in case of $k = n, n - 1$

(continued)

Table 1 (continued)

Sl. no.	Author & Year	Scheme	Pixel expansion	Methods used	Decryption	Size of the reconstructed image	Merits/Demerits
19	Singh et al. (2017)	(n, n)	m = 1	Proposed three algorithms for basis matrices, random shares, and its conversion	XOR	Not Varied	Eliminates the problems related to image contrast, pixel expansion, explicit codebook generation, participant number restriction, share alignment problem, secret recovery loss
20	Yang et al. (2017)	(k, n)	m = 16 for k = 3,4 and m = 2^k-1 for k >=5	Reversible AMBTC- visual cryptography scheme (RAVCS)	OR or XOR	Varied	Same secret image can be shared into multiple groups of shares by considering multiple images
21	Guo and Zhou (2018)	(2, 2) and (3, 3)	Optimal – $2^{|c|}$	By hyper graph decomposition	Without computation	Varied	Prove that pixel expansion is optimal

(continued)

Table 1 (continued)

Sl. no.	Author & Year	Scheme	Pixel expansion	Methods used	Decryption	Size of the reconstructed image	Merits/Demerits
22	Jia et al. (2018)	(k1, n1) and (k2, n2)	m = 2 and m = 3 (minimal pixel expansion)	Using a general access structure	Without computation	Varied	The collaboration between more VC schemes and proved the efficacy of CVC scheme with experimental results
23	Shivani and Agarwal (2018)	(4, 4)	m = 1	Verifiable Progressive Visual Cryptography (VPVC)	Without computation	Not Varied	Used for authentication, copyright protection

is done with or without computation, size of the reconstructed image is varied or not, and a brief description of merits and demerits of each scheme. The result of the comparative study is illustrated in Table 1.

From the study, we noticed that most of the VCSs are done by the general access structures and basis matrices. Some of the other methods used in the visual cryptography are random access structures, induced matching of hyper graph of qualified sets, hybrid encryption procedure using codebook, using a predefined codebook, holography, moiré techniques and dynamic VC, rule-based share generation fuzzy logic, transpose of matrices, and by hyper graph decomposition. The different extended schemes available in the VC are probabilistic VC, deterministic VC, size invariant VC, collaborative VC, general threshold PVC, Verifiable Progressive VC (VPVC), and reversible AMBTC- VC.

5 Conclusions

Here, we have presented a comprehensive study of VCSs for black and white images. Also, we compared the recent research advances in visual cryptography for binary images. This study reveals that the pixel expansion, contrast, and security of the reconstructed original secret image, explicit codebook generation, participant number restriction, share alignment problem, secret recovery loss, computational difficulty, storage issues, transmission problems, and distorted reconstructed image are the major barriers in the visual cryptography research community. The implementation of perfect binary VCS is a great concern of modern society. Our future study will focus to develop more sophisticated VCSs for black and white, grayscale, and color images and also to eliminate the barriers mentioned here.

References

1. Naor, M., Shamir, A.: Visual cryptography, advances in cryptology-eurocrypt'94. LNCS **950**, 1–12 (1995)
2. Pandey, D., Kumar, A., Singh, Y.: Feature and future of visual cryptography based schemes. In: Quality, Reliability, Security and Robustness in Heterogeneous Networks, Lecture Notes of the Institute for Computer Sciences, Social Informatics and Telecommunications Engineering, vol. 115, pp. 816–830. Springer, Berlin (2013). https://doi.org/10.1007/978-3-642-37949-9_71
3. Monoth, T., Babu Anto P, "Analysis and design of tamperproof and contrast-enhanced secret sharing based on visual cryptography schemes, Ph.D Thesis, Kannur University, Kerala, India, (2012). (http:// shodhganga.inflibnet.ac.in)
4. Blundo, C., Cimatob, S., De Santisa, A.: Visual cryptography schemes with optimal pixel expansion. Theor. Comput. Sci. **369**(1-3), 169–182 (2006), (Elsevier) (https://doi.org/10.1016/j.tcs.2006.08.008)
5. Ito, R., Kuwakado, H., Tanaka, H.: Image size invariant visual cryptography. IEICE Trans. Fundam. **E82-A**, 10 (1999)
6. Tzeng, W.G., Hu, C.M.: A new approach for visual cryptography. Des. Codes Cryptogr. **27**(3), 207–227 (2002). https://doi.org/10.1023/A:1019939020426

7. Yang, C.N.: New visual secret sharing schemes using probabilistic method. Pattern Recognit. Lett. **25**(4), 481–494 (2004), (Elsevier)
8. Blundo, Carlo, Cimato, Stelvio, De Santis, Alfredo: Visual cryptography schemes with optimal pixel expansion. Theoret. Comput. Sci. **369**, 169–182 (2006). https://doi.org/10.1016/j.tcs. 2006.08.008. (Elsevier)
9. Klein, A., Wessler, M.: Extended visual cryptography schemes. Inf. Comput. **205**(5), 716–732 (2007). https://doi.org/10.1016/j.ic.2006.12.005. (Elsevier)
10. Hajiabolhassan, Hossein, Cheraghi, Abbas: Bounds for visual cryptography schemes. Discret. Appl. Math. **158**(6), 659–665 (2010). https://doi.org/10.1016/j.dam.2009.12.005. (Elsevier)
11. Liu, Feng, ChuanKun, Wu, Lin, XiJun: A new definition of the contrast of visual cryptography scheme. Inf. Process. Lett. **110**(7), 241–246 (2010). https://doi.org/10.1016/j.ipl.2010.01.003. (Elsevier)
12. Lee, Kai-Hui, Chiu, Pei-Ling: A high contrast and capacity efficient visual cryptography scheme for the encryption of multiple secret images. Opt. Commun. **284**(12), 2730–2741 (2011). https://doi.org/10.1016/j.optcom.2011.01.077. (Elsevier)
13. Liu, Feng, guo, Teng, Wu, ChuanKun, Qian, Lina: Improving the visual quality of size invariant visual cryptography scheme. J. Vis. Commun. Image Represent. **23**(2), 331–342 (2012). https:// doi.org/10.1016/j.jvcir.2011.11.003. (Elsevier)
14. Petrauskiene, V., Aleksa, A., Fedaravicius, A., Ragulskis, M.: Dynamic visual cryptography for optical control of vibration generation equipment. Opt. Lasers Eng. **50**, 869–876 (2012) (Elsevier)
15. Yan, Xuehu, Wang, Shen, Niu, Xiamu: Threshold construction from specific cases in visual cryptography without the pixel expansion. Sig. Process. **105**, 389–398 (2014). https://doi.org/ 10.1016/j.sigpro.2014.06.011. (Elsevier)
16. D'Arco, P., De Prisco, R., De Santis, A.: Measure-independent characterization of contrast optimal visual cryptography schemes. J. Syst. Softw. **95**, 89–99 (2014). https://doi.org/10. 1016/j.jss.2014.03.079. (Elsevier)
17. Chiu, Pei-Ling, Lee, Kai-Hui: User-friendly threshold visual cryptography with complementary cover images. Sig. Process. **108**, 476–488 (2015). https://doi.org/10.1016/j.sigpro.2014.09.032. (Elsevier)
18. Lee, Jung-San, Chang, Chin-Chen, Huynh, Ngoc-Tu, Tsai, Hsin-Yi: Preserving user-friendly shadow and high-contrast quality for multiple visual secret sharing technique. Digit. Signal Proc. **40**, 131–139 (2015). https://doi.org/10.1016/j.dsp.2015.02.012. (Elsevier)
19. Duanhao, Ou, Sun, Wei, Xiaotian, Wu: Non-expansible XOR-based visual cryptography scheme with meaningful shares. Sig. Process. **108**, 604–621 (2015). https://doi.org/10.1016/j. sigpro.2014.10.011. (Elsevier)
20. Palevicius, Paulius, Ragulskis, Minvydas: Image communication scheme based on dynamic visual cryptography and computer generated holography. Opt. Commun. **335**, 161–167 (2015). https://doi.org/10.1016/j.optcom.2014.09.041. (Elsevier)
21. Miss, H.M., Miss, P.V.: Fuzzy logic based image encryption for confidential data transfer using (2, 2) secret sharing scheme. Procedia Comput. Sci. **78**, 632–639. (2016). https://doi.org/10. 1016/j.procs.2016.02.110 (Elsevier)
22. Hodeish, Mahmoud E., Bukauskas, Linas, Humbe, Vikas T.: An optimal (k, n) visual secret sharing scheme for information security. Procedia Comput. Sci. **93**, 760–767 (2016). https:// doi.org/10.1016/j.procs.2016.07.288. (Elsevier)
23. Lakshmanan, R., Arumugam, S.: Des. Codes Cryptogr. **82**(3), 629–645 (2017). https://doi.org/ 10.1007/s10623-016-0181-z. (Springer)
24. Singh, Priyanka, Raman, Balasubramanian, Misra, Manoj: A (n, n) threshold non-expansible XOR based visual cryptography with unique meaningful shares. Sig. Process. (2017). https:// doi.org/10.1016/j.sigpro.2017.06.015
25. Yang, Ching-Nung, Xiaotian, Wu, Chou, Yung-Chien, Zhangjie, Fu: Constructions of general (k, n) reversible AMBTC-based visual cryptography with two decryption options. J. Vis. Commun. Image Represent. **48**, 182–194 (2017). https://doi.org/10.1016/j.jvcir.2017.06.012

26. Guo, Teng, Zhou, LinNa: Constructing visual cryptography scheme by hypergraph decomposition. Procedia Comput. Sci. **131**, 336–343 (2018). https://doi.org/10.1016/j.procs.2018.04.172. (Elsevier)
27. Hua, Hao, Liu, Yuling, Wang, Yongwei, Chang, Dexian, Leng, Qiang: Visual cryptography based multilevel protection scheme for visualization of network security situation. Procedia Comput. Sci. **131**, 204–212 (2018). https://doi.org/10.1016/j.procs.2018.04.204. (Elsevier)
28. Jia, X., Wang, D., Nie, D., Zhang, C.: Collaborative visual cryptography schemes. IEEE Trans. Circuits Syst. Video Technol. **28**(5), 1056–1070 (2018). https://doi.org/10.1109/tcsvt.2016. 2631404. (IEEE)
29. Shivani, S., Agarwal, S.: VPVC: verifiable progressive visual cryptography. Pattern Anal Appl. **21**(1), 139–166 (2018). https://doi.org/10.1007/s10044-016-0571-x. (SpringerLink)

An N-Puzzle Solver Using Tissue P System with Evolutional Symport/Antiport Rules and Cell Division

Resmi RamachandranPillai and Michael Arock

Abstract An N-puzzle is a sliding blocks game that takes place on a grid with tiles each numbered from 1 to N. Many strategies like branch and bound and iterative deepening are exist in the literature to find the solution of the puzzle. But, here, a different membrane computing algorithm also called P systems, motivated from the structure and working of the living cell has been used to obtain the solution. A variation of P system, called tissue P system with evolutional symport/antiport (TPSESA) rules, is used to solve the puzzle. It has been proved that the power of computation of TPSESA rules with membrane division is universal Turing computable. In this paper, a concept of dynamic membrane division is also considered so that it completely simulates the behavior of a living cell. On the basis of experiments performed on a sample of different instances, the proposed algorithm is very efficient and reliable. As far as the author is concerned, this is the first time, an N-Puzzle problem is solved using the framework of membrane computing.

Keywords Membrane computing · Symport/antiport · Active membranes

1 Introduction

Membrane computing is a form of natural computing which mimics the computational processes inside a living cell. Traditional systems have considered only some abstract machines like Turing machines, which proceed with sequential execution model of computation. That means, only one operation can be performed at a particular time. But, all real-time problems need concurrent execution of processes so that it can improve the performance measurements such as throughput and efficiency. Earlier, many researchers described different models which permit multiple processes execute in parallel. But the main limitation imposed in these computing models was

R. RamachandranPillai (✉) · M. Arock
National Institute of Technology, Tiruchirappalli, Tamil Nadu, India
e-mail: resmiramachandranpillai@gmail.com

M. Arock
e-mail: michaelarock@gmail.com

© Springer Nature Singapore Pte Ltd. 2020
K. N. Das et al. (eds.), *Soft Computing for Problem Solving*,
Advances in Intelligent Systems and Computing 1048,
https://doi.org/10.1007/978-981-15-0035-0_40

that it purely depends on the architectural (structural) view of the particular machine and thereby hinders the programmer to program effectively.

Membrane computing is a parallel and distributed framework which involves non-determinism and maximal parallel executions. There are many variants of membrane computing (call it as P systems) including cell-like P system, in which membranes are arranged in hierarchical order that could be visualized by representing it as a tree. In tissue-like P system and spiking neural P systems, the membranes are elementary and can be visualized by representing the membranes, by the nodes of a digraph which clearly depicts the intra- and intermembrane communications through the directed arrows.

An overview of cell-like P systems was presented in [1]. It is mentioned that the system consists of a hierarchical ordering of membranes which have clear boundaries, and in each membrane, multisets of objects are placed. All of these multisets of objects can participate in a manner which is defined by the set of rules. Since the multiplicity of object matters in the computation, it is usually denoted as string objects. There are rules for destroying, creating, dividing, and merging of membranes and of transmission of objects. After the application of those rules, the configuration of the system may change and produce a set of new states and new rules can be applicable to the current state of the system.

Another variation of P system is the P systems with active membranes which was discussed in [2] for solving N-queens problem which is typically known as an NP-Hard problem. Active membranes inspired from biological cell division mainly involve six rules including (a) multiset rewriting rules, (b) rules for passing objects into the particular membranes, (c) rules for sending objects out from a particular membrane, (d) rules for the dissolution of membranes, (e) rules for dividing membranes into two or more, and (f) rules for the division of nonelementary membranes that can be used for processing of objects. Always there exists a trade-off between time and space in computational models. So, with the use of active membranes, one can benefit from the use of exponential workspace in solving problems which are computationally hard and complete. Polarization and polarization less [3, 4] P system can also be employed to solve some problems which are computationally dependant. Job shop scheduling [5] and subset sum problem [6] have been solved using P systems with active membranes.

In this paper, a membrane algorithm using TPSESA rules and simple heuristics has been used. As far as the author is concerned, this is the first time, an N-puzzle is solved using the framework of membrane computing.

The paper is presented as follows: Sect. 2 briefly describes the concept of membrane structure, symport/antiport rules of P system, TPSESA rules and cell division. Section 2 also presents a brief literature survey on the topic and a brief introduction of N-Puzzles and simple heuristics. Section 3 describes the proposed algorithm for solving N-Puzzles. Section 4 presents a complexity analysis of the proposed algorithm. Conclusions of the work done and the future aspects have been considered in Sect. 5.

2 Background

2.1 The Membrane Structure

The membrane structure consists of an outer membrane called skin membrane which contains many inner membranes. The elementary membrane is the one which contains no inner membranes and it may be treated as the output membrane. The membranes can pass objects into other membranes and can evolve into other objects using the rules of the membrane structure. A typical membrane structure [1] is given (Fig. 1) and the environment is responsible for getting the output of the computation through the skin membrane.

2.2 P System with Symport/Antiport Rules and Cell Division

P system with symport/antiport has been employed for many problems because of the coordinated movement of objects from one compartment to another. In uniport communication, only a single kind of molecule can take part in the transmission process. In symport communication, two types of objects take place but in the same direction only. This is used when two objects have to send to a particular membrane for the computation. The antiport communication involves the communication of two different objects in opposite direction. It means that when one object enters into a compartment another object is sent out from the membrane using antiport rules.

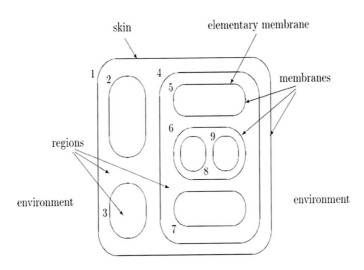

Fig. 1 The membrane structure

Symport/Antiport Rules:

1. (y, in)l, the multiset y moves into the compartment or membrane which surrounds.
2. (y, out)l, the multiset y moves from the cell to the membrane surrounding it.
3. (w, in; x, out), multiset w goes into the membrane from the compartment enveloping it, subsequently multiset x goes in the opposite direction. The *weight* of a particular rule is given by |x| for symport communication, and *max* {|w|, |x|} for antiport communication.

Till now, no proof has been made for P = NP or P ≠ NP. But one thing regarding membrane computing paradigm is that NP-complete and NP-hard problems cannot be solved in polynomial time without an exponential workspace. So, here membrane division has an important role in reducing time complexities.

In [6] it is proved that the power of TPSESA rules has got its superiority over standard P system with symport/antiport rules of same weight.

2.3   TPSESA Rules and Dynamic Cell Division

This section is meant for describing the notions and notations in automata for a better understanding of the problem definition.

In formal language theory, an alphabet τ is denoted as a collection of non-empty set of strings, where string is a collection of symbols basically of letters from the English alphabet. Suppose the string is denoted as V, then the length of the string is denoted by |V| and is defined as the total number of occurrences of every symbol in V. The empty string is denoted by λ.

A TPSESA (with degree k ≥ 1) with membrane division can be defined by

$$\Pi = (\tau, \Sigma, H, M_1 \dots M_k, R, I_{out})$$

where

1. τ is a set of objects.
2. $\Sigma \leq \tau$ is a set of alphabets which are placed in the environment.
3. $M_i, 1 \leq i \leq k$ are multisets over τ.
4. R is a set of the rules of the following forms:

     (a) Evolution rules: $[d \to c]_h$, for h ∈ H, d ∈ τ, cτ*
     (b) Communication rules: $d[]_h \to [c]_h$, for h ∈ H,   d, c ∈ τ
     (c) Communication rules: $[d]_h \to []_h c$, for h ∈ H,   d, c ∈ τ
     (d) Evolutional communication rules:
         (d.1) Evolutional symport rules:

$$[p]_i[q]_j \to []_i[q']_j,$$

where, $1 \leq i \leq j, 0 \leq j \leq k, i \neq j, p \in \tau+, q' \in \tau^*$ or $i = 0, 1 \leq j \leq k, p \in \tau+, q' \in \tau^*$, and there exists at least one object $a, \in p$, such that $p \in \tau \setminus \Sigma$

(d.2) Evolutional antiport rules:

$$[p]_i[q]_j \rightarrow [q']_i[p']_j,$$

where, $0 \leq i \neq j \leq k, p, v \in \tau+, q', q' \in \tau^*$

(e) Division rules: $[p]_i \rightarrow [q]_i[r]_{i+1}[s]_{i+2}[t]_{i+3}$
where $i \neq i_{out}$ and $i = 1$ at the initial stage and p, q, r, s, t $e \in \tau$, $i_{out} \in \{0, 1, \ldots k\}$.

Here four membranes are constructed by membrane division. The number of membranes needed is dependent on the problem configuration and is explained in Sect. 3.

A simple TPSESA with degree $k > 1$ rules does not have cell division rules. Here membrane division on the basis of certain conditions is employed every time. So, instead of dividing the membrane in fixed numbers, it is done dynamically based on certain constraints. This TPSESA can be represented by the nodes of a digraph with nodes depicting the membranes and arrows depicting the transitions and the labels on the arrows depicting the constraints on the evolution.

The above standards of TPSESA principles and cell division in the proposed scheme are connected in a maximally parallel way: at each progression, the framework employs a multiset of rules which is maximal; no additional guideline can be included being pertinent. It has just a single confinement that when a membrane is partitioned, the division rule is the special case which is connected to that membrane at that progression, the items inside that cell does not develop by methods for communication rules [6]. In the event that the recently produced cells do not partition at the subsequent stage, they could take an interest in the association with different cells or with the surroundings by methods for communication rules.

2.4 Overview of N-Puzzle Problem and Simple Heuristics

The N-puzzle problem is a famous NP-complete problem that is also called sliding block puzzle which consists of $k*k$ grid with $((k*k)-1)$ tiles placed in a square matrix of order k. The numbers inside the tiles can range from 1 to N, where N can be any natural number. Given an initial configuration, one can arrive at the goal state by sliding the tiles one position to left, right, top, and bottom in every stage. The empty tile is treated as tile 0. Normally, the standard solution is to slide the empty tile to left, right, top, and bottom in such a way that only one movement can be considered in one chance.

An N-puzzle can be formally defined by
P- (Q, q_0, F, f, C), where

Table 1 Comparison of 8-puzzle and 15-puzzle

	8-puzzle	15-puzzle
Number of states	9!/2	16!/2
Average number of solutions per problem	2.76	17
Maximum number of solutions per problem	64	153
Optimal number of moves	31	80

Q- Set of states of the system.

q_0 - Initial state or initial partial multiset.

F- Final state or goal state.

f- Successor function which includes move left (), move right (), move top (), and move bottom ().

C- Path cost, which is unity for every branch in the path of the state space tree.

For solving N-puzzle problem, one has to apply some heuristics to obtain the solution fast. In this paper, a simple heuristic of counting the misplaced blocks as compared to the goal state has been used. And, then, one has to move the empty tiles to the above-said positions to reach the goal state. So, the heuristic value of the goal state is zero, since there are no misplaced tiles in the block. One can think that it can be done easily by just moving the empty tiles to arrive at the goal state. But finding the solution of N-puzzles using minimum possible movement is an NP-hard problem. For finding the solution of 8 puzzles and 15 puzzles one can easily find the solution in no more than 31 single tile movement. Table 1 shows the comparison of 8-puzzle and 15-puzzle, respectively. Figure 2 shows a 3-puzzle state space.

2.5 Literature Review

A fair quantity of research work has been done in this area and many papers were analyzed. An overview of 15 puzzles and the solution is given by Harsh and Bhasin in [7]. After, Korf proposed a new heuristic function depending on the evaluation of subcodes. As far as the author is concerned, the net cost is lesser compared to existing, but a high degree of candidate pruning is needed in order to avoid duplicate candidates [8]. Another work used Manhattan pair distance heuristics which is a combination of pair distance and Manhattan distance [9]. The algorithm used was iterative deepening A * which is a modified version of the previous work. Another work by Calabro provides $O(n^2)$ algorithm to decide the solvability and $O(n^3)$ moves to solve the problem. The paper primarily dwells on solvability part and not on the solution part [10]. The work by Pizlo develops a cognitive model and analyzes human behavior by solving the problem. It is a cognitive science model and not a computational model [11]. Some of the works like that of Felner uses pattern databases which have not been used in presented approach [12, 13]. A distributed approach was presented by Drogoul by using an eco-problem solving model. The

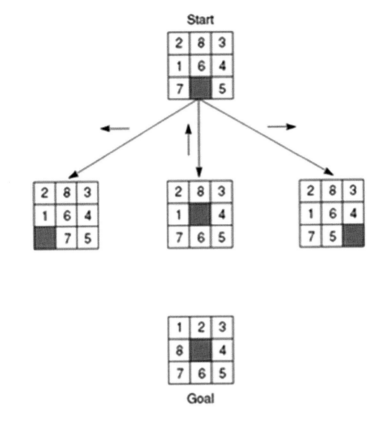

Fig. 2 An example of 3-puzzle state space

strategy is novel but it is meant more for theoretical analysis [14]. Other strategies for N-puzzles like genetic algorithm has also been proposed [15].

3 Proposed Algorithm for N-Puzzle Problem Using TPSESA Rules and Cell Division

In this section, a study of TPSESA with cell division model {Π (n) where n ε N) is proposed to solve N-puzzle problem. The initial partial solution and goal state of the N-puzzle is provided as input to the P system. Then in each membrane, rules are applied to evolve, dissolve, and to communicate, on the objects present in each membrane and final output multiset is produced. Instead of applying the common brute force method as in almost all membrane computing solutions, some heuristic function is provided, in order to reduce the number of probable candidates for a particular cell, thus by eliminating unwanted states which were the main limitations in branch and bound strategies. After applying all the rules, if any membrane satisfies the output criteria or there is no rule remaining, then it sends "YES" to the environment saying that there exists a solution for the particular instance of the problem. Otherwise it sends "NO" to the environment, saying that, with the problem instance given, there is no possible solution, which can be obtained using this type of P systems or the solution does not exist for the particular problem instance.

3.1 Approach: The Model

The input to the system is modeled as an object with initial state and goal state and is given to the system. Then the rules are used to find the solution by moving the empty tiles to left, right, top, and bottom. The proposed algorithm is given as follows:

Algorithm 1 Proposed Algorithm for N-Puzzle solver.

---

1. *Initialize the membrane with initial partial solution and the goal structure.*
2. *Compute the fitness score P_n for each membrane. The fitness score is a measure of the misplaced tiles compared to goal state.*
3. *Generate new membranes by division; each one depicts the arrangement of empty tiles on top, left, right and the bottom.*
4. *Calculate the P_n of each membrane.*
5. *Apply symport/antiport communication rules and find the P_{nbest}.*
6. *Dissolve all other membranes using membrane dissolution rule.*
7. *Repeat Step 3 until all choices have made.*
8. *Repeat Step2 until satisfying solution is obtained.*
9. *Send YES to the environment with the encoded object.*
10. *Otherwise send NO to the environment as the proposed algorithm could not find solution for the instance given.*

Modulo operation is used to implement the Step 3 of the proposed algorithm. There are at most three child nodes at every computation, since a tile is not moved to a position where it has occupied previously. So, at normal cases, there will be three child nodes and is implemented by a modulo-3 operation. If the empty tile is at the edge of the grid, then there will be two cases and perform a modulo-2 operation. And if the empty tile is on one corner of the grid, there will be at most one case and perform a modulo-1 operation to get the child node.

The fitness function is calculated for each membrane and finds the best fitness value P_{nbest}. Then, membrane dissolution operator is applied to dissolve unwanted membranes from the system and is marked by the object "support". Finally, when the fitness value becomes zero or equal to the fitness value of goal state, the computation stops and send "YES" to the environment. Otherwise, it sends "NO" to the environment.

The partial multiset is encoded in an object Z_{ijx} which is treated as input to the problem.

The rules used in the problem are generalized as follows:

1. $[Z_{ijx} \rightarrow S_{ijx}p]_1$
   where $i, j \in$ K, and x represents the item placed in the square (i, j) and p $\in \{1....n\}$, where p is the number of wrongly placed tiles in the initial partial multiset.

2. $[\sim d \sim r_{ij}x w_{ij}k_1 \rightarrow r_{ij}{}^n d\alpha]_1$
   where $i, j \in$ K, and x represents the item placed in the square (i, j) and the object d is produced in this rule. We introduced n copies of objects r_{ij} into the membrane here $n \leq 4$

3. $[\alpha r_{ij}{}^n d]_1 \rightarrow [d r_{1ij}p_1]_1 [d r_{2ij}p_2]_2 [d r_{3ij}p_3]_3$
   This rule is used to divide the membranes based on the problem we are considering. In this particular case, the membrane is divided into three membranes; each one depicts the arrangement of empty tiles in the top, left, right, and bottom positions, and the value of p_n indicates the number of wrongly placed tiles compared to the goal state. This step is carried out by performing modulo operation.

4. $[dr_{1ij}p_1]_1 \rightarrow [p_1']_5$
   In this, the objects d, r_{1ij}, and p_1 are consumed and evolved into p_1' in membrane 5.

5. $[dr_2ijp_2]_1 \rightarrow [p_2']_5$
   In this, the objects d, r_2ij, and p_2 are consumed and evolved into p_2' in membrane 5.

6. $[dr_3ijp_3]_1 \rightarrow [p_3']_5$
   In this, the objects d, r_3ij, and p_3 are consumed and evolved into p_3' in membrane 5.

7. $[dr_3ijp_3]1 \rightarrow [p_3']_5$
   In this, the objects d, r_{4ij}, and p_4 are consumed and evolved into p_4' in membrane 5.

8. $[p_1' \rightarrow ep_{11}'out]_5$

9. $[p_2' \rightarrow ep_{22}out]_5$

10. $[p_3' \rightarrow ep_{33}'out]_5$
    Rules 8, 9, and 10 are used to find out the minimum of p_1', p_2', p_3', p_4' using the modulus operator and is encoded as another object 'c'.

11. $[ep_1']_5 [dp_1]_1 \rightarrow [support]_1$ where $p_1'=c$.

12. $[ep_2']_5[dp_2]_2 \rightarrow [support]_2$ where $p_2'=c$

13. $[ep_3']_5[dp_3]_3 \rightarrow [support]_3$ where $p_3'=c$

14. $[ep_1']_5 [dp_1]_1 \rightarrow [\sim support]_1$ where $p_1' \neq c$

15. $[ep_2']_5[dp_2]_2 \rightarrow [support]_2$ where $p_2' \neq c$

16. $[ep_3']_5[dp_3]_3 \rightarrow [support]_3$ where $p_3' \neq c$
    Rules 11–16 are used for those membranes which satisfy the criteria of membrane dissolution. It sends an object "support" to the membranes to indicate that these membranes should remain in the system for further processing. Otherwise,

sends "~support" to the membranes, which indicates that it will not participate in the successive decision-making.

The same rules can be applied to those which are not participating in the decision-making replacing support with ~support in the corresponding membranes.

17. $[r_{2ij} \text{ support}]_1 \rightarrow [d_{a1ij}p_1]_1 [d_{a2ij}p_2]_2 [d_{a3ij}p_3]_3$
18. $[k_1 \rightarrow k_0]$
19. $[r_{kij} \sim \text{support}]_m \rightarrow \delta$
      where k and m $\in$. ...3}. This rule dissolves the membrane with label where δ indicates the dissolution operator.
20. $[\text{YES}]_m \rightarrow [\ ]_m \text{ YES}$
21. $[\text{NO}]_m \rightarrow [\ ]_m \text{ NO}$

Whenever a membrane produces the value of p_n, which is equal to zero which means that the computation stops and it sends "YES" to the environment. Otherwise, it sends "NO" to the environment indicates that the P system modeled above, cannot find the solution for the particular instant given.

4   Results

4.1   Experimental Setup

To analyze the behavior of proposed algorithm, the experiments are carried out by simulating in P Lingua Simulator 2.0. Since P system contains membranes, processing in parallel, different membranes are represented by a processor connected in parallel.

4.2   Performance Analysis

All the conventional brute force approaches take O (3^n) in the worst case, where, "n" is the level of the state space tree. The reason behind this exponential complexity is that at each level, it will produce at most three nodes. The proposed algorithm is better than the existing ones in terms of time complexity and correctness. Every time, the membrane is divided and the algorithm selects at most one membrane, having the minimum value of Pn. The time complexity of the proposed algorithm is O (h), where "h" is the height of the state space game tree. As the membrane algorithms work in maximally parallel way, it can find solutions in polynomial time. So the time complexity of the proposed algorithm is uniformly polynomial time. An analysis of 8 puzzle solutions has been done here. A graph (Fig. 3) is drawn taking the average number of generations against average number of solutions.

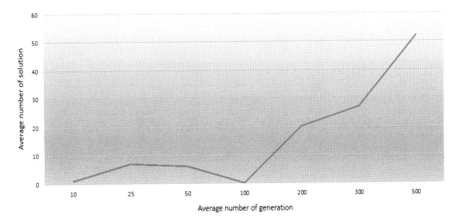

Fig. 3 An analysis of the solutions of the 8-puzzle

5 Conclusion and Future Scope

A new strategy for solving the N-puzzle problem using TPSESA rules and cell division is proposed in this paper. Instead of approaching the problem in brute force manner, some concepts of artificial intelligence are used to get the solution fast. All the computational things are designed in membranes only. Since it uses the concept of heuristic function with the framework of membrane computing, the solution described here yields better performance compared to traditional brute force methods.

References

1. Paun, G.: Introduction to Membrane. Computing (2006). https://doi.org/10.1007/978-3-64256196-2
2. Maroosi, A., Muniyandi, R.C.: Accelerated execution of P system with active membranes to solve the N-queens problem. Theor. Comput. Sci. **551**, 39–54 (2014)
3. Alhazov, A., Pan, L., Paun, G.: Trading polarizations for labels in P systems with active membranes. IJISCE **3** (2010)
4. Alhazov, A., Freund, R.: Polarization less P systems with one active membrane. In: International Conference on Membrane Computing. Springer, Berlin (2015)
5. Xiang, L., Xue, J.: Solving job shop scheduling problems by P system with active membranes. In: 7th International Conference on Intelligent Human-Machine Systems and Cybernetics. 978-1-4799-8646-0/15 $ 31.00. IEEE (2015)
6. Song, B., Zhang, C., Pan, L.: Tissue -like P systems with evolutional symport/antiport rules. Elsevier **378**, 177–193 (2017)
7. Bhasin, H., Singla, N.: Modified genetic algorithms based solution to subset sum problem. IJARAI **1** (2012)
8. Hayes, R.: The Sam Loyd 15-Puzzle. Dublin, Trinity College Dublin, Department of Computer Science, TCD-CS-2001-24, p. 28 (2001)

9. Bauer, B.: The Manhattan Pair Distance Heuristic for the 15 Puzzle. Paderborn, Germany (1994)
10. Calabro, C.: Solving the 15-Puzzle (2005). http://cseweb.ucsd.edu/~ccalabro/essays/15_puzzle.pdf
11. Pizlo, Z., Li, Z.: Solving combinatorial problems: The 15-puzzle. Mem. Cogn. **33**(6), 1069 (2005)
12. Felner, A., Adler, A.: Solving the 24 puzzle with instance dependent pattern databases. In: Proceedings of the Sixth International Symposium on Abstraction, Reformulation and Approximation (SARA05), pp. 248–260 (2005)
13. Felner, A., Korf, R.E., Hanan, S.: Additive Pattern Database Heuristics. J. Artif. Intell. Res. (JAIR) **22**, 279–318 (2004)
14. Drogoul, A., Dubreuil, C.: A distributed approach to N-puzzle solving. In: Proceedings of the Distributed Artificial Intelligence, pp. 95–108 (1993)
15. Bhasin, H., Singla, N.: Genetic based algorithm for N-Puzzle problem. Int. J. Comput. Appl. (0975-8887), **51**(22) (2012)

Renewable Energy Management and Implementation in Indian Engineering Institutions—A Case Study

Shekhar Nair, Senthil Prabu Ramalingam, Prabhakar Karthikeyan Shanmugam and C. Rani

Abstract Renewable energy has slowly come to the forefront of energy production techniques in the world with many countries investing more and more into the renewable sector. Developing country like India with 1.324 billion population as on 2016 has been investing huge amount of money on the renewable sources for meeting their demand. The aim of this paper is to understand how educational institutions are utilizing renewable energy sources and managing energy in meeting their loads. VIT University, Vellore Campus, Tamil Nadu is taken for case study in studying the technical and economic aspects involved in the effective usage of renewable sources. Various challenges involved with respect to educational institutions and energy saving mechanisms are also discussed in this paper.

Keywords Renewable energy sources · Educational institutions · Solar · Wind · Biomass · Energy policy

1 Introduction

1.1 Renewable Energy-INDIA

Electricity is the key factor for industrialization, urbanization, economic growth and improvement of quality of life in society. As of 2014, renewable energy provides an estimated 19.2% of global final energy consumption [1]. India is the world's fifth

S. Nair · S. P. Ramalingam · P. K. Shanmugam (✉) · C. Rani
School of Electrical Engineering, Vellore Institute of Technology, Vellore, India
e-mail: spk25in@yahoo.co.in

S. Nair
e-mail: Shekhar.Nair@hotmail.com

S. P. Ramalingam
e-mail: rsenthilprabu77@gmail.com

C. Rani
e-mail: crani@vit.ac.in

© Springer Nature Singapore Pte Ltd. 2020
K. N. Das et al. (eds.), *Soft Computing for Problem Solving*,
Advances in Intelligent Systems and Computing 1048,
https://doi.org/10.1007/978-981-15-0035-0_41

prevalent country in the electricity sector [2]. It has steadily made large progress in its renewable energy generation as can be seen from Fig. 1. It is also observed that in a span of two years, the generation from renewable [3] has increased by 142.27% (Fig. 1) which is a monumental increase in energy production. This falls in line with India's commitment towards its 2020 goal of 175 GW of production through renewables.

With recent advents in renewables notably solar, the overall conventional cost of producing electricity is seemingly more expensive than producing solar. Figure 2 shows how the overall cost of production (Indian rupees/unit) from solar has seemingly expected to become cheaper than producing from conventional sources.

But at the same time from Fig. 2b, we can also infer that the trend over the capital cost is different from what it is inferred from the overall production cost. Solar PV stands high in capital cost per MW compared to any other sources while the wind takes the second position.

Most parts of India receive 4–7 kWh of solar radiation per square metre per day with 250–300 sunny days in a year. The highest annual radiation energy is received in Western Rajasthan while the North-Eastern region of the country receives the lowest annual radiation. India has a good level of solar radiation, receiving the solar energy equivalent of more than 5000 trillion kWh/yr. Depending on the location, the daily incidence ranges from 4 to 7 kWh/m^2, with the hours of sunshine ranging from 2300 to 3200 per year [4].

With an energy efficiency of the grid becoming an important issue to be addressed about, universities as microgrids are becoming increasingly important in this discussion. As for example, a university campus, with its own distributed energy resources can be configured to operate as microgrid to optimize and improve the use of energy [5]. This paper aims at the Techno-economic challenges on what educational institutions undergo in reaching self-sustainable by adding more renewables into their micro-grid.

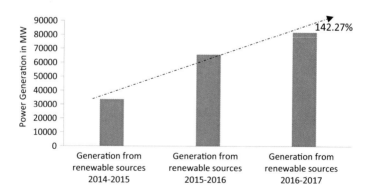

Fig. 1 Generation from renewable sources for the past three years (2014–2017) [4]

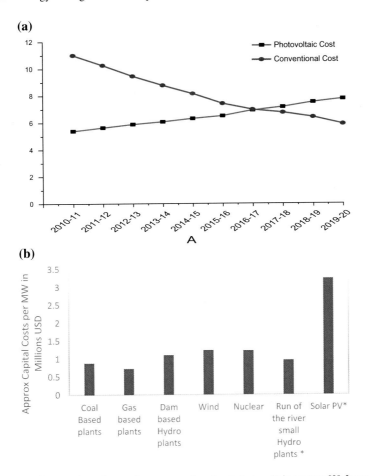

Fig. 2 **a** Cost of generation of power from conventional and photovoltaic sources [2], **b** comparison of capital cost per MW of power plants in India (Million USD) (2012) [20]

2 Renewable Energy Policy in India

As can be seen from Fig. 3, with regards to the installed capacity, renewables constitute 17% of the total energy production. Among them wind continues to be the largest provider with solar right behind it (Fig. 4).

The key legislation which guides the development of renewable energy in India is the Electricity Act, 2003. The Electricity Act 2003 mandates the State Electricity Regulatory Commissions (SERCs) to promote the generation of electricity from renewable sources of energy by providing suitable measures for connectivity with the grid and sale of electricity to any person. The National Tariff Policy, 2006, directs SERC to fix certain minimum percentages for purchase of renewable power [6]. These two policies combined together have incentivized states to increase their renewable

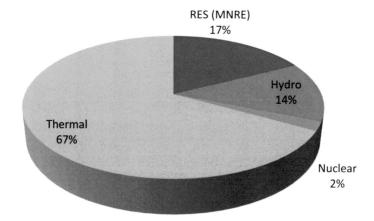

Fig. 3 All India installed capacity (in MW) of power generation in India (as on 30.04.2017) [21]

Fig. 4 Break up of renewable energy sources (RES) all India as on 31.03.2017 is given below (in MW) [21]

▢ Small hydro power

▪ Wind power

▪ Bio- power(both cogen and waste to energy)

▪ solar power

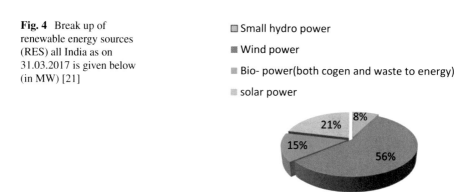

energy production level. Even still it is with the Energy Conservation Act of 2001, An Act to provide for efficient use of energy and its conservation that brings back our focus into educational institutions. As with this policy, large energy consumers, who have a connected load of 500 kW or contract demand of 600 kVA and above or who are intending to use it for commercial purposes have to abide by various energy standards. Some of these are

- furnishing information with regard to the energy consumed by the large consumer
- action was taken on the recommendation of the accredited energy auditor
- appointing an energy manager and submitting a report to the designated agency at the end of every financial year
- to face penalties for not abiding by their energy conservation recommendations [7].

In line with this, even large Universities which consume megawatts of power need to do the same and adopt various methods to reduce their dependence on the grid.

3   VIT University, Tamil Nadu, India as a Case Study

The Vellore Institute of Technology (Henceforth referred to as VIT) is a private University founded in 1984 and its main branch is situated in Vellore, Tamil Nadu, India. VIT is currently having the connected load of 12 MW supplied by the 33 kV line to the main powerhouse situated on campus which is then split from there to six other power stations. As can be seen from Fig. 5, the power is stepped down from a 33 to 11 kV which is then parallelly fed to both the hostel and campus. At the campus, 3000 kVA is provided to the Silver Jubilee Tower (SJT) through a line. Another line is fed to the Technology Tower (TT) station of 1600 kVA and is connected from there to the Main building power station of 2000 kVA. At the hostel end, 'K' block is provided with 3340 kVA, which is then split to 'A' block and 'N' block of 2000 kVA each. VIT also has three other modes of power generation, namely solar panels installed, diesel generators to provide electricity when there are sudden power outages and a Biodiesel plant. Each powerhouse has its own set of diesel generators. VIT as a whole consumes 4,93,212 units per month on an average from the grid which leads to a yearly consumption of 59,18,544 units. As can be seen from the tariff structure given by Tamil Nadu Generation and Distribution Corporation Limited (TANGEDCO) shown in Table 1 [8], VIT being a private educational institution running its own

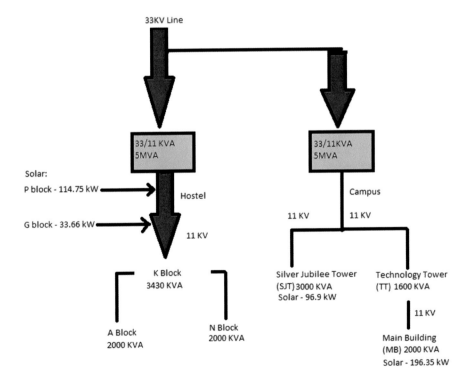

Fig. 5 Electric connection diagram of VIT Vellore campus

Table 1 Tariff structure of different consumers according to TANGEDCO Limited

Tariff	Category of consumers	Energy charges (Indian Rs/unit)	Demand charges (Indian Rs/kVA/month)
I-A	Industries, Registered factories, Textiles, Tea estates, IT services etc.	6.35	350
I-B	Railway traction	6.35	300
II-A	Govt. and Govt. aided Educational Institutions and hostels, Government Hospitals, Public Lighting and Water supply, Actual places of public worship etc.	6.35	350
II-B	Private Educational Institutions & Hostels	6.35	350
III	All other categories of consumers not covered under High Tension-I-A, I-B, II-A, II-B, IV and V	8.00	350
IV	Lift Irrigation societies for Agriculture registered under Co-op societies or under any other Act (Fully subsidised by the Govt.)	6.35	0
V	High tension temporary supply for construction and other temporary purposes and start up power provided to generators	11.00	350

hostel block, gets charged at Rs 6.35 per unit with Rs 350 being charged for every kVA demand per month with a total kVA demand of 8000 kVA.

VIT also has five Solar panels set up on five of its building and has steadily been increasing its solar usage. Table 2 shows the installed capacity of solar panels in VIT as of May 2017 with a total installed capacity of 500 kW. There have been 1295 MWh of energy generated so far as of June 2017 working through 657 days with an average of 1980 kWh per day generation. VIT's Biodiesel plant with a 135 kg/hour reactor capacity and a 100 kW diesel capacity engine producing an average of 1295 units per day. It uses 0.98 kgs of wood to create one unit of electricity and requires

Table 2 Installed solar capacity in various buildings in VIT

Building	Capacity in kW
Sundar Patel Block	114.75
Silver Jubilee Tower	96.9
Ramanujam Block (Men's hostel)	58.65
GDN	196.35
G block (Men's hostel)	33.66

maintenance every 500 h. It uses a particular wood called *Prosopis Juliflora* which is abundant in Tamil Nadu presently costing about Indian Rs of 3800 per tonne.

This paper aims to bring light to the challenges that educational institutions face in line with reducing and conserving their power usage. The challenges are as follows:

3.1 Peak Demand

During peak demand time, priority loading is used to avoid exceeding the maximum demand allowed, in which only the lighting loads are permitted. All AC loads are shut down. As Vellore has an annual mean temp of 29° with a maximum of 39° and minimum of 21° [9], ACs are constantly needed as a form of cooling and constitutes the maximum loads. Due to the restraint of permitted power allowed to avoid transformer overloading and due to the transformer shut down in the Katpadi area where VIT is located due to maintenance, it is for this reason VIT is slowly moving to be connected into the 110 kV line directly to avoid these issues. It is also worth to note that the University's load is increasing rapidly which is eligible for receiving power from the said voltage level. There are six transformers as shown in the electric diagram for managing peak loads in VIT. When any of the rated outputs cross 90% threshold limit, a breaker automatically trips and shuts down AC loads. One initiative is to increase the solar capability in the campus to compensate for the peak demand. The energy produced per month throughout VIT is an average of 60,000 units. If we calculate with the same tariff structure allotted to VIT, the institution saves 3.18 million Indian rupees per month and also reduces peak demand dependability on the grid. There are currently negotiations in place to increase VIT's solar capability by another 620 kW. There are also shading issues to be considered as solar capacity cannot be simply increased due to the positions of the buildings and the shadows that fall due to it.

The Biodiesel plant is also another initiative used to reduce peak demand. It produces 38,850 units per month on an average and saves VIT 3.05 million rupees per month. Though it uses 38,073 kgs of wood, it only costs 144,677.4 Indian rupees. Therefore, the net savings are 2,905,322.6 Indian rupees.

Exploring other technologies as potential sources of electricity generation as shown in Table 3 [10], we find that no other sources can be used other than the already existing ones on campus or that sustainability is too low to be exploited.

As no other source of electricity can be used to help peak demand loads, an initiative that could be used is a solar hybrid radiant cooling system which can be used for high cooling load demand in hot and humid climate. As the majority of the load in VIT is due to air-conditioning, installing a solar hybrid radiant cooling system could reduce its peak loads by not having the need of switching on the AC's itself. The solar hybrid cooling system is expected to effectively handle the sensible and latent cooling loads through the system integration of absorption refrigeration, desiccant dehumidification and radiant cooling in a renewable way, in order to reduce the energy consumption, hence the carbon footprint of the air-conditioning [11].

Table 3 Generation technologies used in VIT campus

Generation technology	Fuel type	Fuel availability	Existence on campus	Overall sustainability
Diesel generator	Diesel	High	Yes	High
Biodiesel	Biofuel	High	Yes	Medium
Photovoltaic	Solar	High	Yes	High
Wind	Wind	Low	No	Very low
Hydroelectric	Water flow	Low	No	Very low
Fuel cell	various	None	No	No

Peak shifting can also be used, by which the consumption profile is smoothed by moving loads to different time periods, can reduce peak load, maximize asset life, and facilitate the increased penetration of on-site renewable energy generation. This can be achieved with electricity storage or demand response (DR) strategies [12]. Electricity storage can be achieved using battery banks. A battery bank can be used to store electricity when electricity prices are cheap and can be used to provide electricity when demand is high by reducing peak demand on transformers or to supplement the use when electricity is more expensive. As battery prices decrease and electricity prices increase, it would become more feasible to use this technology. Demand response strategies can be used to take advantages of energy production on campus to incorporate them onto when necessary. By coupling them with battery storage techniques, this energy produced locally can be used in tangent with the grid supplied electricity to reduce peak demand.

With the help of the optimization technique, the loads can be made ON and OFF from ON peak to OFF Peak period with various constraints [13, 14]. Based on the idea of the Dijkstra's algorithm [13] the DijCostMin algorithm schedules the load in various time durations to avoid the peak demand problem. The Minimum Cost Maximum Power (MCMP) [14] technique is also for peak demand issue and this technique is made ON and OFF from ON peak to OFF peak periods with the help of binary variables. Both papers have suggested if the loads are scheduling along with renewable energy the economic energy management can be achieved and it is the solution for this peak demand problem.

3.2 Power Factor

To maintain power factor, VIT has a capacitor bank attached to each of its transformers. Whenever the power factor increases, the power drawn increases as well and leads to losses. It is for this purpose that for every 25 kW of power increase, a capacitor is turned on to keep the power factor maintained to the nominal value of 50 Hz. It is Technology Tower alone that 22 numbers of 25 kVAR capacitive banks are used to maintain power factor to a near-unity level to avoid losses. They are used for reactive power support, improved voltage control and power factor; reduced system losses and reactive power requirements at generators; and increased steady state stability limits [15]. They have been installed in the distribution transformers to reduce losses, maintain unity power factor and thereby help control and maintain voltage for the area there installed in.

3.3 Line Losses

The primary reason for line losses is due to the length of the wire. The simplest solution for this is to place the transformers with a capacitor in the buildings which consume the power itself so as to not create distance between them and hence increase line losses. As this would be too expensive and increase maintenance charges, VIT has built six powerhouses through which the power is distributed.

3.4 Power Outages

Due to ever-increasing loads which exceed 10 MW, VIT is in the process of shifting to the 110 kV high voltage line which would give them uninterrupted electricity. The reason for interruption of the electricity flow currently is due to the VIT load supply being cut down in half whenever the district power station was overloaded. This will help them reduce the amount of time they run their diesel generators which would save the costs as well as diesel is expensive. Currently whenever there is a loss of power supply from the grid, VIT runs its own Diesel Generators that take a total of two minutes to replace the electricity supply. One to Auto on the DG set and a minute for the changeover. The number of units used on average is given in Table 4.

So, a total of 28,458 units of electricity have been produced on a monthly basis using diesel generators to supply the main campus excluding the hostel requirements due to power outages. The Diesel Generators are switched on for a minimum of 12 h per month.

Table 4 Monthly usage of diesel generator sets in various stations located along VIT

Power house name	Units used per month
TT	7071
Main building	9877
SJT	11,510

3.5   Power Saving and Awareness-Students Contribution

Energy conservation can be achieved through both, the adoption of technical measures and the changing of the behaviour of consumers. It was found that the most committed environmentalists are the most highly educated [16]. In terms of technical measures, there have been numerous initiatives taken by VIT students to reduce the electricity costs and save power; one of them was implementing energy-efficient light automation system [17] as shown in the Fig. 6a, b, which was able to save 71% percent energy in a bathroom fitted with the light automation device compared to an

(a)

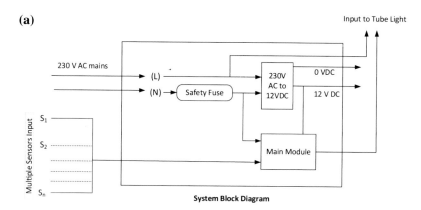

(a)

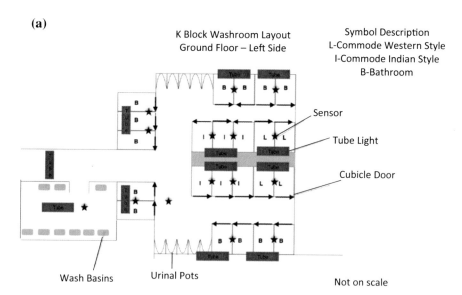

Fig. 6 **a** Block diagram—efficient light automation system implemented in VIT by the students, **b** layout of efficient light automation system

adjacent one without. It was able to save Rs 2000 per month for the electricity used in the washroom. This device operates on sensing movement and keeps the light ON inside the washroom as long as there is someone inside of it and activating only the necessary lights at that time. In terms of educating the students, there are various project-based learning (PBL) courses on renewable energy and awareness created through cultural clubs on campus and competitions organised on the same. VIT students also get involved with the nearby primary/secondary schools in imparting awareness on energy conservation and the significance of renewable sources through their clubs and National Service Schemes (NSS).

Unavailability of human resource with required knowledge and skills is often identified as one of the key reasons for poor dissemination of renewable energy technologies. For a balanced and accelerated diffusion of various renewable energy technologies, adequate number of competent and well-trained professionals are needed [18]. It becomes necessary to seek inputs from the industry on the lack of skilled professionals and what skills are needed for those individuals. Universities can help to bridge that gap and increase the avenues for renewable technology adoption.

3.5.1 Government Projects and VIT, Vellore Campus-(Ongoing)

Table 5 shows some of the relevant projects carried out in the area of renewable energies sanctioned by various agencies of the Government of India. Faculties, staffs and students are involved in the successful completion of projects which will directly or indirectly pave way in increasing the utilisation of renewable energy.

Table 5 List of ongoing projects in VIT Vellore campus in the area of renewable energy sources [20]

S. No.	Project tittle	Issuing agencies
1	Modelling, Analysis and implementation of multilevel multi-string grid Tie Inverter for medium scale Grid connected Photovoltaic system	Technology System Development Programme (TSDP) of Department of Science and Technology (DST)
2	Development and Installation of Micro Thruster augmented Wind power Generator using a 200 kW MICON Power Plant at C-WET facility, Kayathar.	Ministry of New and Renewable Energy (MNRE)
3	Experimental study on Kalina Cycle Cogeneration System.	DST
4	Sustainability and Composition Based Evaluation of Biofuel Energy Resources.	DST—Science and Engg. Research Board (SERB)
5	Synthesis, Characterization, and Studies of Novel Poly[Thieno (Indenoindole)] Based Low Band Gap Polymers for Orlganic Solar Cell Applications	DST—Solar Energy Research Initiative (SERI)

3.5.2 Participation in Open Market

As per Indian market policy, any company/institution can participate in open market trading by buying/selling power if they have a load/source capacity of more than 1 MW. Having two online power trading exchange namely, Indian energy exchange (IEX) and Power Exchange India (PEX), companies/institutions can register and participate in buying/selling the energy. On this context, having approximately 9000 Engineering institutions [19], India can make sufficient contribution from these institutions either through government policies or voluntary means by the institutions.

4 Conclusions

Renewable energy sources are slowly becoming a mainstream way of generating energy. With the cost of production through solar becoming far cheaper than conventional sources and the increase of RES in India, educational institutions can provide a solid infrastructure to show the adaption of RES. To understand how renewable energy production can be implemented on university campuses, VIT has been used a case study to show how solar and biodiesel can be used a regular alternative to diesel generators and relying on mainstream electricity production.

There are various challenges that still prevent us from fully relying on renewables such as size of productions, inability to meet loads with renewable and functionality of solar with respect to VIT such as shade on buildings and building architecture that doesn't allow solar to be implemented everywhere but if used sufficiently can reduce the dependence on grid.

There exist other issues such as peak demand, power factor maintenance, line losses and power outages which can be regulated as mentioned before. These issues are of prominent importance as to how other educational institutions could adopt similar mechanisms to regulate and reduce their energy usage. One important metric is to constantly adapt to newer technologies and to implement energy saving mechanisms. The future work includes other innovative energy saving mechanisms and the development of smart algorithms and optimization techniques to reduce the dependence on the electric grid.

Acknowledgements The authors sincerely acknowledge the management of VIT University, Vellore, Tamil Nadu, India for their constant support during the preparation of this manuscript and sharing of valuable information/data. The authors also thank Mr. S. Umashankar, Supervisor, Power house department, VIT University for the technical support rendered during the research period.

References

1. http://www.ren21.net/wp-content/uploads/2016/05/GSR_2016_Full_Report_lowres.pdf, p. 28
2. Khare, V., Nema, S., Baredar, P.: Status of solar wind renewable energy in India. Renew. Sustain. Energy Rev. **27**, 1–10 (2013)
3. https://renewablesnow.com/news/india-reached-22-of-175-gw-renewable-energy-goal-490964/
4. Kumar, A., Kumar, K., Kaushik, N., Sharma, S., Mishra, S.: Renewable energy in India: Current status and future potentials. Renew. Sustain. Energy Rev. **14**, 2434–2442 (2010)
5. Talei, H., Zizi, B., Abid, M.R., Essaaidi, M., Benhaddou, D., Khalil, N.: Smart campus micro-grid: advantages and the main architectural components. In: 2015 3rd International Renewable and Sustainable Energy Conference (IRSEC)
6. http://www.teriin.org/projects/nfa/pdf/working-paper-14-Governance-of-renewable-energy-in-India-Issues-challenges.pdf
7. http://powermin.nic.in/sites/default/files/uploads/ecact2001.pdf
8. http://www.tangedco.gov.in/linkpdf/ONE_PAGE_STATEMENT.pdf
9. https://www.timeanddate.com/weather/india/vellore/climate
10. Purser, M.S., Kalaani, Y., Haddad, R.J.: A Technical and economical study of implementing a micro-grid system at an educational institution. In: 2015 IEEE Power & Energy Society, Innovative Smart Grid Technologies Conference (ISGT)
11. Fonge, K.F., Chow, T.T., Lee, C.K., Lin, Z., Chan, l.S.: Solar hybrid cooling system for high-tech offices in subtropical climate—radiant cooling by absorption refrigeration and desiccant dehumidification. Energy Convers. Manage. **52**(8–9), 2883–2894
12. Morales Gonz´alez, R., Aslam, M.F., van Goch, T.A.J., Blanch, A., Ribeiro, P.F.: Microgrid design considerations for a Smart-Energy University Campus. In: 2014 5th IEEE PES Innovative Smart Grid Technologies Europe
13. Basit, A., Sardar Sidhu, G.A., Mahmood, A., Gao, F.: Efficient and autonomous energy management techniques for the future smart homes. IEEE Trans. Smart Grid **PP**(2), 1–10 (2015)
14. Singaravelan, A., Kowsalya, M.: A novel minimum cost maximum power algorithm for future smart home energy management. J. Adv. Res. **8**, 731–741 (2017)
15. http://upcommons.upc.edu/handle/2099.1/18886
16. Al-Shemmeri, T., Naylor, L.: Energy saving in UK FE colleges: the relative importance of the socio-economic groups and environmental attitudes of employees. Renew. Sustain. Energy Rev. **68**, 1130–1143
17. Kotawala, S., Das, R.R.: Implementation and Analysis of Energy Efficient Light Automation System, Project Report Submitted to VIT University (2016)
18. Kandpal, T.C., Broman, L.: Renewable energy education: a global status review. Renew. Sustain. Energy Rev. **34**, 300–324 (2014)
19. http://www.aicte-india.org/downloads/List_of_Approved_Institutes_2015-2016.pdf
20. www.vit.ac.in
21. Central energy authority of India report for April 2017, Ministry of Power, Government of India

Inverse Kinematics Analysis of Serial Manipulators Using Genetic Algorithms

Satyendra Jaladi, T. E. Rao and A. Srinath

Abstract This paper gives the insight into inverse kinematics analysis of industrial serial manipulators of different degree of freedom, utilizing homogeneous transformation matrices, Denavit–Hartenberg (D–H) parameters, and the forward kinematics. For a robot with 2/3/4 degrees of freedom, inverse kinematics can be determined by using simple geometry with (sin, cos, tan) trigonometry functions, first by solving the forward kinematics, and then creating simultaneous equations solving those for the joint angles. For a robot with six joints (spherical wrist) or more joints (redundant robots) the inverse kinematics becomes complex to solve manually, in such cases the inverse kinematics will be treated as an optimization problem which can be solved numerically. In the proposed approach, MATLAB's 'ga' solver was used to solve the inverse kinematics of serial manipulators. The joint angles obtained from optimization methods (ga), (pso), and algebraic methods are compared for two degrees of freedom R-R manipulator. A case study on ABB IRB 1600-1.45 serial manipulator with 6-DOF is presented in this paper.

Keywords Genetic algorithm · ga · Inverse kinematics · Robotics toolbox for MATLAB · Optimization · Serial manipulators

1 Introduction

Kinematics is a branch of mathematics that deals with positions and orientations (pose) of bodies relative to base first. This is essential for position analysis. In position analysis the relation between joint space and Cartesian space (the position of end

S. Jaladi (✉) · T. E. Rao · A. Srinath
Mechanical Engineering Department (DST-FIST Sponsored), Robotics & Mechatronics
Engineering, K L E F Deemed to Be University, Vaddeswaram, Guntur, AP, India
e-mail: sam.satya38@gmail.com

T. E. Rao
e-mail: terao1@kluniversity.in

A. Srinath
e-mail: srinath@kluniversity.in

© Springer Nature Singapore Pte Ltd. 2020
K. N. Das et al. (eds.), *Soft Computing for Problem Solving*,
Advances in Intelligent Systems and Computing 1048,
https://doi.org/10.1007/978-981-15-0035-0_42

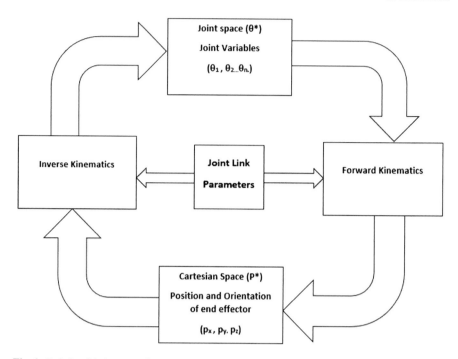

Fig. 1 Relationship between forward and inverse kinematics

effector) and the joint angles and orientation is to be found. This study gives rise to two problems which are forward or direct kinematics and inverse kinematics, as illustrated in Fig. 1. The position and orientation of end effector is collectively termed as configuration of the end effector. In forward kinematics problem, the joint variables are given, and the end-effector configuration must be found. In inverse kinematics, the end-effector configuration is given, and then joint variables are to be found [1]. When these relations are differentiated once velocity analysis arise which is required for smooth motion control of the end effector and differentiated again acceleration analysis arise which is required for the dynamic analysis of the robot. For many serial manipulators, closed-form solutions of the IK exist (e.g., PUMA, FANUC.) [2, 3]. For many other serial manipulators, with higher DOF the IK analytical solution becomes difficult as there is more than one number of possible solutions. To solve IK for a redundant robot, how a genetic algorithm (GA) was used can be found in [4]. In the present approach, MATLAB's 'ga' is proposed for solving joint angles of 2-DOF and 6-DOF robot arms. Traveling salesman problem(TSP) optimization is adopted, and positioning error is the Euclidean distance between the desired and the current known locations in the 2-D/3-D space. When the square root of their difference becomes zero, the corresponding value is global optimum which represents accurate joint angles, and search space for joint angles solution vector lies in the predefined upper and lower bounds range.

This study only deals with the position of bodies and angles between them, not considering the forces and moments acting on the bodies which will cause translational or angular motion.

1.1 Degrees of Freedom

A Robot's Degree of Freedom is defined as the number of independent or minimum coordinates required to fully describe its pose. A simplified form of Grubler–Kutzbach criterion for revolute or prismatic joints is given as follows:

$$N = 3(r-1)-2p \quad \text{(Planar)} \tag{1}$$

$$N = 6(r-1)-5p \quad \text{(Spatial)} \tag{2}$$

where

N = Degree of Freedom of whole system.
r = Number of links in the system.
p = Number of joints in the system.

1.2 Homogeneous Transformation Matrix (HTM)

In [5] describes Homogeneous transformations as the relationships between Cartesian coordinate frames in terms of translation and orientation. HTM describes both the position and orientation of any frame with reference to any other frame, and product of all the coordinate frame transformations matrices from 0 to n links gives the HTM for any given manipulator.

$$^{0}T_{n} = {}^{0}A_{1} * {}^{1}A_{2} * {}^{2}A_{3} \ldots {}^{n-1}A_{n} \tag{3}$$

$$T = \begin{bmatrix} Rotation\ Matrix\ (3 \times 3) & Translation\ Vector\ (3 \times 1) \\ Perspective\ Transformation\ Matrix\ (1 \times 3) & Scale\ Factor\ (1 \times 1) \end{bmatrix}$$

Last row of the HTM (both Perspective matrix and Scale factor) is useful in vision systems. In robotics study last row elements are all zeroes, except scale factor which is 1, for describing the position and orientation of frame {2} with respect to frame {1}, T takes the form $^{1}T_{2}$.

$$^{1}T_{2} = \begin{bmatrix} {}^{1}_{2}R & {}^{1}_{2}P \\ 0\ 0\ 0 & 1 \end{bmatrix}$$

1.3 Denavit–Hartenberg Criterion

Denavit–Hartenberg notation describes the structure of a serial-link manipulator. D–H notation uses Four parameters to describe the relationship between 3D coordinate frames attached to two successive links. four elementary transformations, which include two translations and two rotations are required for describing the pose of the frame J with respect to the frame $J - 1$. In D–H criterion each joint in the robot is described simply by four parameters. four parameters of D–H notation are given below (Table 1).

In Robotics, Vision and Control [6], coordinate transformations can be expressed in the following way:

$$^{J-1}T_J = \text{trotz}(theta\_j) * \text{transl}(0, 0, d\_j) * \text{transl}(a\_j, 0, 0) * \text{trotx}(alpha\_j) \tag{4}$$

The following are the essential transformations for D–H criterion:

- trotz(theta_j)—A rotation around the Z-axis by an angle theta_j.
- transl(0, 0, d_j)—A translation along the Z-axis by a length d_j.
- transl(a_j, 0, 0)—A translation along the X-axis by a length a_j.
- trotx(alpha_j)—A rotation around the X-axis by an angle alpha _j.

2 Forward Kinematics

The mathematical form of forward kinematics can be obtained from the relationship between the robot joint angles and the pose of the robot end effector.

$$\xi_N = \kappa\left(q_j\right)$$
$$q_j = \left\{q_{j}, j\varepsilon[1 \ldots n]\right\} \tag{5}$$

Pose of the robot end effector is the function of robot joint angles.
Here

ξ_N = Pose of the Robot end effector.

Table 1 D–H parameters—theta, d, a, and alpha	S. No	D–H parameter	Symbol	Type
	1	Joint angle	θ_j	Revolute joint variable
	2	Link offset	d_j	Prismatic joint variable
	3	Link length	a_j	Constant
	4	Link twist	α_j	Constant

q_j = Vector of Robot joint angles.

E is the pose of single link 1-DOF planar robot, (Fig. 2). This can be obtained by rotating the coordinate frame by the angle q_1 then by a translation by a length of a_1 along its axis.

$$E = \text{rot2}(q1) * \text{transl2}(a1);$$

$$^0T_1 = \begin{bmatrix} c(q_1) & -s(q_1) & 0 & c(q_1) \times a_1 \\ s(q_1) & c(q_1) & 0 & s(q_1) \times a_1 \\ 0 & 0 & 1 & 0 \\ 0 & 0 & 0 & 1 \end{bmatrix}$$

When joint angle q_1 is 70° and the link length a_1 is 10 m, then the location of End Effector (E) is at the position (X, Y) given as,

$$X = \cos(q_1) * a_1 = \cos(70) * 10 = 3.4202$$
$$Y = \sin(q_1) * a_1 = \sin(70) * 10 = 9.3969$$

Similarly, a 2-DOF R-R planar robot manipulator can be visualized (Fig. 3).

Let a_1 and a_2 are lengths of link 1 and link 2, and θ_1 and θ_2 are the angles between the links with respect to the origin, its joint link parameters are tabulated (Table 2).

Homogeneous transformation matrix for R-R planar manipulator

$$^0T_2 = \begin{bmatrix} c(\theta_1 + \theta_2) & -s(\theta_1 + \theta_2) & 0 & c(\theta_1) \times a_1 + c(\theta_1 + \theta_2) \times a_2) \\ s(\theta_1 + \theta_2) & c(\theta_1 + \theta_2) & 0 & s(\theta_1) \times a_1 + s(\theta_1 + \theta_2) \times a_2) \\ 0 & 0 & 1 & 0 \\ 0 & 0 & 0 & 1 \end{bmatrix}$$

The location of End Effector (E) is at the position (X, Y) given below.

$$\begin{bmatrix} X \\ Y \end{bmatrix} = \begin{bmatrix} c(\theta_1) \times a_1 + c(\theta_1 + \theta_2) \times a_2) \\ s(\theta_1) \times a_1 + s(\theta_1 + \theta_2) \times a_2) \end{bmatrix}$$

Fig. 2 A single link planar robot

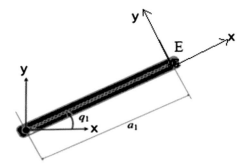

Fig. 3 Two link R-R planar
robot

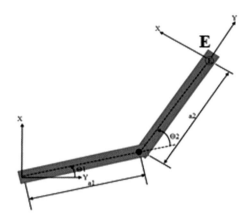

Table 2 D–H parameters of
R-R two-link planar robot

θ_j	D_j	A_j	α_j
θ_1	0	a_1	0
θ_2	0	a_2	0

3 Inverse Kinematics

Inverse kinematics is very important for arm type serial manipulators when com-
parted to parallel manipulators where forward kinematics is of importance. Inverse
kinematics can be described as follows: given robot's end effector at a particular pose
in 2-D/3-D space, finding the joint angles such that the robot end-effector achieves
that pose is inverse kinematics.

$$q_j = \kappa^{-1}(\xi_N) \tag{6}$$

Here

ξ_N = Pose of the Robot end effector.
q_j = Vector of Robot joint angles
    $j \in [1 \dots n]$.

3.1 Trigonometric Equations

Inverse kinematics equations for a serial robotic manipulator can be obtained from
analytical and numerical methods. Trigonometric equations can be used for solving
Inverse kinematics. This is analytic approach of solving inverse kinematics. For a
robotic manipulator with any degrees of freedom, this approach is suitable. Inverse

tangent function is very much useful in calculating the joint angle values of the serial manipulators. TANGENT of Y divided by TANGENT of X is written using atan2 as

$$theta = atan2(y, x);$$

But as the complexity of the robot increases solving algebraic equations manually becomes difficult.

4 Genetic Algorithm

The genetic algorithm [7, 8] is a powerful optimization technique which imitates natural genetics. Genetic algorithm representation of a given problem involves this sequence of steps.

1. Production of initial chromosomes/population (set of individuals),
2. Calculation of fitness of individual in the population,
3. Production of individuals.
4. Selection (stochastic universal sampling)
5. Crossover and mutation. (modification of parents == children)
6. If termination conditions are met, then finish (present case—avg change in fitness function, predefined number of generations), Else
7. Continue to step 2.

$$l_b <= q <= u_b$$

q = joint angles;
l_b = lower bounds;
u_b = upper bounds

formulation basis for obtaining global optimum,

$$q* = \min|\text{Desired Pose} - \text{Current Pose}| \tag{7}$$

q^* = Joint angles obtained from this minimum.

Current pose of the end effector is already known, and the desired pose can be obtained from forward kinematics equations, the Euclidian distance between these two points in 3-D space is position error function, which is applicable for all serial manipulators in this paper.

$$q^* = \sqrt{\begin{array}{c}\left(px_{desired} - px_{current}\right)^2 + \left(py_{desired} - py_{current}\right)^2 \\ + \left(pz_{desired} - pz_{current}\right)^2\end{array}} \tag{8}$$

$(px_{current})$, $(py_{current})$, $(pz_{current})$ are the coordinates of the current known location, and $(px_{desired})$, $(py_{desired})$, $(pz_{desired})$ are points obtained by calculating forward kinematics. MATLAB's 'ga' solver will minimize this criterion to return the accurate joint angles resulting in better positioning accuracy of the robot end effector. Robotics toolbox for MATLAB (RVC) is used along with MATLAB's GUIDE to render a graphical robot, such that a 3-D Pose of robotic manipulator can be visualized.

Forward equations and position error code written in MATLAB,

```
xt = cos(min(1)) * l1 + cos(min(1) + min(2)) * l2;
yt = sin(min(1)) * l1 + sin(min(1) + min(2)) * l2;
min_dist_two_dof = sqrt((xt - px)^2 + (yt - py)^2);
```

For achieving vector of joint angles [20 30], the LB and UB are constrained as (Fig. 4),

```
LB = [pi/10; pi/10]; UB = [pi/4; pi/4];
```

Initial ranges of the joint angle values will largely influence the output solution vector. LB and UB values are passed as input arguments to the 'ga', to ensure that function search space is constrained, and will not assume any other range of values.

Syntax of 'ga' solver written in MATLAB with required parameters.

```
[theta,fval] = ga(FtFn,n,[],[],[],[],LB,UB,[],options);
```

Syntax of 'particle swarm' solver written in MATLAB with required parameters.

```
[theta,fval] = particleswarm(Fitfcn,nvars,LB,UB);
```

R-R type 2-DOF manipulator with link lengths $l1 = 0.2$ m; $l2 = 0.2$ m; input coordinates px and py are read from GUI. px = 0.3164; py = 0.221613. When the results of 'ga' and particle swarm (PSO) are compared, PSO provides better response compared to that of 'ga', in terms of computational time, stable convergence. Joint angle values returned from algebraic method and PSO are same (Table 3).

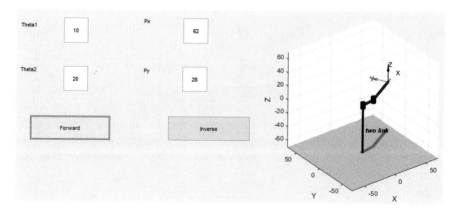

Fig. 4 Input for direct/inverse kinematics for R-R type 2-DOF manipulator

Table 3 R-R type manipulator—comparison of algebraic, ga and PSO outputs

S. No	$\theta°$ (deg)	Algebraic	'ga'	'PSO'
1	θ_1	19.9766	20.0016	19.9766
2	θ_2	30.0698	29.9970	30.0698
Elapsed time (s)		0.002336	0.683423	0.267473

5 Case Study

The genetic algorithm approach is also verified with 6-DOF ABB IRB 1600 (1.45 m reach) robot. A 3-D simulation model of the industrial robot ABB IRB 1600-1.45 designed in MSC ADAMS [9] was used for the study.

The robot is defined by arm lengths $a_1 = 0.448$ m; $a_2 = 1.066$ m; $a_3 = 0.114$ m; D–H parameters are tabulated for this robot. Forward kinematics equations needed for fitness function of 'ga' solver are written as upper and lower bounds on joint angles.

```
LB = [pi/20;pi/10;pi/9;0;0;pi/4];
UB = [pi/18;pi/6;pi/5;0;0;pi/3];
```

'ga' solver with required parameters.

```
ga(Ftfcn,nvars,[],[],[],[],LB,UB,[],options);
```

Position error of 6-DOF ABB robot written in fitness function

```
min = sqrt((xt - px)^2 +(yt - py)^2 +(zt - pz)^2);
```

The above minimization criteria will maximize the position accuracy of the 6-DOF robot (Fig. 5).

$$xt = (\cos(\min(1)) * \cos(\min(2)) * \cos(\min(4)) * \sin(\min(5)) * l6)$$
$$- (\sin(\min(1)) * \sin(\min(4)) * \sin(\min(5)) * l6)$$
$$+ (\cos(\min(1)) * \sin(\min(2)) * \cos(\min(5)) * l6)$$
$$+ (\cos(\min(1)) * \sin(\min(2)) * d3) - (\sin(\min(1)) * l2);$$

$$yt = (\sin(\min(1)) * \cos(\min(2)) * \cos(\min(4)) * \sin(\min(5)) * l6)$$
$$+ (\cos(\min(1)) * \sin(\min(4)) * \sin(\min(5)) * l6)$$
$$+ (\sin(\min(1)) * \sin(\min(2)) * \cos(\min(5)) * l6)$$
$$+ (\sin(\min(1)) * \sin(\min(2)) * l3) + (\cos(\min(1)) * l2);$$

$$zt = (((-\sin(\min(2))) * \cos(\min(4)) * \sin(\min(5)) * l6))$$
$$+ ((\cos(\min(2)) * \cos(\min(5)) * l6)) + (\cos(\min(2)) * d3);$$

Fig. 5 A 6DOF ABB IRB
1600-X/1.45 Robot CAD
model (MSC ADAMS)

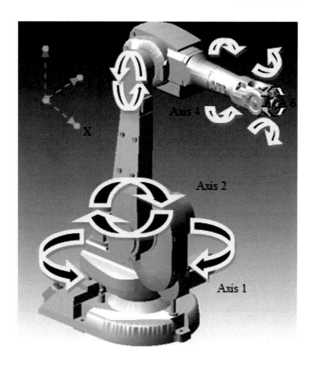

For six joint variables with a population size of 200, the function tolerance is
1.0000e–06, in this case optimization of 6-DOF robot's fitness function is stopped
when average change in the fitness value becomes less than function tolerance
(Tables 4 and 5).

$$\text{Six Operating points } [X, Y, Z] \text{ and their corresponding joint angles.} \qquad (9)$$

$P_1 = [1.533; 0.270; -0.4035]; P_2 = [1.400; 0.3512; -0.1089]; P_3 = [-0.8502; 1.1421; 2.392];$
$P_4 = [1.712; 1.25; 1.17]; P_5 = [1.8; 1.25; 1.17]; \text{ and } P_6 = [-1.25; -1.1; 0.25];$

Table 4 The D–H parameters for a 6-DOF ABB IRB 1600-X/1.45 m (reach) robot

Link [j]	$\theta°$ Initial; Final (deg)		D (m)	a (m)	α (deg)
1	360	− 180; +180	0.72	0.448	− $\pi/2$
2	240	− 90; +150	0	1.066	0
3	310	− 245; +65	0	0.114	− $\pi/2$
4	400	− 200; +200	1.002	0	$\pi/2$
5	130	− 115; +115	0	0	− $\pi/2$
6	800	− 400; +400	0.25	0	0

Table 5 Joint angles obtained after position optimization of given points (P_1–P_6)

P	$\theta_1°$ (deg)	$\theta_2°$ (deg)	$\theta_3°$ (deg)	$\theta_4°$ (deg)	$\theta_5°$ (deg)	$\theta_6°$ (deg)
P_1	9.0000	19.0000	0	0	0	27.5511
P_2	8.6904	19.0003	21.8123	39.2933	39.2933	29.7278
P_3	− 7.0030	18.0000	0	0	0	30.0001
P_4	12.0231	19.0003	29.9970	30.0698	19.0003	59.2111
P_5	12.0231	19.9766	20.0016	19.9766	21.2931	29.0221
P_6	− 8.2033	30.0698	29.9970	30.0698	21.2123	21.122

6 Conclusions

From the above results it is concluded that genetic algorithm (ga) handles all types of constraints. Although 'ga' solver results are accurate, they are stochastic, its results vary every time, setting Initial range of population, upper and lower bounds resulted in an accurate solution vector, like those that are obtained by the algebraic solution. This has been observed and proved for 2-DOF and 6-DOF serial manipulators. Optimizing time taken for solving the joint angles and trajectory generation is in further scope of this work.

Acknowledgements The authors are grateful to the support of K L E F (Deemed to be University), Dept. of Mechanical Engineering, FIST sponsored Advance Prototyping and Manufacturing Lab (SR/FST/ETI—317/2012(C)) for supporting us throughout the project.

References

1. Saha, S.K.: Introduction to Robotics. Tata McGraw-Hill Education (2014)
2. Paul, R.P.: Robot Manipulators: Mathematics, Programming, and Control: the Computer Control of Robot Manipulators. Richard Paul (1981)
3. Tsai, L-W., Morgan, A.P.: Solving the kinematics of the most general six-and five-degree-of-freedom manipulators by continuation methods. J. Mech. Trans. Autom. Des. **107**(2), 189–200 (1985)
4. Garg, D.P., Kumar, M.: Optimization techniques applied to multiple manipulators for path planning and torque minimization. Eng. Appl. Artif. Intell. **15**(3–4), 241–252 (2002)
5. Mittal, R.K., Nagrath, I.J.: Robotics and Control. Tata McGraw-Hill (2003)
6. Corke, P.: Robotics, Vision and Control: Fundamental Algorithms In MATLAB® Second, Completely Revised, vol. 118. Springer (2017)
7. Chipperfield, A.J., Fleming, P.: The MATLAB Genetic Algorithm Toolbox (1995)
8. Goldberg, D.E., Holland, J.H.: Genetic algorithms and machine learning. Mach. Learn. **3**, 95–99 (1988)
9. Adams, M.S.C., Documentation, C.: Msc. Software Corporation (2005)

Deep Learning for People Counting Model

T. Revathi and T. M. Rajalaxmi

Abstract Modeling of automatic people detection and counting in a real-time video is an important feature in a smart surveillance system for safety and security management, marketing research, etc. Face recognition is one of the methods which is used for people detection. In this paper, a real-time automated model is designed using deep learning algorithm such as convolutional neural network which is computationally efficient. The face detected using the proposed algorithm exploits the challenges such as variations in size and shape of the head region to achieve robust detection of a human, even under partial occlusion, dynamically changing background, and varying illumination condition. Here, we have used WIDER face dataset and FDDB dataset to show the results of the proposed method.

Keywords Face detection · Surveillance · GMM · Foreground detection · CNN

1 Introduction

Real-time people detection and counting system are very useful to many applications like security and people management application. In the past years, video cameras were used to track and count the people. Nowadays, the amount of video data tends to be increasing; to process and manage this data manually is difficult. So the automatic system is needed to process the huge data. Motion-based people detection and counting are one of the critical tasks in the current field of computer vision. Motion-based people analysis consists of detection, tracking, and recognition. The group-based people counting detects and tracks the people, but does not recognize the individual human. It is necessary to make the group-based people counting sys-

T. Revathi (✉)
Department of Computer Science, SSN College of Engineering, Chennai, India
e-mail: revathit@ssn.edu.in; revvlr@yahoo.com

T. M. Rajalaxmi
Department of Mathematics, SSN College of Engineering, Chennai, India
e-mail: laxmiraji18@gmail.com

© Springer Nature Singapore Pte Ltd. 2020
K. N. Das et al. (eds.), *Soft Computing for Problem Solving*,
Advances in Intelligent Systems and Computing 1048,
https://doi.org/10.1007/978-981-15-0035-0_43

531

tem to act smart. Therefore, we proposed a novel method to recognize an individual in a group. The proposed system consists of foreground extraction, face detection, region-based segmentation, and counting system.

2 Related Works

The existing human count system counts humans using the following: (a) Component-based people detection (b) Shape-based people detection, and (c) Skin-based people detection. The component-based people detection [1] system uses head and face as the components to count the human in real time. The shape of the human [2] is considered for counting. In skin-based detection, the color of the human skin is concentrated for the counting system. The existing system detects the people-based on the trained features and also maintains individual database to store the human faces. The counting system did not concentrate on individual human. The Omega model [3] for detecting and counting human beings is present in the scene using the features of the head, neck, and shoulder regions of a person.

There are many works related to the problem of multi-face detection over the past two decades. The current face detections work is summarized under the following categories. (i) Cascading (ii) DPM-based [4, 5], and (iii) Neural networks. In [6], people detection is based on the neural network, which trains to classify each pixel in the image as head and non-head. Paper [7] reviews some of the well-known face recognition techniques based on the benefits and drawbacks. Rotation invariant multi-view face detection [8] detects the face based on the rotation in-plane and rotation off-plane. The neural network architecture [9] is designed to detect facial features. This paper [10] estimates the people counting and identification done by the density estimation methods. It [11] detects the people based on the depth information taken from a ToF camera. The people are detected based on deconvolutional networks [12].

3 Proposed System

The proposed system "Real-Time human detection and counting system based on face recognition and region based segmentation" is developed with aim of solving the security-related issue which has been by our society in various places. The system mainly detects the humans based on the video cameras which are present in the entry and exit place in the building. The system not only detects the front face of the human but it also automatically counts when a human enters the system and it reduces the count when the human exits from the system.

The working of the proposed system begins capturing the videos in the entry and exit system. The full-step process of the proposed system consists of the following steps. (i) Automatic human-face detection [13] is considered as the initial process of a fully automated system. The face can be detected by modified CNN with the convolution, pooling, and fully connected layer [14]. (ii) Region-based segmentation is applied to the face-detected image to extract and store each face as an individual image. (iii) Every extracted face was compared with other stored face image before storing to eliminate the face repetition. (iv) Each time when the face is storing, the count gets incremented. (v) The total count gives the total current human count in the respective area. And also it displays the current human faces of the system at any time.

The technical process of the proposed system is discussed here. The input videos are converted to individual frames. The background is modeled using Gaussian Mixture Model. The foreground extraction [15] is done by the background subtraction algorithm. The face is detected using the convolutional neural networks. After the detection of the face, the face is extracted individually using region segmentation. In region segmentation, the connected component labeling is applied. Each region is treated as an individual face image. In the counting system, when anyone enters the system, the count is incremented. Then the exit subsystem reduces the entry count. Then the total count gives the total humans who have been entering into the system (Fig. 1).

There are four steps followed in the proposed system. They are (1) Foreground extraction, (2) Face detection (3) Region-based segmentation, and (4) Counting system.

3.1 Foreground Extraction

Identifying moving objects from a video sequence is a fundamental and critical task in many computer-vision applications. Foreground object extraction is also known as background subtraction. The foreground extraction is very important to concentrate on the moving objects rather the nonmoving objects. The nonmoving and moving objects are considered as the background frame and mo foreground frame, respectively. The foreground extraction is actually the subtraction of the foreground frame from the background frame. The conventional approaches for the foreground object detection are background subtraction, temporal differencing, correlation, color-based segmentation, and optical flow. Out of this, background subtraction is an easy way of localizing moving objects. Here, the foreground is extracted based on the Gaussian Mixture Model (GMM). GMM is used to update the background frame in a timely manner.

The summary of the foreground extraction process is given below. The input image is the current image which is denoted by I. For preprocessing the image, the input RGB image has to be converted to gray-level image. The histogram is applied to the gray-level image to find out the peaks of the image. The output of the peaks is given to the GMM algorithm as the initial peaks to find the background frame (BF). The foreground extraction is done by subtracting the background frame (BF) from the gray-level frame (G). The foreground extraction (FE) gives the binary foreground image, and it is mapped to RGB to get RGB foreground extracted image.

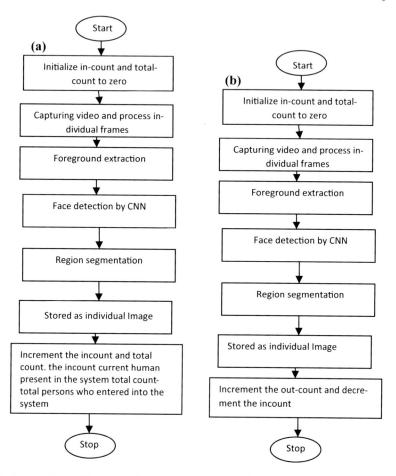

Fig. 1 Process flow of the proposed system: **a** entry system **b** exit system

3.2 Face Detection

The face detection module uses layered CNN for detecting the face. Recently, CNN is applied in various fields for feature extraction. Our work does not rely on hand-crafted features; rather, it uses the image directly. CNN concentrates from low-level feature to high-level feature to classify the data. CNN has different layers of processing from the input image to the output label. They are convolution layer, pooling layer, normalization layer, and fully connected layer.

The face images are taken from the WIDER face dataset to train the face images. For training the dataset, the pretrained ImageNet parameter values are given to train the WIDER face dataset to reduce the computational complexity. The test image is the foreground extracted image which is given as the input to the first layer of CNN. The convolution is performed on the input image with the kernel size of which is the same size of the image to reduce the window sliding so as to reduce the complexity of processing the large dataset. After convolution, the pooling operation is done to down-sample the image to reduce the number of parameters. The fully connected layer gives the one-dimensional feature vector, then the softmax function is applied to the output matrix. The output matrix of the text image is compared with the trained image to determine the face or non-face object.

3.3 Region-Based Segmentation

The region-based segmentation is one type of image segmentation based on regions. The region segmentation can be applied to the color image, gray image or binary image. In the proposed system, the binary image is considered and the region-based segmentation is done by the following steps: Take the binary-detected face image as input. Apply connected component labeling on it.

- Labeling identifies individual regions.
- Each region is considered as the face.
- The individual face is extracted and stored individually.

3.4 Counting System

The counting system is designed to count the human automatically. There are two video cameras to be fixed at the entry and exit place of the system to detect the human. This system integrates all other subsystems. The counting system consists of two subsystems which are (i) Entry system and (ii) Exit system. The two subsystems have to work together to give the efficient count of the humans. The entry system and the exit system incorporate all the subsystems like foreground extraction, face detection, and region segmentation. If the human is entered through the entry system, then the current and total count is added. If the human exits from the system, then the current count is subtracted.

4 Experimental Results

The proposed system is implemented using Python programming in Ubuntu. The video frames are taken from the WIDER face dataset.

Figures 2, 3, 4, and 5 shows the results of the proposed method. Figure 2 shows the results of the face extracted using CNN for WIDER face dataset and FDDB dataset images. Figures 3, 4, and 5 show the process of the proposed system. The proposed system takes the image (Fig. 3a) and gives the foreground extracted image (Fig. 3b).

Fig. 2 Face extracted image **a** WIDER face dataset and **b** FDDB dataset

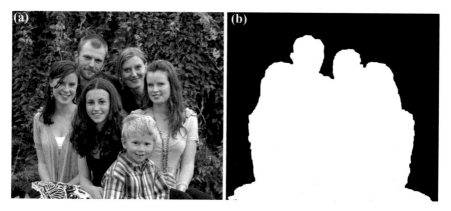

Fig. 3 **a** Input image and **b** Foreground extracted image

Then the foreground extracted RGB image is given as input to the CNN for face extracted image (Fig. 4a). All the extracted faces are segmented as individual faces (Fig. 4b) by region segmentation. Figure 5 gives the current person's faces present in the system.

5 Conclusion

In the literature, the human is detected based on the size and shape of the head region. Our proposed method overcomes the challenges such as variations in size and shape of the head region to achieve robust detection of human beings even under partial occlusion, dynamically changing background, and varying illumination condition. Deep learning for face detection achieves better results without the pose annotation

(a)   **(b)**

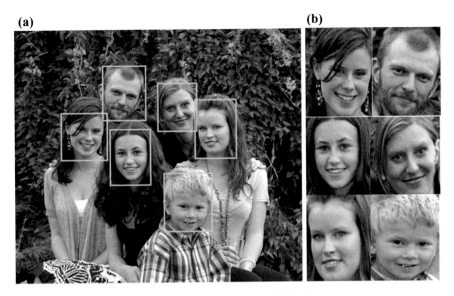

Fig. 4 **a** Face-detected image and **b** Region segmented individual face images

Fig. 5 Displays the current faces present in the system

or predefined facial features. The proposed work is found feasible and is believed that this system can overcome the security-related issues than in the existing systems. To handle a large amount of data, the system can be ported to big data environment.

References

1. Georgino, S.: Automated people counting from video. Int. J. Control Autom. Syst. (2010)
2. Leo, M., Spagnolo, P., Attolico, G., Distante, A.: Shape based people detection for visual surveillance systems. In: International Conference on Audio-and Video-Based Biometric Person Authentication, Germany (2003)
3. Mukherjee, S., Das, K.: Omega model for human detection and counting for application in smart surveillance system. Int. J. Adv. Comput. Sci. Appl. **4**(2), 167–172 (2013). arXiv:1303.0633

4. Mathias, M., Benenson, R., Pedersoli, M., VanGool, L.: Face detection without bells and whistles. In: Proceedings of ECCV (2014)
5. Felzenszwalb, P., McAllester, D., Ramanan, D.: A discriminatively trained, multiscale, deformable part model. In: Proceedings of CVPR (2008)
6. Jiang, H.: E-learned Miller, face detection with the faster R-CNN. In: IEEE International Conference on Automatic Face and Gesture Recognition (2017)
7. Pandya, J.M., Rathod, D., Jadav, J.J.: A survey of face recognition approach. Int. J. Eng. Res. Appl. (IJERA) (2013)
8. Huang, C., Ai, C., Li, Y., Lao, S.: High-performance rotation invariant multiview face detection. IEEE Trans. Pattern Anal. Mach. Intell. (2007)
9. Garcia, C., Delakis, M.: A neural architecture for fast and robust face detection. In: Proceedings of IEEE-IAPR International Conference on Pattern Recognition, Aug 2002
10. Kang, D., Ma, Z., Chan, A.B.: Beyond counting: comparisons of density maps for crowd analysis tasks-counting, detection, and tracking. IEEE Trans. Circuits Syst. Video Technol. (2018)
11. Luna, C.A., Losada-Gutierrez, C., Fuentes-Jimenez, D., Fernandez-Rincon, A., Mazo, M., Macias-Guarasa, J.: Robust people detection using depth information from an overhead time-of-flight camera. Expert Syst. Appl. **71**, 240–256 (2017)
12. YunFei, Z., Zhang, X., Feng, W., Cao, T., Sun, M., Xiaobing, W.: Detection of people with camouflage pattern via dense deconvolution network. IEEE Signal Process. Lett. (2018)
13. Viola, P., Jones, M.J.: Robust real-time face detection. Int. J. Comput. Vis. (2004)
14. Sun, Y., Chen, Y., Wang, X., Tang, X.: Deep learning face representation by joint identification-verification. In: Proceedings of NIPS (2014)
15. Akilan, T., Wu, Q.J., Yang, Y.: Fusion-based foreground enhancement for background subtraction using multivariate multi-model Gaussian distribution. Inf. Sci. **430**, 414–431 (2018)

Hybrid Variable Length Partial Pulse Modulation for Visible Light Communication

Jyothi and Ponnambalam Pathipooranam

Abstract Visible light communication (VLC) is one of the recent technology which is green friendly, efficient, and user affable which bypasses the RF technology. VLC is not regulated by radio regulation law. The technology concentrates on low cost, low power consumption, and also creates a hazardless environment. The light is mainly considered as source, air is the medium, and a photodetector or a phototransistor acts as a receiver. It utilizes LED which is reliable, consumes low power, and efficient instead of incandescent lamps and fluorescent lamps. This paper describes a novel method to transmit data higher than a bit per cycle of the carrier signal. The data transmission is carried during the OFF period of the carrier signal. The method of transmission of data is highly suitable for visible light communication.

Keywords VLC · LED · Illumination and communication

1 Introduction

The usage of electromagnetic (EM) waves for transfer of data is renowned. The transverse waves are measured by their amplitude and wavelength can be divided into a range of frequencies known as the electromagnetic spectrum [1]. Electromagnetic waves responsible for transmitting energy in three the form of microwaves, infrared radiation (IR), visible light (VL), and ultraviolet light (UV), X-rays, and gamma rays. Using basic principles of AM and FM, many other modulation methods or their combinations have been developed over the years. These include Phase Modulation, Quadrature Amplitude Modulation, Space Modulation, Single Sideband Modulation, etc., for analog signals and Amplitude-Shift Keying, Continuous Phase Modulation and Frequency Shift Keying etc., for digital signals. Pulse-width modulation (PWM), or pulse-duration modulation (PDM), is another modulation method used to encode

Jyothi · P. Pathipooranam (✉)
VIT University, Vellore, India
e-mail: ponnambalam.p@vit.ac.in

Jyothi
e-mail: mjsuvarna@gmail.com

© Springer Nature Singapore Pte Ltd. 2020
K. N. Das et al. (eds.), *Soft Computing for Problem Solving*,
Advances in Intelligent Systems and Computing 1048,
https://doi.org/10.1007/978-981-15-0035-0_44

information into a pulsing signal such as a periodic waveform. As applied to square waveforms, PWM is very frequently used in control power supplied to electrical devices [2, 3]. Four PWMs have also been used in certain communication systems where its duty cycle has been used to convey information over a communications channel. However, all present modulation methods can enable only one bit of data transfer per cycle. Further, they use both ON and OFF time period of the cycle and do not allow for variation of any of the time periods according to application needs for more efficient utilization of available cycles [4–12]. Further, one modulation method cannot be combined with another to make a hybrid modulation method with more advantages. Hence, there is a need in the art for a modulation method that allows for data transfer higher than a bit per cycle, uses only OFF period of the cycle and can combine with itself other modulation methods as required.

The method uses a cyclic electromagnetic (EM) signal with an on time (Ton) for a cycle during which the signal is of positive amplitude and an off time (Toff) for the cycle during which the signal is of zero or negative amplitude [13]. Hybrid variable length partial pulse modulation is a technique in which data transmission takes place during the Toff, which can be varied to vary the amount of data transmitted in the cycle. The OFF time period is used for modulation or any encoding by keeping the positive portion of PWM wave as constant. The signal may be visible light, radio waves, or infrared waves and may be pulse-width modulated (PWM) signal. The technique enables multiple bits of data to be transmitted per cycle of the signal and may enable the data transmission only when at least 1 byte of the data is transmitted during Toff. The time taken to transfer 1 byte (T1byte) can be 10/baud rate, and wherein the baud rate can be so selected that the T1byte is less than the Toff.

2 Proposed Method

The proposed method uses visible light as generated by an LED to create a PWM carrier signal. Figure 1 represents the functional block diagram of variable length partial pulse modulation technique. A cyclic light wave can be created by switching the LED "on" or "off" in a regular pattern. Percentage time the LED remains on during one cycle can be termed as duty cycle of the PWM carrier signal. Any other electromagnetic wave can be similarly used along with appropriate switches instead of LEDs. The waveform generated by switching of the LED is the carrier signal which can be represented is shown in Fig. 2.

The proposed method uses time period during which the LED is off (Toff) to transmit data. The data signal can be in the form of bits of data, with a start bit and a stop bit. The data signal can comprise 14 encoded bytes of data, 1 byte having 8 bits of data, one start bit and another stop bit where 1 byte may be made of 10 bits of data. Toff is required to be at least sufficient to accommodate time taken to transfer the 1 byte (i.e., T1byte) of data signal. T1byte, in turn, depends upon the frequency of the data signal, higher the frequency, lower the T1byte. Hence, if Toff is reduced, to transfer same 1 byte of data, a data signal of higher frequency may be required. The

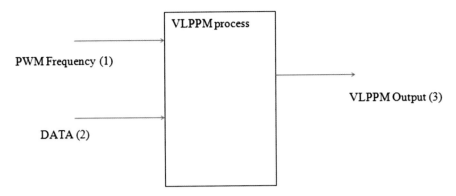

Fig. 1 Functional block diagram of VLPPM

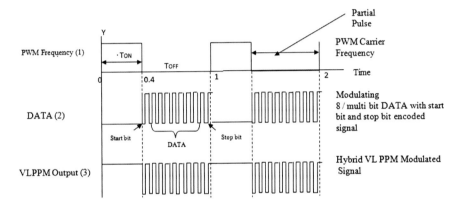

Fig. 2 Variable length partial pulse modulation waveform

modulated signal formed by combination of the carrier signal and the data signal. The hybrid variable length partial pulse modulated signal is formed using a variable length partial pulse modulation (VLPPM) method wherein variable length signifies that "length" of Toff can be varied by varying duty cycle of the carrier signal and partial pulse signifies that the method uses partially only the pulse generated by the carrier signal. Further, other modulation methods can be used to modulate waveform during the period Toff and so, the method can be completely named as hybrid variable length partial pulse modulation (HVLPPM).

For instance, one cycle can be of a total duration of 1 s as shown in Fig. 3. During this time period, electric supply to the LED can be so switched ON or OFF such that during that period of 1 s, the LED is ON for 40% of the time. Hence, for 0.4 s, such duration is termed as Ton and OFF for the remaining which is 0.6 s termed as Toff. The pattern can repeat for next one second time periods in a similar manner. Such a duty cycle can be termed as 40% duty cycle.

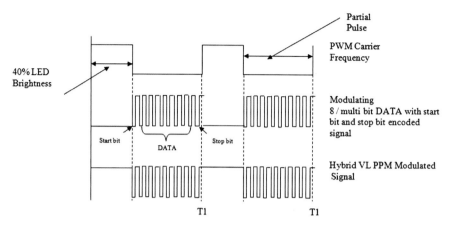

Fig. 3 40% duty cycle variable length partial pulse modulation

At time "0" the LED can be switched on to its full brightness along Y axis which represents amplitude of signal being described, while time may be plotted along X axis. After remaining ON for 0.4 s, the LED can switch OFF. For next period of 0.6 s, the LED can remain off and then be switched on again. Hence an ON period, shown as Ton and an OFF period, shown as Toff of the carrier signal can be achieved in every cycle of the carrier signal.

Also, instead of a cycle of 1 s, the cycle can be configured to be, for instance, 1 μs or any other time period required and so. As the duration of a cycle is lessened, switching of LED will occur faster and maybe ultimately indistinguishable to the human eye, with the difference in different cycles being noticed as variation in LED brightness.

Figure 4 illustrates the effect of varying the duty cycle upon the data signal that can be passed. If the LED is on for 60% of total cycle time, i.e., PWM carrier signal created

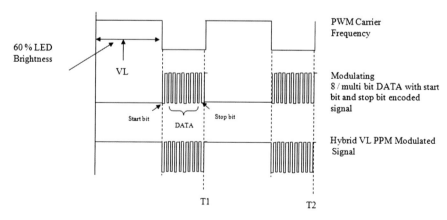

Fig. 4 60% duty cycle variable length partial pulse modulation

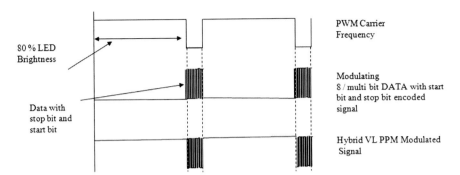

Fig. 5 80% duty cycle variable length partial pulse modulation

is of 60% duty cycle, Toff reduces, as shown, as against Toff for 40% duty cycle. As described above, Toff is required to be at least sufficient to accommodate time taken to transmit the 1 byte, i.e., T1byte of data signal and T1byte, in turn, depends upon the frequency of the data signal, higher the frequency, lower the T1byte. Hence, upon reduction of Toff, frequency of data signal that can pass 1 byte of data may increase as shown, as against frequency of data signal.

Similarly, as shown in Fig. 5, if the LED is on for 80% of total cycle time, i.e., PWM carrier signal created is of 80% duty cycle, Toff further reduces, as shown. As described above, Toff is required to be at least sufficient to accommodate time taken to transmit the 1 byte, i.e., T1byte of data signal and T1byte, in turn, depends upon the frequency of the data signal, higher the frequency, lower the T1byte. Hence, upon reduction of Toff frequency of data signal that can pass 1 byte of data may increase as against frequency of data.

3   Results

Figure 6 elaborates upon a method to determine feasible baud rate for data transfer using the proposed method.

The following steps can be taken to find a feasible baud rate for data transfer using the method proposed herein.

Step 1 can include determination of off time (Toff) of the PWM carrier signal based upon duty cycle proposed of the carrier signal.

For instance, for a PWM carrier signal of frequency 10 kHz total period (Ttotal) of one cycle of the wave is 100 μs (1/10,000 = 100 μs).

If the duty cycle proposed of the PWM carrier is 50%, and it has a frequency of 10 kHz,

$$\text{Toff} = \text{Ttotal}/2$$

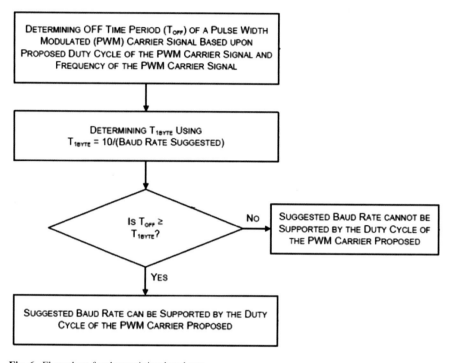

Fig. 6 Flow chart for determining baud rate

$$\text{Toff} = 100/2 = 50\,\mu\text{s}$$

A byte may include 8 bits of data, one start bit and another stop bit, i.e., total 10 bits. Therefore, time taken to transfer 1 byte, T1byte $= 10/$baud rate, where baud rate could be any supported baud rate.

In telecommunication and electronics, baud is a common measure of the speed of communication over a data channel. It is the unit for symbol rate or modulation rate in symbols per second or pulses per second. It is the number of distinct symbol changes made to the transmission medium per second in a digitally modulated signal or a line code.

If baud rate is taken as "b", time taken to transmit 1 byte:

$$\text{T1byte} = 10/\text{b}. \tag{1}$$

It can be readily understood that for at least 1 byte to be transmitted, Toff must be greater or equal to T1byte.

That is, Toff must be at least equal to T1byte and such a substitution can be made in Eq. (1) above.

So, taking Eq. (1) as Toff $= 10/\text{b} \rightarrow \text{b} = 10/\text{Toff} = 10/50/10^{6} = 10^{7}/50 = 200{,}000$ can be the minimum baud rate in this example. As can be explained from above, when

Toff is reduced by increasing Ton of a PWM carrier wave, minimum supported baud rate increases. Toff must be at least equal to T1byte for at least 1-byte data to be transmitted.

Step 2 includes determining T1byte using T1byte = 10/baud rate.

Step 3 includes comparing Toff and T1byte to determine if a suggested baud rate can be supported by duty cycle of the PWM carrier proposed.

If Toff is greater than or equal to T1byte, the suggested baud rate can be supported by duty cycle of the PWM carrier proposed. However, if Toff is less than T1byte the suggested baud rate cannot be supported by duty cycle of the PWM carrier proposed. As can be readily understood, duty cycle of the PWM carrier signal can be varied to accommodate higher/lower baud rates as required.

3.1 Illustrations

For example, if the duty cycle is set at 90%, Toff will reduce and therefore the minimum baud rate supported can be higher. If duty cycle is set at 10%, Toff will increase and consequently minimum baud rate supported will be lower.

In an exemplary embodiment, if baud rate is 921,600, time taken to transmit 1 byte will be

$$T1\,byte = 10/921,600 = 10.8\,\mu s,$$

this shows that with a 921,600 baud rate signal, 1 byte takes 10.8 μs to get transmitted.

Since Toff is 50 μs, it is more than T1byte and so, data transmission is possible.

Further, total number of bytes (Nbytes) transferred in one off period (Toff) can be

$$Nbytes = Toff/T1byte = 50\,\mu s/10.8\,\mu s = 4.629$$

As partial bytes cannot be transmitted, the above figure has to be rounded to the lower numeral and so, it can be determined that four bytes can be sent.

Table 1 shows the total data that can be transmitted per cycle for a PWM carrier signal of frequency 10 kHz, assuming a 50% duty cycle, different transmitting baud rates, and corresponding receiving bandwidths. In an aspect, for a PWM wave of

Table 1 Data transmission for PWM carrier signal

Transmitting baud rate	Maximum time required to transfer 1 byte (μs)	No. of byte sent/cycle	Receiving bandwidth (kbs)
230,400	43.4	1	80
460,800	21.7	2	160
921,600	10.8	4	320

frequency 10 kHz total period (Ttotal) of one cycle of the wave is 100 μs (1/10,000 = 100 μs).

Since the wave is configured with a 50% duty cycle, Toff can be 100/2 = 50 μs.

As shown, a transmitting wave (data signal) can have a baud rate 230,400 and so, time taken to transmit 1 byte (T1byte) can be = 10/230,400 = 43.4 μs.

Since this is less than Toff, data transmission at this baud rate is possible. However, in one cycle 17 of the carrier signal, number of bytes that can be sent can only be 1 (Toff/T1byte, rounded to the lower number), as shown.

As shown an equipment/channel of bandwidth 80 kbs can receive such a signal. Similarly, a transmitting wave can have a baud rate 460,800 and so, time taken to transmit 1 byte (T1byte) can be = 10/460,800 = 21.7 μs, as shown.

Since this is less than Toff, data transmission at this baud rate is possible. However, in one cycle of the carrier signal, number of bytes that can be sent can only be 2 (Toff/T1byte, rounded to the lower number).

Receiving bandwidth can be 160 kbs. Similarly, a transmitting wave can have a baud rate 921,600 and so, time taken to transmit 1 byte (T1byte) can be = 10/921,600 = 10.8 μs.

Since this is less than Toff, data transmission at this baud rate is possible. However, in one cycle of the carrier signal, number of bytes that can be sent can only be 4 (Toff/T1byte, rounded to the lower number).

Receiving bandwidth can be 320 kbs.

4   High Data Transmission Method

Figure 7 illustrates how a higher data transmission can be achieved using the proposed method.

A data stream can be subjected to a hybrid variable length partial pulse modulation (HVLPPM) method. Data stream output of HVLPPM can be provided to a constant current driver and can, in turn, be used to drive a red LED, a green LED, a blue LED, and a white LED, as shown.

Further, an RGBW CCT (correlated color temperature) combination data encoding table as shown in Table 2 can be used in the proposed method.

The gauge of color appearance of white light source is defined by correlated color temperature (CCT) is measured on the Kelvin absolute temperature scale. The most common range is available from 2700 K (warm white) to 7000 K (cold white).

HVLPPM method enables multiple bits of data to be carried in a cycle. Each color temperature can carry 3 bit of information during multiple bit transmission using HVLPPM method proposed.

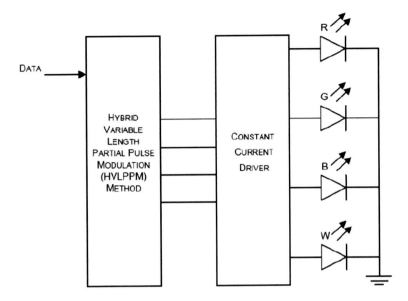

Fig. 7 HVLPPM circuit

Table 2 Data encoding

CCT	DATA
3000	000
3500	001
4000	010
4500	011
5000	100
5500	101
6000	110
7000	111
2000	0
2700	1

4.1 Transmitter Configuration

As an example, at Step 1, a 16-bit data input as under can be provided

$$0000\ 0101\ 0011\ 1001$$

At Step 2, a 3-bit group as under can be made out of above 16-bit data as under

$$1\ 000\ 001\ 010\ 011\ 1001\ 1$$

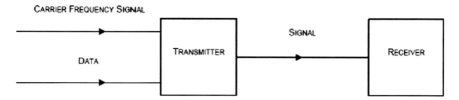

CARRIER FREQUENCY SIGNAL

DATA

TRANSMITTER

SIGNAL

RECEIVER

Fig. 8 Transmitter configuration

At Step 3, using Fig. 7, LED color temperature can be varied using constant current driver and HVLPPM method for multiple LEDs.

For instance, a 3-bit group 000 can be associated with a CCT of 3000, 001 can be associated with 3500, 010 with 4000, 011 with 4500, 100 with 5000, and 1 with 2700. In this manner, 16-bit data (in Step 1) can be compressed to 6 bits of information (the six CCT values derived above).

At a receiver end, these CCT values can be converted back to original 16-bit data by referring table 4B above and hence generating:

$$0000\ 0101\ 0011\ 1001$$

Hence, the HVLPPM method described above increases bandwidth by three times. This is achieved by transmitting 3 bits of data in one cycle, as against one bit of data per cycle permitted by other methods.

Figure 8 illustrates a system proposed which consists of a transmitter configured to transmit data to a receiver. Transmitter can be configured to incorporate a cyclic electromagnetic (EM) signal with an ON time (Ton) for a cycle during which the signal is of a positive amplitude, and an OFF time (Toff) for the cycle during which the signal is of zero or negative amplitude; and transmit the data during said Toff, wherein the Toff can be varied so as to vary the amount of data transmitted in the cycle. The signal can include visible light (VL), radio waves, or infrared waves. Transmitter can implement hybrid variable length partial pulse modulation method described above on signal and data to generate a modulated signal. The modulated signal can be received by receiver which can extract data from the modulated signal. The transmitter itself can generate cyclic signal of appropriate baud rate as required. The signal can be a PWM signal.

5 Conclusions

The proposed system provides a method to transmit data that enables data transfer higher than a bit per cycle of the carrier signal that uses only OFF period of the carrier signal. This method to transmit data that is highly suitable for visible light (VL) communication.

The method includes using a cyclic electromagnetic (EM) signal with an on time (Ton) for a cycle during which the signal is of a positive amplitude and an off time (Toff) for the cycle during which the signal is of zero or negative amplitude; and transmitting the data during the Toff, wherein the Toff can be varied to vary amount of the data transmitted in the cycle. The signal may be visible light, radio waves, or infrared waves and maybe pulse-width modulated (PWM) signal. The method enables multiple bits of data to be transmitted per cycle of the signal, and enable the data transmission only when at least 1 byte of the data is transmitted in the Toff. The method may be combined with other modulation methods.

References

1. Cook, B.: New developments and future trends in high efficiency lighting. Eng. Sci. Educ. J. **A247**, 207–217 (2000)
2. Linnartz, J.P.M.G., Feri, L., Yang, H., Colak, S.B., Schenk, T.C.W.: Communications and sensing of illumination contributions in a power led lighting system. In: 2008 IEEE International Conference on Communications, ICC'08, May 2008, pp. 5396–5400
3. O'brien, D., et al.: Visible light communications: challenges and possibilities. In: Proceedings of the IEEE 19th International Symposium on PIMRC, Sep. 15–18, pp. 1–5 (2008)
4. Gacio, D., et al.: PWM series dimming for slow-dynamics HPF LED drivers: the high-frequency approach. IEEE Trans. Ind. Electron. **59**(4), 1717–1727 (2012)
5. Rajagopal, S., Roberts, R.D., Lim, S.-K.: IEEE 802.15.7 visible light communication: modulation schemes and dimming support. EEE Commun. Mag. **50**(3), 72–82 (2012)
6. Perz, M., Vogels, I., Sekulovski, D., Wang, L., Tu, Y., Heynderickx, I.: Modelling the visibility of the stroboscopic effect occurring in temporally modulated light systems. Light. Res. Technol. **47**(3), 281–300 (2015)
7. Deng, X., Linnartz, J.P.M.G.: Poster: model of extra power in the transmitter for high-speed visible light communication. In: 2016 Symposium on Communications and Vehicular Technologies (SCVT), Nov 2016, pp. 1–5
8. Deng, X., Wu, Y., Khalid, A.M., Long, X., Linnartz, J.-P.M.G.: Led power consumption in joint illumination and communication system. Opt. Express **25**(16), 18,990–19,003 (2017)
9. O'Brien, D.: Visible light communications: challenges and potential. In: IEEE Photonics Conference, pp. 365–366 (2011)
10. Ahmed, F., Ali, S., Jawaid, M.: A review of modulation schemes for visible light communication. Int. J. Comput. Sci. Netw. Secur. **18**, 117–125 (2018)
11. Jovicic, A., Li, J., Richardson, T.: Visible light communication: opportunities, challenges and the path to market. IEEE Commun. Mag. **51**, 26–32 (2013)
12. Rajbhandari, S., et al.: High-Speed integrated visible light communication system: device constraints and design considerations. IEEE J. Sel. Areas Commun. **33**(9), 1750–1757 (2015)
13. Schmid, S., Corbellini, G., Mangold, S., Gross, T.R.: LED-to-led visible light communication networks. In: Proceedings of the Fourteenth ACM International Symposium on Mobile Ad Hoc Networking and Computing, MobiHoc '13, pp. 1–10. ACM, New York, NY, USA (2013)
14. Loo, K., Lun, W.-K., Tan, S.C., Lai, Y., Tse, C.: On the driving techniques for high-brightness LEDs. In: 2009 Energy Conversion Congress and Exposition (ECCE), pp. 2059–2064. IEEE (2009)
15. Beczkowski, S., Munk-Nielsen, S.: LED spectral and power characteristics under hybrid PWM/AM dimming strategy. In: 2010 Energy Conversion Congress and Exposition (ECCE), pp. 731–735. IEEE (2010)

An Efficient Dynamic Background Subtraction Algorithm for Vehicle Detection Tracking System

Rashmita Khilar, Sarat Kumar Sahoo, C. Rani and Prabhakar Karthikeyan Shanmugam

Abstract Background subtraction is an important role in video surveillance system in ITS, yet in complex scenes, it is still a challenging problem; hence, it is required to model the background before subtraction. Various illumination changes and dynamic backgrounds form the major key aspects for background modeling. In this paper, an algorithm (TCO-DBS) is proposed to develop an efficient background subtraction framework to solve the above problems. Here, texture and color features are considered for background modeling, thereby separating the foreground and background video frames. The texture features mainly depend on scale values used, i.e., number of neighboring pixels used for describing local texture description. Among this, local binary pattern (LBP) is mostly used in computer vision applications. LBP texture features along with color feature give a promising result when compared to other methods.

Keywords Background modeling · Texture · Color · LBP · TCO-DBS

1 Introduction

Background subtraction is the first and foremost step for object detection in a video. Handling videos are often affected by various illumination changes such as day, night, or cloud passing over the Sun, camera jitter, and changes in the background geometry.

R. Khilar
Panimalar Engineering College, Chennai, India
e-mail: rashmita.khilar@gmail.com

S. K. Sahoo (✉)
Parala Maharaja Engineering College, Luhajhara, Odisha, India
e-mail: sarat1@rediffmail.com

C. Rani · P. K. Shanmugam
School of Electrical Engineering, VIT University, Vellore, India
e-mail: crani@vit.ac.in

P. K. Shanmugam
e-mail: sprabhakarkarthikeyan@vit.ac.in

© Springer Nature Singapore Pte Ltd. 2020
K. N. Das et al. (eds.), *Soft Computing for Problem Solving*,
Advances in Intelligent Systems and Computing 1048,
https://doi.org/10.1007/978-981-15-0035-0_45

551

Background subtraction plays an efficient method for detecting foreground objects from a dynamic background in surveillance video. First, background modeling has to be done to subtract the foreground and background. Background model facilitates high-level tasks such as intelligent interface, video surveillance, traffic monitoring, and automated event detection.

The above problem has created an overriding need to provide an efficient background and foreground subtraction algorithm. Foreground subtraction from a dynamic background such as swaying of trees, changing billboards, and moving leaves are some of the challenging tasks in background subtraction from a video.

Literature studies so far addressed two different approaches for background subtraction. They are (1) parametric and (2) nonparametric model. In both the cases, background subtraction is done using frame differencing method. In parametric method, background subtraction is represented using normal Gaussian distribution method, but this method fails in the case of complex background such as dynamic background. In nonparametric method, adaptive thresholding technique is used to accurately model the background and then, frame differencing is applied. Various methodologies are applied under this, but all the methods face various problems with dynamic background in terms with computational time. Therefore, an efficient method is required to subtract the background from foreground to handle dynamic background with reduced computational time with various kinds of issues.

2 Related Works

Frame differencing is the common method for background subtraction method. Gaussian mixtures (GM) are commonly used for parametric method. Gaussian mixture model [1] is combined with oriented gradient pixel model to describe the dynamic variations in various regions. Similarity measures [2] and mean shift methods [3] are the commonly used methods for nonparametric model.

Real-time traffic videos are prone to contain shadow and various illumination effects. The shadow and illumination effect has to be removed for better background subtraction process. Many works related to illumination changes [4] are a slow process and takes more computational time [5]. Along with illumination changes, the shadow also has to be removed and it is categorized into two different types, they are 1. Cast shadow which falls on the ground and 2. Self-shadow which falls on the object. Shading differentiation is done for various surface orientation changes due to different types of cast shadow [6]. It is difficult to detect and track a vehicle with the shadow; hence, the shadow has to be removed before a vehicle is detected or tracked after background subtraction because there is a chance of false detection of shadow as a vehicle. Three steps are carried out for shadow detection and removal. First, the difference between the current frame and the reference frame is calculated for detecting the shadow. This assumption will remove the shadow in the current frame. Many of the works published so far exploits pixel appearance and change due to cast shadows [7]. Hence, an effective shadow removal algorithm needs to be implemented for efficient detection and removal of shadow.

Background subtraction plays an important role in vehicle detection and tracking. Foreground, which contains the vehicle, is obtained by subtracting the background from the foreground. Many algorithms work well with static surveillance video, i.e., static background but there is degradation in the performance and accuracy with deep dark moving objects [8] while handling moving videos. Background subtraction method is divided into two major types: (1) Parametric and (2) nonparametric methods. In both the cases, inter-frame differencing and threshold approaches are used for background subtraction. In parametric method, the background subtraction is simple to implement and is represented using normal Gaussian distribution but it fails to mitigate the problem with dynamic background. The adoptive nature of nonparametric model is used accurately to model the background with greater sensitivity and it is designed for dynamic background method. In order to handle dynamic background, the background has to be modeled first. Dynamic background modeling is done using various methods such as 1. Frame differencing method, 2. Gaussian mixture model method, 3. Approximated median filter method, 4. Eigen background method, and 5. Approximated median filter method [8]. Local binary pattern (LBP) texture [9] adopts the idea of Gaussian mixture model and thus, it uses various multiple modes to represent background. In order to reduce the matching complexity and satisfy the real-time scenario, the LBP texture feature vectors are simplified further.

The above-mentioned approaches focused on dynamic background subtraction thereby removing the cast type shadow. Moreover, the dynamic background subtraction is affected by handling multiple vehicles, variation in the background, and illumination changes such as day, night, or cloud passing over the Sun, camera jitter, and changes in the background geometry. These variations make the design of texture and color orientation based dynamic background subtraction (TCO-DBS) from various kinds of recorded videos. This method shows better performance and results when compared with the available methods.

3 Approaches

The goal is to design of texture and color orientation based dynamic background subtraction (TCO-DBS) from various kinds of recorded videos to focus on various illumination changes and suppression of self-shadow. Robustness of dynamic background subtraction from foreground is done with different illumination changes by handling uniform, rotational invariant texture, and color features. But the dynamic changes are very sensitive to color gradient, to overcome this, texture features along with color features are used. Two types of texture features are used in combination with color gradient. They are

- Uniform texture features
- Rotational invariant features.

The novel contributions of the proposed dynamic background subtraction system are listed below:

- Subtracts dynamic background from foreground. Thereby detecting only the vehicle.
- Color entropy features along with textures give a better dynamic background subtraction method.
- Suppress shadow, which gives a false detection.
- Takes care of variation in illuminations changes since rotational invariant textures in eight possible ways.

The statistical texture and color orientation based approach for dynamic background subtraction is proposed in this paper which works across various kind of dynamic background, various illumination changes such as day video and night video. Here, the idea is to apply Gaussian mixture method (GMM) [10] (Sen-Ching et al. 2012) on the frames to capture the texture features (Uniform and Rotational invariant texture features) and color gradient features so as to model the background in order to handle dynamic background subtraction. This paper is motivated to design **TCO-DBS** technique since the color entropy details in the RGB value are used in combination with texture features to subtract the dynamic background and foreground. **TCO-DBS** removes different illumination changes and thus, reduces the computational cost and increases the accuracy with the selection of uniform, rotational invariant texture features, and color feature in eight different possible ways as shown in Fig. 1.

Proposed TCO-DBS technique consists of the following process: Selecting ROI, background modeling, and post-processing. The Architecture diagram describes the above process.

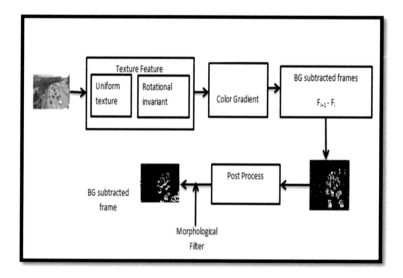

Fig. 1 Architecture diagram for TCO-DBS

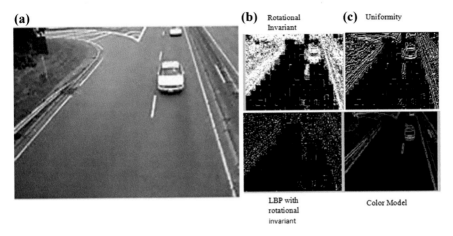

Fig. 2 **a** Input frame, **b** LBP with rotational invariant, **c** uniformity-color model

3.1 Dynamic Background Subtraction

First, the background model needs to be modeled because the intensity values of local pixels remain unstable always. This can also be caused by variation with illumination changes and poor visual appearance in the input given. In order to make the system robust with respect to wide variations, the background needs to be modeled by selecting the texture features (Rotational Invariant and Uniform) and color features in combination with local binary pattern (LBP).

3.2 Uniform and Rotational Invariant Texture Features

Even small change in intensity values (due to illuminations or dynamic changes) of the pixel, produce different LBP code. On the other hand, invariant and uniform textures are efficient against any dynamic changes. Color information is highly sensitive to dynamic changes, to overcome this uniform and rotational invariant texture, features are selected for subtracting the dynamic background as shown in Fig. 2. Gradient calculation is adoptive based on variance value given arbitrarily for an input. The background subtraction gives better performance with gradient information and user-defined threshold and variance values. The algorithm used to model the texture features are as follows:

Step 1: 3 × 3 macroblock conversion are done for each frame.
Step 2: LBP computations are done for each pixel.
Step 3: Periodic shuffling is carried out from LSB to MSB.
Step 4: Find min value among eight possible outcomes from shuffling (invariant texture).

Step 5: carry out XOR operation over successive bit sequence.
Step 6: Find the bit sequence.
Step 7: find number of bit transition from LSB to MSB.
Step 8: Isolate the pixel values based on amount of transitions.

3.3 Background Modeling

The background is modeled by modeling the LBP pixel patterns and the patterns are XOR-ed successively to find the numbers of bits transitioned which yields a result in uniformity of local texture features on the pixels. LBP bit values are rearranged again in eight different possible ways and the minimum and maximum values are taken for calculating the rotational invariant texture model. The above extensive calculations are carried over various frames in the input video to compute the adaptive threshold which is used in subtracting the dynamic background. The obtained foreground is post-processed using a spatial median filter with an adoptive windowing length to remove isolated false positive and false negative responses to get an optimal foreground result shown in Fig. 3.

After LBP extraction, spatial correlation is rechecked with user-defined parameter set. The more the number of ones in the output pattern, there is more possibility for the moving vehicles, lesser this transitions there arises background changes.

Figure 4 represents the calculation carried out for modeling the pixel values to model the background.

Fig. 3 Color gradient variance

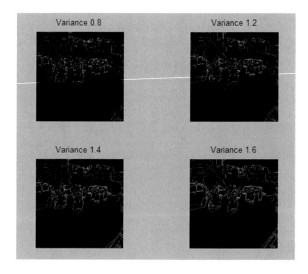

Fig. 4 LBP pixel modeling calculation

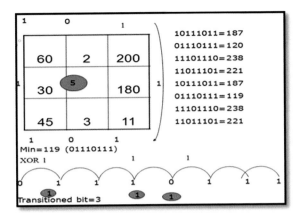

Advantages over Background Modeling:

- Computationally simple.
- Handles various illumination and dynamic changes.
- A better performance is obtained since color gradient along with texture Features.
- Selected threshold values give resistant to dynamic changes.
- Background and foreground separated pixels should be maximum correlated with moving objects in the incoming frames.

3.4 Shadow Suppression

After subtracting the background, the foreground object (Vehicle) is obtained which contains shadow and it leads to false detection of vehicle. Therefore, the shadow has to be suppressed before vehicle detection and tracking process. The saturation components are lower in the region of shadow and are independent of the brightness; hence, HSV color space is selected for suppressing the shadow color models. The incoming frames in RGB which contain both luminance and chrominance value, where chroma value has no or less effect with the shadow, therefore the luminance and chrominance value are separated by using L*a*b conversion method and then, HSV color space is applied for suppressing the shadow. The algorithm used for shadow suppression is as follows:

Algorithm: Shadow Suppression:
For i=1:no.of frames;
Pre_comp(I:,)=L*a*b(i);
Color_model(i:,)=HSV (Pre_comp(I:,));
End

(a)   **(b)**   **(c)**

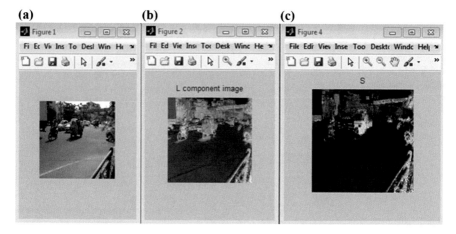

Fig. 5 **a** Input frame, **b** L component image, **c** shadow suppression image

It is required to produce equal performance for removing shadow as it is present along with the foreground object. The shadow has got the same features that of an object; hence, it is difficult to remove shadow while subtracting the background from foreground. Consequently, all these observations, drawbacks, and advantages were analyzed for the above method and in the subsequent section, a more complex method is proposed for vehicle detection and tracking in order to meet the requirements which are shown in Fig. 5.

4 Performance and Experimental Result

The performance of the proposed method has been evaluated based on precision, recall, and F score obtained. In order to have a common technology to evaluate and compare the results from various algorithm, in this project, each text pixel in ground truth and output image is considered in the calculation of precision and recall rate. Precision and recall rates are calculated as follows:

CDP (Correctly Detected Pixels)—the number of pixels matching between output mask image (O)

Ground truth image (GT)—the number of pixels matching between output mask image (O) and ground truth image (GT).

- CDP $= O \cap GT$
- Precision rate (PR)$=$ CDP/(CDP $+$ FP)
- Recall rate (RR)$=$ CDP/(CDP $+$ FN)

In this paper, **TCO-DBS** was attempted over various inter frames with different set of recorded traffic videos with various groups (dynamic background, various illumination changes such as day video, night video, background with high frequency) for carrying out dynamic background subtraction. These dataset videos have been gathered from WSDOT dataset. From the experiment of the above data, it is noted that there is better accuracy rate, F_score value and precision-recall value for the dynamic background subtraction system with various illumination changes when compared with other LBP methods.

Even though multimodal GMM (Yunsheng Zhang et al. 2015) is designed for multimodal background subtraction and LBP-HOG (Hossein Tehrani Niknejad et al. 2012) is designed for histogram-based background subtraction, proposed method is compared along with the above two methods to show the ability of the proposed method subtracts dynamic background from various kinds of recorded traffic video in a better way. The result indicates that TCO-DBS methodology has the efficiency to separate foreground from dynamic background and shows a clear improvement over GMM and LBP methods. During the experiment, it was observed that

- **TCO-DBS** technique of recorded traffic videos.
- The techniques work with various illumination changes since rotational invariant textures in eight possible ways are used for dynamic background subtraction.
- The pixels are invariant over the various image transformations and scaling.
- Color entropy features along with the textures provide a better dynamic background subtraction method.

The Figs. 6, 7 and 8 show the output frames for various inputs.

The comparison results with various existing algorithms are discussed in Table 1 and Fig. 9.

Fig. 6 a Input frame, **b** multimodal GMM, **c** LBP-HOG, **d** proposed TCO-DBS

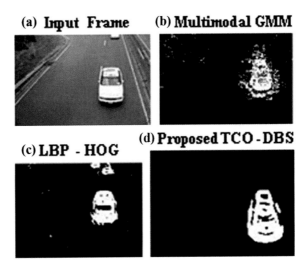

(a) Input Frame   (b) Multimodal GMM

(c) LBP - HOG   (d) Proposed TCO - DBS

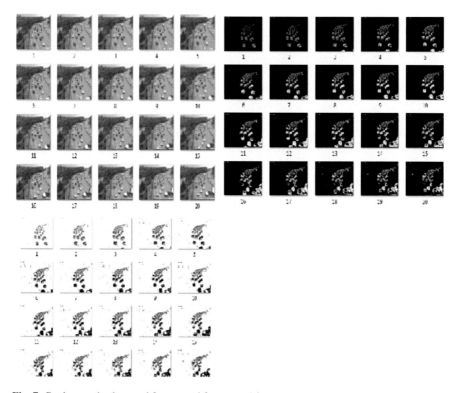

Fig. 7 Background subtracted frames and foreground frames

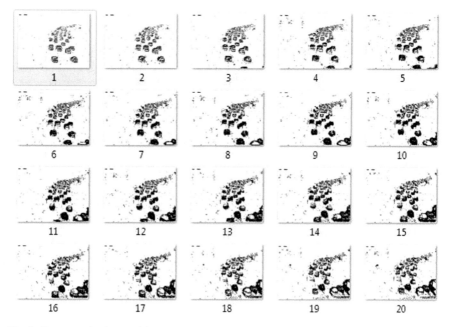

Fig. 8 Foreground subtracted frame

Table 1 Performance comparison with the existing method

Metrics	Multimodal GMM	LBP-HOG	TCO-DBS (proposed method)
Recall	86.12	96.01	98.62
Precision	76.12	90.11	95.78
F-measure	81.83	82.48	84.88
Accuracy	83.26	94.62	96.89

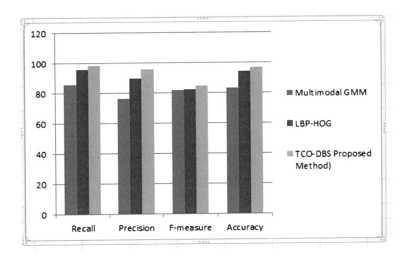

Fig. 9 Comparison graph

4.1 Performance Analysis for Proposed TCO-DBS System

5 Conclusion and Future Enhancement

Thus the TCO-DBS method works across different kind of traffic videos such as video with high density, various illumination and environmental changes, and night videos with complex backgrounds. The texture models are invariant over any of the image transformations and scaling. Color gradient information's along with invariant texture model provides a better dynamic background subtraction. This method can be extended to traffic videos with various other backgrounds and also this method can be used to subtract the foreground and background in biomedical images. The proposed method also reduces the computational complexity and thereby reducing the number of steps while finding the subtracted frames.

References

1. Niknejad, H.T., Kawano, T., Shimizu, M., Mita, S.: Vehicle detection using discriminatively trained part templates with variable size. In: Intelligent Vehicles Symposium (IV). IEEE, pp. 766–771 (2012)
2. Heikkila, M., Pietikainen, M.: A texture-based method for modeling the background and detecting moving objects. IEEE Trans. pattern Anal. Mach. Intell. **28**(4), 657–662 (2006)
3. Zhang, Y., Zhao, C., He, J., Chen, A.: Vehicles detection in complex urban traffic scenes using Gaussian mixture model with confidence measurement. IET Intel. Transp. Syst. **10**(6), 445–452 (2016)
4. Iwasaki, Y., Itoyama, H.: Real-time vehicle detection using information of shadows underneath vehicles. In: Advances in Computer, Information, and Systems Sciences, and Engineering, pp. 94–98 (2006)
5. Guo, Z., Zhang, L., Zhang, D., Zhang, S.: Rotation invariant texture classification using adaptive LBP with directional statistical features. IEEE Int. Conf. Image Process. **7**(3), 285–288 (2010)
6. Zhou, W., Liu, Y., Zhang, W., Zhuang, L., Yu, N.: Dynamic background subtraction using spatial-color binary patterns. In: Sixth International Conference on Image and Graphics (ICIG), pp. 314–319 (2011)
7. Zhu, M., Martinez, A.M.: Subclass discriminant analysis. IEEE Trans. Pattern Anal. Mach. Intell. **28**(8), 1274–1286 (2006)
8. Zhou, X.S., Huang, T.S.: Small sample learning during multimedia retrieval using biasmap. In: Proceedings of the IEEE Computer Society Conference on Computer Vision and Pattern Recognition, CVPR, vol. 1, pp. I–I (2001)
9. Wu, Y., Zeng, D., Li, H.: Layered video objects detection based on LBP and codebook. In: First International Workshop on Education Technology and Computer Science, ETCS'09, vol. 1, pp. 207–213 (2009)
10. Wei, Z., Qiu, X., Sun, Z., Tan, T.: Counterfeit iris detection based on texture analysis. In: 19th International Conference on Pattern Recognition, ICPR, pp. 1–4 (2008)
11. Niknejad, H.T., Takeuchi, A., Mita, S.: On-road multivehicle tracking using deformable object model and particle filter with improved likelihood estimation. IEEE Trans. Intell. Transp. Syst. **13**(2), 748–758 (2012)
12. Wang, W., Chen, D., Gao, W., Yang, J.: Modeling background and segmenting moving objects from compressed video. IEEE Trans. Circ. Syst. Video Technol. **18**(5), 670–681 (2008)
13. Liao, S., Zhao, G., Kellokumpu, V., Pietikäinen, M., Li, S.Z.: Modeling pixel process with scale invariant local patterns for background subtraction in complex scenes. In: IEEE Conference on Computer Vision and Pattern Recognition (CVPR), pp. 1301–1306 (2010)
14. Zhang, S., Yao, H., Liu, S.: Dynamic background modeling and subtraction using spatio-temporal local binary patterns. In: 15th IEEE International Conference on Image Processing, ICIP, pp. 1556–1559 (2008)
15. McHugh, J.M., Konrad, J., Saligrama, V., Jodoin, P.M.: Foreground-adaptive background subtraction. IEEE Signal Process. Lett. **16**(5), 390–393 (2009)
16. Li, L., Huang, W., Gu, I.Y., Tian, Q.: Statistical modeling of complex backgrounds for foreground object detection. IEEE Trans. Image Process. **13**, 1459–1472 (2004)
17. Mandal, M., Nanda, P.K.: Embedded local feature based background modeling for video object detection. In: IEEE Conference on Power, Communication and Information Technology Conference (PCITC) (2015)
18. Kim, K., Chalidabhongse, T.H., Harwood, D., Davis, L.: Real-time foreground–background segmentation using codebook model. Real-Time Imaging **11**(3), 172–185 (2005)
19. Zhong, J., Sclaroff, S.: Segmenting foreground objects from a dynamic textured background via a robust kalman filter. In: Proceedings Ninth IEEE International Conference on Computer Vision, pp. 44–50 (2003)
20. Jacques, J.C.S., Jung, C.R., Musse, S.R.: Background subtraction and shadow detection in grayscale video sequences. In: 18th Brazilian Symposium on Computer Graphics and Image Processing, SIBGRAPI, pp. 189–196 (2005)

PV-Based High-Gain Boost Converter

Ritanjali Behera, Sarat Kumar Sahoo, M. Balamurugan,
Prabhakar Karthikeyan Shanmugam and C. Rani

Abstract This paper presents a high-gain boost DC–DC converter with a photo-voltaic system, which gives high gain at low duty cycle. The Particle Swarm Optimization (PSO) technique with MPPT controller is implemented on the converter, which determines the duty cycle of the converter to maintain constant output voltage at a given input voltage. The working principle of the converter, PSO algorithm with MPPT controller, and the simulation results are presented.

Keywords PV system · Boost converter · Particle Swarm Optimization ·
Maximum power point tracking

1 Introduction

Nowadays, due to increase in the application of electrical energy such as electric vehicles and electric cooking, more electrical energy is required. To fulfill the energy demand, it should be pollution-free, eco-friendly with the environment, focus on different energy sources, i.e., renewable energy sources such as PV, wind, and fuel. Among all the renewable energy sources, PV is more efficient and reliable because of nonradioactive nature, no fuel cost, and noiseless operation. The major limitations of

R. Behera · S. K. Sahoo (✉)
Parala Maharaja Engineering College, Berhampur, Odisha, India
e-mail: sksahoo.ee@pmec.ac.in

R. Behera
e-mail: ritanjalipmecee@gmail.com

M. Balamurugan · P. K. Shanmugam · C. Rani
School of Electrical Engineering, VIT University, Vellore, India
e-mail: balamurugan.m27@gmail.com

P. K. Shanmugam
e-mail: sprabhakarkarthikeya@vit.ac.in

C. Rani
e-mail: crani@vit.ac.in

© Springer Nature Singapore Pte Ltd. 2020
K. N. Das et al. (eds.), *Soft Computing for Problem Solving*,
Advances in Intelligent Systems and Computing 1048,
https://doi.org/10.1007/978-981-15-0035-0_46

PV sources depend on the weather conditions and low output voltage. The output of PV depends on solar irradiance and cell temperature, so the MPPT technique should be implementing in PV application [1–6]. The output voltage of the PV system improves MPPT technique and also uses high-gain DC–DC boost converter [7, 8].

Conventional boost-type DC–DC converter provides a gain of 2–3 times at higher duty ratio which increases the semiconductor losses, EMI problems, parasitic elements, and decreases the efficiency with increasing gain. To avoid these limitations, high-gain boost DC–DC converter is preferred which increases the voltage gain at lower duty ratio and also improves the efficiency [9, 10].

This paper presents a novel system and Particle Swarm Optimization (PSO) algorithms with MPPT controller.

2 Photovoltaic System

The block diagram of a PV system is shown in Fig. 1, which consists of a PV panel, high-gain boost-type DC–DC converter, PSO-to-MPPT technique, and load.

2.1 PV Panel

It is a semiconductor material which consists of both P-type and N-type materials and forms a p–n-junction diode. Figure 2 represents the equivalent circuit of a PV solar panel [11–13]. The relation between the output voltage and current is given in Eq. (1). The parameters are shown in Table 1.

$$I = I_{ph} - I_d \left(\exp \frac{q(V + R_s I)}{NkT} - 1 \right) - \frac{(V + R_s I)}{R_{sh}} \tag{1}$$

where

I $\qquad$ = Load current

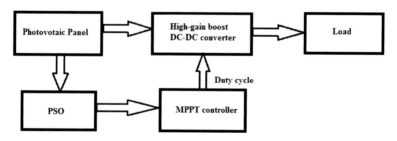

Fig. 1 PV system with PSO MPPT controller

Fig. 2 PV system with PSO
MPPT controller

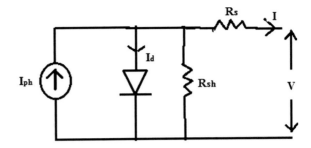

Table 1 PV Panel
parameters

Parameter	Specification
Maximum power (P_{max})	164 W
Voltage at MPP (V_{mp})	22 V
Current at MPP (I_{mp})	7.47 A
Open-circuit voltage (V_{oc})	25 V
Short-circuit current (I_{sc})	7.97 A
Temperature coefficient of V_{oc}	$-$ 0.36101%/°C
Temperature coefficient of I_{sc}	0.10199%/°C

I_{ph} = Photo current
I_d = Reverse saturation diode current
q = Electric charge = 1.6×10^{-19} C
V = Output voltage across diode (V)
k   = Boltzmann constant (J/K)
T   = Cell temperature (K)
N   = Ideality factor of the diode
R_s and R_{sh} = Series and Shunt resistance (Ohm)

3   High-Gain Boost-Type DC–DC Converter

The novel converter as shown in Fig. 3 consists of a coupled inductor, switched capacitor, one power semiconductor switch, three power diodes with parasitic elements, and resistive load, and the input is a PV panel. The working principle of the novel converter is based on the coupled inductor and switched capacitor technique. Figure 4 shows the working principle of the novel converter.

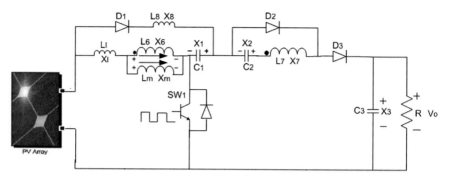

Fig. 3 Circuit diagram of the novel converter

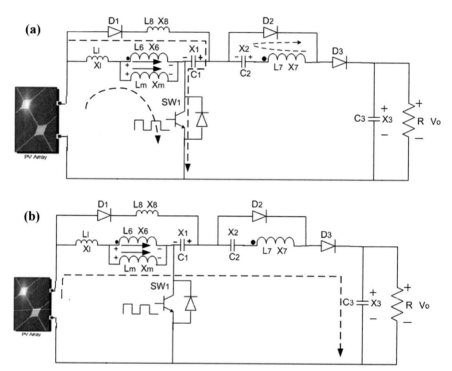

Fig. 4 **a** Operation of the novel converter during Turn-On, **b** Operation of the novel converter during Turn-Off

3.1 Mode-I or Turn-On Mode

When the switch is turned ON, all passive elements get charged with input voltage either directly or indirectly as shown in Fig. 4a. In this mode, a small inductor L_8 in series with a PV solar cell limits the inrush current during Turn-On.

3.2 Mode-II or Turn-Off Mode

During this mode, all passive elements are discharged in series with the input voltage and transfer energy into the output load as shown in Fig. 4b.

3.3 Gain-Analysis of the Novel Converter

During Turn-On, voltage across the magnetizing winding is given by

$$(V_{LM})_{on} = kV_g \tag{2}$$

$$V_{L7} = n(V_{LM})_{on} = nkV_g \tag{3}$$

where

k $\;=$ coupling coefficient $= L_m/(L_k + L_m)$
n $\;=$ turns ratio $= N_2/N_1$

During Turn-Off, voltage across the magnetizing winding is given by

$$(V_{LM})_{off} = V_g + V_{C1} + V_{C2} - V_o \tag{4}$$

Applying volt-second balance in Eqs. (2) and (4), we get

$$\{(V_{LM})_{on} \times D \times T_s\} + \{(V_{LM})_{off} \times (1 - D) \times T_s\} = 0 \tag{5}$$

where D and T_s represent duty cycle and time period of the given converter.
By solving Eq. (5), the gain formula is given by (6)

$$Gain(G_v) = \frac{V_o}{V_g} = \frac{[(kD) + \{(1 - D) \times (2 + nk)\}]}{(1 - D)} \tag{6}$$

4 Application of PSO to MPPT

In this section, the PSO method is implemented including the MPPT controller in the PV system [14–18]. Figure 5 shows the novel flowchart of the PSO with the MPPT controller algorithm. Steps of the novel algorithm are discussed as follows.

Step-1 (Selection of Parameter):

Duty cycle and the output power of the novel converter represent the particle position and fitness value evaluation function. The position and initial velocity of each particle are initialized randomly over the search space in a uniform distribution manner.

Step-2 (Fitness Evaluation):

The particle "i" represents the fitness value which is calculated after the MPPT controller sends the duty cycle, which indicates the position of the particle "i".

Step-3 (Upgradation of Individual and Global Best Data):

The position of each particle of fitness values, i.e., individual best data (P_{best}) and global best data (G_{best}) are updated by comparing previous values and obtaining the new fitness values corresponding to their positions.

Step-4 (Upgradation of Velocity and Position of Each Particle):

The velocities and positions of each particle in the swarm are updated by using the PSO equations, Eqs. (1) and (6).

Step-5 (Determination of Convergence):

The convergence criterion indicates the maximum number of iterations. If the convergence criterion is satisfied, the process can be terminated; otherwise, return to Step-2.

Step-6 (Reinitialization):

Fitness value in a PV system is not constant that depends on the weather conditions and load. So, the PSO method must be reinitialized and the new MPP must be found when the output of the PV is changed. So the novel PSO algorithm is reinitialized when the following functions are met.

$$\frac{P_i(k+1) - P_i(k)}{P_i(k)} > \Delta P$$

where $i = 1, 2, \ldots N$, k = number of iterations.

5 Simulation Results

The novel converter is simulated by using PSpice software. Details of the elements and components used in the simulation are given in Table 2.

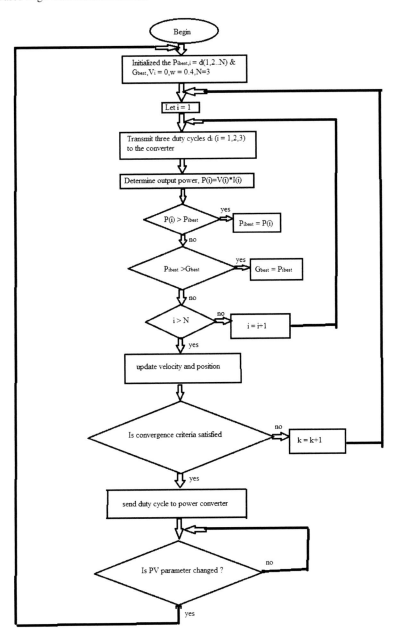

Fig. 5 Flowchart of the PSO method

Table 2 PV panel parameters

Parameter	Specification/rating
MOSFET (SW$_1$)	IRF150
Diode (D_1, D_2, D_3)	HFA25TB60
Inductor L_1	37 μH
Inductor L_2	56 μH
Current-limiting inductor L_3	15 μH
K (coupling coefficient)	0.8
Capacitor C_1	30 μF
Capacitor C_2	30 μF
Capacitor C_3	440 μF
Resistor R	50 Ω
Input voltage V_{in}	25 V
Duty cycle	50%
Switching frequency	13 kHz

Figure 6 shows the simulation of the novel converter, output voltage content less ripple. It is observed that voltage across the switched capacitor, gives low voltage rating of capacitor and also indicates current through the coupled inductor and switched capacitor. Voltage spike of the measured waveforms occurs due to leakage inductance of the coupled inductor when the switch is turned off.

6 Conclusion

High-gain boost converter with a photovoltaic system is proposed and the Particle Swarm Optimization algorithm with maximum power point tracking controller in the PV system is implemented. The novel converter gives fewer ripples in the output voltage waveform. The simulation result of the PSO with the MPPT controller of the converter is presented.

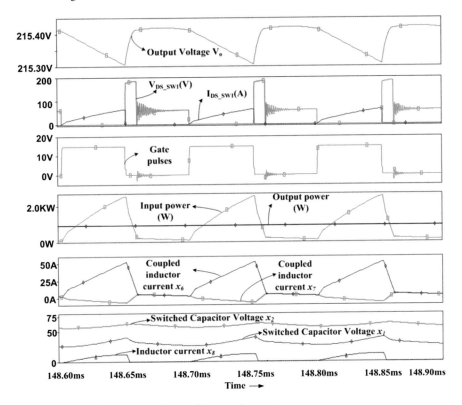

Fig. 6 Voltage and current waveforms of the novel converter

References

1. Zhao, Q., Lee, F.C.: High performance coupled-inductor DC–DC converters, in Conf. Rec. IEEE Appl. Power Electron. Conf. **1**, 109–113 (2003)
2. Tseng, K.C., Liang, T.J.: Novel high-efficiency step-up converter. Proc. Inst. Elect. Eng. Elect. Power Appl. **151**(2), 182–190 (2004)
3. Wai, R.J., Duan, R.Y.: High-efficiency DC/DC converter with high voltage gain. IEEE Proc. Electr. Power Appl. **152**(4), 793–802 (2005)
4. Li, W., Lv, X., Deng, Y., Liu, J., He, X.: A Review of Non-Isolated High Step-Up DC/DC Converters in Renewable Energy Applications. IEEE, pp. 364–369 (2009)
5. Dawidziuk, J.: Review and comparison of high efficiency high power boost DC/DC converters for photovoltaic applications. Bull. Polish Acad. Sci. Tech. Sci. **59**(4), 499–506 (2011)
6. Hsieh, Y.P., Chen, J.F., Liang, T.J., Yang, L.S.: Novel high step-up DC–DC converter with coupled-inductor and switched-capacitor techniques. IEEE Trans. Ind. Electron. **59**(2), 998–1007 (2012)
7. Gu, B., Dominic, J., Lai, J.S., Zhao, Z., Liu, C.: High Boost Ratio Hybrid Transformer DC–DC Converter for Photovoltaic Module Applications, pp. 598–606. IEEE (2012)
8. Laird, I., Lu, D.D.C.: High step-up DC/DC topology and MPPT algorithm for use with a thermoelectric generator. IEEE Trans. Power Electron. **28**(7), 3147–3157 (2013)

9. Kode, R.B.A., Zobaa, A.F.: Comparision between the conventional methods and PSO based MPPT algorithem for photovoltaic systems. World Acad. Sci. Eng. Technol. Int. J. Energy Power Eng. **8**(4), 697–702 (2014)

10. Prakash, J., Sahoo, S.K.: Design of soft switching interleaved boost converter for photovoltaic application. Res. J. Appl. Sci. Eng. Technol. **9**(4), 296–308 (2015)

11. Balamurugan, M., Narendiran, S., Sahoo, S.K., Das, R., Sahoo, A.K.: Application of particle swarm optimization for maximum power point tracking in PV system. In: 3rd Intenational Conference on Electrical Energy Systems, pp. 35–38 (2016)

12. Ishaque, K., Salam, Z., Amjad, M., Mekhilef, S.: An improved particle swarm optimization (PSO)-based MPPT for PV with reduced steady state oscillation. IEEE Trans. Power Electron. **27**(8), 3627–3638 (2012)

13. Esram, T., Chapman, P.L.: Comparision of photovoltaic array maximum power point tracking techniques. Energy Convers. IEEE Tras. **22**, 439–449 (2007)

14. Miyatake, M., Veerachary, M., Toriumi, F., Fujii, N., Ko, H.: Maximum power point tracking of multiple photovoltaic arrays: A PSO approach. IEEE Trans. Aerosp. Electron. Syst. **47**(1), 367–380 (2011)

15. Zhu, Y., Shi, X., Dan, Y., et al.: Application of PSO algorithm in global MPPT for PV array. In: Proceedings of the CSEE, pp. 42–49 (2012)

16. Kenndey, J., Eberhart, R.: Particle swarm optimization. In: Proceedings of IEEE Conference on Neural Networks, pp. 1942–1948 (1995)

17. Trelea, I.C.: The particle swarm optimization algorithm convergence analysis and parameter selection. Inf. Process. Lett. **85**(6), 317–325 (2003)

18. Gayathri, K., Gomathi, S., Suganya, T.: Design of intelligent solar power system using PSO based MPPT with automatic switching between ON grid and OFF grid connections. Int. J. Emerg. Trends Electr. Electron. **1**, 95–97 (2013)

Indoor Object Classification Using Higher Dimensional MPEG Features

Dibyendu Roy Chaudhuri, Dhairya Chandra and Ankush Mittal

Abstract We propose a generic model to classify a given image by detecting a specific image patch. Employing MPEG-7 features along with feature selection populates the feature space which is used to train using SVM classier. Our work target toward classifying objects in an unclassified image. We propose a model that can detect objects through generic framework for larger classes. Our model gives an overall accuracy of over 97%.

Keywords Image classification · Object detection · MPEG · SVM

1 Introduction

Today on the web, there are billions of images. It is not possible to tag all objects in those images, especially when one is a nontechnical person. However, tagging will not only be useful for searching but also helpful for enhancing security, pattern matching, locating people, and several other applications. Here, we propose an automated model which can classify an image using a generic framework for larger classes. The features are extracted and machine learning framework learns the pattern of the object classes through the ground truth.

In this paper, we propose a new approach to classify images using MPEG-7 [1], a feature extraction tool and rank-based features selection algorithm for getting better results by removing irrelevant features. Feature extraction may face many problems such as sometimes images are affected by noise like Salt-and-Pepper noise, Poisson noise, and Gaussian noise. These noises distort regular images which are later difficult to classify. We have used rank-based features selection to remove these distorted features. MPEG-7 is used to extract features and train our model. It is implemented in C++ and has a command-line interface. It can generate six feature classes of an

D. R. Chaudhuri (✉) · D. Chandra · A. Mittal
Department of Computer Science, Graphic Era Deemed to Be University, Dehradun, India
e-mail: dibyendu830711@gmail.com

D. Chandra
e-mail: dhairyachandra@outlook.com

© Springer Nature Singapore Pte Ltd. 2020
K. N. Das et al. (eds.), *Soft Computing for Problem Solving*,
Advances in Intelligent Systems and Computing 1048,
https://doi.org/10.1007/978-981-15-0035-0_47

Fig. 1 Dataset image
contains images of doors,
sign boards, and stairs

image. Out of them, the features like "Color Structure Descriptor" (CSD), "Scalable Color Descriptor" (SCD), "Color Layout Descriptor" (CLD), "Homogeneous Texture Descriptor" (HTD) and "Edge Histogram Descriptor" (EHD) are the relevant feature classes for our classification model. "Color Structure Descriptor" (CSD) is used for feature extraction purpose and "Color Layout Descriptor" (CLD) for block-to-block image comparing. "Edge Histogram Descriptor" (EHD) is used for detecting figures, "Homogeneous Texture Descriptor" (HTD) is used for texture-based image-to-image matching. Later these are passed through "Support Vector Machine" (SVM) for further classification (Fig. 1).

The paper is organized as follows: Sect. 2 describes the literature and overview of our paper. Section 3 introduces methodology, which includes description of MPEG-7 and its features, SVM, feature selection and their formula. Section 4 shows comparison between Receiver Operating Characteristics (ROC) curve and multiple confusion matrix on our final results. Finally, Sect. 5 presents our conclusion.

2 Literature and Overview

Previously, a lot of work and research have been done on image classification such as the work of Ranzato [2] on detection and recognition of generic framework using SVM classifier on six categories (human figures, four-legged animals, airplanes, trucks, cars, and "none of the above").

Another work is of recognizing handwritten character using CNN over a large dataset [3]. The robustness of the model is verified over 78000 different 7-net committees. It reports an error rate of only 0.27%. Similarly, classifying human face is done by using some "face" and "non-face" based model cluster [4]. Model is trained over 4000 human face and non-human face images using both single-layer and multilayer perceptrons and is verified on more than 1000 images. Image classification also gives best results in disease detection [5]. The model is tested on 2000 medical images using CNN and SVM classifiers to classify different categories of lung patterns.

Image classification of indoor–outdoor images [6] is performed by studying image features like histograms in Ohta color space, multiresolution autoregressive model parameters, etc. Model is trained over 1300 consumer images using K-nearest neighbor (KNN) classifies on computed features to give an accuracy of over 90%. Some other models of image classifications are remote sensing image classification [7] of land use and land cover maps, and traffic sign classification as shown in [8] with an accuracy over 95%, histogram-based image classification [9].

Image Classification can be done using random forests and ferns [10]. This model is proposed for categorizing different image patches by applying various object classes which contains a lot amount of object categories. Results are obtained by using SVM classifier. The result is about a 10% improvement over the state of the art for Caltech-256 [11]. Main drawbacks which are present in these works are that they assume the image set will be in perfect condition and free from different noises but in real-life scenario, images are sometimes affected by noises like Salt-and-Pepper, Gaussian, etc. which makes difficulties for a system to classify them correctly. Here, we have used MPEG-7 and rank-based feature selection algorithm for reducing these ambiguous information from the images.

3 Methodology

We first used MPEG-7 [1], feature extraction tool and it generates five features descriptors CSD, SCD, CLD, HTD, and EHD of our training images. Later these are passed into Support Vector Machine (SVM) for further classification which provides an overall accuracy of 97% (Fig. 2).

3.1 MPEG

"MPEG" is a multimedium-based kit which is used to draw out different rudimentary features of an image. It is implemented in C++ platform and used to generate features like "homogeneous texture", "edge histogram", and "texture browsing". Regions for extraction are Globally, Binary mask, Region map, Bounding boxes, etc. We have used MPEG to extract features like "Dominant color", "Color structure", "Color

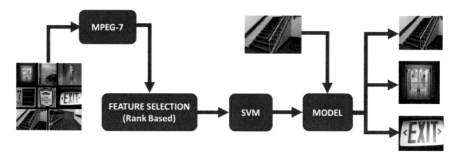

Fig. 2 Block diagram of our approach to classify object using MPEG-7 and SVM

layout", "Scalable color", "Homogeneous Texture", "Edge Histogram", "Region-Based Shape", "Contour-Based figures", etc. "MPEG" provides a lot of components that can help us in originating, packing, and explaining of seven relevant components which are generated earlier. Definition of an "MPEG" starts from a primary node which identifies whether the illustration is perfect or restricted.

Visual illustration of "MPEG" kit containing primary shapes and attributes illustrator that cover some rudimentary optical attributes: tint, patterns, structures, shifting, and face detection. Every classes contain a primary and sophisticated illustrators.

Edge Histogram Descriptor (EHD): EHD is basically used for recognizing structures. Image is splitting into four-cross-four image patches. "Edge histogram" produced of five categories: "45-degree diagonal", "135-degree diagonal", "vertical, horizontal, and non-directional edges". The image patches are splitting into non-intersecting subblocks of little area. The patches blocks generate an "80-bins edge descriptor".

Color Layout Descriptor (CLD): The main functionality of "color layout descriptor" is to denoting spatial scattering of tints. It is primarily applied for block-to-block image comparing. When image is processed, it is sent to image partitioning stage where image gets splitted into eight-cross-eight patches blocks. Then the classification of color selection is performed on the image and tiny image icons are generated in YCbCr color space.

$$D = \sqrt{\sum_i \omega_{yi}\left(DY_i - DY_i'^2\right)} + \sqrt{\sum_i \omega_{bi}\left(DCb_i - DCb_i'^2\right)} + \sqrt{\sum_i \omega_{ri}\left(DCr_i - DCr_i'^2\right)} \quad (1)$$

Then it is sent for DCT transformation which generates (8×8) matrix of 64 Discrete Cosine Transform (DCT) coefficients, i.e., DCTY, DCTCb, and DCTCr. The last stage is zigzag scanning where zigzag matrix matching is done. It generates three-zigzag scanned matrices, i.e., DY, DCb, and DCr which result in CLD of input image.

Color Structure Descriptor (CSD): The main functionality of "color structure descriptor" is extracting image patches. Image contains multiple rectangular shapes or arbitrarily frames. It is calculated by covering every image patch of an image and extracting tints C{0-7} from all eight-cross-eight pixels. This illustrator can classify between two patches where both images have various shapes but the pixel group can have the same tints. Color values are defined in "double-cone" ("Hue-Max-Min-Diff") HMMD color space.

Scalable Color Descriptor (SCD): SCD is in the form of a "color histogram" in the "Hue Saturation and Value" (HSV) color space encoded using "Haar transform". The description may be cached in various forms, starting from 256 bottom to sixteen coefficients in a histogram and description of bit's precision over a long scales of rate of data.

Homogeneous Texture Descriptor (HTD): "Homogeneous texture descriptor" is applied for block-to-block patch comparing using texture-based matches. Importance of HTD is classification of similar text-based areas. It is combined value of the SD, median values of a patch of image, spark, and spark values of Fourier converts of an image. These are drawn out from frequency channels. The distance between two HTDs is computed by

$$d\left(TD_{query}, TD_{db}\right) = \sum_{k}\left|\frac{TD_{query}(k) - TD_{db}(k)}{a(k)}\right| \qquad (2)$$

where normalization value a(k) is the standard deviation of $TD_{db}(k)$ for a given database.

3.2 Support Vector Machine

Support vector machine is one of the machine learning algorithms which is used as a pattern recognizer. The algorithm was first proposed by Vapnik and Cortes [12]. It has two parts—(1) "Kernel Trick" (2) "Support Vectors".

Kernel Trick: "Kernel" methods are pattern analyzer algorithm. It is one of the main members in case of SVM. Pattern annotation is applied to figure out the matching among groups, gradings, main attributes, associations, categories, etc. as shown in Fig. 4. "Kernel methods" owned the name by applying "kernel" methodology. It makes it to operate in a high dimensional and implicit feature space without calculating it's coordinates. This approach by which SVM compute operation on input data is known as "Kernel Trick" [13]. The kernel methods use "kernel trick" for computation as shown in Fig. 3. The points can be classified by

$$\vec{z} \to sgn\left(\vec{\omega} \cdot \varphi\left(\vec{z} - b\right)\right) = sgn\left(\left[\sum_{k=1}^{n} c_i y_i k\left(\vec{x_i}, \vec{z}\right)\right] - b\right) \qquad (3)$$

Fig. 3 Kernel Trick is
separating two different
classes

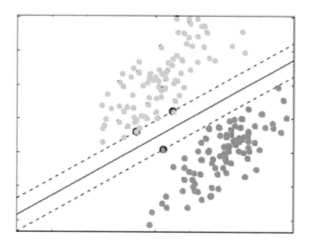

We have used predefined libraries of SVM in Python, written in "sklearn library". This library provides us four types of kernel tricks: (1) "RBF" (2) "Linear" (3) "Poly" (4) "Sigmoid Kernel" tricks transform data from lower to higher dimension.

Support Vectors. SVM is used to find the hyperplane which differentiate between the two classes in the datasets. Vectors which define these hyperplane are known as "Support Vectors". They are also an important part of SVM like kernel trick. 3D model of SVM is shown in Fig. 4.

Fig. 4 Support vectors on
3D modules

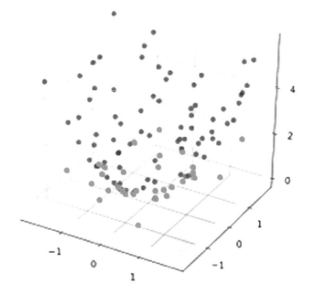

$$\left[\frac{1}{n}\sum_{i=1}^{n}\max(0, 1 - y_i\left(\vec{w}\cdot\vec{x_i} - b\right)\right] + \lambda||\omega||^2 \qquad (4)$$

where parameter lambda determines the tradeoff between increasing the margin size and ensures that xi lies on correct side of the margin.

3.3 Feature Selection

Feature selection is used to remove irrelevant or low information gaining features present in a dataset and extract important features for classification. This is used to reduce training time. Feature selection algorithm are of three types: (1) "Filter Methods" (2) "Wrapper Methods" (3) "Embedded Methods."

Filter Method. In filter method, features are ranked by score and it helps to choose whether a feature should be selected or removed from the dataset. It checks the error rate in the dataset.

Wrapper Method. Wrapper method works as a search problem where different combinations are prepared, evaluated, and compared to other combinations. Combination of dataset is evaluated using a predictive model which assigns a score based on model accuracy.

Embedded Methods. During model creation, embedded methods learn the best attributes that can be applied to the precision of the model. A trained approach takes lead of its own inconstant choosing procedure and executes attributes selection and categorizing simultaneously.

4 Experimental Results

We have used standard test dataset of MCIndoor20000 [14] which have three classes (doors, stairs, sign) of over 2000 images and further we downloaded more images from Internet to increase the test dataset. The following Table 1 summarizes the dataset.

Table 1 Test dataset

Class	Number of images
Doors	754
Signs	702
Stairs	600

4.1 Implementation Framework

The implementation was done in Python 3 using "sklearn" library and Orange 3 tools, an open-source data visualization, machine learning, and data mining toolkit. The code was run on a Quad Core machine.

4.2 Comparison

We have applied different machine learning algorithms like KNN, Logistic Regression, Neural Network, Random Forest, and Naives Bayes. We have also used feature selection algorithm to remove low information gaining features from our model. Rank-based feature selection algorithm is used. Initially, we have around 391 features and applying machine learning algorithm on those features together is computationally more costly. We have selected top 40 features based on different properties like information gain, information gain rate, and gain decreases. We have observed the following Receiver Operating Characteristics (ROC) curves before and after applying feature selection algorithm shown in Figs. 5 and 6, respectively.

After observing these ROC curves, we are quite sure that after applying feature selection algorithm, our model increases Area Under curve (AUC) which is directly proportional to accuracy of a system. It also helps us in reducing our model complexity.

We have observed the following experimental results while cross validation of our model using different machine learning algorithms as shown in Table 2.

For the above experimental result and ROC curve refer to Fig. 7. It is obvious that SVM-based algorithm (Kernel Trick used RBF) is best suited to our model.

Fig. 5 ROC curve is comparing true positive and false positive rates of different ML algorithms

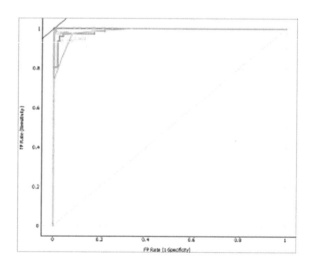

Fig. 6 ROC curve after feature selections is given

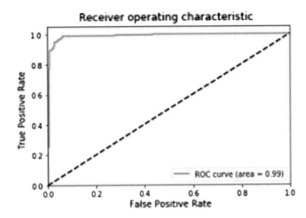

Table 2 70% are train and 30% are test datasets

ML algorithm	Accuracy (%)	Precision (%)	Recall (%)
SVM	98.20	98.20	98.20
Neural network	97.80	97.90	97.80
Random forest	95.40	95.50	95.40
KNN	88.30	89.00	88.30
Logistic regression	97.30	97.30	97.30
Naïve Bayes	90.80	90.80	90.80

Fig. 7 ROC curve is showing true positive and false positive rate of the machine

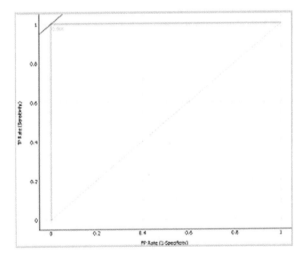

Table 3 Confusion matrix

	Doors	Signs	Stairs
Doors	187	1	1
Signs	6	169	4
Stairs	2	1	143

4.3 Results

We have tested our model on different training and testing ratio and we have come across the following results. Here, we are varying our training percentage and our model gives us these results.

75% Testing and 25% Training. It took 0.57830 s to compute training and 0.06741 s for testing features. It gives an accuracy of 97.10% as shown in Table 3.

From the above confusion matrix, we can see images of two doors are misclassified. One of them is classified as signs and other as stairs. Similarly 10 signs are wrongly classified. Six of them are classified as door and rest as stairs. Total three stairs are misclassified, two of them as doors, and rest as signs.

80% Testing and 20% Training. It took 0.8513 s to compute training and 0.04988 s for testing features. It gives an accuracy of 97.56% as shown in Table 4.

From the following confusion matrix, we can see four doors images are misclassified. Three of them are classified as signs and one as stairs. Similarly five signs are wrongly classified. Three of them are classified as doors and rest as stairs. At last we find that one stair is misclassified as sign.

90% Testing and 10% Training. It took 1.2218 s to compute training and 0.04139 s for testing features. It gives an accuracy of 96.11% as shown in Table 5.

From the above confusion matrix, we can see three doors images are misclassified. Two of them are classified as a sign and stair, respectively. Similarly four signs are wrongly classified. Two of them are classified as door and rest are classified as stairs. At last we find that one stair is misclassified as sign.

Table 4 Confusion matrix

	Doors	Signs	Stairs
Doors	146	3	1
Signs	3	145	2
Stairs	0	1	110

Table 5 Confusion matrix

	Doors	Signs	Stairs
Doors	72	2	1
Signs	2	84	2
Stairs	0	1	43

5 Conclusion

In this paper, we proposed a novel generic framework which can detect objects, matching pattern, and tags images which will help in searching, enhancing security, locating people, and several objects. It will help a non technical person to classify images into different categories. It is an automated framework where features will be extracted and machine learning framework will learn the pattern of the object classes through ground truth.

References

1. Bastan, M., Cam, H., Gudukbay, U., Ulusoy, O.: BilVideo-7: an MPEG-7 compatible video indexing and retrieval system. IEEE MultiMedia **17**(3), 62–73 (2010)
2. Marc'Aurelio Ranzato, F.J.H., Boureau, Y.-L., LeCun, Y.: Unsupervised learning of invariant feature hierarchies with applications to object recognition. In: IEEE Conference on Computer Vision and Pattern Recognition, 2007
3. Ciresan, D.C., Meier, U., Gambardella, L.M., Schmidhuber, J.: Convolutional neural network committees for handwritten character classification. In: 2011 International Conference on Document Analysis and Recognition, 03 Nov 2011
4. Sung, K.K., Poggio, T.: Example-based learning for view-based human face detection. IEEE Trans. Pattern Anal. Mach. Intell. **20**(1) (1998)
5. Li, Q., Cai, W., Wang, X., Zhou, Y., Feng, D.D., Chen, M.: Medical image classification with convolutional neural network. In: 2014 13th International Conference on Control, Automation, Robotics Vision, Marina Bay Sands, Singapore, 10–12th Dec 2014 (ICARCV 2014)
6. Szummer, M., Picard, R.W.: Indoor-outdoor image classification. IEEE Compute Society
7. Al-doski, J., Mansor1, S.B., Shafri, H.Z.M.: Image classification in remote sensing. J. Environ. Earth Sci. **3**(10) (2013). ISSN 2224-3216 (Paper) ISSN 2225-0948 (Online)
8. Gavrila, D.M., Philomin, V.: Real-time object detection for "smart" vehicles. In: Proceedings of the Seventh IEEE International Conference on Computer Vision
9. Chapelle, O., Haffner, P., Vapnik, V.N.: Support vector machines for histogram-based image classification. IEEE Trans. Neural Netw. **10**(5), 1055–1064 (1999)
10. Bosch, A., Zisserman, A., Munoz, X.: Image classification using random forests and ferns. In: IEEE 11th International Conference on Computer Vision, 2007. ICCV 2007, pp. 1–8, Oct 2007. IEEE
11. Griffin, G., Holub, A., Perona. P.: Caltech 256 object category dataset. Technical Report UCB/CSD-04-1366, California Institute of Technology (2007)
12. Cortes, C., Vapnik, V.: Support-vector networks. Mach. Learn. **20**, 273–297 (1995)
13. Theodoridis, S.: Pattern recognition, p. 203. Elsevier B. V (2008). ISBN 9780080949123
14. Data set: MCIndoor20000 .https://github.com/bircatmcri/MCIndoor20000

Lung Nodule Segmentation Using 3-Dimensional Convolutional Neural Networks

Subham Kumar and Sundaresan Raman

Abstract Lung cancer is one of the most deadly diseases in the world today, the annual number of deaths more than the next three cancers combined. Even with our advancement in medical science, the problem still persists. It can be addressed effectively at earlier stages, but most cases are detected at stages 3 or 4, where it is too late to be addressed properly. The objective of this paper is to design an effective Computer Aided Diagnosis (CAD) system which can segment of the CT scan of the lung, and help radiologists identify and diagnose this issue at an early stage. A novel 3-dimensional CNN is used to segment the nodules present in the CT scan, which will help classify the nodules with better accuracy. Various optimizations have been carried out to ensure that the convergence is quick and fast, while yielding the best possible accuracy. The proposed architecture achieves a Dice coefficient of 0.9615, on the LUNA16 dataset.

Keywords Deep learning · 3-D convolutional neural networks · Lung cancer · Nodule segmentation

1 Introduction

Cancer is one of the most deadly diseases on the planet today, killing more than two million people annually with more than hundred million people today are affected by it in some stage or the other. The most common and deadly type of cancer is lung cancer, causing more deaths than the next three, breast, colon and prostate cancer combined [13]. It has a survival rate of just 17%, and this number has not changed significantly, while for other types of cancers there has been significant improvement, the survival rate for breast cancer is 89%. Almost 80% of lung cancer cases can be

S. Kumar (✉) · S. Raman
Birla Institute of Technology and Science Pilani, Pilani, India
e-mail: h20180123@pilani.bits-pilani.ac.in

S. Raman
e-mail: sundaresan.raman@pilani.bits-pilani.ac.in

© Springer Nature Singapore Pte Ltd. 2020
K. N. Das et al. (eds.), *Soft Computing for Problem Solving*,
Advances in Intelligent Systems and Computing 1048,
https://doi.org/10.1007/978-981-15-0035-0_48

Fig. 1 Distribution of various subtypes of lung cancer

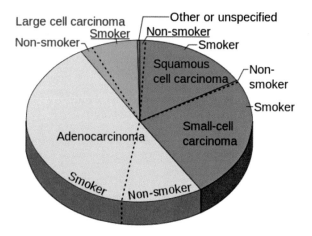

cured if detected at an early stage, but often they are detected only when it reaches an advanced stage. Detecting lung nodules and classifying them as malignant or benign is a very difficult task, because of the particular intensity range and the sizes of the nodules compared to the huge image, making them hard to detect and analyze. To tackle this challenge, Computer Aided Detection (CAD) systems have been created to help radiologists in early diagnosis. A CAD system is designed to help radiologists identify these tumours promptly, as they can leverage the vast knowledge base of all the CT scans collected so far to create an effective system.

Even when the symptoms of lung cancer manifest, they are mistaken for the signs of other diseases, like infections or long-term effects due to smoking, as shown in Fig. 1. This delays the correct diagnosis, and usually reaches stage 3 or 4 by the time it has been detected and confirmed. Tests for problems like pneumonia, heart disease and other lung conditions may result in the discovery of lung cancer. But only a few people among them get cured of it. For many years a lot of effort has been put to construct a good screening test for lung cancer, but only recently studies have proven that Low-Dose CT (LDCT) scans can help lower the risk of deaths due to this disease.

The main objective of this paper is to segment the lung nodules which are present in the CT scan of the lungs. A publicly available dataset is used for developing and testing the proposed solution. Most recent papers to detect lung nodules use Image Processing techniques like graph cuts [2] or basic Machine Learning techniques like SVM [4], and handcrafted features [20]. The handcrafted features require domain knowledge which most of the people who apply Image processing techniques may not know about it. Handcrafted features are not robust, they do not scale well to all the different types of CT scans. Lack of labelled data makes most of the existing methods insufficient. Handcrafted features do not help in the development of a robust CAD system for detecting lung nodules.

Convolution Neural Networks (CNNs) have been used with great success in the classification of natural images. Architectures like AlexNet, GoogleNet [15] and

ResNet [3] have achieved very high accuracies, some of which are higher than the average human being. This is an attempt to explore the vast untapped potential that these networks provide for detecting even the smallest of features which humans cannot perceive.

2 Convolutional Neural Networks in Medical Image Analysis

AlexNet [5] was the breaking ground for the use of CNNs to analyze images. CNNs are comprised of primarily four components: Convolutional Layers, Pooling Layers, Activation Layers and Fully Connected layers. Medical Image Analysis has a lot of challenges which is not present in natural images. Aside from being difficult to interpret for a non-specialist, the databases available are quite small compared to the size of datasets available for other tasks. This section is going to describe how CNNs have performed in this domain so far.

2.1 U-Nets

U-Nets are the first in a class of CNNs designed specifically for medical image analysis, dealing with challenges like small datasets and localization of regions rather than classification [12]. A contracting network is supplemented by successive layers, with upsampling operators coming in place of pooling layers. The resolution of the output increases as a consequence. The contracting path yields high resolution features which are combined with the output of these up-sampled layers for localization. Outputs with more precision can be assembled by a convolution layer. Context information is propagated to the higher-resolution layers because of the large number of feature channels. This results in the network having a U-shaped architecture as shown in Fig. 2, because the expansive path is symmetric to the contracting path. Fully connected layers are absent in this architecture.

3 Segmentation of Lung Nodules

In the field of computer vision, segmentation is the process of generating multiple components by partitioning an image (which can be 2-dimensional or 3-dimensional). The purpose of segmentation is to simplify and modify an image representation into another form which is much more useful and easier to scrutinize, for both experts and common people alike. Image segmentation is used to locate objects and boundaries like lines and curves in images. More precisely, in this case of 3-dimensional data,

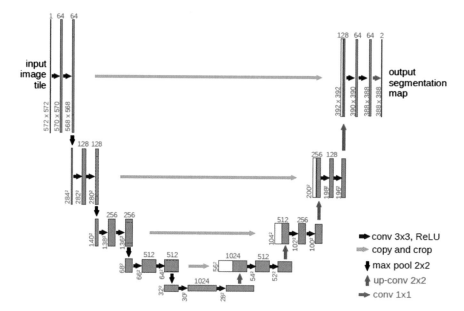

Fig. 2 U-Net architecture [12]

every voxel in an image which is assigned a label such that voxels with the same label share certain characteristics.

A set of segment is generated as a result, collectively covering the whole image, or a particular set of contours are extracted from it. All the voxels in a particular region have similar characteristics or computed properties specified by the user such as colour, intensity, or texture. Adjacent regions are significantly different with respect to the same characteristics. 3-D reconstructions can be created using these generated contours using various interpolation algorithms like Marching Cubes [6].

Segmentation helps us to identify where exactly the problem is. For proper diagnosis, the segmented nodule scan can be examined by the radiologists and diagnosed effectively. When the network is just detecting as in the previous case, it remains like a black box, where the basis for prediction is not known to anyone. The concept of segmentation helps us to address the sensitivity of the issue. This way the CAD system will actually be beneficial to both the patients and the radiologists.

3.1 Dataset

The Lung Image Database Consortium image collection (LIDC-IDRI) [1] is composed of thoracic computed tomography (CT) scans with marked-up annotated lesions for diagnosing and screening lung cancer. In the first phase, each of the four radiologists independently reviewed each CT scan and classified each of the nodules into one of the three categories (nodule <3 mm, nodule ≥3 mm and non-nodule ≥3 mm). In the next phase, each radiologist reviewed their own markings independently, along with the anonymized marks of the three other radiologists to reach a decisive conclusion.

LUNA16 is a smaller dataset of 888 scans from the LIDC dataset, excluding scans with slice thickness greater than 2.5 mm. This dataset contains the metadata which includes lot of information about the particular image, the world and voxel coordinates of the nodule in the image. There is a lot of variety in the dataset, the scans range from 100 units in thickness to 500 units. They contain varying intensities, and contain multiple nodules, both benign and malignant.

3.2   Methodology

For our experiment, we employed V-Net [8] which is a 3-D CNN model that was designed to segment prostate MRI volumes. The prostate can assume a wide range of appearances in different scans because of deformations and variations in intensity. Field inhomogeneity in these images causes artifacts and distortions. It has clinical relevance during treatment, where the anatomical boundary estimated by the network must be accurate. This architecture uses a Dice coefficient-maximization based novel objective function which is optimized while training.

This model is specifically designed for medical image analysis, making certain tweaks in the hyperparameters and the network architecture to obtain the best results on segmenting out the nodules in a reasonable amount of time. It uses many novel features, using only fully convolutional layers, PReLu layers [21] and using residual connections which ensure that the depth of the network does not compromise the convergence time. In each stage the convolution operations use volumetric kernels having size $5 \times 5 \times 5$ voxels. The resolution of the image reduces as the image proceeds through subsequent stages in the compression path. This is achieved through convolution with $2 \times 2 \times 2$ voxels-wide filters which are applied with a stride value of 2. The next operation extracts features by using non-overlapping $2 \times 2 \times 2$ volume patches, which result in feature maps whose size is halved. As a result, no pooling layers are required.

When the network becomes deep, they cannot guarantee the effective forward propagants of the gradient and updation during the back-propagation. ReLU layers are activation functions which are used to effectively train neural networks and alleviate problems like the Dying ReLU, with far better results than the ones used previously like sigmoid, etc. This inadequacy can be addressed effectively by using Parametric ReLU layers or Leaky ReLUs. The proposed model uses Leaky ReLUs which assign a value of 0.01 as the parameter α in the equation $y = \alpha x$ for negative values.

Pooling operations are replaced with convolution operations, which results in a smaller memory footprint while the model is being trained, because no switches are required for the corresponding mapping of the output of the pooling layers to

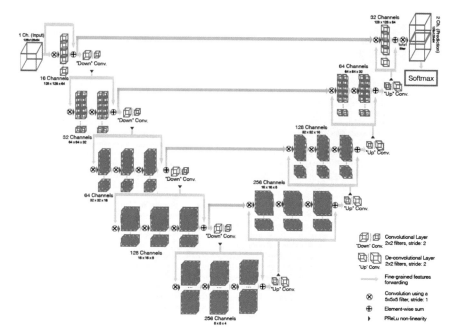

Fig. 3 V-Net architecture [8]

their inputs. This is shown in Fig. 3. Downsampling allows reduction in the size of the input signal and increases the receptive field of the features being computed in successive layers. In each stage present in the left portion of the model, the number of features are doubled compared to that of the previous stage. Feature extraction and expansion of the spatial support of the lower resolution feature maps are done by the right part of the network in order to assimilate the information required to generate the output. The final result is a two channel volumetric segmentation. The two 3-dimensional volumes generated as output are of the same size as the input 3-D image, where each voxel is assigned a probability of being part of the foreground, the nodule in this case and background regions, the lungs.

At every stage in the left portion of the CNN, the features are forwarded to the right part as indicated by the yellow arrows. This way fine-grained details can be assimilated that would be lost in the depth of the compression path, resulting in an improved quality of the final contour prediction. The innermost part of the CNN captures the input volume's complete content. This characteristic is important during segmentation of anatomy which is not clearly visible, and this information along with the fine-grained information coming from the forwarded features gives us a comprehensive solution.

$$D = \frac{2.\sum_i^N p_i g_i}{\sum_i^N p_i^2 + \sum_i^N g_i^2} \tag{1}$$

Dice coefficient is used to design the objective function. It is a quantity ranging between 0 and 1 which is to be maximized as shown above. The Dice coefficient, which is also known as the overlap index, is the most commonly used performance metric in validating medical volume segmentations [16]. It is used extensively to compare generated segmentations with the ground truth and also used to measure reproducibility (repeatability) as shown by [23].

$$\frac{\partial D}{\partial p_j} = 2 \cdot \left[\frac{g_j \left(\sum_i^N p_i^2 + \sum_i^N g_i^2 \right) - 2p_j \left(\sum_i^N p_i g_i \right)}{\left(\sum_i^N p_i^2 + \sum_i^N g_i^2 \right)^2} \right] \tag{2}$$

The gradient is computed with respect to the jth voxel as shown by the above equation. Dice coefficient is a very useful and powerful evaluation metric which compares the entire 3-D volume effectively, while not over penalizing false positives and negatives.

3.3 Preprocessing and Training

Various preprocessing techniques are used to process the 3-dimensional CT scans of LUNA16. The images cannot be fed because they contain noise and lot of details which are unnecessary for segmentation. First, the voxel values are converted to Houndsfield units (HU). This is followed by thresholding and using Marker-driven Watershed algorithm [11] to segment out the lung tissue, while removing all the noise and other unnecessary components like blood, etc. The main area of interest is lung tissue which is around 500 HU. The other regions like air which is around 1000 HU, blood, water and other tissues which are around 0 HU and bone which is 700 HU are masked out. Because CT scans are taken by a variety of equipment, there are scans of different intensities, which means that after preprocessing the output need not have the same texture.

A 3-dimensional architecture is employed, using 3-dimensional filters that perform 3-dimensional convolutions on 3-dimensional images, hence the computation power required is huge. Training requires a GPU, without which it is not possible to train the network in a reasonable amount of time. Therefore, the entire network was trained on Google Colaboratory which is a Google research project. It was started to help disseminate machine learning research and education. The entire dataset is mounted onto Google Drive, and then mounted onto the virtual machine. It is not possible to use the training data all at once because of linking and time constraints. Therefore training is carried out in manual batches, deriving from the concept of transfer learning. There are 888 images in the dataset, out of which 800 has been used for training and validation while 88 images are used for testing. The 800 images are divided into 9 subsets, 89 images in 8 of them and 88 in the last.

Each subset is taken one at a time and used for training the network, using a random initialization of weights, and after the entire subset has been propagated through and back, the weights are updated and saved. This process continues for the rest of the subsets as well. Initially there is a lot of overfitting, but gradually as the training progresses with each successive batch the network architecture generates better segmentations on the training data. PyTorch [10] was the library used for training the network, and SimpleITK [7] was used for visualization of the segmented outputs.

3.4 Optimization

Data augmentation is not required as there are a large number of training images, compared to what is generally available in the medical domain. For the LUNA and LIDC datasets, B-spline interpolation [22] and control points is the most frequently used technique to inflate the dataset. Residual connections are used to help gauge the difference between the input and the output of the layers, to see what those intermediate layers are adding to the learning ability of the model.

4   Results and Discussion

We obtained a Dice coefficient of 0.9615 on the test dataset, which comprised of 88 images from LUNA16 which have varying intensity and resolution values. The network was generic enough to correctly identify and segment out the nodules correctly in a variety of different CT scans. This is a relatively new field where CNNs are being applied, and their enormous potential can be seen. The test images taken from the same dataset, but were not shown to the network in the training phase. The outputs of the two segmentation maps (background and foreground) has been merged, so that the context of the nodules, i.e. the position with respect to the lungs can be seen and is not lost.

The model is able to process different types of scans as shown in Fig. 4. The CT scans are varying in quality, depending on the machine with which it has been taken, but the network architecture is robust enough to accurately segment them out. It is not necessary that there will be only one nodule in a lung. In cases where there are multiple nodules as is the case in Fig. 4f, the model is able to segment them out properly as well, and in this case all three nodules are segmented out.

V-Net outperformed current image processing techniques like using watershed and Markov random field by a significant margin, as shown in [17]. Its performance was much better than the performance of 2-Dimensional architectures, as illustrated in the use of Multiview CNNs [18] and central focused CNNs [19]. This compares favourably to the performance of other deep learning approaches as well. In contrast U-Net was used for segmenting lungs [14] and achieved a Dice coefficient of 0.93, and when used for segmenting nodules achieved a Dice coefficient of 0.79 [9]. The results are summarized in Table 1 and Fig. 5.

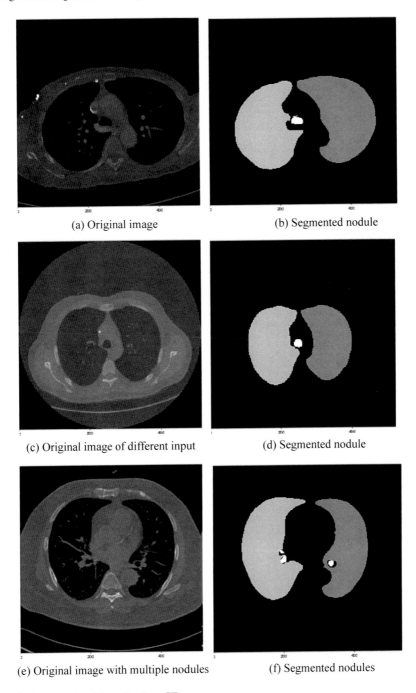

(a) Original image

(b) Segmented nodule

(c) Original image of different input

(d) Segmented nodule

(e) Original image with multiple nodules

(f) Segmented nodules

Fig. 4 Segmented nodules of various CT scans

Table 1 Comparison of results

Model	Dice coefficent
Central focused convolution neural networks [19]	0.81
U-Net with simple diameter information [9]	0.79
Multiview convolution neural networks [18]	0.78
Graph cuts [2]	0.68
Watershed, active contours and Markov random field [17]	0.67
Handcrafted features [20]	0.74
Proposed model	0.96

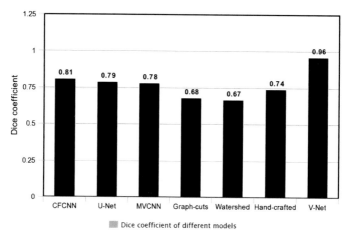

Fig. 5 Graph comparing the performances of various methods on Lung nodule segmentation

Implementing the network was quite challenging because of the computation cost and time constraints involved. Even though hardware is evolving at a very fast pace, some architectures go so deep that out of memory errors are generated in the best of resources available to us. To optimize the code in such a way it can run with the available resources which was a great learning experience.

Thus we achieved a Dice coefficient of 0.96 on the segmentation of the lung nodules by employing our model. This can help in better classification as well as visualization of lung nodules. As deep learning evolves at a rapid pace, with the advent of newer and computationally powerful GPUs, the accuracy of diagnosis can become even better. The positive results of V-Net on segmentation of the nodules provides a viable direction towards diagnosing lung cancer.

References

1. Armato III, S.G., McLennan, G., Bidaut, L., McNitt-Gray, M.F., Meyer, C.R., Reeves, A.P., Zhao, B., Aberle, D.R., Henschke, C.I., Hoffman, E.A., et al.: The lung image database consortium (LIDC) and image database resource initiative (IDRI): a completed reference database of lung nodules on CT scans. Med. Phys. **38**(2), 915–931 (2011)
2. Cha, J., Farhangi, M.M., Dunlap, N., Amini, A.A.: Segmentation and tracking of lung nodules via graph-cuts incorporating shape prior and motion from 4D CT. Med. Phys. **45**(1), 297–306 (2018)
3. He, K., Zhang, X., Ren, S., Sun, J.: Deep residual learning for image recognition. In: Proceedings of the IEEE conference on computer vision and pattern recognition, pp. 770–778 (2016)
4. Keshani, M., Azimifar, Z., Tajeripour, F., Boostani, R.: Lung nodule segmentation and recognition using SVM classifier and active contour modeling: a complete intelligent system. Comput. Biol. Med. **43**(4), 287–300 (2013)
5. Krizhevsky, A., Sutskever, I., Hinton, G.E.: Imagenet classification with deep convolutional neural networks. In: Advances in neural information processing systems, pp. 1097–1105 (2012)
6. Lorensen, W.E., Cline, H.E.: Marching cubes: a high resolution 3D surface construction algorithm. In: ACM siggraph computer graphics, vol. 21, pp. 163–169. ACM (1987)
7. Lowekamp, B.C., Chen, D.T., Ibáñez, L., Blezek, D.: The design of simpleTK. Front. Neuroinform. **7**, 45 (2013)
8. Milletari, F., Navab, N., Ahmadi, S.-A.: V-net: fully convolutional neural networks for volumetric medical image segmentation. In: 3D Vision (3DV), 2016 Fourth International Conference on, pp. 565–571. IEEE (2016)
9. Nam, C.-M., Kim, J., Lee, K.J.: Lung nodule segmentation with convolutional neural network trained by simple diameter information (2018)
10. Paszke, A., Gross, S., Chintala, S., Chanan, G., Yang, E., DeVito, Z., Lin, Z., Desmaison, A., Antiga, L., Lerer, A.: Automatic differentiation in PyTorch. In: NIPS-W (2017)
11. Roerdink, J.B., Meijster, A.: The watershed transform: definitions, algorithms and parallelization strategies. Fundamenta Informaticae **41**(12), 187–228 (2000)
12. Ronneberger, O., Fischer, P., Brox, T.: U-net: convolutional networks for biomedical image segmentation. In: International Conference on Medical image computing and computer-assisted intervention, pp. 234–241. Springer (2015)
13. Siegel, R., Miller, K., Jemal, A.: Cancer statistics. CA: Cancer J. Clin. **68**, 7–30 (2017)
14. Skourt, B.A., El Hassani, A., Majda, A.: Lung CT image segmentation using deep neural networks. Procedia Comput. Sci. **127**, 109–113 (2018)
15. Szegedy, C., Liu, W., Jia, Y., Sermanet, P., Reed, S., Anguelov, D., Erhan, D., Vanhoucke, V., Rabinovich, A.: Going deeper with convolutions. In: Proceedings of the IEEE conference on computer vision and pattern recognition, pp. 1–9 (2015)
16. Taha, A.A., Hanbury, A.: Metrics for evaluating 3D medical image segmentation: analysis, selection, and tool. BMC Med. Imaging **15**(1), 29 (2015)
17. Tan, Y., Schwartz, L.H., Zhao, B.: Segmentation of lung lesions on CT scans using watershed, active contours, and markov random field. Med. Phys. **40**(4) (2013)
18. Wang, S., Zhou, M., Gevaert, O., Tang, Z., Dong, D., Liu, Z., Tian, J.: A multi-view deep convolutional neural networks for lung nodule segmentation. In: Engineering in Medicine and Biology Society (EMBC), 2017 39th Annual International Conference of the IEEE, pp. 1752–1755. IEEE (2017)
19. Wang, S., Zhou, M., Liu, Z., Liu, Z., Gu, D., Zang, Y., Dong, D., Gevaert, O., Tian, J.: Central focused convolutional neural networks: developing a data-driven model for lung nodule segmentation. Med. Image Anal. **40**, 172–183 (2017)
20. Wu, B., Zhou, Z., Wang, J., Wang, Y.: Joint learning for pulmonary nodule segmentation, attributes and malignancy prediction. In: 2018 IEEE 15th International Symposium on Biomedical Imaging (ISBI 2018), pp. 1109–1113. IEEE (2018)

21. Xu, B., Wang, N., Chen, T., Li, M.: Empirical evaluation of rectified activations in convolutional network (2015). arXiv:1505.00853
22. Yin, Y., Hoffman, E.A., Ding, K., Reinhardt, J.M., Lin, C.-L.: A cubic b-spline-based hybrid registration of lung CT images for a dynamic airway geometric model with large deformation. Phys. Med. Biol. **56**(1), 203 (2010)
23. Zou, K.H., Warfield, S.K., Bharatha, A., Tempany, C.M., Kaus, M.R., Haker, S.J., Wells III, W.M., Jolesz, F.A., Kikinis, R.: Statistical validation of image segmentation quality based on a spatial overlap index1: scientific reports. Acad. Radiol. **11**(2), 178–189 (2004)

Improved Performance and Execution Time of Face Recognition Using MRSRC

Jitendra Madarkar, Poonam Sharma and Rimjhim Singh

Abstract Face recognition accuracy is vulnerable to environmental noise, low-resolution images, and other variations such as illumination, pose, and expression. The accuracy of the face recognition mostly relying on the features of training samples and testing samples. Recently, sparse representation based classification (SRC) has shown state-of-the-art results in face recognition and developed several extended versions of SRC methods to improve the performance. The time complexity of the SRC is depended on the size of the dictionary. In this paper, a new fusion approach MRSRC (Multi-resolution sparse representation based classification) is developed by incorporating the wavelet compressed features into the dictionary. MRSRC has shown better performance than an existing algorithm and also reduces the time complexity. The experimentation is carried out on benchmarking databases such as LFW and ORL.

Keywords Face recognition · Sparse representation · Wavelet transformation · Multi-resolution

1 Introduction

Face recognition (FR) is a crucial application of computer vision. Nowadays, most organizations are deploying surveillance cameras and biometric systems for security concerns. Many researchers have developed several methods for face recognition (FR) to achieve better performance such as PCA [1], LBP [2], LDA [3], ICA [4], SVM [5], wavelet [6], HMM, and curvelet [7]. The face recognition process is

J. Madarkar (✉) · P. Sharma · R. Singh
Visvesvaraya National Institute of Technology, Nagpur, India
e-mail: jitendramadarkar475@gmail.com

P. Sharma
e-mail: dr.poonamasharma@gmail.com

R. Singh
e-mail: rimjhimsingh1012@gmail.com

© Springer Nature Singapore Pte Ltd. 2020
K. N. Das et al. (eds.), *Soft Computing for Problem Solving*,
Advances in Intelligent Systems and Computing 1048,
https://doi.org/10.1007/978-981-15-0035-0_49

carried out in 4 steps: 1. Face detection, 2. Preprocessing, 3. Feature extraction, and 4. Classification. Several researchers are mostly focused on feature extraction and classification steps of the face recognition process. A feature extraction methods extracts the discriminative features from the face image such as PCA, LDA, LBP, ICA, etc. A classification is a process that predicts a correct class label for a test image from the training dataset such as nearest neighbor (NN) [8], support vector machine SVM [5], sparse representation-based classification (SRC) [9], and Linear Regression [10]. Even, some researchers have applied preprocessing step on face images to neutralize the effect of noise, illumination, and occlusion.

There are many challenges in face recognition such as linear variation, nonlinear variation, pose variation, undersampled, rotation, scaling, and low-resolution. These challenges affect the performance of face recognition. All these issues of face recognition are not fully explored by any single method.

Recently SRC [9] has shown state-of-the-art results in the face recognition field on the occluded and corrupted face images. The very first time, SRC was introduced in face recognition by J Wright et al. in 2009 and since onwards this method has attracted much attention in this field. SRC has two main disadvantages: first, it requires uncorrupted training samples and second, it requires a large number of atoms in the dictionary. In real-world scenario, collecting a number of samples for each individual is a difficult task. Another challenge is that different resolution images are captured by different capturing devices. Due to high-resolution images, the time and space complexity of an algorithm increases. Face recognition needs a compressed image to reduce time and space complexity. To solve these issues, researchers need a compressed method to compress the image.

The main intention of the researchers is to sustain discriminative information at low-resolution. Existing methods used interpolation technique to rescale the images from one resolution to another. This article proposed a MRSRC method, to improve the performance and reduce the time complexity of face recognition. The MRSRC method has taken the benefits of SRC and wavelet transformation. Wavelet transform had been used in face recognition but did not explore all the ability in the field of face recognition. It reduces the noise and illumination effects of face image to improve the accuracy. Multi-resolution analysis decomposes an image into a different resolution without losing the discriminative feature of the original image. Discrete wavelet transform (DWT) is used to reconstruct the noiseless image and also normalize the illumination effect. SRC represents the test image by a linear selection of the dictionary atoms, which is incorporated by the wavelet feature. The experimental result has shown better performance than the existing SRC method.

The rest of the paper is organized as follows: Sect. 2, describe related work of the face recognition. Section 3, elaborate on the proposed method. Section 4, presents the experimental result. Conclusion introduced in Sect. 5.

2 Related work

The dimension of the images can affect space and time complexity. There are many algorithms that are used for dimension reduction in face recognition such as PCA, FDA, and ICA. These dimension reduction algorithms result gets deflected due to occlusion, pose variation, and an illumination effect. The aforementioned dimension reduction algorithm able to solve the linear data but fails to solve the nonlinear data. To resolve this problem first need to change the nonlinear distribution to linear distribution using kernel function. Kernel function implicitly changes the space from the original dimension to high dimensional. KPCA [11, 12] and KFDA [12] have shown significant performance in face recognition than linear dimensional reduction method. In some cases, images are getting properly analyzed in the space domain, then several transformations have been used in face recognition like Fourier transform, discrete cosine transform, wavelet transform, curvelet [7], and ridge transform. Recently SRC has shown the state of the art result in face recognition. SRC is robust to noise and occlusion but degrades the performance when the training samples have nonlinear variation, pose variation, and low-resolution. Several researchers have developed state of the art methods based on SRC [9] to resolve the issue of linear variation, nonlinear variation, undersampled, pose variation, and low-resolution. Sparse coefficient (code vector) and dictionary play a significant role in SRC, by modifying these two things researchers have improved the performance of SRC. SRC did not depend on the feature of samples but if the sample contains nonlinear variation then it needs a discriminative feature or learning mechanism to improve the performance. SRC [9], WSRC [13], KSRC [14], SNRC [15], SoC SRC [16], Molecular [17], WKSRC [18], and WGSR[19] are methods able to solve only linear issue and GSRC [20], ESRC [21], SSRC [22], SR-RLS [23], SDR [24] and S^3RC [25] are able to solve linear as well as nonlinear issue but failed to overcome the issue of low-resolution and pose variation. ESRC, SSRC, and S^3RC resolve the issues of undersampled. ESRC and SSRC construct new extravariant dictionary to capture all nonlinear variation of the training samples. Due to less number of training samples, S^3RC uses both labeled samples and unlabeled samples to learn proper prototype for each class using GMM and EM algorithm in an unsupervised way. Du et al. [26] uses wavelet in face recognition to neutralize illumination effect, Foon et al. [27] have achieved great performance in face recognition by combining wavelet transform and Zernike moments. Nie [28] combines wavelet transform and kernel PCA to improve the performance of face recognition. In [29], preprocessing done by wavelet transform and then applies KPCA to extract the features of the training sample. Emadi et al. [29] proposed illumination normalization method for face verification using 2D-wavelet.

3   Methodology

This section describes the MRSRC method in detail. The MRSRC method is developed by taking advantage of wavelet transform and SRC to improve the accuracy and execution time of face recognition. SRC uses the dictionary to classify the test sample but time and space complexity depends on the size of the dictionary. Wavelet transform has its own advantage to produce the discriminative feature at different scale and resolution. The MRSRC method is developed by incorporating the wavelet feature into the dictionary as an atom.

SRC represents a test image by a linear combination of the dictionary atoms that is built by the samples of training dataset and test image classifies to the class which produces minimum residual. Let $A = [A_1, A_2, \ldots, A_c]$ ε $\Re^{m*n}$ be the set of training samples of all the classes. A is also called a dictionary which consists n numbers of samples that are belongs to all the classes and m is a dimension of each sample. Each dictionary column is referred to as a dictionary atom. c represents a number of classes and A_i represents the set of samples belonging to the class i. x ε $\Re^n$ is a sparse vector which has very few non zero elements. The test sample y can be represented by Eq. 1.

$$y = Ax_0 \quad \varepsilon \quad \Re^m \tag{1}$$

To find sparsest representation of SRC, linear equation is solved by using ℓ^0-minimization.

$$\hat{x} = \arg \ \min_0 \ \|x\|_0 \ s.t \ y = Ax \tag{2}$$

$\|.\|_0$ refer to as ℓ^0-norm which counts non zero entries in a vector. To solve linear Eq. 2 by ℓ^0-minimization is NP-Hard problem where, it is difficult to find an approximation. If a solution of x_0 is sparse enough, then it can be solved by using ℓ^1-minimization in polynomial time [30]. Then Eq. 2 can be modified with the help of ℓ^1-minimization as shown in Eq. 3.

$$\hat{x} = \arg \ \min_x \ \|x\|_1 \ s.t \ y = Ax$$

or $\tag{3}$

$$\hat{x} = \arg \ \min_x \ \|x\|_1 \ s.t \ \|y = Ax\|_2 \leq \varepsilon$$

The purpose of the SRC is to find more approximate atoms from the dictionary that can sparsely represent the test image y and its sparse vector is obtained by solving the convex problem of the linear equation using (3). Wavelet has a significant advantage over signal and image processing such as noise reduction, illumination normalization, decomposition, and image reconstruction. Wavelet application: DWT and Multi-resolution help to analyze the images. DWT uses two functions namely wavelet and scaling [31] are to be used to analyze the image at different scale using low pass and high pass filter with different cut off frequencies. In 2D-DWT scaling and wavelet function [31] can be mathematically expressed by Eqs. 4 and 5.

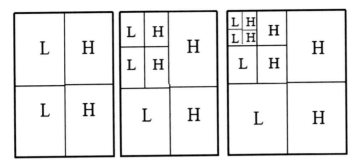

Fig. 1 Multi-resolution decomposition levels of wavelet transform

$$\phi_{j,m,n}(x, y) = 2^{j/2}\ \phi(2^j x - m,\ 2^j y - n) \tag{4}$$

$$\psi_{j,m,n}(x, y) = 2^{j/2}\ \psi(2^j x - m,\ 2^j y - n) \tag{5}$$

Image reconstruction is a preprocessing task of face recognition for a better quality of the image. Multi-resolution (MR) analyzes the frequency of an image at a different resolution and position. It decomposes an image into four different subbands such as approximation, horizontal, vertical, diagonal and each subband carries specific information of the image. The decomposition process can apply recursively on approximation subband as shown in Fig. 1. The approximation subband carries more detail information than other subbands. The successive level contain more discriminative features with less noise. The decomposition process of an image reduced its size(resolution) by (1/4)th at each level. The main intention of the Multi-resolution decomposition is to analyze the image in different resolution because different level has different detail information of the image. The horizontal subband focuses attention on vertical edges. The vertical subband focus attention on horizontal edges and The diagonal subband focus attention on diagonal edges. The approximation (LL) subband is a coarse approximation of the image and it represents low-frequency information which is invariant to small occlusion and facial expression. This information helps to reconstruct the noiseless and quality images at low-resolution.

The main moto of the proposed MRSRC method is to attain better accuracy and reduce time complexity. The face images decomposed and normalize the environmental effects of a face image by discrete wavelet transform. And then apply SRC to represents test sample y by the linear representation of atoms of the dictionary. Overcomplete dictionary $n \gg m$, n is a number of total atoms in the dictionary and m is the dimension of the face image. If the dimension of the face image is high then the overcomplete dictionary required a large number of atoms and the dictionary size increases which is directly proportional to the time and space complexity. If the dimension of an image reduces then, less number of atoms are required to construct a dictionary. After applying a wavelet transform on training images the size of the dictionary becomes $\Re^{p*k}$ and $p \ll m$, $k \ll n$.

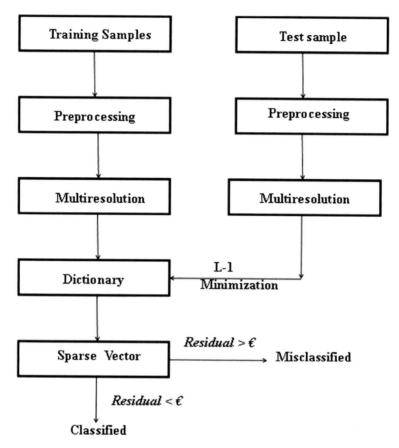

Fig. 2 MRSRC Model

The MRSRC model has shown in Fig. 2, initially, preprocess the all training and testing samples with the help of DWT to remove environmental effects of the image. Multi-resolution reconstructs the training images at different resolutions and then incorporates all training reconstructed images into dictionary. MRSRC uses ℓ^1-minimization to represents a test image by using Eq. 6.

$$\hat{x} = \arg \min_x \ \|x\|_1 \quad s.t \quad \|\phi(y) = \phi(A)x\|_2 \le \varepsilon \tag{6}$$

where dictionary, $\phi(y) \ \varepsilon \ \mathfrak{R}^p$. The test sample is classified based on the residual and it belongs to the class which produces minimum residual. If the residual is greater than the threshold value or sparse coefficient spread over across the dictionary then the test image is misclassified or assume that its class is not available in training dataset.

4 Experiments and Results

The experimentation has done on benchmarking databases like LFW [32] and ORL. For an experimentation, daubechies wavelet has been used to decompose the face images. The experimental results are compared on three features: PCA, Wavelet, and image data itself. In result, SRC has shown better performance when dictionary incorporates with compress feature of wavelet than PCA, and an image sample itself.

4.1 Experiment Setup

In this subsection, the experimental detail of MRSRC method has been discussed. The regularized parameter was set 0.001 for all methods and used homotomy algorithm to represent the test image in the sparse vector. MATLAB 2016 was used for experimentation on the machine having the configuration of 8GB RAM, 3.6 GH $i - 7$ processor. The face images decompose into three levels: level 1 (resolution reduced by the (1/4)th size of the original image), level 2 (resolution reduced by the (1/4)th size of level 2), and level 3 (resolution reduced by the (1/4)th size of level 2 image) using wavelet.

4.2 LFW

The LFW [32] dataset has taken images from the web of 1680 individual. These images are varying in illumination, pose, expression, and different background. LFW database is a benchmark for face recognition application. For experimentation, 360 samples of 40 individual are taken with 95×95 size. The face images cropped by using Voila-Jones face detection algorithm. Each individual has 9 samples and decomposes all the samples into three levels. Experimentation is carried out with different number of training and testing samples as shown in (Figs. 3, 4, 5, 6).

Fig. 3 LFW face dataset samples of 1 individual

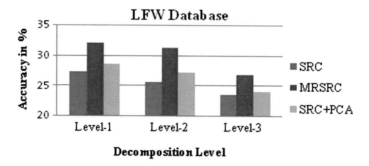

Fig. 4 LFW result (LFW result 2 samples for training and 7 samples for testing)

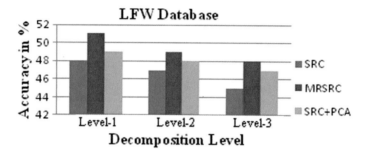

Fig. 5 LFW result (5 samples for training and 4 samples for testing)

Fig. 6 LFW face dataset samples of 1 individual

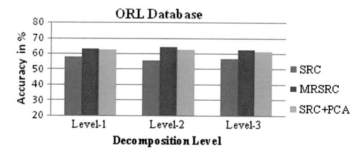

Fig. 7 ORL result (2 samples for training and 8 samples for testing)

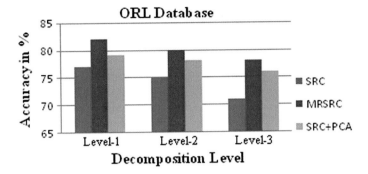

Fig. 8 ORL result (5 samples for training and 5 samples for testing)

4.3 ORL

ORL face database contains a total of 400 face images of 40 individual. Each individual samples capture at a different time, expression, and lighting condition in the same background. For experimentation, taken 112×92 size of the image. Each individual has 10 samples and decomposes all samples into three levels using Multi-resolution wavelet. Experimentation is carried out with a different number of training and testing samples (Figs. 7, and 8).

4.4 Result Analysis

The observation has been observed from the experimental results. For experimentation, we have taken a different number of training and testing samples, By observing the experimental results, we concluded that the performance of the MRSRC method increases as we increase the number of training samples in each class. The proposed approach works better on linear(illumination, occlusion) and nonlinear variation (local expression, deformation) of training samples. Dataset LFW and ORL have more nonlinear variation and results have shown better accuracy than $SRC + PCA$ and SRC. The time and space complexity of SRC based methods depends on the size of the dictionary, as dictionary size reduces then time and space complexity also reduces.

5 Conclusion

In this paper, MRSRC method developed by fusing of two methods: SRC and wavelet transform. The discrete wavelet transform is used to remove noise and illumination effects and Multi-resolution decomposed an image at a different resolution with proper detail. These features incorporated into the dictionary and apply ℓ^1-minimization to represent the test image. The experimental result shows that the MRSRC method performs better than SRC+PCA and SRC. The proposed method reduces the size of the dictionary, hence reduces the execution time of the algorithm.

In the future, the MRSRC method can work on a different type of wavelet, pose variation, and camera captured low-resolution images to improve the accuracy of the FR.

References

1. Turk, M., Pentland, A.: Eigenfaces for recognition. J. Cogn. Neurosci **3**(1), 71–86 (1991)
2. Ahonen, T., Hadid, A., Pietikäinen, M.: Face description with local binary patterns: application to face recognition. IEEE Trans. Pattern Anal. Mach. Intell. **28**, 2037–2041 (2006)
3. Belhumeur, P.N., Hespanha, J.P., Kriegman, D.J.: Eigenfaces vs. fisherfaces: recognition using class specific linear projection. IEEE Trans. Pattern Anal. Mach. Intell. **19**(7), 711–720 (1997)
4. Bartlett, M.S., Movellan, J.R., Sejnowski, T.J.: Face recognition by independent component analysis. IEEE Trans Neural Netw. **13**(6), 1450–1464 (2002)
5. Phillips, P.J.: Support vector machines applied to face recognition. In: Proceedings of the 1998 Conference on Advances in Neural Information Processing Systems II, pp. 803–809. MIT Press, Cambridge, MA (1999)
6. Chien, J.-T., Chia-Chen, W.: Discriminant waveletfaces and nearest feature classifiers for face recognition. IEEE Trans. Pattern Anal. Mach. Intell. **24**(12), 1644–1649 (2002)
7. Sharma, P., Yadav, R.N., Arya, K.V.: Pose-invariant face recognition using curvelet neural network. IET Biom. **3**(3), 128–138 (2014)
8. Cover, T., Hart, P.: Nearest neighbor pattern classification. IEEE Trans. Inf. Theory **13**(1), 21–27 (1967)
9. Wright, J., Yang, A.Y., Ganesh, A., Sastry, S.S., Ma, Y.: Robust face recognition via sparse representation. IEEE Trans. Pattern Anal. Mach. Intell. **31**(2), 210–227 (2009)
10. Naseem, I., Togneri, R., Bennamoun, M.: Linear regression for face recognition. IEEE Trans. Pattern Anal. Mach. Intell. **32**(11), 2106–2112 (2010)
11. Ebied, H.M.: Feature extraction using PCA and Kernel-PCA for face recognition. In: 2012 8th International Conference on Informatics and Systems (INFOS), pp. MM–72–MM–77 (2012)
12. Yang, M.-H.: Face recognition using kernel methods. In: Dietterich, T.G., Becker, S., Ghahramani, Z. (eds.) Advances in Neural Information Processing Systems 14, pp. 1457–1464. MIT Press (2002)
13. Lu, C.-Y., Min, H., Gui, J., Zhu, L., Lei, Y.-K.: Face recognition via weighted sparse representation. J. Vis. Commun. Image Represent. **24**(2), 111–116 (2013). Sparse Representations for Image and Video Analysis
14. Yin, J., Liu, Z., Jin, Z., Yang, W.: Kernel sparse representation based classification. Neurocomputing **77**(1), 120–128 (2012)
15. Hui, K.H., Li, C.L., Zhang, L.: Sparse neighbor representation for classification. Pattern Recogn. Lett. **33**(5), 661–669 (2012)
16. Li, J., Lu, C.-Y.: A new decision rule for sparse representation based classification for face recognition. Neurocomputing **116**, 265–271 (2013). Advanced Theory and Methodology in Intelligent Computing
17. Hu, Z.P., Bai, F., Zhao, S.H., Wang, M., Sun, Z.: Extended common molecular and discriminative atom dictionary based sparse representation for face recognition. J. Vis. Commun. Image Represent. **40**, 42–50 (2016)

18. Liu, X., Lu, L., Shen, Z., Lu, K.: A novel face recognition algorithm via weighted kernel sparse representation. Future Gener. Comput. Syst. **80**, 653–663 (2018)
19. Lai, J., Jiang, X.: Modular weighted global sparse representation for robust face recognition. IEEE Signal Process. Lett. **19**(9), 571–574 (2012)
20. Yang, M., Zhang, L.: Gabor feature based sparse representation for face recognition with gabor occlusion dictionary. In: Proceedings of the 11th European Conference on Computer Vision: Part VI, ECCV'10, pp. 448–461, Berlin, 2010. Springer
21. Deng, W., Hu, J., Guo, J.: Extended src: undersampled face recognition via intraclass variant dictionary. IEEE Trans. Pattern Anal. Mach. Intell. **34**(9), 1864–1870 (2012)
22. Deng, W., Hu, J., Guo, J.: In defense of sparsity based face recognition. In: 2013 IEEE Conference on Computer Vision and Pattern Recognition, pp. 399–406 (2013)
23. Iliadis, M., Spinoulas, L., Berahas, A.S., Wang, H., Katsaggelos, A.K.: Sparse representation and least squares-based classification in face recognition. In: 2014 22nd European Signal Processing Conference (EUSIPCO), pp. 526–530 (2014)
24. Jiang, X., Lai, J.: Sparse and dense hybrid representation via dictionary decomposition for face recognition. IEEE Trans. Pattern Anal. Mach. Intell. **37**(5), 1067–1079 (2015)
25. Gao, Y., Ma, J., Yuille, A.L.: Semi-supervised sparse representation based classification for face recognition with insufficient labeled samples. IEEE Trans. Image Process. **26**(5), 2545–2560 (2017)
26. Du, S., Ward, R.: Wavelet-based illumination normalization for face recognition. In: IEEE International Conference on Image Processing 2005, vol. 2, pp. II–954 (2005)
27. Foon, N.H., Pang, Y.-H., Jin, A.T.B., Ling, D.N.C.: An efficient method for human face recognition using wavelet transform and zernike moments. In: Proceedings of International Conference on Computer Graphics, Imaging and Visualization, CGIV 2004, pp. 65–69 (2004)
28. Nie, X.F.: Face recognition using wavelet transform and kernel principal component analysis. In: 2010 International Conference on Future Information Technology and Management Engineering, vol. 3, pp. 186–189 (2010)
29. Emadi, M., Khalid, M., Yusof, R., Navabifar, F.: Illumination normalization using 2D wavelet. Procedia Eng. **41**, 854–859. International Symposium on Robotics and Intelligent Sensors (IRIS 2012)
30. Donoho, D.L., Tsaig, Y.: Fast solution of ℓ_1-norm minimization problems when the solution may be sparse. IEEE Trans. Inform. Theory **54**(11), 4789–4812 (2008)
31. Ravichandran, D., Nimmatoori, R., Gulam Ahamad, M.: Mathematical representations of 1D, 2D and 3D wavelet transform for image coding, vol. 5, pp. 1457–1464 (2016)
32. Huang, G.B., Ramesh, M., Berg, T., Learned-Miller, E.: Labeled faces in the wild: a database for studying face recognition in unconstrained environments. Technical Report 07-49, University of Massachusetts, Amherst, Oct (2007)

Motion Detection Using a Hybrid Texture-Based Approach

Rimjhim Padam Singh, Poonam Sharma and Jitendra Madarkar

Abstract Motion analysis plays an important role in various real-time applications like object detection, human–computer interaction, surveillance systems, human detection and tracking, event monitoring, etc. Background subtraction that aims at separating the motion regions from the static portions lays the foundation of all such applications. Most of the background subtraction techniques developed to date explore colour features of pixels, either individually or in a spatio-temporal manner. Many other techniques exploit texture characteristics of pixels, while a few have been developed that employ a combination of both texture and colour characteristics for extracting motion-related information from frames. But most of the efficient background modelling techniques demand extensive use of hardware and computation. In this paper, we propose a hybrid sample consensus-based foreground segmentation technique that fuses similarity-based binary patterns of pixels with YCbCr colour space. The core of a pixel-based technique has been reconstructed to obtain drastically refined results.

Keywords Motion detection · Background subtraction · Texture features

1 Introduction

Knowledge extraction from raw videos and images lays the foundation of several research areas in the field of image processing. For instance, security breaches and human impersonation have always been some of the serious threats posed to various organizations, making human identification and authorization, activity monitoring

R. P. Singh (✉) · P. Sharma · J. Madarkar
Visvesvaraya National Institute of Technology, Nagpur, India
e-mail: rimjhimsingh1012@gmail.com

P. Sharma
e-mail: dr.poonamasharma@gmail.com

J. Madarkar
e-mail: jitendramadarkar475@gmail.com

© Springer Nature Singapore Pte Ltd. 2020
K. N. Das et al. (eds.), *Soft Computing for Problem Solving*,
Advances in Intelligent Systems and Computing 1048,
https://doi.org/10.1007/978-981-15-0035-0_50

609

and event detection unavoidable. This has led to burdensome responsibilities of continuously monitoring the areas manually. Also, several human–computer interaction technologies developed to help specially abled persons need automatic extraction of comprehensive information from the videos and images given as input to them. Nowadays, many such systems have moderately replaced their manual counterparts, but tend to generate enormous video data that is impractical to analyse in real-time scenarios. Hence, need arises to model an automated system that is, able to detect, identify and track individual human beings, monitor events, raise alarms in case of unusual activities, etc., in real time. Developing such automated systems for video-analysis involve background subtraction as an initial step to detect and identify motion.

Background subtraction aims at training a model with background scenes capable of detecting unexpected changes (i.e. foreground pixels) in future frames. A powerful and robust background modelling is necessitated as the information provided by it is given as input to high-end models based upon the application. Several unnatural reasons hindering efficient background modelling are improper illumination, camera jitter, noise spikes, etc. Natural reasons like winds, night darkness, etc., can also have an adverse effect on motion analysis. Several background subtraction techniques striving to achieve the best possible results in open and closed environments, complex and dynamic scenes, moving cameras, etc., have been proposed by several researchers. A review of all such techniques can be studied by referring to [1, 2]. In this paper, a motion detection technique inspired by Visual Background Extractor (Vibe) has been developed. This sample consensus-based strategy is deployed to exploit both YCbCr intensities and binary similarity features of pixels for faster ghost absorption into the background. Reliable results can be achieved at very early stages.

2   Related Work

Initially, the very basic background models assumed background scenes to be completely static and subtracted the incoming frames to get motion-related information. This assumption does not hold true in real-time scenarios, thereby leading to corrupted foreground results. Later, background models computed the pixel-wise mean, median, variance and several other statistical features from a set of previous frames to construct the background. Most of these algorithms were pixel-based and utilized pixel intensity or colour information for background construction while some others explored the texture information for the same.

One of the very popular colour-based techniques models each pixel in a frame with the normal distribution. A unimodal Gaussian model proposed by Wren et al. [3] assumed pixel intensities exhibit properties similar to that of a Gaussian model and is incapable of handling dynamic backgrounds with changes, which was later improved by Kim et al. [4] by handling its strict constraints. Later, the Gaussian mixture model capable of handling dynamic backgrounds and illumination conditions was introduced in [5] that modelled pixels using multiple Gaussians in colour space.

Afterwards many improvements were incorporated into mixture of Gaussians model to make it more robust to noise, sudden illumination changes, etc. But all of these models suffered due to a serious problem of parameter estimation.

Pixels in the video have also been modelled using Principal Component Analysis (PCA) technique [6]. In this background model comprising of N eigenvectors is compared to incoming frames for classification. PCA-based background subtraction models [7] have many limitations like, foreground objects must keep continuously moving and must be of comparatively smaller size. Background models employing Support Vector Machine (SVM) [8] and Support Vector Regression (SVR) [9] have also been developed where each class has its own decision boundary (support vector) that aids in the classification of incoming pixels.

The problem of parameter estimation in the above-mentioned techniques led to the development of several non-parametric techniques [10]. Authors used Kernel Density Estimators (KDE) to model the background [11]. Lee and Park [12] later proposed an adaptive version of KDE which was capable of adapting the learning rate parameter according to different environments. But all these techniques incur high memory and computational costs as they require numerous frames for training the model.

Cluster-based models represent altogether different class of background subtraction techniques and commonly include k-means [13], codebook [14] and some other sequential clustering models. Codebook models [15] are more popular and analyse colour distortion of pixels. Each pixel is modelled using a codebook containing one or more codewords that represent a series of important colour values obtained during training. These models proved robust against dynamic backgrounds and illumination changes but are inefficient when both background and foreground pixels have similar colours.

Olivier et al. [16] proposed a pixel-based background subtraction technique, named Visual Background Extractor (Vibe), assuming that neighbouring pixels exhibit similar spatial-temporal information. Hence, they not only updated the corresponding pixel in the background model but also updated the value of its neighbouring pixel. Although the technique dealt with noisy pixels, it produced high false positives. Several modifications to Vibe have been proposed in [17, 18].

Techniques operating on pixels in colour space face serious drawbacks in presence of camouflage occurrences, shadows and illumination changes. Hence, in 2002, Ojala et al. [19] introduced the very basic texture-based Local Binary Patterns (LBP), that encoded a pixel using a series of binary code based upon the relative values of its circular neighbours. But it is quite sensitive to noise as a minor change in the centre pixel changes the complete code. Later, the authors in [13] improved the original LBP descriptors in areas where grey levels of centre and its neighbouring pixels are similar. εLBP [20] and adaptive εLBP [21], the two improvements to LBP, are robust to illumination changes and similarities in colour information with the neighbours.

Later, more compact Centre-Symmetric LBP (CS-LBP) [22] technique was proposed that operated on centre-symmetric pairs only. It was later extended as Spatial Extended Center-Symmetric (SCS-LBP) [23], that made CS-LBP more discriminative by considering its gradient information. It produced relatively a shorter histogram that further required lesser computation. Stereo LBP on Appearance and Motion (SLBP-AM) [24] utilized frame information in three different planes. It is capable of adapting to sudden and large light changes and is more robust to noise.

Bilodeau et al. [25] proposed a Local Binary Similarity Pattern (LBSP) descriptor, that computed binary patterns not only within the frame (intra-LBSP) but also computed LBSP features between incoming frames and background model (inter-LBSP). The technique then obtained Hamming Distance between the two LBSP codes for motion detection. The technique is able to detect both texture and intensity variations in the pixels but absorbs ghosts into the foreground in dynamic scenes.

Authors in [26] modified the core of Vibe by incorporating modified LBSP features into it along with RGB colour space and produced an enhanced version on LBSP features but could not handle dynamic backgrounds, illumination variations efficiently. It also produced incorrect foreground masks for scenarios where the object of interest becomes static for a longer duration.

3 Methodology

As per the drawbacks of several techniques mentioned in the previous section, we prefer to work with a compute-extensive non-parametric approach (Vibe) offering feasible hardware and memory demands. Instead of operating on individual pixels, it considers the similar properties shared by the whole spatio-temporal neighbourhood, efficiently handling scenarios where a whole small region exhibits transition due to noise. But due to its random and selective update policy, pixels getting misclassified as foreground never get absorbed into the background leading to corrupted segmentation results, providing researchers with enough scope to enhance it.

Colour features of pixels fail to differentiate regions containing shadows and improper illumination from the object of interest, leading to incorrect results. Moreover, when a group of pixels has intensity close to the background, applying thresholds on Euclidean distances calculated between them absorbs it into the background. Hence, projecting a new frame and background model in feature space (texture model) adds robustness to the algorithm by accounting for transitions affecting the set of pixels. Hence, in this paper Vibe's core functionality is integrated with a feature space model of pixels where N samples per pixel contained in the background vote for the classification of a pixel.

3.1 Pixel Model

The efficiency of a background subtraction technique critically depends upon the colour space chosen with RGB, HSV, YCbCr and HSI being at the topmost positions. RGB is a widely chosen colour space due to equal distribution of colour and brightness information in all the channels. Moreover, it is robust to camera noise and environmental changes and is the output format of cameras. Though colour information makes background subtraction robust against illumination changes, shadows, etc., but segregation of brightness information from it enhances the efficiency of such techniques.

YCbCr colour space provides such segregation of brightness information from that of chromatic information. Hence, YCbCr is claimed to outperform RGB, HSI and HSV colour spaces. Independent channels make it least sensitive to shadows, noise and illumination changes and a favourable option for our work. Hence, the projection of pixel information into feature space is done using YCbCr colour space instead of RGB colour space. To the best of our knowledge, exploiting YCbCr colour space for generating texture patterns to integrate them further with Vibe has not been done by any researcher to date.

The foremost LBP descriptor proposed by Ojala et al. [19] generates a binary code for a centre pixel ($p = x_p, y_p$) by thresholding its entire circular neighbourhood composed of K pixels. The LBP operator is applied for obtaining binary codes for the incoming frame as well as the background model for their representation in texture space by using Eq. 1.

$$LBP_K(p) = \sum_{i=0}^{K-1} f(t_i - t_p)2^i \qquad (1)$$

where t_p represents the intensity of centre pixel p while t_i represents intensity of each neighbour. Here, f is a thresholding function as defined in Eq. 2

$$f(x) = \begin{cases} 1 & x \geq 0 \\ 0 & \text{otherwise.} \end{cases} \qquad (2)$$

But as LBP descriptors are generated within an image, they are capable of uncovering only texture changes. Say if a group of pixels undergoes intensity change at the same time, LBP descriptors will fail to correctly classify them.

Hence, in 2013 the authors in [25] proposed LBSP features, a novel technique for capturing texture related information of pixels. They generated binary codes for pixels inside the background model (Intra-image LBSP) and between a new frame and background image (Inter-image LBSP) for foreground segmentation. These codes are generated by binary thresholding the absolute difference between pixel intensities over a small area. Calculating the Inter-image LBSP features between a new image and background image uncovers the intensity changes affecting the group of pixels in an area, thereby making LBSP features robust to noise contained either in an individual pixel or a neighbouring group of pixels. LBSP descriptor over a centre pixel (x_c, y_c) and its K circular neighbours can be obtained using Eq. 3.

$$LBSP(x_c, y_c) = \sum_{p=0}^{K-1} f(t(x_p, y_p) - t(x_c, y_c))2^i \qquad (3)$$

For generating Intra-image LBSP features, $t(x_p, y_p)$ represents neighbouring pixel's intensity. $t(x_c, y_c)$ represents intensity of central pixel (x_c, y_c), from background, while for Inter-image LBSP features it represents intensity of central pixel (x_c, y_c) from a reference frame. f is a thresholding function which can be defined as

$$f(t_p, t_c) = \begin{cases} 1 & |t_p - t_c| \geq T \\ 0 & \text{otherwise.} \end{cases} \tag{4}$$

3.2 Pixel Classification and Update

Instead of thresholding distance between LBSP features obtained for a pixel, the proposed approach checks the similarity between the Inter-imageLBSP and Intra-imageLBSP feature descriptors by calculating Hamming Distance (HD) between them. LBSP descriptor is applied on 3×3 area, producing 8-bit binary code per channel. Class C_n for central pixel (x_c, y_c), from new frame n is obtained by thresholding HD between Intra-imageLBSP binary code $((LBSP_M (x_c, y_c)))$ for background model M and Inter-imageLBSP binary code $(LBSP_n(x_c, y_c))$ for new central pixel (x_c, y_c) with respect to background model using Eqs. 5 and 6.

$$C_n(x_c, y_c) = \begin{cases} 0 & simFact_n(x_c, y_c) \leq T_{LBSP} \\ 1 & simFact_n(x_c, y_c) > T_{LBSP} \end{cases} \tag{5}$$

$$simFact_n(x_c, y_c) = HD(LBSP_n(x_c, y_c), LBSP_M(x_c, y_c)) \tag{6}$$

Though LBSP features provide efficient foreground segmentation, they many times fail to detect foregrounds when Intra-image and Inter-image patterns obtained are similar, thereby necessitating the chromatic information for better results. Moreover, Euclidean distance not only demands complex computations but also degrades overall performance as compared to city block distance, making city block distance an obvious choice.

The proposed approach chooses conditional update policy for maintaining the background model, where pixels classified as background are only allowed to update the model, resulting in higher false positives. These false positives are due to ghosts developed from improper illuminations, moving shadows, etc. and the model is never trained with such misclassifications. Hence, a pixel comparison with the average intensity of the background is also integrated into the core algorithm.

Y channel in YCbCr holds purely intensity-related information, hence, we drop Cb and Cr channels for checking average intensity ratio. $AInt_{Ratio}(x_c, y_c)$ provides the ratio of centre pixel's intensity $(Int_Y(x_c, y_c))$ to average intensity of initial background image $(AInt_Y (n = 1))$, that is further thresholded to obtain its new label $L_n(x_c, y_c)$. This two-step comparison technique suppresses the falsely detected ghosts regions and absorbs them into the background model enabling it to provide reliable results.

$$AInt_{Ratio}(x_c, y_c) = Int_Y(x_c, y_c)/AInt_Y(n = 1) \tag{7}$$

$$L_n(x_c, y_c) = \begin{cases} 0 & T1_{IntY} \leq AInt_{Ratio}(x_c, y_c) \leq T2_{IntY} \\ 1 & \text{otherwise} \end{cases} \quad (8)$$

Background model update is also time subsampled, with ϕ governing the probability of a pixel updating its background model. A higher ϕ incorporates changes into the model at a slower rate, neglecting the sudden changes while a lower ϕ updates the model frequently, corrupting the moving objects in case they become stationary for some duration. Detailed analysis of the proposed technique can be done by referring to Algorithm 1.

Input:
    $LBSP_M(x_c, y_c)$ Intra-image LBSP features for centre pixel.
    $LBSP_n(x_c, y_c)$ Inter-image LBSP features for centre pixel.
    $AInt_Y(n = 1)$ Mean of pixel intensities of first frame.
Output:
    $L_n(x_c, y_c)$ Final class label of pixel (x_c, y_c).
begin
    $totMatch = 0$; $nsamples = 1$; $tFlag = 0$
    while $totMatch <= match_{min}$ **or** $nsamples <= N$ **do**
    $totDist = 0$; $tHDist = 0$
      while $ch \leq nbOfChannels$ **and** $Tflag == 0$ **do**
      $HDist(x_c, y_c) = HD(LBSP_n(x_c, y_c)(ch), LBSP_M(x_c, y_c)(ch))$
      $cDist = |Int(x_c, y_c)(ch) - MInt(x_c, y_c)(ch)|$
      if $(HDist < K * T_{LBSP})$ **and** $(cDist < K * T_{int})$ **then**
        $totDist = totDist + cDist$
        $tHDist = tHDist + HDist$
      else
        $tFlag = 1$
      end if
      end while
    if $(tFlag == 0)$ **and** $(totDist < T_{int})$ **and** $(tHDist < T_{LBSP})$ **then**
      Increment $totMatch$
    end if
    Increment $nSamples$
    end while
    if $totMatch \geq match_{min}$ **then**
      $AInt_{Ratio}(x_c, y_c) = Int_Y(x_c, y_c)/AInt_Y(n = 1)$
      if $(AInt_{Ratio}(x_c, y_c)$ **in range** $(T1_{IntY} : T2_{IntY}))$ **then**
        $L_n(x_c, y_c) = 0$
        return and updateBackgroundModel()
      else
        $L_n(x_c, y_c) = 1$
      end if
    else
      return and updateBackgroundModel()
    end if
    end algorithm.

Algorithm 1: Classification of pixel

4 Experiments and Results

4.1 Datasets

For justifying the efficiency of the proposed approach, five sample videos have been chosen from 2014 Change Detection Dataset (2014 CDNet) [27], each presenting a different challenge to the proposed approach. The 2014 CDNet dataset has total 11 categories of videos, each having 4–6 video sequences. Each video sequence contains minimum 1100 video frames and all are annotated with respective ground truths, making it easier to evaluate performance metrics on these sample videos. Videos selected have been shot both in open and closed room environments having shadows, waving and dynamic backgrounds, turbulence situations, etc.

Video sequence 'copyMachine' from *shadow* category and 'office' from *baseline* present a typical close room environment having improper illumination and shadows. Video 'corridor' from *thermal* category has been captured using a thermal camera and again provide an environment with shadows and improper lighting conditions. 'Highway' video sequence contains noisy and dynamic background containing waving leaves and trees along with their shadows. It also contains a moving object in the initial frame which itself is a challenge to background subtraction process. Lastly, video sequence 'turbulence3' from *turbulence* category presents bad weather conditions with poor visibility and continuous noise in the background.

4.2 Evaluation Metrics

Pixel classification has been studied as a binary classification problem, where '0' represents a background pixel and '1' represents a foreground pixel value. We calculated Recall (Re), Precision (Pre) and F-measure (FM) for the segmentation results obtained. Our comparisons mainly rely on *FM* obtained due to the trade-off between Recall and Precision.

$$Recall\ (Re) = \frac{tp}{tp + fn} \tag{9}$$

$$Precision\ (Pre) = \frac{tp}{tp + fp} \tag{10}$$

$$F\text{-}measure\ (FM) = 2 \times \frac{Re \times Pre}{Pre + Re} \tag{11}$$

Here, true positive (tp) holds the correctly classified positive pixels, true negative (tn) holds the correctly classified negative pixels, false positive (fp) represents the incorrectly classified positive pixels and false negative (fn) represents incorrectly classified negative pixels.

Table 1 Comparative results on 2014 CDnet dataset

Video	Metric (%)	GMM	KDE	Vibe	Spectral 360	Ivibe	Ours
Office	Recall	49.04	90.54	94.44	**97.63**	90.46	97.08
	Precision	74.62	96.75	88.8	93.84	98.32	**98.36**
	F-measure	59.18	93.5	91.55	95.69	94.23	**97.72**
Highway	Recall	91.82	93.79	86.31	96.04	79.91	**97.32**
	Precision	92.98	**93.27**	82.51	92.68	84.35	91.62
	F-measure	92.39	93.53	84.37	94.33	82.07	**94.38**
CopyMachine	Recall	53.91	88.55	89.52	90.04	84.8	**90.36**
	Precision	79.41	83.05	85.58	87.54	**90.77**	87.61
	F-measure	64.23	85.71	87.51	88.77	87.68	**88.96**
Corridor	Recall	82.51	83.19	85.21	88.64	88.6	**88.7**
	Precision	80.74	88.06	89.04	90.86	90.4	**91.96**
	F-measure	81.62	85.56	88.62	89.74	89.49	**90.3**
Turbulence3	Recall	74.79	77.4	83.04	**94.16**	78.67	79.67
	Precision	68.66	64.63	70.98	77.25	91.29	**91.59**
	F-measure	70.44	70.44	76.54	84.87	84.51	**84.94**

[a]Best results are marked with bold font

4.3 Discussions

In order to justify the correctness of proposed approach, the paper presents its comparisons with five state-of-the-art approaches, namely, Gaussian Mixture Model (GMM) [5], Kernel Density Estimators (KDE) [11], Vibe [16], Spectral-360 [28] and IVibe [18]. Results for mentioned state-of-the-art approaches have been taken from ChangeDetection.net website (www.changedetection.net) and Ivibe has been implemented specifically with the parameters as suggested by its authors. Table 1 presents its comparisons with other approaches.

It is clearly indicated that $FM \geq 85\%$ for all the sample videos, and $FM \geq 80\%$ is stated as acceptable results in general. The precision achieved is the highest among all other approaches for most of the videos, depicting the clear suppression of false positives due to ghosts getting absorbed into the foreground. It is due to the second verification where pixel's intensity is matched against average intensity of background model, which assumed that initial frame does not contain motion at all and areas affected by poor illumination are much lesser as compared to that of normally illuminated ones (for example, dark corners of a room, small amount of light coming from a direction). So the average obtained from such frames lies somewhere between the intensity values of these two sets of pixels. Hence, this average value is able to differentiate between the actual motion pixels and other static darker regions.

Hence, the proposed approach allows two types of pixels to update the background model, i.e. pixels voted as background by LBSP features and Vibe algorithm and pixels voted as foreground by them but as background by $AInt_{Ratio}(x_c, y_c)$. By doing

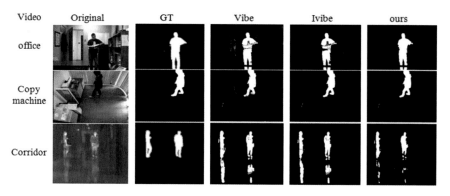

Fig. 1 Comaparative foreground segmentation masks obtained

so, many of the static pixels originally belonging to the background but classified as foreground by Vibe are also classified as background pixels with a chance of updating the background model, thereby suppressing ghost regions. In this way, the model gets trained with all such values in comparatively lesser amount of time.

Recall value got affected in some videos due to strict conditions applied as it may sometimes increase false negatives. Lower *Recall* for 'Turbulence3' video is due to very less proportion of foreground regions in the video. Even a slight increase in false negatives drastically decreased the value of Recall. Except for *Spectral-360*, all other approaches also obtained lower Recall, but the proposed approach obtained highest $FM = 84.94\%$ and $Pre = 91.59\%$.

The efficiency of an approach can be better represented with the F-measure obtained as it is not biased towards more precise strategies. Also, the ranking on ChangeDetection.net website also ranks the algorithms based upon their F-measure. Hence, we can claim our model to be efficient as it obtained the highest F-measure for all the videos. Also, the overall results for selected videos are competitive enough with average $FM = 91.26\%$ and $Pre = 92.228\%$ and $Re = 90.626\%$.

Figure 1 represents graphical output for the above-mentioned three sample videos. Results obtained for 'copyMachine' and 'office' videos clearly represent the better suppression of false positives due to improper illuminations while results for 'corridor' video depict efficient suppression of ghost regions due to hard shadows. The proposed approach produced the best segmentation results for *shadow* category videos.

5 Conclusion

This paper presents a texture-based background subtraction technique, where pixels are classified based on the count of nearest background samples. The technique projects a pixel in both LBSP feature space and YCbCr colour space model, where a single-channel LBSP feature has power for classifying a pixel at the initial step, to maintain lesser time and memory requirements. The two-step verification procedure based on the intensity of Y channel allows faster suppression of ghost regions due to improper illumination conditions achieved. The technique can further be combined with human activity recognition model or event detection model to detect unusual activities occurring in a video, as a direct application to different real-time scenarios.

References

1. Choudhury, S.K., Sa, P.K., Bakshi, S., Majhi,B.: An evaluation of background subtraction for object detection vis-a-vis mitigating challenging scenarios. IEEE Access **4**, 6133–6150 (2016)
2. Bouwmans, T.: Traditional and recent approaches in background modeling for foreground detection: an overview. Comput. Sci. Rev. **11**, 31–66 (2014)
3. Wren, C.R., Azarbayejani, A., Darrell, T., Pentland, A.P.: Pfinder: real-time tracking of the human body. IEEE Trans. Pattern Anal. Mach. Intell. **19**(7), 780–785 (1997)
4. Kim, H., Sakamoto, R., Kitahara, I., Toriyama, T., Kogure, K.: Robust foreground extraction technique using Gaussian family model and multiple thresholds. In: Asian Conference on Computer Vision, pp. 758–768. Springer (2007)
5. Stauffer, C., Grimson, W.E.L.: Adaptive background mixture models for real-time tracking. In: IEEE Conference on Computer Vision and Pattern Recognition, CVPR, pp. 246–252. IEEE (1999)
6. Oliver, N.M., Rosario, B., Pentland, A.P.: A Bayesian computer vision system for modeling human interactions. IEEE Trans. Pattern Anal. Mach. Intell. **22**(8), 831–843 (2000)
7. Xu, Z., Gu, I.Y.H., Shi, P.: Recursive error-compensated dynamic eigenbackground learning and adaptive background subtraction in video. Opt. Eng. **47**(5), 057001 (2008)
8. Lin, H.-H., Liu, T.-L., Chuang, J.-H.: A probabilistic SVM approach for background scene initialization. In: International Conference on Image Processing, ICIP 2002, vol. 3, pp. 893–896. IEEE (2002)
9. Wang, J., Bebis, G., Miller, R.: Robust video-based surveillance by integrating target detection with tracking. In: Conference on Computer Vision and Pattern Recognition Workshop, CVPRW 2006, pp. 137–137. IEEE (2006)
10. Zhu, T., Zeng, P.: Background subtraction based on non-parametric model. In: 4th International Conference on Computer Science and Network Technology ICCSNT, 2015, vol. 1, pp. 1379–1382. IEEE (2015)
11. Elgammal, A., Harwood, D., Davis, L.: Non-parametric model for background subtraction. In: European Conference on Computer Vision, ECCV 2000, pp. 751–767. Springer (2000)
12. Lee, J., Park, M.: An adaptive background subtraction method based on kernel density estimation. Sensors **12**(9), 12279–12300 (2012)
13. Heikkila, M., Pietikainen, M.: A texture-based method for modeling the background and detecting moving objects. IEEE Trans. Pattern Anal. Mach. Intell. **28**(4), 657–662 (2006)
14. Kim, K., Chalidabhongse, T.H., Harwood, D., Davis, L.: Background modeling and subtraction by codebook construction. In: International Conference on Image Processing, ICIP 2004, vol. 5, pp. 3061–3064. IEEE (2004)
15. Kim, K., Chalidabhongse, T.H., Harwood, D., Davis, L.: Real-time foreground–background segmentation using codebook model. Real-Time Imaging **11**(3), 172–185 (2005)
16. Barnich, O., Van Droogenbroeck, M.: Vibe: a universal background subtraction algorithm for video sequences. IEEE Trans. Image Process. **20**(6), 1709–1724 (2011)
17. Droogenbroeck, M.V., Paquot, O.: Background subtraction: experiments and improvements for vibe. In: IEEE Computer Society Conference on Computer Vision and Pattern Recognition Workshops, CVPRW 2012, pp. 32–37. IEEE (2012)

18. Yang, S., Hao, K., Ding, Y., Liu, J.: Improved visual background extractor with adaptive range change. Memet. Comput. **10**(1), 53–61 (2018)
19. Ojala, T., Pietikainen, M., Maenpaa, T.: Multiresolution gray-scale and rotation invariant texture classification with local binary patterns. IEEE Trans. Pattern Anal. Mach. Intell. **24**(7), 971–987 (2002)
20. Wang, L., Pan, C.: Fast and effective background subtraction based on εLBP. In: International Conference on Acoustics, Speech, and Signal Processing, ICASSP 2010, March 2010
21. Wang, L.F., Wu, H.Y., Pan, C.H.: Adaptive εLBP for background subtraction. In: Asian Conference on Computer Vision, ACCV 2010, pp. 560–571. Springer (2010)
22. Heikkilä, M., Pietikäinen, M., Schmid, C.: Description of interest regions with local binary patterns. Pattern Recognit. **42**(3), 425–436 (2009)
23. Xue, G., Sun, J., Song, L.: Dynamic background subtraction based on spatial extended center-symmetric local binary pattern. In: 2010 IEEE International Conference on Multimedia and Expo, ICME 2010, pp. 1050–1054. IEEE (2010)
24. Yin, H., Yang, H., Su, H., Zhang, C., et al.: Dynamic background subtraction based on appearance and motion pattern. In: IEEE International Conference on Multimedia and Expo Workshops, ICMEW 2013, pp. 1–6. IEEE (2013)
25. Bilodeau, G.-A., Jodoin, J.-P., Saunier, N.: Change detection in feature space using local binary similarity patterns. In: International Conference on Computer and Robot Vision, CRV 2013, pp. 106–112. IEEE (2013)
26. St-Charles, P.-L., Bilodeau, G.-A.: Improving background subtraction using local binary similarity patterns. In: IEEE Winter Conference on Applications of Computer Vision WACV, 2014, pp. 509–515. IEEE (2014)
27. Wang, B., Dudek, P.: A fast self-tuning background subtraction algorithm. In: Proceedings of the IEEE Conference on Computer Vision and Pattern Recognition Workshops, CVPRW 2014, pp. 395–398. IEEE (2014)
28. Sedky, M., Moniri, M., Chibelushi, C.C.: Spectral-360: a physics-based technique for change detection. In: Proceedings of the IEEE Conference on Computer Vision and Pattern Recognition Workshops, CVPRW 2014, pp. 399–402 (2014)

Selection of Television Channels for Product Promotion: A Fuzzy-TOPSIS Approach

Arshia Kaul, Sugandha Aggarwal and P. C. Jha

Abstract Today, there are innumerable media that are available to the firm which is advertising their product. The decision that needs to be made is regarding the choice of the media for advertising. There are many traditional and new media which are available for advertising. In different scenarios, different media can be used for advertising and for creating a different impact. In this research paper, we present a case of a firm which advertises its product through television. The decision that needs to be made is for the selection of the appropriate channels in television for advertising. This particular decision of choice of appropriate channels is under the influence of many factors. Thus, it is a multi-criteria decision-making problem. The firm needs to evaluate different potential channels and make the decision of optimum combination of channels as per their product. This decision is dependent on multiple criteria. We have used Fuzzy-Technique for Order Preference by Similarity to Ideal Solution (F-TOPSIS) to make the real-life decision of choosing set of appropriate channels for a firm advertising for their men's product. F-TOPSIS helps to give a quantitative value to the performance of each channel with respect to the different criteria.

Keywords Television · Fuzzy-TOPSIS · Selection

A. Kaul (✉)
ASMSOC NMIMS University, Mumbai, India
e-mail: kaularshia25@gmail.com

S. Aggarwal
LBSIM, 11/07, Dwarka Sector 11, Near Dwarka Sector 11 Metro Station, New Delhi 110075, Delhi, India
e-mail: sugandha_or@yahoo.com

P. C. Jha
Department of Operational Research, University of Delhi, Delhi, India
e-mail: jhapc@yahoo.com

© Springer Nature Singapore Pte Ltd. 2020
K. N. Das et al. (eds.), *Soft Computing for Problem Solving*,
Advances in Intelligent Systems and Computing 1048,
https://doi.org/10.1007/978-981-15-0035-0_51

621

1 Introduction

A very important decision in front of firms which are advertising for their products is in which one particular medium or combination of media is suitable for advertising their product(s). There are many media which are available in front of the firm. There are many media under different classifications such as traditional and new media. Further, each day there is some new version of different media being added and which are available for advertising. The question that arises in front of firms which are advertising for their product, that which is the best media that can be used for advertising or what combination of media must be used for advertising. Here, in particular, we are considering the case of firm advertising their men's product and wanting to make an appropriate choice of the set of channels in television for advertising.

To assist real-life decisions in advertising, many researchers have extensively worked on the television advertisement allocation problem [1–3].

It is important to propose a methodology for the selection of television channel for advertising a firm's advertising for its men's product because:

- The television industry has grown superlatively over the years and the stakes for advertising are very high and there is limited scope for making any error in the decision.
- In the existing literature selection of appropriate television channels prior to placement of advertisement in them has been considered sparsely (refer Sect. 2).
- Moreover, the subjectivity in real-life decision-making needs to be considered quantitatively.

The methodology presented is designed to select the appropriate television channels which will be used for placement of advertisements over a planning horizon in advertising breaks of programs aired on the multiple channels for a variety of products. The case has been applied to a real-life case to show the operational ease of the methodology. To the best of our knowledge there has been limited research which has considered selection of television channels for placement of advertisements in television. Moreover, to consider the subjectivity in real life we have considered the decision-making to be fuzzy in nature. Here, in our paper, we proposed a Multi-Attribute Decision-Making (MADM) technique, namely, Fuzzy-Technique for Order Preference by Similarity to Ideal Solution (F-TOPSIS) to select the appropriate set of channels to be used for advertising.

The remaining paper is presented in the following sections: Sect. 2 presents the literature review. Section 3 gives the problem definition. Section 4 presents the solution methodology. Section 5 presents the case study. Section 6 concludes the paper and provides the future scope.

2 Literature Review

There has been substantial research that has been done in television advertising. Most of them have concentrated on the allocation of advertisements in the slots available keeping different objectives in mind [1–3]. There is only limited literature that has considered the case of selection of either media from those available for advertising or selecting a specific component of a particular medium. The selection is carried out based on multiple criteria, thus it becomes a multiple criteria decision-making.

Ngai [4] discussed the AHP technique for selection of the best website for advertising online. The decision was based on multiple criteria such as impression rate, monthly cost, audience fit, content quality, and look and feel of the websites. A case was presented for the proposed methodology and would be beneficial for the managers in this field. Lin and Hsu [5] in their study discussed model for selection of Internet advertising networks. Here, in this case analytic hierarchy process is used for finding the weights for the criteria and the final ranking for the alternatives is done using the grey relational analysis.

For instance, [6] proposed Analytic Network Process (ANP) technique selection of medium for advertising, incorporating both qualitative and quantitative criteria for selection. A multi-level network is developed wherein each of the criteria is composed of several sub-criteria for selection.

Liao and Chang [7] developed an Analytic Network Process (ANP) selection technique for selection of televised sportscasters (commentator) for the Olympic games. This selection was based on 12 appropriate criteria which were considered on the basis of the discussion with a group of senior executives. Liao and Chang [8] discussed the selection of best capabilities out of those available from those available in case of a TV-shopping company. They proposed ANP for selection wherein multiple financial and non-financial criteria are taken into consideration while making the decisions. The real-life case study is presented for a Taiwan TV-shopping company.

Jeng and Chiu [9] in their research discussed a procedure which had selection of television channels through AHP based on multiple criteria in the first step and placement of advertisements in multiple channels in the second step. The model was developed for maximizing the opportunity to be seen. This methodology was developed for advertising different sections of the theme park in Taiwan and was solved using excel solver. Liao et al. [10] proposed an ANP methodology for selecting the program suppliers in television given the interdependence in criteria. A case for the Taiwan TV industry was presented to show the application of the proposed methodology.

Hsu and Kuo [11] in their research first considered the Nominal Group Technique (NGT) for the selection of criteria for evaluation of the advertising agency. Subsequently, ANP is considered for the selection of an advertising agency.

Hsu [12] in their paper discussed the selection of media agency in a two-step procedure. In the first step, the analytic network process is applied for determining the relative weights of the criteria. In the second step, the Grey Relational Analysis

(GRA) is used for the selection of the media agency. The application of the selection is demonstrated for a Taiwanese Beverage Company.

Tavana et al. [13] discussed an integrated Fuzzy-ANP and COPRAS-G for the selection of social media networks for advertising. Here, Fuzzy-ANP is used for setting weights for each of the criteria for selection. Subsequently, COPRAS-G is used to determine the final ranking for social media websites. The case is for the largest airlines in the Middle-East trans-Gulf Airline for their advertising.

From the extant literature, it can be observed that there is research that has considered multi-criteria decision-making problem related to marketing. There are various different decisions that have been considered but there is limited research that has been considered for the selection of television channels for the advertising a product of a firm. Moreover, the vagueness in real-life decision-making has been considered in fuzzy nature which has not been considered in the extant literature. The proposed research, therefore, has the following distinct features: (i) it considers the selection of channels for advertising from the perspective of the advertising firm which wants to advertise their product (ii) the vagueness in the decision-making is considered by considering the fuzzy environment in the decision-making (iii) the television channel selection is carried out using F-TOPSIS, a method that is easy to understand and incorporates real-life subjectivity in decision-making. Keeping this research gap in mind we have presented this case.

3 Problem Definition

The problem under consideration is the selection of television channels for the advertisement placement for a firm advertising its product. The appropriate selection of channels from those available is based on multiple criteria. The question that is posed in front of the Decision-Makers (DMs) is how the selection decision is made for particular television channels for advertising, such that there is maximum return on investment, i.e., maximum benefits and minimum cost, i.e., minimum negative for the firm. This decision is a complicated one with conflicting preferences and imprecision in the nature of decisions of the DMs. Hence a multi-criteria decision-making scenario is developed in this paper with the following details:

- What are the criteria on the basis of which the selection of channels for advertisements will be made?
- How will the performance of the channel alternatives be done?
- What is the scale on which the performance of alternatives be measured?
- What are the important weights given to each criteria?

The objective is to evaluate the television channel based on the multiple criteria for selection. For initial screening 35 television channels were selected. The potential television channels for advertising are channels from different genres: (i) news, (ii) general entertainment, (iii) sports, (iv) knowledge based, and (v) spiritual. The problem faced by the firm is to determine how to evaluate the television channels quantitatively. This is the particular problem addressed in this study and the research methodology followed is presented in the section that follows.

4   Research Methodology

For selecting the television channel for a firm advertising its product, incorporating the various factors under imprecise environment a multi-criteria decision-making problem is proposed. In this paper, the Fuzzy-Technique for Order Preference by Similarity to Ideal Solution (F-TOPSIS) [14] approach has been applied. The methodology adopted broadly has the following steps:

1. Identifying the criteria which would be used in assessment of the television channels.
2. Calculation of weights of the criteria
3. Evaluation of the available alternatives of television channels
4. Final ranking of the television channels based on closeness coefficient.

4.1   Identifying the Criteria for the Assessment of the Television Channels

The first step in the proposed methodology is to determine the criteria for assessment of television channels for advertising.

As per the discussion with the decision-makers the following are the criteria for selection of television channels: (i) Reputation of Channel (C1), (ii) Coverage of the Channel (C2), (iii) Reach to the target audience (C3), (iv) Repetitions of the advertisements (C4), (v) Less Clutter of same product Advertisements (C5), (vi) Less Clutter of ad space (C6), (vii) Less Length of advertisement breaks (C7), (viii) Quality of Program Content (C8), (ix) Cost effectiveness (C9), (x) Gender Appropriateness (C10), (xi) Less Time duration of production (C11), (xii) Quality of TV channel personnel (C12), (xiii) Financial Status of the channel (C13), (xiv) Ability to complete orders of advertisers (C14), and (xv) Ability to complete orders timely (C15).

4.2   Application of F-TOPSIS Methodology

For evaluation of the television channel alternatives the F-TOPSIS methodology is used. The decision-makers give the weights of the criteria based on their judgment based on the linguistic scale. There are advantages of using the fuzzy nature of the problem, as it helps to incorporate the real-life vagueness. The fuzzy numbers have been used for representing the weights of each criterion and also the performance value. The linguistic words are converted into the triangular fuzzy numbers and are used for analyzing. In the final ranking, the television channels are ranked on the basis of the values in the decision matrix of F-TOPSIS. A per F-TOPSIS the best television channels are the one which has the highest closeness coefficient.

5 Numerical Validations

The F-TOPSIS methodology has been applied for the selection of a set of top 5 television channels from the 35 available for advertising of men's product by the firm based on 15 criteria. In the first step, the weights are computed for the 15 criteria given the linguistic scale [14].

Next, the fuzzy decision matrix is constructed, which defines the performance of each channel alternative with respect to each criterion. DMs fill up the decision matrix based on the fuzzy linguistic scale which is in terms of triangular fuzzy numbers. The decision matrix has not been shown here owing to the lack of readability of the large data tables by user. The authors would be willing to share the details with interested readers.

In the subsequent step, the normalized decision matrix is obtained. For the normalized decision matrix, readers can refer to formula given in [14].

The Positive Ideal Solution(PIS) and the Negative Ideal Solution (NIS) are calculated. In the given problem all are described as benefit criteria. The values of PIS and NIS are (1,1,1) and (0,0,0), respectively.

Then the distance between the value of performance of each of the alternatives from the PIS and NIS is calculated. The distances of the alternatives are given in Table 1.

The Positive Ideal Solution(PIS) and the Negative Ideal Solution (NIS) are calculated. In the given problem all are described as benefit criteria. The values of PIS and NIS are (1,1,1) and (0,0,0), respectively.

Based on distances the closeness coefficient is computed. The top 5 ranked are selected for advertising. The values of the closeness coefficient are given in Table 2. The higher the closeness coefficient value the better is the performance of the alternative.

Table 1 Distance of alternatives from ideal and anti-ideal solution

Distance from FPIS d_i^*				Distance from FPIS d_i^-			
A1	8.12	A18	12.87	A1	8.11	A18	2.86
A2	8.09	A19	12.66	A2	8.14	A19	3.08
A3	7.32	A20	10.95	A3	8.94	A20	5.02
A4	8.41	A21	10.73	A4	7.82	A21	5.26
A5	8.86	A22	9.29	A5	7.33	A22	6.85

(continued)

Table 1 (continued)

Distance from FPIS d_i^*				Distance from FPIS d_i^-			
A6	9.37	A23	9.34	A6	6.76	A23	6.8
A7	10.24	A24	8.67	A7	5.79	A24	7.4
A8	9.1	A25	9.57	A8	7.07	A25	6.55
A9	9.6	A26	9.92	A9	6.52	A26	6.15
A10	9.09	A27	9.59	A10	6.97	A27	6.52
A11	9.22	A28	10.68	A11	6.88	A28	5.34
A12	10.97	A29	9.21	A12	5.02	A29	6.95
A13	8.79	A30	10.16	A13	7.39	A30	5.88
A14	9.82	A31	10.78	A14	6.25	A31	5.19
A15	11.04	A32	8.84	A15	4.92	A32	7.34
A16	10.77	A33	10.72	A16	5.22	A33	5.3
A17	11.04	A34	10.52	A17	4.93	A34	5.52
		A35	10.16			A35	5.88

Table 2 Closeness coefficient of alternatives

Closeness coefficient			
A1	0.4998	A18	0.1821
A2	0.5014	A19	0.1959
A3	0.5499	A20	0.3145
A4	0.4819	A21	0.329
A5	0.453	A22	0.4247
A6	0.4193	A23	0.4213
A7	0.3612	A24	0.4605
A8	0.4372	A25	0.4061
A9	0.4044	A26	0.3828
A10	0.4342	A27	0.405
A11	0.4274	A28	0.3331
A12	0.3141	A29	0.43
A13	0.4565	A30	0.3664
A14	0.3889	A31	0.3251
A15	0.3085	A32	0.4537
A16	0.3265	A33	0.331
A17	0.3088	A34	0.3441
		A35	0.3665

6 Conclusion

In this paper, the selection of television channels for advertisement placement for a firm advertising for its product is done. The selection of television channels depends on multiple criteria. A multi-criteria decision-making technique for taking the decision is used under vague real-life situation. Given there are multiple alternatives of television channels out of which a few are to be chosen based on their evaluation of the multiple criteria, here in this paper, the F-TOPSIS technique is utilized. The set of television which is to be selected for advertisement placement for the case of a men's product is considered. This methodology has been presented for a particular case, in the future it is possible to use the similar methodology for a case similar to the situation. Also in addition to the selection of television channel allocation of advertisements can be carried out through a mathematical model which could be proposed in the future.

References

1. Tiedge, J.T., Ksobiech, K.J.: The "lead-in" strategy for prime-time TV: does it increase the audience? J. Commun. **36**(3), 51–63 (1986)
2. Bollapragada, S., Bussieck, M.R., Mallik, S.: Scheduling commercial videotapes in broadcast television. Oper. Res. **52**(5), 679–689 (2004)
3. Ghassemi Tari, F., Alaei, R.: Scheduling TV commercials using genetic algorithms. Int. J. Prod. Res. **51**(16), 4921–4929 (2013)
4. Ngai, E.W.T.: Selection of web sites for online advertising using the AHP. Inf. Manag. **40**(4), 233–242 (2003)
5. Lin, C.T., Hsu, P.F.: Selection of internet advertising networks using an analytic hierarchy process and grey relational analysis. Int. J. Inf. Manag. Sci. **14**(2), 1–16 (2003)
6. Coulter, K., Sarkis, J.: Development of a media selection model using the analytic network process. Int. J. Adv. **24**(2), 193–215 (2005)
7. Liao, S.K., Chang, K.L.: Select televised sportscasters for Olympic Games by analytic network process. Manag. Decis. **47**(1), 14–23 (2009)
8. Liao, S.K., Chang, K.L.: Selecting key capabilities of TV-shopping companies applying analytic network process. Asia Pacific J. Mark. Logist. **21**(1), 161–173 (2009)
9. Jeng, Y.C., Chiu, F.R.: Allocation model for theme park advertising budget. Qual. Quant. **44**(2), 333–343 (2010)
10. Liao, S.K., Chang, K.L., Tseng, T.W.: Optimal selection of program suppliers for TV companies using an analytic network process (ANP) approach. Asia-Pacific J. Oper. Res. **27**(6), 753–767 (2010)
11. Hsu, P.F., Kuo, M.H.: Applying the ANP model for selecting the optimal full-service advertising agency. Int. J. Oper. Res. **8**(4), 48–58 (2011)
12. Hsu, P.F.: Selection model based on ANP and GRA for independent media agencies. Qual. Quant. **46**(1), 1–17 (2012)
13. Tavana, M., Momeni, E., Rezaeiniya, N., Mirhedayatian, S.M., Rezaeiniya, H.: A novel hybrid social media platform selection model using fuzzy ANP and COPRAS-G. Expert Syst. Appl. **40**(14), 5694–5702 (2013)
14. Chen, C.T.: Extensions of the TOPSIS for group decision-making under fuzzy environment. Fuzzy Sets Syst. **114**(1), 1–9 (2000)

Implementation of ACO Tuned Modified PI-like Position and Speed Control of DC Motor: An Application to Electric Vehicle

Geetha Mani ⓘ

Abstract The modified PI-like control scheme has been derived from classic Internal Model Control (IMC), having two tuning parameters namely controller gain K_c and reset time τ_I. The controller action resembles like PI control action with inherent reset windup protection and dead time compensation, except the inverse of the process model gain is set as controller gain and dominant time constant of the process model set as reset time. The controller structure is feasible enough to implement in hardware. When an accurate process model is not available, the metaheuristic algorithms like Ant Colony Optimization (ACO) is applied to find the optimal tuning parameters. The performance of proposed control scheme for both speed control and position tracking is compared and analyzed under different operating conditions with well-known Proportional-Velocity (PV) control scheme and Two-Degree-of-Freedom (2-DoF) PID structure. The experimental evaluations have been demonstrated via computer-interfaced QUBE-servo DC Motor through Hardware in Loop (HiL) implementation.

Keywords Modified PI-like · 2-DoF · QUBE-servo DC motor · IAE · ISE · ACO

1 Introduction

From past decades, research works in automobile industries focusing on Electrical Vehicles (EVs) increased because of the serious environmental issues emitted by the petroleum product vehicles and statistics shows that the fossil fuels could last for five more decades. We are in a situation to adopt renewable resources than relying on fossil fuels and being renewable resources are clean; it is safer for our ecosystem. Fuel cell-powered vehicles and hybrid EVs are in top priority for automakers

G. Mani (✉)
School of Electrical Engineering, Vellore Institute of Technology, Vellore 632014, Tamil Nadu, India
e-mail: geethamr@gmail.com

© Springer Nature Singapore Pte Ltd. 2020
K. N. Das et al. (eds.), *Soft Computing for Problem Solving*,
Advances in Intelligent Systems and Computing 1048,
https://doi.org/10.1007/978-981-15-0035-0_52

like Tesla, Mercedes-Benz, etc., to lead other companies for the future. In gasoline-powered vehicles components like the combustion engine, carburetor, cooling systems, exhaust systems, etc., are used to run the vehicles. But EV acquires power from the batteries to the motor to run the vehicles. The controller in EVs controls the speed and acceleration of the vehicle as the carburetor does in gasoline-powered vehicles and it can also reverse the direction of the rotor in the motor. When the brakes are applied in EVs, the KE of motion helps to recharge the battery which is called as regenerative brakes where the motor performs as the generator. This method also helps to reduce the wear on brakes, so that it reduces the maintenance cost.

EV requires high dynamic capabilities due to rapid acceleration and ramps. DC brushless motor and induction machines are the main candidates for such applications. AC motors are less expensive with complex structure, but to power the battery, we need inverters which cost higher than the DC setup. Hence, DC motors are preferred more the AC to reduce the overall cost. This inspires the project to work on designing a controller for a DC servo motor.

The well-known PV controller and 2-DoF controller are pointed out by [1–3]. The model-based PI-like control scheme was proposed in [4]. From their investigations, the following constraints are concluded. The proposed controller can be only used for a stable single-loop model. It has only one tuning parameter (controller gain). It provides improved response for higher order systems only. The same structure works well for speed control, because the process model is first-order and type zero system. But the same structure fails in type-1 and second-order system that is position tracking. Hence, the standard structure has been fixed in our proposed work with two tuning parameters; proportional gain and reset time in the closed-loop with positive feedback. Based on the process model, the above-mentioned tuning parameters are computed for speed control. ACO has been implemented to find the optimal tuning values for position tracking.

The primary objective is to design and implement a QUBE-Servo standard modified PI-like controller for both position tracking and speed control in MATLAB-Simulink model and implement the same in DC servo motor through Hardware in Loop (HiL). The end results are compared and analyzed with the performance of well-known 2-DoF PID and PV controller.

The organization of the paper is given below. The glimpse of the mathematical model of DC servo motor details is presented in Sect. 2. Section 3 gives the ACO tuning parameters and cost function details. The design structure of 2-DoF PID, PV, and modified PI-like control is described in Sect. 4. Both simulation results and experimental results were discussed in Sect. 5. Section 6 states the conclusion and future work.

2 Mathematical Model of Dc Servo Motor

The circuit diagram for the servo DC motor with load is shown in Fig. 1. The DC servo motor mathematical model is given in Quanser work manual [5]. The transfer

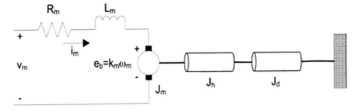

Fig. 1 Circuit diagram of QUBE-servo DC motor

function block diagram of the QUBE DC servo motor for position output and speed output based on first principle modeling is shown in Figs. 2 and 3. These transfer function blocks are considered as a process model in simulation mode.

The QUBE DC servo voltage to speed and position transfer function is given in Eqs. (1) and (2), respectively.

$$P(s) = \frac{K}{(\tau s + 1)} \tag{1}$$

$$P(s) = \frac{K}{s(\tau s + 1)} \tag{2}$$

where

K 23:2 rad/(V-s) is the model steady-state gain
T 0:13 s is the model time constant, obtained from [5]

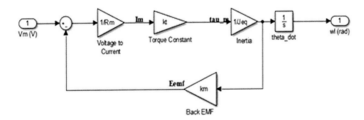

Fig. 2 Block diagram represents the transfer function of servo motor position

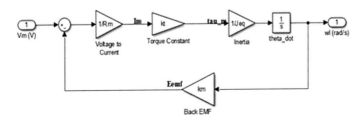

Fig. 3 Block diagram represents the transfer function of servo motor speed

Table 1 DC servo motor specifications

Symbols	Terminology	Values
R_m	Terminal resistance	6.3 Ω
K_t	Torque constant	0.036 N-m/A
K_m	Motor back-emf constant	0.036 V/(rad/s)
J_m	Rotor inertia	4.0×10^{-6} kg-m^2
L_m	Rotor inductance	0.85 mH
m_h	Load hub mass	0.0087 kg
r_h	Load hub mass	0.0111 m
J_h	Load hub inertia	1.07×10^{-6} kg-m^2
m_d	Mass of disk load	0.054 kg
r_d	Mass of disk load	0.0248 m
K	Model steady-state gain	–
τ	Model time constant	–

The DC servo motor parameter specification is provided in Table 1.

3 Ant Colony Optimization (ACO)

An algorithm to solve optimization problems based on the way that ants indirectly communicate directions to each other in search of food. The shortest path is discovered via pheromone trails. Each ant moves at random path, more the number of ants, more the possibility of discovering the shortest path. More pheromone on path increases the probability of path being followed by the rest of the ants, which is the shortest path among all other paths. ACO activities are given below

- Set parameters
- Initialize the pheromonc trails
- Construct the ant solution
- Update the pheromones
- Daemon actions

The probability of transition to j top from I top is expressed in [6]

$$p_{ij}^k = \frac{\tau_{ij}^\alpha \eta_{ij}^\beta}{\sum \tau_{ij}^\alpha \eta_{ij}^\beta}, j \in N_i^k$$
$$P_q^k = 0$$
$$j \notin N_i^k \tag{3}$$

where

Table 2 Fitness function for various ants

No. of ants	No. of iteration	Modified PI-like	2-DoF PID
		Best cost/fitness	Best cost/fitness
10	10	11.3974	14.3225
20	10	12.2738	14.1225
30	10	10.5868	13.6585
40	**10**	**10.4057**	**12.7671**

$\tau_{i,j}$ the amount of phenomena on edge I, j

α parameter to control the influence of $\tau_{i,j}$

     0—greedy algorithm

β parameter to control the influence of $\eta_{i,j}$

     0—the only pheromone is at work → quickly leads to stagnation

$\eta_{i,j}$ the desirability of edge I, j

ACO tuning parameters considered for tuning controllers are given in Table 2. These values are chosen based on optimal fitness function.

- number of iteration $= 10$
- number of ants $= 40$
- alpha $= 0.8$
- beta $= 0.2$
- evaporation rate $= 0.9$
- number of parameters to be tuned $= 5$
- cost function is Integral Square Error (ISE)

4 Design of Controller

4.1 Modified PI-like Control

The unity (single-loop) feedback control block diagram is shown in Fig. 4. Here, $G_P(s)$ is the process transfer function and $G_C(s)$ is the controller's transfer function. Let the model $G_m(s)$ be the model for the process $G_P(s)$, so that a closed-loop transfer function $G_{cl}(s)$ is obtained as in Eq. (4).

$$G_{cl}(s) = \frac{G_c(s)G_m(s)}{1 + G_c(s)G_m(s)} \tag{4}$$

where $G_{cl}(s)$ can also be written as

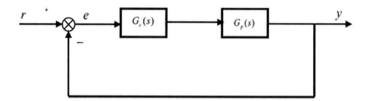

Fig. 4 Block diagram of single-loop negative feedback system

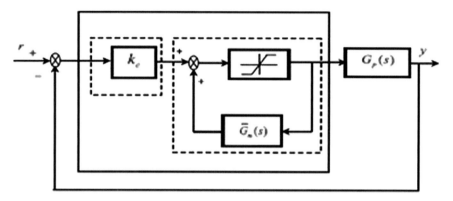

Fig. 5 Block diagram of modified PI-like model

$$G_c(s) = y/r \tag{5}$$

Arun et al. [4] derived the modified PI-like control structure as shown in Fig. 5 from the Direct Synthesis (DS) approach. From that, the equation has written as in Eq. (6)

$$G_c(s) = k_m^{-1} \frac{1}{1 - G_m(s)} \tag{6}$$

And the Eq. (1) can be rewritten as

$$G_c(s) = \frac{u(s)}{e(s)} = kc \frac{1}{1 - G_m(s)} \tag{7}$$

where the k_c is the controller gain, which is the k_{m-1} (inverse of model steady-state gain) and gain free part $\frac{1}{1-G_m(s)}$ is a closed-loop positive feedback, which ensures to eliminate offsets.

The parameter k_c can be tuned by the designer according to their specification of the model. Without any additional tuning parameters in the controller, an offset-free response can be achieved from a closed-loop model. The features of the modified PI-like controller are summarized below based on [4]

- It has only one tuning parameter (k_c) which performs satisfactory and has good robustness from other existing PI-like controllers
- Rather generating a pure integral action, it delays the action of the integral function
- The reset windup protection of the controller makes it easier to realize the reset configuration
- The $G_m(s)$ have to be separated into two parts, steady-state gain and gain free part

The constraints identified are

- The proposed controller can be only used for stable single-loop models
- It can have only one tuning parameter to control.

The Simulink block diagram of a modified PI-like controller is shown in Fig. 6, where K is the inverse of model steady-state gain and "tau" is the time constant of the process model. Arun et al. [4] proposed that inverse of model gain considered has controller gain and gain free part placed in closed-loop positive feedback. In this paper, we proposed the standard structure shown in Fig. 7 is implemented for both speed (first-order system) and position control (second-order system).

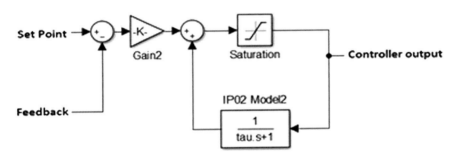

Fig. 6 Block diagram of modified PI-like controller

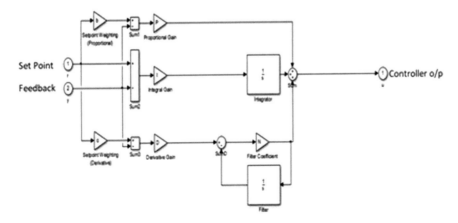

Fig. 7 Block diagram of 2-DoF PID Controller

PV Control design

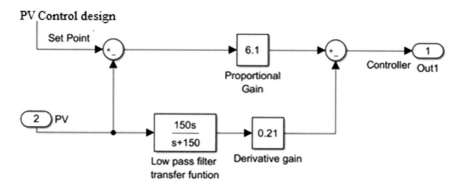

Fig. 8 Block diagram of the PV controller

Table 3 Tuning parameters for speed control

Controller	K_P	K_I	K_d	α	β
Modified PI-like	1/23.2	0.131	–	–	–
2-DOF	0.0308	0.609	0.0003 (N = 30)	0.5	0.5
PV	2.5	–	0.05	–	–

The Simulink block diagram of a 2-DoF PID controller is given in Fig. 7. The optimal tuning parameters are obtained using relay feedback test. The well-known PV controller structure [5] in Simulink is shown in Fig. 8. The proportional gain and derivative gain calculated by desired time domain specifications like peak overshoot and peak time. The tuning parameters used for DC servo speed control is given in Table 3.

Considered desired peak overshoot is 2.5% and the desired peak time is 0.15 s. The overall closed-loop transfer function is given in Eq. (8).

$$\frac{K_{kp}}{\tau_s^2 + (1 + K_{kd})s + K_{kp}} \tag{8}$$

This is a second-order transfer function. Recall the standard second-order transfer function

$$\frac{Y(s)}{R(s)} = \frac{\omega_n^2}{s^2 + 2\xi w_n s + \omega_n^2} \tag{9}$$

The characteristics equation of DC servo motor closed-loop transfer function is

$$\tau_s^2 + (1 + K_d)s + K k_{kd} \tag{10}$$

Equation 10 is restructured into the form in Eq. (11)

$$s^2 + \frac{(1 + K_{kd})s}{\tau} + \frac{Kk_p}{\tau} \tag{11}$$

Equating (11) with the standard second-order transfer function (9), the following equations are obtained:

$$\frac{Kk_p}{\tau} = \omega_n^2 \tag{12}$$

$$\frac{(1 + K_{kd})s}{\tau} = 2\xi w_n \tag{13}$$

By solving the Eqs. (12) and (13), proportional and derivative gain values are obtained

$$k_p = \frac{\tau \omega_n^2}{K} \tag{14}$$

$$K_{kd} = \frac{2\xi w_n \tau - 1}{K} \tag{15}$$

5 Results and Discussion

5.1 DC Servo Motor Speed Control

Simulation Results and Discussion
The comparative servo performance analysis of the aforementioned three controllers for low-speed control in the simulation environment is shown in Fig. 9. Figure 10

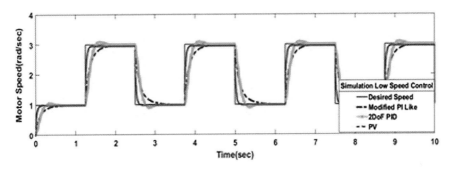

Fig. 9 Comparison of low-speed control in simulation

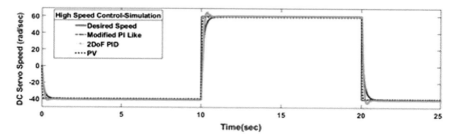

Fig. 10 Comparison of high-speed control in simulation

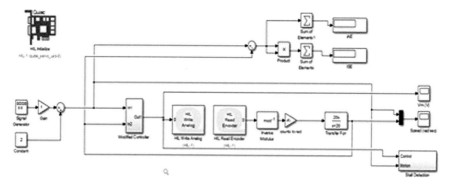

Fig. 11 HiL implementation of speed control using modified PI-like controller

shows the controller's output to achieve the desired speed control in the simulation environment.

Experimental Results and Discussion

The demonstration of both low- and high-speed controller in real time with the respective Simulink model has been implemented through HiL in Quanser QUBE DC Servo system for the Modified PI, 2-DoF PID, and PV controller. Figure 11 shows the overview of HiL implementation in MATLAB environment. Figure 12 shows the comparative low-speed tracking analysis of the aforementioned three controllers in the real-time environment. From the zoom view, it was inferred that PV has offset. Other two controllers provided the cyclic response around the set point.

The comparative of all three controllers is presented in Fig. 13 in real-time environment. PV required more input voltage than other two controllers.

The comparative high-speed tracking analysis of all three controllers in real-time environment is shown in Fig. 14. 2-DoF PID provided maximum peak overshoot compared with the other two controllers. All three controllers track the given set point accurately in real time.

Figure 15 presents the controller's output for achieving high-speed control in HiL. It has inferred that PV spent more input energy to achieve the desired speed compared with both modified PI-like and 2-DoF. IAE and ISE are calculated to measure the

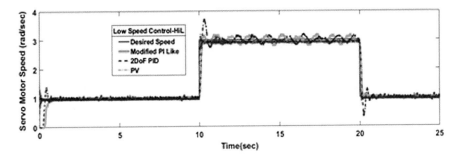

Fig. 12 Comparison of low-speed control in real time

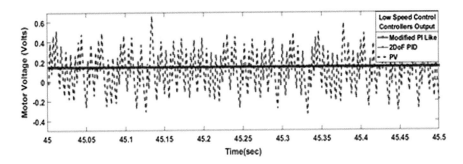

Fig. 13 Comparison of controller's output in low-speed control

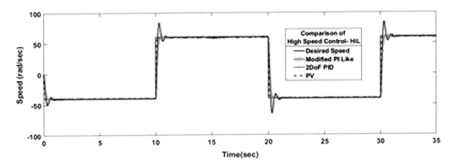

Fig. 14 Comparison of high-speed control in real time

performance of a controller both in simulation and real-time under low-speed and high-speed mode.

Tables 4 and 5 show the quantitative analysis of both low- and high-speed range of control. It has inferred from Table 4 that the modified PI-like controller shows less value of ISE and IAE compared with the other two with respect to real-time implementation. From Table 5, it has been inferred that PV controller shows low ISE and IAE values compared with others.

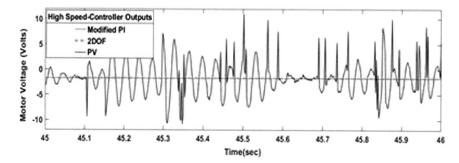

Fig. 15 Comparison of controller's output in high-speed control

Table 4 Quantitative analysis of low-speed control

Controller	IAE	ISE	IAE	ISE
Modified PI-like	2	4.001	2	3.999
2-DOF	2.025	4.102	2	3.999
PV	2.081	4.329	2.002	4.01

Table 5 Quantitative analysis of high-speed control

Controller	IAE	ISE	IAE	ISE
Modified PI-like	100	e+04	100	e+04
2-DOF	99.98	9997	100	e+04
PV	99.35	9870	99.22	9845

5.2 DC Servo Motor Position Tracking

Simulation Results and Discussion

The position tracking response and corresponding controller's output are presented in Figs. 16 and 17 in the simulation environment. 2-DoF PID provided oscillatory response and takes more time to settle. PV required more input voltage to track the desired position.

Experimental Results and Discussion

Figure 18 shows the HiL implementation of position control through MATLAB tool. Figure 19 shows the real-time response of position control of Modified PI-like control structure. The response shows that it fails to track the desired position. Hence, ACO is applied to find optimum tuning parameters of designed controllers.

The comparative position tracking responses and corresponding controller's output are shown in Figs. 20 and 21 in real-time environment. 2-DoF PID is absolutely unstable. Its controller output exceeds the maximum limits compared with the other

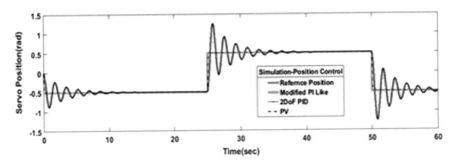

Fig. 16 Comparison of position control in simulation

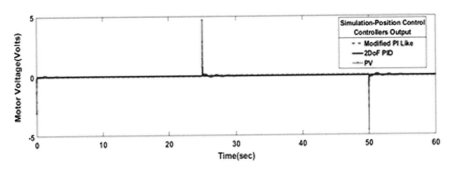

Fig. 17 Comparison of controller output in simulation

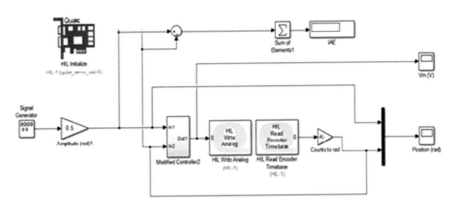

Fig. 18 HiL implementation of position tracking using modified PI-like controller

two controllers. ACO tuned position tracking responses and corresponding con-troller's output of all three controllers are presented in Figs. 22 and 23 in real-time environment. The maximum and minimum range DC servo motor input voltage range is ±15 V. ACO tuned controllers satisfied the imposed input constraints. All

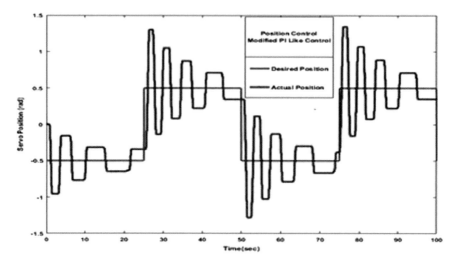

Fig. 19 Position tracking response of modified PI-like controller in real time

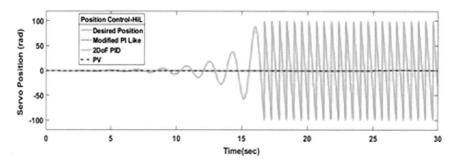

Fig. 20 Comparison of position control in real time

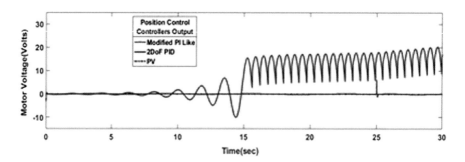

Fig. 21 Comparison of controller output in real time

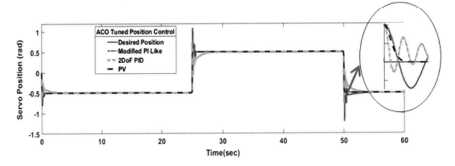

Fig. 22 Comparison of ACO tuned position control in real time

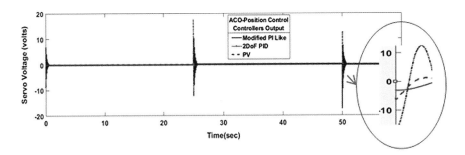

Fig. 23 Comparison of ACO tuned controller output in real time

three controller tracks the desired position quickly. The tuning parameters of the controllers in position tracking are listed in Table 6.

Table 7 shows the ACO tuned parameters of controller structure for position control. From Table 8, it has been inferred that PV controller gives the low value of

Table 6 Tuning parameters of position control

Controller	K_P	K_I	K_d	α	β
Modified PI-like (from model)	1/23.2	0.131	–	–	–
2-DOF (auto tune)	0.030808	0.6097	0.0003 (N = 30)	0.75	0.75
PV (from model)	6.1	–	0.21	–	–

Table 7 ACO applied tuning parameters of position control

Controller	K_P	K_I	K_d	α	β
Modified PI-like	3.0093	1.6626	–	–	–
2-DOF	2.6027	2.1900	0.6205 (N = 30)	0.9	0.5
PV	6.1	–	0.21	–	–

Table 8 Quantitative analysis for position tracking (without ACO)

Controller	Real-time		Simulation	
	IAE	ISE	IAE	ISE
Modified PI-like	-1.071	1.146	-1	0
2-DOF	40.36	1629	-1	1
PV	-1.003	1.006	-0.9984	0.9968

Table 9 Quantitative analysis for ACO tuned position tracking

Controller	Real-time		Simulation	
	IAE	ISE	IAE	ISE
Modified PI-like	-1	1	-0.9996	0
2-DOF	-1	1	-0.9985	0.997
PV	-1.003	1.006	-0.9992	0.9984

ISE and IAE with offset. 2-DoF and modified PI fail to track the desired position. From Table 9, it has inferred that modified PI-like and 2-DOF controller track the desired position with low ISE and IAE.

6 Conclusion

While designing a controller, most of the research works focused on optimizing the process output performance but very few paper works focused on optimizing the controller output's performance which is equally important. This helps to maintain the actuator's condition to perform in its optimum and reliable condition.

The modified PI-like controller provides an optimum controller output for both speed and position control, which makes it unique while comparing it with both 2-DoF PID and PV controller. For low-speed control, modified PI provides oscillatory response around the set point. On the other hand, PV could reach steady state within few seconds. But PV gives offset for both low- and high-speed control. During the change of the set-point period, PV controller output fluctuates a lot and exceeds the maximum limit; it leads to reducing the lifetime of actuators.

Furthermore, the work is extended to validate the response of position tracking for all the three controllers through HiL implementation. Modified PI-like and 2-DoF PID couldn't track the desired position. PV tracks the desired position with minimum offset. Hence, ACO has been applied to find the optimal tuning parameters for the designed controllers. ACO tuned modified PI-like and 2-DoF controllers outperform the classic PV with optimum controller output. Both controller structures provide feasibility for online implementation.

References

1. Chen, D., Seborg, D.E.: PI/PID controller design based on direct synthesis and disturbance rejection. Ind. Eng. Chem. Res. **41**(19), 4807–4822 (2002)
2. Miklosovic, R., Gao, Z.: A robust two-degree-of-freedom control design technique and its practical application. In: Conference Record of the 2004 IEEE Industry Applications Conference, 39th IAS Annual Meeting, pp. 1495–1502. IEEE, Seattle, USA (2004)
3. Araki, M., Taguchi, H.: Two-degree-of freedom PID controllers-their functions and optimal tuning. In: IFAC Workshop on Digital Control: Past, Present and Future of PID Control, Terrassa, Spain 2000, vol. 33, pp. 91–96. Elsevier (2000)
4. Pathiran, A.R., Prakash, J.: Design and implementation of a model-based PI-like control scheme in a reset configuration for stable single-loop systems. Can. J. Chem. Eng. **92**(9), 1651–1660 (2014)
5. Quanser homepage. http://www.quansershare.com
6. Geetha, M., Manikandan, P., Jerome, J.: Soft computing techniques based optimal tuning of virtual feedback PID controller for chemical tank reactor. In: IEEE Congress on Evolutionary Computation (CEC), pp. 1922–1928. IEEE, Beijing, China (2014)

A Bidirectional Converter for Integrating DVR-UCap to Improve Voltage Profile Using Fuzzy Logic Controller

T. Y. Saravanan and Ponnambalam Pathipooranam

Abstract Nowadays the control and automation persuade the severe economic issues on expertise high rating equipments. Especially in distribution sector, it is required to provide elevated quality electrical services to improve the most significant issues of Power Quality (PQ) called Voltage Profile at load end, namely, Voltage Sags and Swells. The drive system failures, tripping of equipment, industrial, and domestic equipment shutdowns are the foremost occurrences by PQ issues. To attain a superlative dynamic capability of a distribution system, a Dynamic Voltage Restorer (DVR) is connected in series to it for voltage reparation of sensitive loads. The voltage sags and swells are mitigated with the help of Ultra-Capacitor (UCap) which will give more superlative characteristics because of low energy density, high power, and an efficiency of 85–90%. The UCap is selected because of the finest energy source device in terms of allowing for long life, safety in operation time, wide temperature range, and less maintenance cost. A bidirectional DC to DC converter is used to integrate the DVR and UCap which results a stiff dc-link voltage and compensates the temporary voltage sag and voltage swell. The Simulink model has been designed for the proposed system with PI Controller and Fuzzy Logic Controller for enhancing power quality in distribution sector and it is compared.

Keywords DC to DC converter · Ultra-Capacitor (UCap) · PI and fuzzy logic controller · Dynamic Voltage Restorer (DVR)

1 Introduction

All over the world, the utilization of energy is increasing day by day because of population growth, technology enhancements, and agriculture, domestic, industrial

T. Y. Saravanan (✉) · P. Pathipooranam
SELECT, VIT, Vellore, India
e-mail: saravanan651988@gmail.com

P. Pathipooranam
e-mail: p.ponnambalam@gmail.com

© Springer Nature Singapore Pte Ltd. 2020
K. N. Das et al. (eds.), *Soft Computing for Problem Solving*,
Advances in Intelligent Systems and Computing 1048,
https://doi.org/10.1007/978-981-15-0035-0_53

loads also due to nonlinear loads invention. So that the disturbances like interruptions, flickering, voltage sags and swells, and power fluctuations are occurred in distribution sectors majorly [1]. Among these, voltage profile problem is the major issue in power quality, i.e., sags and swells caused by nonlinear loads. In general, a momentary decrease in the magnitude of root mean square voltage is called "voltage sag" and momentarily increase in voltage is called "voltage swell". According to the standards of IEEE, i.e., 1159-1195 and 519-1992: if the magnitude of supply voltage is dropped from 10 to 90%, voltage sag is occurred and if voltage rises from 110 to 180%, voltage swell is occurred [2]. In FACTs, there are many classifications under series compensation devices inside that the DVR is the most effective device which stores the energy through injecting voltage for the improvement of voltage profile at load ends. The restoration of voltage and power disruption avoidance by deducting time to few mille seconds are supported by DVR easily. The first DVR for series compensation in voltage source converter is used in North America. In this paper, the bidirectional DC to DC converter is used to integrate the DVR and UCap. Ucap is an Ultra-Capacitor majorly used for rechargeable energy storage at the DC terminal of a DVR [3]. There are many DC to DC converters are available among that buck–boost and CuK converters are considered because of superlative features. PI and fuzzy logic controllers are implemented for both converters by MATLAB/SIMULINK software.

2   DC to DC Converters

The DC to DC converters are proposed for integration and are grouped basically as follows [4].

It is must to follow any one of the above configuration (Table 1) for the all switching converters otherwise switching stages are complicated to take in by a one controlled switch. From Fig. 1, the features of the converters are as follows.

In Table 2, the features of converters are mentioned and therefore buck–boostand CuK converters are considered for proposed integration. The buck converter is shown in Fig. 1, where it acts as step down converter. When switch S_1 is ON, the inductor voltage is difference of input and output voltages; therefore, the current will be difference between V_{in} and voltage across load "R"/L. When switch S_1 is OFF, the inductor L will keep the same current flows through load. The capacitor C is placed

Table 1 Classification of DC to DC converters	Converter configuration	Conversion by seen as
	BUCK converter	Voltage to current (V to I)
	BOOST converter	Current to voltage (I to V)
	BUCK–BOOST converter	Voltage to current to voltage (V to I to V)
	CuK converter	Current to voltage to current (I to V to I)

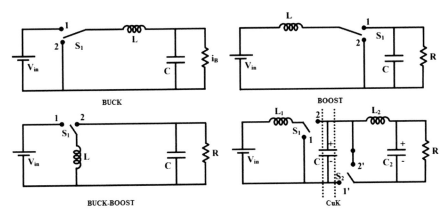

Fig. 1 Schematic circuit of converters

Table 2 Features of DC to DC converters

Converter	Input side	Output side	Voltage
BUCK	Pulsed current, required filter	Continuous current, lower ripple voltage	$V_o < V_{in}$
BOOST	Continuous current, eliminates filter	Pulsed current, higher ripple voltage	$V_o > V_{in}$
BUCK-BOOST	Pulsed current, required filter	Pulsed current, higher ripple voltage	$V_o >$ or $< V_{in}$
CuK	Continuous current	Continuous current	$V_o >$ or $< V_{in}$

across the load R to maintain appreciable change in voltage during switch transition times [4]. The circuit of boost converter is shown in Fig. 1, where it acts as step-up converter. When switch S_1 is ON, the output of inductor is connected between the ground and V_{in} so that the current passes through inductor which increases at the rate of $(V_{in})/L$. When switch S_1 is OFF, the inductor voltage is equal to difference between V_{in} and voltage across load 'R' (V_R) and the current decays at $(V_{in} - V_R)/L$. No boost converter is lossless practically but it achieves efficiency of around 85% [5]. The greater level of competence is achieved by buck–boost converter shown in Fig. 1, than individual buck and boost operations. The main advantage of another converter mentioned, i.e., cuk converter in which the V_R will be either higher or lower voltage.

3 Implementation of DVR and UCap

Energy storage UCAp is connected in one end of DC to DC converter and the other end of converter is connected to DVR shown in Fig. 2 for the sake of (i) increase

(a)

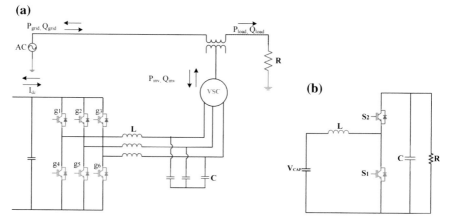

Fig. 2 **a** DVR model, **b** Circuit of DC–DC converter with UCap

in capable of carrying active power in distribution sector and (ii) improving voltage profile through converter control [5].

The injected voltage magnitude and angle of a DVR in compensation are given by

$$V_{DVR} = \sqrt{2} * \left(V_L - V_G^1 \right) \tag{1}$$

$$\angle V_{DVR} = \theta_L \tag{2}$$

where V_L—voltage across the load and V_G^1—voltage across grid during sag

Therefore, the maximum injection of voltage by the DVR is given by [3]

$$V_{DVR\text{-}max} = \frac{V_G^1}{1 - \Delta V_{sag,max}} \sin\theta_L \tag{3}$$

From Fig. 2a

$$I_L = \frac{P_L + j\, Q_L}{V_L} \tag{4}$$

where I_L—current drawn by the load, P_L—real power drawn by the load, and Q_L—reactive power drawn by the load

The model of UCap is shown in Fig. 2b. UCap is the mishmash of correlation pole, electrolyte, electrode, seclusion by membrane, exhaust valve, and seal materials. The concert of UCap can rely on the manufacturing technology, membrane segregation, and electrode– electrolyte composition. Organic polymer electrode ultra-capacitor, double-layer capacitor, and metal oxide electrode supercapacitor are the major types of UCap. Among the three, carbon electrode double-layer capacitor is frequently used.

4 Implementation of Controllers UCap

Due to the voltage profile variation in the DVR system, it is not usually recommended to connect UCap directly like a battery because of the reason is energy discharge in UCap, so that a bidirectional DC to DC converter is connected between UCap and DVR is shown in Fig. 3. The DC-link voltage is maintained at a constant voltage when the voltage of UCap increases for the sake of charging while recharging it can decrease [3].

The gain K_P and K_I are the Proportional and Integral time constants in Fig. 4. Two parameters will do a tuning process. The limits of the PI controller are $+10$ and -10, at zero set point the simulation starts. In this controller, the error deviation is quickly assessed by proportional action function. The offsets between the input and the reference at steady state are used to remove by integral action. It deducts the convolution of DVR system and improves voltage profile.

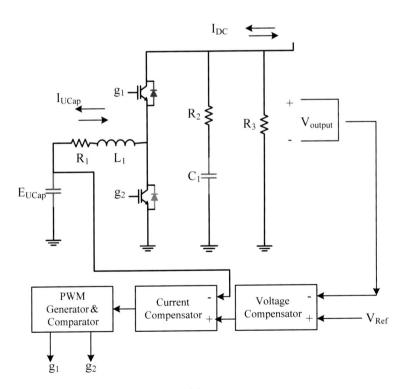

Fig. 3 Controller of DC to DC converter model

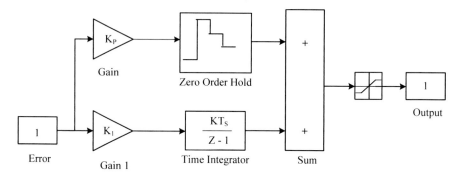

Fig. 4 PI controller block diagram

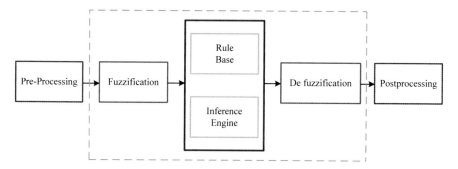

Fig. 5 Fuzzy logic controller block diagram

The fuzzy set theory gives a good number of affluent operations by Fuzzy Logic Controller (FLC) which is shown in Fig. 5. Rather than numerical variables, the FLC utilizes linguistic variables. A human potential behavioral system is constructed for rules by fuzzy logic control technique [4]. The integration of DVR and UCap for voltage profile improvement by FLC through error and change in error are the inputs and duty cycle is the output which is implemented [6].

5 Result/Simulink

For a linear load test system, two parallel feeders are considered in Simulink modeling. The loads are connected to both the feeders and one feeder is connected with DVR. The Simulation diagrams for PI controller of DVR test system and FLC controller of DVR test system as shown in (Figs. 6 and 7) respectively.

The CuK converter integrating DVR and Cuk converter was also simulated as like in Fig. 8. A nonlinear load is considered to test Single Line to Ground Fault (SLG) fault. Between 0.4 and 0.6 s, the fault is created. The load voltage output

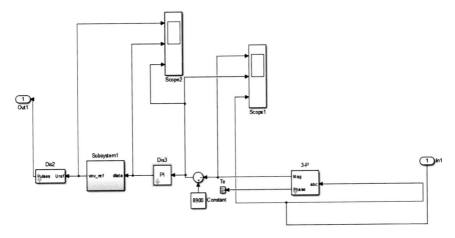

Fig. 6 Simulation diagram for PI controller of DVR test system

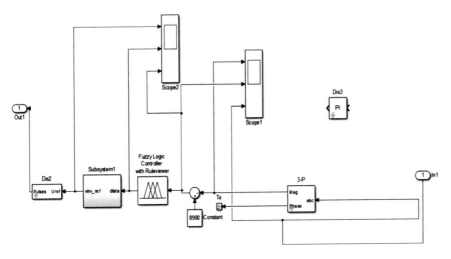

Fig. 7 Simulation diagram for FLC controller of DVR test system

waveforms without and with compensation using PI controller is shown in Figs. 9 and 10. Similarly, the load voltage output waveforms without and with compensation using FLC controller is shown in Figs. 11 and 12. The output waveforms are clearly shaped in phase where fault is cleared.

A SLG fault was created for nonlinear loads and THD level was compared for PI and FLC where it is improved for FLC from 0.61 to 0.47% and finally by integration of DVR and Ucap improved the voltage profile and Cuk (DC to DC converter) gives economic operation than buck–boost converter because of non-isolated type, regulated output DC voltage obtained at higher 'η', reduces ripples in current, and switching losses.

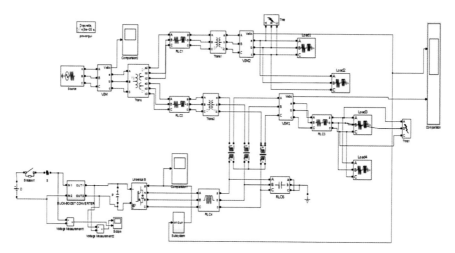

Fig. 8 Simulation diagram for DVR-UCap by buck–boost converter

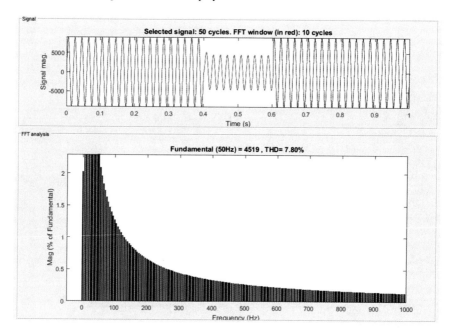

Fig. 9 Simulation output without compensation by PI

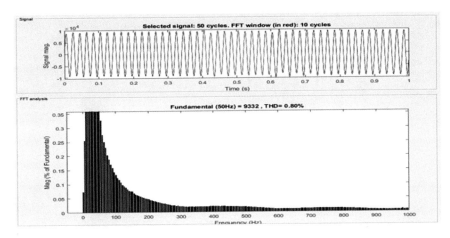

Fig. 10 Simulation output with compensation by PI

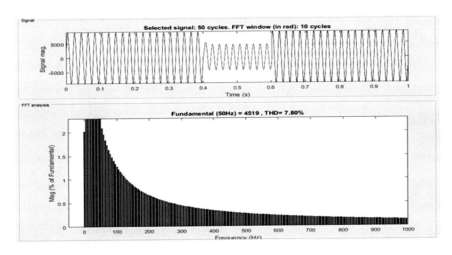

Fig. 11 Simulation output without compensation by FLC

6 Conclusion

It is concluded from the results that the voltage regulation was done under varying load conditions by UCap. Nonlinear loads performance was analyzed using DVR and UCap integration and reduces the harmonics and improves the voltage profile. In future, the UCap is opted for distribution grids from preventing voltage disturbances by sensitive loads and also feasible, economic solution for voltage profile. The inverter inputs, DC to DC converter controllers control actions carried out, found on optimum level integrated controllers. In this paper, a simulation was done using MATLAB by fuzzy logic controller and implementation of integration of DVR and

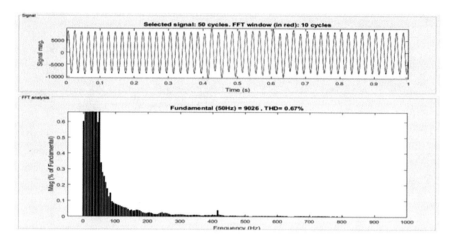

Fig. 12 Simulation output with compensation by PI

UCap. The enhancement of energy security and power reliability in distribution will be improved by UCap because it can act as DC source alternative to investigate the energy storage system and feasibility at a short span of time by providing high power.

References

1. Edomah, N.: Effects of voltage sags, swell and other disturbances on electrical equipment and their economic implications. In: 20th International Conference on Electricity Distribution, Prague (2009)
2. Benachaiba, C., Ferdi, B.: Power quality improvement using DVR. Am. J. Appl. Sci. **6**(3), 396–400 (2009)
3. Abdul Basith, M.B., Sunitha, K.: A novel approach of dynamic voltage restorer integration with ultra capacitor for proper voltage sag compensation. In: 2017 IEEE International Conference on Power, Control, Signals and Instrumentation Engineering (ICPCSI), Chennai, pp. 578–582 (2017)
4. Potnuru, D., Kumar, J.S.V.S.: Design of a front-end DC-DC converter for a permanent magnet DC motor using fuzzy gain scheduling. In: 2017 IEEE International Conference on Power, Control, Signals and Instrumentation Engineering (ICPCSI), Chennai, pp. 1502–1505 (2017)
5. Carlos, G.A.d.A., Jacobina, C.B., dos Santos, E.C.: Investigation on dynamic voltage restorers with two DC links and series converters for three-phase four-wire systems. IEEE Trans. Ind. Appl. **52**(2), 1608–1620 (2016)
6. Chandra, A., Singh, B., Singh, B.N., Al-Haddad, K.: An improved control algorithm of shunt active filter for voltage regulation, harmonic elimination, power-factor correction, and balancing of nonlinear loads. IEEE Trans. Power Electron. **15**(3), 495–507 (2000)

Comparative Study on Histogram Equalization Techniques for Medical Image Enhancement

Sakshi Patel, K. P. Bharath, S. Balaji and Rajesh Kumar Muthu

Abstract Magnetic Resonance Imaging (MRI) is a medical imaging technique used for analyzing and diagnosing diseases such as cancer or tumor in a brain. In order to analyze these diseases, physicians require good contrast scanned images obtained from MRI for better treatment purpose as it contains maximum information of the disease. MRI images are low-contrast images which lead to difficulty in diagnoses, hence better localization of image pixels is required. Histogram equalization techniques help to enhance the image so that it gives an improved visual quality and a well-defined problem. The contrast and brightness are enhanced in such a way that it does not lose its original information and the brightness is preserved. In this paper, we compared the different equalization techniques which are critically studied and elaborated. Various parameters are calculated and tabulated, finally concluded the best among them.

Keywords Histogram equalization · Brain tumor · MRI

1 Introduction

Basically, Digital Image Processing (DIP) is an immense area of research which in turn has various application such as medical imaging, satellite image, and also many industrial applications. Among all these, medical field mainly depends on images such as MRI, X-rays, Ultrasound, and CT to identify the disease for further diagnosis.

S. Patel · K. P. Bharath · S. Balaji · R. K. Muthu (✉)
School of Electronics Engineering, VIT University, Vellore, India
e-mail: mrajeshkumar@vit.ac.in

S. Patel
e-mail: thesakshipatel@gmail.com

K. P. Bharath
e-mail: bharathkp25@gmail.com

S. Balaji
e-mail: sbalaji@vit.ac.in

© Springer Nature Singapore Pte Ltd. 2020
K. N. Das et al. (eds.), *Soft Computing for Problem Solving*,
Advances in Intelligent Systems and Computing 1048,
https://doi.org/10.1007/978-981-15-0035-0_54

These images give a detailed study of various diseases such as brain tumor, cancer, swelling, etc., but it has low contrast. In order to give good treatment, the high-end contrast images are required. Best visual quality and better localization of pixels are carried using different image enhancement techniques. In general, MRI images are low-contrast images. Using various methods of image enhancement techniques leads to improvement in brightness and contrast of the image for practitioners to analyze and treat the infected area [1].

In this paper, we are comparing the different histogram equalization techniques like Typical Histogram Equalization (HE), Global Histogram Equalization (GHE), Local Histogram Equalization (LHE), Dualistic Sub-Image Histogram Equalization (DSIHE),

Brightness Preserving Bi-Histogram Equalization (BBHE), and Recursive Sub-Image Histogram Equalization (RSIHE) using different parametric quantities like Peak Signal-to-Noise Ratio (PSNR), Mean Square Error (MSE), visual quality, complexity, etc.

Rest of the paper is organized as follows, Sect. 2 defines different methodologies, different parameters are discussed in Sect. 3, and simulation results have been discussed in Sect. 4 and concluded in Sect. 5.

2 Methodology

Equalization is a technique used for image enhancement which deals with image histogram which is the plot of number of pixels for each intensity values. Above-mentioned methods equalize the pixels in the input image for each gray level to produce a better resolution.

a. **Typical Histogram Equalization (HE)**:

HE is basically, the mapping of each pixel of the input image to the relating pixels of the output image. This method equalizes the intensity values to full range of the histogram to get an enhanced output image. It enhances the contrast and brightness of the input image by increasing the values of each pixel giving rise to dynamic range expansion [2].

HE can also be defined using

$$P_n = \frac{\text{no. of pixels with intensity } (n_k)}{\text{total no. of pixels } (n)} \tag{1}$$

where $n = 0, 1, \ldots, L - 1$ is the range of gray level values. Probability density function $\text{pdf}(A_k)$ or $p(A_k)$ for an image $A(i, j)$, is given by

$$P(A_k) = \frac{n^k}{n} \tag{2}$$

where n^k is the number of times that the level A_k appears in the input image A. Where $n =$ total number of samples. The number of pixels having specific intensity A_k is represented by $p(A_k)$ which is related to histogram of the image. Now, cumulative density function $c(a)$ is

$$c(a) = \sum_{j=0}^{k} p(A_j) \tag{3}$$

In this method, intensity values are mapped over the dynamic range, (A_0, A_{L-1}), by using $c(a)$ as a transformation function, $f(a)$, i.e., transform function based on $c(a)$ is defined by

$$f(a) = A_0 + (A_{L-1} - A_0)c(a) \tag{4}$$

so, the output image, $B = \{B(i, j)\}$, is given by

$$B = f(a) = \{f(A(i, j)\forall A(i, j) \in A\} \tag{5}$$

Original MRI image of brain along with the histogram plot and resultant image of the HE with its histogram are shown in Fig. 1. The enhanced output is observed but there is unnatural enhancement and excessive increase in brightness when image has high density of pixels for higher gray levels. Here, we can say that major limitation of HE is that it flattens the resultant histogram and also do not consider mean brightness of the input image.

b. **Local Histogram Equalization (LHE):**

LHE implements a block-overlapped technique, which is a subblock implementation to enhance the image. Then, the center pixel value is calculated of subblock, for CDF using typical HE. The subblock is moved one by one pixel and repeated to get the desired output image.

A local transform function for each pixel is calculated based on its neighboring pixels. A small window is defined for the Contextual Region (CR) of that subblock for the center pixel. Pixels falling in that window are considered for CDF calculation. Transform function changes with respect to CR because CDF modifies as the window slides [3]. Local Histogram equalization is as shown in Fig. 2.

LHE gives good contrast enhancement which can sometimes be considered over-enhancement which is the limitation of this method. LHE technique is complex than other algorithms as for every image pixel, local histogram is built and processed.

c. **Brightness Preserving Bi-Histogram Equalization (BBHE):**

BBHE algorithm [4] is basically the next version of typical HE whose main goal is to preserve brightness and avoid false coloring. The block diagram for BBHE is shown in Fig. 3. This technique partitions the input picture histogram into two subparts. The division is carried out using the average intensity of all the pixels which is said to be

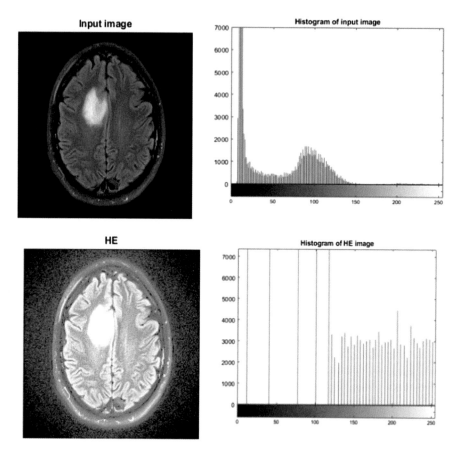

Fig. 1 Histogram equalization for brain tumor

the input mean brightness value of all pixels that is present in the input image. After the division process using mean, these two histograms are equalized independently using the typical histogram equalization method. After performing this step, it is observed that, in the resultant image, mean brightness is present exactly between the input mean and the middle gray level. Then, the two equalized images are combined together to get the resultant image [2]. Mathematically, the gray levels (0 to $L - 1$) present in input image histogram is partitioned into two parts. As a result, we will obtain two histograms, first having range 0 to mean and the other having mean to $L - 1$ range. Let the input image be A and A_m as mean, which can be expressed as E $\in \{A_0, A_1, ..., A_{L-1}\}$.

Decomposition of input into two sub-images A_L and A_U can be given by [4] $A = A_L \cup A_U$, Sub-partition $\mathbf{A}_L$ has $\{A_0, A_1, ..., A_m\}$ levels and $\mathbf{A}_U$ has $\{A_{m+1}, A_{m+2}, ..., A_{L-1}\}$. Now, the probability density functions of $\mathbf{A}_L$ and $\mathbf{A}_U$ can be defined as

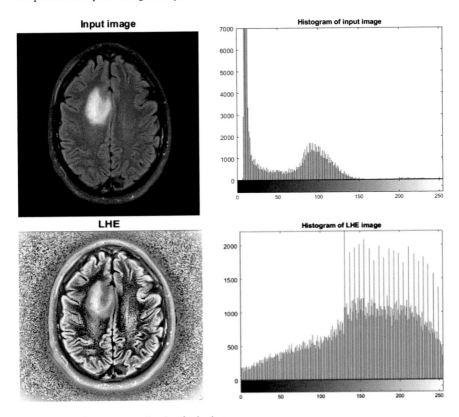

Fig. 2 Local histogram equalization for brain tumor

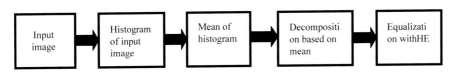

Fig. 3 Block diagram of BBHE

$$p_L(A_k) = \frac{n_L^k}{n_L}, \quad \text{where, } k = 0, 1, \ldots, m, \tag{7}$$

$$p_U(A_k) = \frac{n_U^k}{n_U}, \quad \text{where, } k = m + 1, m + 2, \ldots, L - 1 \tag{8}$$

Then, we define cdf(a) or c(a), i.e., the cumulative density function of the sub-images as

$$c_L(a) = \sum_{j=0}^{k} p_L(a_j), \text{ and } c_U(a) = \sum_{j=m+1}^{k} p_U(a_j), \text{ here, } A_k = a. \tag{9}$$

Here, we take $c_L(A_m) = 1$ and $c_U(A_{L-1}) = 1$ as per the definition. In order to view the proper functioning of the equalization and to cover the entire range, the two partitions of the input are equalized over the ranges $A_0, \ldots, A_m$ and $A_{m+1}, \ldots, A_{L-1}$. The partitioned sub-images are independently equalized using these transform functions, HE and the combination of both equalized sub-images, results in the output of BBHE. So combined resultant image **B** is finally defined as

$$\mathbf{B} = \{B(i, j)\} = f_L(A_L) \cup f_U(A_U) \tag{10}$$

Here, $f_L(\mathbf{A_L}) = \{f_L(A(i,j))| \forall A(i,j) \in \mathbf{A_L}\}$ and, $f_U(\mathbf{A_U}) = \{f_U(A(i,j))| \forall A(i,j) \in \mathbf{A_U}\}$. A is equalized with the methodology that the samples less than the input mean are considered for mapping in the range (A_0, A_m) and for the rest values, which is larger than input mean are then mapped to (A_{m+1}, A_{L-1}) over the entire dynamic range (A_0, A_{L-1}). Therefore, the result from HE, the mean brightness lies in the middle gray level of the output. $E(.) = $ statistical expectation.

$$E(B) = \sum_{A_0}^{A_{L-1}} ap(a)dx = \sum_{A_0}^{A_{L-1}} \frac{a}{A_{L-1} - A_0} \tag{11}$$

In BBHE, if A is considered as a random variable, having symmetric distribution around its mean, A_m. When two partitions are considered for equalization, we can express the mean brightness as

$$E(B|A \le A_m) = \frac{(A_0 + A_m)}{2}, \text{ and}$$

$$E(B|A > A_m) = \frac{(A_m + A_{L-1})}{2}$$

$$E(B) = \frac{(A_m + A_G)}{2} \tag{12}$$

where $A_G = \frac{(A_0 + A_{L-1})}{2}$ is middle gray level, so mean brightness of output from BBHE will lie in the mid of input mean and the middle gray level. The output mean of the resultant image is the function of input mean, A_m. By observing the resultant output from BBHE, we can conclude that brightness of the image is preserved as compared to typical HE, in which mean always lies in middle of the gray level. Resultant image of the BBHE with its histogram is shown in Fig. 4.

d. **Dualistic Sub-Image Histogram Equalization (DSIHE):**

DSIHE algorithm uses the same idea as BBHE technique. Here also the input image histogram is decomposed into two subdivisions but the difference is it takes median under consideration to partition the image being one bright and one dark. Then,

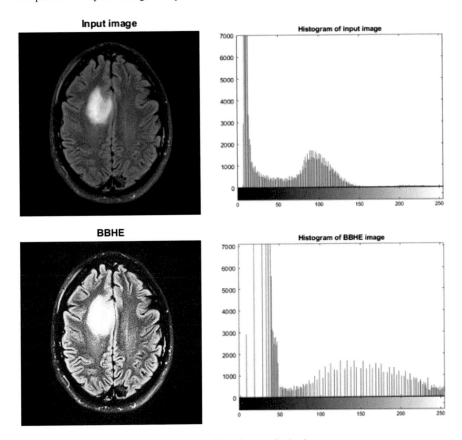

Fig. 4 Brightness Preserving Bi-Histogram Equalization for brain tumor

typical HE is applied separately on both the sub-histograms. After the equalization process, both the parts are combined to give the resultant output [5]. Let input image be A, this image will be partitioned into two parts A_L and A_U the median value A_D, as above it is already defined as

$$A = A_L \cup A_U, \text{ and median,}$$

$$X_D = \text{argMin}_{0 \leq k \leq L-1} \left| \text{cdf}(A_k) - \frac{\text{cdf}(A_0) + \text{cdf}(A_{L-1})}{2} \right| \qquad (13)$$

The final result from the DSIHE algorithm is considered when the two equalized partitions are composed into one image. If $B(i, j)$ denotes the processed image, then:
$$B = \{B(i, j)\} = f_L(A_L) \cup f_U(A_U)$$

$$B(i, j) = \begin{cases} A_0 + (A_m - A_0)c_L(A_k) \\ A_{m+1} + (A_{L-1} - A_{m+1})c_U(A_k) \end{cases} \qquad (14)$$

Also note that sub-image A_L is equalized by the function $f_L(A_L)$ over the range (A_0, A_{D-1}) and A_U *is* done over the range (A_D, A_{L-1}) by function $f_U(A_U)$. The output image of DSIHE algorithm is given in Fig. 5.

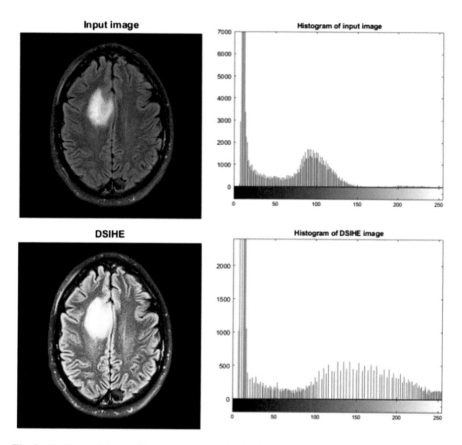

Fig. 5 Dualistic sub-image histogram equalization for brain tumor

e. **Recursive Sub-image Histogram Equalization (RSIHE):**

The generalized version of DSIHE is RSIHE algorithm. This method divides the input image histogram into sub-histograms using Cumulative Density Function (CDF). The image is recursively divided into equal parts for number of recursion levels using the CDF value equal to 0.5. The subdivisions A_i will have equal number of pixels. Scalable brightness preservation is achieved using this recursive process. By using this method, we divide the histogram using median estimation of brightness instead of using input mean brightness. The major task here is to find the ideal estimation of iteration element to get huge enhancement results. But this method does not provide mechanism for modifying the level of improvement. The original MRI image of brain along with the histogram plot and resultant image of the RSIHE with its histogram is shown in Fig. 6.

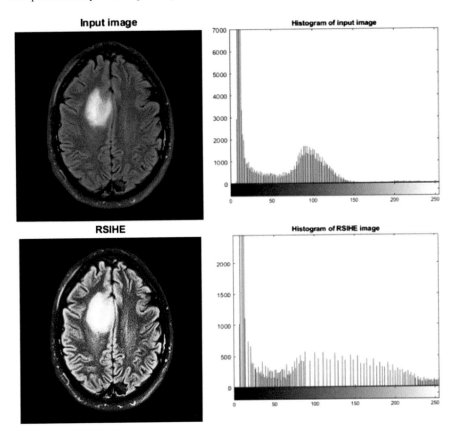

Fig. 6 Recursive sub-image histogram equalization for brain tumor

3   Parameter Measurements

The various parameters using which the study is carried out are [6–8]

a. **Mean Square Error (MSE):**

This parameter calculates the average of square of errors between the input and output image. MSE should be minimum to get image with good quality. It can be calculated as

$$\text{MSE} = \frac{1}{RC} \sum_{i=1}^{R} \sum_{j=1}^{C} \left(A_{i.j} - B_{i,j}\right)^2 \tag{15}$$

A = matrix of input image, B = matrix of output image, R, C = rows and columns of and image, i, j = index of R, C resp.

b. **Peak Signal-to-Noise Ratio (PSNR)**:

PSNR gives the peak error in the output image. The value of this parameter should be large, as it represents the ratio of signal power-to-noise power, noise power should be minimum. It can be represented as

$$PSNR = 10\log\frac{255^2}{MSE} \tag{16}$$

c. **Visual Quality**:

This parameter is measured by the vision of human eye, which can differentiate between the best image and a moderate image. Physicians' needs images with good visual quality to identify the exact texture of the infection, regarding the same quality of the image should be good.

d. **Complexity**:

The complexity of an algorithm is calculated using the number of additions and multiplications that a particular algorithm is performing. More the complexity, more time a method will take to give the output. But, if time is not the matter of constrain, then the best algorithm should be selected regardless of the time taken.

4 Simulation Results

- This section gives the simulation results of different histogram equalization techniques. Here, we have taken five test images from database Brain Tumor Segmentation(BraTS) [9] and correspondingly the output of various techniques like HE, LHE, GHE, BBHE, DSIHE, and RSIHE are shown in Fig. 7. Also, the histograms of all the outputs are also plotted, shown in Fig. 8. Then quality measures are tabulated in Table 1 for each technique to make comparison easy.

a. **Output Images**:

See Fig. 7.

b. **Histogram of output images**:

See Fig. 8.

c. **Quality Parameters**:

From Table 1, we can see that the RSIHE algorithm gives better result than other equalization techniques such as BBHE and DSIHE in terms of MSE, visual quality, and PSNR.

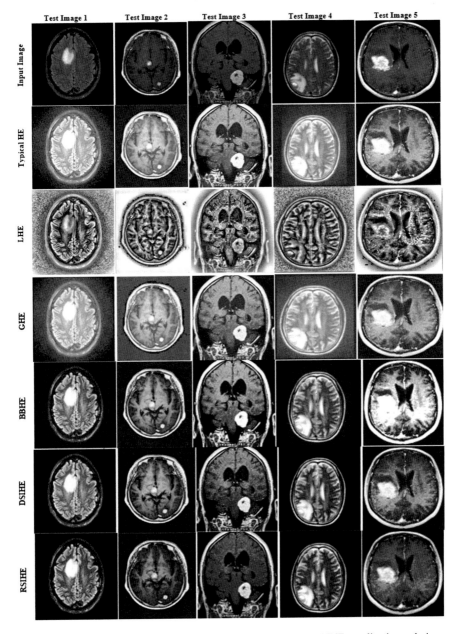

Fig. 7 Output images input image, HE, LHE, GHE, BBHE, DSIHE, RSIHE equalization techniques

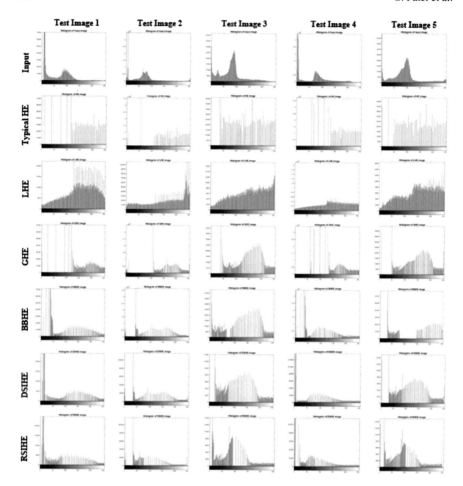

Fig. 8 Histograms of input image of HE, LHE, GHE, BBHE, DSIHE, RSIHE equalization techniques

Table 1 Quality parameters for different histogram equalization techniques

Measures	HE	LHE	GHE	BBHE	DSIHE	RSIHE
PSNR	9.178198	5.193755	9.909195	15.34653	15.95524	17.80369
MSE	−2.38769	9.937468	4.877582	3.780637	4.38935	6.237799
Quality	Moderate	Moderate	Moderate	Good	Better	Better
Complexity	Very low	Low	Low	Moderate	Moderate	Moderate

5 Conclusion

In this paper, a comparative study of various histogram equalization techniques is successfully performed using histogram processing. The major role of these techniques is to enhance the brightness and contrast of the original image. But HE, GHE, and LHE techniques lead to false coloring. Moreover, the image should not lose its important information while performing image enhancement techniques. We observe that BBHE algorithm preserves brightness of the image to a certain extent but when observed on the basis of various factors like PSNR, visual quality, MSE, etc., it was found that DSIHE and RSIHE which are the improved version of BBHE give better results. Hence, these methods can be used for the enhancement of medical images for better diagnosis.

References

1. Gonzalez, R.C., Woods, R.E.: Digital Image Processing, 2nd edn. Prentice Hall (2002)
2. Kim, Y.T.: Contrast enhancement using brightness preserving bi-histogram equalization. IEEE Trans. Consum. Electr. **43**(1), 1–8 (1997)
3. Abdullah-Al-Wadud, M., et al.: A dynamic histogram equalization for image contrast enhancement. IEEE Trans. Consum. Electr. **53**(2), 593–600 (2007)
4. Chen, S.D., Ramli, A.R.: Minimum mean brightness error Bi-histogram equalization in contrast enhancement. IEEE Trans. Consum. Electr. **49**(4), 1310–1319 (2003)
5. Patel, O., Maravi, Y.P.S., Sharma, S.: A comparative study of histogram equalization based image enhancement techniques for brightness preservation and contrast enhancement. Signal Image Process. Int. J. (SIPIJ) **4**(5) (2013)
6. Senthilkumaran, N., Thimmiaraja, J.: A study on Histogram Equalization for MRI Brain Image Enhancement. Association of Computer Electronics and Electrical Engineers (2014). doi: 03.AETS.2014.5.348
7. Garg, R.: Histogram equalization techniques for image enhancement. IJECT **2** (2011)
8. Kaur, M.: Survey of contrast enhancement techniques based on Histogram equalization. Int. J. Adv. Comput. Sci. appl. **2** (2011)
9. https://www.med.upenn.edu/sbia/brats2017/data.html

A Fuzzy Multi-criteria Decision Model for Analysis of Socio-ecological Performance Key Factors of Supply Chain

Rahul Solanki, Jyoti Dhingra Darbari, Vernika Agarwal and P. C. Jha

Abstract Amidst stiff competition, Small and Medium scale Enterprises (SMEs) need to continuously improve their supply chain sustainable performance. Due to customer pressure and government interventions, the focus has also shifted to enhancing the social and environmental performance of the SC. However, the challenge faced by SMEs is to identify the necessary socio-ecological key factors (KF)s required for a better sustainable performance and how to constantly assess the KFs for improving the SC performance. Hence, the research objective of the study is development of a decision model for assessing the social and environmental key factors or Indian SMEs and analyzing their interrelationships. To attain the research objective, the KFs for social and environmental performance are first identified through an extensive literature survey. Further, the KFs are evaluated using fuzzy-DEMATEL technique. The fuzzy-DEMATEL method is used to draw a directed graph for visibly capturing the causal relationships between the multiple KFs. The structural graph helps to portray a contextual relationship among 12 identified KFs including green market trends, business risk protection, and green collaborations. The result reveals that product responsibility and stakeholders' involvement are both significant factors affecting other factors and play key roles in overall sustainable enhancement of SC for SMEs. The research is based on data collected from an electronics firm, analyzed

R. Solanki (✉) · P. C. Jha
Department of Operational Research, University of Delhi, New Delhi, India
e-mail: solanki.rahul1470@gmail.com

P. C. Jha
e-mail: jhapc@yahoo.com

J. D. Darbari
Department of Mathematics, Lady Shri Ram College, University of Delhi, New Delhi, India
e-mail: jydbr@hotmail.com

V. Agarwal
Amity International Business School, Amity University, Noida, Uttar Pradesh, India
e-mail: vernika.agarwal@gmail.com

© Springer Nature Singapore Pte Ltd. 2020
K. N. Das et al. (eds.), *Soft Computing for Problem Solving*,
Advances in Intelligent Systems and Computing 1048,
https://doi.org/10.1007/978-981-15-0035-0_55

to extract cause factors, which foster sustainability satisfaction from the SME's perspective. The results of the study provide a benchmark for decision-makers to estimate the uncertain environment for developing a socio-ecological responsive and performance-driven SC.

Keywords Supply chain performance · Fuzzy-DEMATEL · Small and Medium Enterprises (SMEs) · Socio-ecological sustainability

1 Introduction

Contemporarily, the Supply Chain (SC) management operations are concerned only with the economic stability, where stakeholders and managers are not motivated enough to pay attention to ecological or social concerns of a firm. It was only in late 1980s ecological consciousness emerged among the world [1]. Since then a large number of related definitions and regulations have come up to make sustainability more actionable and concrete with participation at various levels from groups, societies, and organizations. This led to the realization among organizations that traditional business practices are merely not enough for long-term efficient performance and sustainability. Therefore, motivating researchers and stakeholders toward sustainability not only helps to achieve an advantage over competitors but also to perform better in the long run, as the firm can develop specific ecological or social competencies that are not easy to imitate [2]. The key components of sustainability are economic, ecological, and social developments. Adoption of ecological practices not only improves the environment but it also has a meaningful impact on economic performance [3]. On the other hand, managing social issues are equally significant as they directly or indirectly impact the people related to SC. Practicing social sustainability tends to increase SC performance as an effect of improved competitive edge, customer satisfaction, and reputation [4], which also maximizes its market value. Social sustainability deals with issues that can affect the stakeholders involved in both upstream and downstream SC [5]. Therefore, ignoring the social issues in the SC can result in backlash from customers, penalties, and reduction in their market shares.

For large scale industries, SMEs are the major suppliers, and hence sustainability adoption of both levels of industries hold equal importance. But implementation of sustainability practices for SMEs is quite a difficult task due to lack of resources, expertise, and information. Particularly in the Indian scenario, sustainable operations are not widely incorporated by firms [6, 7]. Due to immense pressure from the government and customers, SMEs in India have now gradually started to incorporate ecological sustainability, but still lack in exploring social sustainability. A limitation in implementing sustainable practices is that the companies do not have an illustrated view and understanding about the relationships between ecological and social aspects. Balancing ecological issues and sound social practices in the complex and uncertain environment setting is a big challenge. Identification and effective implementation of the right social and ecological sustainability Key Factors (KFs) is still a challenging task.

This leads us to the understanding that there is an urgent need for Indian SMEs to identify and evaluate environmental and social sustainability KFs, which can improve the overall performance of the firm. To address this major issue, the present research gives an in-depth analysis using a case example of an Indian manufacturing SME. The case is considered to explore the following research questions: (i) what are the socio-ecological KFs, on the basis of which the firm's overall performance can be measured? (ii) How to deal with the vague judgments of a group of experts or decision-makers and model the casual relationships of the identified KFs? (iii) What is the strength and influencing capability of the identified KFs on the overall system? To achieve the aforementioned objectives, the study presents an integration of Group Decision-Making (GDM), fuzzy logic, and Decision-Making Trial and Evaluation laboratory (DEMATEL) technique, collectively called fuzzy-DEMATEL. The purpose of GDM is to arrive at an acceptable judgment from among the multiple conflicting judgments. To achieve a desirable solution, GDM is an essential part. At first, a group of relevant decision-makers (DMs) is identified for the evaluation process, and then FST is used to incorporate the uncertainties or vagueness in human judgments and decision-making. Further, the fuzzy data is converted into crisp values. Finally, DEMATEL identifies the strength and casual relationships among the shortlisted KFs and simultaneously builds a structural model to visualize the relationships.

The rest of the paper is outlined as follows: Sect. 2 describes the problem and a case study to support the problem. In Sect. 3, the proposed solution methodology, fuzzy-DEMATEL is discussed. Section 4 presents the numerical illustration. Finally, Sect. 5 describes the conclusion, findings, and future suggestions.

2   Case Study

Managing environmental, ecological, and societal issues are significant because they directly or indirectly enhance a firm's SC performance. Extensive research has been undertaken that describes the importance of economic sustainability of Indian SMEs. However, scholars and researchers find it difficult to address the issues pertaining to ecological and societal development that can help strengthening the overall SC performance of the firm. For this purpose, the study under consideration aims to understand the relationship between the key factors of ecological and social sustainability. This is illustrated in the following case study:

A case company XYZ is an Indian electronics SME with an annual turnover of 63 crore rupees. The company has four manufacturing facilities and products are distributed to various retailers. The current focus of XYZ is only economical sustainability but in order to remain competitive and perform better, company seeks the underlying objectives:

1. Determine ecological and social sustainability KFs that can measure SC performance.
2. Recognize and analyze the contextual relationship among the identified KFs from the DMs' perspective.
3. To build a structural model that can define the strength of KFs and understand how they can help in enhancing the overall firm's SC performance.

In order to tackle the above-stated objectives, firm XYZ considers formation of a committee for the identification of KFs, i.e., comprising of a group of seven DMs, including directors, consultants, managers, investors, and partners. Inputs from the relevant DMs were gathered with the help of e-mail, telephonic, and direct interviews. A total of 12 evaluation KFs were shortlisted by the GDMs, which are defined as follows: (i) Protection a Business from Risk of Environment and Safety Factors, (ii) Economical and Environmental Literacy among Supply Chain Partners, (iii) Unethical Practices, (iv) Product Responsibility, (v) Public Perception, (vi) Cultural Heritage, (vii) Eco-Design of a Product, (viii) Green Collaborations, (ix) Green Market Trends, (x) Government Management, Rewards, Support and Initiatives, (xi) Stakeholders and Environmental Advocacy Group Involvement, and (xii) Employment Practices and Stability. The useful information extracted is thus analyzed using fuzzy-DEMATEL.

3 Research Methodology

To capture the contextual relationship and build a structural model for the purpose of analyzing the influence relation among the complex KFs, the fuzzy-DEMATEL a comprehensive method is utilized [8]. The extension of DEMATEL for the purpose of decision-making along with fuzzy set theory is discussed below.

3.1 Fuzzy Set Theory

Due to various conditions, modeling real-life situations are not possible as human judgments are subjective, ambiguous, and not easy to estimate with the help of numerical values or data. To solve the uncertainties because of human perception and reasoning, the linguistic variables approach is one of the best ways to express the DM's judgments [9]. The linguistic variables are represented using fuzzy numbers, among which the most commonly used number is Triangular Fuzzy Number (TFN). These numbers assess the DM's preferences. A TFN can be defined as a number A with an associated triplet (p, q, r), where $p \leq q \leq r$ and a membership function $\alpha_{\tilde{A}}(x)$ defined as below:

$$\alpha_{\tilde{A}}(x) = \begin{cases} 0, & x < p \\ \frac{(x-p)}{(q-p)}, & p \le x \le q \\ \frac{(r-x)}{(r-q)}, & q \le x \le r \\ 0, & x > r \end{cases},$$

where p, q, and r are real numbers.

A fuzzy process always includes a defuzzification method. Here, Converting Fuzzy Data into Crisp Scores (CFCS) method is used for defuzzification, as it gives better crisp values compared to the centroid method. The method is explained below:

Let $a_{ij}^k = \left(p_{ij}^k, q_{ij}^k, r_{ij}^k \right)$ be the fuzzy judgment of kth decision-maker which represent the degree of impact of criterion 'i' on criterion 'j'

Step (1) Normalization:

$$xp_{ij}^k = \left(p_{ij}^k - \min p_{ij}^k \right) / \Delta_{min}^{max}, \ xq_{ij}^k = \left(q_{ij}^k - \min q_{ij}^k \right) / \Delta_{min}^{max}, \ xr_{ij}^k = \left(r_{ij}^k - \min r_{ij}^k \right) / \Delta_{min}^{max},$$

where $\Delta_{min}^{max} = \max r_{ij}^k - \min p_{ij}^k$.

Step (2) Left and right normalization:

$$xLs_{ij}^k = xq_{ij}^k / \left(1 + xq_{ij}^k - xp_{ij}^k \right) \ \& \ xRs_{ij}^k = xr_{ij}^k / \left(1 + xr_{ij}^k - xp_{ij}^k \right).$$

Step (3) Total normalized crisp value conversion:

$$x_{ij}^k = \left[xLs_{ij}^k \left(1 - xLs_{ij}^k \right) + xRs_{ij}^k x Rs_{ij}^k \right] / \left(1 - xLs_{ij}^k + xRs_{ij}^k \right).$$

Step (4) Integrated crisp values:

$$zs_{ij}^k = \min p_{ij}^k + x_{ij}^k \Delta_{min}^{max} \ \& \ zs_{ij} = \frac{1}{v} \left(zs_{ij}^1 + zs_{ij}^2 + \cdots + zs_{ij}^v \right).$$

3.2 Integrating FST and DEMATEL Method

The DEMATEL method obtains the solution to fragmented and ambiguous problems. This methodology not only structures the relations among the KFs considered in the system but also analyzes the strength of the interrelationships. The DEMATEL methodology identifies the most influential KFs [10]. For the evaluation of KFs efficaciousness, this technique reduces the number of KFs for system under consideration. Alongside, a firm can simultaneously work on improving the effectiveness of other KFs based on the impact diagram map. Further, this method transforms the interrelationships into intangible source of information as a structural model. The methodological steps of fuzzy-DEMATEL are briefly explained below:

Step 1: *Formation of a group decision-making body and identifying the decision goals*: For the purpose of problem-solving, acquiring the relevant knowledge using GDM constitutes is an important requirement. This helps in accumulating the relevant information for the system, generating the alternatives, evaluating these alternatives for advantage or disadvantage and obtaining the best possible and optimal alternatives to achieve the desired decision goals [9].

Step 2: *To identify and evaluate the KFs*: This step comprises establishing the KFs that are to be evaluated. A pairwise comparison scale for the DMs is designated at five levels, where the following scores 0, 1, 2, 3, and 4 represent different levels of influence, respectively. Then, the following fuzzy linguistic scale is used: No Influence- N_0, Extremely Low Influence-EL, Low Influence-L, High Influence-H, and Extreme High Influence-EH, expressed in terms of FTNs as (k_{ij}, l_{ij}, m_{ij}).

Step 3: *Forming initial direct relation matrix*: The GDM involves assessment of the influence of each criterion on the other, for measuring the relationship among all listed KFs. Using the pairwise comparisons based on KFs influences and directions as explained in Step 2, a direct relation matrix $A = [A_{ij}]$ is obtained, where A_{ij} denotes the judgment representing the degree of influence of KF 'i' on KF 'j'. Further, these fuzzy evaluations are defuzzified into crisp scores, using CFCS method. Hence, the fuzzy initial direct relational matrix is converted to initial direct relational matrix.

Step 4: *Constructing the structural model*: Next, a defuzzified normalized matrix is obtained, using the following formula:

$$D = [d_{ij}]\max = \frac{A}{s} \quad \& \quad s = \max \left(\max_{1 \le i \le n} \sum_{j=1}^{n} a_{ij}, \ \max_{1 \le j \le n} \sum_{j=1}^{n} a_{ij} \right)$$

Next, matrix T which represents the total relations among the criteria is constructed using the following formula:

$$T = [r_i]_{n \times 1} = \lim_{m \to \infty} (D + D^2 + D^3 + \cdots + D^\infty) = D(I - D)^{-1}$$

After which column and row sum of T are evaluated and presented under 'C' and 'R', respectively.

$$R = [r_i]_{n \times 1} = \left[\sum_j t_{ij} \right]_{n \times 1} \quad \& \quad C = [c_{ij}]_{1 \times n} = \left[\sum_j t_{ij} \right]_{1 \times n}$$

Further, the values of $R + C$ (horizontal axis) and $R - C$ (vertical axis) are obtained that help to build a casual diagraph. The cause group consists of all criteria with positive '$R - C$'. The remaining constitutes the effect group. Therefore, the complex relationships among the criteria are well represented by cause and effect diagraph that helps in visualizing the structural model.

4 Case Illustration

The DMs assess each criterion on a scale of 0–4 as explained before, and the influence of each criterion defined is converted into a positive TFN as shown in Table 1. The transformed fuzzy matrix ($\tilde{A}$) is shown in Table 2, with the linguistic values, i.e., ZI (0, 0, 0.25), ELI (0, 0.25, 0.5), LI (0.25, 0.5, 0.75), HI (0.5, 0.75, 1.0), and EHI (0.75, 1.0, 1.0).

Further, the TFNs are defuzzified using CFCS method and these crisp values are divided by total number of DMs, i.e., seven. Further, matrix $\tilde{A}$ is shown in Table 2 and matrix A is Table 3.

The normalized direct relation matrix is shown by Table 4 and total relation matrix T along with their row sum R and column sum C are obtained as represented by Table 5. A threshold value is set up to be 0.40 with the help of DMs.

Further, Fig. 1 depicts the casual diagram obtained by plotting the values of $R + C$ and $R - C$ obtained in Table 6, i.e., values of ($R_i + C_i$), horizontal axis and ($R_i - C_i$), and vertical axis, which gives the degree of relation.

On analyzing the causal diagram, we can observe that the KFs in the cause group are KF1: protecting a business from risks of environmental and safety factors, KF2: economical and environmental literacy among supply chain partners, KF4: product responsibility, KF10: government management, rewards, support, and initiatives, and KF11: stakeholders and environmental advocacy group involvement. The effect group consists of the following KFs: KF3: unethical practices, KF5: public perception, KF6: cultural heritage, KF7: eco-design of a product, KF8: green collaboration, KF9: green market trends, KF12: employment practices and stability.

5 Discussions and Conclusion

The growing environmental deterioration and depletion of natural resources are putting pressure on the companies to incorporate sustainability practices into their workings. In the present empirical study, the company under consideration wants to implement sustainability into its operations, in doing so the first step is to understand the effect of social and environmental KFs on the economic forefront.

To address this problem, the present study develops a decision model based on multi-criteria analysis for (i) identifying the KFs of social and ecological sustainability for electronics firms in India and (ii) categorizing them into cause group (influencing factors) and the effect group (influenced factors). A total of 12 KFs were identified from rigorous literature survey and discussion with the GDMs, which were used for evaluation of sustainability performance of the SC. Fuzzy-DEMATEL

Table 1 Pair wise comparison matrix

KFs	KF1	KF2	KF3	KF4	KF5	KF6	KF7	KF8	KF9	KF10	KF11	KF12
KF1	–	N_0	H	EH	L	L	H	EL	N_0	H	L	H
KF2	L	–	H	N_0	H	N_0	EH	L	N_0	EL	N_0	N_0
KF3	EH	EL	–	N_0	EL	L	EH	EH	EL	H	H	N_0
KF4	H	L	EH	–	L	L	EH	EH	L	EL	N_0	EH
KF5	H	EL	0	L	–	N_0	L	EH	N_0	H	N_0	EL
KF6	L	L	0	EL	L	–	L	N_0	N_0	1	H	EH
KF7	0	EL	0	0	EL	L	–	N_0	H	EH	N_0	EH
KF8	H	EH	EH	EH	N_0	H	L	–	L	EH	EL	N_0
KF9	N_0	L	EL	L	EH	N_0	N_0	L	–	H	EL	N_0
KF10	H	H	EH	H	H	EH	N_0	H	1	–	H	H
KF11	EH	EH	EH	EL	EH	EH	H	EH	L	EL	–	EL
KF12	N_0	EL	H	L	H	EH	H	EL	L	N_0	EH	–

Table 2 Fuzzy initial direct relation matrix ($\tilde{A}$)

KFs	KF1	KF2	KF3	KF4	KF5	KF6	KF7	KF8	KF9	KF10	KF11	KF12
KF1	(0, 0, 0.25)	(0.25, 0.5, 0.75)	(0.5, 0.75, 1.0)	(0.75, 1.0, 1.0)	(0.25, 0.5, 0.75)	(0.25, 0.5, 0.75)	(0.5, 0.75, 1.0)	(0, 0.25, 0.5)	(0, 0, 0.25)	(0.5, 0.75, 1.0)	(0.25, 0.5, 0.75)	(0.5, 0.75, 1.0)
KF2	(0.25, 0.5, 0.75)	(0, 0, 0.25)	(0.5, 0.75, 1.0)	(0, 0, 0.25)	(0.5, 0.75, 1.0)	(0, 0, 0.25)	(0.75, 1.0, 1.0)	(0.25, 0.5, 0.75)	(0, 0, 0.25)	(0, 0.25, 0.5)	(0, 0, 0.25)	(0, 0, 0.25)
KF3	(0.75, 1.0, 1.0)	(0, 0.25, 0.5)	(0, 0, 0.25)	(0, 0, 0.25)	(0, 0.25, 0.5)	(0.25, 0.5, 0.75)	(0.75, 1.0, 1.0)	(0.75, 1.0, 1.0)	(0, 0.25, 0.5)	(0.5, 0.75, 1.0)	(0.5, 0.75, 1.0)	(0, 0, 0.25)
KF4	(0.5, 0.75, 1.0)	(0.25, 0.5, 0.75)	(0.75, 1.0, 1.0)	(0, 0, 0.25)	(0.25, 0.5, 0.75)	(0.25, 0.5, 0.75)	(0.75, 1.0, 1.0)	(0.75, 1.0, 1.0)	(0.25, 0.5, 0.75)	(0, 0.25, 0.5)	(0, 0, 0.25)	(0.75, 1.0, 1.0)
KF5	(0.5, 0.75, 1.0)	(0, 0.25, 0.5)	(0, 0, 0.25)	(0.25, 0.5, 0.75)	(0, 0, 0.25)	(0, 0, 0.25)	(0.25, 0.5, 0.75)	(0.75, 1.0, 1.0)	(0, 0, 0.25)	(0.5, 0.75, 1.0)	(0, 0, 0.25)	(0, 0.25, 0.5)
KF6	(0.25, 0.5, 0.75)	(0.25, 0.5, 0.75)	(0, 0, 0.25)	(0, 0.25, 0.5)	(0.25, 0.5, 0.75)	(0, 0, 0.25)	(0.25, 0.5, 0.75)	(0, 0, 0.25)	(0, 0, 0.25)	(0, 0.25, 0.5)	(0, 0, 0.25)	(0.75, 1.0, 1.0)
KF7	(0, 0, 0.25)	(0, 0.25, 0.5)	(0, 0, 0.25)	(0, 0, 0.25)	(0, 0.25, 0.5)	(0.25, 0.5, 0.75)	(0, 0, 0.25)	(0, 0, 0.25)	(0.5, 0.75, 1.0)	(0.75, 1.0, 1.0)	(0, 0, 0.25)	(0.75, 1.0, 1.0)
KF8	(0.5, 0.75, 1.0)	(0.75, 1.0, 1.0)	(0.75, 1.0, 1.0)	(0.75, 1.0, 1.0)	(0, 0, 0.25)	(0.5, 0.75, 1.0)	(0.25, 0.5, 0.75)	(0, 0, 0.25)	(0.25, 0.5, 0.75)	(0.75, 1.0, 1.0)	(0, 0.25, 0.5)	(0.75, 1.0, 1.0)
KF9	(0, 0, 0.25)	(0.25, 0.5, 0.75)	(0, 0.25, 0.5)	(0.25, 0.5, 0.75)	(0, 0, 0.25)	(0, 0, 0.25)	(0, 0, 0.25)	(0.25, 0.5, 0.75)	(0, 0, 0.25)	(0.5, 0.75, 1.0)	(0, 0.25, 0.5)	(0, 0, 0.25)
KF10	(0.5, 0.75, 1.0)	(0.5, 0.75, 1.0)	(0.75, 1.0, 1.0)	(0.5, 0.75, 1.0)	(0.5, 0.75, 1.0)	(0.75, 1.0, 1.0)	(0, 0, 0.25)	(0.5, 0.75, 1.0)	(0, 0.25, 0.5)	(0, 0, 0.25)	(0.5, 0.75, 1.0)	(0.5, 0.75, 1.0)
KF11	(0.75, 1.0, 1.0)	(0.75, 1.0, 1.0)	(0.75, 1.0, 1.0)	(0, 0.25, 0.5)	(0.75, 1.0, 1.0)	(0.75, 1.0, 1.0)	(0.5, 0.75, 1.0)	(0.75, 1.0, 1.0)	(0.25, 0.5, 0.75)	(0, 0.25, 0.5)	(0, 0, 0.25)	(0, 0.25, 0.5)
KF12	(0, 0, 0.25)	(0, 0.25, 0.5)	(0.5, 0.75, 1.0)	(0.25, 0.5, 0.75)	(0.5, 0.75, 1.0)	(0.75, 1.0, 1.0)	(0.5, 0.75, 1.0)	(0, 0.25, 0.5)	(0.25, 0.5, 0.75)	(0, 0, 0.25)	(0.75, 1.0, 1.0)	(0, 0, 0.25)

Table 3 Direct relation matrix (A)

KFs	KF1	KF2	KF3	KF4	KF5	KF6	KF7	KF8	KF9	KF10	KF11	KF12
KF1	0.00	0.52	0.77	0.96	0.52	0.52	0.77	0.27	0.33	0.77	0.52	0.77
KF2	0.52	0.00	0.77	0.33	0.77	0.33	0.96	0.52	0.33	0.27	0.33	0.33
KF3	0.96	0.27	0.00	0.33	0.27	0.52	0.96	0.96	0.27	0.77	0.77	0.33
KF4	0.77	0.52	0.96	0.00	0.52	0.52	0.96	0.96	0.52	0.27	0.33	0.96
KF5	0.77	0.27	0.33	0.52	0.00	0.33	0.52	0.96	0.33	0.77	0.33	0.27
KF6	0.52	0.52	0.33	0.27	0.52	0.00	0.52	0.33	0.33	0.27	0.77	0.96
KF7	0.33	0.27	0.33	0.33	0.27	0.52	0.00	0.33	0.77	0.96	0.33	0.96
KF8	0.77	0.96	0.96	0.96	0.33	0.77	0.52	0.00	0.52	0.96	0.27	0.33
KF9	0.33	0.52	0.27	0.52	0.96	0.33	0.33	0.52	0.00	0.77	0.27	0.33
KF10	0.77	0.77	0.96	0.77	0.77	0.96	0.33	0.77	0.27	0.00	0.77	0.77
KF11	0.96	0.96	0.96	0.27	0.96	0.96	0.77	0.96	0.52	0.27	0.00	0.27
KF12	0.33	0.27	0.77	0.52	0.77	0.96	0.77	0.27	0.52	0.33	0.96	0.00

Table 4 Normalized direct relation matrix

KFs	KF1	KF2	KF3	KF4	KF5	KF6	KF7	KF8	KF9	KF10	KF11	KF12
KF1	0.00	0.07	0.10	0.12	0.07	0.07	0.10	0.03	0.04	0.10	0.07	0.10
KF2	0.07	0.00	0.10	0.04	0.10	0.04	0.12	0.07	0.04	0.03	0.04	0.04
KF3	0.12	0.03	0.00	0.04	0.03	0.07	0.12	0.12	0.03	0.10	0.10	0.04
KF4	0.10	0.07	0.12	0.00	0.07	0.07	0.12	0.12	0.07	0.03	0.04	0.12
KF5	0.10	0.03	0.04	0.07	0.00	0.04	0.07	0.12	0.04	0.10	0.04	0.03
KF6	0.07	0.07	0.04	0.03	0.07	0.00	0.07	0.04	0.04	0.03	0.10	0.12
KF7	0.04	0.03	0.04	0.04	0.03	0.07	0.00	0.04	0.10	0.12	0.04	0.12
KF8	0.10	0.12	0.12	0.12	0.04	0.10	0.07	0.00	0.07	0.12	0.03	0.04
KF9	0.04	0.07	0.03	0.07	0.12	0.04	0.04	0.07	0.00	0.10	0.03	0.04
KF10	0.10	0.10	0.12	0.10	0.10	0.12	0.04	0.10	0.03	0.00	0.10	0.10
KF11	0.12	0.12	0.12	0.03	0.12	0.12	0.10	0.12	0.07	0.03	0.00	0.03
KF12	0.04	0.03	0.10	0.07	0.10	0.12	0.10	0.03	0.07	0.04	0.12	0.00

Table 5 Total relation matrix

KFs	KF1	KF2	KF3	KF4	KF5	KF6	KF7	KF8	KF9	KF10	KF11	KF12
KF1	0.34	0.34	0.45	0.39	0.37	0.39	0.45	0.36	0.27	0.4	0.34	0.4
KF2	0.33	0.22	0.37	0.27	0.33	0.3	0.4	0.33	0.23	0.29	0.26	0.29
KF3	0.44	0.31	0.35	0.32	0.33	0.38	0.46	0.43	0.26	0.4	0.36	0.35
KF4	0.45	0.36	0.49	0.3	0.39	0.41	0.49	0.46	0.31	0.37	0.34	0.44
KF5	0.37	0.27	0.34	0.3	0.26	0.31	0.36	0.38	0.23	0.36	0.27	0.29
KF6	0.33	0.28	0.32	0.26	0.31	0.26	0.35	0.3	0.23	0.28	0.32	0.36
KF7	0.31	0.26	0.33	0.27	0.29	0.33	0.29	0.3	0.28	0.37	0.27	0.37
KF8	0.46	0.42	0.5	0.42	0.38	0.44	0.45	0.36	0.31	0.45	0.34	0.38
KF9	0.31	0.28	0.31	0.28	0.35	0.29	0.31	0.32	0.18	0.33	0.25	0.28
KF10	0.49	0.42	0.53	0.42	0.45	0.49	0.46	0.47	0.3	0.36	0.42	0.45
KF11	0.49	0.43	0.51	0.35	0.46	0.47	0.49	0.48	0.32	0.39	0.31	0.38
KF12	0.36	0.3	0.42	0.32	0.38	0.42	0.43	0.35	0.28	0.34	0.38	0.29

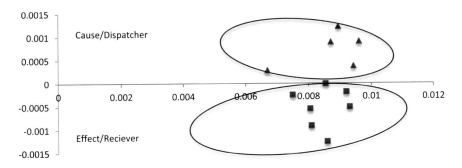

Fig. 1 Cause and effect groups

Table 6 Degree of relation among KFs

KFs	R_i	C_i	$R_i + C_i$	$R_i - C_i$	Group
KF1	3.07528	2.79852	5.87380	**0.27675**	**Cause**
KF2	2.42605	2.38566	4.81170	**0.04039**	**Cause**
KF3	2.55133	3.13884	5.69016	−0.58751	Effect
KF4	3.17564	2.52020	5.69583	**0.65544**	**Cause**
KF5	2.68426	2.72238	5.40664	−0.03812	Effect
KF6	2.47122	2.99955	5.47076	−0.52833	Effect
KF7	2.21966	3.23031	5.44997	−1.01066	Effect
KF8	2.85116	2.92222	5.77339	−0.07106	Effect
KF9	2.03883	2.70597	4.74481	−0.66714	Effect
KF10	3.69571	2.31243	6.00814	**1.38327**	**Cause**
KF11	3.42918	2.37688	5.80606	**1.05231**	**Cause**
KF12	2.45465	2.96000	5.41464	−0.50535	Effect

methodology is used for segmentation of KFs into cause and effect groups, which can aid the DMs in successful implementation of business sustainability.

On analyzing the results, it can be seen that KF1: protecting a business from risks of environmental and safety factors, KF2: economical and environmental literacy among supply chain partners, KF4: product responsibility K10: government management, rewards, support, and initiatives, and KF11: stakeholders and environmental advocacy group involvement, which implies that high level of focus should be done on these KFs. It can also be seen from Table 6 that KF10 and KF11 are the most influencing factors within the cause group. It can be inferred that government support along with stakeholder involvement can aid the SMEs immensely in the social and ecological improvement of the SC. The strength of the cause group KFs is depicted by Fig. 2, stressing on the importance of KF10 and KF11. Hence, a company working toward making its practices more social and environment friendly needs all the aid that the government can provide in terms of subsidies. Analyzing

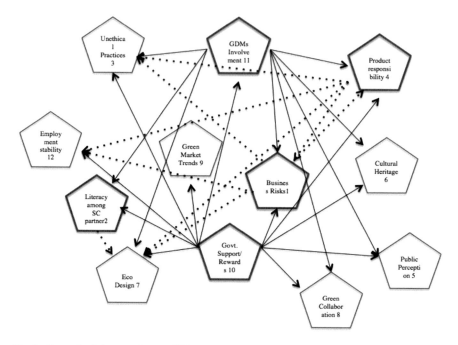

Fig. 2 Strength of the casual group KFs

Table 6 further, depicts that KF4 (Product responsibility) ranks next in terms of level of influence, which shows that it is very important that the SMEs must take responsibility and accountability toward the product's entire life cycle for ensuring social and ecological stability.

This study is applicable to the general electronics industry. Although the decision-makers judgments may vary as per each firm's requirements, however, the assessment and validation of the judgments can be checked using structural equation modeling in the future.

References

1. Keeble, B.R.: The Brundtland report: 'our common future'. Med. War **4**(1), 17–25 (1988)
2. Voegtlin, C., Scherer, A.G.: Responsible innovation and the innovation of responsibility: governing sustainable development in a globalized world. J. Bus. Ethics **143**(2), 227–243 (2017)
3. Golicic, S.L., Smith, C.D.: A meta-analysis of environmentally sustainable supply chain management practices and firm performance. J. Supply Chain Manag. **49**(2), 78–95 (2013)
4. Saeidi, S.P., Sofian, S., Saeidi, P., Saeidi, S.P., Saaeidi, S.A.: How does corporate social responsibility contribute to firm financial performance? The mediating role of competitive advantage, reputation, and customer satisfaction. J. Bus. Res. **68**(2), 341–350 (2015)
5. Mani, V., Agarwal, R., Gunasekaran, A., Papadopoulos, T., Dubey, R., Childe, S.J.: Social sustainability in the supply chain: construct development and measurement validation. Ecol.

Ind. **71**, 270–279 (2016)

6. Govindan, K., Mangla, S.K., Luthra, S.: Prioritising indicators in improving supply chain performance using fuzzy AHP: insights from the case example of four Indian manufacturing companies. Prod. Plan. Control **28**(6–8), 552–573 (2017)

7. Luthra, S., Govindan, K., Mangla, S.K.: Structural model for sustainable consumption and production adoption—a grey-DEMATEL based approach. Resour. Conserv. Recycl. **125**, 198–207 (2017)

8. Author, F.: Contribution title. In: 9th International Proceedings on Proceedings, pp. 1–2. Publisher, Location (2010)

9. Soleimani, H., Govindan, K., Saghafi, H., Jafari, H.: Fuzzy multi-objective sustainable and green closed-loop supply chain network design. Comput. Ind. Eng. **109**, 191–203 (2017)

10. Opricovic, S., Tzeng, G.H.: Compromise solution by MCDM methods: a comparative analysis of VIKOR and TOPSIS. Eur. J. Oper. Res. **156**(2), 445–455 (2004)

A Fuzzy MCDM Model for Facility Location Evaluation Based on Quality of Life

Aditi, Arshia Kaul, Jyoti Dhingra Darbari and P. C. Jha

Abstract The decision for facility location selection is an important one in the context of management of Supply Chain (SC). Facility location decision affects the overall SC performance, since there is an influence on the cost, delivery speed, service levels and effectiveness of SC. Hence it is essential to evaluate the impact of selection of each facility location. In addition to the traditional criteria such as market, labor, transportation, community and climate, Quality of Life (QOL) has also become an important factor for sustainable facility location selection decision. Since QOL includes aspects of social, economic, environmental, and psychological well-being therefore quantification of QOL for facility location evaluation is a challenging task. Due to subjectivity of the decision makers, vagueness sets in the decision making process and hence fuzzy decision making approach is required for handling the vagueness in assessment of QOL factor. Thus the present study considers an effective integrated approach using Fuzzy Analytical Hierarchy Process (FAHP) and Fuzzy Technique for Order Performance by Similarity to Ideal Solution (FTOPSIS) for evaluation of facility locations, considering the criteria of QOL. FAHP is used for calculating the weight of each QOL criterion and FTOPSIS methodology is used for computing the rank of the facility location options under the fuzzy environment. The application of this integrated approach is applied to case of an Indian manufacturing company.

Keywords Facility location selection · Quality of life · FAHP · FTOPSIS

Aditi (✉) · P. C. Jha
Department of Operational Research, University of Delhi, Delhi, India
e-mail: aditibajpai.du.or.16@gmail.com

P. C. Jha
e-mail: jhapc@yahoo.com

A. Kaul
Asia-Pacific Institute of Management, New Delhi, India
e-mail: arshia.kaul@gmail.com

J. D. Darbari
Department of Mathematics, Lady Shri Ram College, University of Delhi, Delhi, India
e-mail: jydbr@hotmail.com

© Springer Nature Singapore Pte Ltd. 2020
K. N. Das et al. (eds.), *Soft Computing for Problem Solving*,
Advances in Intelligent Systems and Computing 1048,
https://doi.org/10.1007/978-981-15-0035-0_56

687

1 Introduction

Globalization of Supply Chains (SCs) is encouraging manufacturing companies to expand their SC network worldwide. To cater to the needs of the expanded SC, many strategic decisions are to be taken and one of the most crucial decisions in this regard is to choose the best suited location for opening new facilities [1]. Recent trends show that many international companies are setting up new facilities in developing nations, particularly in India. However, setting up facilities can only be done in collaboration with the national firms. The decision of choosing the right facility location is therefore faced with the conflicting opinions of the various SC actors of the firm. Broadly, there are two major SC actors involved in the process of decision making who have conflicting objectives: (1) the state/local government who look for overall growth and development of the society and (2) the firm which is concerned mainly with achieving profit. There are many other key players such as suppliers, retailers, customers and employees who are also affected by the location decision of the facility and hence their concerns must also be addressed.

Traditionally, a firm's decision for the facility location has been driven by factors such as ease of transportation, availability of labor, availability of raw materials, community surrounding the facility, firm's interactions within the market (political, economic, competition) and ingress into markets. In recent times, there is further research that has been carried out on the additional factors that may affect the decisions related to facility location [2]. It can be observed that Quality of Life (QOL) is one such factor, which is gaining special significance while taking decisions for modern facility location. The research in the extant literature suggests that nowadays in addition to traditional factors, a lot more importance is been given to QOL, which is concerned with basic needs and general well-being of individuals and society [3]. The reason is attributed to the rising concerns regarding sustainability. The World Commission on Environment and Development defines sustainability as continuation of QOL for upcoming generation [4] and the Committee of the Regions considers QOL as one of the basic principles of sustainable development [5]. Hence QOL must be considered as an important factor while evaluating the locations for setting up of facilities.

In this regard, the present paper contributes to the research area of facility location selection problem by incorporating QOL factor in the evaluation process. In this regard the research problem addressed in the study involves the following concerns: (i) How to evaluate the QOL factor for facility location selection decision? (ii) Which approach is best suitable for assessing QOL based on both subjective and objective data?

The rest of the paper is organized as follows: the literature review is presented in Sect. 2. In Sect. 3, integration of FAHP and FTOPSIS methodology is elaborated. In Sect. 4, a numerical illustration is presented validating the effectiveness of the proposed methodology for the selection of facility location considering QOL. The conclusion and the future scope of the study are provided in Sect. 5.

2 Literature Review

In the context of management of SC, the decision for facility location selection has been given due importance by both researchers and practitioners. Christodoulou et al. [6] discussed that for global manufacturing firms, the designing and establishment of manufacturing plants or facilities is a long term strategic decision. Oshri [7] discussed that shifting the processes and operations to another facility location could be considered as an effort towards finding solution to current manufacturing problems quickly. The decision of selection of facility location depends on many factors such as market, transportation, labor, community etcetera [8]. Although most firms do not take facility selection decisions with regard to sustainable development, however one factor which has found prominence in the decision making is Quality of Life(QOL). In general, QOL has two aspects: subjective and objective. The detailed understanding of the basic concept of QOL is important for efficient decision making in different fields [9–11]. Darbari et al. [12] researched the conceptual content of QOL, environmental quality and sustainability, and noted the lack of a complete methodological framework of analysis in relation to urban QOL and human welfare, as well as the lack of a specific indicator system. A few have considered the concept of the QOL factor in the decision for location of facilities [3], clearly highlighting the need for further research required for detail understanding of the QOL factor while taking facility location decision.

This leads us to the motivation of our research. The present study aims to analyze the QOL factor in detail for selection of facility location so that the decision has a positive impact on the wellbeing of the community. Since the selection of facility location is a Multi Criteria Decision Making (MCDM) problem and also involves uncertainty in handling the subjectivity of decision makers, therefore many applications of the MCDM techniques integrated with fuzzy logic for facility location selection are found in the literature [13, 14]. Although integrated approach of FAHP and FTOPSIS has been used for solving many problems [15] but the advantages of this approach have not been utilized in quantification of QOL for facility location selection. This paper highlights the efficiency of this integrated approach. The case example and the methodology adopted in the study are discussed below.

3 Problem Definition

The problem addressed in the study is of an Indian manufacturing company wanting to select the best facility location for setting up a new manufacturing plant. Since the firm is inclined towards achieving a sustainable SC, therefore in addition to traditional evaluation factors such as geographical location, transportation, availability of resources, setting up cost and operational cost, the firm also wants to take QOL factor into consideration. The challenge faced by the decision makers (DMs) is how to assess a location based on the QOL factor so that the decision leads to maximum benefit to society but has minimum negative financial impact for the firm. This decision making process is also complicated by the fact that there are conflicting preferences of various stakeholders in the SC and imprecise nature of DMs'judgments. Hence a decision hierarchy is developed in this paper for the DMs of the firm which has the following levels in order: (i) identification of the alternative locations (ii) identification of sub factors of QOL based on which facilities are evaluated (iii)choosing an effective technique for evaluation and selection of facilities.

For the initial screening of the locations an inspection team was formed consisting of experts, academicians and industry associates. The team shortlists six facility location options out of which the firm needs to now select the best facility location. The potential six locations are: Jhilmil (A_1), Kirti Nagar (A_2), Mongolpuri (A_3), Rewari Line (A_4), Wazirpur (A_5) and Sawan Park (A_6). However, the problem still faced by the firm is how to evaluate the six locations based on the QOL criteria. This particular problem is addressed in this study. The research methodology adopted for achieving the purpose is presented below.

4 Research Methodology

In this paper, the integrated FAHP- FTOPSIS approach [16] has been applied for evaluation of facility locations based on the QOL factor. The methodology adopted is broadly discussed below:

1. Identification of QOL criteria for evaluation of facility locations.
2. Determining weights of QOL criteria using FAHP [17].
3. Evaluation of facility locations based on the QOL criteria using FTOPSIS [18].

The schematic representation of the proposed methodology for selection of facility location on the basis of QOL factor is shown in Fig. 1.

4.1 Identification of Criteria for Assessment of QOL

First phase of the proposed methodology is to identify the criteria for QOL assessment for facility location selection.

As per the report of the Economist Intelligence Unit (EIU), nine criteria are used for assessing the QOL which are: (i) healthiness, (ii) gender equality, (iii) family life, (iv) political stability and security, (v) political freedom, (vi) community life, (vii) climate and geography, (viii) material well-being, (ix) job security. Out of the nine criteria, six factors of QOL considered by the decision making team are: (i) health and safety (C_1), (ii) community life (C_2), (iii) geography and climate (C_3), (iv) job security (C_4), (v) infrastructure and transport (C_5) and(vi) gender equality (C_6).

The hierarchy for the problem of selection of facility location is depicted by Fig. 2.

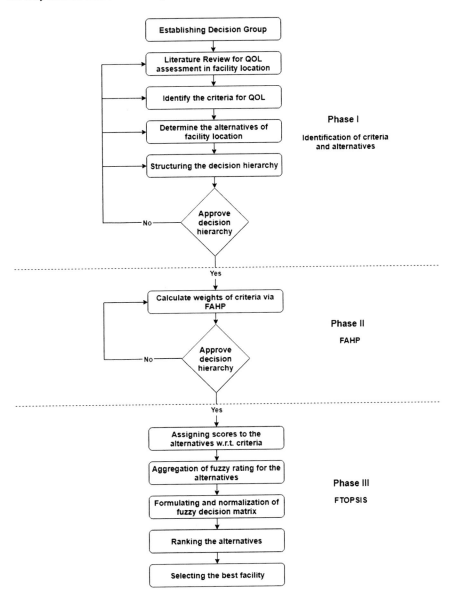

Fig. 1 The schematic representation of the proposed method

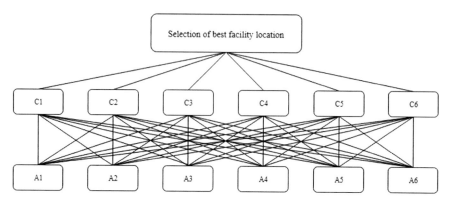

Fig. 2 The decision hierarchy of facility location problem

4.2 Application of FAHP-FTOPSIS Methodology

In phase II, FAHP method is applied for computing the weights of QOL criteria. The AHP methodology proposed by Saaty [19], is a multi- criteria decision making technique. For tackling the unstructured and complex problems, it can be used by forming a structured hierarchy. The main idea behind AHP is to prioritize the preference of each criterion by using numerical values. It can be done for evaluating the importance of each criterion. Due to vagueness in decision making process, fuzzy numbers are used for representing the relative importance of one criterion over the other criterion and obtaining the judgment vectors for each criterion. The steps of FAHP are as follows:

(a) Development of fuzzy pair wise comparison matrix using triangular fuzzy number.
(b) Calculation of geometric mean for each criterion.
(c) Evaluate the fuzzy weights of each criterion.
(d) Defuzzifying the fuzzy weights of each criteria by Centre of area method proposed by Chou and Chang [20].

In the final phase of proposed methodology, ranking of the facility location options is computed using FTOPSIS method. FTOPSIS is one of the multi-criteria decision making technique based on Euclidean distance. The steps of FTOPSIS are as follows:

(a) Development of a fuzzy comparison matrix by assigning the score to alternatives with respect to each criterion.
(b) Aggregation of fuzzy rating for the alternatives.
(c) Formulation and normalization of fuzzy decision matrix.
(d) Develop ranking preference for alternatives.

5 Numerical Validation

The FAHP-FTOPSIS methodology is applied in this section for evaluation of the six alternative locations based on the six quality criteria.

To begin with, the weights for the shortlisted six criteria are computed using FAHP. Table 1 gives the final pair wise comparison matrix for criteria.

The weights of shortlisted criteria C_1, C_2, C_3, C_4, C_5 and C_6 are 0.427, 0.241, 0.024, 0.072, 0.080 and 0.156 respectively.

It can be observed that criterion C_1 has obtained the highest weight and thus it is the most important criterion for facility location selection problem.

Next, we construct the initial FTOPSIS decision matrix. To obtain the decision matrix, DMs are asked to give their opinions using linguistic term and fuzzy triangular scale and a fuzzy decision matrix is constructed. In the decision matrix, the alternatives are compared with respect to each criterion. The fuzzy decision matrices are aggregated to obtain the final decision matrix for each alternative as shown by Table 2.

Table 1 Final pair wise comparison matrix for criteria

	C_1	C_2	C_3	C_4	C_5	C_6
C_1	(1.00, 1.00, 1.00)	(2.50, 3.50, 4.50)	(8.00, 8.50, 9.00)	(4.50, 5.50, 6.50)	(3.50, 4.50, 5.50)	(3.50, 4.50, 5.50)
C_2	(0.24, 0.29, 0.42)	(1.00, 1.00, 1.00)	(5.50, 6.50, 7.50)	(3.50, 4.50, 5.50)	(3.50, 4.50, 5.50)	(1.50, 2.50, 3.50)
C_3	(0.11, 0.12, 0.13)	(0.14, 0.16, 0.19)	(1.00, 1.00, 1.00)	(0.16, 0.19, 0.24)	(0.16, 0.19, 0.24)	(0.12, 0.14, 0.16)
C_4	(0.16, 0.19, 0.24)	(0.19, 0.23, 0.29)	(4.50, 5.50, 6.50)	(1.00, 1.00, 1.00)	(1.00, 1.00, 1.00)	(0.23, 0.29, 0.42)
C_5	(0.19, 0.23, 0.29)	(0.19, 0.23, 0.29)	(4.50, 5.50, 6.50)	(1.00, 1.00, 1.00)	(1.00, 1.00, 1.00)	(0.25, 0.42, 0.75)
C_6	(0.19, 0.23, 0.29)	(0.29, 0.42, 0.75)	(6.50, 7.50, 8.50)	(2.50, 3.50, 4.50)	(1.50, 2.50, 3.50)	(1.00, 1.00, 1.00)

Table 2 Decision matrix for the facility location selection

	C_1	C_2	C_3	C_4	C_5	C_6
A_1	(5, 7.67, 11)	(1, 3, 7)	(1, 4.33, 7)	(5, 8.33, 11)	(1, 5, 11)	(3, 5.67, 9)
A_2	(3, 5.67, 9)	(1, 4.33, 7)	(1, 5.67, 7)	(1, 2.33, 5)	(1, 4.33, 7)	(1, 2.33, 7)
A_3	(3, 7.67, 11)	(3, 7, 11)	(1, 3.67, 9)	(3, 5.67, 9)	(1, 3, 7)	(1, 5, 11)
A_4	(1, 2.33, 5)	(1, 3, 7)	(1, 3, 7)	(1, 2.33, 5)	(1, 4.33, 11)	(1, 4.33, 9)
A_5	(3, 5.67, 9)	(1, 5, 9)	(1, 3, 7)	(5, 7.67, 11)	(1, 4.33, 7)	(1, 2.33, 5)
A_6	(1, 3, 7)	(5, 7.67, 11)	(3, 7, 11)	(3, 5.67, 9)	(5, 8.33, 11)	(1, 3.67, 9)

Subsequently, the normalized decision matrix is obtained as shown in Table 3. After the computation of the normalized decision matrix, the weighted normalized decision matrix is obtained as given in Table 4. Criteria weights which were calculated

Table 3 The resulted normalized decision matrix

	C_1	C_2	C_3	C_4	C_5	C_6
A_1	(0.455, 0.697, 1)	(0.091, 0.273, 0.636)	(0.091, 0.394, 0.636)	(0.455, 0.757, 1)	(0.091, 0.455, 1)	(0.272, 0.515, 0.818)
A_2	(0.273, 0.515, 0.818)	(0.091, 0.394, 0.636)	(0.091, 0.515, 0.636)	(0.091, 0.212, 0.455)	(0.091, 0.394, 0.636)	(0.091, 0.211, 0.636)
A_3	(0.273, 0.697, 1)	(0.273, 0.636, 1)	(0.091, 0.334, 0.818)	(0.273, 0.515, 0.818)	(0.091, 0.273, 0.636)	(0.091, 0.455, 1)
A_4	(0.091, 0.212, 0.455)	(0.091, 0.273, 0.636)	(0.091, 0.273, 0.636)	(0.091, 0.212, 0.455)	(0.091, 0.394, 1)	(0.091, 0.393, 0.818)
A_5	(0.273, 0.515, 0.818)	(0.091, 0.455, 0.818)	(0.091, 0.273, 0.636)	(0.455, 0.697, 1)	(0.091, 0.394, 0.636)	(0.091, 0.212, 0.455)
A_6	(0.091, 0.273, 0.636)	(0.455, 0.697, 1)	(0.273, 0.636, 1)	(0.273, 0.515, 0.818)	(0.455, 0.757, 1)	(0.091, 0.334, 0.818)

Table 4 The weighted normalized decision matrix

	C_1	C_2	C_3	C_4	C_5	C_6
A_1	(0.194, 0.298, 0.427)	(0.022, 0.066, 0.153)	(0.002, 0.009, 0.015)	(0.036, 0.061, 0.080)	(0.014, 0.071, 0.156)	(0.020, 0.037, 0.059)
A_2	(0.116, 0.220, 0.349)	(0.022, 0.095, 0.153)	(0.002, 0.012, 0.015)	(0.007, 0.017, 0.036)	(0.014, 0.061, 0.099)	(0.007, 0.015, 0.046)
A_3	(0.116, 0.298, 0.427)	(0.066, 0.153, 0.241)	(0.002, 0.008, 0.020)	(0.022, 0.041, 0.065)	(0.014, 0.042, 0.099)	(0.007, 0.033, 0.072)
A_4	(0.039, 0.090, 0.194)	(0.022, 0.066, 0.153)	(0.002, 0.007, 0.015)	(0.007, 0.017, 0.036)	(0.014, 0.061, 0.156)	(0.007, 0.029, 0.059)
A_5	(0.116, 0.220, 0.349)	(0.022, 0.110, 0.197)	(0.002, 0.007, 0.015)	(0.036, 0.056, 0.080)	(0.014, 0.061, 0.099)	(0.007, 0.015, 0.033)
A_6	(0.039, 0.116, 0.272)	(0.110, 0.168, 0.241)	(0.007, 0.015, 0.024)	(0.22, 0.041, 0.065)	(0.071, 0.118, 0.156)	(0.007, 0.024, 0.059)

Table 5 Positive ideal solution and negative ideal solution

	C_1	C_2	C_3	C_4	C_5	C_6
A^+	(0.427, 0.427, 0.427)	(0.241, 0.241, 0.241)	(0.024, 0.024, 0.024)	(0.080, 0.080, 0.080)	(0.156, 0.156, 0.156)	(0.072, 0.072, 0.072)
A^-	(0.039, 0.039, 0.039)	(0.022, 0.022, 0.022)	(0.002, 0.002, 0.002)	(0.007, 0.007, 0.007)	(0.014, 0.014, 0.014)	(0.007, 0.007, 0.007)

Table 6 Separation measures, closeness coefficient and rank

	D_i^+	D_i^-	CC_i	Rank
A_1	0.500	0.551	0.525	1
A_2	0.612	0.406	0.399	5
A_3	0.518	0.566	0.522	2
A_4	0.718	0.319	0.308	6
A_5	0.574	0.459	0.444	4
A_6	0.541	0.495	0.478	3

via FAHP are used in this stage for determining the weighted normalized decision matrix.

Next, Positive Ideal Solution (PIS) and Negative Ideal Solution (NIS) are calculated [14]. In our problem, all criteria are benefit criteria. The respective values of PIS and NIS are given in Table 5.

Next, distances of each alternative from PIS and NIS are calculated and the closeness coefficient (CC_i) for each alternative with respect to PIS is computed.

The separation measures, the closeness coefficient and the final ranking of facility locations based on the value of CC_i are shown in Table 6.

The ranking of the alternatives are A_1, A_3, A_6, A_5, A_2 and A_4 in descending order based on the values of closeness coefficient. Therefore, A_1 is the best possible alternative of the facility location for manufacturing taken into consideration quality of life.

6 Conclusion

Facility location selection is an important strategic decision as the operational and tactical decisions are strongly affected by the facility location and thus is crucial decision for the business continuity. Right facility location can also ensure success of the firm in the current global competitive environment. The decision options for DMs while expanding their SC are either to expand the existing facilities or add new facilities while retaining the existing ones or move to a new facility. In case of a new facility opening, one of the many complex decisions is to decide the best alternative location for the facility. The facility selection decision making is complicated owing to the conflicting perspectives of the stakeholders and vagueness in the judgments of the DMs. Thus, there is need of an effective and systematic approach for improving the quality of the decision to be taken. Participation of experienced judgment makers from different areas could enhance accuracy and efficiency of the decision. Moreover, since facility location influences the SC performance, hence evaluation of the locations must be done based on various criteria. The manufacturing company considered in the study wants to understand how to evaluate facility locations based on the QOL performance. QOL includes aspects of social, economic, environmental, and psychological well-being, therefore it is closely related to the three pillars of sustainable development. The study proposes a decision making approach for facility location selection problem incorporating QOL factors using an integrated FAHP-FTOPSIS methodology. The considered factors of QOL are: (i) health and safety, (ii) community life, (iii) geography and climate, (iv) job security, (v) infrastructure and transport, (vi) gender equality. For the final selection of the facility location, the evaluation score of facilities based on QOL can be aggregated with other evaluation scores based on the traditional criteria. The current research provides an effective procedure for best facility location selection. In the future, other MCDM methods could be utilized and a comparative analysis could be carried out to assess the best method for the given situation. Also optimization models could be proposed for integrating the evaluation weights for selection of location of facility while minimizing the cost under other constraints of the company.

References

1. Kalantari, A.H.: Facility location selection for global manufacturing. https://dc.uwm.edu/cgi/viewcontent.cgi?referer=https://scholar.google.co.in/&httpsredir=1&article=1238&context=etd. Accessed 29 Nov 2018
2. Chu, T.-C.: Facility location selection using fuzzy TOPSIS under group decisions. Int. J. Uncertain. Fuzziness Knowl. Based Syst. **10**(6), 687–701(2002)
3. Feneri, A.M., Vagiona, D., Karanikolas, N.: Multi-criteria decision making to measure quality of life: an integrated approach for implementation in the urban area of Thessaloniki, Greece. Appl. Res. Qual. Life **10**(4), 573–587 (2015)
4. World commission on Environment and Development: Report of the world commission on environment and development: our common future. http://www.un-documents.net/our-common-future.pdf. Accessed 29 Nov 2018
5. Lambiri, D., Biagi, B., Royuela, V.: Quality of life in the economic and urban economic literature. Soc. Indic. Res. **84**(1), 1–25 (2007)
6. Christodoulou, P., Fleet, D., Hanson, P., Phaal, R., Probert, D., Shi, Y.: Making the right things in the right places: a structured approach to developing and exploiting manufacturing footprint strategy. Institute for Manufacturing, University of Cambridge, Cambridge, UK (2007)
7. Oshri, I.: Offshoring Strategies: Evolving Captive Center Models. MIT Press, Cambridge, Massachusetts. https://mitpress.mit.edu/books/offshoring-strategies (2011)
8. Chadawada, R., Sarfaraz, A., Jenab, K., Pourmohammadi, H.: Integration of AHP-QFD for selecting facility location. Benchmarking Int. J. **22**(3), 411–425 (2015)

9. Tesfazghi, E.S., Martinez, J.A., Verplanke, J.J.: Variability of quality of life at small scales: Addis Ababa, Kirkos Sub-City. Soc. Indic. Res. **98**(1), 73–88 (2010)
10. Das, D.: Urban quality of life: a case study of Guwahati. Soc. Indic. Res. **88**(2), 297–310 (2008)
11. Marans, R.W.: Quality of urban life studies: an overview and implications for environment-behaviour research. Procedia Soc. Behav. Sci. **35**, 9–22 (2012)
12. Darbari, J.D., Agarwal, V., Yadavalli, V.S., Galar, D., Jha, P.C.: A multi-objective fuzzy mathematical approach for sustainable reverse supply chain configuration. J. Transp. Supply Chain Manag. 11(1), 1–12 (2017)
13. Farahani, R.Z., Asgari, N.: Combination of MCDM and covering techniques in a hierarchical model for facility location: a case study. Eur. J. Oper. Res. **176**(3), 1839–1858 (2007)
14. Farahani, R.Z., SteadieSeifi, M., Asgari, N.: Multiple criteria facility location problems: a survey. Appl. Math. Model. **34**(7), 1689–1709 (2010)
15. Shukla, R.K., Garg, D., Agarwal, A.: An integrated approach of Fuzzy AHP and Fuzzy TOPSIS in modeling supply chain coordination. Prod. Manuf. Res. **2**(1), 415–437 (2014)
16. Kishore, P., Padmanabhan, G.: An integrated approach of fuzzy AHP and fuzzy TOPSIS to select logistics service provider. J. Manuf. Sci. Prod. **16**, 51–59 (2016)
17. Ayhan, M.B.: A fuzzy AHP approach for supplier selection problem: a case study in a Gear motor company. arXiv preprint arXiv:1311.2886 (2013)
18. Nazam, M., Xu, J., Tao, Z., Ahmad, J., Hashim, M.: A fuzzy AHP-TOPSIS framework for the risk assessment of green supply chain implementation in the textile industry. Int. J. Supply Oper. Manag. **2**(1), 548 (2015)
19. Saaty, T.L.: The Analytic Hierarchy Process, vol. 137. New York (1980)
20. Chou, S.W., Chang, Y.C.: The implementation factors that influence the ERP (enterprise resource planning) benefits. Decis. Support Syst. **46**(1), 149–157 (2008)

Analytical Structural Model for Implementing Innovation Practices in Sustainable Food Value Chain

Rashi Sharma, Jyoti Dhingra Darbari, Venkata S. S. Yadavalli, Vernika Agarwal and P. C. Jha

Abstract In the current scenario, sustainability is asserting a profound effect on the global Food Supply Chain (FSC). It is driven primarily by growing consciousness of consumers who want healthy food and at the same time, they demand that food production should not harm the environment. However, sustainability cannot be improved in isolation. It has to be a collaborative effort of all the players involved in the supply chain. This study is aimed at exploring the possibilities for agri-food sector of India to sustainably remunerate as good as its potential. From the perspective of a company engaged in production of an agri-food product, it is a challenging area of research to investigate into the decision-making methodologies which suit the requirements of the stakeholders as well as generate a positive sustainable impact on the FSC. In this study, a detailed analysis is done considering the present sustainable practices followed and scope for future strategies which can be adopted by its Production Plant (PP). This is achieved with the aid of Interpretive Structural Modeling (ISM) technique, a multi-criteria decision methodology and fuzzy-MICMAC analysis. A structured framework is obtained which shows the strength of the impact of each practice on the other. Using the result findings, it has been concluded that the PP must prioritize their efforts in taking measures for water reservation, pollution reduction, creating awareness among farmers and traders, and adopting sustainable

R. Sharma (✉) · P. C. Jha
Department of Operational Research, University of Delhi, Delhi, India
e-mail: rashilakhanpal9@gmail.com

P. C. Jha
e-mail: pcjhadu@gmail.com

J. D. Darbari
Department of Mathematics, Lady Shri Ram College, University of Delhi, Delhi, India
e-mail: jydbr@hotmail.com

V. S. S. Yadavalli
Department of Industrial & Systems Engineering, University of Pretoria, Pretoria, South Africa
e-mail: sarma.yadavalli@up.ac.za

V. Agarwal
Amity International Business School, Amity University, Noida, Uttar Pradesh, India
e-mail: vernika.agarwal@gmail.com

© Springer Nature Singapore Pte Ltd. 2020
K. N. Das et al. (eds.), *Soft Computing for Problem Solving*,
Advances in Intelligent Systems and Computing 1048,
https://doi.org/10.1007/978-981-15-0035-0_57

699

employment practices. This research work can, hence, steer the focus of the company in the direction of appropriately prioritizing their sustainability practices for achieving a sustainable supply chain.

Keywords Sustainable practices · Agri-food supply chain · Interpretive structural modeling · Fuzzy-MICMAC analysis

1 Introduction

Recently, there has been a growing concern for implementation of sustainability practices in Supply Chain (SC) management. The focus is drifting toward long-standing sustainable development rather than short-term growth and on balancing profitability and socio-environmental impacts. Some of the major reasons for the same are government regulations, increasing competition, increasing demand for sustainable products, increasing concern for depleting biodiversity, etc. [1]. In case of agri-food industry, the main drivers behind this adoption are industrialization of agriculture, food safety and quality concerns, government intervention, customer concerns, emergence of modern retailer forms, and multinational corporations [2–4]. The changing consumption pattern and demand for food is also putting tremendous pressure on the Food Supply Chain (FSC) members to embrace sustainability in its operations [5]. In general, an agri FSC network is structured in multi-level echelons such as the input suppliers, producers, intermediaries, processors, retailers, and consumers.FSC operations include procurement, production, storage, processing, marketing, distribution, food services, and consumption [6]. The FSC configuration is quite complex due to the inclusion of characteristics such as product perishability, production seasonality and variability in quantity and quality of supply, and traceability [7]. Although the agri-food industry has immense prospective for sustainability in terms of fulfillment of farmer's needs, employment prosperity, local growth, private enterprise, sustainable utilization, and environmental impact, however, encompassing sustainability concerns in a FSC is all the more challenging due the complex and dynamic nature of the product in focus [8]. It is mainly because efficient policies have to be deployed according to the needs of the agri-food industry and subject to the constraints of cost efficiency of operations, logistics infrastructure, access to resources, seasonal variability, and regulatory conditions [9]. A typical wheat SC involves movement of farm produce from farmer to the end consumer and therefore, the role of farmers in the overall sustainability is crucial.

   In the context of this, the research objective of the present study is to develop an integrated sustainable FSC framework that can be adopted in agri-food sector, particularly in the wheat milling sector, in order to cultivate a sustainable culture which is beneficial for all the members of the SC, in particular for the farmers. As the sustainability issues are industry-specific as well as company-specific, hence, case study of Delhi Flour Mill Company (DFM) based in National Capital Region, India has been taken into consideration. Since sustainability can only be attained with the support of all the stakeholders, therefore, a multi-criteria decision-making approach is required for developing the analytical framework. Eight key emerging sustainable practices are identified by the decision body of the company for management of their SC. Adoption of these may result in enhancement of social well-being of farmers along with significant reduction of the total energy use, product waste, greenhouse emissions, environmental impacts, harmonizing of food safety policies, and traceability systems, etc. Understanding the interrelationships between these practices and their mutual impact on each other is essential for their effective implementation. This is the core objective of the present work which has been accomplished through the use of ISM process and fuzzy-MICMAC analysis. Results of the same provide insightful implications for the company.

The rest of the paper is structured as follows: Sect. 2 gives a brief literature review. Section 3 presents an overview of the flour mill company and presents the list of identified sustainable practices. The methodology adopted is presented in Sect. 4, and Sect. 5 elaborates upon the result and draws useful implications. The conclusion emerging out of the study is summarized in Sect. 6.

2　Literature Review

2.1　Sustainable Food Value Chain

The changing consumption pattern and demand for food is putting tremendous pressure on the FSC members to embrace sustainability in its operations [5]. Smith [10] discussed the importance of engaging consumers and highlighted the importance of cooperation among food manufacturers, retailers, NGOs, governmental and farmers' organizations. The study by [11] evaluated the critical factors for sub-supplier management in a sustainable FSC perspective. In case of developing economies, [12] addressed the issue of the impact of sustainable SC practices on food safety assurance in food firms. Irrespective of the above-mentioned research, the dimension considering sustainability initiatives in FSC planning in Indian context still remains untouched which is the focus of the present.

2.2　ISM Methodology

The review of literature shows that ISM is extremely popular choice for identification of significant barriers [13] and enablers [14] in various industries. Darbari and Agarwal [13] employed ISM methodology to derive the dominant enablers for implementing sustainability initiatives in the Indian retail sector. Darbari et al. [15] analyzed the potential barriers which would hinder the manufacturing organizations from embracing Industry 4.0 using ISM and fuzzy-MICMAC. A number of researchers have contributed in the application of ISM methodology to delve into deeper understanding of implementation of sustainability practices in Agro SC. Kamble et al. [16] evaluated the barriers for implementation of sustainability in Indian FSC.

3 Problem Description

3.1 Overview of the Supply Chain of Delhi Flour Mills

The Delhi Flour Mills Company Limited, part of DFM group, is a century-old public limited enterprise. It is classified as nongovernment company and is registered at Registrar of Companies, Delhi. Figure 1 provides the pictorial representation of their SC.

3.2 Sustainability Practices of Delhi Flour Mills

The company in an effort to abide by the sustainability regulations imposed by the government and pressure mounted by the stakeholders has been involved in few sustainability initiatives which are briefly described in Table 1.

3.3 Objective of the Study

The problem faced by the company is how to draw maximum benefits from these sustainable practices given the limitations of the company and the challenges it faces.

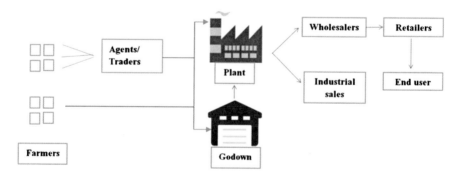

Fig. 1 DFM supply chain

Table 1 Present sustainability practices

S. no.	Practices	Description
1	Alliance with social groups, projects, or institutes for societal development	Increasing sustainability awareness amongst all the members of supply chain
2	Adoption of pollution reduction measures	Reducing pollution caused during all the stages of procurement, production, and distribution
3	Adoption of water conservation measures	Conserving water during all the stages of production
4	Recyclable packaging	Ecofriendly packaging of raw material as well as final products
5	Appropriate quality measures	Strict quality inspection during all the stages of procurement, production, and distribution of wheat
6	Effective measures for attaining a holistic sustainable SC	Proper handling, production and storage of wheat in ways that prevent foodborne illness, flour fortification, reducing waste, feed livestock, etc.
7	Adoption of energy-efficient measures	Providing work from home options to employees, using energy-efficient equipments at the facility, organizing Skype interviews, improving the energy efficiency of the processes or equipments used, etc.
8	Innovative employment practices	Providing opportunities for staff training and education, conducting recreational activities, providing sustainable work environment, etc.

There is scope of improvement in each of these initiatives but it must be done within the economic constraints. Thus, an in-depth analysis of the impact of each of these practices on the overall performance of the SC must be made so that right decisions can be taken at the strategic, tactical, and operational levels. The analysis is carried with the aid of the ISM and fuzzy-MICMAC to attain the following objectives:

- To identify practices which would enable sustainability compliance in the flour mill.
- To rank the identified practices
- To map the relationships between the practices
- To highlight the most imperative practices.

4 Methodology

A prerequisite for ensuring sustainability across the multitier FSC is that the most effective solutions must be recognized by the stakeholders through a multi-criteria decision-making process. The methodology adopted involves data collection through interviews and structured questionnaires and data analysis using ISM methodology and fuzzy-MICMAC.

4.1 Data Collection

To understand and identify the key efforts made by the stakeholders of the mill toward sustaining high performance, a team of six Decision-Makers (DMs) including the senior and junior functionaries were consulted. They were designated as Assistant General Manager (GM)-Wheat Management, GM-Business Operations, GM-Human Resources, GM-works production, Chief Financial Officers, and the Company Secretary. Data collection stressed on the collection of the qualitative and quantitative data relating to the various aspects of sustainability initiatives undertaken by the mill and also the problems faced and prospects that lie ahead. The primary information was gathered through a structured questionnaire for soliciting responses from executives and senior and junior functionaries. The secondary information was collected from the records of the company and also their official website. A list of sustainable practices presented in Table 1 in Sect. 3.2, is prepared based on the information gathered through the data collection process.

Overview of steps of ISM as taken in this study is given below:

Step 1. Identify and define the factors
In this step, factors (sustainable practices) are identified by the team of DMs, as discussed in Sec 4.1, which impact the main objective (of sustainable growth). The identified factors are listed in Table 1.

Step 2. Define the comparative/contextual relationship between these factors
Based on DM's opinions, the contextual relationship is established between the eight factors based on how one influences the other in attaining the objective.

Step 3. Construct the Reachability Matrix
Initial Reachability Matrix is constructed using the binary values' representation of pairwise relationships among factors, where value "1" (or "0") at the (i, j)th entry of the matrix signifies that the ith factor *influences* (*does not influence*) the jth factor. Final Reachability Matrix, displayed in Table 2, is formed by considering all the transitivity's which means if (i, k)th entry and (k, j)th entry is 1 then (i, j)th entry will also be 1.

Step 4. Obtain the Level partitioning
The hierarchical configuration is done by building reachability and antecedent sets from the FRM. Reachability set of practice F_i consists of practices F_j's which are reachable from or are affected by factor F_i, implying that (i, j)th entry of the matrix

Table 2 Final reachability matrix

	F1	F2	F3	F4	F5	F6	F7	F8
F1	1	1	1	1	1	0	1	1
F2	0	1	1	1	1	0	1	1
F3	0	0	1	0	1	0	0	0
F4	0	0	0	1	0	0	0	0
F5	0	0	0	0	1	0	0	0
F6	1	1	1	1	1	1	1	1
F7	0	1	1	1	1	0	1	1
F8	0	0	0	0	0	0	0	1

is "1" Antecedent set consists of practices F_j's which practice F_i gets affected by or is reached from implying that (j, i)th entry of the matrix is "1". At each level m, a practice F_i is allocated level m, if its reachability set is contained in its antecedent set. The corresponding ith row and ith column are eliminated from the matrix for the next level. These iterations are repeated until each practice is allocated a level. The eight sustainable practices considered in the study are allocated levels in four iterations, out of which first and final iterations have been shown in Table 3.

Table 3 Level partitioning

Practices	Reachability set	Antecedent set	Intersection set	Level
		Iteration 1		
1	1, 2, 3, 4, 5, 7	1, 6	1	
2	2, 3, 4, 5, 7, 8	1, 2, 6, 7	2, 7	
3	3, 5	1, 2, 3, 6, 7	3	
4	4	1, 2, 4, 6, 7	4	I
5	5	1, 2, 3, 5, 6	3, 5	I
6	1, 2, 3, 4, 5, 6, 7, 8	6	6	
7	2, 3, 4, 5, 7, 8	1, 2, 6, 7	2, 7	
8	8	1, 2, 6, 7, 8	8	I
		Iteration 4		
1	1	1, 6	1	IV
2	2, 7	1, 2, 6, 7	2, 7	III
3	3	1, 2, 3, 6, 7	3	II
4	4	1, 2, 4, 6, 7	4	I
5	5	1, 2, 3, 5, 6	3, 5	I
6	1, 6	6	6	V
7	2, 7	1, 2, 6, 7	2, 7	III
8	8	1, 2, 6, 7, 8	8	I

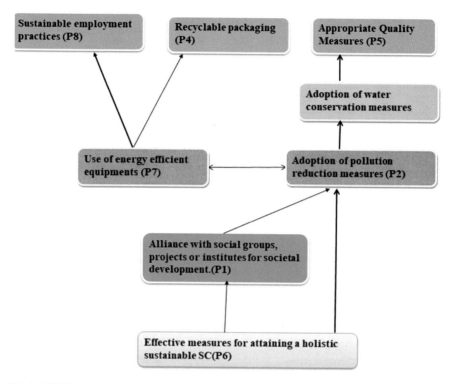

Fig. 2 ISM hierarchal model

Step 5. Developing ISM Model

Diagraph or directed graph is made as given in Fig. 2 in order to represent the hierarchical configuration. Nodes representing the various practices are placed as per the levels along with directed links between the practices.

4.2 Fuzzy-MICMAC Analysis

The major drawback of using ISM is that only binary values of 0 or 1 are considered and there is no option of determining the strength of the relationship between the variables (practices). It is assumed that when there is no linkage between variables, then it is denoted by 0 and when there is linkage, then a value of 1 is used to denote this relationship. The fact that the relationship can be very strong, strong, weak, or very weak is not considered. In order to overcome this limitation of ISM modeling, fuzzy-MICMAC is used for more precise analysis.

The following steps are used to carry out the fuzzy-MICMAC analysis:

Step 1: Construction of Binary Direct Relationship Matrix (BDRM)

The initial reachability matrix is used to construct BDRM. In this step, the diagonal elements are considered as zero.

Table 4 Stabilized matrix

	F1	F2	F3	F4	F5	F6	F7	F8	Dr P
F1	0	0.5	0.3	0.5	0.3	0	0.5	0.5	2.6
F2	0	0.5	0.3	0.5	0.3	0	0	0.4	2
F3	0	0	0	0	0	0	0	0	0
F4	0	0	0	0	0	0	0	0	0
F5	0	0	0	0	0	0	0	0	0
F6	0	0.5	0.2	0.4	0.7	0	0.7	0.7	3.2
F7	0	0	0.3	0.5	0.3	0	0.5	0.5	2.1
F8	0	0	0	0	0	0	0	0	0
De P	0	1.5	1.1	1.9	1.6	0	1.7	2.1	

Step 2: Determination of Fuzzy Direct Relationship Matrix (FDRM)

In this step, fuzzy set theory is utilized to consider the additional possibility of relationship between practices. The possible values of reachability were taken as "0-No", "0.1-Very low", "0.3-Low", "0.5- Medium", "0.7-High", "0.9-Very high", and "1-Full". Based on these values, the opinions of DMs are considered and the BDRM matrix is converted to FDRM.

Step 3: Obtaining fuzzy-MICMAC Stabilized Matrix

The FDRM matrix is multiplied repeatedly until the hierarchies of the driving power and dependence power stabilize [17, 18]. Fuzzy matrix multiplication generates another fuzzy matrix using the given below rule [19].

$$C = A \times B = \max k\left[\min\left(a_{ik}, b_{kj}\right)\right] \text{where } A = [a_{ik}] \text{ and } B = \left[b_{kj}\right]$$

Table 4 gives the stabilized matrix. The driving and dependence power of the practices are derived by summing the entries of the possibilities of interactions in the rows and columns.

Step 4: Classification of factors

In this step, the factors are classified into linkage, driving, autonomous, and dependent factors based on their driving and dependence power.

- **Autonomous factors**: They have less driving as well as dependence power. They are not affected by the other factors nor do they affect the factors.
- **Linkage factors**: They have high driving as well as dependence power. They are not stable as any little variation in them affects the system.
- **Dependent factors**: They have high dependence and less driving power.
- **Driving factors**: They have high driving and less dependence power.

Figure 3 shows the graphical representation of this classification.

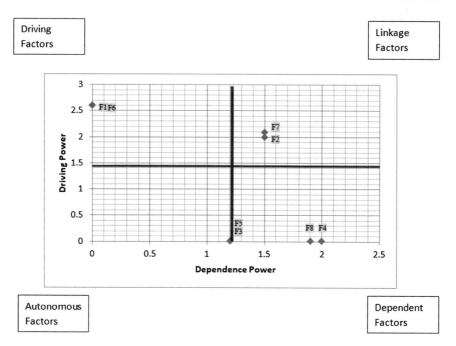

Fig. 3 Classification of factors using fuzzy-MICMAC

5   Result Discussion

In ISM, "Effective measures for attaining holistic sustainable SC" is observed to be at the top level followed by "Alliance with social groups, projects, or institutes to help the firm work toward social sustainability". It is further validated by the fuzzy-MICMAC analysis in which these practices have "Driving" power 3.2 and 2.6, respectively, which is highest among the rest. As observed in Table 3, both these practices also have "0" "Dependence" power. ISM provides the direction and order of the relationship between each practice and fuzzy-MICMAC provides a useful insight for DMs to understand the relationship between "Driving" and "Dependent" factors as classified in Fig. 3. DMs can also comprehend the relative importance and interdependencies among all practices in order to map the detailed framework of all the practices and effect of their implementation on the SC.

Accentuating "Effective measures for attaining holistic sustainable SC", would provide the much needed encouragement to the wheat industry in streamlining the process of sustainable growth of the SC. To implement this practice, the focus needs to be laid on sustainable food safety, consumption, and wastage measures which will automatically take care of related practices like "Adoption of pollution reduction measures". Thus, there will be a significant reduction of the total energy use, product waste, environmental impacts, harmonizing of food safety policies and traceability systems, and enhancement of social well-being. It also implies that for attaining sustainable SC, company will have to do "Alliance with social groups, projects or institutes to help the firm work towards social sustainability". It will help the company to increase sustainability awareness among all the members of the SC, viz, farmers, agents, facility, customer, etc. Creating sustainability awareness at the grassroot level that is the second tier suppliers (or farmers) will have a vertical impact on the whole supply chain.

6 Conclusion

The current study attempts to present a prioritization model for sustainability practices that can be implemented in the wheat mil fuzzy-MICMAC ling sector with a case-based study of DFM Company. The most crucial practices, as identified by the study, will help the company in addressing the sustainability concerns related to farmers and thus, accelerate the overall progress of the wheat SC towards sustainability. A multi-criteria decision-making technique called ISM has been utilized for the purpose of identifying the most imperative practice which helps in making rational decisions in a conflicting environment. It is based on graph theory approach which helps in reducing the complexities involved in the decision-making process. The result obtained from it is reinforced using fuzzy-MICMAC analysis. Therefore, the framework proposed for attaining the desired sustainable FSC for the flour mill is not just a theoretical possibility but is practically feasible.

The inference derived from the study indicates that "Effective measures for attaining holistic SC" is the most pivotal sustainable practice. Company stakeholders must encourage investment in sustainable food safety, consumption, and wastage measures. Also, association with social groups working in the same direction can turn out to be a major step while stepping forward on the ladder of sustainability.

Any new practice introduced by the company impacts all the decisions made at the strategic, operational and tactical level, therefore, the investments and efforts to be made in the redesigned FSC must initially focus on the most significant practices. Infusion of the most impactful practices into the SC network is sure to enhance the sustainable performance of the wheat milling sector. The framework provided by the study highlights the hidden dependencies between all the practices and provides a clear understanding of the areas which require urgent accountability. Accordingly, timely action and emphasis would help in attaining a sustainable value chain from farmers to consumers. The company may face many challenges and discover many new opportunities in their journey toward prudent and effective execution of these practices. A generalized FSC framework can be developed which will overcome the case-dependent nature of the present study and aid in upliftment of farmers from various agro-sectors.

Compliance with Ethical Standards This is an independent and non-funded research study, thus, there are no potential source(s) of conflict of interests. Also, informed consents were taken from all the respondents of the questionnaire.

References

1. Dubey, R., Gunasekaran, A., Childe, S.J., Papadopoulos, T., Luo, Z., Wamba, S.F., Roubaud, D.: Can big data and predictive analytics improve social and environmental sustainability? Technol. Forecast. Soc. Change (2017)
2. Marsden, T., Smith, E.: Ecological entrepreneurship: sustainable development in local communities through quality food production and local branding. Geoforum **36**(4), 440–451 (2005)
3. Maloni, M.J., Brown, M.E.: Corporate social responsibility in the supply chain: an application in the food industry. J. Bus. Ethics **68**(1), 35–52 (2006)
4. Mariani, M.: Sustainable agri-food supply chains and systems. In: Forum China–Europe, Work in Progress, vol. 27, p. 12, June 2007
5. De Haen, H., Réquillart, V.: Linkages between sustainable consumption and sustainable production: some suggestions for foresight work. Food Secur. **6**(1), 87–100 (2014)
6. Jaffee, S., Siegel, P., Andrews, C.: Rapid agricultural supply chain risk assessment: a conceptual framework. Agriculture and Rural Development Discussion Paper, vol. 47 (2010)
7. Kirwan, J., Maye, D., Brunori, G.: Acknowledging complexity in food supply chains when assessing their performance and sustainability. J. Rural Stud. **52**, 21–32 (2017)
8. Chen, K.: Agri-food supply chain management: opportunities, issues, and guidelines. In: Proceedings of the International Conference on Livestock Services, Beijing, People's Republic of China, pp. 16–22, Apr 2006
9. Naik, G., Suresh, D.N.: Challenges of creating sustainable agri-retail supply chains. IIMB Manag. Rev. (2018)
10. Smith, B.G.: Developing sustainable food supply chains. Philos. Trans. R. Soc. Lond. B Biol. Sci. **363**(1492), 849–861 (2008)
11. Grimm, J.H., Hofstetter, J.S., Sarkis, J.: Critical factors for sub-supplier management: a sustainable food supply chains perspective. Int. J. Prod. Econ. **152**, 159–173 (2014)
12. Talib, F., Rahman, Z., Qureshi, M.N.: Analysis of interaction among the barriers to total quality management implementation using interpretive structural modeling approach. Benchmarking Int. J. **18**(4), 563–587 (2011)
13. Darbari, J.D., Agarwal, V.: Identification of key facilitators for adoption of socially sustainable reverse logistics system. BULMIM J. Manag. Res. **1**(2), 112–118 (2016)
14. Darbari, J.D., Agarwal, V.: (2016) An interpretive structural modelling approach for analyzing the enablers towards adoption of sustainability initiatives by retailers. In: Retail Marketing in India: Trends and Future Insights, pp. 103–116. Emerald Group Publishing, India
15. Darbari, J.D., Agarwal, V., Sharma, R., Jha, P.C.: Analysis of impediments to sustainability in the food supply chain: an interpretive structural modeling approach. In: Quality, IT and Business Operations, pp. 57–68. Springer, Singapore (2018)
16. Kamble, S.S., Gunasekaran, A., Sharma, R.: Analysis of the driving and dependence power of barriers to adopt industry 4.0 in Indian manufacturing industry. Comput. Ind. **101**, 107–119 (2018)
17. Gorane, S.J., Kant, R.: Supply chain management: modelling the enablers using ISM and fuzzy MICMAC approach. Int. J. Logist. Syst. Manag. **16**(2), 147–166 (2013)
18. Khan, J., Haleem, A.: An integrated ISM and fuzzy MICMAC approach for modelling of the enablers of technology management. Indian J. Appl. Res. **3**(7), 236–242 (2013)
19. Kandasamy, W.V., Smarandache, F., Ilanthenral, K.: Elementary fuzzy matrix theory and fuzzy models for social scientists. In: Infinite Study (2007)

Mathematical Design and Analysis of Photovoltaic Cell Using MATLAB/Simulink

CH Hussaian Basha, C. Rani, R. M. Brisilla and S. Odofin

Abstract This study explored different models of PV cell, namely, single diode model and double diode models using MATLAB/Simulink Environment. The output power and current characteristics are analyzed for different solar intensity radiations and temperature variations of PV cell. Simulation results are obtained for different atmospheric and temperature conditions. The simulation results reveal that the double diode model generates maximum power and has a higher efficiency compared to single diode model.

Keywords Current and power against voltage curves · Modeling of PV cell · Solar irradiations

Nomenclature

I_{PV}	PV cell current (A)
V_{PV}	PV cell voltage (V)
T_{ref}	Reference temperature of PV cell (Kelvin)
T	Operating temperature of PV cell (Kelvin)
T_n	Nominal temperature of PV cell (Kelvin)
V_{oc}	PV cell open-circuit voltage (V)
K_v	Voltage coefficient (-0.360 V)

CH Hussaian Basha · C. Rani (✉) · R. M. Brisilla
School of Electrical Engineering, VIT University, Vellore, India
e-mail: crani@vit.ac.in

CH Hussaian Basha
e-mail: hussaianbasha.ch@vit.ac.in

R. M. Brisilla
e-mail: brisilla.rm@vit.ac.in

S. Odofin
School of Energy and Environment, University of Derby, Derby, UK
e-mail: s.odofin@derby.ac.uk

© Springer Nature Singapore Pte Ltd. 2020
K. N. Das et al. (eds.), *Soft Computing for Problem Solving*,
Advances in Intelligent Systems and Computing 1048,
https://doi.org/10.1007/978-981-15-0035-0_58

711

K Boltzmann's constant (1.38×10^{-23})
Ns Number of cells in series
Np Number of cells in parallel
G Solar irradiations
G_n Nominal solar irradiations (1000 W/m^2)
K_i Current coefficient (0.06)
q Electrical value (1.6×10^{-19} C)
a_1, a_2 Ideality factor of diode
$I_{0\mathrm{n}}$ Nominal diode saturation current (A)
Ig Photon current of PV (A)
I_sh Shunt resistor current (A)
R_p Shunt resistor (Ω)
R_s Series resistor (Ω)
I_d Current flowing through the diode (A)
$I_{\mathrm{sc\_n}}$ Nominal short-circuit current (A)
V_MPP Maximum peak voltage
I_MPP Maximum peak current
$I_{\mathrm{g\_STC}}$ Photon current at STC

1 Introduction

Fossil fuel consumption leads to environmental pollution and hence nonconventional (or) renewable source of energy plays a major role in world electricity generation. Out of all the renewable sources, solar is the most considerable and useful energy source because of its sustainability, ubiquity, and abundance. Moreover, solar is availing excess in nature and free of cost [1].

At present, PV systems are the most consumed power generation systems and it converts solar energy into useful electrical energy. Moreover, PV is static and quite semiconductor device and it does not consist of any rotating parts. The I-V and P-V curves of a solar cell strongly depend on operating temperature, insolation, and open-circuit voltage of PV. The solar PV operating principle can be explained by the working of a PN diode. When solar photons incident on the P-N semiconductor materials, the electrons in the N-type semiconductor gets energized and moves freely from N-type materials to P-type materials and current flows if the circuit closed as shown in Fig. 1.

The photovoltaic module is modeled by series combination of cells. The series and the parallel combination of modules form a panel. As a result, sunlight incident PV area is increased. Hence the generation of PV power is increased. The series connected modules improve the voltage profile of the PV system and parallel connected PV modules improve the current profile of the PV system [2, 3]. Single and double diode circuit models of a PV cell give the basic power conversion of a PV system. The output power of solar PV and its I-V and P-V characteristics mainly

Fig. 1 Generation of
photovoltaic current

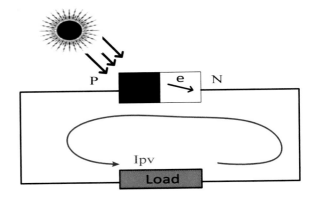

Fig. 1 Generation of
photovoltaic current

depends on solar irradiation incident angle, operating temperature, series resistance
(R_s), parallel resistance (R_p), and diode ideality factor (a) [4, 5]. The continuous
changes in atmospheric conditions, solar PV gives nonlinear I-V and P-V characteristics. So, the maximum power point tracking technique (MPPT) is used to track the
maximum power point of solar PV [5]. The PV cells are mono-, poly-, and thin-film
technology. Mono and polycrystalline PV cells are designed using microelectronic
manufacturing technique and its efficiencies are 13% and 10%, respectively. Thin-
film PV cells are designed using CdTe, a-Si, and CuInse2 semiconductor materials
and its efficiency is 9–10 %. Hence, in this paper, high-efficiency monocrystalline
silicon material is used [6].

Many researchers are updating the modeling of PV system for easy understanding.
The implementation of both diode models of solar PV and its power against voltage
curves are explained at different atmospheric conditions in Sects. 2 and 3. Section 4
gives the step by step PV array design and its nonlinear P-V and I-V characteristics.
Finally, Sect. 5 gives the conclusion of the article.

2 Mathematical Modeling of PV Cell

Basically, PV is a PN diode semiconductor device. The voltage generation of PV cell
is in between 0.5 and 0.8. It is not useful and insufficient for practical use. To achieve
high voltages, a number of PV cells are connected in series to form a module. The
electric current flows in a PN diode based on electromagnetic solar radiations [6].
The series connection of cells and Simulink modeling is given in this section.

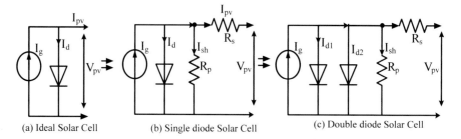

(a) Ideal Solar Cell   (b) Single diode Solar Cell   (c) Double diode Solar Cell

Fig. 2 The equivalent circuit of, **a** ideal diode, **b** single diode and **c** double diode PV cell

2.1 Ideal PV Cell Modeling

The ideal PV cell is given in Fig. 2a and it consists of a diode connected in parallel with the photon current. The model Eq. 1 describing the ideal PV cell.

$$I = I_g - I_0 \left(e^{\left(\frac{qV_d}{nKT} \right)} - 1 \right) \tag{1}$$

2.2 Mathematical Modeling of Single Diode PV Cell

The single diode PV cell includes a parallel resistance (R_p) and series resistance (R_s) along with the diode. The single diode model equivalent circuit of PV is given in Fig. 2b and its extraction parameters are taken from the article [5] and it is given in Table 1. From Fig. 2b, the photon current is written as

$$I_{pv} = I_g - I_d - I_{sh} \tag{2}$$

and

$$I_g = \left(I_{g\_STC} + K_i \Delta T \right) * \frac{G}{G_{STC}} \tag{3}$$

Table 1 Single diode PV cell operating parameters

Parameters	Values	Parameters	Values
N_p	1	I_{sc_n}	4.252 A
N_s	36	V_{oc}	20.359 V
a	1.3	I_{g_STC}	5.243 A
R_s	0.392 Ω	I_{MPPT}	4.8 A
R_p	149.361 Ω	V_{MPPT}	15.12 V

The current flowing through the diode and shunt resistor is derived as

$$I_d = I_o\left(e^{\left(\frac{V_{pv}+I_{pv}R_s}{aV_t}\right)} - 1\right) \tag{4}$$

where $V_t = \frac{N_s K T}{q}$

$$I_{sh} = \frac{V_{pv} + I_{pv}R_s}{R_p} \tag{5}$$

The diode reverse saturation current is derived as

$$I_o = \frac{I_{sc\_STC} + K_i\Delta T}{e^{\left(\frac{V_{oc\_STC}+K_v\Delta T}{aV_t}\right)} - 1} \tag{6}$$

2.3 Mathematical Modeling of Double Diode PV Cell

The double diode solar cell is modeled by adding an additional diode in parallel with the shunt connected resistor of the single diode PV cell [7–9]. The double diode equivalent circuit model of PV is illustrated in Fig. 2c. The modeling of single and double circuit model is explained by using global PV cell given in Fig. 3.

From Fig. 2c, the output current of PV is derived as

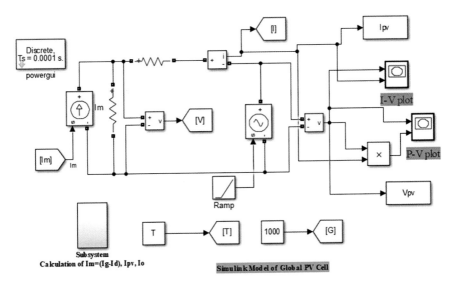

Fig. 3 Simulink model of global PV cell

Table 2 Double diode operating parameters

Parameters	Values	Parameters	Values
N_p	1	R_p	103.326 Ω
N_s	36	I_{sc_n}	4.2 A
A	1.0	$I_{01} = I_{02}$	$8.234 \times e^{-10}$ A
$a1$	1.21	V_{oc}	20.359 V
R_s	0.5 Ω	I_{g_STC}	5.432 A
I_{MPPT}	4.78 A	V_{MPPT}	15.10 V

$$I_{pv} = I_g + I_{d1} - I_{d2} - \frac{V_{pv} + I_{pv} R_s}{R_p} \tag{7}$$

where

$$I_{d1} - I_{d2} = I_{01}\left(e^{\left(\frac{V_{pv}+I_{pv} R_s}{a_1 V_{t1}}\right)} - 1\right) - I_{02}\left(e^{\left(\frac{V_{pv}+I_{pv} R_s}{a_2 V_{t2}}\right)} - 1\right) \tag{8}$$

and

$$V_{t1} = V_{t2} = \frac{N_s K T}{q}$$

The double diode model design parameters are taken from the article [5] and it is given in Table 2.

3 Performance Analysis of Single and Double Diode PV Cells

The Simulink model of single and double diode cell is given in Fig. 3 which is used to plot the I-V and P-V characteristics given in Fig. 4, 5, 6, 7, 8, 9 and 10. The physical working behavior of both PV systems is related to I_g, I_d, R_p, and R_s. At standard test condition (1000 W/m^2), temperature ($T = 25$ °C), series resistance ($R_s = 0.39$), and parallel resistances ($R_p = 149.39$), both the diode models of the I-V and P-V curves are shown in Fig. 4a, b. For modeling of single diode circuit PV type, there are four parameters required which are short-circuit current (I_{sc}), open-circuit voltage (V_{oc}), a peak current of PV (I_{MPPT}), and peak voltage (V_{MPPT}) [3]. The double diode circuit PV cell modeling is done by adding an additional diode parallel with the shunt resistance in a single diode PV array. The double diode circuit PV cell modeling requires two more factors compared to a single diode circuit, diode ideality factor 'a', and reverse saturation current 'Io'.

At static irradiation condition, the current against voltage, and power curves parameters are investigated and it is given in Table 3 and it is observed that the

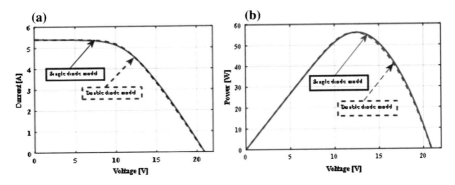

Fig. 4 **a** Current against voltage and, **b** P-V curves at 1000 W/m^2

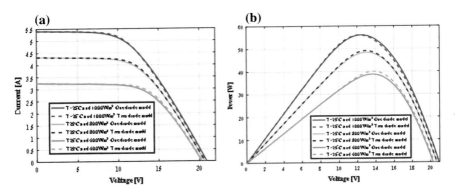

Fig. 5 **a** I-V and, **b** P-V characteristics of PV cell at (1000, 800 and 600 W/m^2)

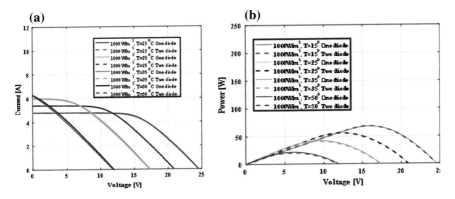

Fig. 6 **a** I-V curve and, **b** P-V curves at different temperatures

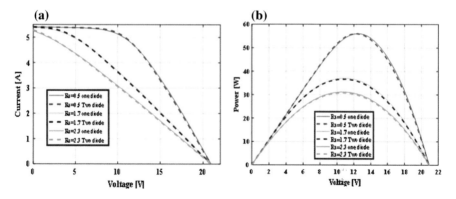

Fig. 7 **a** I and V relationship and, **b** P and V relationship for different series resistances

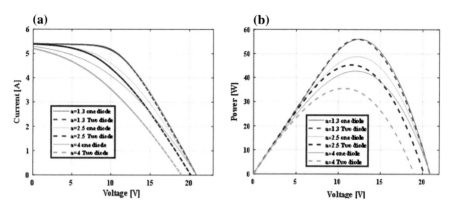

Fig. 8 **a** I-V and, **b** P-V curves at different ideality factors condition

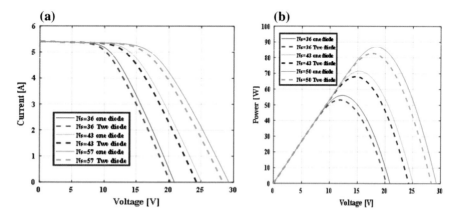

Fig. 9 **a** I and V and, **b** P-V curves for different series connected cells

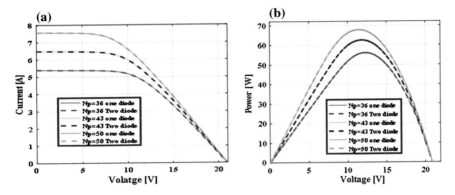

Fig. 10 a I-V curves and, b P-V curves for different parallel connected cells

double diode PV panel power generation is slightly higher than the single diode. In addition, from the I-V curves, the current is decreasing in opposite direction to the power in the P-V curve. At one point in the PV curve, power reaches the maximum point. That is called a maximum power point of the PV cell.

3.1 Variation of Irradiances (G)

From Eqs. (1) and (3), the power generation of the PV is majorly depending on the solar irradiances. The dynamic characteristics of the PV cell under different irradiation conditions (1000, 800, and 600 W/m^2) are shown in Fig. 5a, b. However, at constant temperature ($T = 25$ °C), when the solar irradiances step down from 1000 to 800 W/m^2 the power generation of the PV cell gets reduced to 6 to 7 W approximately.

3.2 Variation of Temperature (T)

From Eq. (6), it is noted that the reverse saturation current of PV cell depends on the operating temperature of the PV. At different atmospheric temperature conditions, the PV array I-V and P-V characteristics are analyzed and given in Fig. 6a, b.

From Fig. 6a, b, at static irradiation condition (1000 W/m^2), the step by step temperature of the cell varies. As a result, the PV cell current varies in ascending manner and its corresponding open-circuit voltage varies in descending way. For different ascending temperature conditions, the maximum power point voltage of PV is measured and it is recorded in Table 3.

Table 3 Performance parameters of single and double diode PV cell

Changing parameter	Single diode model PV cell				Double diode model PV cell			
Irradiance (W/m^2)	V_{MPP} (V)	I_{MPP} (A)	P_{MPP} (W)	FF	V_{MPP} (V)	I_{MPP} (A)	P_{MPP} (W)	FF
1000	12.41	4.52	56.10874	0.648	13.55	4.284	58.0597	0.670
800	14.18	3.544	50.2790	0.580	13.80	3.525	48.6562	0.562
600	14.40	2.792	40.21563	0.464	14.84	2.657	39.4516	0.455
400	14.98	2.560	38.3518	0.443	14.90	2.027	30.21399	0.349
200	14.32	1.932	29.7025	0.343	15.76	1.275	20.112	0.232
Temperature (°C)								
15	17.02	3.272	55.78	0.771	17.20	3.384	58.22	0.788
25	13.41	4.152	55.710	0.643	13.60	4.176	56.822	0.656
35	8.466	5.30628	39.2832	0.45377	8.560	4.7875	40.981	0.4733
50	4.0418	6.4506	22.8501	0.2639	3.5313	6.4932	22.963	0.2652
55	2.803	6.501	18.2201	0.2104	2.8301	6.7206	19.02	0.21970
Resistance (R_s)								
0.5	13.28	4.216	56	0.6394	13.27	4.218	56.83	0.6489
1.7	12.02	3.105	37.27	0.4256	12.82	2.9641	38.00	0.4339
2.3	11.36	2.799	31.80	0.3632	12.120	2.656	32.20	0.3677
2.9	11.12	2.401	26.7	0.3048	11.500	2.408	27.70	0.3163
3.5	10.02	2.1397	21.44	0.24437	11.102	2.0689	22.80	0.2603
Ideality factor (a)								
1.3	13.8	3.8637	53.32	0.6088	14.02	3.923	55.01	0.6280
2.5	13.02	3.4562	45.001	0.5138	13.6	3.529	48	0.5481
4	11.82	2.893	34.200	0.3905	12.83	2.805	36	0.41109
5.5	9.20	2.888	26.00	0.2969	11.02	2.196	24.2	0.2763
Series cells (N_s)								
36	21.22	4.147	88.2	0.7769	22.02	4.055	89.3	0.7866
43	17.52	4.1381	72.5	0.6386	18.23	4.0482	73.8	0.6501
50	15.38	3.8036	58.5	0.5153	16.01	3.70	59.2	0.5214
57	12.07	3.55426	42.9	0.3779	13.026	3.3225	43.28	0.3812
Parallel cells (N_p)								
36	13.2	5.1515	68.00	0.599	13.24	5.2288	69.23	0.609
43	12.92	4.953	64.00	0.5637	13.12	4.974	65.26	0.5748
50	11.83	4.6491	55.12	0.4855	11.89	4.3460	56.02	0.4933
57	10.22	4.1291	42.20	0.3717	11.22	3.8413	43.10	0.3796

3.3 Variation of the Series Resistor (R_s)

From Eqs. (4) and (5), the output current of a PV cell is directly depended on the series resistance of a PV cell. For easy understanding of the PV systems performance, the series resistance of PV cell is neglected and sometimes it is taken as '0'. However, by varying the series resistance of PV the I-V and P-V characteristics are analyzed and are shown in Fig. 7a, b. For different series resistances, the slope of I-V and P-V characteristics vary. Hence, the maximum power point of PV gets deviated. Moreover, for different series resistances, the maximum voltage and current parameters are calculated and are mentioned in Table 3.

From the maximum voltage and power parameters, it is observed that the MPP of PV is varying inversely proportional to the series resistance of the PV cell.

3.4 Variation of Diode Ideality Factor (a)

Ideality factor of the diode depends on P-N junction operating temperature and type of semiconductor material used. Ideality factor of a single diode is assumed to be constant. But in practice, it is a function of the voltage across the P-N diode and high voltage application it is taken as one. Here, the diode ideality factor of both the diode models are $a_1 = 1$ and $a_2 = 1.3$, respectively. For different ideality factors, the 'I' against 'V' curves are given in Fig. 8a, b.

3.5 Variation of Series Connected Cells

By connecting additional cells in series to the existing cells, the power generation of PV system increases because the voltage of the PV increases while the output current of PV cell remains constant as shown in Fig. 9a, b. If any mismatch in series connected cells, the open-circuit voltage, and short-circuit current create a heating effect on power generation of PV cell. As a result, the efficiency of the PV cell gradually decreases. However, in series connection uniform current flow is required in all the cells.

3.6 Variation of Parallel Connected Cells

The cells are connected in parallel for the high PV current application. The mismatch effect in parallel connection reduces the power generation of the PV cell. Hence, for connecting cells in parallel, the current rating of all the cells must be constant. For different parallel connected cells, the I and V, P and V characteristics are investigated and it is shown in Fig. 10a, b.

3.7 Fill Factor

The ratio of peak power to V_{oc} and I_{sc} is known as fill factor. From Fig. 11, The fill factor of PV cell is derived as

$$FF = \frac{V_m I_m}{V_{oc} I_{sc}} \tag{9}$$

where V_m = Maximum voltage of PV and I_m = Maximum current of PV.

Fill factor is used to measure the quality of the PV cell. For different radiances, temperatures, ideality factors, and series resistances, the fill factor of the PV cell is calculated and it is given in Table 3.

4 Mathematical Modeling of PV Array

Combinations of hundreds of solar cells create an array [10]. MATLAB/Simulink model of both diode models-based global solar arrays can be explained using Fig. 3. Here, a single and double diode-based PV array is designed by multiplying the number of series and parallel modules ((N_{ss} = 20) * (N_{pp} = 10)). The mathematical equations of a single and double diode array-based PV arrays are explained in this section.

Fig. 11 Evaluation of fill factor

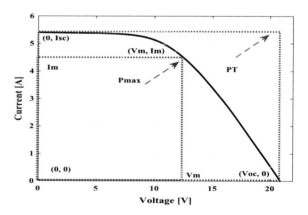

4.1 Mathematical Modeling of Single Diode-Based PV Array

The output current of PV array directly depends on the ratio of series and parallel connection cells. When the ratio increases the corresponding power generation of PV array also increases.

$$I_{pv} = I_g N_{pp} + I_0 N_{pp} \left(e^{\left(\frac{V_{pv} + I_{pv} R_s N}{a V_t N_{ss}} \right)} - 1 \right) - I_{sh} \tag{10}$$

$$I_{sh} = \frac{V_{pv} + I_{pv} R_s N}{R_p N} \tag{11}$$

$$I_g = \left(I_{g\_STC} + K_i \Delta T \right) * \frac{G}{G_n} \tag{12}$$

where $N = \frac{N_{ss}}{N_{pp}}$.

4.2 Mathematical Modeling of Double Diode-Based Array

The current in double diode array is given by (Fig. 12)

$$I_{pv} = I_g N_{pp} + I_{d1} - I_{d2} - I_{sh} \tag{13}$$

$$I_{d1} = I_{01} N_{pp} \left(e^{\left(\frac{V_{pv} + I_{pv} R_s N}{a_1 V_{t1} N_{ss}} \right)} - 1 \right) \tag{14}$$

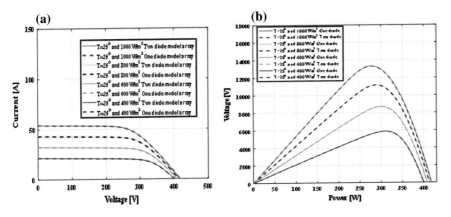

Fig. 12 **a** I and V and, **b** P and V curves of one diode array at 25 °C

$$I_{d2} = I_{02} N_{pp} \left(e^{\left(\frac{V_{pv} + I_{pv} R_s N}{a_2 V_{t2} N_{ss}} \right)} - 1 \right) \tag{15}$$

The shunt current through the resistor is the same as Eq. (10). The I-V and P-V characteristics of the array are analyzed at different irradiations and temperature conditions as shown in Figs. 13a, b and 14a, b. For different irradiances, temperature conditions, the PV output maximum power (P_{mpp}), voltage (V_{mpp}), current (I_{mpp}), and fill factor (FF) parameters are evaluated and it is given in Table 4.

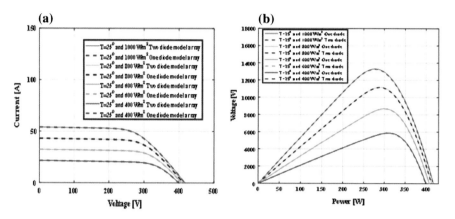

Fig. 13 **a** I and V and, **b** P-V curves of double diode array at 25 °C

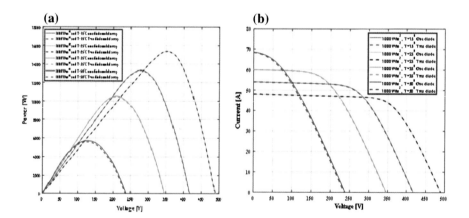

Fig. 14 **a** I and V and, **b** P-V curves of double diode array at 1000 W/m^2

Table 4 Investigation of PV array for different atmospheric conditions

Changing parameters	One diode PV				Two diode model PV cell			
Irradiance (W/m^2)	V_{MPP} (V)	I_{MPP} (A)	P_{MPP} (W)	FF	V_{MPP} (V)	I_{MPP} (A)	P_{MPP} (W)	FF
1000	322	40.3726	13,000	0.866	323	40.00	13,120	0.8746
800	310	40.322	12,500	0.833	315	39.9365	12,580	0.8386
600	285	41.402	11,800	0.7866	300	39.366	11,810	0.78733
400	270	41.84	11,296	0.7530	292	39.0344	11,320	0.75466
200	260	41.92	10,900	0.7266	280.5	38.5964	11,000	0.7333
Temperature (°C)								
15	323.52	40.8	13,200	0.88	330.45	40.98	13,300	0.8862
25	312.18	40.722	12705.2	0.8466	314.15	40.681	12,780	0.852
35	289.827	41.9	11900.27	0.7933	299.98	39.733	11,920	0.79466
50	271.88	41.922	11,398	0.7598	298.2	38.255	11,400	0.76
55	244.56	41.98	10,267	0.6844	280.01	36.792	10,302	0.6868

5 Conclusion

The equivalent circuits of single and double diode models of PV cell and its corresponding arrays are modeled using MATLAB/Simulink. The performance of these PV models is analyzed for different solar intensity radiations and temperature conditions by plotting I and V and P and V curves. From this analysis, we conclude that, the double diode PV system generates accurate maximum power and has a greater efficiency as compared to a single diode PV array.

Acknowledgements I would like to thank the **University Grants Commission (Govt. of India)** for funding my research program and I especially thank VIT University management for providing all the facilities to carry out my research work.

References

1. Tsai, H.-L.: Insolation-oriented model of photovoltaic module using Matlab/Simulink. Sol. Energy **84**(7), 1318–1326 (2010)
2. Chin, V.J., Salam, Z., Ishaque, K.: An accurate and fast computational algorithm for the two-diode model of PV module based on a hybrid method. IEEE Trans. Ind. Electron. **64**(8), 6212–6222 (2017)
3. Ishaque, K., Salam, Z.: A comprehensive MATLAB Simulink PV system simulator with partial shading capability based on the two-diode model. Sol. Energy **85**(9), 2217–2227 (2011)
4. Ram, J.P., Sudhakar Babu, T., Rajasekar, N.: A comprehensive review on solar PV maximum power point tracking techniques. Renew. Sustain. Energy Rev. **67**, 826–847 (2017)

5. Rani, C., Hussaian Basha, C.H., Odofin, S.: Analysis and comparison of SEPIC, Landsman and Zeta converters for PV fed induction motor drive applications. In: 2018 International Conference on Computation of Power, Energy, Information and Communication (ICCPEIC). IEEE (2018)
6. Rani, C., Hussaian Basha, C.H.: A review on non-isolated inductor coupled DC-DC converter for photovoltaic grid-connected applications. Int. J. Renew. Energy Res. (IJRER) 7(4), 1570–1585 (2017)
7. Taheri, H., Taheri, S.: Two-diode model-based nonlinear MPPT controller for PV systems. Can. J. Electr. Comput. Eng. 40(2), 74–82 (2017)
8. Cárdenas, A.A., et al.: Experimental parameter extraction in the single-diode photovoltaic model via a reduced-space search. IEEE Trans. Ind. Electron. 64(2), 1468–1476 (2017)
9. Boutana, N., et al.: An explicit IV model for photovoltaic module technologies. Energy Convers. Manag. 138, 400–412 (2017)
10. Rani, C., Hussaian Basha, C.H., Odofin, S.: Design and switching loss calculation of single leg 3-level 3-phase VSI. In: 2018 International Conference on Computation of Power, Energy, Information and Communication (ICCPEIC). IEEE (2018)

Development of Cuckoo Search MPPT Algorithm for Partially Shaded Solar PV SEPIC Converter

CH Hussaian Basha, Viraj Bansal, C. Rani, R. M. Brisilla and S. Odofin

Abstract Photovoltaic (PV) power generation is playing a prominent role in rural power generation systems due to its low operating and maintenance cost. The output properties of solar PV mainly depend on solar irradiation, temperature, and load impedance. Hence, the operating point of solar PV oscillates. Due to the oscillatory behavior of operating point, it is difficult to transform maximum power from the source to load. To maintain the operating point constant at the maximum power point (MPP) without oscillations, a maximum power point tracking (MPPT) technique is used. Under partial shading condition, the nonlinear characteristics of PV comprise of multiple maximum power points (MPPs). As a result, discovering true MPP is difficult. The traditional and neural network MPPT methods are not suitable to track the MPP because of oscillations around MPP and impreciseness in tracking under partial shading (PS) condition. Therefore, in this article, a biological intelligence cuckoo search optimization (CSO) technique is utilized to track and extract the maximum power of the solar PV at two PS patterns. MATLAB/Simulink is used to demonstrate the CSO MPPT operation on SEPIC converter.

Keywords CS MPPT · Duty cycle · PV cell and · Partial shading

CH Hussaian Basha · V. Bansal · C. Rani (✉) · R. M. Brisilla
School of Electrical Engineering, VIT University, Vellore, India
e-mail: crani@vit.ac.in

CH Hussaian Basha
e-mail: hussaianbasha.ch@vit.ac.in

V. Bansal
e-mail: viraj12800@gmail.com

R. M. Brisilla
e-mail: brisilla.rm@vit.ac.in

S. Odofin
School of Energy and Environment, University of Derby, Derby, UK
e-mail: s.odofin@derby.ac.uk

© Springer Nature Singapore Pte Ltd. 2020
K. N. Das et al. (eds.), *Soft Computing for Problem Solving*,
Advances in Intelligent Systems and Computing 1048,
https://doi.org/10.1007/978-981-15-0035-0_59

727

1 Introduction

In the past few decades, the use of solar electricity is increasing for household, industries, vehicles, satellite systems, solar updraft tower, fuel production, and water pumping systems [1–3]. The incidence of solar radiation on PV is converted to heat and used to generate electrical energy. PV effect is the property used to convert solar energy into direct current. When solar radiation falls on the active surfaces of the PV panel, it starts absorbing heat energy resulting in the recombination of electrons and holes. Solar technology is very useful and popular these days because additional sources such as water, fuel, and transportation are not required, and it does not have any rotating parts [4–6].

The advantage of the solar power plant is that it has low operating and maintenance cost. But, the cost of installation is very high, which can be minimized by using different PV manufacturing technologies. One of the major problems we are facing in power generation industry at present is the irregularity of the power supply from the central grid. During this situation, to overcome this discontinuity, diesel generators are used for backup supply. But it produces hazardous inflammable gases and has a direct impact on the environment. Moreover, the cost of generation increases [7]. Hence, the most popular renewable and eco-friendly source is solar energy.

Due to the stochastic nature of atmospheric light conditions, the output characteristics of the PV is nonlinear. To extract the maximum power from the solar PV, MPPT methods are applied to move the operating point of PV closer to MPP [8]. Solar power systems are usually installed in the shadow-free rural regions, unlike urban areas where tall trees and high-rise buildings cause PS on PV system. Because of this PS condition, multiple MPP peaks are generated on Power against Voltage, and Current against Voltage graphs [9]. In Fig. 1, two local peaks ("2" and "3") and one global MPP are observed. We neglect the local peaks and use the global peak to extract the maximum power from solar PV.

The traditional MPPT methods are utilized in constant irradiation conditions to improve the efficiency of a solar PV. Such methods are incremental resistance (IR),

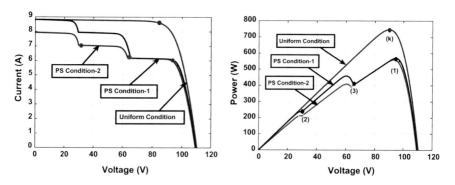

Fig. 1 I-V and P-V characteristics at different partial shading patterns

perturb and observe (P&O), incremental conductance (IC), and hill climb (HC). These methods track MPP efficiently with high convergence under constant irradiation conditions. The major drawback of these traditional MPPT methods is the disability to suppress the operating point of PV oscillations around MPP. Due to the oscillatory behavior of operating point, power loss occurs in the PV system at steady-state condition and they don't give accurate MPP for shading [10].

To overcome the disadvantages of classical MPPT techniques, artificial intelligence methods are used to track MPP of solar PV under partial shading. Such algorithms are neural network, genetic algorithm, ANFIS, and fuzzy logic. Though the flexibility of ANFIS techniques is poor, they give high convergence speed and accuracy in MPP tracking. For instance, the fuzzy MPP controller mainly depends on the user's knowledge and experience of having applied fuzzy in a PV system and selecting a membership function [11, 12].

Because of its high training time, the accuracy in MPP tracking is poor but artificial neural network gives good convergence speed. ANN is implemented using costly microprocessors. Hence, the implementation cost of the ANN-based MPPT controller is high. The genetic algorithm optimization technique is inspired by biological evolution. The major drawback of GA is the difficulty in obeying inequality constraints, sensitivity in the initialization of population, and deterioration of solution quality with an increase in problem size [13, 14].

The biological algorithms are applied to solve the linear and nonlinear problems and the most popular MPPT method is cuckoo search. Cuckoo search gives less oscillations around MPP and its implementation is simple [15, 16]. The diagram of SEPIC converter fed PV under PS condition is given in Fig. 2.

The partial shading behavior of PV array is explained in Sect. 2 and the importance of MPPT under partial shading is explained in Sect. 3. Section 4 gives the CSO MPPT algorithm operation. Finally, Sects. 5 and 6 give the simulation results and conclusion of the article.

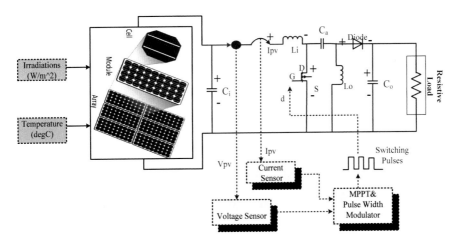

Fig. 2 Block diagram of an MPPT controlled PV fed SEPIC converter system

2 Modeling of PV Cell

The modeling of PV parameters has taken from article [17]. In this, the modeling parameters depend on the solar irradiation's incident angle with the PV array and atmospheric temperature conditions. In this work, a single diode circuit model PV array is considered to implement the PV panel because of the advantages being simple in design and easy to analyze the PV performance. There are four parameters required to model the single diode circuit PV array. Those are open-circuit voltage (Voc), series resistance (Rs), short circuit current (Isc), and shunt resistance. The MATLAB Simulink is used to model the PV cell. The diode (d) is connected is across the current source (Ipv) to model PV cell which is given in Fig. 3 and its I-V and P-V characteristics are drawn at different irradiations and it is given in Fig. 1.

$$I = I_{ph}N_{pp} - I_0 N_{pp}\left(\exp\left(\frac{V + I R_s(N_{ss}/N_{pp})}{a * V_t * N_{ss}}\right) - 1\right) - \frac{V + I R_s(N_{ss}/N_{pp})}{R_p * (N_{ss}/N_{pp})} \quad (1)$$

where I is the output current of PV cell, I_{ph} is the photon current, and I_0 is the reverse saturation current. The power rating of the PV panel is improved by increasing the series (N_{ss}) and parallel (N_{pp}) connection of cells.

$$I_{ph} = \left(I_{ph\_STC} + k_I \Delta T\right) * G/G_{STC} \quad (2)$$

From Eq. (2), the generated photocurrent varies in direct proportion to the change in atmospheric temperature and irradiation conditions. The diode reverse saturation current is given in Eq. (3) and it depends on the diode manufacturing material property and operating temperature.

$$I_0 = \frac{I_{sc\_STC} + K_i \Delta T}{\exp\left((V_{oc\_STC} + K_v \Delta T)/\{a/p\}V_t\right) - 1} \quad (3)$$

The thermal voltage of PV is derived from the Boltzmann constant (k) and the temperature across PV-junction and it is given in Eq. (4).

Fig. 3 Equivalent circuit of PV cell

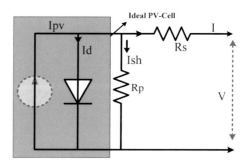

$$V_t = \frac{N_s * KT}{q} \tag{4}$$

3 Partial Shading of PV Array

In urban centers, the PV systems are installed on the rooftop of the buildings. The shading effect occurs on the panel due to the shadows of high-rise buildings and tall trees. Since the power output from a solar PV depends on solar irradiation and temperature, the performance in urban areas is affected [15]. Additionally, the high-irradiated PV panels lose their power to low-irradiated panels, and the energy is dissipated in the form of heat. As a consequence, the overall power output decreases. To overcome this drawback, a metaheuristic CSO MPPT technique is implemented in this work [16]. Parallel diodes are connected across each panel to limit the current passing to the shaded panel. There are three cells connected in series to increase the power rating of the PV. For each shading pattern, different MPP peaks occur on power against voltage curve and it is shown in Fig. 1. There are two shading patterns considered in this work to analyze the operation of partial shaded solar PV. In first pattern, solar irradiations are considered as 900, 800, and 700 W/m² and its peak powers are 200, 415, and 574 W, respectively. The second pattern of PV system is 1000, 900, and 700 W/m² and its corresponding MPP values are 270, 475, and 580 W, respectively. Hence, from the performance of I-V curves, it is observed that the partial shading behavior creates multiple MPPs and it is difficult to transfer the maximum power from source to resistive load.

4 Cuckoo Search MPPT Tracking

The importance of MPPT in PV operation is to extract maximum power at different weather conditions and improving PV system efficiency by evaluating the maximum peak voltage and current. There are different MPPs on power against voltage curves under partial shading. There are many MPPT techniques used to track the MPP under different bad weather conditions. But some techniques track MPP with high speed at constant uniform irradiation conditions and some other techniques gives accurate results in tracking MPP. To overcome this issue, a recently existed metaheuristic CS technique is used in this work to track global MPP with high accuracy and speed [18]. The CSO consists of many advantages because of its easy implementation, simplicity, and it can solve the real-time nonlinear problems effectively and efficiently with less periodic tuning constraints. Hence, the convergence speed of CSO is high.

4.1 Behavior of Cuckoo Birds

The CSO MPPT technique convergence is mainly depending on the Levy flight function and it is inspired by a parasitic swarm intelligence of cuckoo birds. The broad parasitic will be performed by a different type of cuckoo birds. The parasitic search involved three steps: intraspecific, takeover, and cooperative. Some cuckoo birds mimic the color and shape of the host birds to increase its production capability. Most of the birds lay on their own eggs up to certain time. As a result, chances of the host birds occupying other birds eggs are less. The implementation of CS is given in Fig. 4.

4.2 Leavy Flight

Finding a particular nest of host birds is an important criterion in cuckoo search. The finding of host birds' nest is same as birds searching for food. The birds will search the food in a random manner. The similar concept is applied to control the duty ratio of SEPIC converter. The searching for food in a particular direction can be derived as a mathematical equitation's. Levy is used to search the random step size of cuckoo birds. The cuckoo eggs break by host birds by using three steps. In that, every bird gives one egg at every time. In first iteration, the cuckoo gives quality one then only it will go for next step and final one is the host nest birds constant value. The implementation of cuckoo is given in Fig. 4. Different duty cycles are initiated randomly at first, after that, each one is given to the PV system, and the generated voltage and currents are used to estimate the power. The duty cycle is updated until the best optimum duty cycle and fitness function.

5 Simulation Results of CS MPPT Technique

Three series connected panels are used to perform the CS operation for the control of output voltage of SEPIC converter. The CS optimization technique performance is analyzed at two different irradiation patterns. The nonlinear characteristics of PV and multiple MPPs exist on current against power curves. The tracking of true MPP is difficult. Hence, the transferring of maximum power from source to load is difficult. The SEPIC converter is placed in between the source and resistive load and its duty cycle is controlled by using CS technique.

The SEPIC converter design parameters are input capacitor ($C_i = 10.22$ µF), resistor ($R_0 = 115$ Ω), intermediate capacitor ($C_a = 0.6$ mF), output capacitor ($C_0 = 2$ mF), and an inductor ($L_i = 0.5$ µH). The intermediate capacitor is used to minimize the switching losses across the switch and output inductor is used to improve the voltage gain of the converter. The input and output capacitors are used to suppress

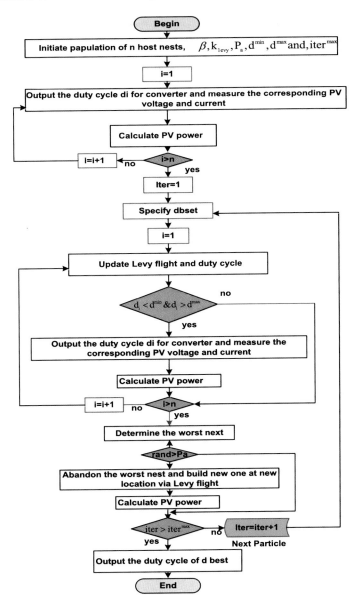

Fig. 4 Operation of cuckoo search technique

ripples in voltage and current waveforms. High switching frequency metal oxide semiconductor switch is used in converter to improve the power handling capability of the converter.

From Fig. 5a, it is observed that the CS MPPT controller uses instantaneous changes of PV voltage and current parameters to achieve the optimum duty of SEPIC converter. As a result, the extreme power is transferred from PV system to load. The first irradiation pattern used to analyze the PV performance is 900, 800, and 700 W/m². At this pattern, the SEPIC output voltage is fluctuating up to 0.65 s. After that it is stabilized with the voltage magnitude of 230 V. The duty ratio is varying in between 0, 0.3, 0.57, and 0.9, respectively, and is given in Fig. 5b. As a result, the cuckoo is searching an entire space region of nonlinear characteristics and the obtained duty ratio is given to SEPIC converter to control the PV output voltage. The SEPIC output current and voltage waveforms are shown in Fig. 5c, d.

For the second pattern, the PV output voltage, converter current, and voltage waveforms are shown in Fig. 6. From Fig. 6a, the voltage of SEPIC converter is oscillating with the time duration of 0.4 s. After that, it is settled and the magnitude of voltage is 250 V. The duty of converter is 0, 0.3, 0.53, and 0.95, respectively. Hence, the MPPT controller covers entire I and V and Power and Voltage curves. Hence, the CS is tracking a true MPP to transfer extreme power from PV array to load at different irradiation shading conditions.

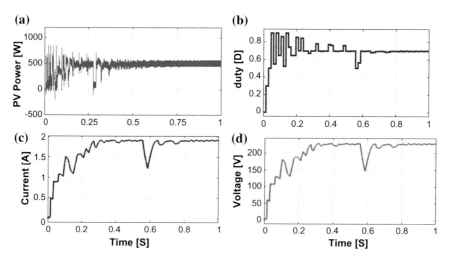

Fig. 5 **a** PV output voltage, **b** converter duty, **c** output current of SEPIC, and **d** SEPIC output voltage at 900, 800, and 700 W/m²

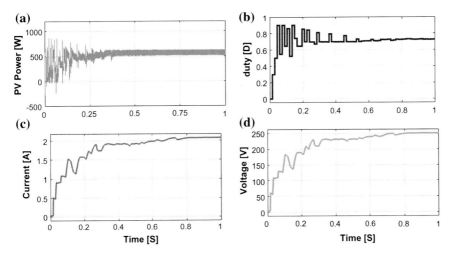

Fig. 6 **a** PV output voltage, **b** converter duty, **c** output current of SEPIC, and **d** SEPIC output voltage at 1000, 900, and 700 W/m^2

6 Conclusion

The simulation of CS optimization MPPT technique is implemented by using MAT-LAB/Simulink software under partial shading condition. Due to the shading, there are two local MPPs and one global MPP which is required to get the maximum power of the solar system. The MPPT tracking of CS is performed at two different irradiation patterns. From the simulation results, the CSO tracking technique is giving accurate MPP with high convergence speed by utilizing entire search space of current against voltage characteristics. In addition, the development of CS is simple and has less complexity.

Acknowledgements I would like to thank to the **University Grants Commission (Govt. of India)** for funding my research program and I especially thank VIT University management for providing all the facilities to carry out my research work.

References

1. López, J.M.G., et al.: Smart residential load simulator for energy management in smart grids. IEEE Trans. Ind. Electron. **66**(2), 1443–1452 (2019)
2. Charuchittipan, D., et al.: A semi-empirical model for estimating diffuse solar near infrared radiation in Thailand using ground-and satellite-based data for mapping applications. Renew. Energy **117**, 175–183 (2018)
3. Aliyu, M., et al.: A review of solar-powered water pumping systems. Renew. Sustain. Energy Rev. **87**, 61–76 (2018)

4. Woodruff, D.L., et al.: Constructing probabilistic scenarios for wide-area solar power generation. Solar Energy **160**, 153–167 (2018)
5. Rani, C., Hussaian Basha, C.H., Odofin, S.: Analysis and comparison of SEPIC, Landsman and Zeta converters for PV fed induction motor drive applications. In: 2018 International Conference on Computation of Power, Energy, Information and Communication (ICCPEIC). IEEE (2018)
6. Rani, C., Hussaian Basha, C.H.: A review on non-isolated inductor coupled DC-DC converter for photovoltaic grid-connected applications. Int. J. Renew. Energy Res. (IJRER) **7**(4), 1570–1585 (2017)
7. Yuan, J., et al.: Coal use for power generation in China. Resour. Conserv. Recycl. **129**, 443–453 (2018)
8. Singh, N.K., Badge, S.S., Salimath, G.F.: Solar tracking for optimizing conversion efficiency using ANN. In: Intelligent Engineering Informatics, pp. 551–559. Springer, Singapore (2018)
9. Tey, K.S., et al.: Improved differential evolution-based MPPT algorithm using SEPIC for PV systems under partial shading conditions and load variation. IEEE Trans. Ind. Inf. (2018)
10. Saravanan, S., Ramesh Babu, N., Sanjeevikumar, P.: Comparative analysis of DC/DC converters with MPPT techniques based PV system. In: Advances in Power Systems and Energy Management, pp. 275–284. Springer, Singapore (2018)
11. Harrag, A., Messalti, S.: How fuzzy logic can improve PEM fuel cell MPPT performances. Int. J. Hydrogen Energy **43**(1), 537–550 (2018)
12. Farayola, A.M., et al.: Distributive MPPT approach using ANFIS and perturb & observe techniques under uniform and partial shading conditions. In: Artificial Intelligence and Evolutionary Computations in Engineering Systems, pp. 27–37. Springer, Singapore (2018)
13. Lee, C.-T., et al.: Application of the hybrid Taguchi genetic algorithm to maximum power point tracking of photovoltaic system. In: 2018 IEEE International Conference on Applied System Invention (ICASI). IEEE (2018)
14. Ebrahim, A.F., et al.: Vector decoupling control design based on genetic algorithm for a residential microgrid system for future city houses at islanding operation. In: SoutheastCon 2018. IEEE (2018)
15. Nguyen, T.T., Vo, D.N., Dinh, B.H.: An effectively adaptive selective cuckoo search algorithm for solving three complicated short-term hydrothermal scheduling problems. Energy **155**, 930–956 (2018)
16. Peng, B.-R., Ho, K.-C., Liu, Y.-H.: A novel and fast MPPT method suitable for both fast changing and partially shaded conditions. IEEE Trans. Ind. Electron. **65**(4), 3240–3251 (2018)
17. Rani, C., Hussaian Basha, C.H., Odofin, S.: Design and switching loss calculation of single leg 3-level 3-phase VSI. In: 2018 International Conference on Computation of Power, Energy, Information and Communication (ICCPEIC). IEEE (2018)
18. Ahmed, J., Salam, Z.: A maximum power point tracking (MPPT) for PV system using Cuckoo search with partial shading capability. Appl. Energy **119**, 118–130 (2014)

An Improved Fuzzy Clustering Segmentation Algorithm Based on Animal Behavior Global Optimization

A. Absara⊙, S. N. Kumar⊙, A. Lenin Fred⊙, H. Ajay Kumar⊙
and V. Suresh⊙

Abstract The bio-inspired optimization algorithms play vital role in many research domains and this work analyzes animal behavior optimization algorithm. Medical image segmentation helps the physicians for disease diagnosis and treatment planning. This work incorporates ABO algorithm for cluster centroid selection in Fuzzy C-means clustering segmentation algorithm. The Animal Behavior Optimization (ABO) algorithm was developed based on the group behavior and was validated on 13 benchmark functions. The dominant nature of an animal species decides the fitness function value and each solution in problem space depicts the animal position. The ABO algorithm was coupled with the classical FCM for the analysis of region of interest in abdomen CT and brain MR datasets. The results were found to be efficient when compared with the FCM coupled with artificial bee colony (ABC), firefly, and cuckoo optimization algorithms. The promising results generated by ABC makes it an efficient one for real-world problems.

Keywords Segmentation · Animal behavior optimization · Fuzzy C-means · Artificial bee colony · Firefly optimization · Cuckoo optimization

A. Absara (✉) · S. N. Kumar · H. Ajay Kumar · V. Suresh
School of Electronics and Communication Engineering, Mar Ephraem College of Engineering and Technology, Marthandam, Kanyakumari, India
e-mail: alexmaryalison@gmail.com

S. N. Kumar
e-mail: appu123kumar@gmail.com

H. Ajay Kumar
e-mail: ajayhakkumar@gmail.com

A. Lenin Fred
School of Computer Science and Engineering, Mar Ephraem College of Engineering and Technology, Marthandam, Kanyakumari, India
e-mail: leninfred.a@gmail.com

© Springer Nature Singapore Pte Ltd. 2020
K. N. Das et al. (eds.), *Soft Computing for Problem Solving*,
Advances in Intelligent Systems and Computing 1048,
https://doi.org/10.1007/978-981-15-0035-0_60

1 Introduction

Global optimization algorithms are broadly classified into deterministic and meta-heuristic techniques and find its applications in engineering, science, and mathematics [1, 2]. A detailed analysis of metaheuristic algorithms for partition clustering was proposed in [3]. In [4], improved gray wolf optimization approach was used for solving the benchmark functions and biomedical applications. The optimization algorithm was tested on 23 standard benchmark function and generates improved performance when compared with classical gray wolf optimization and particle swarm optimization. The improved Bee Colony Optimization (BCO) termed as patch levy-based bees algorithm outperforms the classical BCO and was validated on benchmark functions [5]. The biological traits of lion pave a way toward the generation of Lion Pride Optimization (LPO) algorithm; efficient results were produced when compared with classical optimization algorithms for both unimodal and multimodal benchmark functions [6]. A detailed study of metaheuristic algorithms based on the group behavior of animals was proposed for the traveling salesman problem [7]. The Social Emotional Optimization Algorithm (SEOA) was developed for solving the real-world problems; SEOA with gauss distribution was found to be efficient [8].

The optimization algorithms gain important role in image processing applications like preprocessing, segmentation, and compression [9]. The improved quantum behaved PSO generates efficient multilevel thresholding segmentation results for benchmark images [10]. The adaptive bacterial foraging algorithm was employed for the optimization of fitness function in fuzzy entropy segmentation for images [11]. The artificial fish swarm optimization algorithm when coupled with fuzzy C-means clustering generates efficient segmentation results when compared with the classical FCM algorithm [12]. Section 2 discusses the characteristics of animal behavior optimization and the role of optimization in FCM algorithm for medical image segmentation. Section 3 depicts the result and discussion with performance metrics and finally, conclusion is drawn in Sect. 4.

2 Materials and Methods

2.1 Data Acquisition

The validation of optimization algorithm was done on benchmark functions and for the application, the algorithm was analyzed on abdomen CT and brain MR datasets for extraction of region of interest. The DICOM abdomen CT and brain MR datasets are obtained from the Metro Scans and Research Laboratory. A normal and a benign abdomen CT dataset and two MR datasets of malignant lesion are used in this research work.

2.2 Animal Behavior Optimization Algorithm

The animal behavior optimization algorithm relies on the biological behavior of animals. Each solution in the problem space depicts the position of animal. The animal dominant position is determined from the fitness value. This research work focusses on the validation of ABO algorithm on standard benchmark functions and coupled with the classical FCM algorithm for the segmentation of medical images. Based on the biological characteristics of animals, the memory is classified into two types; best position at each generation (m_u) and other for storing the best historical positions during the complete evolutionary process (m_h).

The population is initialized randomly and optimization is an iterative process and it comprises four stages that repeat until convergence is reached.

- Maintain the location of best individuals
- Animals can move with in the neighborhood
- Random movement is also possible
- Memory is updated and compete for space within a determined distance.

Stage 1: Maintain the location of best individuals
The first stage of ABO algorithm is the initialization of population of animal positions.

Let N_p represents the animal positions such that $s = \{S_1, S_2, \ldots, S_{N_p}\}$ denote the set.

A lower bound S_j^{low} and upper bound S_j^{high} value are specified and the animal positions are uniformly and randomly distributed.

$$S_{i,j} = S_j^{\text{low}} + \text{rand}(0, 1).\left[S_j^{\text{high}} - S_j^{\text{low}}\right] \quad \begin{matrix} j = 1, 2, \ldots, N \\ i = 1, 2, \ldots, N_p \end{matrix} \quad (1)$$

where S_{ij} is the jth parameter of the ith individual.

The fitness function reflects the dominant property of the animals and the animal positions are sorted in accordance with it to form a set $Y = \{Y_1, Y_2, \ldots, Y_{N_p}\}$. The best positions (B) are determined from the set "Y" and stored in memory elements (m_h, m_g). In the initial stage, the memory elements have same value.

Stage 2: Animals can move within the neighborhood
The ABO is an evolutionary process and the first B elements $s = \{S_1, S_2, \ldots, S_B\}$ of the new animal position set are created.

The "B" positions are estimated from the historical memory (m_h) with a slight random perturbation around them. The expression for the best animal positions is expressed as follows.

$$s_l = m_h^l + u \quad (2)$$

where $l \in \{1, 2, 3, \ldots, B\}$, m_h^l depicts the lth element of m_h, u is a random movement with a small length.

Stage 3: Random movement is also possible

The biological characteristics reveal that there is a random attraction or repulsion between the animals. The random attraction or repulsion is represented as follows:

$$S_i = \begin{cases} y_i \pm r.\left(m_h^{near-y_i}\right) & \text{with probability H} \\ y_i \pm r.\left(m_g^{near-y_i}\right) & \text{with probability } (1-H) \end{cases} \tag{3}$$

where $i \in \{B+1, B+2, \ldots, N_P\}$, m_h^{near} and m_g^{near} represents the nearest elements of m_h and m_g to y_i.

```
if (r > 0)
    Position  y_i  is attracted to  m_h^near or  m_g^near
else
    Repulsion between positions
```

r is a random number between $[-1 \ 1]$.

Stage 4: Memory is updated and competed for space within a determined distance

The biological trait reveals that under same probability P, random change in position of animal takes place

$$S_i = \begin{cases} r & \text{with probability P} \\ y_i & \text{with probability } (1-P) \end{cases} \tag{4}$$

where $i \in \{B+1, B+2, \ldots, N_P\}$, R is a random vector in the problem space.

Once the above four stages are over, the memory m_h has to be updated. The dominant property of animal decides the value of m_h. Animals of different species maintain a minimum distance among them and that distance is termed as "d" in the ABO algorithm. The distance "d" depends upon the aggressive nature of the animals, when two animals confront each other, the dominant one exist in the same location, while the weaker one tries to leave from the place.

The historical memory m_h is updated as follows.

The components of m_h and m_g are grouped into m_u.

$$m_u = m_h \cup m_g \tag{5}$$

Each component m_u is compared pairwise with the other elements. When the distance between the two elements is less than "d", the component has good fitness value and it exists, otherwise it will not be considered. The best "B" positions are framed from the above step to build the new m_h value. The value of "d" depends on the problem space size. The larger value of "d" increases the exploration capacity, however, the convergence rate decreases.

The empirical model for the estimation of "d" is expressed as follows:

$$d = \frac{\prod_{j=1}^{D} \left(S_j^{\text{high}} - S_j^{\text{low}} \right)}{10 \cdot D} \tag{6}$$

where S_j^{low} and S_j^{high} represent the lower and upper bound of the jth parameter, respectively.

The steps of ABO algorithm is summarized as follows:

Step 1: Initialize the parameters N_P, B, H, P and number of iteration.

Step 2: Random generations of animal positions $S = \{S_1, S_2, \ldots, S_{N_p}\}$.

Step 3: Sort the elements of S in accordance with the objective function based on the dominant nature of the animal $y = \{y_1, y_2, \ldots, y_{N_p}\}$.

Step 4: Select the first B positions of Y and place in the memory m_g.

Step 5: Update m_h.

Step 6: The first "B" positions are determined with a slight random perturbation $S_l = m_h^l + u$, where u is a random vector of small length.

Step 7: The rest of elements of S are generated through the attraction, repulsion, and random movements

```
for  i = B+1 : N
     {
     if (r₁ > P) then
        Movement is local attraction/repulsion
if (r₂ > H) then
```
$$S_i = y_i \pm r.\left(m_h^{\text{nearest}} - y_i \right)$$
```
else if
```
$$S_i = y_i \pm r.\left(m_g^{\text{nearest}} - y_i \right)$$
```
     }
else
     movement is random
     {
```
$$S_i = r$$
```
     }
end
```

where r_1, r_2 are the random variables.

Step 8: When the number of iterations is completed proceed to Step 3.

The best value m_h depicts the global solution for the optimization problem. The flow diagram of the ABO algorithm is depicted in Fig. 1.

Fig. 1 Flow diagram of the
ABO algorithm

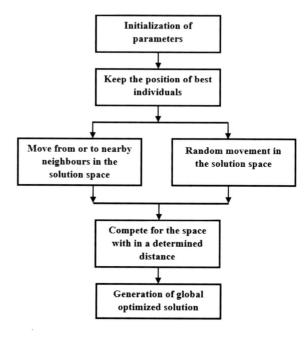

2.3   *Improved FCM Based on ABO Algorithm*

Clustering algorithms are mainly classified into hard clustering and soft clustering.
The K-means is a soft clustering algorithm and fuzzy C-means (FCM) is a soft
clustering algorithm. In K-means algorithm, a pixel can accommodate in one of the K
classes, however in many cases, hard partition will not be good, since a pixel can have
degree of participation in more than one class. The fuzzy clustering incorporates this
by a fuzzy membership function whose value lies in the range of 0–1. The Euclidean
distance is a widely used distance metric for the generation of optimum number of
clusters.

The random centroid initialization in classical FCM was carried out by ABO
algorithm and the membership function is reformulated as follows:

$$U_{ij} = \frac{1}{\sum_{k=1}^{s} \left(\frac{\|y_i - f_j\|}{\|y_i - f_k\|} \right)^{\frac{2}{f-1}}}; \quad 1 \le i \le N \tag{7}$$

where U_{ij} represents the degree of membership of best animal position S_j, f represents the measure of fuzziness and it is usually taken as 2.

$$T_j = \frac{\sum_{i=1}^{N} u_{ij}^{f} \cdot x_i}{\sum_{i=1}^{N} u_{ij}^{f}} \tag{8}$$

T_j represents the solution after applying FCM in the animal best position j. The best position of the animals is updated and the membership function is also updated. The algorithm executes for fixed number of iterations and until the best solution was determined.

3 Results and Discussion

The algorithms are developed in Matlab 2015a and tested on benchmark functions and real-time CT/MR datasets. The system specification is as follows Intel core i3 @3.30 GHz processor with 4 GB RAM. The benchmark functions are depicted in Table 1. The first four benchmark functions are two variable function and other benchmark functions are multivariable functions. The parameter tuning plays vital role and here the parameters to be tuned are few and are as follows.

N_p Population of animals (Here the value is chosen as 40 and depending upon the application, it can be changed)

Iter   Number of iterations and here 100 is chosen

Reps   Number of times the optimization function has to repeat, here 10 is chosen

H Probability value associated with the random movement, here $H = 0.5$ is chosen

P Probability value associated with the memory updation and competes for space, here $P = 0.8$ is chosen.

The number of best B positions are determined as follows: $B = N_p/10$ here, $B = 4$.

Table 1 Performance measures of animal behavior optimization corresponding to objective functions ($f1, f2, f3, f4, f5, f6$)

Objective function	Performance measure	Number of iterations					
		100	200	400	600	800	1000
Beale (f1)	P1	2.140540e−04	4.714259e−05	5.688110e−06	2.336961e−06	3.830721e−06	1.054360e−06
	P2	2.579139e−04	5.143372e−05	1.068128e−05	2.047707e−06	6.722768e−06	1.200402e−06
	P3	3.653766e−06	1.315522e−06	7.022577e−07	1.110983e−07	2.591679e−08	5.785720e−10
Matya (f2)	P1	2.737188e−06	6.714207e−07	4.035586e−07	1.821945e−08	6.768208e−08	4.348249e−09
	P2	2.753945e−06	6.609876e−07	4.888349e−07	2.870525e−08	1.661016e−07	6.440013e−09
	P3	1.387531e−07	5.184708e−09	2.547884e−08	8.497423e−10	3.160188e−10	4.389758e−11
Booth (f3)	P1	1.300434e−04	2.037454e−05	9.755533e−06	1.663407e−06	9.612183e−07	2.655221e−07
	P2	1.563943e−04	1.904203e−05	1.841700e−05	2.408378e−06	1.329177e−06	4.474700e−07
	P3	4.960031e−06	2.772613e−08	5.675163e−09	1.497018e−08	4.048310e−08	2.017842e−09
Boha1 (f4)	P1	3.032690e−03	9.401993e−04	1.861723e−04	4.012270e−05	1.154513e−04	2.077159e−05
	P2	1.347487e−03	5.457137e−04	2.382839e−04	3.217678e−05	1.404227e−04	3.482000e−05
	P3	8.319221e−05	1.118719e−04	7.418531e−06	1.812253e−07	1.455357e−07	4.115917e−07
Rosen (f5)	P1	1.835114e+00	1.379002e+00	1.793025e+00	8.831297e−01	1.476727e+00	2.183196e+00
	P2	1.017505e+00	1.802674e+00	1.871631e+00	1.054613e+00	2.040373e+00	1.821430e+00
	P3	3.443427e−01	3.575594e−02	8.084439e−02	6.923606e−02	8.182475e−02	3.680042e−02
Sumpow (f6)	P1	5.581083e−06	2.309346e−06	3.805674e−07	1.404919e−07	1.179774e−07	9.466422e−08
	P2	5.887690e−06	1.774666e−06	3.904235e−07	1.582076e−07	7.353890e−08	1.218066e−07
	P3	2.126650e−07	1.935900e−07	2.947472e−08	1.698599e−09	4.115459e−09	1.323485e−09

For the validation of clustering algorithm the following metrics are used: Partition Coefficient (PC), Partition Entropy (PE), Davies Bouldin Index (DBI), Xie and Beni Index (XBI), and WL Index (WLI). High values of PC, DBI, XBI, and low values of PE, WLI represent the efficiency of clustering algorithm.

$$PC = \frac{1}{N} \sum_{p=1}^{P} \sum_{i=1}^{N} \mu_{ip}^2 \tag{9}$$

$$PE(K) = \frac{1}{N} \sum_{p=1}^{P} \sum_{i=1}^{N} \mu_{ip} \log_2(\mu_{ip}) \tag{10}$$

$$DBI = \frac{1}{P} \sum_{P=1}^{P} \max \frac{S_j + S_p}{\|V_j - V_p\|} \tag{11}$$

$$XBI(K) = \frac{\sum_{p=1}^{P} \sum_{i=1}^{N} \mu_{ip}^2 \|u_i - v_p\|^2}{N \cdot \min_{i \neq j} \|v_i - v_j\|^2} \tag{12}$$

$$WLI = \sum_{P=1}^{P} \sum_{i=1}^{N} \frac{\mu_{ilP}^2 \|U_i - V_P\|^2}{\sum_{i=1}^{N} \mu_{iP}} \tag{13}$$

The results are depicted in Tables 1 and 2 and benchmark functions are depicted in Table 3. The ABO algorithm was validated by changing the number of iterations from 100 to 1000 and results are depicted here. Figure 2 depicts the input MR images of brain and CT abdomen images. The DICOM input images are subjected to FCM clustering by coupling with ABC, Cuckoo, Firefly, and ABO algorithms. Figure 3 depicts the FCM-ABO results of MR brain datasets and Fig. 4 depicts the CT abdomen datasets.

Table 2 Performance measures of animal behavior optimization corresponding to objective functions $(f7, f8, f9, f10, f11, f12, f13)$

Objective function	Performance measure	Number of iterations					
		100	200	400	600	800	1000
Sumsqu (f7)	P1	1.393985e−02	4.860044e−03	9.753996e−04	2.323271e−04	2.051227e−04	1.107736e−04
	P2	8.256243e−03	2.953751e−03	6.216143e−04	1.537582e−04	1.968874e−04	7.563357e−05
	P3	2.022864e−03	1.208871e−03	1.345399e−04	3.941030e−05	2.235706e−05	1.747270e−05
Spheref (f8)	P1	2.967168e−03	8.715806e−04	5.591253e−05	9.778561e−05	2.888412e−05	7.621305e−06
	P2	1.830729e−03	5.271283e−04	3.611820e−05	1.175138e−04	3.125284e−05	7.924703e−06
	P3	8.929385e−04	5.162595e−05	6.360145e−06	8.772174e−06	4.480987e−06	3.951219e−07
Schwef (f9)	P1	1.627430e+08	1.391807e+13	4.700266e+25	1.605331e+39	2.292618e+55	2.697597e+66
	P2	1.809706e+08	2.105598e+13	1.325501e+26	3.769610e+39	6.273692e+55	8.454647e+66
	P3	4.801480e+08	6.159042e+13	4.228109e+26	1.204817e+40	1.994805e+56	2.675960e+67
Rastr (f10)	P1	7.657927e+00	8.022748e+00	6.036737e+00	6.042415e+00	5.782710e+00	7.707868e+00
	P2	4.357540e+00	4.400296e+00	5.093117e+00	3.173727e+00	3.298049e+00	3.990239e+00
	P3	1.946097e+00	2.166299e+00	1.020513e+00	2.023023e+00	1.908598e+00	2.997675e+00
Levy (f11)	P1	1.161348e−01	3.496050e−02	9.498232e−03	9.223335e−03	9.020384e−03	5.462139e−05
	P2	1.588299e−01	4.639834e−02	2.817187e−02	2.823783e−02	2.830913e−02	5.607592e−05
	P3	4.479096e−03	5.068268e−04	6.953183e−05	3.929830e−06	1.974751e−06	1.626435e−06

(continued)

Table 2 (continued)

Objective function	Performance measure	Number of iterations					
		100	200	400	600	800	1000
Griewank (f12)	P1	6.245766e−01	4.266366e−01	3.637815e−01	3.159239e−01	2.714446e−01	3.437938e−01
	P2	2.922388e−01	1.932306e−01	1.719354e−01	1.182561e−01	1.413093e−01	1.676683e−01
	P3	3.261874e−01	2.453368e−01	1.037132e−01	1.532196e−01	8.425464e−02	1.257945e−01
Ackley (f13)	P1	1.715168e+00	9.996519e−01	4.823678e−01	3.973626e−01	3.289433e−01	1.876420e−01
	P2	9.683803e−01	1.084070e+00	8.662691e−01	7.617482e−01	6.201746e−01	4.625219e−01
	P3	2.573226e−01	3.578799e−02	2.295142e−02	2.327714e−02	3.855257e−03	6.944255e−03

Table 3 Benchmark functions used for the validation of optimization algorithm

Function	Formula
Beale function	$Y = \left\{ (1.5 - x_1 + x_1 x_2)^2 + \left(2.25 - x_1 + x_1 x_2^2\right)^2 + \left(2.625 - x_1 + x_1 x_2^3\right)^2 \right\}$
Matyas function	$Y = \left\{ \left[\left(0.26 - \left(x_1^2 + x_2^2\right)\right)^2 \right] + \left[(-0.48 * x_1 * x_2) \right] \right\}$
Booth function	$f(x) = (x_1 + 2x_2 - 7)^2 + (2x_1 + x_2 - 5)^2$
Bohachevsky function	$Y = x_1^2 + \left(2 * x_2^2\right) + (-0.3 * \cos(3 * pi * x_1)) + (-0.4 * \cos(4 * pi * x_2)) + 0.7$
Rosenbrock function	$f(x, y) = (a - x)^2 + b\left(y - x^2\right)^2$
Sum of different powers function	$f(x, y) = \sum_{i=1}^{d} \|x_i\|^{i+1}$
Sum squares function	$r_k(n) = \left\| \left\{ (a_1, a_2, \ldots, a_k) \in Z^k : n = a_1^2 + a_2^2 + \cdots + a_k^2 \right\} \right\|$
Sphere function	$f(x) = \frac{1}{899} \left(\sum_{i=1}^{6} x_i^2 2^i - 1745 \right)$
Ackley function	$f(x) = -a . \exp\left(-b \sqrt{\frac{1}{d} \sum_{i=1}^{d} x_i^2} \right) - \exp\left(\frac{1}{d} \sqrt{\sum_{i=1}^{d} \cos(cx_i)} \right) + a + \exp(1)$
Rastrigin function	$f(x) = 10d + \sum_{i=1}^{d} \left[x_i^2 - 10\cos(2\pi x_i) \right]$
Levy function	$f(x) = \sin^2(\pi \omega_i) + \sum_{i=1}^{d-1} (\omega_i - 1)^2 \left[1 + 10\sin^2(\pi \omega_i + 1) \right] + (\omega_d + 1)^2 \left[1 + \sin^2(2\pi \omega_d) \right]$ where, $\omega_i = 1 + \frac{x_i - 1}{4}$ for all i = 1, 2, …, d
Griewank function	$f(x) = \sum_{i=1}^{d} \frac{x_i^2}{4000} - \prod_{i=1}^{d} \cos\left(\frac{x_i}{\sqrt{i}} \right) + 1$
Schwef function	$f(x) = 418.9829d - \sum_{i=1}^{d} x_i \sin\left(\sqrt{\|x_i\|} \right)$

(a)　　　　**(b)**　　　　**(c)**　　　　**(d)**

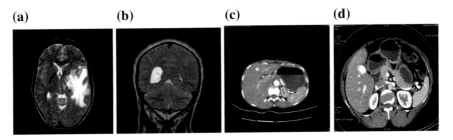

Fig. 2 **a**, **b** MR brain DICOM images, **c**, **d** CT abdomen DICOM images

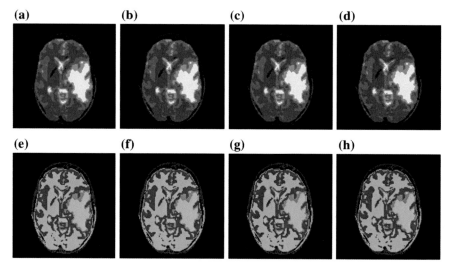

Fig. 3 Segmentation results **a**, **e** FCM-ABC, **b**, **f** FCM-Cuckoo, **c**, **g** FCM-Firefly, **d**, **h** FCM-ABO

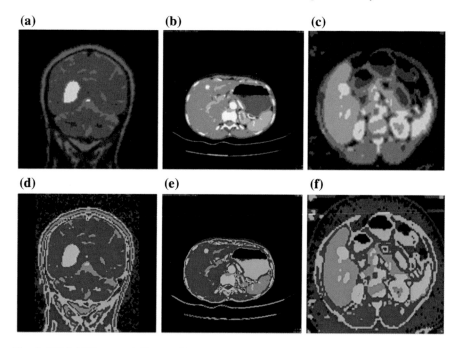

Fig. 4 FCM-ABO segmentation results

Table 4 Performance metric evaluation of clustering algorithms for Dataset 1

ID1	FCM-ABC	FCM-Cuckoo	FCM-Firefly	FCM-ABO
PC +	0.9028	0.9248	0.92638	0.9568
PE −	0.2369	0.2368	0.2161	0.2065
DBI +	0.1459	0.1452	0.1458	0.1952
XBI +	0.1152	0.1155	0.1155	0.1254
WLI −	0.0142	0.0124	0.0109	0.0110

Table 5 Performance metric evaluation of clustering algorithms for Dataset 2

ID2	FCM-ABC	FCM-Cuckoo	FCM-Firefly	FCM-ABO
PC +	0.9058	0.9059	0.9063	0.9069
PE −	0.2362	0.2361	0.2352	0.2338
DBI +	0.1843	0.1858	0.1834	0.1860
XBI +	0.0101	0.0102	0.0102	0.0103
WLI −	0.0096	0.0096	0.0095	0.0094

The performance metrics of the clustering algorithms when coupled with optimization techniques are depicted in Tables 4, 5, 6, and 7.

Table 6 Performance metric evaluation of clustering algorithms for Dataset 3

ID3	FCM-ABC	FCM-Cuckoo	FCM-Firefly	FCM-ABO
PC +	0.9536	0.9136	0.9236	0.9556
PE −	0.1866	0.1766	0.1866	0.1766
DBI +	0.1478	0.1488	0.1498	0.2498
XBI +	0.0120	0.0121	0.0123	0.0126
WLI −	0.0102	0.0102	0.0103	0.0104

Table 7 Performance metric evaluation of clustering algorithms for Dataset 4

ID4	FCM-ABC	FCM-Cuckoo	FCM-Firefly	FCM-ABO
PC +	0.9068	0.9170	0.9174	0.9176
PE −	0.2305	0.2090	0.2098	0.2089
DBI +	0.1590	0.1597	0.1601	0.1661
XBI +	0.0109	0.0112	0.0119	0.0120
WLI −	0.0124	0.0120	0.0114	0.0116

4 Conclusion

This research work analyzes animal behavior optimization based on the collective behavior of animal species and their interactions. The algorithm was validated on benchmark functions and for real-time application, FCM algorithm was coupled with ABO for the extraction of region of interest in CT/MR images. The FCM-ABO algorithm generates robust segmentation results when compared with the classical optimization algorithms. The future work will be the application of ABO in other image processing domains like classification and compression.

Acknowledgements The authors would like to acknowledge the support provided by DST under IDP scheme (No: IDP/MED/03/2015).

References

1. Pardalos, P.M., Romeijn, H.E., Tuy, H.: Recent developments and trends in global optimization. J. Comput. Appl. Math. **124**(1–2), 209–228 (2000). https://doi.org/10.1016/S0377-0427(00)00425-8
2. Floudas, C.A., Akrotirianakis, I.G., Caratzoulas, S.: Global optimization in the 21st century: advances and challenges. Comput. Chem. Eng. **29**(6), 1185–1202 (2005). https://doi.org/10.1016/j.compchemeng.2005.02.006
3. Nanda, S.J., Panda, G.: A survey on nature inspired metaheuristic algorithms for partitional clustering. Swarm Evol. Comput. **16**, 1–18 (2014). https://doi.org/10.1016/j.swevo.2013.11.003
4. Singh, N., Singh, S.B.: A modified mean gray wolf optimization approach for benchmark and biomedical problems. Evol. Bioinform. **13** (2017). https://doi.org/10.1177/2F1176934317729413
5. Hussein, W.A., Sahran, S., Sheikh Abdullah, S.N.H.: An improved Bees algorithm for real parameter optimization. Int. J. Adv. Comput. Sci. Appl. **6**, 23–39 (2015)
6. Wang, B., Jin, X., Cheng, B.: Lion pride optimizer: an optimization algorithm inspired by lion pride behavior. Sci. China Inf. Sci. **55**(10), 2369–2389 (2012). https://doi.org/10.1007/s11432-012-4548-0
7. Ruiz-Vanoye, J.A., Díaz-Parra, O., Cocón, F., Soto, A., Buenabad Arias, M.D.L.Á., Verduzco-Reyes, G., Alberto-Lira, R.: Meta-heuristics algorithms based on the grouping of animals by social behaviour for the traveling salesman problem. Int. J. Comb. Optim. Probl. Inf. **3**(3), 104–123 (2012)
8. Cui, Z., Xu, Y., Zeng, J.: Social emotional optimization algorithm with random emotional selection strategy. In: Theory and New Applications of Swarm Intelligence. InTech. vol. 3, pp. 33–50 (2012)
9. Qin, Z.T.: Optimization Algorithms for Structured Machine Learning and Image Processing Problems. Columbia University (Thesis) (2013). https://doi.org/10.7916/D8JH3TDM
10. Gao, H., Xu, W.: Multilevel thresholding for image segmentation through an improved quantum-behaved particle swarm algorithm. IEEE Trans. Instrum. Meas. **59**(4), 934–946 (2010). https://doi.org/10.1109/TIM.2009.2030931
11. Sanyal, N., Chatterjee, A., Munshi, S.: An adaptive bacterial foraging algorithm for fuzzy entropy based image segmentation. Expert Syst. Appl. **38**(12), 15489 (2011). https://doi.org/10.1016/j.eswa.2011.06.011
12. Chu, X., Zhu, Y., Shi, J., Song, J.: Method of image segmentation based on fuzzy C-means clustering algorithm and artificial fish swarm algorithm. In: 2010 International Conference on Intelligent Computing and Integrated Systems (ICISS), pp. 254–257. IEEE (2010). https://doi.org/10.1109/ICISS.2010.5657199

Fitness-Based Controlled Movements in Artificial Bee Colony Algorithm

Harish Sharma, Kritika Sharma, Nirmala Sharma, Assif Assad and Jagdish Chand Bansal

Abstract Artificial Bee Colony (ABC) is an efficient metaheuristic algorithm is used for solving various complex optimization problems. A new variant of ABC, namely, fitness-based controlled movements in ABC (ConABC) is presented here. In ConABC, an Intelligent Term (IT) is introduced in the employed bee stage, which enhances the solution search ability of the ABC algorithm. The IT is actually controlling the step size of an individual according to its fitness. The presented algorithm is extensively inferred to 12 benchmark functions. It is then compared with ABC, its two recent variants, titled Best-So-Far ABC (BSFABC), Modified ABC (MABC) and some more state-of-the-art algorithms. The observational outcomes unfold that ConABC has potential to solve the problems in a better way than ABC algorithm.

Keywords Nature inspired algorithms · Collective behaviour · Guided search · Artificial bee colony

1 Introduction

Nature is a very rich source of inspiration for researchers in various ways [9]. It offers some of the efficient ways for solving various real-world problems [15]. There are various algorithms inspired by processes, observed from nature. These are termed as Nature Inspired Algorithms (NIAs). The purpose of developing such algorithms is to solve various optimization problems. NIAs are further mainly categorized as Swarm Intelligence (SI) based algorithms and Evolutionary Algorithms (EAs) [9]. EAs are grounded on the Darwins theory of the survival of the fittest [14]. Some of

H. Sharma (✉) · K. Sharma · N. Sharma
Rajasthan Technical University (R.T.U.), Kota, India
e-mail: harish.sharma0107@gmail.com

A. Assad
IUST, Awantipora, India

J. C. Bansal
South Asian University, New Delhi, India

© Springer Nature Singapore Pte Ltd. 2020
K. N. Das et al. (eds.), *Soft Computing for Problem Solving*,
Advances in Intelligent Systems and Computing 1048,
https://doi.org/10.1007/978-981-15-0035-0_61

the EAs are Genetic Programming (GP) [13], Differential Evolution (DE) [7], etc. SI-based algorithms emphasize the collective intelligent behaviour of social creatures like termites, bees, ants, etc., that coordinate with each other using decentralized control and self-organization [11]. Some popular SI-based algorithms are Artificial Bee Colony Optimization (ABC) [10], Bacterial Foraging Optimization Algorithm (BFOA) [6], Spider Monkey Optimization (SMO), [4] etc.

The ABC algorithm was given by Karaboga [10] in 2005. It is influenced by the intelligent food-seeking behaviour of the honey bees present in nature. Here, the population comprises possible solutions regarding the food sources for honey bees. The fitness is resolved in terms of the quality of the food source. It contains three classes of honey bees named as employed bees, onlooker bees and scout bees. As we know, the performance of any optimization algorithm basically relies on two factors. One is the exploration and the other one is exploitation of the search space. Exploration defines the capability to search the entire search space for discovering the true solution of the optimization problem, whereas exploitation is defined as the potential to refine the previously explored search space. Both the properties are contradictory in nature. An optimization algorithm that always maintains a proper harmony between exploration and exploitation is considered to be an effective algorithm. It should be kept in mind that the ABC algorithm's solution search equation is fine at exploration yet bad at exploitation [3]. So, it is required to retain an appropriate balance between exploration and exploitation ability of ABC algorithm.

In the above context, this article proposed a fitness-based controlled movement in ABC algorithm (ConABC) Algorithm. In this proposed strategy, an Intelligent Term (IT) is introduced to control the step size of an individual. By this IT, the higher fit solution will move with lesser step size which helps in the exploitation and the low fit solutions will move with greater step size which results in exploring the search domain. The presented method is then compared with ABC, its two variants, named, Best-So-Far Artificial Bee Colony (BSFABC) [2], Modified Artificial Bee Colony (MABC) [1] and also with Shuffled Frog Leaping Algorithm (SFLA) [8], Particle Swarm Optimization (PSO) [12], Gravitational Search Algorithm (GSA) [16] and DE [7].

Further, the article is structured as: the ABC algorithm is briefly illustrated in Sect. 2. The presented algorithm ConABC is narrated in Sect. 3. Outcomes and discussions are analysed in Sect. 4. Finally, the work is concluded in Sect. 5.

2　Artificial Bee Colony (ABC) Algorithm

The ABC algorithm has four stages. These are Initialization stage, Employed bees stage, Onlooker bees stage and Scout bees stage. All of them are illustrated below.

2.1 Initialization Stage

Each food source represented as x_i where $(i = 1, \ldots, S)$, having D-dimensional vector is generated by the Eq. 1. D represents the number of optimization variables.

$$x_{ij} = x_{lowj} + random[0, 1] * (x_{uppj} - x_{lowj}), \forall j = 1, 2, \ldots, D \qquad (1)$$

where random[0, 1] is a uniformly distributed random number in the [0, 1] range. x_{uppj} and x_{lowj} are maximum and minimum values of x_i in jth direction, respectively.

2.2 Employed Bees Stage

The position of the ith candidate for the jth dimension is updated in this stage as follows:

$$v'_{ij} = x_{ij} + \phi_{ij} * (x_{ij} - x_{lj}) \qquad (2)$$

where the term $\phi_{ij} * (x_{ij} - x_{lj})$ represents the step size, $j \in \{1, \ldots, D\}, l \in \{1, \ldots, SN\}$ be two randomly selected tokens. l should be different from i and ϕ_{ij} is a random number having the value ranging from -1 to 1. Now, the greedy selection is applied here. If the robustness of the currently generated food source is greater than the former one, the bee in this stage renews her position with the current position.

2.3 Onlooker Bees Stage

Here, every employed bee provides the quality and location of the updated solutions to the onlooker bees. Onlooker bees analyse the accessible data regarding the food source. After that, it scrutinizes the feasible data. Then, the selection of a solution as per the probability $prob_i$ is done which is computed as follows:

$$prob_i = \frac{0.9 \times Fit_i}{max Fit} + 0.1. \qquad (3)$$

Here, maxFit represents the maximum fitness amidst each of the solutions and Fit_i depicts the fitness of a particular ith solution. The position update process is alike to employed bee stage. Again, a greedy selection is applied. If the calculated fitness of the new position is greater than that of the old position, the bee retains the new position and obliterate the previous one.

2.4 Scout Bees Stage

In this stage, if a solution is not updating itself upto the predefined number of iterations, called *Limit*, then the solution is believed to be stuck and that solution is randomly initialized in the given search space.

3 Proposed ConABC Algorithm

In view of balancing the exploration and exploitation abilities of SI-based algorithms, it is necessary that the good solution should exploit the specified search space, whilst the bad ones should explore it. For this, a novel variant of ABC algorithm is presented in this article, where the position update procedure of the employed bee stage of the traditional ABC algorithm is altered. The proposed position update equation is given by Eq. 4.

$$v'_i = x_{ij} + \phi_{ij} * (x_{Gbestj} - x_{ij}) + IT \qquad (4)$$

The term v'_i represents the updated position of the ith solution. The term x_{ij} is the current position of the solution. Here, the term IT is named as *Intelligent Term*. The term x_{Gbestj} represents the solution in the swarm having best fitness. The Intelligent Term IT is calculated by Eq. 5.

$$IT = (1.1 - prob_i) * \phi_{ij} \qquad (5)$$

In the Eq. 5, $prob_i$ is a function of the fitness for an individual and computed as described in Eq. 3. It is evident from Eq. 3 that the range of $prob_i$ is from $[0.1, 1]$, according to the fitness of the solution. So the range of IT will be $[-1, 1]$ as per the fitness of the individual. As shown in Eq. 4, this Intelligent Term controls the movement of the individuals in given search space specified by its fitness. For the high fit solutions, the value of IT will be low because of a higher value of $prob_i$. Thus it's step size is intended to be low. So the higher fit solution will proceed with lesser step size which assists in the exploitation of the nearby search space. Whereas the low fit solutions will have a higher magnitude of IT. This leads to a greater value of the step size, which results in exploring the search space.

So the fundamental contribution of the presented work is to control the step size of an individual throughout the solution search procedure in the employed bee stage of ABC algorithm. The individuals possessing higher probability (higher fitness) will proceed steadily and exploit the search space, whereas the lower probability individuals having the higher value of IT will explore the search space. Therefore, a proper balance between exploration and exploitation is established. Now, since the exploration is better taken care of by the term IT; therefore, to obtain faster convergence x_{Gbest} is used in the place of any arbitrary selection solution in the position update process. This replacement will ensure that the updated position will be inclined in the direction of best of the swarm.

4   Outcomes and Discussions

To check the performance of the presented ConABC, we assessed it on 12 distinct global optimization problems (f_1 to f_{12}) indexed in Table 1.

To prove the significance of ConABC, a comparative study is done amongst ConABC, ABC, its two modifications, namely, Modified ABC (MABC) [1], Best-so-far ABC (BSFABC) [2] and also with other four algorithms, namely, Shuffled Frog Leaping Algorithm (SFLA) [8], Particle Swarm Optimization (PSO) [12], Gravitational Search Algorithm (GSA) [16] and DE [7].

The following experimental setting is adopted to test ConABC, ABC, BSFABC, MABC, PSO, GSA, SFLA and DE:

- Total experimental runs = 100,
- Swarm size $S = 25$,
- $\phi_{ij} =$ is a random number between -1 to 1 and $Limit = D \times S$ where D denotes the Dimension,
- The setting of parameter for all the other mentioned algorithms, namely, BSFABC, MABC, PSO, GSA, SFLA and DE is same as their original research paper,
- Termination criteria is the maximal number of function evaluation which is fixed to be 200,000 for all the considered algorithms or achieving the acceptable error.

Table 2 depicts the obtained results regarding four parameters that are standard deviation (SD), mean error (ME), average number of function evaluations ($AFEs$) and success rate (SR). It is clearly demonstrated that ConABC outperforms other mentioned algorithms in terms of reliability, efficiency and accuracy.

Additionally, boxplots are designed using $AFEs$. The boxplots for ConABC, ABC, MABC, BSFABC, PSO, GSA, SFLA and DE are depicted in Fig. 1. It clearly shows that the interquartile range and the median of ConABC are comparatively lower than other examined algorithms.

Further, the Acceleration Rate (AR) [17] is calculated using the AFEs of the algorithms as stated below:

$$AR = \frac{AFE_{ALGO}}{AFE_{ConABC}}, \tag{6}$$

Here, ALGO $\in$ ABC, MABC, BSFABC, PSO, GSA, SFLA and DE. Using Eq. 6, AR is calculated between ConABC and ABC, ConABC and MABC, ConABC and BSFABC, ConABC and PSO, ConABC and GSA, ConABC and SFLA, ConABC and DE. This comparison is listed in Table 3 which shows that ConABC is fastest amongst all the observed algorithms for almost all the test functions. Also, a test, namely, Mann–Whitney U Rank Sum (MWURS) test [5] is performed on function evaluations to examine whether these outcomes are with a significant difference or not. The level of significance is taken as 5%, i.e. $\alpha = 0.05$ for comparison between ConABC–ABC, ConABC–MABC, ConABC–BSFABC, ConABC–PSO, ConABC–GSA, ConABC–SFLA and ConABC–DE.

Table 1 Test problems (D: Dimension, AE: Acceptable Error)

S. no.	Test problem	Objective function	Search range	Optimum value	D	AE
1	Michalewicz	$f_1(x) = -\sum_{i=1}^{D}\sin x_i(\sin(\frac{i x_i^2}{\pi}))^{20}$	$[0, \pi]$	$f_{min} = -9.66015$	10	1.0E-05
2	Inverted cosine wave	$f_2(x) = -\sum_{i=1}^{D-1}\left(\exp\left(\frac{-(x_i^2+x_{i+1}^2+0.5x_ix_{i+1})}{8}\right) \times I\right)$	$[-5, 5]$	$f(\mathbf{0}) = -D+1$	10	1.0E-05
3	Neumaier 3 problem (NF3)	$f_3(x) = \sum_{i=1}^{D}(x_i-1)^2 - \sum_{i=2}^{D}x_ix_{i-1}$	$[-D^2, D^2]$	$f_{min} = -\frac{(D(D+4)(D-1))}{6}$	10	1.0E-01
4	Branin function	$f_4(x) = a(x_2 - bx_1^2 + cx_1 - d)^2 + e(1-f)\cos x_1 + e$	$x_1 = [-5, 10],$ $x_2 = [0, 15]$	$f(-\pi, 12.275) = 0.3979$	2	1.0E-05
5	2D Tripod function	$f_5(x_2) = p(x_2)(1 + p(x_1)) + |(x_1 + 50p(x_2)(1 - 2p(x_1)))| + |(x_2 + 50(1 - 2p(x_2)))|$	$[-100, 100]$	$f(0, -50) = 0$	2	1.0E-04
6	Shifted Rosenbrock	$f_6(x) = \sum_{i=1}^{D-1}[100(z_i^2 - z_{i+1})^2 + (z_i - 1)^2] + f_{bias},$ $z = x - o + 1, x = [x_1, x_2, \ldots, x_D], o = [o_1, o_2, \ldots, o_D]$	$[-100, 100]$	$f(o) = f_{bias} = 390$	10	1.0E-01
7	Goldstein–Price	$f_7(x) = (1 + (x_1 + x_2 + 1)^2 \cdot (19 - 14x_1 + 3x_1^2 - 14x_2 + 6x_1x_2 + 3x_2^2)) \cdot (30 + (2x_1 - 3x_2)^2 \cdot (18 - 32x_1 + 12x_1^2 + 48x_2 - 36x_1x_2 + 27x_2^2))$	$[-2, 2]$	$f(0, -1) = 3$	2	1.0E-14
8	Six-hump camel back	$f_8(x) = (4 - 2.1x_1^2 + x_1^4/3)x_1^2 + x_1x_2 + (-4 + 4x_2^2)x_2^2$	$[-5, 5]$	$f(-0.0898, 0.7126) = -1.0316$	2	1.0E-05
9	Hosaki problem	$f_9(x) = (1 - 8x_1 + 7x_1^2 - 7/3x_1^3 + 1/4x_1^4)x_2^2\exp(-x_2),$ subject to $0 \le x_1 \le 5.0, 0 \le x_2 \le 6$	$x_1 = [0, 5],$ $x_2 = [0, 6]$	-2.3458	2	1.0E-6
10	McCormick	$f_{10}(x) = \sin(x_1 + x_2) + (x_1 - x_2)^2 - \frac{3}{2}x_1 + \frac{5}{2}x_2 + 1$	$-1.5 \le x_1 \le 4, -3 \le x_2 \le 3$	$f(-0.547, -1.547) = -1.9133$	2	1.0E-04
11	Meyer and Roth	$f_{11}(x) = \sum_{i=1}^{5}\left(\frac{x_1x_3t_i}{1+x_1t_i+x_2v_i} - y_i\right)^2$	$[-10, 10]$	$f(3.13, 15.16, 0.78) = 0.4E-04$	3	1.0E-03
12	Sinusoidal	$f_{12}(x) = -[A\prod_{i=1}^{D}\sin(x_i - z) + \prod_{i=1}^{D}\sin(B(x_i - z))],$ $A = 2.5, B = 5, z = 30$	$[0, 180]$	$f(90 + z) = -(A + 1)$	10	1.0E-02

Table 2 Analysis of the results for test problems (T.P.)

TP	Algorithm	SD	ME	AFE	SR
f_1	ConABC	3.43E–06	4.52E–06	32494.12	100
	ABC	3.74E–06	3.66E–06	27213.73	100
	MABC	3.28E–06	6.70v06	36367.70	100
	BSFABC	3.76E–06	3.63E–06	46436.86	100
	PSO	3.46E–01	3.71E–01	198,447	1
	GSA	5.59E–07	9.30E–06	154,615	100
	SFLA	4.49E–01	1.10E+00	200,000	0
	DE	4.26E–02	4.21E–02	171,411	20
f_2	ConABC	2.46E–06	7.08E–06	76826.33	100
	ABC	7.36E–02	1.13E–02	87518.62	90
	MABC	1.47E–06	8.30E–06	68454.01	100
	BSFABC	2.00E–01	5.37E–02	122447.64	85
	PSO	6.17E–01	1.37E+00	197,401	4
	GSA	3.91E–03	8.60E–02	200,000	0
	SFLA	6.70E–01	2.82E+00	200,000	0
	DE	6.30E–01	9.56E–01	184,340	11
f_3	ConABC	1.06E–02	9.93E–02	95947.03	99
	ABC	7.71E–01	8.85E–01	196800.17	3
	MABC	1.12E–01	1.17E–01	131102.02	97
	BSFABC	5.59E+00	4.77E+00	200024.69	0
	PSO	6.89E–03	9.42E–02	35997.50	100
	GSA	2.56E–01	2.48E+00	200,000	0
	SFLA	2.57E–07	9.67E–06	40275.77	100
	DE	1.56E–02	7.97E–02	10885.5	100
f_4	ConABC	6.27E–06	5.40E–06	1548.99	100
	ABC	6.26E–06	5.44E–06	2051.31	100
	MABC	6.22E–06	5.23E–06	16990.25	93
	BSFABC	6.59E–06	5.74E–06	21587.17	90
	PSO	3.52E–06	6.12E–06	26572.50	88
	GSA	5.23E–02	1.88E–02	142,220	86.67
	SFLA	6.40E–01	2.72E+00	200,000	0
	DE	7.46E–06	6.50E–06	35696.50	83
f_5	ConABC	2.24E–05	7.24E–05	6253.98	100
	ABC	2.37E–05	6.71E–05	11791.41	100
	MABC	7.14E–04	1.91E–04	56423.43	95
	BSFABC	3.52E–04	1.25E–04	11196.38	96
	PSO	3.36E–01	1.30E–01	38,554	87
	GSA	1.07E–04	2.23E–04	75	100
	SFLA	2.80E–07	5.31E–07	1269.20	100
	DE	2.55E–01	7.01E–02	17,166	93

(continued)

Table 2 (continued)

TP	Algorithm	SD	ME	AFE	SR
f_6	ConABC	5.40E–02	1.01E–01	100973.51	90
	ABC	2.67E+00	1.33E+00	182516.38	15
	MABC	9.66E–01	6.10E–01	170082.71	29
	BSFABC	3.92E+00	2.17E+00	178501.97	22
	PSO	6.74E+00	2.03E+00	186800.50	48
	GSA	2.23E–07	7.07E–07	145708.33	100
	SFLA	1.93E+00	1.78E+00	135338.73	57
	DE	1.71E+00	2.27E+00	192,987	4
f_7	ConABC	3.12E–14	2.00E–14	55103.53	98
	ABC	7.67E–05	1.18E–05	113701.83	53
	MABC	4.85E–14	5.49E–14	116054.78	45
	BSFABC	6.14E–03	2.91E–02	197487.95	2
	PSO	4.57E–14	4.94E–14	102,817	51
	GSA	8.88E–13	1.85E–12	21,290	100
	SFLA	2.77E+04	5.76E+04	145.20	100
	DE	4.83E–14	5.33E–14	107778.50	47
f_8	ConABC	1.02E–05	1.19E–05	889.51	100
	ABC	1.11E–05	1.19E–05	1015.02	100
	MABC	1.52E–05	1.56E–05	88816.08	56
	BSFABC	4.66E–14	3.73E–14	78238.27	65
	PSO	1.19E–05	1.57E–05	91685.50	55
	GSA	2.19E+04	3.46E+04	50	100
	SFLA	6.80E–01	2.92E+00	200,000	0
	DE	1.42E–05	1.86E–05	118652.50	41
f_9	ConABC	6.21E–06	5.37E–06	551	100
	ABC	6.68E–06	6.21E–06	672.50	100
	MABC	6.52E–06	5.69E–06	16939.22	92
	BSFABC	5.32E–03	4.91E–01	2805.73	100
	PSO	3.22E–06	5.56E–E–06	17,325	92
	GSA	6.09E+03	9.34E+03	631.67	100
	SFLA	6.60E–01	2.83E+00	200,000	0
	DE	6.16E–06	5.65E–06	20857.50	90
f_{10}	ConABC	5.93E–06	8.81E–05	901.50	100
	ABC	6.06E–06	8.80E–05	1243.55	100
	MABC	6.57E–06	8.91E–05	1764.36	100
	BSFABC	6.40E–06	5.86E–06	16628.28	92
	PSO	6.85E–06	8.80E–05	1526.50	100
	GSA	5.95E–06	5.50E–06	39686.67	100
	SFLA	6.50E–01	2.81E+00	200,000	0
	DE	6.87E–06	8.72E–05	982.50	100

(continued)

Table 2 (continued)

TP	Algorithm	SD	ME	AFE	SR
f_{11}	ConABC	2.89E–06	1.95E–03	7483.80	100
	ABC	2.90E–06	1.95E–03	22086.89	99
	MABC	2.91E–06	1.95E–03	8830.62	100
	BSFABC	6.68E–06	8.83E–05	1013.53	100
	PSO	2.80E–06	1.95E–03	3348.50	100
	GSA	6.00E–06	8.96E–05	43526.67	100
	SFLA	5.20E–18	1.26E–02	200,000	0
	DE	1.86E–05	1.95E–03	3830.50	99
f_{12}	ConABC	2.03E–03	7.77E–03	10627.58	100
	ABC	2.09E–03	7.61E–03	53121.45	100
	MABC	9.71E–02	6.01E–01	200035.80	0
	BSFABC	5.21E–06	4.44E–06	8951.57	100
	PSO	2.89E–01	4.48E–01	183,011	19
	GSA	2.88E–19	1.87E+02	200,000	0
	SFLA	4.86E–17	9.30E–16	37419.80	100
	DE	2.31E–01	5.33E–01	199,685	1

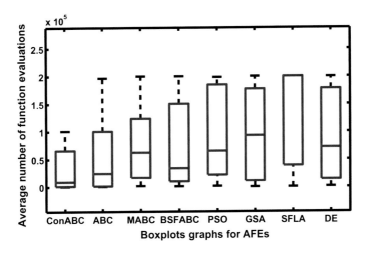

Fig. 1 Boxplots graphs for AFEs

Table 3 Acceleration Rate (AR) of ConABC compared to the basic ABC, MABC, BSFABC, PSO, GSA, SFLA and DE

Test problems	ABC	MABC	BSFABC	PSO	GSA	SFLA	DE
f_1	0.84	1.12	1.43	6.11	4.76	6.15	5.28
f_2	1.14	0.89	1.59	2.57	2.60	2.60	2.40
f_3	2.05	1.37	2.08	0.38	2.08	0.42	0.11
f_4	1.32	10.97	13.94	17.15	91.81	129.12	23.05
f_5	1.89	9.02	1.79	6.16	0.01	0.20	2.74
f_6	1.81	1.68	1.77	1.85	1.44	1.34	1.91
f_7	2.06	2.11	3.58	1.87	0.39	0.00	1.96
f_8	1.14	99.85	87.96	103.07	0.06	224.84	133.39
f_9	1.22	30.74	5.09	31.44	1.15	362.98	37.85
f_{10}	1.38	1.96	18.45	1.69	44.02	221.85	1.09
f_{11}	2.95	1.18	0.14	0.45	5.82	26.72	0.51
f_{12}	5.00	18.82	0.84	17.22	18.82	3.52	18.79

Table 4 Comparison based on MWURS Test at a $\alpha = 0.05$ significance level ('+' shows ConABC is cost effective, '−' shows ConABC is costlier, whilst = shows the difference is not significant)

Test Problems	ConABC versus ABC	ConABC versus MABC	ConABC versus BSFABC	ConABC versus PSO	ConABC versus GSA	ConABC versus SFLA	ConABC versus DE
f_1	−	+	+	+	+	+	+
f_2	+	−	+	+	+	+	+
f_3	+	+	+	−	+	−	−
f_4	+	+	+	+	+	+	+
f_5	+	+	+	+	−	−	+
f_6	+	+	+	+	+	+	+
f_7	+	+	+	+	−	−	+
f_8	+	+	+	+	−	+	+
f_9	+	+	+	+	+	+	+
f_{10}	+	+	+	+	+	+	+
f_{11}	+	+	−	−	+	+	−
f_{12}	+	+	−	+	+	+	+

The end results of the MWURS test for the AFEs on 100 runs are being shown in table 4. Furthermore, the conditions where $ConABC$ performed well are denoted by '+' sign whereas '−' sign denotes that it is not performing well. The Table 4, contains 70 '+' signs from 84 comparisons. Thus, it can be deduced that the outcomes of ConABC are profitable in comparison with ABC, MABC, BSFABC, PSO, GSA, SFLA and DE for most of the test problems.

5 Conclusion

This article aims at developing a novel variant of ABC algorithm, titled fitness-based controlled movements in ABC algorithm (ConABC). In this variant, the modification is done within the employed bee stage of ABC algorithm by introducing an Intelligent Term (IT), which guides the individuals during solution search process. The IT controls the step size of an individual as per the fitness of it in the swarm and thus balances the exploration and exploitation abilities of ABC. For evaluating the presented algorithm, it is assessed on 12 benchmark functions. The statistical investigation over test problems reveals that ConABC will be a preferable option to solve various continuous optimization problems.

References

1. Akay, B., Karaboga, D.: A modified artificial bee colony algorithm for real-parameter optimization. Inf. Sci. **192**, 120–142 (2012)
2. Banharnsakun, A., Achalakul, T., Sirinaovakul, B.: The best-so-far selection in artificial bee colony algorithm. Appl. Soft Comput. **11**(2), 2888–2901 (2011)
3. Bansal, J.C., Sharma, H., Arya, K.V., Deep, K., Pant, M.: Self-adaptive artificial bee colony. Optimization **63**(10), 1513–1532 (2014)
4. Bansal, J.C., Sharma, H., Jadon, S.S., Clerc, M.: Spider monkey optimization algorithm for numerical optimization. Memet. Comput. **6**(1), 31–47 (2014)
5. Bansal, J.C., Sharma, H., Nagar, A., Arya, K.V.: Balanced artificial bee colony algorithm. Int. J. Artif. Intell. Soft Comput. **3**(3), 222–243 (2013)
6. Das, S., Biswas, A., Dasgupta, S., Abraham, A.: Bacterial foraging optimization algorithm: theoretical foundations, analysis, and applications. In: Foundations of Computational Intelligence, vol. 3, pp. 23–55. Springer (2009)
7. Das, S., Suganthan, P.N.: Differential evolution: a survey of the state-of-the-art. IEEE Trans. Evol. Comput. **15**(1), 4–31 (2011)
8. Eusuff, M., Lansey, K., Pasha, F.: Shuffled frog-leaping algorithm: a memetic meta-heuristic for discrete optimization. Eng. Optim. **38**(2), 129–154 (2006)
9. Fister Jr, I., Yang, X.-S., Fister, I., Brest, J., Fister, D.: A brief review of nature-inspired algorithms for optimization (2013). arXiv:1307.4186
10. Karaboga, D.: An idea based on honey bee swarm for numerical optimization. Technical report, Technical report-tr06, Erciyes University, Engineering Faculty, Computer Engineering Department (2005)
11. Karaboga, D., Basturk, B.: On the performance of artificial bee colony (ABC) algorithm. Appl. Soft Comput. **8**(1), 687–697 (2008)
12. Kennedy, J.: Particle swarm optimization. In: Encyclopedia of Machine Learning, pp. 760–766. Springer (2011)
13. Koza, J.R.: Genetic programming as a means for programming computers by natural selection. Stat. Comput. **4**(2), 87–112 (1994)
14. Luthra, I., Chaturvedi, S.K., Upadhyay, D., Gupta, R.: Comparative study on nature inspired algorithms for optimization problem. In: 2017 International Conference of Electronics, Communication and Aerospace Technology (ICECA), vol. 2, pp. 143–147. IEEE (2017)
15. Marrow, P.: Nature-inspired computing technology and applications. BT Technol. J. **18**(4), 13–23 (2000)
16. Rashedi, E., Nezamabadi-Pour, H., Saryazdi, S.: Gsa: a gravitational search algorithm. Inf. Sci. **179**(13), 2232–2248 (2009)

17. Sharma, H., Sharma, S., Kumar. S.: Lbest gbest artificial bee colony algorithm. In: 2016 International Conference on Advances in Computing, Communications and Informatics (ICACCI), pp. 893–898. IEEE (2016)

Herbal Plant Classification and Leaf Disease Identification Using MPEG-7 Feature Descriptor and Logistic Regression

Ajay Rana and Ankush Mittal

Abstract Plant disease classification, especially herbal plant disease classification is a prominent problem in the field of botany. It is compelling problem due to the heterogeneity among the plants of the same category and dearth of awareness about the immense medicinal properties of herbal leaf. By not only classifying herbal plant but also identifying the diseased and non-diseased traits among herbal plants will facilitate the naive population as well as herbal product manufacturing industry and pharmaceutical industry to enrich the global economy. In this paper, we have presented how MPEG-7 color and texture feature descriptors are incorporated with the traditional classifiers (for example, Logistic regression and Support Vector Machine, etc.) to yield very impressive results on wide range of classes. A total of two datasets: herbal plant dataset and leaf disease dataset are used to evaluate the results. This classification strategy is not only accurate but also very efficient in terms of number of computations needed and overall performance of the system. Comparison with other traditional features indicates the potential of MPEG-7 feature descriptors.

Keywords MPEG-7 · Black rot · Logistic regression · Bacterial spot · Cross entropy · Haar transform

1 Introduction

Food and crops are copiously available across the globe, but only an insignificant amount of it embodied herbal properties into it. Food security remained a threatening factor due to the existence of pathogens and lack of pollinators. The herbal plants are endangered species and further being threatened by the plant diseases, despite having so much improvement in the food preservation techniques. The small farm holders, naive population, herbal product manufacturing industries, and pharmaceutical industry are drastically affected by this menace, leading a great loss to the global economy every year.

A. Rana (✉) · A. Mittal
Department of Computer Science, Graphic Era Deemed To Be University, Dehradun, India
e-mail: ajaysinghrana16@gmail.com

© Springer Nature Singapore Pte Ltd. 2020
K. N. Das et al. (eds.), *Soft Computing for Problem Solving*,
Advances in Intelligent Systems and Computing 1048,
https://doi.org/10.1007/978-981-15-0035-0_62

The traditional methods used for plant classification includes manual monitoring, which requires inherent knowledge that is only acquired by an expert botanist. The availability of skilled botanist is significantly small in remote areas where there is no way to provide such facility. Even if availability is there, the process of manual monitoring is laborious, time taking, economically extravagant, and less accurate. But if computer vision and machine learning algorithms are employed, it will be very easy to classify plant categories within no time, more accurately and less economically.

This paper merges the ideas of herbal plant classification and plant disease classification together to form herbal plant disease classification using various combinations of MPEG-7 color and texture feature descriptors [1] and by incorporating various machine learning algorithms and computer vision. First, herbal plant classification is done to find out the herbal plant category. Then, further classification is done to find out whether the classified herbal plant has some disease or not, if yes then from which disease that herbal plant is suffering.

The challenges inherent with the problem of herbal leaf disease classification includes: (1) Data acquisition of herbal plants, (2) large image resolution resulting in high computational cost, and (3) different leaf orientations among the same category of the dataset. This paper is organized as follows: Sect. 2 represents the literature review. Section 3 presents the methodology and theory behind the MPEG-7 feature descriptors and other machine learning algorithms used. Result and Discussion are discussed in Sect. 4. Future perspective and conclusion are discussed in Sect. 5.

2 Literature Review

Several studies were conducted in the field of plant classification in the last few years. One of the earliest work had been done by Wu et al. [2]. They extracted 12 morphological features and 5 geometrical features. Then apply dimensionality reduction algorithm, Principal Component Analysis (PCA), and feed it to the Probabilistic Neural Network (PNN). They conducted this research on Flavia plant dataset, which is their own creation and was able to get a classification accuracy of 90.35%.

By employing the same technique as used by Wu et al. [2] and different datasets, Amin and Hossain [3] in 2010 achieved a classification accuracy of 93% using K-nearest neighbor classifier. Using Isomap distance measure, Du et al. [4] attained an accuracy of 92.3% on 20 categories and 2000 plant images in total. By using only shape and color as features and K-nearest classifier as classifier, Munisami et al. [5] got an accuracy of 87.3% on the dataset containing 640 images distributed among 32 plant categories with each image having resolution of 1980 * 1024.

Hernandez-Serna and Jimenez-Segura [6] used eight texture, two morphological, and six geometrical features, fed them to an artificial neural network to yield an accuracy of 92.9% on Flavia plant dataset. In 2017, A. Begue et al. [7] extracted some basic features such as width, length, perimeter, area of white space, and area of hull. (40 in total) and fed it into random forest classifier to get an accuracy of 90.1% on 24 categories of herbal plants containing 720 images in total (30 images per category). The dataset was developed on the tropical island of Mauritius using a smartphone.

A substantial amount of work has been done in the field of leaf disease classification and detection. Some of them are as follows:

Mohanty et al. [8] in 2016, conducted an experiment over 54306 images belonging to 14 crop species and 26 diseased (a total of 38 plant categories) categories using AlexNet and GoogLeNet deep learning architecture. After multiple hours of training, the result they found was very astonishing. Dhaygude et al. [9] presented four steps process, first a color transformation is created for RGB images. Then by setting a threshold, green pixels are removed. Then masking of useful regions is done using precomputed threshold level. And at last segmentation is done to detect the plant disease.

Badnakhe and Deshmukh [10] presented the relevance of machine learning algorithms in India, which can save effort, time, and money. They used color co-occurrence method for feature extraction and neural network for training and testing, resulting in very accurate plant disease detection. Arivazhagan et al. [11] introduced four steps method as follows: First color transformation is done on RGB, then masking of green pixels is performed using a specific threshold value, then segmentation is performed and texture statistics are computed for getting useful segment. At the end, they fed the extracted features into the classifier to yield a classification accuracy of 94%.

Naikwadi and Amoda [12] proposed how histogram matching is performed to identify plant disease. Plant disease appears on the leaf area of a plant; hence, histogram matching is performed on the basis of color feature and edge detection technique.

Patil et al. [13] in 2011, presented triangle threshold for lesion region area and simple threshold method used to segment leaf area. At last, classification was done by computing the quotient of leaf and lesion area. For measuring the leaf disease severity, this is fast and accurate. It used threshold segmentation to compute the leaf area. Singh and Misra [14] proposed segmentation of leaf components using genetic algorithm. The average accuracy achieved by employing the proposed work is 97.6% on a dataset containing five classes of leaf disease.

3 Methodology

3.1 Proposed Framework

First of all, a total of five MPEG-7 feature set including three color feature set and two texture feature set are extracted. The extracted features are then passed to classifier to determine the herbal plant type. Then classification is performed one more time to find out whether the classified herbal plant has any disease or not, if yes then from which disease the herbal plant is suffering. In doing so, different combinations

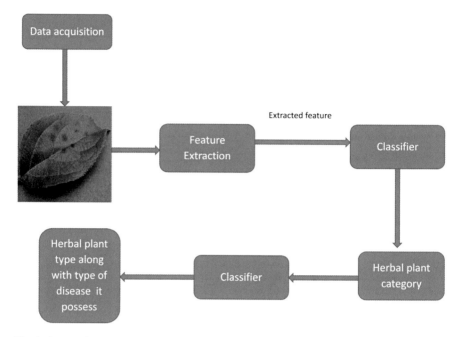

Fig. 1 System figure

of features are used in order to observe the effect of these feature descriptors on the performance of the system. No preprocessing or post-processing is done on the datasets. The system figure is presented in Fig. 1.

3.2 Feature Descriptor

The primary goal of feature extraction is to draw indispensable attributes from the multimedia and facilitate the classification and retrieval process. Feature extraction is done by computing descriptors from the pixels of multimedia around the point of interest. A descriptor is a vector representation of an image in which each value conveys a special meaning. Based on what these values signify, they are broadly classified into two categories [15]: global and local descriptors. Global descriptors describe the multimedia using fewer vector values, ignoring the minute details so as to represent a generalized view of the object.

In contrast with global descriptors, local descriptors identify the keypoints of an object and then compute a descriptor for each one of them. Local descriptor is a compact representation of point's local neighborhood. Unlike global descriptors, local descriptors try to match shape only in the local neighborhood around the point. Working with local descriptors leads to increased dimension of feature space, hence ignored for the main result section of this paper. However, a comparison has been

Table 1 Features along with number of dimensions

Feature name	Number of dimensions
Edge histogram descriptor	80
Homogenous texture descriptor	62
Scalable color descriptor	128
Color structure descriptor	32
Color layout descriptor	120

made using global features alone and global features along with local feature named Speeded Up Robust Features (SURF).

The feature descriptors which we employ in our work are given in Table 1. The detailed description of the MPEG-7 features can be found in MPEG-7 overview [1], but to elucidate the key concept behind MPEG-7 feature descriptors, a brief discussion is presented.

Color Layout Descriptor (CLD)

Color layout descriptor is used to represent the spatial dispersion of color in an image. The process involved is depicted in four phases: (a) Partitioning of image into 64 nonoverlapping blocks to ensure resolution invariance, (b) Average color of each block is computed, (c) Discrete Cosine Transformation (DCT) applied on 64 average colors, so as to obtain three sets of 64 DCT coefficients, and (d) Finally, zigzag scanning is performed on these three sets of 64 DCT coefficients to filter out low-frequency coefficient [1]. CLD uses a combination of grid-wise dominant color and grid structure descriptor. It computes the matching between two CLD's using distance measure which is described as follows:

Suppose, we have two CLD's $\{DY, DCr, DCb\}$ and $\{DY', DCr', DCb'\}$. Then the distance measure

$$D = \sqrt{\sum_k w_{yk}\left(DY_k - DY'_k\right)^2} + \sqrt{\sum_k w_{rk}\left(DCr_k - DCr'_k\right)^2} + \sqrt{\sum_k w_{bk}\left(DCb_k - DCb'_k\right)^2}$$

where $(DY_k, DCr_k,$ and $DCb_k)$ depict the kth coefficients of the respective color components.

CLD is a very robust representation of color that is compact, resolution invariant, and computationally inexpensive.

Color Structure Descriptor (CSD)

The color structure descriptor captures both color and structure content of the image (or frame). It is ideally used for static image retrieval system. This descriptor can differentiate between two images, where the amount of each color is similar but group of pixels lying in that color is different. It uses 8×8 structuring element in order to compute CSD. The number of samples is kept as 64, while the spatial extent and subsampling factor vary with the size of image. The subsampling is calculated as follows:

$$p = \max\{0, \text{round}(0.5 * \log_2 WH - 8)\}$$
$$K = 2^p$$
$$E = 8K$$

where

H and W are height and width of the image, respectively;
K is the subsampling factor;
$E \times E$ is the spatial extent of structuring element.

Color values are represented in HMMD color space, which is nonuniformly quantized into 32, 64, 128, and 256 number of bins. It provides improved similar-image retrieval, especially in case of natural images.

Scalable Color Descriptor (SCD)

It is a compound descriptor containing color quantization, color space, and histogram descriptor. It is a color histogram represented in HSV color space that is uniformly quantized into 256 bins. The histogram values are abbreviated using 11-bit integer representation. In order to achieve more efficient encoding, mapping is done to obtain a 4-bit nonlinear representation.

It requires 1024 bits/histogram for 4-bit representation of 256 bin HSV histogram. In order to truncate the number of bits/histogram, Haar transform is used. It is a very useful descriptor for image-based retrieval depending on the color features.

Edge Histogram Descriptor (EHD)

It represents the spatial distribution of five types of edge orientations, classified as one nondirectional and four directional edges. Edge is an indispensable trait of an object, and using it in classification can give significantly good results even in the case of natural images, where edge distribution is nonuniform. The performance of EHD can be substantially increased by combining it with a color descriptor (e.g., CSD). The reason behind this drastic performance boost is the non-dependency of texture descriptors (EHD) to the color descriptors (CSD).

Logistic Regression (LR)

It establishes the relationship between dependent (class/labels) and independent (features) variables, by using the logistic function. These probabilities are transformed into binary representation using sigmoid function. Sigmoid function normalizes the real-valued number in the range between 0 and 1. This transformed value is then transformed to binary label 0 or 1, using classifier threshold.

The cross-entropy loss function used for logistic regression is given as follows:

$$\text{Loss} = \frac{-1}{N} \sum_i y_i \log a_i + (1 - y_i) \log(1 - a_i)$$

where

$$z_i = W x_i \text{ and } a_i = \sigma(z_i)$$

Computing the derivative of loss with respect to W, we get

$$\frac{\text{dLoss}}{\text{dW}} = \frac{1}{N} \sum_i -(y_i - a_i) * x_i$$

In order to get the optimum value we can set the derivative, $\frac{\text{dLoss}}{\text{dW}} = 0$.

Liblinear optimization algorithm is ideally used for small dataset, and since in our approach the feature space is very small, hence liblinear algorithm is the optimal choice for our work.

4 Experiment and Result

4.1 Implementation Framework

The implementation has been done in Python 3.6 using scikit-learn library. It took 2.8 and 12.1 s to perform training and testing on herbal plant dataset and leaf disease dataset, respectively. We have used Intel core i3 processor.

4.2 Dataset Description

Herbal plant dataset contains 1043 herbal leaf images, split across 24 categories. We have used the same dataset used by [7]. Few categories are displayed in Fig. 2.

Using a public dataset provided by PlantVillage, a subset of 20 categories are extracted containing 7 healthy and 13 diseased plants, containing 100 images each to make the leaf disease dataset. Sample leaf disease categories from leaf disease dataset are displayed in Fig. 3.

(a)   **(b)**   **(c)**

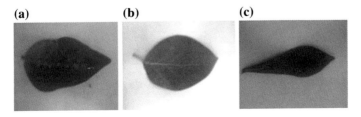

Fig. 2 Sample image categories from herbal plant dataset: **a** Antidesma, **b** Avocado, **c** Ayapana

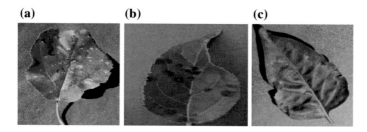

Fig. 3 Some of the leaf disease type: **a** Apple scab, **b** Apple black rot, **c** Pepper bell bacterial spot

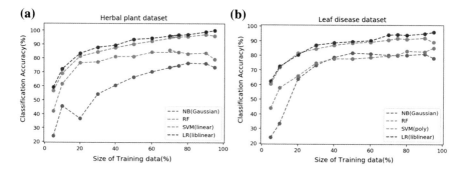

Fig. 4 Experiment 1: performance of classifiers on varying training size. **a** for herbal plant dataset, **b** for leaf disease dataset

4.3 Result and Discussion

In this section, experiments using herbal plant dataset and leaf disease dataset are presented. A comparison has been performed using four different classifiers incorporated with the combination of global and local feature descriptors. Global feature descriptors include CSD, CLD, SCD, Homogenous Texture Descriptor (HTD), EHD; while local descriptors include SURF.

Now we are going to demonstrate a comparison between the performance of Naive Bayes (NB), Support vector machine (SVM), Random forest (RF), and Logistic regression (LR). For the purpose of performance comparison, we undertake two types of experiments: (1) with varying training data size, (2) with varying number of dimensions. Experiment 1 evaluates the performance of different types of classifiers on different training sizes, it signifies how well the system is able to generalize on even small number of training examples, while Experiment 2 depicts the effect of dimensional space size on the system's performance.

For Experiment 1, depicted in Fig. 4, number of dimensions is taken as 360 (including CSD, CLD, SCD, and EHD for both the datasets) and various training sizes. Logistic regression performs the best on both the datasets. It is relevant to look at the performance of logistic regression classifier at 20% training and 80% testing data, which crosses 80% accuracy on both the datasets. This depicts how well logistic

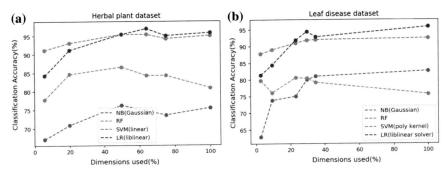

Fig. 5 Experiment 2: performance of classifiers on variable dimensions. **a** for herbal leaf dataset, **b** for leaf disease dataset

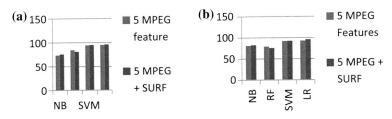

Fig. 6 Represent the effect of adding SURF descriptor to the MPEG-7 features on the performance, **a** for herbal plant dataset, **b** for leaf disease dataset

regression generalizes on small training examples. After that SVM remains as the second-best classifier, achieving an accuracy of 76.64% and 81.5% on herbal plant dataset and leaf disease dataset, respectively, on 20% training data.

Experiment 2 presented in Fig. 5, shows the performance of different classifiers on varying number of dimensions. Figure 5 shows how we varied the dimensions of feature space and compute results on each one of them for both the datasets. It is interesting to note that by employing SVM and only 32 dimensions of CSD, an accuracy of 91.19% and 87.80% is achieved on herbal leaf plant dataset and leaf disease dataset, respectively. This experiment shows the relevance of CSD features.

On the other side, if we take a look at the performance obtained by incorporating SURF feature along with 5 MPEG-7 feature descriptors (CSD, CLD, SCD, EHD, and HTD) as shown in Fig. 6. It is clearly shown that SURF feature increases the feature space by large extant, providing a very small boost in performance of classifier on both the datasets and even deteriorate the performance in case of random forest classifier on both the dataset (as indicated in Fig. 6a, b). Hence, it is being inferred that the local descriptors will always increase the feature space, but it does not always ensure boost in accuracy.

Adding Homogenous Texture Descriptor (HTD) with CSD, CLD, SCD, EHD results in decreased performance on all the classifiers in case of herbal plant dataset (Refer to Fig. 7 to see the effect of HTD).

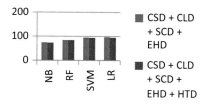

Fig. 7 Effect of adding HTD to combination of CSD, CLD, SCD, and EHD on the performance using herbal plant dataset

Table 2 Performance comparison between using EHD alone and EHD, CSD together on both the datasets

Classifier	Herbal plant dataset (accuracy %)			Leaf disease dataset (accuracy %)		
	EHD	CSD	EHD + CSD	EHD	CSD	EHD + CSD
Naive Bayes	36.40	67.05	70.88	43.80	62.80	73.80
RF	47.89	77.78	84.67	32.60	79.80	76.00
SVM	61.30	91.19	93.10	53.20	87.80	89.00
LR	62.84	84.29	91.19	44.40	81.40	84.40

Table 3 Performance of various classifiers on: (a) leaf disease dataset, (b) herbal plant dataset

Classifier	Accuracy
(a)	
Naïve Bayes	80.00
Random forest	80.40
SVM (poly kernel)	91.80
Logistic regression (liblinear)	94.40
(b)	
Naïve Bayes	74.71
Random forest	84.29
SVM (linear kernel)	95.40
Logistic regression (liblinear)	96.93

For another experiment, if we look at the classifiers performance considering only one texture feature (EHD), the results are quite disappointing. But on adding a color descriptor for example CSD, which is independent of the texture descriptor increases the performance drastically [1] (as shown in Table 2). It signifies that it is fruitful to combine two independent feature descriptors together.

The final result calculations are carried out using 360 dimensional vector (including CSD, CLD, SCD, and EHD), 75% training and a set of classifiers. The results obtained are shown in Table 3a, b.

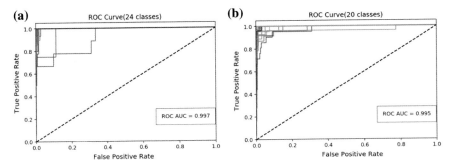

Fig. 8 ROC curve of: **a** Herbal plant dataset, **b** leaf disease dataset

Table 4 Performance in terms of precision, recall and ROC Area Under Curve (AUC)

Dataset	Precision	Recall	Average ROC AUC
Herbal plant	0.97	0.97	0.997
Leaf disease	0.95	0.94	0.995

The best results are provided by logistic regression classifier, achieving an accuracy of 96.93% on herbal plant dataset, and 94.40% on leaf disease dataset. The previous results on the herbal plant dataset which we have used in our work were calculated by [7], they achieved an accuracy of 90.1% using Random Forest classifier. Though the accuracy achieved by our model is comparatively more impressive than previous methods, our methodology still suffers from certain limitations: (a) We have used only leaf for classification purpose, not other parts of the plant, (b) we are currently constrained to the classification of single leaf, not much variations in orientation, and on homogenous background.

Receiver Operating Characteristic (ROC) curve has been calculated for both the datasets and shown in Fig. 8. It shows the relationship between true positive rate and false positive rate provided by the classifier.

Average precision, recall value, and ROC Area Under Curve (AUC) on both the datasets are presented in Table 4. AUC is the area enclosed under the ROC curve. Since we have a multiclass problem here, we have taken the average AUC of 24 classes in case of herbal plant dataset and 20 classes of leaf disease dataset.

5   Conclusion and Future Work

In this paper, we have developed a generic framework using MPEG-7 feature descriptors incorporated with traditional machine learning algorithms in order to identify the herbal plant category of the plant along with the disease possessed by it. Result and Discussion section (Experiment 2) shows how the used framework is able to generalize on very few dimensions. The methodology that we propose is very efficient in terms of dimensionality space size and the number of computations required to train the model. This automated system will not only help people in the field of taxonomy, pharmaceutical industry, and herbal product manufacturing industry but also enables naive people to draw distinction between the plant categories within no time and more accurately.

For future perspective, a more sophisticated dataset can be developed and tested on our model in order to serve the purpose of early plant disease detection. We can also calculate the degree of deterioration of a plant on the basis of color and texture. For color and texture extraction, MPEG-7 would be a great choice. We can further employ deep learning models on a more robust dataset to achieve even more impressive results.

References

1. Martınez, J.M.: MPEG-7 overview (version 10), vol. 3752. Technical report (2004)
2. Wu, S.G., Bao, F.S., Xu, E.Y., Wang, Y.-X., Chang, Y.-F., Xiang, Q.-L.: A leaf recognition algorithm for plant classification using probabilistic neural network. In: 2007 IEEE International Symposium on Signal Processing and Information Technology, pp. 11–16. IEEE (2007)
3. Hossain, J., Amin, M.A.: Leaf shape identification based plant biometrics. In: 2010 13th International Conference on Computer and Information Technology (ICCIT), pp. 458–463. IEEE (2010)
4. Du, J.-X., Wang, X.-F., Zhang, G.-J.: Leaf shape based plant species recognition. Appl. Math. Comput. **185**(2), 883–893 (2007)
5. Munisami, T., Ramsurn, M., Kishnah, S., Pudaruth, S.: Plant leaf recognition using shape features and colour histogram with K-nearest neighbour classifiers. Procedia Comput. Sci. **58**, 740–747 (2015)
6. Hernández-Serna, A., Jiménez-Segura, L.F.: Automatic identification of species with neural networks. PeerJ **2**, e563 (2014)
7. Begue, A., Kowlessur, V., Singh, U., Mahomoodally, F., Pudaruth, S.: Automatic recognition of medicinal plants using machine learning techniques. Int. J. Adv. Comput. Sci. Appl. **8**(4), 166–175 (2017)
8. Mohanty, S.P., Hughes, D.P., Salathé, M.: Using deep learning for image-based plant disease detection. Front. Plant Sci. **7**, 1419 (2016)
9. Dhaygude, S.B., Kumbhar, N.P.: Agricultural plant leaf disease detection using image processing. Int. J. Adv. Res. Electr. Electron. Instrum. Eng. **2**(1), 599–602 (2013)
10. Badnakhe, M.R., Deshmukh, P.R.: An application of K-means clustering and artificial intelligence in pattern recognition for crop diseases. In: International Conference on Advancements in Information Technology (2011)
11. Arivazhagan, S., Newlin Shebiah, R., Ananthi, S., Vishnu Varthini, S.: Detection of unhealthy region of plant leaves and classification of plant leaf diseases using texture features. Agric. Eng. Int. CIGR J. **15**(1), 211–217 (2013)
12. Naikwadi, S., Amoda, N.: Advances in image processing for detection of plant diseases. Int. J. Appl. Innov. Eng. Manag. (IJAIEM) **2**(11) (2013)
13. Patil, S.B., Bodhe, S.K.: Leaf disease severity measurement using image processing. Int. J. Eng. Technol. **3**(5), 297–301 (2011)
14. Singh, V., Misra, A.K.: Detection of plant leaf diseases using image segmentation and soft computing techniques. Inf. Process. Agric. **4**(1), 41–49 (2017)
15. Mittal, A., Cheong, L.-H.: Addressing the problems of Bayesian network classification of video using high-dimensional features. IEEE Trans. Knowl. Data Eng. **16**(2), 230–244 (2004)

Simulation of Metaheuristic Intelligence MPPT Techniques for Solar PV Under Partial Shading Condition

CH Hussaian Basha, C. Rani, R. M. Brisilla and S. Odofin

Abstract The nonlinear characteristics of solar PV consist of different MPPs under the partial shading condition. Hence, it is difficult to find out true MPP. The conventional MPPT methods are not giving an accurate position of MPP. In this work, two global metaheuristic optimization techniques are simulated and the comparative analysis is carried out in terms of tracking speed, steady-state oscillations, algorithm complexity, periodic tuning, and dynamic response. Those are the Cuckoo Search Optimization (CSO) and Particle Swarm Optimization (PSO) MPPT methods used to extract the maximum power of solar PV under partial shading condition. The Matlab/Simulink is used to evaluate performance results of CSA and PSO MPPT techniques.

Keywords Boost converter · CSO · Duty cycle · I-V and P-V curves · Partially shaded solar PV · PSO

1 Introduction

Due to the reduction of conventional energy sources and environmental pollution, the world is focusing on renewable energy sources. Out of all the renewable energy sources, solar is playing a major role because of maintenance-free, environmentally friendly nature, and excess availability in nature [1–3]. The installation and operating cost of solar PV is reduced by improving the manufacturing technology of

CH Hussaian Basha · C. Rani (✉) · R. M. Brisilla
School of Electrical Engineering, VIT University, Vellore, India
e-mail: crani@vit.ac.in

CH Hussaian Basha
e-mail: hussaianbasha.ch@vit.ac.in

R. M. Brisilla
e-mail: brisilla.rm@vit.ac.in

S. Odofin
School of Energy and Environment, University of Derby, Derby, UK
e-mail: s.odofin@derby.ac.uk

© Springer Nature Singapore Pte Ltd. 2020
K. N. Das et al. (eds.), *Soft Computing for Problem Solving*,
Advances in Intelligent Systems and Computing 1048,
https://doi.org/10.1007/978-981-15-0035-0_63

773

semiconductor material [4]. The PV semiconductor materials are mono, poly, and thin-film crystalline. In that most of the manufacturers focusing on monocrystalline silicon technology because of higher efficiency between 10 and 13% compared to other technologies [5, 6].

The PV cell converts sunlight energy into electrical energy using a photovoltaic effect. The solar cell working behavior is similar to the P–N diode operation. Whenever the solar radiations incident on the P–N junction, the electrons in the N-type gets energized and moves freely from N-type materials to P-type materials and current form a closed path of the circuit called photovoltaic effect [7–9]. The voltage generation of single PV cell lies between 0.55 and 0.88. It is not useful and insufficient for practical use. To achieve high voltage and currents, a number of PV cells are connected in series and parallel to form a module. The series-connected modules from a string and the multiple combinations of strings form an array [10].

The PV systems are can be designed by using single or double diode circuit topologies. From article [11], most of the research scholars interested on double diode circuit model of solar PV compared to single diode circuit topology because the single diode is useful for low-power applications and it does not suit for different irradiation conditions. The solar PV characteristics are nonlinear because of continuous changes in atmospheric conditions. In order to extract the maximum power of the solar PV, MPPT technique is used to run the operating point of PV closer to the MPP [12].

Most of the solar power plants installed in the shadow-free region but in urban areas tree falling and high height buildings are causes of Partial Shading (PS) on PV system. Due to this PS condition, multiple MPP peaks exist on power against voltage and current against voltage characteristics [13, 14]. From Fig. 1a, b, there are two local peaks ("a" and "b") and one global MPP is "c" which is useful to extract the maximum power from solar PV. The classical MPPT techniques are used under constant irradiation condition to improve the efficiency of a solar PV. Such techniques are Perturb and Observe (P&O), Hill Climb (HC), Incremental Resistance (IR), and Incremental Conductance (IC), etc. [2, 15]. These techniques track MPP

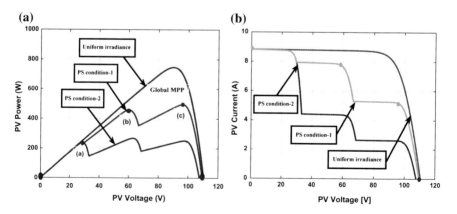

Fig. 1 **a** Power against voltage and, **b** current against voltage curves

effectively with high convergence speed under uniform irradiation condition. The major drawback of classical MPPT methods is unable to suppress the operating point of PV oscillations around MPP. Due to the oscillatory behavior of operating point, the power loss occurs in the PV system at steady-state condition and they are not giving accurate MPP for shading [11].

In order to overcome the drawbacks of classical MPPT techniques, Artificial Intelligence techniques are used to track MPP for partial shaded solar PV. Such algorithms are a genetic algorithm, Fuzzy logic, and Neural network, etc. Excluding flexibility, artificial intelligence techniques give less convergence speed to track MPP and have high implementation complexity compared to classical methods. For example, the fuzzy controller MPP tracking is mainly depended on user knowledge and experience of the application of fuzzy in a PV system and selection of membership function [16].

The artificial neural network gives excellent convergence speed but the accuracy in MPP tracking is less due to its high training time. ANN is implemented using an expensive microcontroller. As a result, the implantation cost ANN-based MPPT controller is high. Genetic algorithms are the computational probability search techniques inspired by is a biological evaluation. The PID controller is used in GA's to track MPP of solar PV and the controller constraints (Kp, Ki, Kd) are encoded as binary strings. These strings are called chromosomes and its combination form a mating pool. The fitness rate of each chromosome is determined from the PID controller proportionality constants [17]. The disadvantages of GA are difficulty in obeying inequality constraints and sensitive in the initialization of population. The limitation of GA is that and the solution quality deteriorates with increase in problem size and it cannot guarantee to find the true MPP.

The biological algorithms are used for solving all complex linear and nonlinear problems with less mathematical computations because it requires optimal searching. These techniques are easy to implement and very attractive to track MPP under shaded condition. The biological algorithms are Ant colony, Firefly algorithm, CSO, and PSO. The most effective and simple in design CSO and PSO techniques are considered in this work to track MPP under PS solar PV application [18, 19]. These algorithms are used for many industrial applications where multiple local peaks exist and it gives tremendous advantages compared to other techniques such as fast MPP tracking and high convergence speed without identifying the shading pattern of solar PV [20].

The mathematical modeling of double diode model of solar PV is given in Sect. 2 and the partial shading behavior is explained in Sect. 2.1. Section 3 gives the CS and PSO MPPT algorithms operation. Finally, Sects. 4 and 5 give the simulation results and conclusion of this chapter.

2 Modeling of Solar PV

The perfect model of solar PV gives high efficiency at all weather conditions and it utilizes all solar irradiations effectively. Hence, the PV system generates accurate IV and PV curves. The modeling of solar PV requires four major parameters such as short circuit current (Isc), open circuit voltage (Voc), maximum peak voltage (Vmpp), and maximum peak current (Impp). In addition, PV cell design requires diode ideality factor (a), photon current (Ipv), shunt resistance (Rsh), and series resistance (Rs) [2]. These parameters are calculated from any analytical method or optimization technique. In this work, double diode circuit topology is used to model the PV array because of its high accuracy and efficiency compared to single diode circuit topology.

The double diode circuit PV array can be modeled by the use of the following equation [2],

$$I = I_{ph}N_{pp} + X - Y - \frac{V + IR_s(N_{ss}/N_{pp})}{R_p * (N_{ss}/N_{pp})} \tag{1}$$

$$X = I_{01}N_{pp}\left(\exp\left(\frac{V + IR_s(N_{ss}/N_{pp})}{a_1 * V_{t1} * N_{ss}}\right) - 1\right) \tag{2}$$

$$Y = I_{02}N_{pp}\left(\exp\left(\frac{V + IR_s(N_{ss}/N_{pp})}{a_2 * V_{t2} * N_{ss}}\right) - 1\right) \tag{3}$$

where

$$I_{ph} = \left(I_{ph\_STC} + k_I\Delta T\right) * G/G_{STC} \tag{4}$$

From Eq. (1), the photon current is directly depending on the solar irradiations. Whenever the solar irradiations changes in ascending way the generated PV current steps up from one level to another level as shown in Fig. 1.

$$I_{01} = I_{02} = I_0 = \frac{I_{sc\_STC} + K_i\Delta T}{\exp\left((V_{oc_{STC}} + K_v\Delta T)/\{a_1 + a_2/p\}V_t\right) - 1} \tag{5}$$

where I_{01} and I_{02} are the single and double diode reverse saturation currents and it purely depends on the PV cell semiconductor material property and operating temperature. From Eq. (6), when the PV module series connected cells are increased, the thermal voltage of PV is increased and it is equal for both circuit models.

$$V_t = V_{t1} = V_{t2} = \frac{N_s * KT}{q} \tag{6}$$

2.1 Partial Shading of Solar PV

The most important task of MPPT controller in PV system is tracking of MPP under partial shaded condition. The multiple strings are connected in series and parallel to form an array PS which occurs in PV system because of nonuniform solar irradiation distribution on panels. The nonuniform solar irradiation distribution on panels because of birds falling or building shadows. In order to show the multiple MPP points on I-V and P-V curves, three series-connected panels are utilized under uniform and PS condition [1].

The partially shaded panel consumes the power of unshaded panel and it dissipates in the form heat. In series-connected panels, the uniform irradiation panels currents are limited from the shaded panel. To eliminate this issue a P–N junction diode is connected across each panel and it is forward biased to limit current of shaded panels [18]. The power against voltage characteristics of the series-connected panel is shown in Fig. 1 under uniform and two partial shading conditions. A unique MPP exist for each irradiation and temperate condition and it keeps on changing at diverse weather conditions. As a result, per unit generation cost of PV is increased. Hence, the operating point of PV forcibly coincides with true MPP. In this article, metaheuristic MPPT controllers are developed by interfacing a boost converter in between PV system and load.

3 Metaheuristic MPPT Algorithms

3.1 Particle Swarm Optimization

The PSO is one of the biological intelligence optimization technique introduced by Eberhart [20]. The operating principle of PSO is taken from the hunting action of birds and fish school system. In this work, each bird is considered as one particle. In the search space, each particle shares their previous and present data with other particles for achieving the optimal solution of solar PV. In PSO, the particle movement in search space is mainly depended on two variables P_{best} and G_{best}. The P_{best} is used to find out each particle best position in search space and G_{best} is helpful to compare all particle positions and stored the best position of swarm [21]. The swarm utilizes this process until obtaining the best position of particles. The implementation of PSO is given in Fig. 2.

The position of velocity and duty cycle of boost converter is updated by using Eqs. (7) and (8).

$$i^{n+1} = d_i^n + V_i^{n+1} \tag{7}$$

$$i^{n+1} = W V_i^n + C_a r_a \left(P_{best} - d_i^n\right) + C_b r_b \left(G_{best} - d_i^n\right) \tag{8}$$

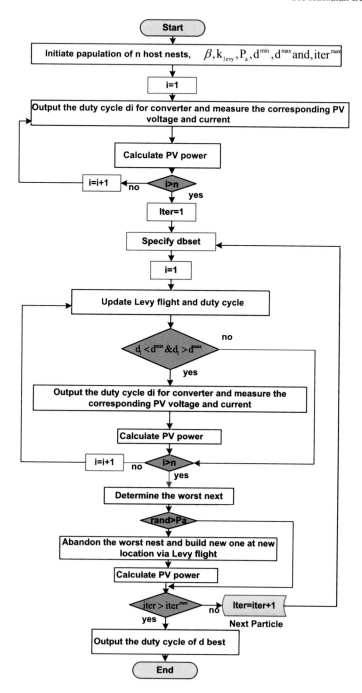

Fig. 2 Flow chart of PSO

where "V_i" is the step size of iteration "$n + 1$" and "W" are the inertial weight, "C_a" and "C_b" are the acceleration factors, "r_a" and "r_b" are the random variables. The implementation of the PSO algorithm is given in Fig. 2. At first iteration, different duty values are assigned randomly to all particles that are applied to the DC–DC converter. The sensors are used to measure PV voltage and current parameters and the calculated power is used as a fitness function of the particle "i". The new fitness value is compared with previous fitness values and the resultant value is considered as "P_{best}".

3.2 Cuckoo Search Algorithm

Among all MPPT algorithms, Cuckoo is the recent and best optimization technique for fast-tracking MPP and it extracts the maximum power of a solar PV. The CS technique is explained by the use of brood parasitism which is classified as interspecific and cooperation of birds [22]. The flow chart of CSA is given in Fig. 3.

The host bird breaks the CS eggs and nests after that it creates a new nest and this algorithm works based on three rules. First one, at a time each cuckoo should generate only one egg and second is the generated eggs in the first iteration should be good quality then only it will go to the next iteration. Finally, host nets are considered a fixed value. Levy flights are used to generate new eggs and a step size of the Cuckoo is determined by using the Levy function as follows [2].

$$\text{Levy}(\beta) = L^{-\beta} \tag{9}$$

where "L" is the length of flight and "β" is a constant. The value of "β" is in between "1" and "3". The optimizing flight size coefficient "α" is considered as a fraction of discharged eggs (P_a).

In this work, the main aim of the cuckoo search is to find an optimum duty cycle boost converter to extract the maximum power of solar PV. So, a duty cycle "d" is considered as an optimum function and the PV power is selected as a fitness function. The duty of boost converter is updated by using the Levy function is taken from the article [3],

$$d_i^{k+1} = d_i^k + \alpha \oplus \text{Levy}(\beta) \tag{10}$$

$$d_i^{k+1} = d_i^k + k_{\text{levy}} * \left(\frac{u}{\lfloor v \rfloor^{1/\beta}} \right) * \left(d_{\text{best}}^k - d_i^k \right) \tag{11}$$

where β is considered as 1.5 and "k_{levy}" is a multiplication coefficient. The functions "u" and "v" are determined by the use of distribution function:

$$u = N\left(0, \sigma_0^2\right), v = N\left(0, \sigma_v^2\right) \tag{12}$$

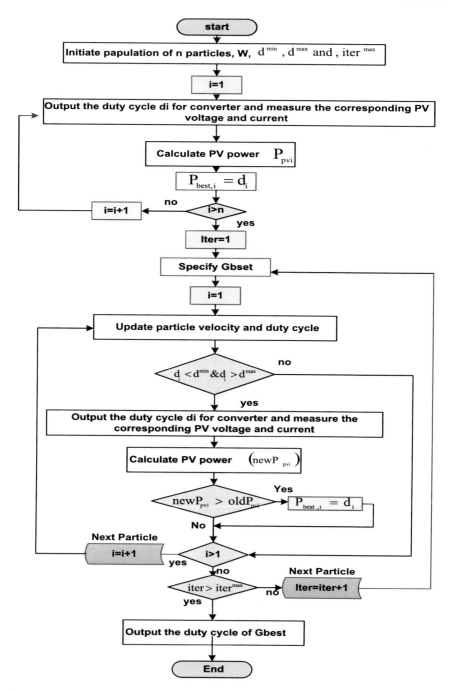

Fig. 3 Flow chart of CSA

where σ_u and σ_v functions are as follows:

$$\sigma_u = \left(\frac{r(1+\beta) * \sin(\pi\beta/2)}{r((1+\beta)/2) * \beta * 2^{(\beta-1/2)}} \right)^{1/\beta} \text{ and } \sigma_v = 1 \qquad (13)$$

The optimization of the boost converter duty cycle is given in Fig. 3. Different duty cycles are initiated randomly at first, after that each one is given to the PV system and the generated voltage and currents are used to estimate the power. The duty cycle is updated until the best optimum duty cycle and fitness function.

4 Discussion of Simulation Results

The CS and PSO optimization techniques performance is analyzed with the help of three series-connected PV panels and this is simulated at three different irradiation patterns. Due to the nonlinear characteristics of solar PV and multiple MPP peaks on PV curves it is difficult to transfer the maximum power from source to load. Hence a boost converter is interfaced in between PV array and load whose duty cycle is controlled by CS and PSO MPPT algorithms. The simulated diagram of PV fed boost converter under partial shading condition is given in Fig. 4.

The parameters considered for designing boost converter are input DC-link capacitor ($C_i = 10 \, \mu F$), an inductor ($L_i = 0.5 \, \mu H$), an output capacitor ($C_0 = 2 \, mF$), and a load resistor ($R_0 = 100 \, \Omega$). The input DC-link capacitor is useful to suppress the PV voltage and current ripples and output capacitor stabilize the load voltage. The Metal Oxide Semiconductor Field Effect Transistor is used as a switch because of its superior characteristics such as high switching speed, less power consumption,

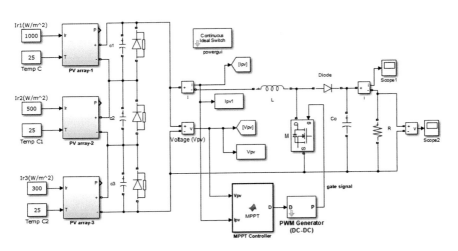

Fig. 4 Simulink model of photovoltaic power system

Table 1 Comparison parameters of CSA and PSO

Shading condition	MPPT controller	Peak power (W)	Tracking time		Global power (W)	Efficiency (%)
			Seconds	Iterations		
First condition	CS	513.93	0.7	9	515	99.792
	PSO	512.98	1.5	4		99.607
Second condition	CS	515.27	0.6	7	520	99.09
	PSO	515.03	1.8	4		99.04
Third condition	CS	265.51	0.85	6	280	94.82
	PSO	265.22	1.6	4		94.721

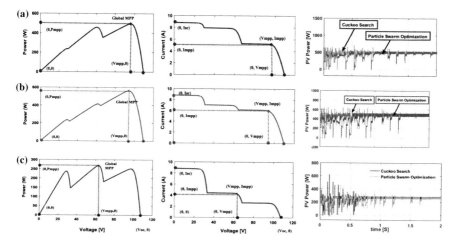

Fig. 5 Simulated, **a** PV, **b** IV and, **c** PV output power at three different irradiation patterns

and dissipation, high power control capability and temperature withstand capacity. The comparison results of CSO and PSO is given in Table 1.

From Fig. 4, the MPPT controller takes the changes in PV voltage and current and instantaneous voltage and the current rate for obtaining the suitable duty cycle of the boost converter. The metaheuristic MPPT techniques are utilizing the PV voltage and current parameters for achieving the optimum duty cycle of boost converter thereby extracting the maximum PV and converter voltages and currents. These MPPT techniques tracking performance is evaluated at each pattern of irradiation. For the first pattern, solar irradiations incident each PV are 1000, 900, and 600 W/m^2.

Similarly, the second pattern of irradiations is 1000, 800, and 700 W/m^2. Final shading pattern of PV power system is 1000, 500, and 300 W/m^2. For three shading patterns the generated power and current against voltage characteristics are shown in Fig. 5a–c. From Fig. 5a it is observed that there are two local MPPs and one global MPP which is required to track for extracting maximum power of solar PV. The generated PV peak power for three shading patterns are 513, 515, and 280 W,

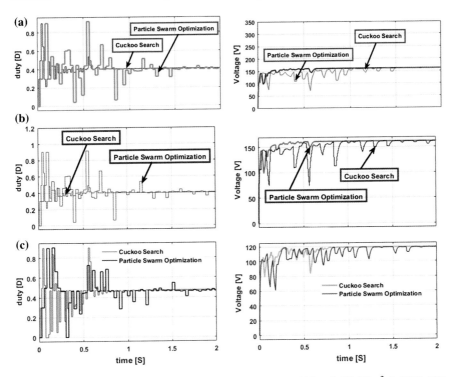

Fig. 6 Duty cycle and boost converter output voltage, **a** 1000, 900 and 600 W/m², **b** 1000, 800, and 700 W/m²˙ and **c** 1000, 500 and 300 W/m²

respectively. Hence, the PSO and CSA are extracting maximum output power of PV with high tracking speed.

The duty cycle and converter output voltages are plotted at three different shading patterns by using two successive MPPT techniques and it is given in Fig. 6. From Fig. 6a, the CS and PSO algorithms initial duty values are 0, 0.3, 0.5, and 0.9, respectively. Hence, both MPPT techniques covering the entire search space of PV curve. The duty cycle of boost converter is updated until achieving the optimum duty cycle of boost converter.

The generated switching pulse from CSO and PSO MPPT controller are given to the MOSFET to step the PV voltage in between 115 and 158 V. The MPP tracking time of CS and PSO controllers are 0.8 and 1.4 s which is given in Fig. 6a. Hence the MPP tracking time of CSO is 57% less than the PSO and it gives less oscillation around the MPP of solar PV. The CSO-based MPPT controller boost converter output voltage 152 V which is higher than the PSO. For the remaining two shading patterns the CSA and PSO controller-based converter and PV output voltages are plotted in Figs. 5 and 6b, c. The CSO and PSO algorithms MPP tracking time, number of iterations and peak power are evaluated at all shading pattern conditions and it is given in Table 1.

5 Conclusion

The CSA and PSO algorithms are implemented successfully by using Mat-lab/Simulink software. The conventional MPPT methods are not suitable to track the MPP and it creates high oscillations around MPP. The CSA and PSO optimization techniques are simulated and the comparative analysis is carried out. Under partial shading condition, the metaheuristic techniques tracking the global MPP with less oscillations. From the simulation results, the MPP tracking speed of CSA is high and its algorithm complexity is less. In addition, CSO technique gives high efficiency compared to PSO.

Acknowledgements I would like to thank the **University Grants Commission (Govt. of India)** for funding my research program and I especially thank VIT University management for providing all the facilities to carry out my research work.

References

1. Ram, P.J., Sudhakar Sudhakar, T., Rajasekar, N.: A comprehensive review on solar PV maximum power point tracking techniques. Renew. Sustain. Energy Rev. **67**, 826–847 (2017)
2. Rezk, H., Fathy, A., Abdelaziz, A.Y.: A comparison of different global MPPT techniques based on meta-heuristic algorithms for photovoltaic system subjected to partial shading conditions. Renew. Sustain. Energy Rev. **74**, 377–386 (2017)
3. Nasir, M., et al.: Solar PV-based scalable DC microgrid for rural electrification in developing regions. IEEE Trans. Sustain. Energy **9**(1), 390–399 (2018)
4. Capizzi, G., et al.: Optimizing the organic solar cell manufacturing process by means of AFM measurements and neural networks. Energies **11**(5), 1221 (2018)
5. Lee, H.K.H., et al.: Organic photovoltaic cells–promising indoor light harvesters for self-sustainable electronics. J. Mater. Chem. A **6**(14), 5618–5626 (2018)
6. Mathew, M., Kumar, N.M., Ponmiler i Koroth, R.: Outdoor measurement of mono and poly c-Si PV modules and array characteristics under varying load in hot-humid tropical climate. Mater. Today Proc. **5**(2), 3456–3464 (2018)
7. Rani, C., Hussaian Basha, C.H., Odofin, S.: Design and switching loss calculation of single leg 3-level 3-phase VSI. In: 2018 International Conference on Computation of Power, Energy, Information and Communication (ICCPEIC). IEEE (2018)
8. Segev, G., et al.: The spatial collection efficiency of charge carriers in photovoltaic and photo-electrochemical cells. Joule (2018)
9. Lloyd, J., et al.: Performance of a prototype stationary catadioptric concentrating photovoltaic module. Opt. Express **26**(10), A413–A419 (2018)
10. Rani, C., Hussain Basha, C.H.: A review on non-isolated inductor coupled DC–DC converter for photovoltaic grid-connected applications. Int. J. Renew. Energy Res. (IJRER) **7**(4), 1570–1585 (2017)
11. Laudani, A., et al.: Irradiance intensity dependence of the lumped parameters of the three-diodes model for organic solar cells. Sol. Energy **163**, 526–536 (2018)
12. Rani, C., Hussaian Basha, C.H., Odofin, S.: Analysis and comparison of SEPIC, Landsman and Zeta converters for PV fed induction motor drive applications. In: 2018 International Conference on Computation of Power, Energy, Information and Communication (ICCPEIC). IEEE (2018)

13. Guichi, A., et al.: A new method for intermediate power point tracking for PV generator under partially shaded conditions in hybrid system. Sol. Energy **170**, 974–987 (2018)
14. Gallardo-Saavedra, S., Karlsson, B.: Simulation, validation and analysis of shading effects on a PV system. Sol. Energy **170**, 828–839 (2018)
15. Dolara, A., et al.: An evolutionary-based MPPT algorithm for photovoltaic systems under dynamic partial shading. Appl. Sci. **8**(4), 558 (2018)
16. Samal, S., Barik, P.K., Sahu, S.K.: Extraction of maximum power from a solar PV system using fuzzy controller based MPPT technique. In: Technologies for Smart-City Energy Security and Power (ICSESP). IEEE (2018)
17. Tey, K.S., et al.: Improved differential evolution-based MPPT algorithm using SEPIC for PV systems under partial shading conditions and load variation. IEEE Trans. Ind. Inf. (2018)
18. Dawson, F.H., Kern-Hansen, U.: Aquatic weed management in natural streams: the effect of shade by the marginal vegetation: with 4 figures and 2 tables in the text. Int. Ver. Theor. Angew. Limnol. Verh. **20**(2), 1451–1456 (1978)
19. Li, G., et al.: Application of bio-inspired algorithms in maximum power point tracking for PV systems under partial shading conditions—a review. Renew. Sustain. Energy Rev. **81**, 840–873 (2018)
20. Soufyane Benyoucef, A., et al.: Artificial bee colony based algorithm for maximum power point tracking (MPPT) for PV systems operating under partial shaded conditions. Appl. Soft Comput. **32**, 38–48 (2015)
21. Babu, T.S., et al.: Particle swarm optimization based solar PV array reconfiguration of the maximum power extraction under partial shading conditions. IEEE Trans. Sustain. Energy **9**(1), 74–85 (2018)
22. García, J., et al.: A binary cuckoo search big data algorithm applied to large-scale crew scheduling problems. Complexity **2018** (2018)

Closed Loop Control of Diode Clamped Multilevel Inverter Using Fuzzy Logic Controller

K. Muralikumar and Ponnambalam Pathipooranam

Abstract This paper propose the closed loop control of multilevel inverter. Multilevel inverter (MLI) used is five-level neutral point clamped inverter. The switches of inverter are controlled by multicarrier phase opposition disposition SPWM technique. This type of multilevel configuration is deployed to reduce total harmonics distortion compared to two-level conventional PWM inverter. Fuzzy logic controller is designed for closed loop control of RMS voltage of multilevel inverter. The performance of fuzzy logic-based closed loop control of multilevel inverter is studied using Simulink tool of MATLAB.

Keywords Multilevel inverter · Multicarrier phase disposition · Rule-based fuzzy logic controller · Closed loop control

1 Introduction

Multilevel inverter (MLI) has gained much popularity in power electronics. It produces output in different voltage levels which leads to smoother waveform [1, 2]. Conventional two-level inverter only has two voltage level +Vdc and −Vdc which leads to output voltage full of harmonics distortion. This results in increased switching loss, electromagnetic interference, and low efficiency. Filter circuit must be implemented to get smooth output waveform with less harmonics content which is costly. The appropriate solution to this problem is implementation of multilevel inverter. Increase in number of voltage levels in output waveform makes the waveform smoother and leads to reduction in total harmonics distortion (THD) without implementation of LC and EMI filters. Increase in voltage level output waveform of inverter leads to increase in components and complexity of circuit. Increase in components also increases the cost. Smoother waveform with low THD obtained

K. Muralikumar (✉) · P. Pathipooranam
Vellore Institute of Technology, Vellore 632014, Tamilnadu, India
e-mail: kolamuralikumar@gmail.com

P. Pathipooranam
e-mail: p.ponnambalam@gmail.com

© Springer Nature Singapore Pte Ltd. 2020
K. N. Das et al. (eds.), *Soft Computing for Problem Solving*,
Advances in Intelligent Systems and Computing 1048,
https://doi.org/10.1007/978-981-15-0035-0_64

from implementation of multilevel inverter is trade-off for the increased cost of component. The output of waveform with low THD reduces the loss and increases the efficiency of the system. MLI can be realized from various topologies. One of the typical topologies is diode clamped multilevel inverter.

2 Diode Clamped MLI

Diode clamped MLI is also called neutral point clamped inverter. It makes use of clamping diode for realization of multilevel inverter. The use of diode in neutral point inverter is mainly to reduce the voltage stress over the switch. For m level diode clamped multilevel inverter, it requires $(m-1)$ voltage levels and $2(m-1)$ number of switches. Figure 1 shows one leg of five-level diode clamped inverter. It contains eight switches (S1, S2, …S8) and four voltage level each of Vdc/4 V with dc-link voltage Vdc [3–5].

The switches in upper part are complementary to switches in lower part. Switch S1 is complementary pair of switch S5, similarly S2 of S6, S3 of S7, and S4 of S8. Hence four triggering pulses generated for upper switches can be complemented for low switches. Four switches operate at a time to get desired voltage level. Table 1 shows the switching pattern of switches to get different voltage levels. The switches in upper part are complementary to switches in lower part. Switch S1 is complementary pair of switch S5, similarly S2 of S6, S3 of S7, and S4 of S8. Hence, four triggering pulses generated for upper switches can be complemented for low switches. Four switches

Fig. 1 Five-level diode clamped inverter

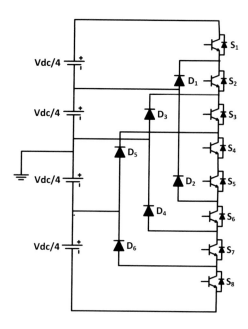

Table 1 Switching states

V_{out}	S1	S2	S3	S4	S5	S6	S7	S8
V/2	1	1	1	1	0	0	0	0
V/4	0	1	1	1	1	0	0	0
0	0	0	1	1	1	1	0	0
−V/4	0	0	0	1	1	1	1	0
−V/2	0	0	0	0	1	1	1	1

Fig. 2 Phase opposition
disposition technique

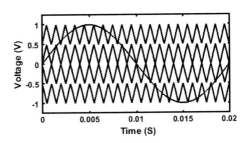

operate at a time to get desired voltage level. Figure 2 shows the switching pattern of switches to get different voltage levels. From Table 1, the maximum voltage level is Vdc/2 which is obtained when S1, S2, S3, and S4 is turned on. Similarly, other four voltage levels Vdc/4, 0, −Vdc/4, and −Vdc/2 can also be obtained from switching pattern shon in the below Table 1.

3 Phase Opposition Disposition (POD) PWM Method

PWM is obtained by comparison between carrier signal and reference signal. Sinusoidal pulse width modulation (SPWM) is obtained by comparison between triangular carrier and sinusoidal reference. When triangular carrier is less than sinusoidal reference, the PWM output is high. When triangular carrier is less than sinusoidal reference the PWM output is low. Phase disposition method of carrier-based PWM is used. For five-level inverter, four carrier waveforms are generated. For phase disposition method, all the carriers are in same phase with its neighboring waveforms. The waveform of phase disposition of five-level MLI is shown in Fig. 2 [6].

4 Fuzzy Logic Controller (FLC)

Fuzzy logic is a method that is based on degree of membership. Fuzzy logic uses the way which is similar to decision-making in humans that involves all intermediate states between two fixed states [6–10]. Fuzzy logic controller contains three basic blocks (1) Fuzzifier, (2) Inference engine, and (3) Defuzzifier. Fuzzy logic controller first takes in crisp input. These crisp inputs are converted in various fuzzy sets by fuzzifier. The knowledge base contains all the rules provided by expert in form of If–Then rules. The inference engine determines the degree of matching of fuzzy input based on rules in knowledge base. The fuzzy output is transformed into crisp output by defuzzifier [3].

4.1 Fuzzy Logic Implementation in Diode Clamped Multilevel Inverter

A fuzzy logic controller is designed for five-level DCMLI. Block diagram of designed fuzzy logic controller for closed loop system is shown in Fig. 3. The Figs. 4 and 5 show the membership function of input variable and output variable. In Fig. 5,

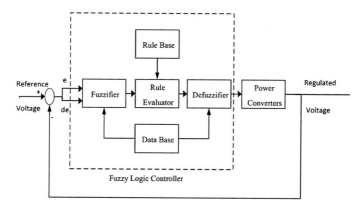

Fig. 3 FLC closed loop system for DCMLI [2]

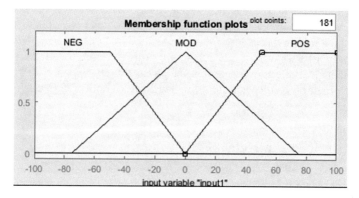

Fig. 4 Input variable of FLC

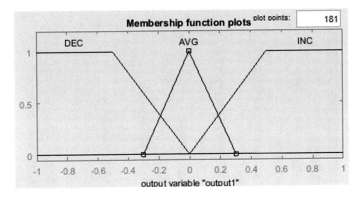

Fig. 5 Output variable of FLC

abscissa represents the amplitude of output variable whereas ordinate presents the membership. The fuzzy logic controller was used for closed loop control of five-level diode clamp inverter.

FLC is used to control the RMS voltage output of MLI. The RMS voltage of MLI is taken as feedback. This feedback signal is compared to reference and fed to fuzzy logic controller. Using the rule from rule base, the controller adjusts the parameter and tries to equalize the output voltage and input reference voltage.

5 Simulation Results

First of all, conventional SPWM inverter was simulated in Simulink. Fast Fourier transform analysis was performed and total harmonics distortion was found to be 68.96%. Figure 6 shows the FFT analysis of conventional SPWM inverter.

Under same condition of conventional SPWM inverter, multilevel inverter was simulated. FFT analysis of five-level diode clamped inverter showed that the THD was reduced to 34.66%. Figure 7 shows output waveform of five-level neutral clamped inverter and Fig. 8 shows its FFT analysis.

The reference voltage is 50 V in the beginning and after 0.2 s it is raised to 70 V. The output of closed loop system is shown in Fig. 9.

Figure 10 shows the response of the closed loop system when the input reference is fixed 70 V. It can be seen that the system is second order system with rise time 3.6345×10^3 s, overshoot of 3.4634 s, settling time of 1.054×10^4 s, and peak value of 72.4403 V.

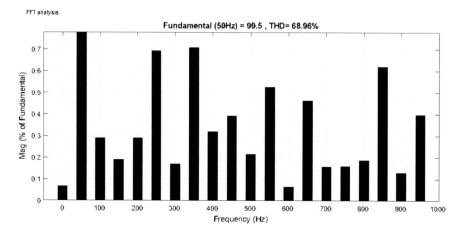

Fig. 6 Fourier analysis of conventional SPWM inverter

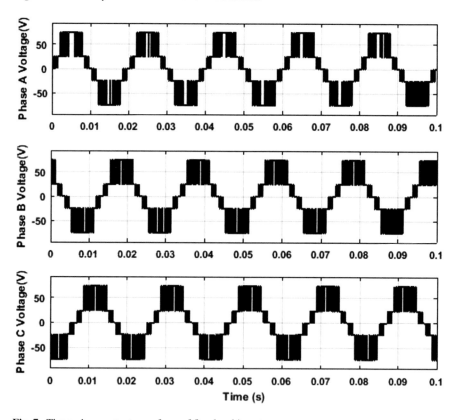

Fig. 7 Three-phase output waveform of five-level inverter

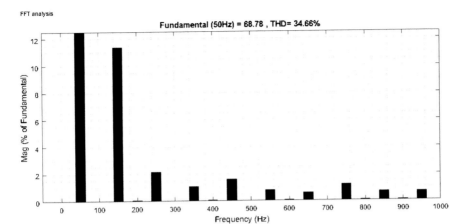

Fig. 8 Fourier analysis of five-level inverter

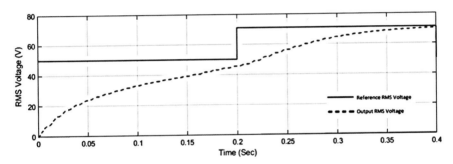

Fig. 9 System response for 50 V reference and 70 V reference

Fig. 10 Response of system for fixed reference of 70 V

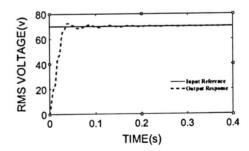

6 Conclusion

Multilevel diode clamped inverter can be used in place of conventional two-level inverter for harmonics sensitive application. It was found that there was drastic reduction of harmonics content from 68 to 34.66% when FFT analysis was performed. In order to get desired output voltage, closed loop control of inverter using fuzzy logic was performed and desired output was obtained.

References

1. Rodriguez, J., Lai, J.S., Peng, F.Z.: Multilevel inverters: a survey of topologies, controls, and applications. IEEE Trans. Industr. Electron. **49**(4), 724–738 (2002)
2. Colak, I., Kabalci, E., Bayindir, R.: Review of multilevel voltage source inverter topologies and control schemes. Energy Convers. Manag. **52**(2), 1114–1128 (2011)
3. Newton, C., Sumner, M.: Neutral point control for multi-level inverters: theory, design and operational limitations. In: Industry applications conference. Thirty-Second IAS Annual Meeting, IAS'97., Conference Record of the 1997 IEEE 1997 Oct 5 (vol. 2, pp. 1336–1343). IEEE (1997)
4. Franquelo, L.G., Rodriguez, J., Leon, J.I., Kouro, S., Portillo, R., Prats, M.A.: The age of multilevel converters arrives. IEEE Ind. Electron. Mag. **2**(2), 2008
5. Kumar, P.A., Kartheek, K., Madhukar, B.: Simulation of five level diode clamped multilevel inverter. Int. J. Recent Innov. Trends Comput. Commun. **5**(7), 19–24 (2017)
6. Ponnambalam, P., Aroul, K., Reddy, P.P., Muralikumar, K.: Analysis of fuzzy controller for H-bridge flying capacitor multilevel converter. In: Proceedings of sixth international conference on soft computing for problem solving, pp. 307–317. Springer, Singapore (2017)
7. Reddy, V.P., Muralikumar, K., Ponnambalam, P., Mahapatra, A.: Asymmetric 15-level multilevel inverter with fuzzy controller using super imposed carrier PWM. In: Power and advanced computing technologies (i-PACT), 2017 Innovations in 2017 Apr 21, pp. 1–6. IEEE (2017)
8. Ponnambalam, P., Muralikumar, K., Vasundhara, P., Sreejith, S., Challa, B.: Fuzzy controlled switched capacitor boost inverter. Energy Procedia. **1**(117), 909–916 (2017)
9. Viswanath, Y., Muralikumar, K., Ponnambalam, P., Kumar, M.P.: Symmetrical cascaded switched-diode multilevel inverter with fuzzy controller. In: Soft computing for problem solving 2019, pp. 121–137. Springer, Singapore
10. Madhav, K.L., Babu, C., Ponnambalam, P., Mahapatra, A.: Fuzzy logic controller for nine level multi-level inverter with reduced number of switches. In: Power and advanced computing technologies (i-PACT), 2017 innovations in 2017 Apr 21, pp. 1–7. IEEE (2017)

Prediction of California Bearing Ratio Using Particle Swarm Optimization

T. Vamsi Nagaraju, Ch. Durga Prasad and M. Jagapathi Raju

Abstract California bearing ratio (CBR) is one of the soul parameters for the pavement designs. CBR value can assess the stiffness and strength of the subgrade. Moreover, it was influenced by various properties such as index properties and compaction characteristics of the soils. The paper aims the viability of the swarm assisted particle optimization (PSO) for estimation or prediction of subgrade CBR. CBR estimation model equations by using PSO have been developed by considering index properties and compaction characteristics. The results show that the developed model equations are satisfactory agreement with the test data.

1 Introduction

Recent years, the Indian government has been giving prime importance to road network to connect each and every nook by Pradhan Mantri Gram Sadak Yojana scheme (PMGSY). However, most of the soil deposits covered in the India were problematic soils such as expansive clays, marine clays, collapsible soils, and loose sands. These soils possess big challenges for geotechnical engineers to involve doing rigorous experimental testing to assess or understand the soil behavior.

To counteract aforementioned problem, one should cope with the advanced tools to predict or assess the soil behavior. Especially in flexible pavements, stiffness and strength of the subgrade of soil is one which decides the overall thickness and cost of the pavement. The strength and stiffness modulus of subgrade is indirectly measured

T. V. Nagaraju (✉) · M. J. Raju
Department of Civil Engineering, S. R. K. R. Engineering College,
Bhimavaram 534204, India
e-mail: Varshith.varma@gmail.com

M. J. Raju
e-mail: profmjraju99@gmail.com

Ch. D. Prasad
Department of Electrical and Electronics Engineering, S. R. K. R. Engineering College,
Bhimavaram, India

© Springer Nature Singapore Pte Ltd. 2020
K. N. Das et al. (eds.), *Soft Computing for Problem Solving*,
Advances in Intelligent Systems and Computing 1048,
https://doi.org/10.1007/978-981-15-0035-0_65

795

by the CBR test. However, subgrade CBR values were influenced by various factors such as water content, plasticity characteristics, and compaction characteristics [2, 7, 10, 11]. And also, chemical additives influence the soil structure and further effects CBR value [4, 11]. While adopting optimization techniques to predict or estimate the CBR values the index properties and compaction characteristics were taken into consideration as input variables [4, 9]. Among all the variables, maximum dry density is the most effective variable on stiffness modulus nothing but CBR value [8]. In this regard, some research has been adopted new computational prediction models to predict the California bearing ratio of soil, and a brief account of the results is presented below.

Patel et al. [6] adapted regression analysis for developing correlations between CBR values and the index and compaction properties. And also comparison is done between the test data and calculated data. Taskiran [8] reported that the artificial neural networks (ANN) and gene expression programming (GEP) were successfully developed models for prediction of CBR. Jyoti et al. [4] adapted a genetic algorithm for estimating CBR values for the fly ash-soil blends. CBR values were effect with the variation of fly ash content in the blends. Valentine et al. [9] reported the relatively fair coefficients of R^2 determination between California bearing ratio values and maximum dry density using single linear regression analysis (SLRA) and multiple linear regression analysis (MLRA). This paper investigates the prediction of California bearing ratio (CBR) of subgrade soil using particle swarm optimization (PSO). The predicted model equations have been developed, uses five input parameters, namely, liquid limit, plasticity index, optimum moisture content (OMC), and maximum dry density (MDD). This paper also explores the importance of PSO application in geotechnical engineering parameter prediction studies.

2 Data Base Compilation

The collected test data was from the various road construction projects ongoing in the tracts of coastal region of Andhra Pradesh. The data set includes index properties as liquid limit (LL), plastic limit (PL) and particle sizes; and compaction data as optimum moisture content (OMC) and maximum dry density (MDD), and California bearing ratio (CBR) of soils. The test data shows Appendix A.

3 View of the Particle Swarm Optimization (PSO)

Traditional methods for prediction or estimation models involve mathematical regression analysis, which are manually calculated and time-consuming applications. As a result, new optimization techniques are emerging for better convergent results and easy operated tools within a limited time. In particular the study of the prediction of a give variable with a limited data, metaheuristic particle swarm optimization (PSO) method has been receiving attention as a potential method for prediction of a given variable with limited data set [3, 5]. PSO has predominantly applied to continuous discrete heterogeneous strongly nonlinear numerical optimization and it thus used in every field of engineering [1, 5, 12, 13]. In PSO, the optimization mechanism can be stated that a group of particles (in general fishes) that searching for food in a pond (constraint based optimization) with random manner. The location of each fish can be treated as a solution to given problem. Initially these locations (solutions) are randomly generated. These current positions (current solution) of a particle may not consist food hence the bird will move to next position (best solution) by update its velocity. To get closer to the location of the food (considered as best solution), all of the fishes follow the nearest fish that is closer to the food place. Each particle/fish is having two vectors x_i, v_i representing the position and velocity of the particle at every iteration. All of the particles regulate their route for the next iteration based on their experience and other particles experiences at the present iteration. Let x_i^k denotes the particle position vector in search of space of solutions and the new position for each particle is taken place randomly as

$$x_i^{k+1} = x_i^k + v_i^{k+1} \tag{1}$$

where v_i^{k+1} is the velocity vector of particle that drives the process toward optimization. The position of every individual bird is p_i^k and the velocity is v_i^k. The best position of each particle is p_{best} and the overall optimized position is g_{best}. Then the velocity of each organism in the swarm is updated as

$$V_i^{k+1} = \omega V_i^k + c_1 r_1 (p_{best_i} - x_i^k) + c_2 r_2 (g_{best_i} - x_i^k) \tag{2}$$

In Eq. (2), c_1 and c_2 are acceleration constants, r_1 and r_2 are random numbers varies between 0 and 1, and ω is the inertia weight factor. PSO technique has two type parameters known as common parameters and control parameters. Every meta heuristic optimization algorithm needs common parameters, such as population size, number of iterations, etc., but control parameters vary from one algorithm to other. After initialization and selection of both common and control parameters, iterative procedure will start by random initial solution matrix. For each solution, the function value is calculated known as fitness of the particles (solutions). Based local best and global best concepts (information with a particular particle experience and other particle experiences), updating of position and velocity will take place. This updating procedure will continue until either optimization value is arrived or iteration criteria are met. In this paper, a constant inertia weight strategy is applied for estimation. The acceptable range of ω for engineering problem is lie in between [0.4 and 0.9]. Even identification of suitable value of ω is difficult in its wide range; several investigations are carried out in this paper for selecting an optimal ω to get close results.

4 Multi-Variable Linear Equation Models for Prediction of California Bearing Ratio (CBR)

The data information of liquid limit (LL), plastic limit (PL), gravel content in percentage (G), sand content in percentage (S), fines (clay and silt) content in percentage (F), optimum moisture content (OMC), and maximum dry density (MDD) are considered as input variables to predict California bearing ratio (CBR). Linear regression models were used to estimate or predict unknown variables with known variables are in vogue [2, 6]. In this paper, the multi-linear model equations were developed for predicting California bearing ratio (CBR) instead of taking direct regression models. For this developed equation, each variable such as gravel content, sand content, fines content, LL, PL, OMC, and MDD was multiplied by respective coefficients such as a_1, a_2, a_3, a_4, a_5, a_6, and a_7; and additional coefficient a_0 was considered. The developed multi-variable equation is shown in Eq. 3.

$$\text{CBR}_{(est)} = a_1.G + a_2.S + a_3.F + a_4.LL + a_5.PL + a_6.OMC + a_7.MDD + a_o \tag{3}$$

where a_1, a_2, a_3, a_4, a_5, a_6, and a_7 are coefficients need to be estimated properly to calculate final output variable CBR. On the other hand, an additional constant is used to convert it into multiple linear regression models. For the entire data samples, CBR is estimated and an error is evaluated from actual data of C_c. For evaluation of error, the following formula is used.

$$E(k) = \sum_{i=1}^{N} \left(\text{CBR}_{(act)}(i, k) - \text{CBR}_{(est)}(i, k) \right)^2 \tag{4}$$

In the above equation, k represents number of iterations. N is total number of test data samples, CBR is actual value of California bearing ratio, and $\text{CBR}_{(est)}$ estimated value of California bearing ratio. The error calculated from both estimated and actual values of CBR for an iteration 'k' is represented by E. From calculated error values, the objective function (J) is formed as shown in Eq. (5)

$$J = \min(E) \tag{5}$$

The above objective function depends on the prediction of coefficients estimated by optimization techniques. Limits of these variables depend on the user requirement and/or data set availability. For this problem, the limits are fixed in between $-10{,}000$ and $10{,}000$ obviously large solution space.

5 Results and Discussion

Input variables considered for developing model equations for estimation of CBR were particle size distribution (gravel content, sand content, and fines content), plasticity characteristics (liquid limit and plastic limit) and compaction characteristics (optimum moisture content and maximum dry density). For predicting CBR estimation, metaheuristic swarm assisted particle optimization method was used. The following equations were formulated for estimation CBR with varying inertia weights 0.4, 0.5, 0.6, and 0.7, respectively.

$$
\begin{aligned}
CBR_{(est)} = {} & -72.2516.G - 72.5434.S - 72.8630.F - 0.4137.LL \\
& + 0.7671.PL + 2.2575.OMC + 11.5871.MDD + 7219.16
\end{aligned} \tag{6}
$$

$$
\begin{aligned}
CBR_{(est)} = {} & 8.8974.G + 8.5993.S + 8.5815.F - 0.0030.LL \\
& - 0.1672.PL - 0.2932.OMC + 9.6051.MDD - 865.1109
\end{aligned} \tag{7}
$$

$$
\begin{aligned}
CBR_{(est)} = {} & 54.6347.G + 54.3368.S + 54.4862.F + 0.2226.LL \\
& - 0.6953.PL - 1.6878.OMC + 8.5816.MDD - 5422.2847
\end{aligned} \tag{8}
$$

$$
\begin{aligned}
CBR_{(est)} = {} & -9.2243.G - 9.6661.S + 9.7267.F - 0.2162.LL \\
& + 0.6969.PL + 1.8356.OMC - 26.3831.MDD + 1038.078
\end{aligned} \tag{9}
$$

The following equations were best trails of the respective inertia weights varying 0.4, 0.5, 0.6, and 0.7. Among all the inertia weights, the best estimation was obtained at the 0.5 inertia weight (vide Fig. 2). This can be attributed from the overall errors obtained from all available 134 samples data. Table 1 shows the coefficients for the formulated equations with varying inertia weights. For all cases, the estimated and actual values comparisons plots are presented from Figs. 1, 2, 3, and 4.

From all PSO estimates, errors are calculated using actual samples and estimated samples. Figure 5 shows the errors between actual data and predicted data. The error was formulated by using the Eq. (10).

$$
Error = Abs\big((CBR_{act} - CBR_{predicted})/CBR_{act}\big) \tag{10}
$$

Table 1 Coefficients of model equations with varying inertia weights using PSO

Inertia weight in PSO	a_0	a_1	a_2	a_3	a_4	a_5	a_6	a_7
0.4	−72.251	−72.543	−72.863	−0.413	0.767	2.257	11.587	7219.16
0.5	8.8974	8.5993	8.5815	−0.003	−0.167	−0.293	9.6051	−865.110
0.6	54.634	54.336	54.486	0.222	−0.695	−1.687	8.5816	−5422.28
0.7	−9.2243	−9.6661	9.7267	−0.2162	0.696	1.835	−26.383	1038.07

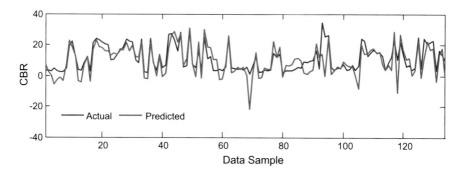

Fig. 1 Predicted CBR values using PSO with inertia weight 0.4

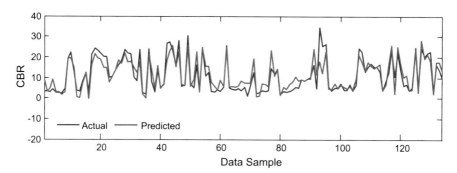

Fig. 2 Predicted CBR values using PSO with inertia weight 0.5

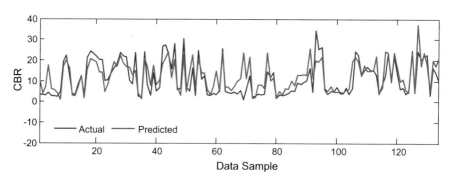

Fig. 3 Predicted CBR values using PSO with inertia weight 0.6

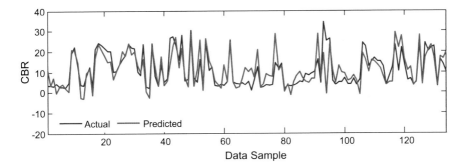

Fig. 4 Predicted CBR values using PSO with inertia weight 0.7

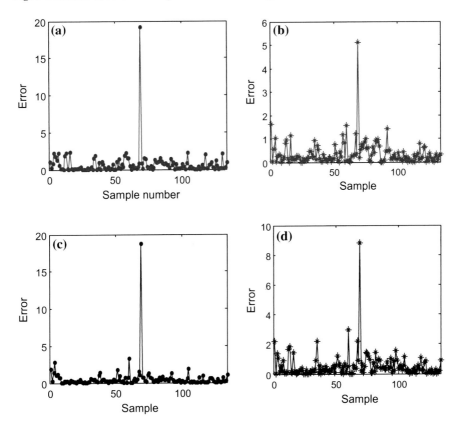

Fig. 5 Overall errors of all 134 samples using PSO estimates

6 Conclusions

The following conclusions can be drawn from the foregoing study:

1. The California bearing ration (CBR) is one of the soul parameters for pavement designs. CBR of subgrade pavement is influenced by various factors particle size, plasticity characteristics, and compaction characteristics.
2. Soil investigation for a highway pavement construction, CBR values for every 250 m is necessary. In this context, in order to go for laborious testing, one should aim to cope with to assess or guess the stiffness or strength of the subgrade soil nothing but CBR. As a reason, prediction of CBR value is necessary.
3. This paper employs a swarm assisted particle multi-linear regression model which enables adequately predicts the CBR value. And also, with varying inertia weights the models were varied. The best convergent results for the predicting CBR value are found at 0.5 inertia weight.
4. Primarily emphasis on the swarm assisted particle optimization for predicting geotechnical parameters.

References

1. Ahmed, A.A.E., Germano, L.T.: Application of particle swarm optimization to optimal power system. Int. J. Innovative Comput. Inf. Control. **8**(3(A)), 1705–1716 (2012)
2. Isabel, G.F., William, A., Gaby, R.: Prediction of California bearing ratio from index properties of soils using parametric and non-parametric models. Geotech. Geol. Eng. **36**(6), 3485–3498 (2018)
3. Kennedy, J., Eberhart, R.: Particle swarm optimization, IEEE, pp 1942–1948 (1995). ISBN 0-7803-2768-3/95
4. Jyoti, S.T., Sandeep, N., Chakradhar, I.: Optimum utilization of fly ash for stabilization of sub grade soil using Genetic algorithm. Procedia Eng. **51**, 250–258 (2013)
5. Ninad, K.K., Sujata, P., Trupti, B., Omkar, K., Kakandikar, Nandedkar, V.M.: Particle swarm optimization to mechanical engineering—a review. Mater. Today Proc. **2**(4–5), 2631–2639 (2015)
6. Patel, R.S., Desai, M.D.: CBR predicted by index properties for alluvial soils of South Gujarat. In: Indian Geotechnical Conference, GEOtrendz, pp. 79–82 (2010)
7. Phanikumar, B.R., Nagaraju, T.V.: Effect of fly ash and rice husk ash on index and engineering properties of expansive clays. Geotech. Geol. Eng. **36**(6), 3425–3436 (2018)
8. Taskiran, T.: Prediction of California bearing ratio (CBR) of fine grained soils by AI methods. Adv. Eng. Softw. **41**, 886–892 (2010)
9. Valentine, Y.K., Souleyman, M.M., Bertille, M., Armand, S.L.W., Lawrence, A.B.: Correlation of California bearing ratio (CBR) value with soil properties of road subgrade soil. In: Geotechnical and Geological Engineering, pp. 1–18 (2018). ISSN 1573-1529
10. Nagaraju, T.V., Satyanarayana, P.V.V.: Geotechnical aspects of various constructions along the canal embankments. In: Tyagaraj, T. (ed) Ground Improvement Techniques and Geosynthetics. Chapter 16, pp. 143–150. Springer Nature (2019). ISBN: 978-981-13-0558-0
11. Tan, Yunzhi, Mozhen, Hu, Li, Dianqing: Efeects of agglomerate size on California bearing ratio of lime treated lateritic soils. Int. J. Sustain. Built Environ. **5**, 168–175 (2016)

12. Sung, Yu-Chi, Wang, Chun-Ying, Teo, Eng-Huat: Application of particle swarm optimization to construction planning for cable-stayed bridges by the cantilever erection method. J. Struct. Infrastruct. Eng. **12**(2), 208–222 (2016)
13. Prasad, C.D., Nayak, P.K.: Performance assessment of swarm-assisted mean error estimation-based fault detection technique for transmission line protection. Comput. Electr. Eng. **71**, 115–128 (2018)

Adaptive Multi-swarm Bat Algorithm (AMBA)

Reshu Chaudhary and Hema Banati

Abstract Modified shuffled multi-population bat algorithm (MSMPBat) is a recently proposed swarm algorithm. It divides its population into multiple sub-populations (SPs), each of which uses different parameter settings and evolves independently using an enhanced search mechanism. For information exchange among these SPs, a solution from one SP is copied to the next after every generation. This process leads to duplication of solutions over time. To overcome this drawback, different techniques are introduced. Opposition-based learning is used to generate a diverse starting population. For information exchange, if a solution comes too close to the swarm best, only then it is sent (moved, not copied) to another swarm. Four techniques are proposed to select this second swarm. Initially, the selection probability of each technique is same. The algorithm adaptively updates these probabilities based on their success rate. The swarm which gave up the solution uses a modified opposition-based learning technique to generate a new solution. These changes help to maintain the overall diversity of the population. The proposed approach, namely, adaptive multi-swarm bat algorithm (AMBA), is compared to six algorithms over 20 benchmark functions. Results establish the superiority of adaptive multi-swarm bat algorithm.

Keywords Bat algorithm · Multi-swarm optimization · Adaptive algorithm · Numerical optimization

R. Chaudhary (✉)
Department of Computer Science, University of Delhi, Delhi 110007, India
e-mail: reshu.dumca@gmail.com

H. Banati
Dyal Singh College, Lodhi Road, Delhi 110003, India
e-mail: banatihema@hotmail.com

© Springer Nature Singapore Pte Ltd. 2020
K. N. Das et al. (eds.), *Soft Computing for Problem Solving*,
Advances in Intelligent Systems and Computing 1048,
https://doi.org/10.1007/978-981-15-0035-0_66

805

1 Introduction

Real-world optimization problems are usually complex, having multiple variables with intricate constraints. Optimization algorithms need to find a set of values for the variables, such that they satisfy the given constraints while minimizing or maximizing the function value. Minimization problems are targeted at minimizing the consumption of limited resources like time and money, whereas maximization problems target the maximization of performance and profit. Traditional optimization approaches do not fare well for complex problems; hence, the research community has shifted emphasis on nature-inspired algorithms. Nature-inspired algorithms [1] are heuristic algorithms which ape their model from some natural phenomenon occurring in nature. Eberhart and Kennedy proposed particle swarm optimization [2] algorithm which is based on the foraging behavior of birds. John Holland proposed genetic algorithm (GA) [3] based on Charles Darwin's survival of the fittest theory. It models mutation and crossover to produce better solutions. The firefly algorithm [4] proposed by Yang is based on the luminescence characteristics of fireflies to attract potential mates.

Bat algorithm (BA) [5] proposed by Yang is a nature-inspired algorithm which apes the echolocation characteristics of bats to formulate an efficient optimization algorithm. BA is a popular algorithm frequently used for various real-world problems [6–9], hence to improve its efficiency, many improvements have been recently proposed [10–18]. Multi-modal bat algorithm with improved search (MMBAIS) [10] is one of the recently proposed variants of BA. It comprises two modifications: MMBA and BAIS. Multi-modal bat algorithm (MMBA) enhances the efficiency of BA by incorporating the foraging behavior of bats into BA. Bat algorithm with improved search (BAIS) enhances the convergence capabilities by a specially introduced parameter which reduces the range of random movement as the algorithm progresses. Modified shuffled multi-population bat algorithm (MSMPBat) [12] is a variant of BA which partitions its population into independent sub-populations (SPs). Each of these SPs evolves using BAIS and have independent parameter sets. After evolution, a master SP is formed which evolves using parameter setting of the SP which was able to produce the global best solution. Also, the SPs exchange information by copying their best solutions to the next SP in a directed ring manner.

Due to the manner in which information is exchanged in MSMPBat, it leads to duplicity of solutions resulting in poor population diversity. This paper proposes an adaptive variant of MSMPBat, namely, adaptive multi-swarm bat algorithm (AMBA). AMBA employs various techniques to enhance and maintain population diversity. It uses opposition-based learning to generate a diverse starting population. AMBA moves a solution from one swarm to another only if it comes close to the swarm best solution. Four techniques are used to select this second swarm. The algorithm adaptively updates the selection probabilities of these techniques based on their success rate. Also a new solution is added to the swarm which gave up a solution using a modified opposition-based learning technique. These techniques help impart AMBA with a diverse population, helping it avoid local optima. Sections 2 and 3 present MSMPBat and AMBA, respectively. Experimentation is discussed in Sect. 4. The chapter concludes with a summary of the proposed work in Sect. 5.

2 Modified Shuffled Multi-population Bat Algorithm

Banati and Chaudhary proposed MSMPBat in 2018 [12]. It is an improved version of the bat algorithm (BA) [5]. MSMPBat enhances the exploration and exploitation capabilities of BA using different techniques. It partitions its population into multiple sub-populations (SPs) which can simultaneously explore different areas of the search space. It further enhances on the exploration capabilities by assigning independent parameter sets per SP. It also uses the parameter proposed in bat algorithm with improved search (BAIS) [10] to enhance its convergence capabilities. The algorithm works as follows.

MSMPBat starts with initializing a population of N bats. Every bat represents a possible solution to the optimization problem. Each bat x_i, $(i = 1, 2, \ldots, N)$, is randomly initialized within a range [LB, UB]. Subsequently, the fitness of every bat is computed, and the fittest bat is the global best solution X_{GB}.

Once initialized, the population is partitioned into S independent sub-populations (SPs). The best solution is assigned to SP 1, next solution is assigned to SP 2, Sth best solution is assigned to Sth SP, $(S + 1)$th solution is assigned to SP 1, and so on. After SPs are formed, every SP identifies the best solution X_{SB} within itself. Every bat x_i moves toward X_{SB} using

$$f_i = f_{min} + (f_{max} - f_{min})\beta \tag{1}$$

$$v_i^t = v_i^{t-1} + (x_i^{t-1} - X_{SB}^t)f_i \tag{2}$$

$$x_i^t = x_i^{t-1} + v_i^t \tag{3}$$

where $f_i \in [f_{min}, f_{max}]$ is the frequency, $\beta \in [0, 1]$ is a random number, v_i^{t-1} and x_i^{t-1} are the velocity and position of bat x_i at generation t $- 1$.

Bat algorithm uses two parameters, pulse rate r and loudness A, to balance the exploration and exploitation of the algorithm. As the algorithm progresses, value of r increases and value of A decreases. These parameters are retained as is in MSMPBat. Based on the value of r, a given bat x_i may be replaced by a new solution. BA uses only the best solution to generate the new solution, while MSMPBat computes the new solution by taking a weighted sum of the swarm best X_{SB} and solution x_i itself. The following equation is used

$$x_i = \varphi X_{lb} + (1 - \varphi)x_i + random * \epsilon * A^t \tag{4}$$

where ϵ is the scaling factor (value used 0.1), φ is the weight (randomly selected to not favor any 1 solution) and A^t is the average loudness of all the bats in the population.

Once the final position is computed, every bat performs random movement around itself. BA has a range of $[-1, 1]$ for this movement. MSMPBat adopts a step-size parameter δ from BAIS [10], which reduces the range of random movement as the algorithm progresses. Using rand $\in [-1, 1]$, the following equation is employed:

$$x_i = x_i + \delta * rand \tag{5}$$

The change in the position of a bat is saved back into the population if its fitness is improved, with a probability of its loudness A. If saved, pulse rate value and loudness value of a bat are also updated using the following:

$$A_i^{t+1} = \alpha A_i^t \tag{6}$$

$$r_i^t = r_i^0 [1 - exp(-\gamma t)] \tag{7}$$

α and γ are constants which influence the rate of change of A and r, and r_i^0 is initial pulse rate.

Once all SPs have evolved, best bats from the sub-populations are combined to form a master population. The master population evolves the same as all the SPs. The global best solution X_{GB} is updated irrespective of the value of A. After SPs and master population have evolved, best solutions from one SP are copied to the next in a directed ring direction.

If the global best solution is not improved for G number of generations, two things are changed. First, all the SPs are combined and fresh SPs are formed using the same technique as explained above. Secondly, the value of step-size δ is decreased using

$$\delta = \delta * \omega \tag{8}$$

where ω is the search weight factor by which value of step-size is decreased. The algorithm keeps executing till the termination criterion is not met, which is usually the number of function evaluations.

3 Adaptive Multi-swarm Bat Algorithm (AMBA)

This paper proposes an adaptive multi-swarm bat algorithm (AMBA). As the name suggests, it is an adaptive algorithm which uses multiple swarms. AMBA is an extension to MSMPBat [12] explained above. For information exchange, MSMPBat copies best solutions from one swarm/sub-population to the next in a directed ring direction. This is done at the end of every generation. Copying solutions, that too very frequently may lead to loss in diversity of the population. This can easily lead to the algorithm being caught in local optima. To ensure population is diverse, AMBA uses different techniques. First, AMBA uses opposition-based learning technique [19] to get a diverse starting population. Then, unlike MSMPBat, in AMBA, if a solution comes close to the swarm best solution, only then it is moved to another swarm. The solution is moved and not copied, hence, eliminating duplicity. The selection of the second swarm can be done by four different techniques: randomly, swarm with best average fitness, swarm with worst average fitness, and swarm with farthest swarm best. Every technique has its own advantage. Initially, the probability of all these techniques is equal. As the algorithm progresses, these probabilities are updated based on their success to improve the fitness of the solution. The swarm which gave up a solution uses a modified opposition-based learning technique to produce a new solution. This fresh solution helps to maintain population diversity. These modifications impart AMBA with better efficiency.

The population in AMBA is generated using the same method as in MSMPBat. After the N solutions have been computed, opposition-based learning (OBL) [19] is applied to form N more solutions. Hence for every solution x_i, another solution ox_i is computed using

$$\text{ox}_i = \text{UB} + \text{LB} - x_i \tag{9}$$

The best N solutions from these $2N$ solutions are selected as the starting population for AMBA. OBL is found to be an efficient technique to cover diverse areas of the search space. Once population is initialized, swarms are formed same as in MSMPBat, with each swarm having its own set of parameters. Swarms evolve independently using BAIS. Every swarm identifies the best solution X_{SB} and evolve using Eqs. (1) to (8) of MSMPBat.

3.1 Swarm Communication

The directed ring exchange of MSMPBat is dropped in AMBA. In AMBA, when a particle comes very close to the swarm centre, it is sent to another swarm and the current swarm gets a new solution in its place. Bat x_i is considered close to swarm best X_{SB} if

$$(\|x_i - X_{SB}\| \leq (\text{Swarm Best Error/Swarm Error})) \text{ and} \\ (|\text{Fitness}_{X_{SB}} - \text{Fitness}_{x_i}| \leq E - 10) \tag{10}$$

where $\|x_i - X_{SB}\|$ is the Euclidian distance between x_i and X_{SB}. Selection of the swarm and the technique to generate new solution helps to enhance the overall efficiency of the algorithm.

The other swarm can be selected in four ways:

- *Random selection*: This is the simplest form of selection, wherein a swarm is randomly selected and the bat is sent to this swarm.
- *Swarm with worst average fitness*: This technique might generate a performance enhancement as the swarm with worst average fitness is short of good solutions and can hence benefit from a good incoming solution.
- *Swarm with best average fitness*: This technique might generate a performance enhancement as the particle gained as much enhancement as it could from its previous swarm. Sending it to a good swarm gives it more chances of further enhancement as the bat will get better learning opportunity.
- *Swarm with farthest centre*: This technique might generate a performance enhancement as selecting a swarm with centre farthest from this bat (in terms of Euclidian distance) will allow it to traverse more area of the search space.

3.2   Computing Selection Probability

Initially, the probability of selection (P_i) of each of these techniques is set to 0.25. As bats are transferred using these approaches, their probabilities of selection are updated after G generations. After every transfer, it is recorded whether or not the bat improved its fitness value. If technique i is used to transfer total T_i^t bats in G generations, of which T_i^s bats were able to gain fitness, then improvement percentage T_i^{IP} for this technique can be computed as

$$T_i^{IP} = T_i^s / T_i^t \tag{11}$$

None of the techniques can ever have a selection percentage 0. If a technique was not selected for transferring a bat, then its T_i^{IP} cannot be computed, as both T_i^s and T_i^t are 0. Such techniques are assigned a minimum selection probability of P_{Min}, kept at 0.1 in our implementation. In order to do so, they are assigned a T_i^{IP} value computed using the following:

$$T_i^{IP} = \frac{\left(\sum_{i=1 \text{ and } T_i^t \neq 0}^{4} T_i^{IP}\right) * P_{Min}}{(1 - \text{count} * P_{Min})} \tag{12}$$

where count is the number of techniques which had $T_i^t = 0$. To compute updated selection percentage for the four techniques, the following equation is used

$$P_i = T_i^{\text{IP}} / \sum_{i=1}^{4} T_i^{\text{IP}} \tag{13}$$

Hence, the selection technique resulting in more cases of improvement will have better selection probability.

3.3 Generating Replacement Bat

The swarm which lost a bat that was close to the swarm best needs to generate a replacement solution. The bat being replaced, x_i, is used to generate the new solution. A modified opposition-based learning method is introduced, which is

$$x_{\text{newMOBL}} = (x_{\text{UB}} * \text{rand1}) + (x_{\text{LB}} * (1 - \text{rand1})) - x_i \tag{14}$$

The above equation still uses the solution x_i but generates a solution that is not at the same Euclidian distance from the swarm best.

3.4 No Improvement Scenario

Similar to MSMPBat, if the global best is not improved for consecutive G generations, all the populations are combined together and fresh swarms are formed. The swarms will retain their parameter values, though they will have different bats now. This technique will help escape any possible local optima as different bats are learning from each other and are now using different parameter values to do so. The step-sizes of the swarms are also updated. Unlike MSMPBat, AMBA does not form the master population/swarm.

4 Experimentation

To compare the performance of the proposed work, it is compared to BA [5], GA [2], PSO [2], MSMPBat [12], BAIS [10], and directional bat algorithm (dBA) [18]. A set of 20 benchmark functions is employed, taken from [18]. Table 1 presents the list of benchmark function, while the parameter setting employed is listed in Table 2. The number of bats is kept 30, with number of generations at 500, same as dBA so as to keep the number of function evaluations same. The population size remains constant.

Table 1 List of benchmark functions

$F^{\#}$	Name	F	Name	F	Name	F	Name	F	Name
1	Sphere	5	Trid	9	Schwefel	13	Michalewicz	17	Weierstrass
2	Sum of different powers	6	Rastrigin	10	Rosenbrock	14	Powell singular	18	Styblinski Tang
3	Rotated	7	Levy	11	Zakharov	15	Bent Cigar	19	Salomon
4	Griewank	8	Ackley	12	Dixon and price	16	Alpine	20	Schaffer

All definitions can be seen in [18]
F function number

Table 2 Parameter setting

Parameter	Values				
	Swarm 1	Swarm 2	Swarm 3	Swarm 4	Swarm 5
fMin*	0.7	0	1	0.55	0.85
fMax*	1	0.5	1.5	0.85	1.15
Loudness A	0.85	0.5	0.95	0.75	0.9
Pulse r	0.85	0.5	0.95	0.75	0.9
Alpha α	0.5	0.1	0.9	0.25	0.75
Gamma γ	0.5	0.1	0.9	0.25	0.75
Step-size β	0.5	0.1	1	0.35	0.65
Weight ω	0.5	0.1	0.95	0.35	0.65

fMin minimum frequency; *fMax* maximum frequency

4.1 Numerical Comparison

BA, BAIS, and MSMPBat are executed 50 times each. Results for GA, PSO, and dBA have been taken directly from [18]. The best, average, and worst fitness computed by these algorithms over the 50 executions are selected as the comparison criteria. Table 3 presents the comparative results. To compare the performance, each algorithm is given a rank. The best performing algorithm in a given criterion, for a given function, is given rank 1, next algorithm is given rank 2 and so on. The average rank of an algorithm over all the 20 functions, for a criterion is hence computed which reflects the overall performance of the algorithm.

It can be seen from the results that AMBA is able to produce best results for most functions. Considering the best fitness, AMBA produces average rank of 1.95, followed by MSMPBat and dBA with ranks 2.8 and 3.45, respectively. For average fitness, AMBA got average rank of 1.6, followed by MSMPBat and dBA with ranks 2.65 and 3.45, respectively. Similar results can be seen for worst fitness, with AMBA coming in first place with average rank of 1.7. MSMPBat and dBA follow with ranks 2.6 and 3.55, respectively. Taking an average of the ranks overall three criteria, the ranks hence computed are presented in Table 4. It can be seen that AMBA comes in first place with overall rank of 1.75.

4.2 Statistical Test

To ascertain the superiority of AMBA over the other algorithms, Friedman's test [20] is employed. The computed F-value is 54.425, which is more than the critical F-value of 1.318. Hence, the null hypothesis which states that all algorithms have comparable performance is rejected. Since AMBA got the least rank, it is used as the control algorithm and p-values are computed. The values are adjusted using

Table 3 Comparative results

Algorithm	$F^{\#}$	Best	Average	Worst	F	Best	Average	Worst
BA	1	2.05E+02	2.56E+02	2.80E+02	2	7.69E+02	1.22E+03	1.48E+03
PSO$		1.11E+03	2.85E+03	5.62E+03		1.60E+20	1.04E+33	1.72E+34
GA$		5.51E+00	1.67E+03	7.96E+03		7.48E+04	1.04E+40	2.390E+41
BAIS		2.90E−01	2.75E+01	6.03E+01		9.88E+01	2.57E+02	6.30E+02
dBA$		1.92E−03	2.25E−01	2.23E+00		7.53E+01	1.51E+02	2.50E+02
MSMPBat		1.14E−03	6.48E−03	**1.20E−02**		7.15E−04	1.69E−02	5.67E−02
AMBA		**1.12E−03**	**1.35E−03**	6.51E−02		**2.66E−05**	**7.03E−03**	**5.14E−02**
BA	3	2.72E+06	4.38E+06	5.30E+06	4	1.99E+01	2.00E+01	2.03E+01
PSO		4.82E+03	1.56E+04	3.41E+04		3.04E+01	7.48E+01	1.68E+02
GA		8.28E+01	8.13E+03	3.29E+04		1.08E−01	1.90E+01	5.57E+01
BAIS		1.41E+01	2.08E+04	1.78E+05		1.23E+01	1.43E+01	1.70E+01
dBA		1.01E+6	1.36E+12	1.71E+13		3.21E+00	5.83E+00	8.80E+00
MSMPBat		9.06E−10	2.93E−07	1.63E−06		5.17E−04	1.19E−02	4.94E−02
AMBA		**1.28E−12**	**1.30E−07**	**1.27E−06**		**2.16E−04**	**1.08E−02**	**3.90E−02**
BA	5	5.14E+01	6.16E+01	7.25E+01	6	4.63E+05	7.08E+05	8.68E+05
PSO		3.07E+05	6.20E+05	1.22E+06		1.70E+02	2.59E+02	3.45E+02
GA		6.32E+03	3.19E+05	7.00E+05		**2.99E+01**	**5.74E+01**	**9.91E+01**
BAIS		1.72E+00	7.27E+00	1.09E+01		2.49E+05	7.23E+05	1.89E+06
dBA		3.46E−02	3.71E+00	2.04E+01		1.68E+03	3.42E+04	9.70E+04
MSMPBat		**2.81E−04**	2.53E−03	1.36E−02		4.73E+03	4.80E+03	4.88E+03
AMBA		4.62E−04	**2.21E−03**	**4.03E−03**		4.16E+03	4.78E+03	4.91E+03

(continued)

Table 3 (continued)

Algorithm	F#	Best	Average	Worst	F	Best	Average	Worst
BA	7	4.19E+02	5.90E+02	6.73E+02	8	3.97E+02	4.43E+02	4.93E+02
PSO		2.12E+01	3.97E+01	8.05E+01		1.25E+01	1.47E+01	1.73E+01
GA		1.09E+00	5.67E+00	1.56E+01		2.59E+00	5.92E+00	1.14E+01
BAIS		2.62E+01	6.57E+01	1.48E+02		6.84E+01	1.34E+02	1.89E+02
dBA		5.04E−03	1.40E−01	5.63E−01		6.81E+01	1.19E+02	2.47E+02
MSMPBat		3.79E−08	1.30E−06	1.46E−05		7.33E−04	4.62E−02	2.12E−01
AMBA		**1.80E−08**	**1.03E−06**	**7.97E−06**		**2.67E−04**	**9.73E−03**	**2.97E−02**
BA	9	2.13E+06	3.09E+06	4.22E+06	10	3.57E+01	4.65E+01	5.87E+01
PSO		7.29E+03	8.71E+03	9.48E+03		8.56E+03	8.15E+04	2.81E+05
GA		2.73E+03	4.20E+03	5.99E+03		1.04E+02	5.96E+03	4.79E+04
BAIS		2.19E−01	7.80E+01	1.64E+02		1.35E+00	7.20E+00	1.44E+01
dBA		2.91E+01	1.64E+02	1.01E+03		1.51E+00	4.71E+00	9.99E+00
MSMPBat		**2.87E+01**	**2.87E+01**	**2.89E+01**		1.12E+00	1.64E+00	2.24E+00
AMBA		**2.87E+01**	**2.87E+01**	**2.89E+01**		**9.89E−01**	**1.59E+00**	**2.09E+00**
BA	11	5.50E+08	6.31E+08	7.01E+08	12	7.94E+03	1.15E+04	1.51E+04
PSO[a]		5.75E+02	1.05E+03	1.616E+03		6.914E+03	3.864E+04	1.202E+05
GA[a]		1.12E+01	9.88E+07	9.31E+08		**1.20E+01**	6.49E+03	3.63E+04
BAIS		2.50E+02	1.27E+04	7.50E+04		1.70E+02	2.65E+02	3.53E+02
dBA[a]		4.49E+01	4.92E+02	2.51E+03		1.13E+02	**1.95E+02**	**2.68E+02**
MSMPBat		1.92E+00	6.65E+02	3.68E+03		5.18E+02	5.86E+02	6.60E+02
AMBA		**3.38E−02**	**7.85E+01**	**7.51E+02**		4.80E+02	5.28E+02	5.72E+02

(continued)

Table 3 (continued)

Algorithm	$F^{\#}$	Best	Average	Worst	F	Best	Average	Worst
BA	13	−7.14E+00	−6.21E+00	−5.39E+00	14	1.19E+06	1.80E+06	2.34E+06
PSO		−1.30E+01	−9.62E+00	−7.01E+00		2.18E+03	9.52E+03	3.69E+04
GA		**−2.47E+01**	**−2.18E+01**	**−1.87E+01**		7.20E+01	2.09E+03	6.45E+03
BAIS		−1.31E+01	−8.94E+00	−6.55E+00		7.66E−01	3.61E+00	9.04E+00
dBA		−2.09E+01	−1.49E+01	−1.01E+01		**7.44E−01**	1.91E+01	1.04E+02
MSMPBat		−8.65E+00	−6.65E+00	−5.70E+00		9.86E−01	1.02E+00	1.17E+00
AMBA		−1.04E+01	−7.85E+00	−6.70E+00		9.84E−01	**9.96E−01**	**1.00E+00**
BA	15	5.04E+03	7.32E+03	9.16E+03	16	4.58E+01	5.10E+01	5.47E+01
PSO		3.98E+07	8.30E+07	1.55E+08		1.60E+01	2.69E+01	3.97E+01
GA		1.47E+06	1.80E+07	8.33E+07		**1.02E−01**	1.30E+00	4.27E+00
BAIS		2.14E+00	7.93E+00	1.88E+01		4.56E+01	5.10E+01	5.39E+01
dBA		1.34E+00	4.89E+01	1.91E+02		2.71E+01	3.05E+01	3.32E+01
MSMPBat		5.41E−05	4.25E−03	3.66E−02		7.88E−01	2.29E+00	4.23E+00
AMBA		**2.22E−06**	**6.30E−04**	**3.73E−03**		1.86E−01	**1.05E+00**	**2.10E+00**
BA	17	5.65E+03	6.83E+03	7.66E+03	18	9.57E+00	1.07E+01	1.17E+01
PSO		8.76E+00	2.88E+01	3.28E+01		5.20E+02	6.77E+02	9.58E+02
GA		2.53E+00	7.68E+00	1.46E+01		2.61E+02	3.62E+02	6.59E+02
BAIS		4.52E+01	1.96E+02	4.25E+02		4.78E+00	6.16E+00	8.05E+00
dBA		3.55E−01	1.41E+00	2.35E+00		3.86E+00	5.26E+00	6.76E+00
MSMPBat		9.00E−06	4.16E−04	1.34E−03		3.27E−02	6.04E−02	2.15E−01
AMBA		**1.48E−06**	**6.00E−05**	**2.28E−04**		**5.49E−03**	**5.78E−02**	**1.36E−01**

(continued)

Table 3 (continued)

Algorithm	$F^{\#}$	Best	Average	Worst	F	Best	Average	Worst
BA	19	2.60E+05	3.83E+05	4.68E+05	20	8.08E+03	9.79E+03	1.05E+04
PSO		4.30E+02	1.00E+03	2.29E+03		5.73E+00	6.61E+00	7.41E+00
GA		9.30E+00	2.31E+02	7.37E+02		**7.87E−01**	**2.06E+00**	**3.68E+00**
BAIS		6.74E+00	2.38E+03	1.03E+04		5.92E+03	7.40E+03	8.57E+03
dBA		1.63E−02	1.46E+01	1.25E+02		2.89E+03	4.35E+03	5.64E+03
MSMPBat		8.65E−05	1.45E−02	1.72E−01		8.29E+03	9.44E+03	1.01E+04
AMBA		**4.04E−06**	**1.21E−03**	**6.30E−03**		8.14E+03	8.98E+03	9.48E+03

[a] Results for PSO, GA, and dBA are taken from [18]

[#] Function number as listed in Table 1

Table 4 Average rank of algorithms overall three criteria

Algorithm	Rank	Algorithm	Rank	Algorithm	Rank	Algorithm	Rank
AMBA	1.75	dBA	3.483	BAIS	4.417	BA	6.183
MSMPBat	2.683	GA	4.117	PSO	5.3		

Table 5 Adjusted p-values with AMBA as the control algorithm

Algorithm	Unadjusted P	Bonfernni	Holm	Holland	Finner
BA	1.30E−29	7.81E−29	7.81E−29	0.00E+00	0.00E+00
PSO	1.12E−19	6.71E−19	5.60E−19	0.00E+00	0.00E+00
BAIS	6.80E−12	4.08E−11	2.72E−11	2.72E−11	1.36E−11
GA	9.78E−10	5.87E−09	2.93E−09	2.93E−09	1.47E−09
dBA	5.57E−06	3.34E−05	1.11E−05	1.11E−05	6.68E−06
MSMPBat	9.00E−03	5.40E−02	9.00E−03	9.00E−03	9.00E−03

different methods [20]. The corresponding results are given in Table 5. It can be seen that all values are less than 0.5, implying that AMBA significantly outperformed them. Results hence establish AMBA as a strong optimization algorithm.

4.3 Convergence Comparison

The above computations reflect upon the ability of these algorithms to produce good solutions. Convergence comparison is conducted to measure the efficiency of an algorithm to produce good solutions in few iterations. Convergence comparison is done among AMBA, BA, BAIS, and MSMPBat over Himmelblau's function. Every algorithm is executed 50 times independently, and the average fitness at regular intervals is plotted. The corresponding diagram is presented in Fig. 1. It can be seen that AMBA is able to reach near-optimal solution in the least number of generations.

5 Conclusion

Modified shuffled multi-population bat algorithm (MSMPBat) is a variant of bat algorithm (BA). It improves the exploration and exploitation capabilities of BA. It uses multiple sub-populations, which evolve using different parameter settings. They also employ a parameter to refine the random movement of bats. Further, MSMPBat uses a weighted sum of the best and a solution while performing search around the best solution. After every generation, the best solutions are picked from each sub-population to form a master sub-population. Also, the best solution from each

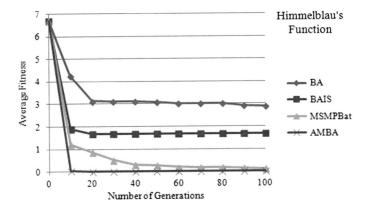

Fig. 1 Convergence comparison

sub-population is copied into the next sub-population using directed ring. If global best solution does not improve for few generations, all sub-populations are combined and fresh sub-populations are formed.

The proposed algorithm, adaptive multi-swarm bat algorithm (AMBA) is an adaptive variant of MSMPBat. Copying solutions at the end of every generation leads to duplicity of solutions in MSMPBat. AMAB uses multiple techniques to maintain population diversity. First of all it uses opposition-based learning to generate a richer starting population. It then partitions its population into swarms which evolve same as in MSMPBat. That is, different swarms use different parameter settings, while also employing refined random movement. In AMBA, swarms move a solution from one to another if a solution becomes too close to the swarm best. If so, four different techniques are used to select the swarm where this solution should be sent. The applied techniques are random selection, swarm with best average fitness, swarm with worst average fitness, and swarm with best farthest from the solution. Initially the selection probabilities of these techniques are same. The algorithm adaptively updates these probabilities based on their success in improving fitness of a solution. The swarm which gave up a solution close to swarm best employs a modified opposition-based learning technique to generate a new solution. All these changes help AMBA maintain a diverse population which aid in escaping any local optima.

AMBA is compared to bat algorithm (BA), bat algorithm with improved search (BAIS), MSMPBat, genetic algorithm, particle swarm optimization algorithm, and directional bat algorithm (dBA). A set of 20 multi-dimensional benchmark optimization functions has been picked from dBA. AMBA obtained best overall rank of 1.75 over different comparison criteria. Friedman's test was also employed to compare the performance of the proposed work. The results show that AMBA significantly outperforms BA, GA, PSO, BAIS, MSMPBat, and dBA. To check whether AMBA can provide good solutions using few iterations, convergence comparison is performed.

AMBA is compared to BA, BAIS, and MSMPBat. Results show that AMBA outperforms these algorithms and reaches a good solution using very few iterations. The cumulative results hence establish AMBA as a strong optimization algorithm.

References

1. Yang, X.S.: Nature-Inspired Metaheuristic Algorithms, 2nd edn. Luniver Press, UK (2010)
2. Kennedy, J., Eberhart, R.C.: Particle swarm optimization. In: Proceedings of the IEEE International Conference on Neural Networks, pp. 1942–1948. Australia (1995)
3. Haupt, R.L., Haupt, S.E.: Practical Genetic Algorithms, 2nd edn. John Wiley and Sons, USA (2004)
4. Yang, X.S.: Firefly algorithms for multimodal optimization. In: Watanabe, O., Zeugmann, T. (eds.) Stochastic Algorithms: Foundations and Appplications, SAGA 2009. LNCS, vol. 5792, pp. 169–178. Springer-Verlag, Berlin (2009)
5. Yang, X.S.: A new metaheuristic bat-inspired algorithm. In: Nature Inspired Cooperative Strategies for Optimization (NISCO 2010). In: Gonzalez, J.R. et al. (eds.) Studies in Computational Intelligence, vol. 284, pp. 65 –74, Springer, Berlin (2010)
6. Alihodzic, A., Tuba, M.: Improved bat algorithm applied to multilevel image thresholding. Sci. World J. **2014**, 16 (2014), Article ID 176718
7. Xiao, L., Qian, F., Shao, W.: Multi-step wind speed forecasting based on a hybrid forecasting architecture and an improved bat algorithm. Energy Convers. Manag. **143**, 410–430 (2017)
8. Naderi, M., Khamehchi, E.: Well placement optimization using metaheuristic bat algorithm. J. Petrol. Sci. Eng. **150**, 348–354 (2017)
9. Rahmani, M., Ghanbari, A., Ettefagh, M.M.: Robust adaptive control of a bio-inspired robot manipulator using bat algorithm. Expert Syst. Appl. **56**, 164–176 (2016)
10. Banati, H., Chaudhary, R.: Multi-Modal bat algorithm with improved search (MMBAIS). J. Comput. Sci. **23**, 130–144 (2017)
11. Chaudhary, R., Banati, H.: Shuffled multi-population bat algorithm (SMPBat). In: 2017 International Conference on Advances in Computing, Communications and Informatics (ICACCI), pp. 541–547. IEEE, Udupi (2017)
12. Chaudhary, R., Banati, H.: Modified shuffled multi-population bat algorithm. In: 2018 International Conference on Advances in Computing, Communications and Informatics (ICACCI), pp. 943–951. IEEE, Bangalore (2018)
13. Al-Betar, M.A., Awadallah, M.A.: Island bat algorithm for optimization. Expert Syst. Appl. (2018). https://doi.org/10.1016/j.eswa.2018.04.024
14. Al-Betar, M.A., Awadallah, M.A., Faris, H., Yang, X.S., Khader, A.T., Alomari, O.A.: Bat-inspired algorithms with natural selection mechanisms for global optimization. Neurocomputing **273**, 448–465 (2018)
15. Meng, X.-B., Gao, X.Z., Liu, Y., Zhang, H.: A novel bat algorithm with habitat selection and Doppler effect in echoes for optimization. Expert Syst. Appl. **42**, 6350–6364 (2015)
16. Topal, A.O., Altun, O.: A meta-heuristic algorithm: dynamic virtual bats algorithm. Inf. Sci. **354**, 222–235 (2016)
17. Banati, H., Chaudhary, R.: Enhanced shuffled bat algorithm (EShBAT). In: 2016 International Conference on Advances in Computing, Communications and Informatics (ICACCI), pp. 731–738. IEEE, Jaipur (2016)
18. Chakri, A., Khelif, R., Benouaret, M., Yang, X.S.: New directional bat algorithm for continuous optimization problems. Expert Syst. Appl. **69**, 159–175 (2017)

19. Ahandani, M.A., Alavi-Rad, H.: Opposition-based learning in the shuffled differential evolution algorithm. Soft. Comput. **16**, 1303–1337 (2012)
20. Derrac, J., Garcia, S., Molina, D., Herrera, F.: A practical tutorial on the use of nonparametric statistical tests as a methodology for comparing evolutionary and swarm intelligence algorithms. Swarm Evol. Comput. **1**, 3–18 (2011)

Fuzzy-Based-Cascaded-Multilevel Inverter Topology with Galvanic Isolation

K. Muralikumar, C. Sivakumar, Ankit Rautela and Ponnambalam Pathipooranam

Abstract The key issues regarding the "multilevel inverters (MLI)" is as follows: reduction of number of "switches" that are used in structure, on the other hand increasing the levels that are in output, multiple DC supplies that are being used for the MLI's. "This paper gives latest method of providing, the device count reduction technique and also overcomes the disadvantage different DC power supplies". This method provides a "high-frequency magnetic link" which is used to be provide galvanic isolation at output and input parts of inverters. It is needed in various values in grid-connected and industrial applications. "This method is needed to transfer for present one-phase cascaded-multilevel inverter to three-phase inverter without using more number of electronic components like MOSFET, IGBT". In our topology, we use "asymmetrical mode" mode in our project which represents "3-phase inverter and cascaded h-bridge inverters" in order to produce "high-frequency magnetic link" which is necessary, for (toroidal-core in our method). "As a result we can decrease the size and increase the power density". "This proposed method can be presented by the help of simulation".

Keywords Cascaded-multilevel inverter (CMLI) · Three-phase inverter (TPI) · High-frequency magnetic link · Phase generator (PG) · Bidirectional switches (BD) · Fuzzy logic controller

K. Muralikumar (✉) · C. Sivakumar · A. Rautela · P. Pathipooranam
Vellore Institute of Technology, Vellore 632014, Tamilnadu, India
e-mail: kolamuralikumar@gmail.com

C. Sivakumar
e-mail: c.sivakumar2018@vitstudent.ac.in

A. Rautela
e-mail: ankit.rautela2018@vistudent.ac.in

P. Pathipooranam
e-mail: p.ponnambalam@gmail.com

© Springer Nature Singapore Pte Ltd. 2020
K. N. Das et al. (eds.), *Soft Computing for Problem Solving*,
Advances in Intelligent Systems and Computing 1048,
https://doi.org/10.1007/978-981-15-0035-0_67

1 Introduction

The important points in our paper is to select the cascaded-multilevel inverter, so it will obtain large number of voltages by the help of reduced power components [1]. Cascaded-multilevel inverters are more useful in our method which has an application in tidal power station, solar power station, charge-storage, etc. "As well as it is also beneficial in industrial applications such as motor drives". In our method we are connecting more cells in series, so we can increase the levels in voltages with dc power supply. "There are two modes in cascaded-multilevel inverters that can categorized as symmetric and asymmetric modes". "Asymmetrical mode of operation has property to generate large number of levels at the output voltages which is more applicable as compared with counter mode of operation", i.e., symmetrical mode by utilizing existing power switches and supply. And in our method we use high-frequency transformer that can isolate multilevel inverters to grid in order to get the galvanic isolation. "We need to consider size and weight of transformer which is an important parameter in multilevel inverter application" [1–3].

A multilevel inverter with high frequency consists of toroid-core, "magnetic-core materials such as finnet" which speeds up the total performance and the high "frequency magnetic link" which is completely depended on cells that are arranged in series that work as cascaded inverters used to produce various levels of voltage waveform in output that uses one DC supply in primary and many windings in secondary side winding in the "high-frequency magnetic link" in this topology. "The important objective of our methods is to implement from one-phase inverter to three-phase inverter without using more number of switching components" ("HFML makes CMLI work in a single dc supply"), the limitations of electronic components in three-phase structure can be avoided [4, 5].

2 Proposed Topology of CMLI

Our present topic "CMLI" has the following components such as: "HMLI, AC to DC converters, simple structured MLI which will be connected in the cascaded type. The square wave voltage generator which generates the square wave at high-frequency on the output side." The SWV generator is fed from the fuzzy logic controller for pulse generating. We simulated the same SWV generator by feeding it with SPWM technique of switching [1–3].

On the output side of multi-winding transformer, the DC to AC converters are connected which serves the purpose of establishing the DC supplies that are required by the cascaded cell inverters and also the three-phase inverter. Here the output three-level voltages are shifted to the three-phase inverter at the junction points, namely A, B, and C.

The DC supply that is fed to the cascaded connection inverters are modified by "varying the winding turns ratio of MWT". This ease of adjustment is very beneficial for this project. Also, another helpful factor in this topology is that we use different

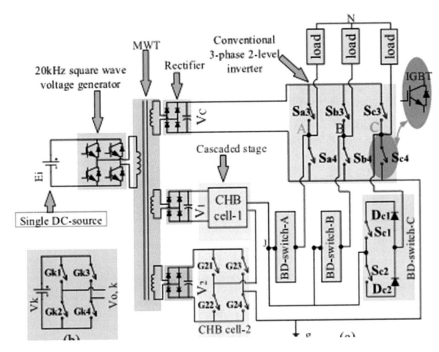

Fig. 1 Implementation structure of H-bridge cells in a cascade type connection [1]

number of "cascaded H-bridge cells". And these can be replaced by any other per-sisting "cascaded stage inverter cells". The "TPI" will have three legs and each leg comprises two switches. These switches operate on the counter operation and will give the voltage levels of "Vc and 0". The same is repeated for rest of the legs but the only difference that "120 degree delay".

2.1 Established Notion for Three-Phase CMLI

The pictorial representation is shown in Fig. 1 which accommodates a "cascaded stage" inhered with "the polarity generator stage". This "cascaded structure" is not constant, it is used with accessible "multilevel inverter structure". The polarity generator is finished and utilizing the "bi-directional (BD) switches, three-phase inverter (TPI)".

3 Fuzzy Logic Controller

A fuzzy controller is based on fuzzy code designed to control mechanical as well as electrical systems. They can be in the form of software or the hardware and we can use them in anything from the small circuits to large scale of mainframe circuits.

Currently, the fuzzy controllers are used in flight control system, anti-lock braking systems, and camcorder stabilization lifting bodies, etc. We use fuzzy controllers in order to acquire high robustness, usage of multiple inputs and outputs. Also they can be easily modified. They are much simpler than linear algebraic equations. On the other hand, they are very quick and cheaper to implement. The procedure involved in constructing the fuzzy controller are as follows: [6–10].

(1) Creating membership values, i.e., fuzzification.
(2) Specifying the rule tables.
(3) Determination of procedure for defuzzification.

At first, we should create the membership values for the data and place them into the fuzzy sets. Then divide them into the required ranges. The Y axis value must be 0–1 that means 0–100%. The X axis will be an arbitrary range which we determined. We should now create the rules table in order to specify the output ranges which are used. Now, we should specify what to do with the result that we got from the rules. This can be done by using Mamdani's Center of Gravity method. And the membership functions are shown in Figs. 2 and 3.

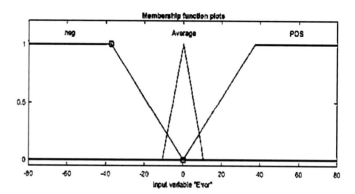

Fig. 2 Input (error) membership function for the fuzzy controller

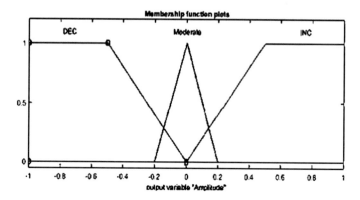

Fig. 3 Output (amplitude) "membership function for the designed fuzzy controller"

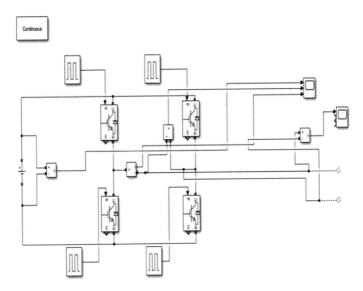

Fig. 4 Simulation diagram square wave generator

4 Square Wave Voltage Generator

The SWV generator simulation circuit determined in Fig. 4 is worked with 4 "(IGBT) insulated-gate bipolar transistors". "The SWV generator converts the input DC supply (240 V) into a square wave voltage with a very high frequency which is headed to primary winding in multi-winding transformer. The pattern of switches, i.e., [S1 and S3] and [S2 and S4] which are available in SWV generator that operates on flipping condition to generate 240 V with 25 kHz SWV". The input and output voltages of SWV generator are shown in Fig. 5.

5 High Frequency Magnetic Link

Here the "multiple-winding transformer" consists two coils of "secondary side" and fundamental structure of multi-winding is shown in Figs. 6 and 7. Here, the multi-winding transformer is utilized for confining the source voltages. "The whole coils which are available of secondary side could be balanced that totally rely upon a required power supply voltage. Here we utilized two coils that lie opposite side of primary side of transformer which is being connected to three-phase inverter along with the cascaded cells. "The assumed value Vc, i.e., a DC source must be "2 V". The feeding voltage given to cascaded cells must match the value of "v" that could be transformed by changing multi-winding transformer turns proportion of 2:2:1". The "diode-bridges" are associated with the "transformer's secondary side" with the end

Fig. 5 Input voltage and output voltage of SWV generator

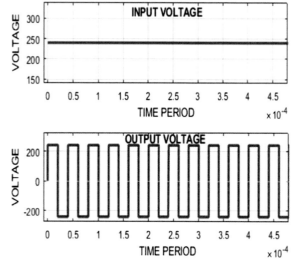

Fig. 6 Multi-winding transformer has two windings in it's secondary side

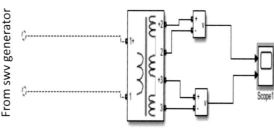

Fig. 7 Basic structure of MW transformer [1]

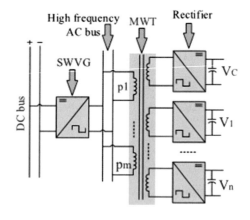

goal to deliver the DC source power for "CHB-cells" and the "CTPTL-inverters". With the aim to eliminate or decrement the ripples that occur in output voltage we are introducing the capacitors in parallel to "diode-bridges".

5.1 Bidirectional Switches

A bidirectional switch is working with the arrangement of two "insulated-gate bipolar transistors (IGBT)" in one arm and two diodes that are connected across them. A "BD-switch" will "TURN-ON" if the two IGBTs in the same arm are in ON condition. The steps in the output voltage that are obtained within the regions V_c, zero (0) are due to the operation of "Cascaded H-bridge" only during the on-state of "BD-switches". The three "BD-switches" transfer the generated voltages for junction points A-B-C. They are working with 120° difference in each phase. The IGB's mounted on first leg in the "TPI" and the "BD-switch" shall be working with "toggle mode" of operation.

5.2 Three-Phase Inverter (TPI)

See Figs. 8 and 9.

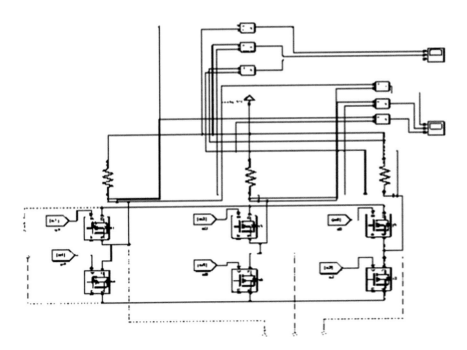

Fig. 8 Simulation diagram of three-phase inverter

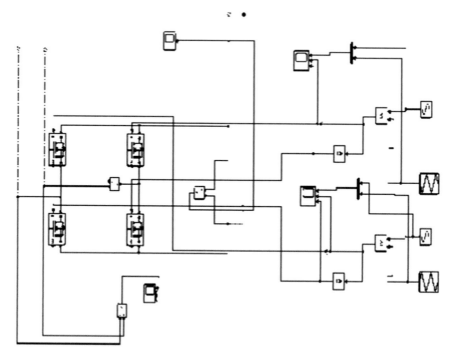

Fig. 9 Simulation of H-bridge cell

5.3 "H-Bridge Topology" Used as Cascaded Stage

"The cascaded H-Bridge (CHB) MLI" the most famous topology that can be applied for one-phase and also three-phase. Any cascaded-multilevel inverter, i.e., one-phase and three-phase can be treated as cascaded cell. The simulation diagrams of three-phase inverter and "H-bridge" cell is shown in Figs. 8 and 9. The conventional "CHBMLI" topology that is described in the diagram for different "cascaded H-bridge cells" used to check this topology. The new "three-phase MLI structure" is utilizing one-H-bride cell that is lying in cascaded stage. If we desire to achieve large voltage levels in output side then we should increase the series connection, i.e., "cascaded stage". Every cascaded cell will have four switches, for example, MOSFET and IGBTs. And they will operate in opposingly and give us the voltage waveform of positive of supply and negative of supply as well as zero level, since asymmetrical methods are more preferable than the symmetrical mode in various output stage levels. Our DC voltages that are used were adjusted in binary form such as "(1:2:3...2n)". The used voltage of Vc must be higher than the summation of all cascaded stage cells. Let us assume that the input voltages be: "$V1, V2,...Vn$". Also here we consider all the voltages are same, i.e., '$V1 = V2 = v$'. The voltage supply given to inverter cells are as follows:

$$Vc = \sum_{n}^{n} Vx + V1 = (V1 + V2 + \dots Vn) + v$$

Here the magnitudes given to H-bridge cell input voltage is maintained in a symmetric, i.e., $V1 = v$, $Vc = 2v$. Here two switches that are present in "TPI" will work for producing voltage level of Vc and 0. The next level of voltage ($V1$) that lies in between the 0 and Vc is produced by the usage of various control logical gates to CHB-cells, "when the BD switches are turned on".

6 "Sinusoidal Wave Pulse Width Modulation"

"The Sinusoidal Wave Pulse Width Modulation (SPWM)" is habitually useful to voltage source inverter. The triggering pulse patterns will be achieved by comparing the triangular wave with a sinusoidal wave, here the triangular wave is considered as carrier wave and the sine wave is treated as a reference wave. Reference signal frequency will be equal to the frequency that is obtained "inverter output". The "frequency" that is given to the carrier wave, i.e., triangular wave in this topology explains how many pulse signals in one-half cycle. "The carrier signal's frequency is very higher compared to reference signals. In "the" implemented structure "sine wave" is having a "phase-delay" of 120 between one to the other sine wave and compared with the triangular wave. The pulses generated are "shown in Figs. 10 and 11".

7 Results

The Corresponding results obtained from simulation, the output voltage from the H-bridge cell, the Output voltage of multi-winding transformer and Output voltage of bridge rectifier given to three-phase inverter is as shown in Figs.12, 13 and 14. The three-phase inverter of phase voltage and line voltage is as shown in Figs. 15 and 16.

7.1 Fuzzy-Based-Output for SWV Generator

See Figs. 17 and 18.

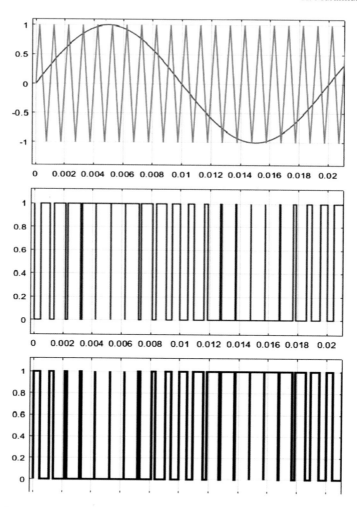

Fig. 10 "SPWM pulses for the G11 and G12 switches"

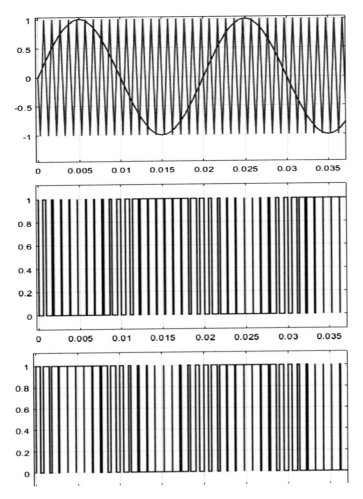

Fig. 11 "SPWM pulses for the G13 and G14 switches"

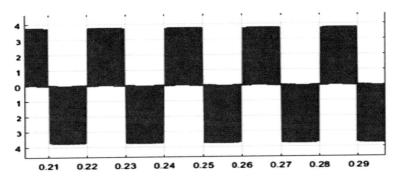

Fig. 12 Obtained output voltage from the H-bridge cell

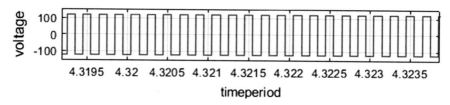

Fig. 13 Output voltage of multi-winding transformer

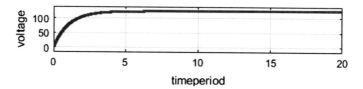

Fig. 14 Output voltage of bridge rectifier given to three-phase inverter

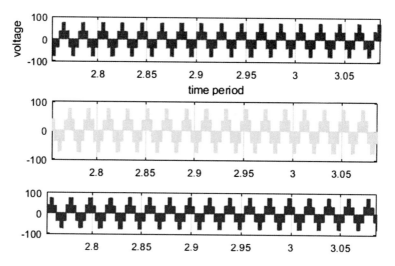

Fig. 15 "Phase voltages of three-phase inverter"

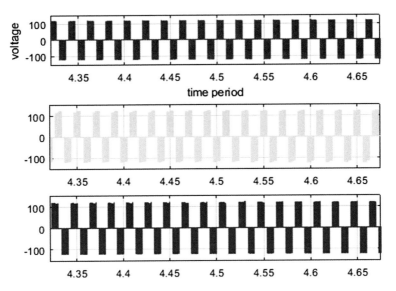

Fig. 16 Line voltages from three-phase inverter

Fig. 17 Fuzzy output
voltage from SWV generator

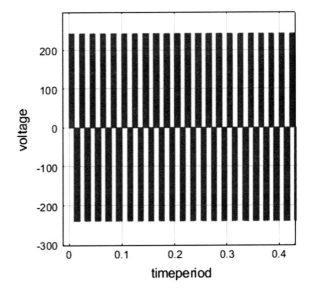

Fig. 18 Fuzzy reference voltage for SWV generator

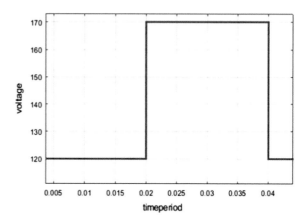

8 Conclusion

A Fuzzy-based-magnetic link usage for CMLI is used in the present paper that employees a high frequency. This paper shows a few focal points when compared with other topologies. The implementation of this topology "from one-phase to three-phase" is done rather than maximizing the used switching devices. The Fuzzy based "Cascaded-multilevel inverter with galvanic isolation" has the capability to reduce the size, increase power-density, and the performance with fuzzy logic controller.

References

1. Hasan, M.M., Abu-Siada, A., Islam, S., Dahidah, M.S.: A new cascaded multilevel inverter topology with galvanic isolation. IEEE Trans. Ind. Appl. (2018)
2. Hasan, M.M., Abu-Siada, A., Islam, S.M., Muyeen, S.M.: A novel concept for three-phase cascaded multilevel inverter topologies. Energies **11**(2), 268 (2018)
3. Rodriguez, J., Lai, J.S., Peng, F.Z.: Multilevel inverters: a survey of topologies, controls, and applications. IEEE Trans. Ind. Electron. **49**(4), 724–38 (2002)
4. Franquelo, L.G., Rodriguez, J., Leon, J.I., Kouro, S., Portillo, R., Prats, M.A.: The age of multilevel converters arrives. IEEE Ind. Electron. Mag. **2**(2), (2008)
5. Patel, J.V., Mishra, N.G.: Analysis of five-level cascade multilevel inverter using SPWM Technique
6. Ponnambalam, P., Aroul, K., Reddy, P.P., Muralikumar, K.: Analysis of fuzzy controller for H-bridge flying capacitor multilevel converter. In: Proceedings of sixth international conference on soft computing for problem solving 2017, pp. 307–317. Springer, Singapore
7. Reddy, V.P., Muralikumar, K., Ponnambalam, P., Mahapatra, A.: Asymmetric 15-level multi-level inverter with fuzzy controller using super imposed carrier PWM. In: Power and advanced computing technologies (i-PACT), 2017 innovations in 21 Apr 2017, pp. 1–6. IEEE (2017)
8. Ponnambalam, P., Muralikumar, K., Vasundhara, P., Sreejith, S., Challa, B.: Fuzzy controlled switched capacitor boost inverter. Energy Procedia. **117**, 909–16 (2017)

9. Viswanath, Y., Muralikumar, K., Ponnambalam, P., Kumar, M.P.: Symmetrical cascaded switched-diode multilevel inverter with fuzzy controller. In: Soft computing for problem solving, pp. 121–137. Springer, Singapore (2019)
10. Madhav, K.L., Babu, C., Ponnambalam, P., Mahapatra, A.: Fuzzy logic controller for nine level multi-level inverter with reduced number of switches. In: Power and advanced computing technologies (i-PACT), 2017 innovations in 21 Apr 2017, pp. 1–7. IEEE (2017)

Modeling the Efficacy of Geopolymer Mosquito Repellent Strips Leachate Distribution Using Meta-heuristic Optimization

D. K. D. B. Rupini and T. Vamsi Nagaraju

Abstract Many mosquito repellents were available in the markets in various forms such as coils, plug-in repellents, papers, creams, and other repellent imparted synthetics and fibers are in vogue. Moreover, all the aforementioned repellents were used in in-doors, applying for human bodies and some of them are imparted in clothes and accessories. However, one should give prime importance to control mosquitoes at their breeding stage itself in stagnant waters or drain waters. In this context, geopolymer soils imparted with mosquito repellent was developed to eradicate mosquitoes at their breeding stage itself. This paper presents the leachate distribution assessment of VR-geo mosquito repellent strip using swarm-assisted multi-linear regression. A model equation has been developed for the prediction of leachate distribution in terms of pH using input parameters like volume of the geopolymer repellent strip, molarity of NaOH, Na_2SiO_3/NaOH ratio, and alkali-activator content.

Keywords Geopolymerization · Leachate assessment · VR-geo mosquito repellent · PSO

1 Introduction

In India, most of the rural villages and tribal areas are covered with open drainages or drainage pits due to these conditions many viral diseases are spreading. Dengue is one of the major mosquito-borne acute viral diseases increasing annually and severe health threat globally reported by World health Organization "Impact of Dengue". A major issue with mosquito-borne diseases was it causes illness, vomiting, abdominal pain, and skin rashes for humans. For instance, approximately 2.5 billion people are at risk for dengue fever globally. Furthermore, 22,000 people were dead with this dengue virus annually. Many mosquito repellents are in vogue in various forms

D. K. D. B. Rupini (✉) · T. V. Nagaraju
S. R. K. R. Engineering College, Bhimavaram 534204, India
e-mail: rupini.52@gmail.com

T. V. Nagaraju
e-mail: varshith.varma@gmail.com

© Springer Nature Singapore Pte Ltd. 2020
K. N. Das et al. (eds.), *Soft Computing for Problem Solving*,
Advances in Intelligent Systems and Computing 1048,
https://doi.org/10.1007/978-981-15-0035-0_68

such as plug-in repellents, coils, creams, and other imparted repellent synthetics. However, all the aforementioned repellents were concentrating only on adult mosquitoes. So, one should cope with to control mosquitoes at their early stages (larvae and pupa). Naturally, larvae and pupa stages were experienced in the stagnant pools; their survivability depends on many factors such as pH of water, salinity, alkaline environment, and temperature [4].

In this regard, some research has been reported that the survival of mosquitoes depends upon the temperature, alkaline environment, salanity, and humidity [4, 6, 10].

To optimize on survival of mosquitoes, prediction models are necessary. This study aims at the applicability of computations applications in the discipline of biosciences, especially mosquito studies.

To know the knowledge of the authors, very limited studies have been carried on the computational applications on the mosquito studies. One of the few studies carried out by Keun et al. [8] developed models for predicting mosquito abundances in urban areas. Most of the existing research focuses on mathematical modeling and social media [1, 13]. On the other hand, Agriculture Research Service, Florida has developed computer models to assess mosquito repellent compounds. Recently, Lenus et al. [9] explore the viability of machine learning techniques for potential distribution models for an invasive mosquito species.

This paper aims the importance of particle swarm optimization for the prediction of leachate parameters of VR-Geo mosquito repellent strips. The input variables such as volume of the geopolymer repellent strip, molarity of NaOH, Na_2SiO_3/NaOH ratio, and alkali-activator content were considered for the prediction of pH of the leachate mosquito repellent strips. This paper also explores the importance of soft computing approach for the mosquito-related studies.

2 Materials and Methodology

The main material used in this study was geopolymer soil blends imparted with mosquito repellent (VR-Geo mosquito repellent an innovative product). Geopolymers are globally recognized binders, having three-dimensional tetrahedral Al and Si frameworks [3, 12].

In order to develop simulation models, we briefly recall the VR-geo mosquito repellent strips. These strips were made by using geopolymer soil blended with mosquito repellent. In this study, mosquito repellent strips were made by varying molarity of NaOH, Na_2SiO_3/NaOH ratio, and alkali-activator content to assess the leachate distribution for 1 cubic meter water tank. The samples were placed into the water tank and the variation of pH values of water was observed using pH meter. Further, with the help of data set obtained, the pH values were estimated by using particle swarm optimization using volume of the geopolymer repellent strip, molarity of NaOH, Na_2SiO_3/NaOH ratio, and alkali-activator content as input variables. The test data was tabulated in Appendix.

3 Particle Swarm Optimization (PSO)

Many traditional approaches for prediction models like mathematical regression analysis are quite well-known from long time, which are time-consuming and manually tabulated or calculated. As a result, emerging consortium of techniques like soft computing is desirable for all disciplines for better convergent results. In particular, the study of the prediction of a given variable with a limited data, meta-heuristic Particle Swarm Optimization (PSO) method has been receiving attention as a potential method for prediction of a given variable with limited data set [7]. On the other hand, PSO can be used in various disciplines due to its simple and general principles [2]. In PSO, the optimization mechanism can be stated that a group of particles (in general fishes) that searching for food in a pond (constraint-based optimization) with random manner. The location of each fish can be treated as a solution to given problem. Initially, these locations (solutions) are randomly generated. These current position (current solution) of a particle may not consist food; hence, the bird will move to next position (best solution) by updating its velocity. To get closer to the location of the food (considered as best solution), all of the fishes follow the nearest fish that is closer to the food place. Each particle/fish is having two vectors x_i, v_i represent the position and velocity of the particle at every iteration. All of the particles regulate their route for the next iteration based on their experience and other particles experiences at the present iteration. Let x_i^k denotes the particle position vector in search of space of solutions and the new position for each particle is taken place randomly as

$$x_i^{k+1} = x_i^k + v_i^{k+1} \tag{1}$$

where v_i^{k+1} is the velocity vector of particle that drives the process toward optimization. The position of every individual bird is p_i^k and the velocity is v_i^k. The best position of each particle is p_{best} and the overall optimized position is g_{best}. Then the velocity of each organism in the swarm is updated as

$$V_i^{k+1} = \omega V_i^k + c_1 r_1 (p_{\text{best}_i} - x_i^k) + c_2 r_2 (g_{\text{best}_i} - x_i^k). \tag{2}$$

In Eq. (2), c_1 and c_2 are acceleration constants r_1 and r_2 are random numbers which vary between 0 and 1 and ω is the inertia weight factor. PSO technique has two type parameters known as common parameters and control parameters. Every meta-heuristic optimization algorithm needs common parameters such as population size and number of iterations but control parameters vary from one algorithm to other. After initialization and selection of both common and control parameters, iterative procedure will start by random initial solution matrix. For each solution, the function value is calculated known as fitness of the particles (solutions). Based on local best and global best concepts (information with a particular particle experience and other particle experiences), updating of position and velocity will takes place. This updating procedure will continue until either optimization value is arrived or iteration criteria are met. In this paper, a constant inertia weight strategy is applied for estimation. The acceptable range of ω for engineering problem lies in between [0.4 and 0.9]. Even identification of suitable value of ω is difficult in its wide range; several investigations are carried out in this paper for selecting an optimal ω to get close results.

4 PSO-Based Prediction Models for Leachate Distribution

Estimation of pH of the leachate distribution using Particle Swarm Optimization (PSO). The data information of volume of the geopolymer repellent strip (Vol), molarity of NaOH (M), Na_2SiO_3/NaOH ratio (R), and alkali-activator content (AAC) are considered as input variables to predict pH of the leachate distribution. The multi-linear model equations were developed for predicting pH of leachate distribution. For this formulated equations, two models were considered based on the additional coefficient. In the case Model A, each parameter such as Vol, M, R, and AAC was multiplied by respective coefficients such as x_1, x_2, x_3, and x_4. The developed multivariable equation is shown in Eq. 3.

$$pH_{(est)} = Vol.x_1 + M.x_2 + R.x_3 + AAC.X_4 \tag{3}$$

where a_1, a_2, a_3, a_4, a_5, a_6, and a_7 are coefficients need to be estimated properly to calculate final output variable pH of the leachate. On the other case Model B, an additional constant is used to convert it into swarm-assisted multiple linear regression models. For the total data set, pH is estimated and an error is evaluated from actual data of pH. For evaluation of error, the following formula is used.

$$pH(est) = Vol.x_1 + M.x_2 + R.x_3 + AAC.X_4 + X_0 \tag{4}$$

In the above equation, k represents number of iterations. N is totalling number of test data samples, pH is actual value, and $pH_{(est)}$ estimated value of pH. The error calculated from both estimated and actual values of pH for an iteration "k" is represented by E. From calculated error values, the objective function (J) is formed as shown in Eq. (5)

$$J = \min(E) \tag{5}$$

The above objective function depends on the prediction of coefficients estimated by optimization techniques. Limits of these variables depend on the user requirement and/or data set availability. For this problem, the limits are fixed in between $-10,000$ and $10,000$ obviously large solution space.

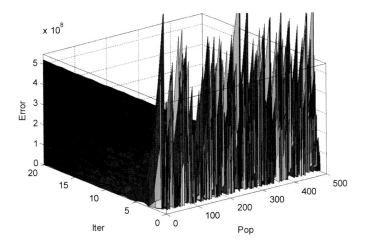

Fig. 1 Surface of the predicted pH (Model A)

Fig. 2 Estimation of pH
(Model A)

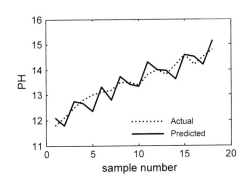

Model A:

In the case of Model A, for formulating model equation of pH estimation the following Eq. (6) was developed using the PSO (Figs. 1 and 2).

$$pH_{(est)} = 0.001559.Vol - 0.10528.M + 0.5620.R + 0.63658.AAC \qquad (6)$$

Model B:

As discussed in section Model A, the simulation models do not yield close prediction; hence, model B is selected and the coefficients are obtained by using PSO. Equation (7) is the final model equation to predict pH using PSO with additional coefficient taken into consideration to minimize local struck (Figs. 3 and 4).

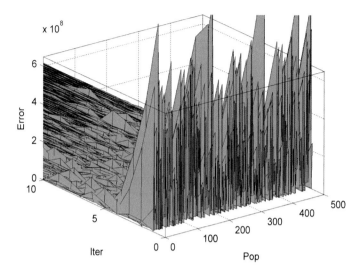

Fig. 3 Surface of the predicted pH (Model B)

Fig. 4 Estimation of pH
(Model B)

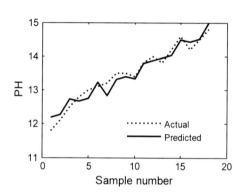

$$pH_{(est)} = 0.00148.Vol + 0.02896.M + 0.4888.R + 0.1881.AAC + 7.3516 \quad (7)$$

Figure 5 shows, by comparison, variation of percentage of error with the involvement of additional coefficient. This can be attributed due to the minimization of local struck. The error was formulated by using the following Eq. (8). Table 1 shows the error for both cases (Model A and Model B).

$$Error = Abs\left((pH_{act} - pH_{predicted})/pH_{act}\right) \quad (8)$$

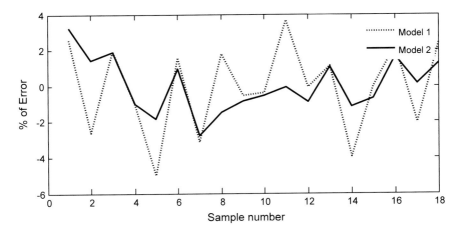

Fig. 5 Comparison of percentage of error in Model 1 and Model 2

Table 1 Comparison of error with actual and predicted

pH (actual)	Model 1 with four variables		Model 2 with five variables	
	pH (predicted)	Error	pH (predicted)	Error
11.8000	12.1025	2.5634	12.1855	3.2667
12.1000	11.7866	−2.5900	12.2724	1.4244
12.5000	12.7439	1.9512	12.7355	1.8837
12.8000	12.6645	−1.0586	12.6744	−0.9816
13.0000	12.3486	−5.0105	12.7612	−1.8366
13.1000	13.3059	1.5719	13.2244	0.9493
13.2000	12.7886	−3.1163	12.8381	−2.7419
13.5000	13.7459	1.8218	13.3012	−1.4727
13.5000	13.4301	−0.5180	13.3881	−0.8291
13.4000	13.3507	−0.3681	13.3270	−0.5451
13.8000	14.3080	3.6809	13.7901	−0.0719
14.0000	13.9921	−0.0565	13.8770	−0.8789
13.8000	13.9551	1.1242	13.9475	1.0688
14.2000	13.6393	−3.9488	14.0344	−1.1663
14.6000	14.5966	−0.0236	14.4975	−0.7021
14.2000	14.5172	2.2336	14.4364	1.6647
14.5000	14.2013	−2.0600	14.5233	0.1605
14.8000	15.1586	2.4229	14.9864	1.2594

5 Conclusions

This paper presents a mosquito repellent leachate distribution assessment using meta-heuristic particle swarm optimization. The leachate distribution is influenced by many factors such as volume of the geopolymer repellent strip, molarity of NaOH, Na_2SiO_3/NaOH ratio, and alkali-activator content.

Different experiments were conducted to predict the pH value, in the case Model A, no additional coefficient was taken, and due to this large error is witnessed in the clusters. On the other hand, the developed model B with additional coefficient demonstrated that proposed condition enhanced the convergent results of the PSO for prediction of pH value. It was evident that additional coefficient consideration can enhance the prediction models by minimizing the local struts.

In summary, PSO was able to predict the given parameter with the limited data set (particularly in few disciplines where limited data set is available).

Appendix: Test Data of the VR-Geo Mosquito Repellent Leachate Distribution

Set I	Volume of strip	345	345	345	345	345	345
	Molarity of NaOH	7	10	13	7	10	13
	Na_2SiO_3/NaOH	1.5	1.5	1.5	2.5	2.5	2.5
	Akali-activator content	18	18	20	18	18	20
	pH	11.8	12.1	12.5	12.8	13	13.1
Set II	Volume of strip	785	785	785	785	785	785
	Molarity of NaOH	7	10	13	7	10	13
	Na_2SiO_3/NaOH	1.5	1.5	1.5	2.5	2.5	2.5
	Akali-activator content	18	20	20	18	20	20
	pH	13.2	13.5	13.5	13.4	13.8	14
Set III	Volume of strip	1533	1533	1533	1533	1533	1533
	Molarity of NaOH	7	10	13	7	10	13
	Na_2SiO_3/NaOH	1.5	1.5	1.5	2.5	2.5	2.5
	Akali-activator content	18	18	20	18	18	20
	pH	13.8	14.2	14.6	14.2	14.5	14.8

References

1. Angelina, M.L., Melissa, A.P., Thomas, S., Nakul, C.: Mathematical modeling of mosquito dispersal in a heterogeneous environment. Math. Biosci. **241**, 198–216 (2012)
2. Khare, A., Rangnekar, S.: A review of particle swarm optimization and its applications in solar photovoltaic system. Appl. Soft Comput. **13**, 2997–3006 (2013)
3. Davidovits, J.: Properties of geopolymer cements. In: First international conference on alkaline cements and concretes (1994)
4. Dmitri, Y.B., Leonid, L.M., Paul, J.L., James, R.T., Peter, J.S.S., William, R.H.: Insitu analysis of pH gradients in mosquito larvae using non-invasive, self-referencing, pH sensitive microelectrodes. J. Exp. Biol. **204**, 691–699 (2001)
5. Impact of Dengue. http://www.who.int/csr/disease/dengue/impact/en/
6. Irish, S.: Effects of different pH levels on the viability, metamorphosis rate and morphology of aedes mosquitoes. In: Central visayas health research and innovation conference, Talamban, Cebu City (2016)
7. James, K., Russell, E.: Particle swarm optimization. IEEE 0-7803-2768-3/95, pp. 1942–1948 (1995)
8. Lee, K.Y., Chung, N., Hwang, S.: Application of an artificial neural networks (ANN) model for predicting mosquito abundances in urban areas. Ecol. Inform. **36**, 172–180 (2015)
9. Linus, F., Helge, K., Antje, K., Gunter, A.S., Doreen, W., Ralf, W.: Modelling the potential distribution of an invasive mosquito species: comparative evaluation of four machine learning methods and their combinations. Ecol. Model. **388**, 136–144 (2018)
10. Pelizza, S.A., Lopez, L.C.C., Becnel, J.J., Bisaro, V., Garcia, J.J.: Effects of temperature, pH and salinity on the infection of leptolegnia chapmanii Seymour (Peronosporomycetes) in mosquito larvae. J. Invertebr. Pathol. **96**(2), 133–137 (2007)
11. USDA/Agricultural Research Service.: Computer model for finding mosquito repellent compounds. ScienceDaily, 12 June 2008
12. Nagaraju, V.T.: Potential of geopolymer technology towards ground improvement. In: 2nd International conference on Advances in concrete, structural and geotechnical engineering. BITS Pilani, Rajasthan (2018)
13. Jain, V.K., Kumar, S.: Effective surveillance and predictive mapping of mosquito-borne diseases using social media. J. Comput. Sci. **25**, 406–415 (2017)

Optimization of Drilling Rig Hydraulics in Drilling Operations Using Soft Computing Techniques

G. Sangeetha, B. Arun kumar, A. Srinivas, A. Siva Krishna, R. Gobinath and P. O. Awoyera

Abstract The primary goal for all oil and gas producing companies is to produce oil as much as possible by optimization with lower cost. One way to increase the productivity index is essentially related to a proper drilling technique. In this study, optimization of drilling rig hydraulics in drilling operations was performed using soft computing techniques. Data such as flow rate, the angle of inclination, yield point, plastic viscosity and depth of the well were used as input. Thus, using the neural network (ANN) approach, five process parameters are the inputs to the model and output from this model is cutting concentration. The best model for the drilling rig contains five input parameters, seven hidden layers and one output parameter. The optimization aids the inherent characteristics of the system as well as the factors like exorbitant surface torque, unexplained drop in the rate of infiltration and a sudden change in surface weight

Keywords Drilling · Drilling fluid carrying capacity · Hydraulics · Optimization · Rate of infiltration · Rate of penetrations

1 General Introduction

Optimization in drilling engineering is one of the most important divisions in petroleum exploration. Majorly, the organizations are looking forward to optimizing in order to enhance their productivity in general. However, if the system is not optimized, both time and cost might be affected.

The optimization of drilling hydraulics is a typical term used to allude to a vast assortment of circumstances utilizing liquids. In boring, power through pressure assumes a key job too. No boring can be completed without using a liquid to carry out various occupations in the meantime. This liquid is constrained through various

G. Sangeetha · B. Arun kumar · A. Srinivas · A. Siva Krishna · R. Gobinath (✉)
Civil Department and Center for Materials and Methods, SR Engineering College, Warangal, India
e-mail: r.gobinath2013@vit.ac.in

P. O. Awoyera
Department of Civil Engineering, Covenant University, Ota PMB 1023, Nigeria

© Springer Nature Singapore Pte Ltd. 2020
K. N. Das et al. (eds.), *Soft Computing for Problem Solving*,
Advances in Intelligent Systems and Computing 1048,
https://doi.org/10.1007/978-981-15-0035-0_69

parts of optimizations such as a penetrate string, at that point comes back from the annulus to surface finishing a cycle. One of its critical capacities is to counter the subsurface development weight. On the off-chance that the counterweight applied is excessive, the liquid can be lost in the development while insufficient can result in an undesirable inundation of arrangement liquids. Inside the penetrate string, distinctive parts are tended to particular needs of the activity, each bit of gear, because of its inward distance across, outside measurement and cosmetics that have particular weight misfortune related with it. Understanding these weight misfortunes is imperative with the end goal to optimize or to comprehend and gauge the correct weight that would be experienced while boring. Realizing the weight, one can even more likely arrangement for the fitting hydrostatic go to bore a particular development. The parametric examination is done to optimize the impact of every parameter on the cutting fixation with a time of the directional boring and the limited distinction strategy for optimizations was appointed to settle the cutting transport condition numerically for the flat wellbore [1].

Application of nanotechnology in drilling fluids for the oil and gas industry has been a focus of several recent studies in well optimizations [2]. The representative of RASL, apparent rock strength log was utilized for the optimization purpose [3, 4]. Different data mining techniques studied for optimizations such as multiple linear regression, principal component register and partial least square have been studied and optimized to obtain accurate output in terms of regression error constant (R), root mean square error (RMSE) and F-test (FF, cost objective (p) [5]). Interaction effects of the process parameters play a significant role in the overall optimizations [6–8]. The systematic study reveals the inherent mechanisms of the process and is useful in the scaling and modelling of the whole system [6, 9–12]. As the outcome of the process is being influenced by each process parameter, the multivariable model needs to be optimized with an objective of reducing the root means square error, which thus results in the higher efficient model [13–16]. The parametric study considered for optimizations were Reynolds number, density, specific heat and thermal conductivity and the flow assumed to be laminar, according to their thermal properties [17]. It was concluded that other properties might be optimized such as the viscosity that may add a significant effect rather than compressibility in the behaviour of what so-called foamy oil compared to the presence or absence of asphaltenes and other polar oil components [18–21]. Oil viscosity affects the process in many ways, and it affects the diffusivity of gas in the oil [22]. A new well configuration was able to significantly optimize the application of steam-assisted gravity drainage process [23, 24].

Since the cutting concentration depends on process parameters like flow rate, the angle of inclination, yield point, plastic viscosity and depth of the well, so to improve the cutting concentration, these parameters should be varied. The interactive effects of these parameters dictate the efficiency of cutting concentration. Therefore, the interactive effects of these parameters should be systematically varied. To limit the number of experimental runs and minimize the cost, a data-driven model should be developed to capture the inherent mechanism. In this regard, a well-known ANN is implemented. Based on the importance of the parameters that affect the accumulation of the cuttings at the horizontal section, a sensitivity analysis was conducted at various specific hole cleaning parameters optimization such as fluid flow rate, the angle of inclination, yield point, plastic viscosity and the rate of penetration. To achieve the proper investigation and optimization stage of cutting concentration at the horizontal section, experimental or real field data were obtained to reach the required objective.

2 Materials and Methods

2.1 Static Condition

Static conditions exist at various events in the well like association time, preceding running packaging times when penetrating is halted for stream check, rig issues causing a close-down of the apparatus, wireline tasks and so forth. Currently, the descending power applied by a liquid segment is given by $F = P \times A$ (where p is weight and the cross-sectional zone), when there is a descending power, there must be an equivalent upward power to make an equalization as an optimization; these shift with profundity.

   In boring, the liquids managed (penetrating mud and saltwater) have insignificant compressibility. For the extent of this module, the rearranged condition for hydrostatic weight for an incompressible liquid in field units is given by $P = 0.00982\rho D + p_o$, where ρ is the density of the fluid in kg/m^3 and p_o is the static surface pressure and D is the depth. As specified before, this optimized weight must be sufficient to apply a positive weight on the arrangement. Generally, the likelihood of breaking the development with an excess of weight can result. On the other side of optimizations, lacking weight can make the development liquids enter the borehole and drive the penetrating liquid out of the wellbore causing a victory. It is conceivable to locate a liquid section of shifting densities in the wellbore and subsequently, to decide the weight applied by such a segment, one needs to include the different modules of liquid optimization as appeared in Fig. 1.

   Another optimization technique for deciding the weight applied by the segment of liquid is to think about just a single liquid segment. The identical mud thickness is optimized and can be figured by Eqs. 1 and 2.

$$p = p_o + 0.00981 \sum_{i=1}^{n} (D_i - D_{i-1}) \tag{1}$$

$$\rho_e = \frac{P}{0.00981 \times D} \tag{2}$$

Fig. 1 A complex liquid column

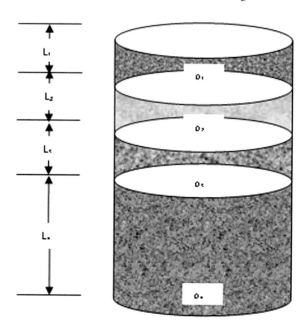

2.2 Equivalent Circulating Density (ECD)

In static conditions, it is protected to accept that the weight at a specific profundity will be a similar weight applied for optimizations on the development. Once the liquid begins pumping, it needs to defeat rubbing in the annulus and the subsequent weight is more noteworthy optimized than the static weight. The viable liquid thickness applied on the arrangement at a specific profundity is known as comparable course thickness (optimized ECD) and is ascertained as follows:

$$\text{ECD} = \frac{\rho + P}{0.00981 \times D} \tag{3}$$

where ρ is the mud weight (kg/m^3), P is the pressure drop in the annulus between depth D and surface (kPa) and D is the true vertical depth (m).

2.3 Optimization of the Dynamic Conditions

Once the boring liquid is in movement, the penetrate string is additionally in movement and henceforth, the elements of the borehole wind up hard to decide the diverse frictional misfortunes that the liquid needs to survive. These are vital to know with the end goal to optimize the ECD's while penetrating/establishing/stumbling and additionally computing the ideal size of bore string and bit spouts. Because of the differing sizes of the penetrate string in the gap, optimizing the speed of the boring liquid will fluctuate as it goes out of the bit spouts and up the annulus. It is important to optimize and have adequate mean speed with the end goal to lift the stone cuttings out of the gap. Two of these basic speeds can be figured with the accompanying optimization conditions:

$$\text{For pipes, } v = \frac{1.273 \times 10^6 q}{d^2} \tag{4}$$

$$\text{For the Annulus, } v = \frac{1.273 \times 10^6 q}{d_2^2 - d_1^2} \tag{5}$$

where v = average velocity m/min

q = flow in m^3/min
d = internal diameter of pipe mm
d^2 = internal diameter of the outer pipe or borehole mm
d^1 = external diameter of inner pipe mm

As the drilling fluid travels down the bottom hole assembly, it finally enters a restriction, smallest in the BHA and the optimized bit nozzles. It is here that the maximum fluid velocity is optimized and hence the most pressure loss of the drill string is observed. This pressure drop can be calculated with the optimization of the following equation:

$$\Delta P_b = \frac{2.48 * 10^5 \rho q^2}{\left(d_1^2 + d_2^2 + d_3^2\right)^2} \text{kpa} \tag{6}$$

where ρ is in kg/m^3, q is in m^3/min and d is the nozzle diameter in mm.

Once Δp_b is known, the nozzle velocity can be calculated using the following equation:

$$V_n = \sqrt{\frac{1996 * \Delta p_b}{\rho}} \tag{7}$$

where v_n is in m/min and ρ in kg/m^3.

2.4 Model Development Using Artificial Neural Network Framework

Conducting these experiments are not only tedious but also expensive. In case, if there is a need to carry few more experiments, where the cutting concentrations have to be validated, then the whole setup has to be assembled with the new well configuration. Therefore, it would be wise, based on the existing data, a model is developed which can be used for prediction at different operating conditions. Many researchers prefer a linear regression, as it is easily evaluated. But drilling hydraulics is not a linear process, but it is highly non-linear. Therefore, instead of a regression model, artificial neural network (ANN) approach will be the best approach, as it not only considers the inherent characteristics of the process but also optimizes the process [11–14, 25–27], as the cutting concentrations depend on flow rate, the angle of inclination, yield point, plastic viscosity and depth of the well. Therefore, these parameters are considered as process parameters in this study.

3 Results and Discussion

3.1 Study of Interaction Effects of Process Parameters

The parameters that are investigated and optimized in the present study are flow rate, the angle of inclination, yield point, plastic viscosity and depth of the well. 'The interaction effects of these process parameters are investigated systematically, and the net outcome of the process model is optimized to result in a low cost of operation. The unsteady cutting transport model is employed and optimized in the present analysis. The finite difference technique was optimized to solve the cutting transport equation numerically. The significant contribution of this study is to optimize the cutting concentrations that are accumulated and hindering at the horizontal section of the wellbore. Based on the importance of the parameters that affect the accumulation of the cuttings at the horizontal section, a sensitivity analysis was conducted at various specific hole cleaning parameters optimization such as fluid flow rate, the angle of inclination, yield point, plastic viscosity and the rate of penetration. To achieve the proper investigation and optimization stage of cutting concentration at the horizontal section, experimental or real field data were obtained to reach the required objective'. Table 1 shows the comparison between an analytical modelling approach for horizontal wellbore and optimized numerical modelling for a vertical wellbore.

3.2 Impact of Flow Rate and Angle of Inclination on the Cutting Concentration

One of the more significant key elements is particularly, understanding the parameters that control hole cleaning for vertical and horizontal wells is the well optimizations. Thus, a specific well with bore size will be obtained. Therefore, optimization models for vertical wells and horizontal wells will be achieved. The result of the cutting concentration against flowrate is shown in Fig. 2. It was observed that the relationship between each other is inversely proportional, whereas the flow rate increases the

Table 1 A comparison of numerical modelling approach vs the respective analytical modelling approach

Numerical approach	Analytical modelling approach
Hole cleaning model was assigned for vertical wellbores (material balance model)	Hole cleaning model was assigned for horizontal and deviated wellbores (material balance model), which does not depend on the location
Parameters like wellbore constituent and individual cutting concentration were numerically calculated	An empirical model was developed, once the experimental runs were run to predict the required fluid velocity to move the cuttings, clean the hole and eventually will have good hydraulics
After that the matrix of equations should be formed, then equations should be ready for a numerical solution and finally can be solved to find the cuttings concentration at each node or element	An experimental study was focused on the minimum transport velocity that is really needed to carry the cuttings or fractures out of the wellbores
The cutting concentration was estimated using the finite difference method.	The minimum transport velocity was calculated using an equation of $V_{\min} = V_{\text{cut}} + V_{\text{slip}}$

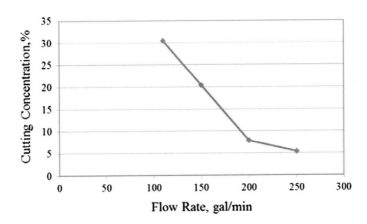

Fig. 2 Variation of cuttings concentration versus flowrate

cutting concentration decreases, therefore, it has to be optimized. The effects of the angle of inclination on the cuttings concentration are shown in Fig. 3. It is apparent that initially, the concentration of cuttings is quite low at 20°, then it optimized to the state as the angle of inclination gets higher due to the nearly horizontal section or bedding section and the issues at the throat. The yield point is one of the more important physical property of mud, where carrying capacity must exist and thereafter should be optimized. It was observed that the cutting concentration increases as the yield point decreases and this is another sign of optimization. The yield point is a kind of force to carry or lift the cuttings from the wellbore to the surface. It was also observed that the cutting concentration increased with an increase in plastic viscosity, which is in direct relation to the yield point.

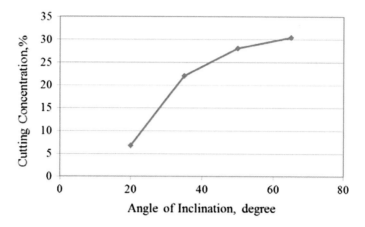

Fig. 3 Variation of cuttings concentration versus flowrate

3.3 Model Development Using ANN

As the cutting concentration depends on the five processes, namely, flow rate, the angle of inclination, yield point, plastic viscosity and depth of the well, so using the neural network architecture in ANN, a data-driven model is developed. So in this ANN framework, the five process parameters are the inputs to the model and the only output from this model is cutting concentration. Since the performance and efficiency of ANN architecture depend on the number of hidden layers, therefore, the efficacy of this model is tested for a different number of neurons. The variation of the correlation coefficient (R^2) and mean squared error (MSE) with respect to a number of neurons in hidden layer in ANN architecture is shown in Fig. 4. Since seven number of neurons in hidden layer results in higher R^2 and lower MSE, therefore the topology of the neural network architecture consider further in this study is 5-7-4 as shown in Fig. 5. The optimum weights and biases for the ANN model for 5-7-1 topology are given in Table 2.

The experimentally obtained data is segregated as a training set, testing set and validating set with 70%, 15% and 15%, respectively. This segregation is done randomly so that there is no bias in model and uncertainty in the process is distributed randomly. Using the optimal 5-7-1 topology obtained, systematically the model is obtained by training using the training set data. It was observed that the trained model resulted in R^2 of 0.9688. Since any value higher than 0.85 is considered good, therefore, this model is used to test its performance. So the model is tested using the testing set data. Interestingly, the model was able to give R^2 of 0.9106 which is a good agreement. With the confidence in the model performance on the testing data, this model is again validated on the validation data. Again, remarkably, the identified model was able to give R^2 of 0.92787. When this model is tested for the whole data

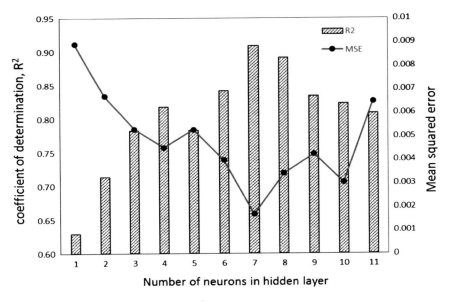

Fig. 4 Variation of correlation coefficient R^2 and mean squared error (MSE) with respect to number of neurons in hidden layer in ANN architecture

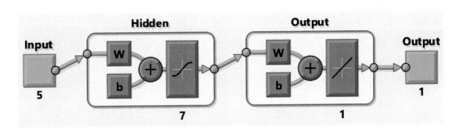

Fig. 5 Topology of the neural network architecture in ANN

Table 2 Optimum weights and biases for ANN model for 5-7-1 topology

Input (5)-Hidden layer (7)				Hidden layer (7)-Output (1)			
Weights			Bias	Weights			Bias
−0.443	−0.875	0.929	1.388	−0.026	−0.301	−1.092	−0.301
1.388	0.625	−0.714	−1.321	0.966	0.281	−0.501	
0.525	0.923	0.954	−0.274	−0.271	−0.927	0.063	
−0.158	1.388	0.224	−1.388	−0.480	−0.301	−0.497	
0.666	1.016	0.267	1.101	1.388	−1.388	0.141	
−0.408	−0.771	−1.101	−1.183	0.582	−1.388	−0.236	

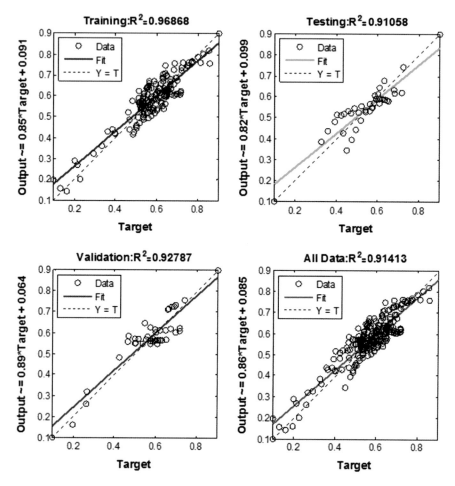

Fig. 6 Plot showing the performance of ANN for training data, testing data, validation data and all data

(training set + testing set + validating set), it was observed that the model was very successful and resulted in $R^2 = 0.91413$. The performance of ANN topology for various scenarios is shown in Fig. 6. These results thus confirm the efficacy of the 5-7-1 ANN topology.

3.4 Optimization of the Cutting Concentration

The numerical result of the cuttings concentrations against depth at a different rate of penetration indicates that they are correlated. It is evident that the optimized rate of penetration increases, as well as the cutting concentration and it, is not a

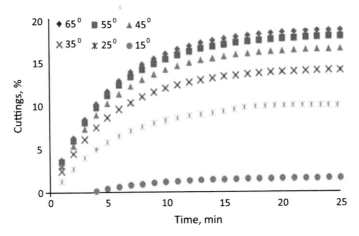

Fig. 7 Optimization of the angle of inclination on the cutting concentrations

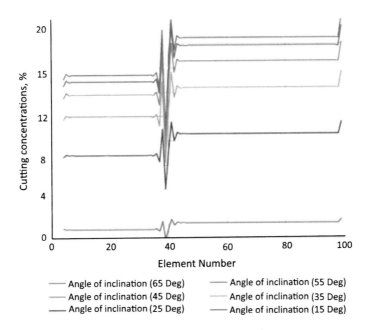

Fig. 8 Optimization of element number on the cutting concentrations

depth-dependent. It was observed that after reaching a peak of 0.2160% cutting concentration could not be optimized further, the cutting concentration rate declines slowly to 23 min then becomes steady till the end.

The variation of cutting concentrations against time at various angles of inclinations is shown in Fig. 7. It is visible that cuttings concentrations are optimized and the angle of inclination increases with time. Figure 8 presents the angle of inclina-

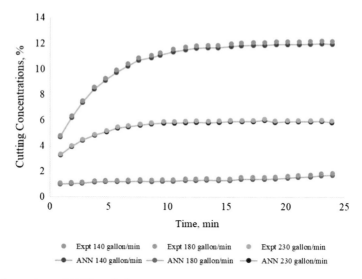

Fig. 9 Comparison of ANN predictions versus experimental values

tion against cutting concentration; it is noted that at the tangent section, the optimum angle required for rotary drilling to best maintain both the angle and the direction is approximately 55°. However, this is a difficult angle to keep the drilling bit.

Based on the optimized values, given by the ANN model, the experimental runs are carried for the same operating conditions. A comparison of ANN model predictions against experimental values is shown in Fig. 9. The model predictions are in good agreement with the experimental runs, and therefore these results thus indicate that the optimized relationship will present the inherent mechanisms of the hole depth and drilling parameters.

4 Conclusion

This study focuses on optimization of drilling rig hydraulics in drilling operations using soft computing techniques. The conclusions drawn from the study are as follows:

i.   The assurance of the correct piece spout sizes is one of the more regular utilizations of the frictional weight misfortune condition in boring. Choosing the correct spouts can result in a critical increment in the rate of infiltration. Also, the streaming activity of spouts helps in cleaning the opening of the cuttings empowering the bit to nibble new development inevitably. In delicate arrangements, the equivalent streaming activity assumes an essential job in breaking the development, so the bit needs to do less as the new opening is made. A considerable measure of variables becomes possibly the most important factor when endeavouring to choose the streamlining of the bit planes.

ii. The numerical results show that at higher flow rates cutting concentration decreases. As the rate of penetration increases the cutting concentration increases as well. Cuttings concentration decreases with an increase in the flow rate of the drilling mud. The present optimized model can be extended to include the bed cutting concentration with the effect of other parameters, such as three-dimensional modelling, the pressure of the drilling mud, drilling string rotation among others which need more and more investigation.

References

1. Busahmin, B., Saed, N.H., Alusta, G., Zahran, M.M.: Review on hole cleaning for horizontal Wells. ARPN J. Eng. Appl. Sci. **12**(16), 4697–4708 (2017)
2. Hoelscher, K.P., Stefano G., Riley, M., Young S.: Application of nanotechnology in drilling fluids. In: SPE international oilfield nanotechnology conference and exhibition. Society of Petroleum Engineers (2012)
3. Rao, K.R., Rao, D.P., Venkateswarlu, C.: Soft sensor based nonlinear control of a chaotic reactor. IFAC Proc. Volumes **42**(19), 537–543 (2009)
4. Abdrazakov, S. et al.: From seismic to ROP and $/m for exploration Wells. In: 45th US rock mechanics/geomechanics symposium (2011)
5. Lanka, S., Radha, M., Abusahmin, B.S., Puvvada, N.: Predictive data mining techniques for management of high dimensional big-data. J. Ind. Pollut. Control **33**, 1430–1436 (2017)
6. Lingamdinne, L.P., et al.: Process optimization and adsorption modeling of Pb(II) on nickel ferrite-reduced graphene oxide nano-composite. J. Mol. Liq. **250**, 202–211 (2018)
7. Gobinath, R., et al.: Characterization of iron nano particles (Fe_2O_3) synthesized through coprecipitation and sol-gel methods. Clay Res. **34**(2), 59–65 (2015)
8. Karri, R.R., Sahu, J.N., Jayakumar, N.S.: Optimal isotherm parameters for phenol adsorption from aqueous solutions onto coconut shell based activated carbon: error analysis of linear and non-linear methods. J. Taiwan Inst. Chem. Eng. **80**, 472–487 (2017)
9. Murthi, P., Awoyera, P.O., Palanisamy S., Dharsana D., Gobinath R.: Using silica mineral waste as aggregate in a green high strength concrete: workability, strength, failure mode, and morphology assessment. Aust. J. Civil Eng. 1–7 (2018)

10. Anandaraj, S., Rooby J., Awoyera, P.O., Gobinath R.: Structural distress in glass fibre-reinforced concrete under loading and exposure to aggressive environments. Constr. Build. Mater. (2018)

11. Lingamdinne, L.P., et al.: Multivariate modeling via artificial neural network applied to enhance methylene blue sorption using graphene-like carbon material prepared from edible sugar. J. Mol. Liq. **265**, 416–427 (2018)

12. Karri, R.R., et al.: Optimization and modeling of methyl orange adsorption onto polyaniline nano-adsorbent through response surface methodology and differential evolution embedded neural network. J. Environ. Manage. **223**, 517–529 (2018)

13. Karri, R.R., Sahu, J.N.: Process optimization and adsorption modeling using activated carbon derived from palm oil kernel shell for Zn(II) disposal from the aqueous environment using differential evolution embedded neural network. J. Mol. Liq. **265**, 592–602 (2018)

14. Karri, R.R., Sahu, J.N.: Modeling and optimization by particle swarm embedded neural network for adsorption of zinc (II) by palm kernel shell based activated carbon from aqueous environment. J. Environ. Manage. **206**, 178–191 (2018)

15. Gopalakrishnan, V., Suji, D., Gobinath, R.: Comparative studies on advanced wastewater pre treatment system for textile dyeing industries: a case study of three plants in Tirupur District, Tamilnadu. Pollut. Res. **34**(1), 179–186 (2015)

16. Karri, R.R., Jayakumar, N.S., Sahu, J.N.: Modelling of fluidised-bed reactor by differential evolution optimization for phenol removal using coconut shells based activated carbon. J. Mol. Liq. **231**, 249–262 (2017)

17. Saeid, N.H., Abusahmin, B.S.: Transient cooling of a cylinder in cross flow bounded by an adiabatic wall. ASEAN J. Chem. Eng. **17**(2), 17–26 (2017)

18. Busahmin, B., et al.: Studies on the stability of the foamy oil in developing heavy oil reservoirs. Defect Diffus. Forum **371**, 111–116 (2016)

19. Busahmin, B., Maini, B.: A potential parameter for a non-darcy form of two-phase flow behaviour, compressibility related. Int. J. Eng. Technol. (UAE) **7**(3), 126–131 (2018)

20. Abusahmin, B.S., Karri, R.R., Maini, B.B.: Influence of fluid and operating parameters on the recovery factors and gas oil ratio in high viscous reservoirs under foamy solution gas drive. Fuel **197**, 497–517 (2017)

21. Busahmin, B.S., Maini, B.B.: Effect of solution-gas-oil-ratio on performance of solution gas drive in foamy heavy oil systems (2010)

22. Maini, B.B., Busahmin, B.: Foamy oil flow and its role in heavy oil production. In: AIP conference proceedings (2010)

23. Tavallali, M. et al.: Assessment of SAGD Well configuration optimization in Lloydminster heavy oil reserve. In: Society of petroleum engineers—SPE/EAGE European unconventional resources conference and exhibition (2012)

24. Elmabrouk, S.K., Mahmud, W.M., Sbiga, H.M.: Calculation of EUR form oil and water production data. In: Proceedings of the international conference on industrial engineering and operations management (2018)

25. Rao, K.R., Srinivasan, T., Venkateswarlu, C.: Mathematical and kinetic modeling of biofilm reactor based on ant colony optimization. Process Biochem. **45**(6), 961–972 (2010)

26. Awoyera, P.O.: Mechanical and microstructural characterization of ceramic-laterized concrete composite. PhD Thesis (2018). Covenant University, Ota, Nigeria

27. Awoyera, P.O.: Predictive models for determination of compressive and split-tensile strengths of steel slag aggregate concrete. Mater. Res. Innovations **22**, 287–293 (2018)

Renewable Energy Harnessing by Implementing a Three-Phase Multilevel Inverter with Fuzzy Controller

K. Muralikumar and Ponnambalam Pathipooranam

Abstract This paper presents a multilevel topology implemented to overcome losses, possessed with the aid of the current renewable energy harnessing techniques, thereby making it more dependable; and the efficient output strength goes with the flow to the grid. Multilevel inverters are used as they offer an extraordinary method in power digital interfacing. They additionally eliminate the cause of the transformer in the tool, thereby disposing off several transformer losses and the location of the setup. This can increase the efficiency of the output strength generated to the grid. The device implements a dc-to-dc enhanced converter to improve the output dc energy acquired from the renewable supply delivered. It additionally implements a flying capacitor which removes voltage spikes generated. This inverter also can be introduced properly into a PV cellular for additional extension. The whole harmonic distortion is measured, and the simulation outcomes are acquired additionally using fuzzy controller by way of a closed-loop system.

Keywords Modified H-bridge (MHB) · Cascaded H-bridge (CHB) · Total harmonic distortion (THD) · Multilevel inverters (MLI) · Maximum power point tracking (MPPT) · Nearest level control (NLC)

1 Introduction

Inside the winning scenario, because of the herbal troubles and constrained fossil stays, the hobby on sustainable energy sources is increasing. To attend to this, growing name for wind turbines and PV power gadgets have modified into the vital essential piece of network-associated sustainable energy supply. Exploiting of electric energy after the PV frameworks presents smooth energy age. This dedication

K. Muralikumar (✉) · P. Pathipooranam
School of Electrical Engineering, VIT, Vellore 632014, Tamilnadu, India
e-mail: kolamuralikumar@gmail.com

P. Pathipooranam
e-mail: p.ponnambalam@gmail.com

© Springer Nature Singapore Pte Ltd. 2020
K. N. Das et al. (eds.), *Soft Computing for Problem Solving*,
Advances in Intelligent Systems and Computing 1048,
https://doi.org/10.1007/978-981-15-0035-0_70

finished it countlessly in the gift day worldwide climatic situations. Extended endura-
tion, excessive efficiency, and infection unrestricted electricity age are the advantages
of PV frameworks [1, 2]. The low-yield voltage PV modules are associated in preparations to gain excessive voltage dc, and the inverter that is implemented is used to
interface with the network. The system goals immoderate rated voltage devices to
the inverter. Consequently, to remedy this trouble, step-up transformer comes into
effect to overcome the trouble. This system empowers to utilize the inverter with
low-voltage devices and later the transformer boosts the voltage. This may lead to
increase in losses and cost of the system. The usage of transformer-less thoughts
allows us to decrease the scale, price, weight, and masses besides functions of the
inverter aside from enhancing performance. The multilevel inverter legitimately utilizes electricity semiconductor gadgets to generate a staircase sinusoidal waveform
from a multiple lower degree dc voltage waveform. The small yield voltage results
in great output voltage, reduced [3–5].

The transformer-less PV system with multilevel inverter is shown in Fig. 1. This
method dreams excessive rated voltage device to the inverter. Consequently, in the
direction of the solution to this problem, the step-up transformer comes into effect
to overcome the problem. This gadget empowers to make use of the inverter with
low-voltage devices and afterward the transformer boosts the voltage. Eventually, it
increases cost and generates more losses. The ides of making use of the transformer

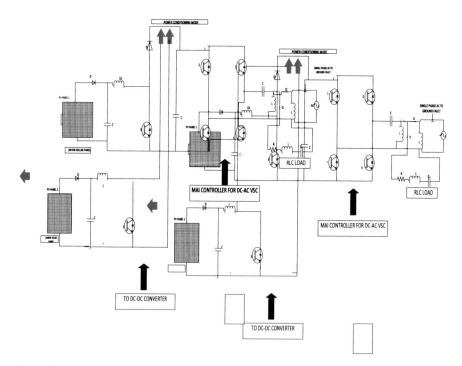

Fig. 1 Transformer-less PV system equipped with MLI

a lot much lesser enables us to cutting down the scale, price, weight, and many-sided traits of the inverter apart from improving the efficiency. The multilevel inverter legitimately makes use of energy semiconductor devices to generate a staircase sinusoidal waveform from a more than one decrease degree dc voltage waveform. The minor yield voltage effects in brilliant output voltage decreased "voltage stress" on switching devices, thereby reducing switching losses and enhancing the enactment. The series H-bridge is supposed regularly. The diode "secured convertor" is used as a collection of collecting capacitors. A later enhancement unfolds the at the wing capacitor style in which the capacitors had been variable rather than series-linked. An assorted shape consists of parallel connotation of electrical converter ranges over inter-phase gadgets. In this fragment, the semiconductors slab the substantial-ranging dc voltage, however, they surpluse the weight current. Numerous combination flairs have added steadily emerged a few with the uncomplicated topologies [2]. These graces may additionally have higher power brilliant for a given choice of semi-conductor gadgets than the uncomplicated topologies. The first MLI changed into applied at some point of the 12 months of 1981 and went directly to enforce Cascaded H-bridge (CHB) in 1995. These favorable situations superior the MLI due to the fact the reasonable choice for grid associated sustainable power deliver in each with and without transformer preparations. The implementation of MLIs results in a reduction of cost and size of filtering requirements in transformer less PV system. Inside the present multilevel configuration, the main demerit is the need for excessive amount of power switches. The reduction of power semiconductor switches in the MLI topologies has results to enormous significance in academia and the industry. The reduction of switches moreover reduces the sort of segments as well as gate drives. In a lengthy time period, the lower in switches balances the performance and reduce the control complexity. The MLI topologies, 1-phase and 3-phase are concentrating takes place reducing the range of switches. Such on its very own diploma topologies are an awful lot much less complex however spreading those to some-degree makes the design and control complex. The losses created due to switching had been decreased by imposing the approach of hybrid management. This prompted the development of hybrid three-phase symmetrical and asymmetrical arrangements.

In this paper, the structure of CHB is exhibited, wherein the development is saved smoothly, a variety of switches are reduced and now there are no longer any capacitor balancing problems. Each element is superior from an MHB. A PV cellular also may be proposed for this MLI circuit for additional extension. The choice of this topology is due to easy manufacturing, use of a smaller variety of electricity digital switches and less necessity of gate driving force circuits.

2 Modular H-Bridge Model

The predicted cascaded hybrid module "MLI" is created after an "MHB" module and "T-type three-leg inverter shape" (TTL).

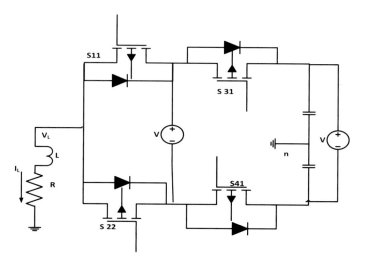

Fig. 2 MHB module of the proposed MLI

2.1 MHB Module

The switching analysis, we ought to definitely discover that transfer pairs in together the arms (S11, S21) and (S31, S41) is complimentary in nature. The module can produce four stages of voltage with proper combination of switches. The levels of voltage are (3 V/2, V/2, −V/2, −3 V/2). Normally, every MHB issue has three sources to yield the output. It is not that powerful to layout an MLI with more sources. With the give up intention to lessen one source, capacitors and one dc supply are employed. The voltage balance among two capacitors is ensured through a charge balance control method. To exploit this manipulating strategy, the switching statuses need to have a recurrence. The proposed MHB module has no reiteration within the switching states in a single module, and through the morals of its shape, unvarying balancing of capacitors is empowered and the MHB module is as shown in Fig. 2.

2.2 TTL Inverter

The TTL inverter diagram is shown in Fig. 3. The "SPWM converter" is the maximum considerable "multilevel converter" used in the nontraditional electricity form for its small swapping loss and compact device voltage strain. However, it suffers from immoderate conduction losses. The SPWM is investigated in the works to overcome the weaknesses of I type inverter. But, the high component and low facet power gadgets of SPWM want to block the entire dc hyperlink voltage. The switching frequency is expressive. The "TTL inverter" assembly consists of six unidirectional switches namely (Ta1, Ta2), (Tb, Tb2), and w22 (Tc1, Tc2) in their stages a, b, and c, respectively.

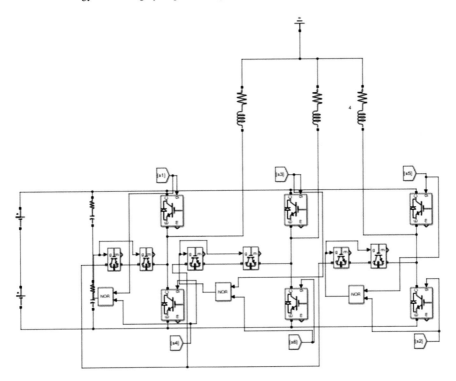

Fig. 3 Inverter shape of TTL

It is mathematically expressed as

$$\int_0^{\theta_1} Il(t)Vl(t)\mathrm{dwt} = \int_0^{\theta_2} Il(t)Vl(t)\mathrm{dwt}$$

$$\int_{\theta_2}^{\pi} Il(t)Vl(t)\mathrm{dwt} = \int_{\theta_4}^{2\pi} Il(t)Vl(t)\mathrm{dwt}$$

The TTL inverter shape is included through three bidirectional switches: P, Q, and R which are prepared by means of again to back arrangement of switches. This inverter can yield three voltage grades (zero, $+V/2$ and $-V/2$) in the respective section. The dc-interface voltage is given with the aid of typically practical capacitors C1 and C2 associated over one deliver V_t.

2.3 Hybrid Modular Inverter Structure

The combination of two structures "MHB" and "TTL" inverters produces Hybrid modular "MLI" is as shown in Fig. 4. This inverter shape can yield nine voltage

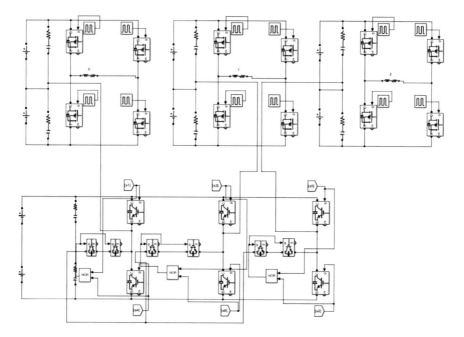

Fig. 4 Cascaded hybrid model

stages in individual tiers. The voltage tiers are +2 V, +3 V/2, +V, +V/2, zero, −
V/2, −V, −3 V/2, and −2 V within the proportioned mode. The switching table for
the two phases is equal, but their switching instants are 120 degree and 240 degree
one after the different from the phase "A". This topology can be clearly prolonged
to any diploma independent through cascading the "MHB modules".

3 Operational Modes

In constituent of this deliberate hybrid dropped modular "MLI" is its capability to
work in both symmetrical and asymmetrical modes. In this component, systematic
inquiry of diverse constrictions is conceded out. A three-phase L-level HCM multilevel inverter structure is shown in Fig. 5.

3.1 Symmetrical Mode

In this grace, the quantity of the "dc voltage" resources in every "MHB module" and
that of "TTL inverter" are set at the same price.

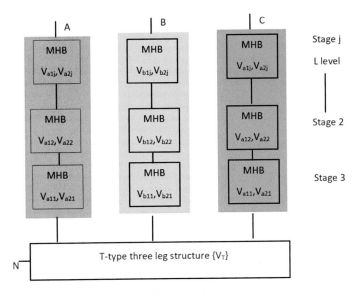

Fig. 5 Proposed three-phase L-level HCM MLI structure

$$V_{a11} = V_{a12} = \ldots\ldots\ldots = V_{a1j} = V$$
$$V_{a21} = V_{a22} = \ldots\ldots\ldots = V_{a2j} = V$$
$$V_{b11} = V_{b12} = \ldots\ldots\ldots = V_{b1j} = V$$
$$V_{b21} = V_{b22} = \ldots\ldots\ldots = V_{b2j} = V$$
$$V_{c11} = V_{c12} = \ldots\ldots\ldots = V_{c1j} = V$$
$$V_{c21} = V_{c22} = \ldots\ldots\ldots = V_{c2j} = V$$
$$V_t = V$$

The numeral of stages "L" engendered in line with "phase voltage" is expressed as

$$L = 6j + 3$$

The intense value of "phase voltage" $V_{A, B, C}$ attained in this mode is

$$V_{A,B,C,\text{MAX}} = \frac{(1+3j)V}{2}$$
$$S = 12(1 + j)$$

The gate drive responsibility (G) is expressed as

$$G = S - 3 = 9 + 12j$$

This voltage disruptive functionality (VBC) is the voltage that appears across the transfer as soon because the device is grew to become off country. It may be said as follows for 3-section symmetrical movement.

In the "TTL inverter" construction,

$$T_{A,B,C,1} = T_{A,B,C,2} = V_P = Q = R = 2V$$
$$V_{BCTTL} = 9V$$

3.2 Asymmetrical Mode

Using numerous dc voltages via suitable ratios can increase the "output voltage primary harmonic distortion (THD)" and from then on the power is remarkable. In this approach, the magnitudes of the dc voltage reason in each MHB module and that of "TTL inverter are regular at numerous values". This increases the volume rely "L" with the equal transfer "S" and gate pressure "G" depends. Case 1: In this example, the importance of the dc voltage motives in every "MHB module" is identical at price and that of "TTL inverter" is about double the price. Equations are comparable as that of a symmetrical gadget. The other vocabularies are modified as

$$V_t = 2V_L = (4j + 1)$$
$$V_{A,B,C,}\text{max} = \frac{(2 + 3j)V}{2}$$
$$V_{BCTTL} = 18V$$
$$V_{B,C} = 18(1 + j)V$$

4 Control System Strategies

The series hybrid sectional "MLI" designed for the "PV nonconventional power" source measures the use of this approach on the age side and on the inverter component. The thoughts of those mechanism structures are discussed in this area. Figure 6 shows the four-level MHB circuit connected with PV nonconventional energy source and Fig. 7 shows the nine-level PV renewable energy source connected HCM MLI.

4.1 Sinusoidal PWM

Industries will use this approach. A sinusoidal orientation sign and a triangular provider wave can be related, and a generated gating signal may be obtained. The size of every pulse is modified proportionally to the breadth of a sine wave expected in the center of the equal pulse; "it is a famous control method" frequently applied

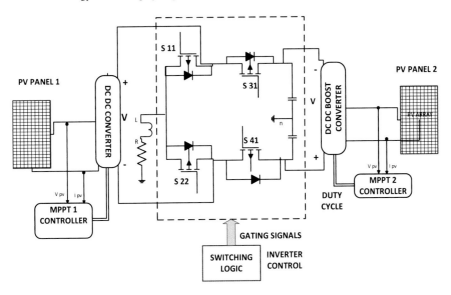

Fig. 6 Four-level MHB circuit connected with PV nonconventional energy source

in an "energy electronics inverter circuit". It is by far having many advantages like conduction losses, fewer disturbances in the output, and ease of technique application.

5 Fuzzy Logic Controller

Block diagram of a fuzzy logic controller is shown in Fig. 8. Fuzzy common sense is a shape of many-valued good judgment; it deals with reasoning that is approximate as opposed to fixed and genuine ones. In contrast with the conventional logic theory, in which binary units have two-valued good judgment, authentic or fake, fuzzy common-sense variables can also have a truth value points to be consider [4, 6–9].

Fuzzy good judgment has long been addressing the concept of partial reality, where the fact price can also vary between absolutely real and absolutely false. Moreover, while linguistic variables are used, those stages can be controlled through unique functions. It started with the 1965 concept of fuzzy set theory via Lotfi Zadeh. Fuzzy logic has been applied to many fields, from control idea to synthetic intelligence.

5.1 Stages of Reality

Fuzzy good judgment and probabilistic common sense are mathematically similar; each have fact values ranging between 0 and 1—but are conceptually wonderful because of special interpretations; see interpretations of opportunity principle. Fuzzy

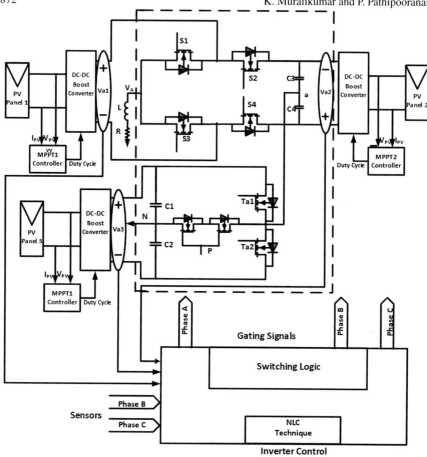

Fig. 7 Nine-level PV renewable energy source connected HCM MLI

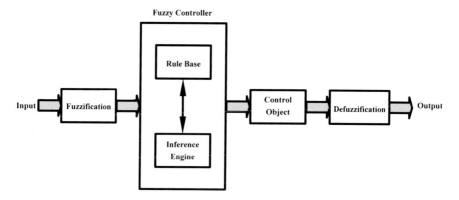

Fig. 8 Block diagram of fuzzy logic controller (FLC)

good judgment corresponds to "stages of fact", while probabilistic good judgment corresponds to "chance, chance"; as these differ, fuzzy common sense and the same actual-global conditions.

6 Simulation Results

6.1 Sinusoidal PWM Input and Output

The Sinusoidal PWM input voltage and Sinusoidal PWM output voltage waveforms are shown in Figs. 9 and 10.

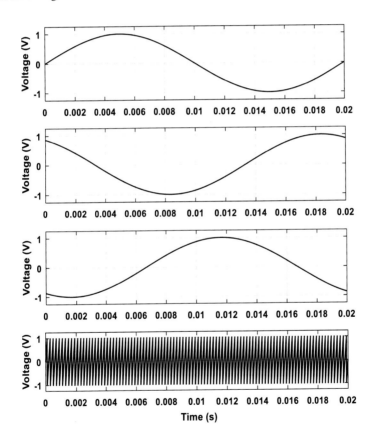

Fig. 9 Sinusoidal PWM input voltage

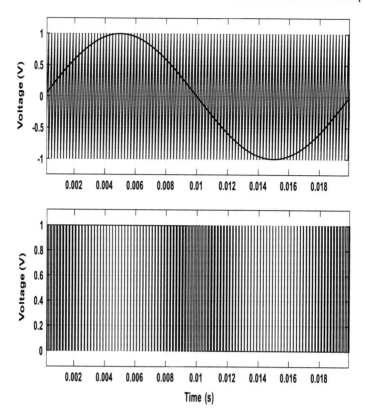

Fig. 10 Sinusoidal PWM output voltage

6.2 Three-Phase MHB MLI Output Voltage and Current

The Three-phase MHB MLI output voltage and three phase MHB MLI output current waveforms are shown in Figs. 11 and 12.

6.3 Three-Phase MHB MLI Output Voltage and RMS Voltage with Fuzzy Logic Controller

The Fuzzy controlled MHB MLI output voltage and Fuzzy controlled MHB MLI RMS output voltage are shown in Figs. 13 and 14.

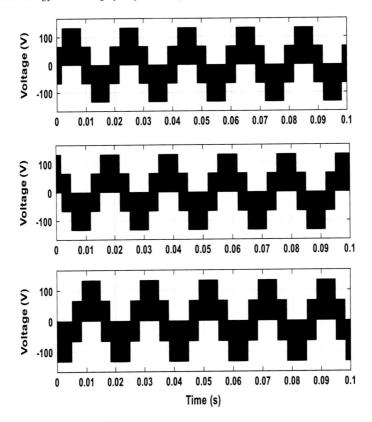

Fig. 11 Three-phase MHB MLI output voltage

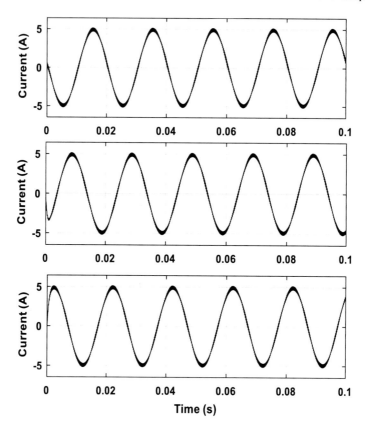

Fig. 12 Three phase MHB MLI output current

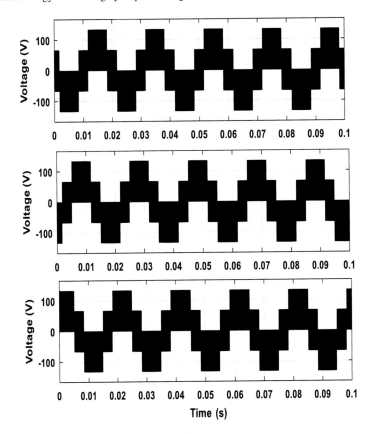

Fig. 13 Fuzzy controlled MHB MLI output voltage

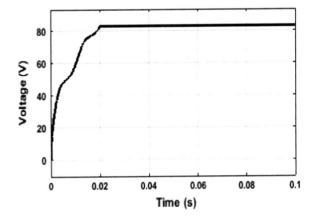

Fig. 14 Fuzzy controlled MHB MLI RMS output voltage

7 Conclusion

In this paper, a "three-segment hybrid cascaded model" of multilevel inverter topology is hooked from the MHB module. The improvement mainly concentrates on the evaluation of cascaded H-bridge multilevel inverter with THD analysis. For this reason, we are growing the number of tiers with a low THD and bases on the output without calculating any issues to the energy circuit. The main space of this topology is to condense the whole Harmonic Distortion (THD), lesser electromagnetic interference technology, and accomplishment of excessive output voltage.

References

1. Hasan, M.M., Abu-Siada, A., Islam, S.M., Dahidah, M.S.: A new cascaded multilevel inverter topology with galvanic isolation. IEEE Trans. Ind. Appl. **54**(4), 3463–3472 (2018)
2. Rodriguez, J., Lai, J.-S., Peng, F.Z.: Multilevel inverters: a survey of topologies, controls, and applications. IEEE Trans. Ind. Electr. **49**(4), 724–738 (2002)
3. Franquelo, L.G., Rodriguez, J., Leon, J.I., Kouro, S., Portillo, R., Prats, M.A.: The age of multilevel converters arrives. IEEE Ind. Electron. Mag. **2**(2), 28–39 (2008)
4. Muralikumar, K.: Analysis of fuzzy controller for H-bridge flying capacitor multilevel converter. In: Proceedings of sixth international conference on soft computing for problem solving: SocProS 2016, vol. **1**, p. 307. Springer (2017)
5. Calais, M., Agelidis, V.G.: Multilevel converters for single-phase grid connected photovoltaic systems-an overview. In: Industrial electronics, 1998. Proceedings. ISIE'98. IEEE international symposium on. Vol. 1. IEEE (1998)
6. Kang, F.-S. et al.: Multilevel PWM inverters suitable for the use of stand-alone photovoltaic power systems. IEEE Trans. Energy Convers. **20**(4), 906–915 (2005)
7. Reddy, V.P., Muralikumar, K., Ponnambalam, P., Mahapatra, A.: Asymmetric 15-level multilevel inverter with fuzzy controller using super imposed carrier PWM. In: Power and advanced computing technologies (i-PACT), 2017 innovations in, pp. 1–6. IEEE (2017)
8. Viswanath, Y., Muralikumar, K., Ponnambalam, P., Kumar, M.P.: Symmetrical cascaded switched-diode multilevel inverter with fuzzy controller. In: Soft computing for problem solving 2019 (pp. 121–137). Springer, Singapore (2019)
9. Ponnambalam, P., Muralikumar, K., Vasundhara, P., Sreejith, S., Challa, B.: Fuzzy controlled switched capacitor boost inverter. Energy Procedia. **1**(117), 909–916 (2017)

Design of SVPWM-Based Two-Leg VSI for Solar PV Grid-Connected Systems

CH Hussaian Basha, V. Govinda Chowdary, C. Rani, R. M. Brisilla and S. Odofin

Abstract In this work, a four-switch Voltage Source Inverter (VSI) is considered for highly efficient and low power solar PV grid-connected applications to optimize the cost and size of the PV system. The Perturb and Observe (P&O) Maximum Power Point Tracking (MPPT) technique is used to track Maximum Power Point (MPP) of solar PV. This technique is simple, easy to design, and less complexity. By using two-leg four-switch inverter (B-4 inverter) the cost of the PV system can be reduced compared to six switch inverters, as the cost of inverter mainly depends on the cost of semiconductor switches. The boost converter is utilized to step-up the PV voltage. This work is to analyze the Space Vector Pulse Width Modulation Technique (SVPWM) in two-leg B-4 inverter topology to reduce the ripples at time of switching thereby reducing the Total Harmonic Distortion (THD) and reducing the inverter switching and conducting losses at high pulse width modulation frequency. Moreover, SVPWM technique improves the utilization factor of B-4 inverter. The results are analyzed by using MATLAB Simulink window.

Keywords B-4 inverter · Design of PV panel · P&O MPPT technique · SVPWM
. generation

CH Hussaian Basha · V. Govinda Chowdary · C. Rani (✉) · R. M. Brisilla
School of Electrical Engineering, VIT University, Vellore, India
e-mail: crani@vit.ac.in

CH Hussaian Basha
e-mail: hussaianbasha.ch@vit.ac.in

V. Govinda Chowdary
e-mail: govindachowdaryvankayalapati@gmail.com

R. M. Brisilla
e-mail: brisilla.rm@vit.ac.in

S. Odofin
School of Energy and Environment, University of Derby, Derby, UK
e-mail: s.odofin@derby.ac.uk

© Springer Nature Singapore Pte Ltd. 2020
K. N. Das et al. (eds.), *Soft Computing for Problem Solving*,
Advances in Intelligent Systems and Computing 1048,
https://doi.org/10.1007/978-981-15-0035-0_71

1 Introduction

In India, the government funding agencies motivates the people to install solar power plant by giving subsidies. Solar energy is an environmental free less maintenance power generation compared to other renewable energy sources. But it is having a drawback of high installation cost. To overcome this drawback many researchers are developing different advanced MPPT technologies and interfacing DC–DC converters [1]. Moreover, the recent focus of the power system researchers is using different convertor topologies to improve the use of solar energy. To achieve the maximum power through a boost and VSI switching control, P&O MPPT technique is employed in this work due to its low cost, high accuracy, and flexibility [2].

The converters are classified based on its application as isolated and non-isolated DC–DC converters. The non-isolated converters are used for low and medium power applications while the isolated converters are used for high power applications. In this work, the PV voltage is improved by a non-isolated boost convertor of high efficiency, less design cost, and minimum use of passive elements. This will enhance the power factor of the grid thereby matching the impedance between the (photovoltaic) PV and the grid [3]. The B-4 inverter is used to control the output voltage of the boost converter and maintaining constant voltage magnitude and frequency at the grid and it gives square or quasi square wave out waveform [4]. This paper optimizes the power circuit of PV system by employing four B-4 topologies other than the conventional six switches three-leg topology, by directly connecting one of the three terminals of grid to the midpoint of the DC-link capacitors. This concept, as shown in Fig. 1, reduces the THD in the system, by minimizing number of switches. The overall switching losses and the size of the B-4 inverter circuit are reduced [5, 6].

The B-4 topology reduces the peak voltage of the system to $V_{dc}/3.464$, which improves the utilization factor of the inverter circuit. The flow of the phase current through the neutral point of the DC-link capacitors will fluctuate the phase voltages of the inverter under steady state. The use of nonidentical DC-link capacitors causes overmodulation of the pulse width modulation process. This topology can

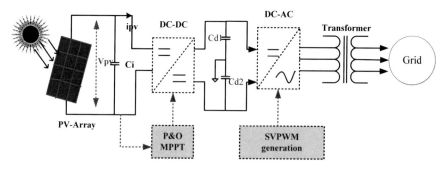

Fig. 1 Block diagram of solar PV fed NPC-VSI

be employed for both single-phase and three-phase, symmetrical as well as asymmetrical grid power application [4, 7]. The balancing of neutral point is done by using PWM technique. Most commonly used PWM technique is Sine Pulse Width Modulation Technique (SPWM). In this technique a sine wave is compared with triangular wave and its resultant pulse generation is given to switching devices in B-4 inverter [8]. But it is having drawback of high switching expensive because of its low turn ON and OFF time duration. Due to its high PWM switching frequency, the B-4 inverter switching conduction losses are high and hence output voltage of the inverter is reduced.

To overcome the drawbacks of the basic PWM techniques, a Space Vector Pulse Width Modulation (SVPWM) technique is proposed in this work. This technique is mainly depending on d, q transformation theory and it gives a high modulation index and fast switching and conduction duration calculation of inverter switches. Moreover, this PWM technique gives 15% higher inverter output voltages compared to SPWM [8, 9]. In SVPWM-based B-4 inverter, the difference of the instantaneous current and the voltages of half-link DC capacitors, measured using sensors, are considered.

The paper is composed as follows, Sect. 2 glances to the equivalent circuit of solar PV cell and the design of boost converter is highlighted in Sect. 3. The two-leg B-4 inverter design is analyzed in Sect. 4, and the results are discussed in Sect. 5.

2 Single Diode Equivalent of Solar PV

The current source connected through a diode using series and parallel resistance gives the equivalent model of the PV cell, as shown in Fig. 2. This panel is designed for B-4 inverter fed PV grid-connected application. The shunt resistance (R_{sh}) and current (I_{out}) are generated by the use photovoltaic effect. The combination of shunt (R_{sh}) and series resistances (R_{se}) gives that the PV panels are manufactured by the series and shunt combination of solar cells [1].

$$I_{out} = I_{PV\_Cell} - I_d \tag{1}$$

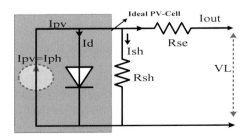

Fig. 2 Single diode equivalent circuit

Table 1 Solar panel measured parameters

Parameters	Values
Maximum wattage (W_p)	100 W
Maximum phase current (I_{Ph})	43.55 A
Phase voltage (V_{ph})	23.50 V
Irradiances (I_i)	1000 W/m^2
Reverse saturation current (I_0)	$2.077 * e^{-0.005A}$
Boltzmann constant	$1.3805 * e^{-23}$

$$I_d = I_{0\_Cell}\left(e^{\left(\frac{qV}{\eta KT}\right)} - 1\right) \tag{2}$$

By substituting Eq. (2) in (1), the PV cell current is derived as

$$I_{out} = I_{PV\_Cell} - I_{0\_Cell}\left(e^{\left(\frac{qV}{\eta KT}\right)} - 1\right) \tag{3}$$

where I_d is the Shockley diode current, I_{PV_Cell} is a PV cell current, q is an electrical charge ($1.602 * 10^{-19}$ C), and T is a temperature of the p–n diode junction temperature.

$$I_{out} = I_{PV\_Cell} - I_{0\_Cell}\left(e^{\left(\frac{V+IR_s}{\eta V_T}\right)} - 1\right) - \frac{V + I_{out}R_{Se}}{R_{Sh}} \tag{4}$$

where $V = \left(\frac{N_S TK}{q}\right)$ is the thermal voltage of panel and $N_S = 12$ is the number of series connected cells and $N_p = 4$ is the number of parallel connected cells. The PV array voltage is increased by increasing the series connection of PV cell and current is increased by increasing the parallel connection of cells. The PV panel design and performance parameters are given in Table 1.

The series and shunt resistance of the PV panel is derived as

$$R_{Se} = \left\{\frac{\frac{N_P}{N_S} * (V_{OC} - V_{MPP})}{I_{SC}}\right\} = 1.54 \text{ m}\Omega \tag{5}$$

$$R_{Pa} = \left\{\frac{\frac{N_S}{N_P}(V_{MPP})}{I_{SC} - I_{MPP}}\right\} = 69.5035 \ \Omega \tag{6}$$

a. **P&O MPPT technique**

In this P&O technique, the PV voltage and current parameters are measured by using iterative process and its corresponding power is evaluated [2]. If the PV panel voltage and power product is more than the final iteration product, then the panel voltage is increased, otherwise it is decreased and its corresponding new values are

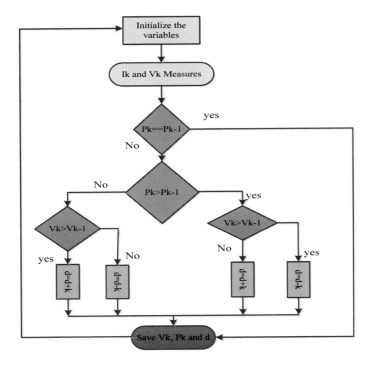

Fig. 3 Switching pulse generation of boost converter

updated. From this technique, it has been observed that the PV voltage is inversely proportional to the duty cycle of the converters and the switching pulses of the converters generate by comparing the sawtooth wave with duty cycle [10]. The block diagram of switching pulse generation of boost converter is given in Fig. 3.

3   Design of Boost Converter

To overcome the low output voltage of the PV cell, a boost converter is used to step-up the PV voltage to meet the load demand. Assuming the semiconductor devices are ideal (Fig. 4a), the MOSFET has a fast transient response and the diode has a zero threshold value. The input and output capacitors of boost converter are used to reduce ripple so as to maintain constant voltage [6].

The boost converter operates in two modes of operation, namely, the conduction state and the blocking state. In conduction state, the MOSFET gets forward biased and diode gets reverse polarized, so there is no supply between the source and load. The conduction state of boost converter is given in Fig. 4b.

The voltage across the inductor and current flowing through the capacitor can be derived as in Eqs. (7) and (8).

Fig. 4 a Basic boost
converter topology,
b conduction state of boost
converter, and **c** blocking
state of boost converter

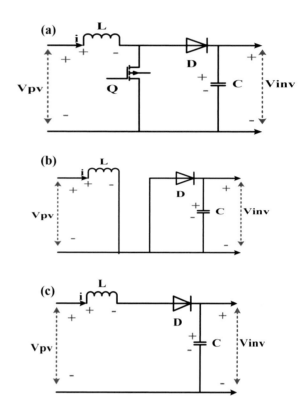

$$V_{pv} = L\frac{di}{dt} \tag{7}$$

$$\frac{V_0}{R} + C\frac{dV_0}{dt} = 0 \tag{8}$$

Similarly, if Switch Q is in blocking state, the diode D gets forward biased and the stored energy in the inductor flows through the inverter to the grid. The blocking state of MOSFET is given in Fig. 4c. The voltage across the input inductor ($L_1 = 2.2$ mH) and current ($C = 0.12$ mF) flowing through the output capacitor is calculated in Eqs. (9) and (10).

$$L\frac{di}{dt} = -V_0 + V_{pv} \tag{9}$$

$$C\frac{dV_0}{dt} = i - \frac{V_0}{R} \tag{10}$$

4 Operation of B-4 Inverter Topology

The boost converter output voltage is given to the two-leg B-4 inverter to convert DC–AC dc-ac, B-4 inverter technology is used, with two driver circuits to trigger the four switches. Also, the common-mode voltage is 2/3 times of B-6 inverter [9]. But an unbalance in DC-link capacitor is caused due to the supply variations and also due to unequal loading of split-link capacitors at low output frequencies.

To ensure the power quality at the output of VSI, the inverter should have the ability to produce balanced output under the practical DC-link conditions. The size and cost will be the consequences faced while improving the capacitance of DC link as a solution for above problem [11, 12]. Real-time compensation SVPWM technique generates control timing signals depending on the space vectors in FSTPI by direct calculation. The block diagram of SVPWM generation B-4 inverter topology is shown in Fig. 5.

a. **Effect of DC-link neutral point unbalance in inverter**

The initial value of the voltage in the capacitor is given by

$$V_0 = \frac{Cd_1}{C_{d1} + C_{d2}} * V_{dc} \tag{11}$$

As C_{d1} is not equal to C_{d2} the asymmetrical voltage in the capacitors circulates a DC current causing voltage unbalance in the circuit, eventually creating voltage

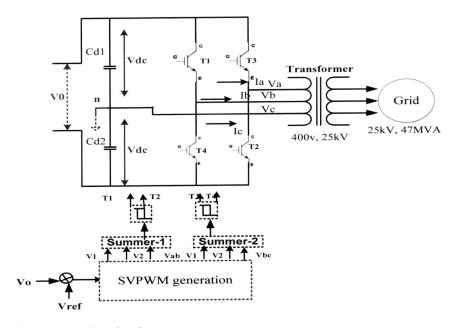

Fig. 5 Two-leg three-phase inverter

unbalance in the grid-connected inverter. The voltage difference in the upper and lower part of DC-link capacitors creates ripples and unbalance in three-phase grid voltage. The unbalance in neutral point at unequal capacitors is overcome by using reasonable mathematical transformation of SVPWM technique [11].

b. Analysis of SVPWM generation for B-4 inverter

In operation of SVPWM generation, the sine voltage amplitude considered as a constant amplitude rotating vector at constant switching frequency and the combination of four B-4 inverter switching vectors give approximate reference vector. In SVPWM each vector is displaced by at an angle 90° [12]. The space vector representation is given in Fig. 6 and its controller is given in Fig. 7. The four vectors combinations and phase, line voltages are given in Table 2 and its d, q axis voltages are given in Table 3.

From Fig. 7 for easy understanding space vector is represented in d, q theory. The a, b, and c coordinates are transformed to d, q by using coordinate transformation technique. The d, q coordinates gives the spatial sum of a, b, and c vectors. The V_α and V_β components are calculated by using DC-link capacitor voltage (V_{dc}) and it is given in Eqs. (12) and (13).

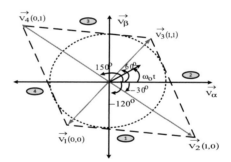

Fig. 6 Block diagram of SVPWM generation

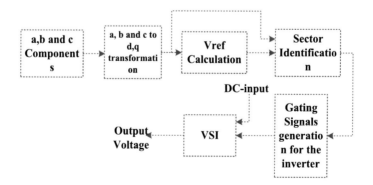

Fig. 7 SVPWM controller for B-4 inverter

Table 2 Combination of space and switching vectors

Space vectors	Switching states		Phase voltages			Line voltages		
	T_a	T_b	V_{an}	V_{bn}	V_{cn}	V_{ab}	V_{bc}	V_{ca}
V_1	0	0	$\frac{V_{dcl}}{3}$	$-\frac{V_{dcl}}{3}$	$-\frac{2V_{dcl}}{3}$	0	$-V_{dcl}$	V_{dcl}
V_2	1	0	V_{dcl}	$-V_{dcl}$	0	$2V_{dcl}$	$-V_{dcl}$	$-V_{dcl}$
V_3	1	1	$\frac{V_{dcl}}{3}$	$\frac{V_{dcl}}{3}$	$-\frac{2V_{dcl}}{3}$	0	V_{dcl}	$-V_{dcl}$
V_4	0	1	$-V_{dcl}$	V_{dcl}	0	$-2V_{dcl}$	V_{dcl}	V_{dcl}

Table 3 Representation of two-phase voltages

Vector	Switching states		Direct voltage vector	Quadrature voltage vector	Resultant vector
	T_a	T_b	V_d	V_q	$V = V_d + V_q$
V_1	0	0	$-\frac{V_{dcl}}{3}$	$-\frac{V_{dcl}}{\sqrt{3}}$	$\frac{2V_{dcl}}{3} * e^{-j\frac{2\pi}{3}}$
V_2	1	0	V_{dcl}	$-\frac{V_{dcl}}{\sqrt{3}}$	$\frac{2V_{dcl}}{\sqrt{3}} * e^{-j\frac{\pi}{6}}$
V_3	1	1	$\frac{V_{dcl}}{3}$	$\frac{V_{dcl}}{\sqrt{3}}$	$\frac{2V_{dcl}}{3} * e^{j\frac{\pi}{3}}$
V_4	0	1	$-V_{dcl}$	$\frac{V_{dcl}}{\sqrt{3}}$	$\frac{2V_{dcl}}{\sqrt{3}} * e^{j\frac{5\pi}{6}}$

$$V_\alpha = \frac{2}{6}V_{dc} - \frac{2}{6}S_a V_{dc} - \frac{2}{6}S_b V_{dc} \tag{12}$$

$$V_\beta = -\frac{2\sqrt{3}}{6}S_a V_{dc} - \frac{2\sqrt{3}}{6}S_b V_{dc} \tag{13}$$

The voltage vectors V_d, V_q, V_{ref}, and α are determined along with the switching time intervals T_0, T_a, T_b to analyze the space vector modulation. From the first sector, the switching time intervals are calculated by the using the reference vectors $\vec{V}_\alpha$ and $\vec{V}_\beta$ it is given in equation's.

$$V_1 T_a \cos(-120°) + V_2 T_b \cos(-30°) = V_s T_s \cos(240° + \theta)$$

$$-\frac{V_1 T_a}{2} - \frac{\sqrt{3}V_2 T_b}{2} = -V_s T_s \cos(60° + \theta) \tag{14}$$

$$V_1 T_a \cos(210°) + V_2 T_b \cos(120°) = V_s T_s \cos(210° - \theta) \tag{15}$$

$$-\frac{\sqrt{3}V_1 T_a}{2} - \frac{V_2 T_b}{2} = -V_s T_s \cos(30° - \theta) \tag{16}$$

By simplifying the above equations, the time intervals T_a and T_b are calculated as

Table 4 Each sector switching time intervals

Sector number	Upper switches	Lower switches
1	$S_1 = T_b + \frac{T_o}{2}$,	$S_2 = T_a + \frac{T_o}{2}$,
	$S_3 = \frac{T_o}{2}$	$S_4 = T_a + T_b + \frac{T_o}{2}$
2	$S_1 = T_a + T_b + \frac{T_o}{2}$,	$S_2 = \frac{T_o}{2}$,
	$S_3 = T_b + \frac{T_o}{2}$	$S_4 = T_a + \frac{T_o}{2}$
3	$S_1 = T_a + T_b + \frac{T_o}{2}$,	$S_2 = T_b + \frac{T_o}{2}$,
	$S_3 = T_a + \frac{T_o}{2}$	$S_4 = \frac{T_o}{2}$
4	$S_1 = \frac{T_o}{2}$,	$S_2 = T_a + T_b + \frac{T_o}{2}$,
	$S_3 = T_a + \frac{T_o}{2}$	$S_4 = T_b + \frac{T_o}{2}$

$$T_a = \frac{V_s T_s \cos(\theta)}{V_1} \tag{17}$$

$$T_b = \frac{V_s T_s \cos(\theta)}{V_2} \tag{18}$$

The common way of calculating four sectors switching time intervals is given in equation

$$T_a = \frac{V_s T_s}{V_1} \cos\left(\theta + \left(\frac{n-1}{2}\right)\pi\right) \tag{19}$$

$$T_b = \frac{V_s T_s}{V_2} \sin\left(\theta - \left(\frac{n-1}{2}\right)\pi\right) \tag{20}$$

The triggering time interval of each IGBT is calculated as (Table 4),

$$T_o = T_s - (T_a + T_b) \tag{21}$$

5   Simulation Results

The parameters used to design PV panel with an output voltage (22 V) and current 46 A are shown in Table 1 at 1000 W/m^2. The PV voltage is given to the boost converter to step-up the voltage to 450 V. The switching time intervals are given in Fig. 8.

In this paper, two-leg boost converter is used to convert DC to AC source. SVPWM generation is used to generate switching pulses for B-4 inverter because it gives less current ripple, better fundamental voltages, and less total harmonic distortion. The

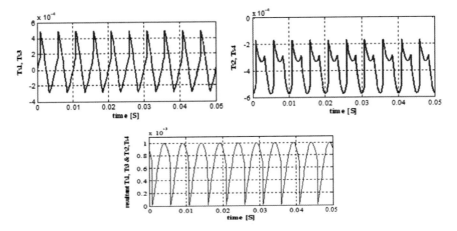

Fig. 8 Switching time intervals of B-4 inverter

switching time duration of each switch and its PWM waveforms of two-leg three-phase VSI is obtained from Fig. 7 and its simulated waveforms are given in Figs. 8 and 9.

From Fig. 8, at 0.01 s switches Ts_1 and Ts_3 are ON condition and Ts_2 and Ts_4 are in off condition. Hence, Phase A and B supply voltage to the grid. At 0.015 s, Switches Ts_2 and Ts_4 are in ON condition and Ts_1 and Ts_3 are in OFF condition. Hence, Phase A and B supply voltage to the grid. The inverter supplies peak–peak rms voltage of 400 V is given to the 400/25 kV transformer to step the voltage and given to the gird of 25 kV, 47 MVA.

From Fig. 9, it is clearly observed that S_1 pulse is a commentary of S_3 and S_2 is commentary of S_4. In this way, the B-4 inverter converts DC–AC and its output line voltages and currents are given in Fig. 10a, b.

The THD analysis, B-4 inverter output voltage consists of less ripple and its line voltage THD is nearly 4.07% and it is given in Fig. 11.

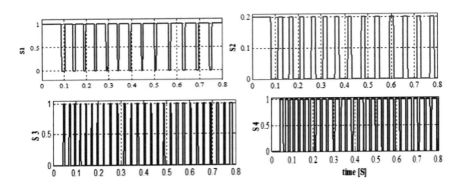

Fig. 9 SVPWM pules B-4 inverter

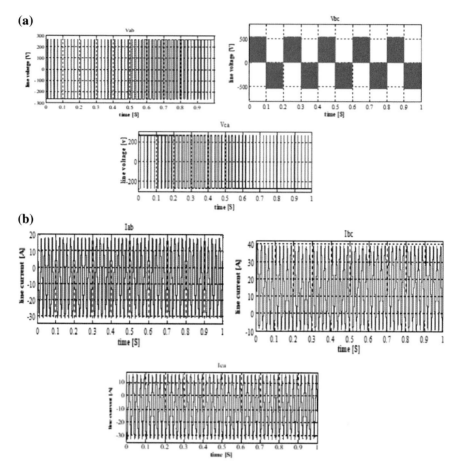

Fig. 10 **a** B-4 inverter output line voltages, **b** B-4 inverter output line currents

6 Conclusion

From the simulation results, the B-4 inverter is the best solution for low power PV grid-connected applications. The B-4 inverter is giving high stability and with less THD output voltage waveforms compared to basic six-switch inverter. In addition, B-4 inverter is applicable for both symmetrical and asymmetrical voltage applications. The SVPWM removes the ripples in inverter output voltage at the time of high switching frequency. Moreover, SVPWM generation reduces the inverter switching and conducting losses at high pulse width modulation frequency. Hence the utilization factor of B-4 inverter is improved.

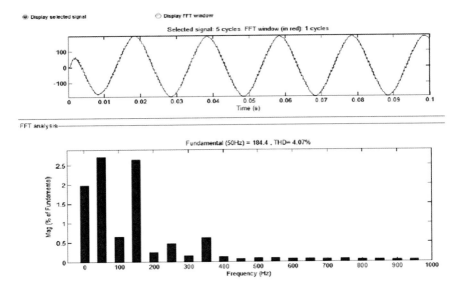

Fig. 11 Analysis of harmonic spectrum at fundamental frequency for line voltage (V_{ab})

Acknowledgements I would like to thank the **University Grants Commission (Govt. of India)** for funding my research program and I especially thank VIT University management for providing all the facilities to carry out my research work.

References

1. Rojas, C.A., et al.: DC–DC MMC for HVdc grid interface of utility-scale photovoltaic conversion systems. IEEE Trans. Ind. Electron. **65**(1), 352–362 (2018)
2. Kamala, D.V., et al.: A modified Perturb & Observe MPPT technique to tackle steady state and rapidly varying atmospheric conditions. Sol. Energy **157**, 419–426 (2017)
3. Rani, C., Basha, Ch H.: A review on non-isolated inductor coupled DC-DC converter for photovoltaic grid-connected applications. Int. J. Renew. Ener. Res. (IJRER) **7**(4), 1570–1585 (2017)
4. Lee, H.H., Dzung, P.Q., Thanh, H.T.: The adaptive space vector PWM for four switch three phase inverter fed induction motor with DC-link voltage imbalance. In: TENCON 2008-2008 IEEE Region 10 Conference. IEEE (2008)
5. Su, J., Sun, D.: Model predictive torque-vector control for four-switch three-phase inverter-fed PMSM with capacitor voltage offset suppression. In: 2017 20th International Conference on Electrical Machines and Systems (ICEMS). IEEE (2017)
6. Rani, C., Basha, Ch H., Odofin, S.: Analysis and comparison of SEPIC, Landsman and Zeta converters for PV fed induction motor drive applications. In: 2018 Internat 2018 International Conference on Computation of Power, Energy, Information and Communication (ICCPEIC) Ional Conference on Computation of Power, Energy, Information and Communication (ICCPEIC). IEEE (2018)

7. Rani, C., Basha, ChH., Odofin, S.: Design and switching loss calculation of single leg 3-level 3-phase VSI. In: 2018 Internat 2018 International Conference on Computation of Power, Energy, Information and Communication (ICCPEIC) Ional Conference on Computation of Power, Energy, Information and Communication (ICCPEIC). IEEE (2018)

8. Lee, H.H., Dzung, P.Q.: Dynamic adaptive space vector PWM for four switch three phase inverter fed induction motor with compensation of DC-link voltage ripple. In: PEDS 2009. International Conference on Power Electronics and Drive Systems, 2009. IEEE (2009)

9. Peng, X., et al.: Opposite vector based phase shift carrier space vector pulse width modulation for extending the voltage balance region in single-phase 3LNPC cascaded rectifier. IEEE Trans. Power Electron. **32**(9), 7381–7393 (2017)

10. Liu, F., et al.: A variable step size INC MPPT method for PV systems. IEEE Trans. Ind. Electron. **55**(7), 2622–2628 (2008)

11. Dasgupta, S., et al.: Application of four-switch-based three-phase grid-connected inverter to connect renewable energy source to a generalized unbalanced microgrid system. IEEE Trans. Ind. Electron. **60**(3), 1204–1215 (2013)

12. ELbarbary, Z.M.S., Hamed, H.A., El-kholy, E.E.: Comment on 'A performance investigation of a four-switch three-phase inverter-fed IM drives at low speeds using fuzzy logic and PI controllers'. IEEE Trans. Power Electron. (2017)

Performance Analysis and Optimization of Process Parameters in WEDM for Inconel 625 Using TLBO Couple with FIS

Anshuman Kumar, Chinmaya P. Mohanty, R. K. Bhuyan and Abdul Munaf Shaik

Abstract The present investigation highlights an experimental study and optimization of machining outcomes characteristics (such as MRR and Ra) during WEDM process of Inconel 625. The present work examined the effects of wire electrode materials, such as Zn-coated brass electrode (ZCBE) and uncoated brass electrode (UBE) on work material during WEDM process. Based on L_{16} orthogonal array, the experiment was performed in consideration with four process factor: spark-on time (S_{on}), flushing pressure (P_f), wire-tension (T_w), and discharge current (D_c), within selected experimental domain. The additional objective of present investigation is to develop a multi-response optimization tool for selection of satisfactory process parameter setting during WEDM of Inconel 625. Nonlinear regression model was applied to formulate statistical models for multi-objective optimization using, fuzzy inference system (FIS) combination with TLBO for fulfill this objective. Finally, the satisfactory process parameter obtained by TLBO was compared with the genetic algorithm (GA) individually and found out that, the TLBO algorithm was found to be simpler, effective, and time-saving approach while solving multi-objective problems.

1 Introduction

Inconel 625 super alloy picked up for extensively used in manufacturing industries for its mechanical, thermal, and corrosion resistance properties. This material has vast applications like in pressure vessels, chemical plant, reactors, etc. When heated, it can form a thick and stable, passivating oxide coating to protecting the surface from further attack. It is a very challenging task to get stringent cutting through

A. Kumar (✉) · R. K. Bhuyan · A. M. Shaik
Department of Mechanical Engineering, Koneru Lakshmaiah Education Foundation, Vaddeswaram, Andhra Pradesh, India
e-mail: anshu.mit06@gmail.com

C. P. Mohanty
School of Mechanical Engineering, Vellore Institute of Technology, Vellore, Tamil Nadu, India

© Springer Nature Singapore Pte Ltd. 2020
K. N. Das et al. (eds.), *Soft Computing for Problem Solving*,
Advances in Intelligent Systems and Computing 1048,
https://doi.org/10.1007/978-981-15-0035-0_72

conventional methods such as turning, grinding, and broaching milling. Low thermal conductivity caused transferring high heat energy during machining to these types conventional tools, which responsible for softening the tool and cause the massive tool wear. Therefore, nonconventional processes are much cost-effective and more suitable for machining for this alloy. For intricate design and stringent cutting, WEDM process is the best option among the other nonconventional machining processes. WEDM is a thermoelectric process used to produce complex shape, through electrically conductive material by using wire. In this process, MRR occurs from initiation of repetitive pulse discharges between the gap of workpiece and the wire electrode connected in an electrical circuit. A distilled water (dielectric) is continuously feed in the spark gap provided by workpiece and the wire electrode. The gap of 0.025–0.05 mm, between the wire electrode and workpiece is maintained constantly by the numerical program [1–3]. The used wire tool cannot reuse due to the variation in dimensional accuracy [4, 5]. Improving the stringent cutting is a significant area of study for all researchers' community and education institution in WEDM. The dimension inaccuracy has been happened not only due to geometrical error along the direction of the wire electrode but also happening due to electrode wire lags (measured between the position of the guides (programed) and the deformed wire (actual), which is strongly dependent upon workpiece thickness, gap force intensity, etc.

In addition, the stringent cutting may be destroyed because of wire vibration, external force on the wire, etc. due to improper selection of process parameter. To requirements of stringent cutting and wire vibration, optimized process parameter maybe work to minimize these defects and improve wire path that CNC programs control. Bobbili et al. [6] studied effect of WEDM process parameters on surface finish on armor steel. In this investigation, found that high S_{on} value, deteriorate the surface finish whereas high-value S_{on} may increase the surface finish. Other process parameters such as wire-tension wire feed and dielectric pressure were found nonsignificant effect on surface roughness.

Nowadays, high-speed growing industries, demanding precise product with high production rate is a real challenging task in WEDM process. Purchase updated machine may be uneconomical for the industry. A small changing may be increase or decrease the cutting speed, quality and tolerances. Among few strategies for improving the fast cutting and quality, changing the wire electrode may be the good option to achieve these machining responses and economical also. The past researchers to improve the cutting speed and surface roughness in Wire EDM with the changing the wire electrode of various materials have conducted noticeable pioneer research. Prohaszka et al. [7] investigate the suitable wire materials and effect of wire material properties for better machinability. Antar et al. [8] discussed the zinc-coated copper wire and ZCBE effects on the productivity. Almost 70% productivity was increased compared to UBE with exact parameter setting. The recast layer of WEDMed surface also decreases from 40 to 25% as compared to UBE to ZCBE for titanium alloy. Kumar et al. [9] investigated a strategy to increase the dimension accuracy with the most popular wire electrode, i.e., ZCBE and UBE.

Kumar et al. [10] determined an optimal setting of WEDM parameter toward achieving satisfactory results in an angular error and surface roughness during WEDM of Inconel 718 work material.

Moreover, most of the past work considered to increase the quality and production rate using WEDM of verity materials, but the potential of high-speed cutting technique with quality is still to be explored for the machining of Inconel 625. Moreover, effective usage of WEDM, also requires the most suitable machining parameter condition to grow the production rate at least preparing cost. Thus, the objective of the present investigation, to obtain a suitable process parameter and compare the performance of wire electrode (UBE and ZCBE) to achieve satisfactory machining yield in terms of MRR and Ra. Moreover, multi-optimization process has been utilized for optimizing the process parameter through hybrid technique using fuzzy inference system (FIS) based TLBO with both the wire electrode.

2 Experimental Details

The controllable parameter has been selected based on machining parameters. The experiments were conducted according to L_{16} orthogonal array. 4-level-4-factor design is selected in the rough machining process for this present study: spark-on time (S_{on}), flushing pressure (P_f), wire-tension (T_w), and discharge current (D) as shown in Table 1.

The experiment has been conducted on WEDM manufacture electronica eNova 1S. A ZCBE and UBE were used as tool electrodes with 0.20 mm diameter and compare to each other with the same design of experiment. The dielectric flow rate and spark-off Time (S_{off}) have been kept constant at 60 μs and 1.4 bars, respectively, throughout the experiments. Deionized water has been used as dielectric fluid. The schematic diagram of the WEDM process has been depicted in Fig. 1. Workpiece material as Inconel 625 with dimensions (50 mm $\times$ 20 mm $\times$ 4 mm) was used. During the experiment, 10 mm length was cut along the length of the workpiece. The MRR calculation has been done by using the following formulas (Eq. 1):

$$MRR = W \times C_s \times \rho \times t \tag{1}$$

Table 1 Process factors and factors levels

Parameter	Notation	Unit	Level of Variations			
			I	II	III	IV
Spark-on time (S_{on})	A	[μs]	95	100	105	110
Flushing Pressure (P_f)	B	[kg/mm^2]	9	12	15	20
Wire-tension (T_w)	C	[N]	6	9	11	14
Discharge Current (D_c)	D	[A]	12	14	16	18

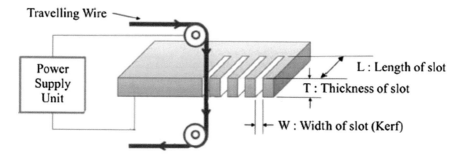

Fig. 1 Schematic diagram of WEDM [11]

Here W is the kerf-width, ρ is the density of the Inconel 625 (8.44 g/cm³) and cutting speed (C_s). Using the Carl Zeiss optical microscope (Manufactured by Germany), the kerf was measured. The kerf measurement is made at five different regions and average of five-time measurement with 3 mm increment along the kerf length. The process factors and factors levels have been shown in Table 1.

The value of Cs is directly noted down from the attached monitor on the machine. In addition to that, the WEDMed surface and the finishing mechanism were measured through Scanning Electron Microscopy (SEM) device (Model: Nova Nano SEM 450). Table 2 shows the experimental domain of the present investigation along with the corresponding outcomes furnished after each trial.

3 Methodologies Explore

3.1 Fuzzy Inference System (FIS)

FIS process is defined as the correlation between input data vector and desired output using application of fuzzy logic rule. Fuzzy logic is a problem-solving methodology system of inexact reasoning, which converts the human reasoning process in linguistic terms in a faster way [12, 13]. Fuzzy controllers and fuzzy reasoning may be applied in a very complicated industries problem, which cannot solve through assumptions. A FIS mainly comprises an inference engine, fuzzifier, a database, and a defuzzifier.

The Mamdani FIS rule used directly in this present study to provides reasonable results for fuzzy inference reasoning.

3.2 Nonlinear Regression

This mathematical method converts experimental data (process parameter and machining responses) into a mathematical equation. This is accomplished using

Table 2 Design of experiment and collected response data [in coded form]

Ex. No	A [μs]	B [kg/mm²]	C [N]	D [A]	MRR [g/min]		Surface roughness [μm]	
					ZCBE	UBE	ZCBE	UBE
1	1	1	1	1	0.0460	0.0272	2.88	3.18
2	1	2	2	2	0.0772	0.0368	3.21	3.63
3	1	3	3	3	0.0952	0.0528	3.41	3.68
4	1	4	4	4	0.1072	0.0788	3.22	3.58
5	2	1	2	3	0.1076	0.0532	3.29	3.88
6	2	2	1	4	0.1332	0.0848	4.45	4.82
7	2	3	4	1	0.0408	0.0248	3.35	3.92
8	2	4	3	2	0.0884	0.0348	3.62	3.87
9	3	1	3	4	0.1364	0.0920	4.12	4.35
10	3	2	4	3	0.1048	0.0604	3.51	3.68
11	3	3	1	2	0.0608	0.0328	3.48	3.97
12	3	4	2	1	0.0372	0.0212	3.72	3.79
13	4	1	4	2	0.0736	0.0548	3.47	3.77
14	4	2	3	1	0.0576	0.0356	3.87	4.18
15	4	3	2	4	0.1380	0.0792	4.51	4.88
16	4	4	1	3	0.0936	0.0548	3.89	4.21

iterative estimation algorithms. The proposed mathematical model of MRR and Ra for each wire electrode is mentioned in Eq. 2:

$$Y_u = x \times A^l \times B^m \times C^n \times D^o \qquad (2)$$

where x denotes the constant, A denotes S_{on}, B denotes P_f, C denotes $W_{t,}$, D denotes D_c and l, m, n, o are estimated coefficients of the nonlinear regression model. Here, SYSTAT 7.0 software package has been used to generate the coefficient, by applying Gauss–Newton algorithm.

3.3 Teaching Learning Based Optimization Algorithm

The TLBO algorithm is nature-based optimization and solves various optimization problems effetely and proposed by Rao et al. [14]. This algorithm based on the behavior of student and the teachers during the classes. Student can be called as learners and quality teaching by the teachers. A quality teacher can motivate the learners (Students) in the class and help to improve the performance of the class. Therefore, learners also follow the quality teaching of the teacher to improve their individual and group performances. This is the basic concept of learner and student concept has been applied in this algorithm [15, 16]. The TLOB algorithm of the present study has been presented in Fig. 2. Figure 3 shows the comparison of performance between the ZCBE and uncoated counterpart. Form the productivity viewpoint MRR should be "higher-is-better criteria". Thus, the percentage improvement in the mean value of MRR is observed to be 41%. This is in support with previous literature, which says that, cutting speed of the ZCBE is almost twice in comparison to UBE [17]. Nevertheless, in the quality viewpoint ZCBE also found an improved WEDMed surface than the uncounted counterpart. Thus, improvement in the mean of Ra from ZCBE to UBE is 09.3%. Figure 3 shows the SEM image revealed machined surface of WEDMed Inconel 625 with both UBE and ZCBE with process parameter conditions (Run no. 9). The spark-on time (105 μs), flushing pressure (8 kg/mm^2) wire-tension (10 N), and discharge current (18 A). During the analysis ZCBE gives better result than the uncoated counterpart. From Fig. 3 it has been observed microholes, melted

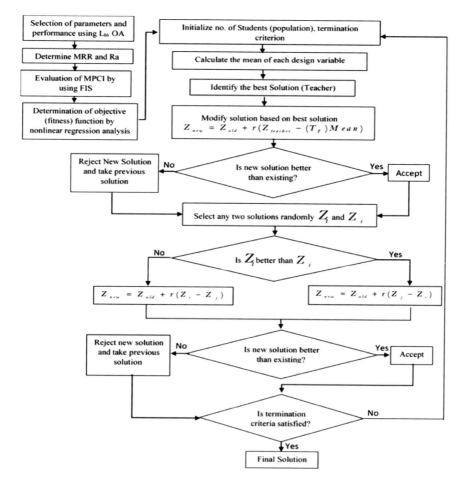

Fig. 2 Flowchart of the proposed optimization route

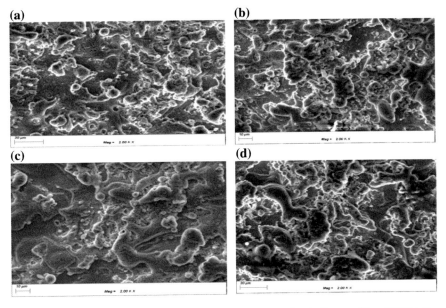

Fig. 3 SEM image of WEDMed surface using ZCBE **a** Run no. 3 and **c** Run no. 15 and UBE **b** Run no. 3 and **d** Run no. 15

material deposited on the WEDMed surface. Additional zinc on the wire electrode helps to control the discharge properties, subsequently enhancing machining performance. Moreover, additional zinc protects the core surface of wire electrode, which controls heat of successive sparks release. Similarly, UBE has the ability to generate stable spark but it has low electrical conductivity.

4 Results and Discussions

4.1 Machining Performance Optimization

In the fast-growing industries, any process or manufactured product performances is characterized by multi-performance features; hence, selecting single response optimization technique may not be always beneficial, because optimal parameter setting may appear to be different for different objective functions. Therefore, it essentially requirement for multi-optimization tool for optimize multiple process characteristics in a single parameter setting. Experimental design and responses, viz., MRR and Ra for both the wire electrodes have been depicted in Table 2. The present work, attempted a satisfaction function approach according by convert individual machining responses into their satisfaction measures. Moreover, satisfaction machining responses have been feed into Fuzzy Inference System (FIS) for Multi-Performance

Table 3 Normalized value of experimental results and MPCI for both the wire electrode

Sl no.	ZCBE			UBE		
	N-MRR	N-RA	MPCI	N-MRR	N-RA	MPCI
1	0.087302	1	0.554	0.084746	1	0.553
2	0.396825	0.797546	0.613	0.220339	0.735294	0.487
3	0.575397	0.674847	0.599	0.446328	0.705882	0.570
4	0.694444	0.791411	0.622	0.813559	0.764706	0.677
5	0.698413	0.748466	0.647	0.451977	0.588235	0.522
6	0.952381	0.036810	0.493	0.898305	0.035294	0.463
7	0.035714	0.711656	0.387	0.050847	0.564706	0.339
8	0.507937	0.546012	0.531	0.192090	0.594118	0.418
9	0.984127	0.239264	0.609	1	0.311765	0.650
10	0.670635	0.613497	0.602	0.553672	0.705882	0.610
11	0.234127	0.631902	0.451	0.163842	0.535294	0.368
12	0	0.484663	0.250	0	0.641176	0.330
13	0.361111	0.638037	0.500	0.474576	0.652941	0.567
14	0.202381	0.392638	0.348	0.203390	0.411765	0.351
15	1	0	0.500	0.819209	0	0.403
16	0.559524	0.380368	0.469	0.474576	0.394118	0.436

Characteristic Index (MPCI) value. The predicted MPCI values have been treated as single objective function. This objective function has been derived using nonlinear regression analysis using SYSTAT 7.0. Hence, the derived mathematical models have been used for optimization individually using TLBO algorithm in the selected parametric search space.

However, collected response data and normalized data (Table 3) to avoid data variation, diverse units along with conflict in criterial requirement. Moreover, machining responses are conflicting in nature, viz., MRR and Ra for both the wire electrode. As MRR corresponds to "Higher-is-Better" (HB) criteria (from productivity view-point), whereas Ra corresponds to "Lower-is-Better" (LB) criteria (from quality view-point). Therefore, the following equation has been used for normalization of collected machining response data.

$$\text{"Higher-is-Better" (HB) criteria} = \frac{X_{ij}}{X_{\max}} \tag{3}$$

$$\text{"Lower-is-Better"(LB) criteria (from quality view-point)} = \frac{X_{\min}}{X_{ij}} \tag{4}$$

where X_{ij} is experimental value whereas, X_{ij} (min) is the maximum value and X_{ij} (max) is minimum value.

The above-mentioned normalized data shown in Table 3 has been used as inputs to the FIS designed (Fig. 4) herein. Above said data has been expressed in linguistic variable using fuzzy membership function via "High", "Medium", and "Low", although, Fuzzy-single output (MPCI) value has been expressed using other membership function design, viz., "Very-High", "High", "Medium", "Low", and "Very-Low" (Fig. 5). Present study, the fuzzy set comprises for each input variable and output responses as symmetric Gaussian membership function as depicted in Fig. 6, the Mamdani's implication method has been implemented for fuzzy reasoning based on the fuzzy rules. To acquire a rule

$$S_i: if\ a_1\ is\ B_{i1}.a_2\ is\ B_{i2},\ and\ a_s\ is$$
$$then\ y_i\ is\ G_i,\ i = 1, 2, \ldots, N.$$

$u_{agg}(y)$ is the output of Mamdani type FIS and it has expressed by a crisp value for the next operation of the fuzzy controller. Centre-of-gravity (CG) technique (Eq. 5) has been chosen for the defuzzification.

$$Y_0 = \frac{\sum_{i=1}^{m} y_i u_{agg}(y_i)}{\sum_{i}^{m} u_{agg}(y_i)} \tag{5}$$

In this section, mathematical model using nonlinear regression analysis based on MPCI value (Table 3) for both the wire electrode, which is shown in Eqs. (6) and (7). The R^2 value of MPCI value for both the wire electrode were 93% for ZCBE wire and 99.1% UBE

$$MPCI\ (ZCBE) = 649.779 \times A^{(-1.927)} \times B^{(-0.244)} \times C^{0.136} \times D^{0.776} \tag{6}$$

Fig. 4 Membership function for MPCI

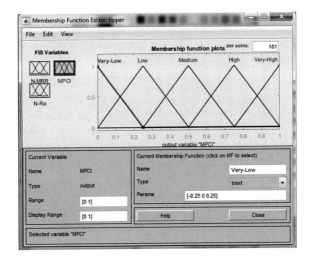

Fig. 5 Fuzzy rule editor

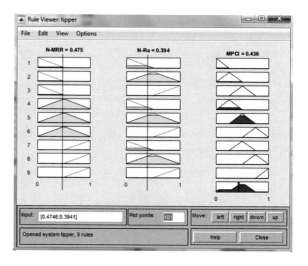

Fig. 6 Fuzzy rule matrix

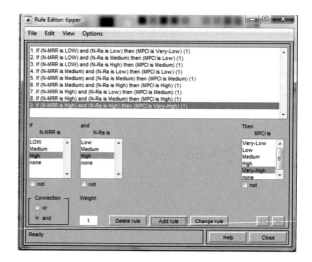

$$\text{MPCI (UBE)} = 99.436 \times A^{(-1.636)} \times B^{(-0.294)} \times C^{0.354} \times D^{0.823} \qquad (7)$$

Hence, it can be say about the aforesaid model is capable enough to select an objective function and can use for multi-objective function using FIS based TLBO and Genetic Algorithm (GA), respectively. The predicted optimal process parameter solution as acquired by maximizing MPCI through TLBO algorithm has shown for ZCBE as A = 95 μs, B = 9 kg/mm², C = 11 N, D = 18 A and for UBE as A = 95 μs, B = 9 kg/mm², C = 11 N, D = 18 A with fitness functions (MPCI) value of 0.797315 and 0.815352. The optimal result as obtained through TLBO has been validated via. evolutionary algorithms GA [18, 19]; it seems a good agreement has been found (Table 4 and Figs. 7, 8).

Table 4 Single and multi-optimization results using TLBO comparison with GA

Wire materials	Algorithm	Responses	Optimal parametric combination				Fitness value
			A	B	C	D	
ZCBE	TLBO	MRR	95	9	6	18	0.1563
		Ra	95	9	11	12	2.8885
		MPCI	**95**	**9**	**11**	**18**	**0.7973**
	GA	MRR	110	9.001	6.032	18	0.1453
		Ra	95	9.001	11	12	2.8885
		MPCI	**95.115**	**9.041**	**10.907**	**17.964**	**0.7975**
UBE	TLBO	MRR	95	9	11	18	0.1020
		Ra	95	9	11	12	3.2387
		MPCI	**95**	**9**	**11**	**18**	**0.8153**
	GA	MRR	95.028	9	10.994	18	0.1020
		Ra	95	9.121	11	12	3.2398
		MPCI	**95**	**9**	**10.948**	**17.999**	**0.8154**

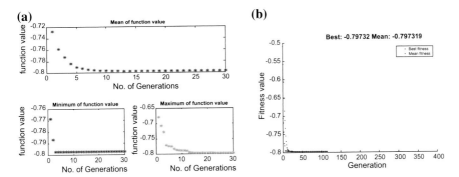

Fig. 7 Convergence plot for optimizing MPCI through the **a** TLBO algorithm, **b** GA for ZCBE

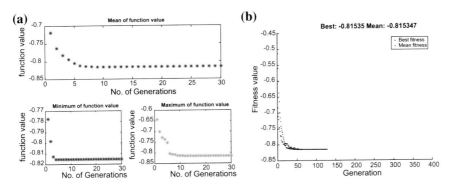

Fig. 8 Convergence plot for optimizing MPCI through the **a** TLBO algorithm, **b** GA for UBE

5 Conclusions

In this experiment, the study focused on multi-objective optimization technique for achieving the good machinability at minimum production cost. MRR and surface roughness have been selected for finding out machinability criteria via WEDM process on cutting the Inconel 625 with ZNCB and UCB. Fuzzy inference system (FIS) based with TLBO for fulfill this objective. The following conclusions are drawn from this investigation:

i. Machining with ZNCB electrode was found the most suitable wire electrode for better machinability as compare with UCB on Inconel 625 using WEDM. Moreover, from the SEM analysis, it is also confirmed that ZNCB produces minimum globules of debris, crater and re-solidify materials on WEDMed surface.
ii. Improvement in the mean of Ra and MRR from ZCBE to UBE is observed 09.3% and 41%.
iii. Spark-on time and discharge-current were found the most significant process parameter influence the MRR and Spark-on time and flushing pressure were found the most affecting parameter on surface roughness.
iv. The predicted mathematical model was found to have better machinability results. The FIS coupled TLBO algorithm was found best suited optimization technique for obtained best machining conditions.
v. The optimal process parameter for high MRR and minimum surface roughness was obtained at A = 95 μs, B = 9 kg/mm^2, C = 11 N, D = 18A for both the wire electrode.

References

1. Tarng, Y., Ma, S., Chung, L.: Determination of optimal cutting parameters in wire electrical discharge machining. Int. J. Mach. Tools Manuf. **35**(12), 1693–1701 (1995)
2. Spedding, T.A., Wang, Z.: Parametric optimization and surface characterization of wire electrical discharge machining process. Precision Eng. **20**(1), 5–15 (1997)
3. Kumar, A., Majumder, H., Vivekananda, K., Maity, K.: NSGA-II approach for multi-objective optimization of wire electrical discharge machining process parameter on inconel 718. Mater. Today Proc. **4**(2), 2194–2202 (2017)
4. Scott, D., Boyina, S., Rajurkar, K.: Analysis and optimization of parameter combinations in wire electrical discharge machining. Int. J. Prod. Res. **29**(11), 2189–2207 (1991)
5. Kumar, A., Abhishek, K.: Influence of process parameters on MRR, kerf width and surface roughness during WEDM on Inconel 718: performance analysis of electrode tool material. Int. J. Ind. Syst. Eng. **30**(3), 298–315 (2018)
6. Bobbili, R., Madhu, V., Gogia, A.: Effect of wire-EDM machining parameters on surface roughness and material removal rate of high strength armor steel. Mater. Manuf. Process. **28**(4), 364–368 (2013)
7. Prohaszka, J., Mamalis, A., Vaxevanidis, N.: The effect of electrode material on machinability in wire electro-discharge machining. J. Mater. Process. Technol. **69**(1), 233–237 (1997)
8. Antar, M., Soo, S., Aspinwall, D., Jones, D., Perez, R.: Productivity and workpiece surface integrity when WEDM aerospace alloys using coated wires. Procedia Eng. **19**, 3–8 (2011)

9. Kumar, A., Abhishek, K., Vivekananda, K., Maity, K.: Effect of wire electrode materials on die-corner accuracy for Wire Electrical Discharge Machining (WEDM) of Inconel 718. Mater. Today Proc. **5**(5), 12641–12648 (2018)

10. Kumar, A., Abhishek, K., Vivekananda, K., Upadhyay, C.: Experimental study and optimization of process parameters during WEDM taper cutting. In: Soft Computing for Problem Solving, pp. 721–736. Springer (2019)

11. Varun, A., Venkaiah, N.: Simultaneous optimization of WEDM responses using grey relational analysis coupled with genetic algorithm while machining EN 353. Int. J. Adv. Manuf. Technol. **76**(1–4), 675–690 (2015)

12. Zadeh, L.A.: A fuzzy-algorithmic approach to the definition of complex or imprecise concepts. Int. J. Man Mach. Stud. **8**(3), 249–291 (1976)

13. Verma, R.K., Abhishek, K., Datta, S., Mahapatra, S.S.: Fuzzy rule based optimization in machining of FRP composites. Turk. J. Fuzzy Syst. **2**(2), 99–121 (2011)

14. Rao, R.V., Savsani, V.J., Vakharia, D.: Teaching–learning-based optimization: a novel method for constrained mechanical design optimization problems. Comput. Aided Des. **43**(3), 303–315 (2011)

15. Rao, R.V., Savsani, V.J., Vakharia, D.: Teaching–learning-based optimization: an optimization method for continuous non-linear large scale problems. Inf. Sci. **183**(1), 1–15 (2012)

16. Rao, R.V.: Parameter optimization of machining processes using TLBO algorithm. In: Teaching Learning Based Optimization Algorithm, pp. 181–190. Springer (2016)

17. Golshan, A., Ghodsiyeh, D., Izman, S.: Multi-objective optimization of wire electrical discharge machining process using evolutionary computation method: effect of cutting variation. Proc. Inst. Mech. Eng. Part B J. Eng. Manuf. 0954405414523593 (2014)

18. Palanikumar, K.: Surface roughness model for machining glass fiber reinforced plastics by PCD tool using fuzzy logics. J. Reinf. Plast. Compos. **28**(18), 2273–2286 (2009)

19. Kaveh, A., Talatahari, S.: Optimum design of skeletal structures using imperialist competitive algorithm. Comput. Struct. **88**(21–22), 1220–1229 (2010)

Application of WDO for Decision-Making in Combined Economic and Emission Dispatch Problem

V. Udhay Sankar, Bhanutej, C. H. Hussaian Basha, Derick Mathew, C. Rani and K. Busawon

Abstract A number of optimization techniques have been used by researchers to solve the combined economic and emission dispatch problem. In this paper, we have applied Wind Driven Optimization (WDO), a heuristic global optimization technique to solve the CEED problem. The technique was applied to three different test systems and the results obtained were compared and analyzed with the results obtained from other techniques. MATLAB R2017a was used for the coding and execution of the algorithm.

Keywords Combined economic emission dispatch · Emission cost · Wind-driven optimization · Fuel cost

Nomenclature

CEED   Combined Economic and Emission Dispatch
WDO   Wind Driven Optimization

V. Udhay Sankar (✉) · Bhanutej · C. H. Hussaian Basha · D. Mathew · C. Rani
School of Electrical Engineering, VIT University, Vellore, India
e-mail: udhaysankarv2014@gmail.com

Bhanutej
e-mail: bhanutej.mtech@gmail.com

C. H. Hussaian Basha
e-mail: sbasha238@gmail.com

D. Mathew
e-mail: derik.mathew@vit.ac.in

C. Rani
e-mail: crani@vit.ac.in

K. Busawon
Faculty of Engineering and Environment, Northumbria University, Newcastle upon Tyne, UK
e-mail: krishna.busawon@northumbria.ac.uk

© Springer Nature Singapore Pte Ltd. 2020
K. N. Das et al. (eds.), *Soft Computing for Problem Solving*,
Advances in Intelligent Systems and Computing 1048,
https://doi.org/10.1007/978-981-15-0035-0_73

907

1 Introduction

Researchers have used various nature-inspired optimization techniques to solve real-world problems. Particle Swarm Optimization (PSO), Ant Colony Optimization (ACO), Flower Pollination Algorithm (FPA), Artificial Bee Colony Algorithm (ABC), and many such optimization techniques have been used to solve various economic dispatch problems over the years [1–4].

With the rising emission of pollutants in air due to fossil fuels, the need to reduce the same has arisen. Global warming and greenhouse gas effects can lead to climate change that might affect our future generations. Carbon dioxide (CO_2), Sulfur oxide (SO_x), Nitrogen oxide (NO_x) are some of the pollutant gases whose emission should be controlled. Thus, the paper has been written to solve the economic emission and dispatch problem by considering fuel cost and emission cost using Wind Driven Optimization (WDO), a population-based iterative optimization technique. Due to the presence of conflicting objectives in economic emission dispatch problem, it is necessary need to find a way to introduce both in our objective function to find the ideal output from each generator that simultaneously satisfies our problem constraints too. The constraints for the problem are the generator capacity limits and power balance constraints [5].

Venkatesh P. et al. applied Evolutionary Programming (EP) to economic emission dispatch problem with tie-line constraints using standard IEEE 14-, 30, and 180-bus test systems [5]. D. N. Jayakumar et al. applied particle swarm optimization to solve various ED problems such as Multi-Area Economic Dispatch (MAED) and CEED [6]. Basu M. (2011) used Multi-Objective Differential Evolution (MODE) technique for the same [7].

Güvenç U. et al. used gravitational search algorithm to solve the multi-objective dispatch problem [8]. Chatterjee A. et al. (2012) used Opposition-based Harmony Search Algorithm (OHS) to deal with CEED [9]. Abdelaziz A. et al (2016) used Flower pollination algorithm (FPA) to solve CEED and its superiority over traditional algorithms was evident [3]. WDO is a new technique that hasn't been applied yet to the domain of economic emission dispatch problems. We tried to gauge the effectiveness of the algorithm in solving economic emission dispatch problems by applying it to four different test systems.

The paper is structured as follows. In Sect. 2, the CEED problem is formulated in detail. The quadratic fuel cost function to be minimized and the set of constraints are explained. Sect. 3 deals with the WDO algorithm and the steps involved in the method are discussed. The method of application of WDO to CEED problem is explained in Sect. 4. The results obtained for the three different test systems are tabulated and analyzed in Sect. 5. Finally, conclusions are drawn in Sect. 6.

2 Problem Formulation

a. Economic Dispatch Problem

The economic dispatch problem is formulated to reduce the fuel cost of a system of generators while supplying the load demand. If F_C is the total fuel cost of the system with n number of generators, it is expressed as

$$F_C = \sum_{i=1}^{n} F_i \tag{1}$$

$$F_i = a_i + b_i P_i + c_i P_i^2 \tag{2}$$

In (2), F_i is the cost function of the ith generator in a system of n generators. a_i, b_i, and c_i are the fuel cost coefficients and P_i is the real output power generated from the ith generator.

In (2), we make an assumption that fuel cost increases in a quadratic manner. But, practically valves are used to control the flow of steam through separate nozzle groups. So, the increase in output is achieved by the sequential opening of the valves, leading to ripples in the input–output curve as shown in Fig. 1. Thus, the cost function can no longer be a quadratic function and the inclusion of value point effect is shown in (3).

$$F_i = a_i + b_i P_i + c_i P_i^2 + \left| g_i \times \sin\left(f_i \times \left(P_i^{\min} - P_i\right)\right) \right| \tag{3}$$

where g_i and f_i are the valve point coefficients of the ith generator.

$$P_i = P_D + P_L \tag{4}$$

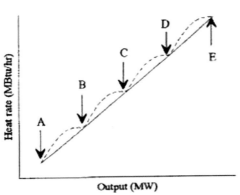

Fig. 1 Input–output curve with five value points

Heat rate (MBtu/hr)

D
E
C
B
A

Output (MW)

A, B, C, D, E: Operating points of admission valves

where P_D is the total power demand from the system and P_L is the power loss in the system. The power loss is calculated by (5).

$$P_L = \sum_{i=1}^{n} \sum_{j=1}^{n} P_i B_{ij} P_j \qquad (5)$$

B_{ij} is the loss coefficient matrix. This constraint is known as the power balance constraint. The generators are also subject to the capacity constraints

$$P_i^{min} \leq P_i \geq P_i^{max} \qquad (6)$$

where P_i^{min} and P_i^{max} are the minimum and maximum power generation capacities of a generator.

b. Emission Dispatch Problem

Emission dispatch problem deals with reducing the pollutants released into the air due to the burning of coal in the generators. The emission function E_i is a quadratic function of the power generated P_i with coefficients d, e, and f represented as

$$E_T = \sum_{i=1}^{n} (E_i) = \sum_{i=1}^{n} \left(d_i + e_i P_i + f_i P_i^2 \right) \qquad (7)$$

where E_T is the total emission in ton and d_i, e_i, and f_i are the emission coefficients from the generators.

c. Combined Economic and Emission Dispatch Problem

CEED combines both the economic and emission problems by using weight factor w. The function to be minimized becomes

$$F = w \times F_C + (1 - w) \times E_T \qquad (8)$$

$$F = \sum_{i=1}^{n} w \times (a_i + b_i + c_i P_i^2) + \left| g_i \times \sin(f_i \times P_i^{min} - P_i) \right| + \sum_{i=1}^{n} (1 - w) \times (d_i + e_i P_i + f_i P_i^2) \qquad (9)$$

The value of w ranges from 0 to 1. The maximum value of $w = 1$, makes the minimizing equation a pure economic dispatch problem and minimum value of $w = 0$, makes it a pure emission dispatch problem.

But, using weight factor w also leads to a mismatch of units in the Eq. (9) as fuel cost is measured in dollars and emission is calculated in kg. So, another way of combining both the equations is by using price penalty factor h [5].

$$F = \sum_{i=1}^{n} (a_i + b_i P_i + c_i P_i^2) + \left| g_i \times \sin(f_i \times P_i^{min} - P_i) \right| + \sum_{i=1}^{n} h_i \times (d_i + e_i P_i + f_i P_i^2) \qquad (10)$$

Equation (10) is the objective function proposed in this problem with h_i the price penalty factor of the ith generator and is given as

$$h_i = \frac{F_i^{\max}}{E_i^{\max}} \tag{11}$$

where $F_i^{\max}$ is the maximum fuel cost of the ith generator, $E_i^{\max}$ is the maximum emission from a generator. This method is better than the weight factor method as the units on both sides of the Eq. (10) are equivalent [5].

3 Wind-Driven Optimization

WDO is a nature-inspired technique based on the earth's atmospheric motion. It is multidimensional and can be applied to a variety of real-world problems. The technique was developed by Zikri Bayraktar et al. and was applied to optimization problems in electromagnetics. Bhandari, Singh, Kumar, and Singh used WDO for image segmentation. Derrick M. et al. used WDO for the estimation of solar photovoltaic parameters. Yet, WDO remains a relatively unexplored technique and the algorithm can be coded easily and is effective for a multitude of problems [10–12].

Wind in the atmosphere blows from a region of high pressure to low pressure, such that the imbalance in air pressure is equalized. The velocity of the wind is proportional to the pressure gradient. Using Newton's second law of motion, the force applied on an air parcel is given by

$$\rho \vec{a} = \sum \vec{F_t} \tag{12}$$

where ρ is the air density of the air parcel and $\vec{a}$ is the acceleration vector. The equation relating air pressure P to its density and temperature is given by the ideal gas law

$$P = \rho RT \tag{13}$$

where R and T are the universal gas constant and temperature respectively.

Different forces such as the pressure Gradient force (F_{PG}), Frictional force (F_F), Gravitational force (F_G), and Coriolis force (F_C) act on the air parcel influencing its movement. The force because of the pressure gradient is

$$\overrightarrow{F_{PG}} = -\overrightarrow{\nabla P} \cdot \delta V \tag{14}$$

The force due to the pressure gradient is opposed by the frictional force F_F and is expressed as

$$\vec{F_F} = -\rho\alpha\vec{u} \tag{15}$$

ρ is the air density, α is the coefficient of friction and $\vec{u}$ is the wind velocity vector. Gravitational force acting on the air parcel is given as

$$\vec{F_G} = \rho \cdot \delta V \cdot \vec{g} \tag{16}$$

where $\vec{g}$ is the gravitation vector.

Due to earth's rotation, Coriolis force acts on the air parcel opposite to the direction of velocity of the air parcel

$$F_C = -2\theta \times \vec{u} \tag{17}$$

θ is the earth's rotation. Cumulating all the forces acting on the air parcel using Newton's second law of motion gives us

$$\rho\frac{\vec{\Delta u}}{\Delta t} = -\vec{\nabla P} \cdot \delta V - \rho\alpha\vec{u} + \rho \cdot \delta V \cdot \vec{g} - 2\theta \times \vec{u} \tag{18}$$

Air parcels moving with the wind have infinitesimal volume. Thus, ignoring δV and equating Δt to 1 for simplicity and the new expression is obtained

$$\vec{\Delta u} = \vec{g} + \left(-\vec{\Delta P} \cdot \frac{RT}{P_c}\right) + (-\alpha\vec{u}) - \left(\frac{2\theta\vec{u}RT}{P_c}\right) \tag{19}$$

In (19), the first and second vectors are broken down into their magnitude and direction. The velocity ($\vec{\Delta u}$ vector is resolved into $\vec{u_{new}}$ and $\vec{u_{old}}$ and the equation is further simplified as

$$\vec{u_{new}} = \left((1-\alpha)\vec{u_{old}}\right) + g\left(-\vec{x_{old}}\right) + \left[\left|\frac{P_{max}}{P_{old}} - 1\right|RT(x_{max} - x_{old})\right] + \left[\frac{-c_{old}^{otherdim}}{P_{old}}\right] \tag{20}$$

Thus, new velocity to be calculated ($\vec{u_{new}}$) depends on the various variables in the current iteration such as the velocity of current iteration $\left(\vec{u_{old}}\right)$, current position of the air parcel $\vec{x_{old}}$, distance from the highest point of pressure x_{max}, maximum pressure P_{max}, pressure of the air parcel in the current location P_{old}, acceleration due to gravity g, temperature T, universal gas constant R, and the constants α and c. But the search swarm moves toward the highest pressure point iteratively, obeying physical equations that govern motion in the atmosphere. After the velocity is updated, the position of the air parcel can be updated using

$$\vec{x_{new}} = \vec{x_{old}} + \left(\vec{u_{new}} \times \Delta t\right) \tag{21}$$

Fig. 2 Flow chart of the
WDO Algorithm

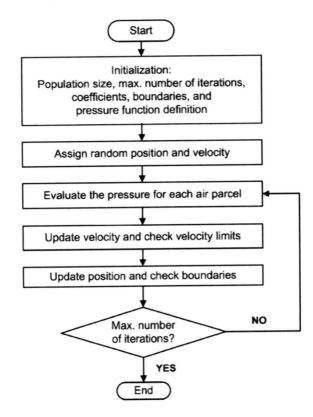

The algorithm is executed for a predetermined number of iterations and the global best is updated for each iteration. The constants play an important role in the convergence of the algorithm. Each air parcel also has a velocity vector which is updated by (20). The position of the air parcels is continuously updated using (21). Figure 2 shows the linear flowchart of the execution of the proposed algorithm.

4 Wind-Driven Optimization and Its Application for Combined Economic and Emission Dispatch

Adapting WDO to our CEED problem involves the following steps such as

Step 1: Declaration of initial parameters: Different parameters of the problem are declared initially such as the population of the air parcels, number of dimensions, number of iterations, and the value of the constants RT, g, α, and c. Each generator is equivalent to a dimension in the problem, with its power generation limits forming the upper and lower limits of the dimension.

Step 2: Calculation of initial pressure and Ranking: A number of random air parcels equal to the population size are created within the generator limits for each generator. The position of the air parcels are created inside the limits $[-1, +1]$, for each dimension using the formula

$$\text{pos} = \left(\frac{(x - \text{gen}_{\text{low}})}{(\text{gen}_{\text{high}} - \text{gen}_{\text{low}})} \times 2 \right) - 1 \qquad (22)$$

where gen_{low} and gen_{high} are the upper and lower power capacity limits of the generator. The velocity of each air parcel is declared randomly for all the dimensions. For each air parcel, the pressure is calculated using (10), the objective function. The air parcels are sorted and ranked based on the values of pressure calculated and their position is noted. The lowest pressure value is set as the global best.

Step 3: Calculation of new position and velocity: For subsequent iterations, the new velocity and position of the air parcel are calculated using Eqs. (16) and (15). In order to prevent the air parcel from taking large steps, the velocity is limited to a predetermined range, and the new velocity calculated is checked in every iteration such that it remains within the same range.

$$u_{\text{new}} = \begin{cases} u\text{max} \ if \ u_{\text{new}} > u\text{max} \\ -u\text{max} \ if \ u_{\text{new}} < u\text{max} \end{cases} \qquad (23)$$

Step 4: Updating the pressure: If the pressure (objective function) of an air parcel is found to be lesser than the global best, it is updated.

Step 5: End of algorithm: The algorithm is terminated when the number of iterations set at the start is completed. The minimization of our objective function is complete.

5 Results and Discussion

a. Six-generator system

It consists of six generators with power loss, and the generator data such as capacity limits and loss coefficients are taken from [6]. The values of RT Coefficient, g, α, and c are set as follows and shown in Table 1.

The system is solved using WDO, PSO, CEP, and FCGA and the results are tabulated in Tables 2 and 3.

The convergence of the solution when solved using WDO is shown in Fig. 2. The demand from the system is set at two different values such as 700 and 1000 MW. In Fig. 3, we see the solution for demand of 700 MW from the system is obtained

Table 1 Value of different constants in the method

Constants	Value of the constant
RT coefficient	3
g (gravitational constant)	0.2
α (update equation constant)	0.4
c (coriolis effect)	0.4
Maximum limiting velocity	0.3

Table 2 Solution for test system 1—demand = 700 MW

Method	CEP [6]	FCGA [6]	PSO [6]	WDO
G1 (MW)	77.274	80.16	77.3421	79.758
G2 (MW)	49.639	53.71	49.9888	47.256
G3 (MW)	48.535	40.93	48.2616	49.823
G4 (MW)	103.525	116.23	103.9630	103.513
G5 (MW)	260.695	251.20	260.1879	258.034
G6 (MW)	191.233	190.62	191.0598	191.416
Fuel cost ($/h)	38216.47	38408.82	38216.52	38172.84
Emission cost (ton)	524.49	527.46	527.166	523.333
Total cost ($/h)	19369.84	NA	19371.84	19349.007

Table 3 Solution for test system 1—demand = 1000 MW

Method	PSO [5]	WDO
P1 (MW)	125.0000	125.0000
P2 (MW)	83.9168	79.0009
P3 (MW)	74.2969	75.0752
P4 (MW)	167.3707	164.5273
P5 (MW)	325.0000	325
P6 (MW)	288.1127	295.3899
Fuel cost ($/h)	55456.387	55339.78
Emission cost (ton)	1012.137	1010.61
Total cost (FC) ($/h)	28234.7372	28174.3667

around the 20th iteration. We see the convergence is much quicker than other swarm optimization methods such as PSO which typically takes longer cycles to reach the optimum.

The output current of the PV array directly depends on the ratio of series and parallel connection cells. When the ratio increases, the corresponding power generation of PV array also increases.

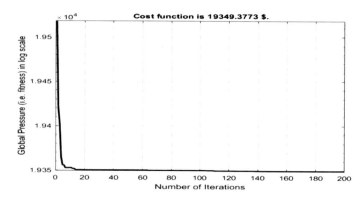

Fig. 3 Convergence of CEED 6—generator problem

The cost function is found to be 19349 $ for the system which was lower than the cost estimated by PSO, CEP, and FCGA.

The population of the air parcels was varied from 25 to 200 in steps of 25. The computation time varied with different values of population. The results were analyzed and tabulated in Table 4. It was inferred from Table 4 that as the population of the air parcels is increased, the computation time also gets higher (Fig. 4).

b. **11-generator testing system**

Test system 2 consists of 11 generators with the generator data and loss coefficients taken from [13]. The total cost of the system was found to be 9204.38 $/h. The fuel cost was found to be 8471.85 $/h and the emission cost 204.98 ton. It was lower than other cost estimations from different methods. The results are tabulated in Table 6. The demand of the system ranged from 1000 to 2000 MW. It is obvious from the table that WDO yields consistently better results than GA for different values of demand. Table 5 gives us the comparison of the results from WDO and other optimization techniques for test system 2 with the demand set at 1000 MW. The fuel cost provided by WDO is approximately 17 $/h lesser than the solutions from other techniques.

Table 4 Solution for test system 1—demand = 700 MW

Population size	Fuel cost ($/h)	Emission cost (ton)	Computation time (s)
25	38173.229	525.59	0.22
50	38174.01	524.34	0.25
75	38172.52	526.32	0.27
100	38174.49	524.05	0.36
125	38171.795	526.622	0.430
150	38173.06	525.49	0.496
175	38174.42	524.029	0.58
200	38173.18	525.57	0.67

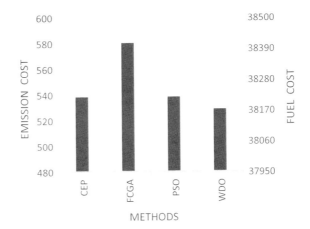

Fig. 4 Comparison of methods for $P_d = 700$ MW in test system 1

Table 5 Comparison of results—test system 2, demand = 1000 MW

Population size	Fuel cost ($/h)	Emission cost (ton)	Computation time (s)
25	38173.229	525.59	0.22
50	38174.01	524.34	0.25
75	38172.52	526.32	0.27
100	38174.49	524.05	0.36
125	38171.795	526.622	0.430
150	38173.06	525.49	0.496
175	38174.42	524.029	0.58
200	38173.18	525.57	0.67

The population size of the air parcels was set at 1000 and the maximum number of iterations was 1000. The average execution time of the program was 0.68 s which is significantly faster than other techniques. But the convergence of the technique was longer than expected, at around 300 iterations as shown in Fig. 5.

a. **20-generator testing system**

Test system 3 consists of 20 generators and with a total demand of 2500 MW. The emission coefficients, loss matrix, and generator data are taken from [9] and [14]. The results are tabulated in Table 7 along with solutions from other optimization techniques such as Hopfield modeling, biogeography-based optimization, and opposition-based harmony search. The Emission cost of the test system was found to be 27287 ton and the total cost was 62295.46 $/h which is lower than the solutions provided by the other techniques (Fig. 6).

Test system 4 consists of 40 generators with value point loading effects. The generator data was taken from [7]. The system demand was $P_d = 10500$ MW. The

Table 6 Solution for test system 2—compared with genetic algorithm (GA) based on similarity crossover

Demand	1000 MW		1250 MW		1500 MW	
Method	GA [13]	WDO	GA [13]	WDO	GA [13]	WDO
G1 (MW)	86.27	86.82	93.93	94.14	106.51	106.26
G2 (MW)	76.97	77.94	83.07	84.19	87.14	89.87
G3 (MW)	85.43	89.29	96.32	101.03	105.36	105.93
G4 (MW)	74.18	76.03	98.48	100.85	128.38	126.74
G5 (MW)	48.46	50.98	64.42	69.90	84.39	79.02
G6 (MW)	82.01	83.03	100.84	101.09	123.97	125.27
G7 (MW)	55.12	50.84	63.77	62.52	76.65	81.08
G8 (MW)	132.00	128.11	167.96	168.11	216.23	203.45
G9 (MW)	118.31	131.44	160.39	164.54	189.76	193.28
G10 (MW)	122.21	119.44	161.13	156.87	193.60	200.46
G11 (MW)	119.02	110.21	159.69	146.71	187.97	188.60
Fuel cost ($/h)	8501.85	**8485.58**	9107.99	**9083.06**	9733.22	**9706.58**
Emission cost (ton)	205.175	**204.98**	339.7063	**336.39**	539.49	**537.03**
Total cost ($/h)	NA	**9204.38**	NA	**10236.63**	NA	**11500.05**
CPU (s)	NA	**0.64**	NA	**0.69**	NA	**0.70**
Demand	1750 MW			2000 MW		
Method	GA [13]		WDO	GA [13]		WDO
G1 (MW)	113.03		113.20	113.03		113.20
G2 (MW)	94.88		95.57	94.88		95.57
G3 (MW)	114.55		116.09	114.55		116.09
G4 (MW)	150.52		151.15	150.52		151.15
G5 (MW)	93.41		95.52	93.41		95.52
G6 (MW)	144.97		149.07	144.97		149.07
G7 (MW)	99.01		93.56	99.01		93.56
G8 (MW)	235.16		236.63	235.16		236.63
G9 (MW)	224.31		226.54	224.31		226.54
G10 (MW)	122.21		119.44	161.13		156.87
G11 (MW)	119.02		110.21	159.69		146.71
Fuel cost ($/h)	8501.85		**8485.58**	9107.99		**9083.06**
Emission cost (ton)	205.175		**204.98**	339.7063		**336.39**
Total cost ($/h)	NA		**9204.38**	NA		**10236.63**
CPU (s)	NA		**0.64**	NA		**0.69**

Fig. 5 Convergence of the CEED 11—generator problem

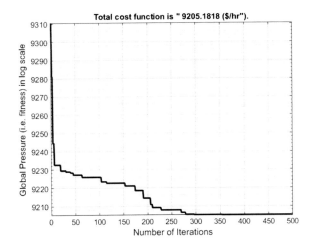

Total cost function is " 9205.1818 ($/hr").

Table 7 Solution for test system 3—demand = 2500

Method	HFM [9]	BBO [9]	OHS [9]	WDO
G1 (MW)	512.7804	513.0892	511.8518	512.345
G2 (MW)	169.1035	173.3533	151.9528	163.23
G3 (MW)	126.8897	126.9231	111.3158	114.15
G4 (MW)	102.8656	103.3292	102.1659	102.45
G5 (MW)	113.6836	113.7741	103.6150	115.64
G6 (MW)	73.5709	73.06694	71.3042	70.065
G7 (MW)	115.2876	114.9843	114.2378	112.456
G8 (MW)	116.3994	116.4238	105.7168	108.45
G9 (MW)	100.4063	100.6948	118.5139	115.30
G10 (MW)	106.0267	99.99979	103.3803	104.45
G11 (MW)	150.2395	148.977	171.4014	160.23
G12 (MW)	292.7647	294.0207	313.3944	300.33
G13 (MW)	119.1155	119.5754	120.2281	119.82
G14 (MW)	30.8342	30.54786	42.3455	36.45
G15 (MW)	115.8056	116.4546	145.3918	130.347
G16 (MW)	36.2545	36.22787	38.7053	36.222
G17 (MW)	66.8590	66.85943	47.4920	54.28
G18 (MW)	87.9720	88.54701	82.3015	83.4561
G19 (MW)	100.8033	100.9802	82.7111	88.234
G20 (MW)	54.3050	54.2725	53.5811	54.163
Emission cost (ton)	NA	NA	27318.00	**27287.13**
Total cost (FC) ($/h)	62456.6341	62456.7926	62340.00	**62295.46**

Fig. 6 Comparison of the results for $P_D = 2500$, for test system 3

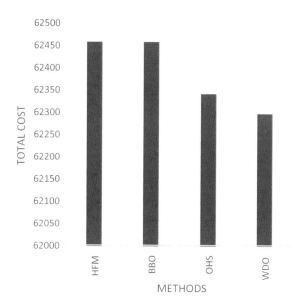

results were compared with other algorithms as shown in Table 8. The convergence of the algorithm for the 40-unit system is shown in Fig. 7.

6 Conclusion

The WDO technique was applied to different test systems and the results were compared with results from other optimization techniques by minimizing the fuel and emission cost. The results show that the proposed technique is competitive with all other existing algorithms. Further research can be done by applying the WDO to other economic dispatch problems such as simple economic dispatch, cubic cost economic dispatch, dynamic economic dispatch, dynamic economic and emission dispatch, and MAED with tie-line constraints.

Table 8 Solution for test system 4—demand = 2500

Method	MODE [9]	PDE [9]	NSGA-2 [16]	SPEA [9]	GSA [8]	WDO
G1 (MW)	113.595	112.1549	113.85	113.96	113.99	113.87
G2 (MW)	114.100	113.9431	113.61	114.00	113.96	113.78
G3 (MW)	120.100	120.0000	120.10	119.87	119.95	119.97
G4 (MW)	179.015	180.2647	180.47	179.92	179.77	180.23
G5 (MW)	96.716	97.0000	97.10	97.00	97.00	97.00
G6 (MW)	139.260	140.01	140.01	139.27	139.01	139.23
G7 (MW)	300.010	299.88	300.01	300.00	299.98	300.00
G8 (MW)	298.93	300.00	299.04	298.27	300.00	300.00
G9 (MW)	290.37	289.89	288.80	290.52	296.20	292.18
G10 (MW)	130.95	130.57	131.62	131.48	130.38	130.52
G11 (MW)	244.79	244.10	246.58	244.67	245.47	245.66
G12 (MW)	317.88	318.28	318.88	317.20	318.21	318.0
G13 (MW)	395.36	394.78	395.74	394.73	394.62	394.28
G14 (MW)	394.42	394.21	394.19	394.62	395.20	394.22
G15 (MW)	305.84	305.96	305.51	304.72	306.00	305.23
G16 (MW)	394.89	394.13	394.68	394.72	395.10	395.22
G17 (MW)	487.92	489.30	489.44	487.98	489.25	488.47
G18 (MW)	489.11	489.64	488.21	488.53	488.75	488.74
G19 (MW)	500.55	499.98	500.79	501.16	499.23	500.18
G20 (MW)	457.02	455.41	455.26	456.43	455.28	455.23
G21 (MW)	434.68	435.28	434.69	434.78	433.45	433.34
G22 (MW)	434.50	433.73	434.10	434.39	433.81	433.12
G23 (MW)	444.62	446.24	445.85	445.07	445.51	445.10
G24 (MW)	452.02	451.88	450.79	451.89	452.05	452.41
G25 (MW)	492.71	493.22	491.25	492.39	492.88	492.29
G26 (MW)	436.37	434.75	436.38	436.99	433.36	433.23
G27 (MW)	10.01	11.80	11.27	10.77	10.00	10.00
G28 (MW)	10.31	10.75	10.10	10.29	10.02	11.73
G29 (MW)	12.39	10.33	12.04	13.70	10.01	12.48
G30 (MW)	96.90	97.01	97.20	96.24	96.91	96.26
G31 (MW)	189.77	190.10	189.46	190.00	189.96	190.00
G32 (MW)	174.24	175.35	174.71	174.21	175.00	175.00
G33 (MW)	190.10	190.10	189.25	190.00	189.01	190.00
G34 (MW)	199.66	200.01	200.50	200.00	200.00	200.00
G35 (MW)	199.82	200.01	199.98	200.00	200.00	200.00
G36 (MW)	200.10	200.10	199.56	200.00	199.99	199.93

(continued)

Table 8 (continued)

Method	MODE [9]	PDE [9]	NSGA-2 [16]	SPEA [9]	GSA [8]	WDO
G37 (MW)	110.10	109.93	108.36	110.00	109.99	110.00
G38 (MW)	109.94	109.86	110.00	109.62	109.01	109.37
G39 (MW)	108.16	108.99	109.79	108.50	109.45	109.62
G40 (MW)	422.02	421.32	421.57	421.85	421.99	421.56
Fuel cost (FC) ($\times 10^5$ \$/h)	1.25	1.26	1.26	1.25	1.25	**1.25**
Emission cost (FC) ($\times 10^5$ ton)	2.11	2.13	2.10	2.11	2.10	**2.10**

Fig. 7 Convergence of WDO test system 4

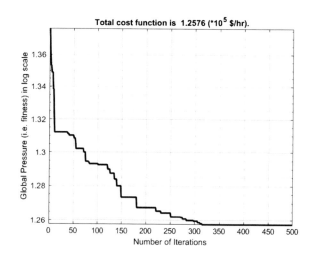

References

1. Kumar, A.I.S., et al.: Particle swarm optimization solution to emission and economic dispatch problem. In: TENCON 2003, Conference on Convergent Technologies for the Asia-Pacific Region, vol. 1. IEEE (2003)
2. Gopalakrishnan, R., Krishnan, A.: An efficient technique to solve combined economic and emission dispatch problem using modified Ant colony optimization. Sadhana 38(4), 545–556 (2013)
3. Abdelaziz, A.Y., Ali, E.S., Abd Elazim, S.M.: Combined economic and emission dispatch solution using flower pollination algorithm. Int. J. Electr. Power Energy Syst. 80, 264–274 (2016)
4. Aydin, D., et al.: Artificial bee colony algorithm with dynamic population size to combined economic and emission dispatch problem. Int. J. Electr. Power Energy Syst. 54, 144–153 (2014)
5. Venkatesh, P., Gnanadass, R., Padhy, N.P.: Comparison and application of evolutionary programming techniques to combined economic emission dispatch with line flow constraints. IEEE Trans. Power Syst. 18(2), 688–697 (2003)
6. Jeyakumar, D.N., Jayabarathi, T., Raghunathan, T.: Particle swarm optimization for various types of economic dispatch problems. Int. J. Electr. Power Energy Syst. 28(1), 36–42 (2006)

7. Basu, M.: Economic environmental dispatch using multi-objective differential evolution. Appl. Soft Comput. **11**(2), 2845–2853 (2011)
8. Güvenç, U., et al.: Combined economic and emission dispatch solution using gravitational search algorithm. Sci. Iranica **19**(6), 1754–1762 (2012)
9. Chatterjee, A., Ghoshal, S.P., Mukherjee, V.: Solution of combined economic and emission dispatch problems of power systems by an opposition-based harmony search algorithm. Int. J. Electr. Power Energy Syst. **39**(1), 9–20 (2012)
10. Bayraktar, Z., Komurcu, M., Bossard, J.A., Werner, D.H.: The wind driven optimization technique and its application in electromagnetics. IEEE Trans. Antennas Propag. **61**(5), 2745–2757 (2013)
11. Mathew, D., Rani, C., Kumar, M.R., Wang, Y., Binns, R., Busawon, K.: Wind-driven optimization technique for estimation of solar photovoltaic parameters. IEEE J, Photovolt. **8**(1), 248–256 (2018)
12. Bhandari, A.K., Singh, V.K., Kumar, A., Singh, G.K.: Cuckoo search algorithm and wind driven optimization based study of satellite image segmentation for multilevel thresholding using Kapur's entropy. Expert Syst. Appl. **41**(7), 3538–3560 (2014)
13. Guuml, U.: Combined economic emission dispatch solution using genetic algorithm based on similarity crossover. Sci. Res. Essays **5**(17), 2451–2456 (2010)
14. dos Santos Coelho, L., Lee, C.-S.: Solving economic load dispatch problems in power systems using chaotic and Gaussian particle swarm optimization approaches. Int. J. Electr. Power Energy Syst. **30**(5), 297–307 (2008)
15. Balamurugan, R., Subramanian, S.: A simplified recursive approach to combined economic emission dispatch. Electr. Power Compon. Syst, **36**(1), 17–27 (2007)
16. Dhanalakshmi, S., Kannan, S., Mahadevan, K., Baskar, S.: Application of modified NSGA-II algorithm to combined economic and emission dispatch problem. Int. J. Electr. Power Energy Syst. **33**(4), 992–1002 (2011)

Application of Wind-Driven Optimization for Decision-Making in Economic Dispatch Problem

V. Udhay Sankar, Bhanutej, C. H. Hussaian Basha, Derick Mathew, C. Rani and K. Busawon

Abstract A number of optimization techniques have been used by researchers to solve the non-convex Economic Dispatch (ED) problem. In this paper, we have applied Wind-Driven Optimization (WDO), a heuristic global optimization technique to solve the ED problem. The technique was applied to three different test systems and the results obtained were compared and analyzed with the results obtained from other techniques. MATLAB R2017a was used for the coding and execution of the algorithm.

Keywords Economic dispatch · Emission cost · Wind-driven optimization · Fuel cost

Nomenclature

ED   Economic Dispatch
WDO   Wind-Driven Optimization

V. Udhay Sankar (✉) · Bhanutej · C. H. Hussaian Basha · D. Mathew · C. Rani
School of Electrical Engineering, VIT University, Vellore, India
e-mail: udhaysankarv2014@gmail.com

Bhanutej
e-mail: bhanutej.mtech@gmail.com

C. H. Hussaian Basha
e-mail: sbasha238@gmail.com

D. Mathew
e-mail: derik.mathew@vit.ac.in

C. Rani
e-mail: crani@vit.ac.in

K. Busawon
Faculty of Engineering and Environment, Northumbria University, Newcastle upon Tyne, UK
e-mail: krishna.busawon@northumbria.ac.uk

© Springer Nature Singapore Pte Ltd. 2020
K. N. Das et al. (eds.), *Soft Computing for Problem Solving*,
Advances in Intelligent Systems and Computing 1048,
https://doi.org/10.1007/978-981-15-0035-0_74

925

1　Introduction

Researchers have used various nature-inspired optimization techniques to solve real-world problems. Particle Swarm Optimization (PSO), Ant Colony Optimization (ACO), Flower Pollination Algorithm (FPA), Artificial Bee Colony (ABC) algorithm, and many such optimization techniques have been used to solve various economic dispatch problems over the years [1–4].

With the rising emission of pollutants in air due to fossil fuels, the need to reduce the same has arisen. Global warming and greenhouse gas effects can lead to climate change that might affect our future generations. Carbon dioxide (CO_2), Sulfur oxide (SO_x), and Nitrogen oxide (NO_x) are some of the pollutant gases whose emission should be controlled. Thus, the paper has been written to solve the economic dispatch problem by considering fuel cost using Wind-Driven Optimization (WDO), a population-based iterative optimization technique [5].

Venkatesh P. et al. applied Evolutionary Programming (EP) to economic emission dispatch problem with tie line constraints using standard IEEE 14-, 30, and 180-bus test systems [5]. D. N. Jayakumar et al. applied Particle Swarm Optimization to solve various ED problems such as Multi-Area Economic Dispatch (MAED) and CEED [6]. Basu M. (2011) used Multi-Objective Differential Evolution (MODE) technique for the same [7].

Güvenç U. et al. used gravitational search algorithm to solve the multi-objective dispatch problem [8]. Chatterjee A. et al. (2012) used Opposition-based Harmony Search algorithm (OHS) to deal with ED problems [9]. Abdelaziz A. et al. (2016) used Flower Pollination Algorithm (FPA) to solve ED and its superiority over traditional algorithms was evident [3]. WDO is a new technique that has not been applied yet to the domain of economic emission dispatch problems. We tried to gauge the effectiveness of the algorithm in solving economic emission dispatch problems by applying it to four different test systems.

The paper is structured as follows. In Sect. 2, the ED problem is formulated briefly. The quadratic fuel cost function to be minimized and the set of constraints are briefly explained. Section 3 deals with the WDO algorithm and the steps involved in the method are discussed. The method of application of WDO to ED problem is explained in Sect. 4. The results obtained for the three different test systems are tabulated and analyzed in Sect. 5. Finally, conclusions are drawn in Sect. 6.

2　Problem Formulation

a.　Economic Dispatch Problem

The economic dispatch problem is formulated to reduce the fuel cost of a system of generators while supplying the load demand. If FC is the total fuel cost of the system with n number of generators, it is expressed as follows:

Fig. 1 Input–output curve with five value points

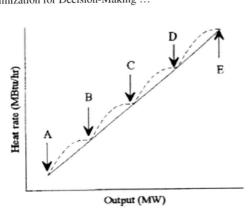

A, B, C, D, E: Operating points of admission valves

$$F_C = \sum_{i=1}^{n} F_i \tag{1}$$

$$F_i = a_i + b_i P_i + c_i P_i^2 \tag{2}$$

In (2), F_i is the cost function of the ith generator in a system of n generators. a_i, b_i, and c_i are the fuel cost coefficients and P_i is the real output power generated from the ith generator.

In (2) we make an assumption that fuel cost increases in a quadratic manner. But, practically valves are used to control the flow of steam through separate nozzle groups. So, the increase in output is achieved by the sequential opening of the valves, leading to ripples in the input–output curve as shown in Fig. 1. Thus, the cost function can no longer be a quadratic function and the inclusion of value point effect is shown in (3).

$$F_i = a_i + b_i P_i + c_i P_i^2 + \left| g_i \times \sin\left(f_i \times \left(P_i^{\min} - P_i\right)\right)\right| \tag{3}$$

where g_i and f_i are the valve point coefficients of the ith generator.

Equation 3 is subjected to constraints as follows:

$$P_i = P_D + P_L \tag{4}$$

where P_D is the total power demand from the system and P_L is the power loss in the system. The power loss is calculated by (5).

$$P_L = \sum_{i=1}^{n} \sum_{j=1}^{n} P_i B_{ij} P_j \tag{5}$$

B_{ij} is the loss coefficient matrix. This constraint is known as the power balance constraint. The generators are also subjected to the capacity constraints

$$P_i^{\min} \leq P_i \geq P_i^{\max} \tag{6}$$

where $P_i^{\min}$ and $P_i^{\max}$ are the minimum and maximum power generation capacities of a generator.

3 Wind-Driven Optimization

WDO is a nature-inspired technique based on the earth's atmospheric motion. It is multidimensional and can be applied to a variety of real-world problems. The technique was developed by Zikri Bayraktar et al. and was applied to optimization problems in electromagnetics. Bhandari, Singh, Kumar, and Singh used WDO for image segmentation. Derrick M. et al. used WDO for the estimation of solar photovoltaic parameters. Yet WDO remains a relatively unexplored technique and the algorithm can be coded easily and is effective for a multitude of problems [10–12].

Wind in the atmosphere blows from a region of high pressure to low pressure, such that the imbalance in air pressure is equalized. The velocity of the wind is proportional to the pressure gradient. Using Newton's second law of motion, the force applied on an air parcel is given by

$$\rho \vec{a} = \sum \vec{F_t} \tag{12}$$

where ρ is the air density of the air parcel and $\vec{a}$ is the acceleration vector. The equation relating air pressure P to its density and temperature is given by the ideal gas law

$$P = \rho RT \tag{13}$$

where R and T are the universal gas constant and temperature, respectively.

Different forces such as the pressure Gradient force (F_{PG}), Frictional force (F_F), Gravitational force (F_G), and Coriolis force (F_C) act on the air parcel influencing its movement. The force because of the pressure gradient is

$$\overrightarrow{F_{PG}} = -\overrightarrow{\nabla P} \cdot \delta V \tag{14}$$

The force due to the pressure gradient is opposed by the frictional force F_F and is expressed as

$$\overrightarrow{F_F} = -\rho \alpha \vec{u} \tag{15}$$

ρ is the air density, α is the coefficient of friction, and $\vec{u}$ is the wind velocity vector.

Gravitational force acting on the air parcel is given as

$$\vec{F_G} = \rho \cdot \delta V \cdot \vec{g} \tag{16}$$

where $\vec{g}$ is the gravitation vector.

Due to earth's rotation, Coriolis force acts on the air parcel opposite to the direction of velocity of the air parcel

$$F_C = -2\theta \times \vec{u} \tag{17}$$

θ is the earth's rotation. Cumulating all the forces acting on the air parcel using Newton's second law of motion gives us

$$\rho \frac{\vec{\Delta u}}{\Delta t} = -\vec{\nabla P} \cdot \delta V - \rho \alpha \vec{u} + \rho \cdot \delta V \cdot \vec{g} - 2\theta \times \vec{u} \tag{18}$$

Air parcels moving with the wind have infinitesimal volume. Thus, ignoring δV and equating Δt to 1 for simplicity and the new expression is obtained as

$$\vec{\Delta u} = \vec{g} + \left(-\vec{\Delta P} \cdot \frac{RT}{P_c}\right) + (-\alpha \vec{u}) - \left(\frac{2\theta \vec{u} RT}{P_c}\right) \tag{19}$$

In (19) the first and second vectors are broken down into their magnitude and direction. The velocity $(\vec{\Delta u})$ vector is resolved into $\vec{u_{new}}$ and $\vec{u_{old}}$ and the equation is further simplified as follows:

$$\vec{u_{new}} = \left((1-\alpha)\vec{u_{old}}\right) + g\left(-\vec{x_{old}}\right) + \left[\left|\frac{P_{max}}{P_{old}} - 1\right| RT (x_{max} - x_{old})\right] + \left[\frac{-c_{old}^{otherdim}}{P_{old}}\right] \tag{20}$$

Thus, new velocity to be calculated $(\vec{u_{new}})$ depends on the various variables in the current iteration such as the velocity of current iteration $(\vec{u_{old}})$, current position of the air parcel $\vec{x_{old}}$, distance from the highest point of pressure x_{max}, maximum pressure P_{max}, pressure of the air parcel in the current location P_{old}, acceleration due to gravity g, temperature T, universal gas constant R, and the constants α and c. But the search swarm moves toward the highest pressure point iteratively, obeying physical equations that govern motion in the atmosphere. After the velocity is updated the position of the air parcel can be updated using

$$\vec{x_{new}} = \vec{x_{old}} + \left(\vec{u_{new}} \times \Delta t\right) \tag{21}$$

Fig. 2 Flow chart of the
WDO algorithm

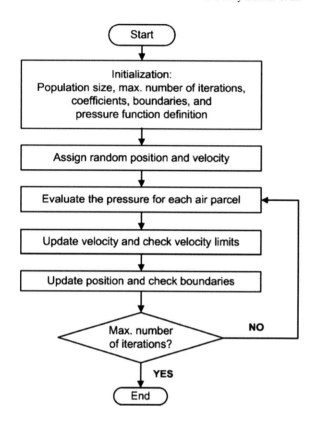

The algorithm is executed for a predetermined number of iterations and the global best is updated for each iteration. The constants play an important role in the convergence of the algorithm. Each air parcel also has a velocity vector which is updated by (20). The position of the air parcels is continuously updated using (21). Figure 2 shows the linear flowchart of the execution of the proposed algorithm.

4   Wind-Driven Optimization and Its Application for Economic Dispatch

Adapting WDO to our CEED problem involves the following steps:

Step 1: Declaration of initial parameters: Different parameters of the problem are declared initially such as population of the air parcels, number of dimensions, number of iterations, and the value of the constants RT, g, α, and c. Each generator is equivalent to a dimension in the problem, with its power generation limits forming the upper and lower limits of the dimension.

Step 2: Calculation of initial pressure and Ranking: A number of random air parcels equal to the population size are created within the generator limits for each generator. The position of the air parcels are created inside the limits $[-1, +1]$, for each dimension using the formula

$$\text{pos} = \left(\frac{(x - \text{gen}_{\text{low}})}{(\text{gen}_{\text{high}} - \text{gen}_{\text{low}})} \times 2 \right) - 1 \qquad (22)$$

where gen_{low} and gen_{high} are the upper and lower power capacity limits of the generator. The velocity of each air parcel is declared randomly for all the dimensions. For each air parcel, the pressure is calculated using (10), the objective function. The air parcels are sorted and ranked based on the values of pressure calculated and their position is noted. The lowest pressure value is set as the global best.

Step 3: Calculation of new position and velocity: For subsequent iterations, the new velocity and position of the air parcel are calculated using Eqs. (16) and (15). In order to prevent the air parcel from taking large steps the velocity is limited to a predetermined range, and the new velocity calculated is checked in every iteration such that it remains within the same range.

$$u_{\text{new}} = \begin{cases} u_{\max} \; if \; u_{\text{new}} > u_{\max} \\ -u_{\max} \; if \; u_{\text{new}} < u_{\max} \end{cases} \qquad (23)$$

Step 4: Updating the pressure: If the pressure (objective function) of an air parcel is found to be lesser than the global best it is updated.

Step 5: End of algorithm: The algorithm is terminated when the number of iterations set at the start is completed. The minimization of our objective function is complete.

5 Results and Discussion

MATLAB 2017a was used for coding and executing the algorithm on an Intel i7 1.80 GHz, 16 gb ram computer. Due to the stochastic nature of WDO, different results are obtained for different trials. So, for each test system, the least value for 1000 trials is considered for comparison with the other algorithms. Also, the average, maximum and least values obtained during the trial for different values are also observed.

There are five different used-defined constants in the WDO algorithm. The ideal values for the constants are selected after testing for various standard functions. The values are tabulated in Table 1.

Table 1 Value of different constants in the method

Constants	Value of the constant
RT coefficient	3
g (gravitational constant)	0.2
α (update equation constant)	0.4
c (coriolis effect)	0.4
Maximum limiting velocity	0.3

Table 2 Solution for test system 1—demand = 850 MW

Generator	P_{min} (MW)	P_{max} (MW)	A	B	C	E	F
1	100	600	0.001562	7.92	561	300	0.0315
2	50	200	0.004820	7.97	78	150	0.063
3	100	400	0.001940	7.85	310	200	0.042

Table 3 Solution for test system 1—demand = 850 MW

Method	Average cost ($/h)	Maximum cost ($/h)	Minimum cost ($/h)
GAB	NA	NA	8234.08
GAF	NA	NA	8234.07
CEP	8235.97	8241.83	8234.07
FEP	8234.24	8241.78	8234.07
MFEP	8234.71	8241.80	8234.08
IFEP	8234.16	8231.54	8234.07
FA	8234.08	8241.23	8234.07
WDO	8233.89	8241.5131	8233.7292

a. **Three generator systems**

It consists of three generators with valve point loading effects. The total demand from the system was 850 MW The value of the cost coefficients of the generators were taken from [17] and is shown in Table 2.

Using the generator data, the WDO algorithm was executed on the system to minimize the cost function. The population size of the air parcels was set initially at 100 and the number of maximum iterations was set at 100. The minimum cost obtained during the trials was found to be 8233.7292 $ and the maximum was found to be 8241.5141 $. The individual output from the three generators for the minimum cost was 300.24 MW, 149.7623 MW, and 399.999 MW, respectively (Table 3).

In Table 2 we can see that WDO gives a lower minimum cost function than other methods, the maximum cost is a few MW's higher than IFEP and FFA (Table 4).

The computation time 0.054 s shows the efficiency of the algorithm. But, it's speed could not be compared with those of other methods used for comparison because of

Table 4 Output power of each generator in the best solution for three-unit system—demand = 850 MW

Unit	FA	WDO
1	300.267	300.24
2	149.733	149.7623
3	400.000	399.999
Total cost ($/h)	8234.074	8233.7292

the different hardware used for the former techniques. The convergence of the WDO algorithm for three-generator system is shown in Fig. 3.

The size of the population makes a significant impact on the execution time of the algorithm as the computation load increases with increase in the population of the algorithm. The population was varied from 25 to 200 and the execution time is tabulated in Table 5.

From Table 5 it is evident that as the populations of air parcels increase the computation time owing to the increased computation load to be executed by the computer. Low population size favors quicker output and higher population size leads to a relative increase in execution time, but the cost is not necessarily low as seen from the table.

b. 13-generator testing system

It consists of 13 generators with valve point loading effects, the cost coefficients and the valve point coefficients are taken from [16] and are tabulated in Table 5. The demand was set at 1800 MW.

Due to the increase in the number of generators this test system represents a more complex problem than the three-generator system. The population size was set at 100, the maximum number of iterations 2000, and the algorithm was run. The maximum cost found during the runs was 18234.4069 $ and the minimum cost was 17963.81 $ which was also lower than the minimum cost from other methods. The

Fig. 3 Convergence of WDO for test system 1

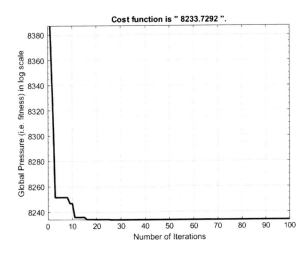

Table 5 Solution for test system 1—demand = 850 MW

Population size	Fuel cost ($)	Computation time (s)
25	8233.7298	0.0176
50	8233.7871	0.0312
75	8233.7446	0.0427
100	8233.7423	0.0762
125	8233.7612	0.0915
150	8233.7362	0.0751
175	8233.7551	0.1122
200	8233.7199	0.1251

Table 6 Coefficients for 13-unit test system—demand = 1800 MW

Generator	a	b	c	e	f
1	0.00028	8.10	550	300	0.035
2	0.00056	8.10	309	200	0.042
3	0.00056	8.10	307	200	0.063
4	0.00324	7.74	240	150	0.063
5	0.00324	7.74	240	150	0.063
6	0.00324	7.74	240	150	0.063
7	0.00324	7.74	240	150	0.063
8	0.00324	7.74	240	150	0.063
9	0.00324	7.74	240	150	0.063
10	0.00284	8.6	126	100	0.084
11	0.00284	8.6	126	100	0.084
12	0.00284	8.6	126	100	0.084
13	0.00284	8.6	126	100	0.084

total number of trials for the evaluation of average, minimum, and maximum cost was 1000. It should also be noted that increasing the number of iterations increased the computation time but the solution was not necessarily better (Table 6).

The execution time for the algorithm was 0.55 s. The individual output from each generator for the minimum cost function obtained from the algorithm is shown in Table 7.

c. **40-generator testing system**.

This test system has 40-generating units with valve point loading effects. The demand for the test system was 10500 MW. Due to the increase in the number of units from 13-generator test system, the computation needed to minimize the objective function increases. The algorithm should not be trapped in a local minimum due to the vast solution space of the algorithm. Thus, the effectiveness of the algorithm cannot be judged over a single trial, several runs are required.

Table 7 Output power of each generator in the best solution for 13-unit system—demand = 1800 MW

Generator	P_{out} (MW)
1	628.3312
2	149.585
3	222.74921
4	109.86354
5	109.86354
6	109.86354
7	109.86354
8	60.0000
9	109.885
10	40.0123
11	40.0123
12	55.01002
13	55.0000

Table 8 Comparison of solution for 13-unit test system, demand = 1800 MW

Method	Average cost ($/h)	Maximum cost ($/h)	Minimum cost ($/h)
CEP	18190.32	18404.04	18048.21
FEP	18200.79	18453.82	18018.00
MFEP	18192.00	18416.89	18028.29
IFEP	18127.06	18267.42	17994.07
EP-SQP	18106.93	NA	17991.03
HDE	18134.8	NA	17975.73
CGA-MU	NA	NA	17975.34
PSO-SQP	18029.99	NA	17969.93
HS	17986.53	18070.176	17965.62
IGA-MU	NA	NA	17963.98
DEC (1)-SQP	17984.81	17984.81	17963.94
ST-HDE	18046.38	NA	17963.89
FA	18029.16	18168.10	17963.83
WDO	18040.23	18234.4069	17963.81

The population for the WDO was set at 100 and the m maximum number of iterations was set at 2000. The cost coefficients and valve point data are taken from [16]. The algorithm is run and the results are tabulated in Table 8, with the output from each generator for the minimum cost over 1000 trials is compared with three PSO methods and Firefly Algorithm.

The average, maximum, and minimum cost for WDO during the trials for the 40-generator test system is tabulated along with various other algorithms in Table 9.

Table 9 Coefficients and solution For 40-unit test system—demand = 10,500 MW

Unit	P_{min}	P_{max}	a	b	c	e	f	Individual output from each unit for different methods				
								PSO-LRS (MW)	NPSO (MW)	NPSO-LRS (MW)	FA (MW)	WDO (MW)
1	36	114	0.00690	6.73	94.705	100	0.084	111.9858	113.9891	113.9761	110.8099	111.0523
2	36	114	0.00690	6.73	94.705	100	0.084	110.6273	113.6334	113.9986	110.8059	110.8045
3	60	120	0.02028	7.07	309.54	100	0.084	98.5560	97.5500	97.4241	97.40230	97.4123
4	80	190	0.00942	8.18	369.03	150	0.063	182.9266	180.0059	179.7327	179.7332	179.7342
5	47	97	0.01142	5.35	148.89	120	0.077	87.7254	97.0000	89.6511	92.70700	92.70650
6	68	140	0.01142	8.05	222.33	100	0.084	139.9933	140.0000	105.4044	140.0000	140.0000
7	110	300	0.00357	8.03	287.71	200	0.042	259.6628	300.000	259.7502	259.6004	259.5999
8	135	300	0.00492	6.99	391.98	200	0.042	297.7912	300.000	288.4534	284.6004	284.5999
9	135	300	0.00573	6.60	455.76	200	0.042	284.8459	284.5797	284.6460	284.6004	284.5999
10	130	300	0.00605	12.9	722.82	200	0.042	130.000	130.0517	204.812	130.0028	130.0026
11	94	375	0.00515	12.9	635.20	200	0.042	94.6741	243.7131	168.8311	168.8008	168.7994
12	94	375	0.00569	12.8	654.69	200	0.042	94.3734	169.0104	94.0000	168.8008	168.7994
13	125	500	0.00421	12.5	913.40	300	0.035	214.7369	125.0000	214.7663	214.7606	214.7645
14	125	500	0.00752	8.84	1760.4	300	0.035	394.1370	393.9662	394.2852	304.5204	304.5214
15	125	500	0.00708	9.15	1728.3	300	0.035	483.1816	304.7586	304.5187	394.2801	394.2794
16	125	500	0.00708	9.15	1728.3	300	0.035	304.5381	304.5120	394.2811	394.2801	394.2794
17	220	500	0.00313	7.97	647.85	300	0.035	489.2139	489.6024	489.807	489.2801	489.2994
18	242	550	0.00313	7.95	649.69	300	0.035	489.6154	489.6087	489.2832	489.2801	489.2994
19	242	550	0.00313	7.97	647.83	300	0.035	511.1782	511.7903	511.2845	511.2817	511.2830
20	254	550	0.00313	7.97	647.81	300	0.035	511.7336	511.2624	511.3049	511.2817	511.2830

(continued)

Table 9 (continued)

Unit	P_{min}	P_{max}	a	b	c	e	f	Individual output from each unit for different methods				
								PSO-LRS (MW)	NPSO (MW)	NPSO-LRS (MW)	FA (MW)	WDO (MW)
21	254	550	0.00298	6.63	785.96	300	0.035	523.4071	523.3274	523.2916	523.2793	523.2790
22	254	550	0.00298	6.63	785.96	300	0.035	523.4599	523.3274	523.2853	523.2793	523.2790
23	254	550	0.00284	6.66	794.53	300	0.035	523.4756	523.4707	523.2797	523.2832	523.2840
24	254	550	0.00284	6.66	794.53	300	0.035	523.7032	523.0661	523.2994	523.2832	523.2840
25	254	550	0.00277	7.10	801.32	300	0.035	523.7854	523.3978	523.2865	523.2793	523.2793
26	254	550	0.00277	7.10	801.32	300	0.035	523.2757	523.2897	523.2936	523.2793	523.2793
27	10	150	0.52124	3.33	1055.1	120	0.077	10.0000	10.0208	10.0000	10.0000	10.0000
28	10	150	0.52124	3.33	1055.1	120	0.077	10.6251	10.0927	10.0001	10.0000	10.0000
29	10	150	0.52124	3.33	1055.1	120	0.077	10.0727	10.0621	10.0000	10.0000	10.0000
30	47	97	0.01140	5.35	148.89	120	0.077	51.3321	88.9456	89.0139	87.8008	87.8013
31	60	190	0.00160	6.43	222.92	150	0.063	189.8048	189.9951	190.0000	189.9989	189.9982
32	60	190	0.00160	6.43	222.92	150	0.063	189.7386	190.0000	190.0000	189.9989	189.9982
33	60	190	0.00160	6.43	222.92	150	0.063	189.9122	190.0000	190.0000	189.9989	189.9982
34	90	200	0.0001	8.95	107.87	200	0.042	199.3258	191.2978	172.0275	164.8036	164.8040
35	90	200	0.0001	8.62	116.58	200	0.042	199.3065	172.4153	165.1397	164.8036	164.8040
36	90	200	0.0001	8.62	116.58	200	0.042	192.8977	191.2978	172.0275	164.8036	164.8040
37	25	110	0.0161	5.88	307.45	80	0.098	110.0000	109.9893	110.0000	110.0000	110.0000
38	25	110	0.0161	5.88	307.45	80	0.098	109.8628	109.9521	110.0000	110.0000	110.0000
39	25	110	0.0161	5.88	307.45	80	0.098	92.8751	109.8733	93.0962	110.0000	110.0000
40	242	550	0.00313	7.97	647.83	300	0.035	511.6883	511.5671	511.2996	511.2794	511.2800

The best solution 121,414.98 $/h is slightly lower than the best solution provided by Firefly Algorithm (FA), is competitive when compared to other solutions. The average solution is 121,430.38 $/h, which is much closer to the best solution meaning the algorithm will consistently produce solutions that are closer to the ideal solution. In practical situations we can not expect a technique to provide the best solution in the first trial. Therefore, the average cost can be used as a parameter to judge the technique's effectiveness. The average solution is very much closer to the best solution meaning the algorithm consistently provides solutions closer to the best solution. The computation time was 6.56 s, and it is greater than the computation time for 3 and 13 generator systems because of the increase in the solution space due to number of generators in the problem (Table 10).

Table 10 Comparision of 13-unit test system

Method	Average cost ($/h)	Maximum cost ($/h)	Minimum cost ($/h)
HGPSO	126,855.70	NA	124,797.13
SPSO	126,074.40	NA	124,350.40
PSO	124,154.49	NA	123,930.45
CEP	124,793.48	126,902.89	123,488.29
HGAPSO	124,575.70	NA	122,780.00
FEP	124,119.37	127,245.59	122,679.71
MFEP	123,489.74	124,356.47	122,647.57
IFEP	123,382.00	125,740.63	122,624.35
TM	123,078.21	124,693.81	122,477.78
EP-SQP	122,379.63	NA	122,323.97
MPSO	NA	NA	122,252.26
ESO	122,558.45	123,143.07	122,122.16
HPSOM	124,350.87	NA	122,112.40
PSO-SQP	122,245.25	NA	122,094.67
PSO-LRS	122,558.45	123,461.67	122,035.79
Improved GA	122,811.41	123,334.00	121,915.93
HPSOWM	122,844.	NA	121,915.30
IGAMU	NA	NA	121,819.25
HDE	122,705.66	NA	121,813.26
DEC (2)-SQP (1)	122,295.12	122,839.29	121,741.97
PSO	122,513.91	123,467.40	121,735.47
APSO (1)	122,221.36	122,995.09	121,704.73
ST-HDE	122,304.30	NA	121,698.51
NPSO-LRS	122,209.31	122,981.59	121,664.43
APSO (2)	122,153.61	122,912.39	121,663.52

(continued)

Table 10 (continued)

Method	Average cost ($/h)	Maximum cost ($/h)	Minimum cost ($/h)
SOHPSO	121,853.57	122,446.30	121,501.14
BBO	121,512.06	121,688.66	121,479.50
BF	121,814.94	NA	121,423.63
GA-PS-SQP	122,039.00	NA	121,458.00
PS	122,332,65	125,486.29	121,415.14
FA	121,416.57	121,424.56	121,415.05
WDO	121,420.23	121,430.38	121,414.98

6 Conclusion

The WDO technique was applied to three different test systems and the results were compared with results from other optimization techniques by minimizing the fuel. The results show that the proposed technique is competitive and robust with all other existing algorithms. Further research can be done by applying the WDO to other Economic dispatch problems such as cubic cost economic dispatch, dynamic economic dispatch, dynamic economic and emission dispatch, and MAED with tie line constraints.

References

1. Kumar, A.I.S., et al.: Particle swarm optimization solution to emission and economic dispatch problem. In: Conference on Convergent Technologies for the Asia-Pacific Region, TENCON 2003, vol. 1. IEEE (2003)
2. Gopalakrishnan, R., Krishnan, A.: An efficient technique to solve combined economic and emission dispatch problem using modified Ant colony optimization. Sadhana **38**(4), 545–556 (2013)
3. Abdelaziz, A.Y., Ali, E.S., Abd Elazim, S.M.: Combined economic and emission dispatch solution using flower pollination algorithm. Int. J. Electr. Power Energy Syst. **80**, 264–274 (2016)
4. Aydin, D., et al.: Artificial bee colony algorithm with dynamic population size to combined economic and emission dispatch problem. Int. J. Electr. Power Energy Syst. **54**, 144–153 (2014)
5. Venkatesh, P., Gnanadass, R., Padhy, N.P.: Comparison and application of evolutionary programming techniques to combined economic emission dispatch with line flow constraints. IEEE Trans. Power Syst. **18**(2), 688–697 (2003)
6. Jeyakumar, D.N., Jayabarathi, T., Raghunathan, T.: Particle swarm optimization for various types of economic dispatch problems. Int. J. Electr. Power Energy Syst. **28**(1), 36–42 (2006)
7. Basu, M.: Economic environmental dispatch using multi-objective differential evolution. Appl. Soft Comput. **11**(2), 2845–2853 (2011)
8. Güvenç, U., et al.: Combined economic and emission dispatch solution using gravitational search algorithm. Sci. Iran. **19**(6), 1754–1762 (2012)
9. Chatterjee, A., Ghoshal, S.P., Mukherjee, V.: Solution of combined economic and emission dispatch problems of power systems by an opposition-based harmony search algorithm. Int. J. Electr. Power Energy Syst. **39**(1), 9–20 (2012)

10. Bayraktar, Z., Komurcu, M., Bossard, J.A., Werner, D.H.: The wind driven optimization technique and its application in electromagnetics. IEEE Trans. Antennas Propag. **61**(5), 2745–2757 (2013)
11. Mathew, D., Rani, C., Kumar, M.R., Wang, Y., Binns, R., Busawon, K.: Wind-driven optimization technique for estimation of solar photovoltaic parameters. IEEE J. Photovolt. **8**(1), 248–256 (2018)
12. Bhandari, A.K., Singh, V.K., Kumar, A., Singh, G.K.: Cuckoo search algorithm and wind driven optimization based study of satellite image segmentation for multilevel thresholding using Kapur's entropy. Expert Syst. Appl. **41**(7), 3538–3560 (2014)
13. Güvenç, U.: Combined economic emission dispatch solution using genetic algorithm based on similarity crossover. Sci. Res. Essays **5**(17), 2451–2456 (2010)
14. dos Santos Coelho, L., Lee, C.-S.: Solving economic load dispatch problems in power systems using chaotic and Gaussian particle swarm optimization approaches. Int. J. Electr. Power Energy Syst. **30**(5), 297–307 (2008)
15. Balamurugan, R., Subramanian, S.: A simplified recursive approach to combined economic emission dispatch. Electr. Power Comp. Syst. **36**(1), 17–27 (2007)
16. Dhanalakshmi, S., Kannan, S., Mahadevan, K., Baskar, S.: Application of modified NSGA-II algorithm to combined economic and emission dispatch problem. Int. J. Electr. Power Energy Syst. **33**(4), 992–1002 (2011)

Reliability–Redundancy Allocation Using Random Walk Gray Wolf Optimizer

Shubham Gupta, Kusum Deep and Assif Assad

Abstract From some past recent years, Swarm Intelligence (SI) based optimization algorithms have shown their impact in finding the efficient solutions of real-life application problems that occur in engineering, science, industry, and in various other fields. Gray Wolf Optimizer (GWO) is an efficient and popular optimizer in the area of SI to solve nonlinear complex optimization problems. GWO mimics the dominant leadership characteristic of gray wolves to catch the prey. But, like other stochastic search algorithms, GWO gets trapped in local optimums in some cases. Therefore in the present study, Random Walk Gray Wolf Optimizer (RW-GWO) is applied to determine—(1) the optimal redundancies to optimize the system reliability with constraints on volume, weight, and system cost in series, series–parallel, and complex bridge systems and (2) the optimum cost of two different types of complex systems with constraints imposed on system reliability. The obtained results are compared with classical GWO and some other optimization algorithms that are used to solve reliability problems in the literature. The comparison shows that the RW-GWO is comparatively an efficient algorithm to solve the reliability engineering problems.

Keywords Gray wolf optimizer · Swarm intelligence · Constraint handling · System reliability

S. Gupta (✉) · K. Deep
Department of Mathematics, Indian Institute of Technology Roorkee, Roorkee 247667, Uttarakhand, India
e-mail: g.shubh93@gmail.com

K. Deep
e-mail: kusumfma@iitr.ac.in

A. Assad
Department of Computer Science and Engineering, Islamic University of Science & Technology, Awantipora 247955, Jammu and Kashmir, India
e-mail: assifassad@gmail.com

© Springer Nature Singapore Pte Ltd. 2020
K. N. Das et al. (eds.), *Soft Computing for Problem Solving*,
Advances in Intelligent Systems and Computing 1048,
https://doi.org/10.1007/978-981-15-0035-0_75

1 Introduction

The importance of system reliability has attracted researchers toward the field of reliability optimization. In the construction of modern engineering design, computer simulation plays an important role but it causes some difficulties in reliability optimization like increasing computational complexity. Sometimes the computer simulation process takes a very long time to design the system. Therefore for the optimization process in reliability systems, an algorithm is required which takes less computational time and provides high accuracy in obtaining the solution. In this direction, nature-inspired optimization algorithms [1] have achieved enormous attention in finding the solution to reliability problems. The probabilistic nature of these techniques helps in avoiding the local optimums and provides an optimum direction toward the global optima. These techniques do not use the gradient or any other information related to the problem and considered the optimization problem as a black box. In the area of nature-inspired algorithms, one of the interesting development was the No Free Lunch Theorem (NFL) [2] which states that there is no single optimization algorithm exist or can be designed which is suitable for all type of problems. SI is a very popular branch in the field of nature-inspired optimization algorithms, in which the social and intelligent behavior of various creatures (like ant, bee, whale, etc.) are mimicked in mathematical form to find the solution of the problem. In SI based algorithms, a swarm of search agents is evolved to progress the search. "Particle Swarm Optimization (PSO)" [3], "Ant Colony Optimization (ACO)" [4], "Artificial Bee Colony (ABC) algorithm" [5], "Whale Optimization Algorithm (WOA)" [6] and "Gray Wolf Optimizer (GWO)" [7] etc., are some popular SI algorithms. In these algorithms, exploitation and exploration are two conflicting operators [8] that works to achieve the optimal solution. Exploration refers to discovering new search regions of search space to escape from stagnation in suboptimal solutions and exploitation refers to visit the promising search regions around previously discovered solutions. For the appropriate working of an optimization algorithm, a suitable balance between these two operators is necessary. In the literature, from the experimental analysis, it has been analyzed that the classical GWO becomes trapped in local solutions due to insufficient diversity. Therefore in the present work, RW-GWO [9, 10] is implemented on reliability problems, which is an enhanced version of classical GWO to prevent the stagnation in local solutions.

The remaining article is organized as—Sect. 2 presents a summary of reliability optimization. In Sect. 3, the mathematical form of the reliability problems is described. Section 4 presents an overview of GWO and RW-GWO. In Sect. 5, a simple constraint handling based on simple natural selection is discussed. In Sect. 6, the experimental results are presented and analyzed. Finally, in Sect. 7, the conclusion of the work is discussed with some future directions.

2 Literature Review

Generally, reliability refers—"the probability that the system will not fail during delivery of service" [11]. A system designer always wants to design a system with minimum cost and maximum reliability with increasing comfort and enhancing the functional safety and this can be achieved by the process of optimization. In [12], first time Coit and Smith applied the Genetic Algorithm (GA) to find solution of reliability problems. Ravi et al. [13] have developed the modified Simulated Annealing (SA) named as INESA to solve the reliability problems. In [14], a variety of complex reliability problems are formulated as fuzzy optimization problems. Ant Colony Optimization (ACO) algorithm has been also implemented on these reliability problems [15]. Deep and Deepti [16] have implemented self-organizing migrating GA to find the efficient solution of these problems. In [17], Mutingi and Kommula have used fuzzy multi-criteria GA on the reliability optimization problems. In [18], Kuo and Prasad presented an overview of methods applied on reliability problems. In [19], Levitin and Lisnianski have proposed a new technique for multistate system problems of reliability. In these techniques, GA and UGF (universal generating system) are hybridized. Kumar et al. [20] have used GWO for only one case of complex bridge system and for life life-support system in space capsule. Majety et al. [21] have used cutting plane method for reliability allocation. In [22], random search technique has been employed on reliability allocation problem. In [23], surrogate constraint algorithm and in [24], a dynamic programming has been used to solve redundancy reliability allocation problem. In [25], Liang and Chen have shown the efficiency of neighborhood search algorithm in obtaining the solution of redundancy allocation in series–parallel systems. In [26], Tabu search algorithm has been implemented to solve redundancy allocation problem. Kuo et al. [27] have critically reviewed heuristics for system reliability. In [28], surrogate constraint algorithm has been used for reliability optimization problem. In [29], an improved version of algorithm proposed by Agarwal et al. [30] is presented for reliability optimization. Xu et al. [31] have proposed an approach to solve system reliability problems. Hsieh et al. [32] and Yokota et al. [33] have used Genetic Algorithm (GA) for reliability design problems. In [34], Chen has used penalty guided artificial immune algorithm for redundancy allocation problems in reliability. Kim et al. [35] have use simulated annealing optimization approach for solving redundancy allocation problems. Yeh et al. [36] have used ABC algorithm and Wu et al. [37] have used improved PSO for solving reliability problems. Dhingra [38] have used sequential minimization techniques with heuristic algorithms to find an optimum solution of reliability problems.

3 Problem Description

In this section, four problems of finding the number of redundancies to maximize the system reliability and two problems of reliability optimization have been considered to analyze the efficiency of random walk gray wolf optimizer. In the reliability design systems, it is assumed that if any component fails to operate, then the entire system will not be damaged. The notations which are used in the paper are as follows:

$q_i = (1 - r_i)$	Unreliability of the subsystem i
y_i	No. of components in ith subsystem
R_S	System reliability
$z_i = 1 - (1 - r_i)^{r_i}$	ith subsystem reliability
w_i	Weight component in an ith subsystem
c_i	Cost of component in an ith subsystem
p_i	Volume of component in an ith subsystem
Z^+	Set of all positive integers

3.1 Reliability–Redundancy Allocation Problem

In this section, four problems concerning the determination of number of redundancies to maximize the system reliability are discussed. In the first three problems, all constraints are nonlinear in nature and are series–parallel, series system, and complex system, respectively. The fourth problem is overspeed protection system problem. The objective in all these four problems is to optimize the reliability of the system with multiple nonlinear constraints. In these problems, y_i represents the number of components in a subsystem i and this should be integer. Therefore, these problems are considered as mixed-integer optimization problems. In these problems, redundancy allocations and component reliabilities are determined simultaneously. The line diagram of these systems is presented in Fig. 1. These problems summarized as follows:

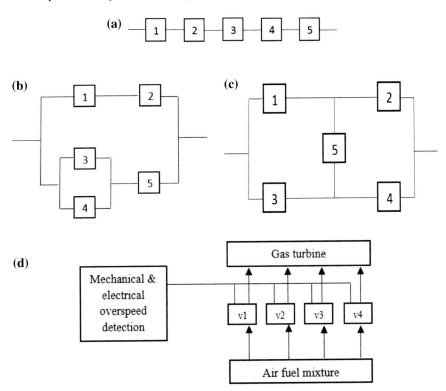

Fig. 1 a Series, b series–parallel, c complex bridge, d overspeed gas turbine systems

1. **Series system problem**

$$\textbf{Max } R_S = \prod_{i=1}^{5} 1 - (1 - r_i)^{y_i}$$

$$\textbf{s.t.} \quad \sum_{i=1}^{5} p_i y_i^2 - 110 \leq 0$$

$$\sum_{i=1}^{5} c_i \left(\frac{-1000}{\log(r_i)} \right)^{\beta_i} \left[y_i + e^{y_i/4} \right] - 175 \leq 0$$

$$\sum_{i=1}^{5} w_i \left[y_i \cdot e^{y_i/4} \right] - 200 \leq 0$$

$$\beta_i = 1.5 \quad \text{for all } i$$

$$0.5 \leq r_i \leq 1 \quad \text{for all } i$$

$$1 \leq y_i \leq 5, \text{ and } y_i \in Z^+ \quad \text{for all } i$$

The values of p_i, c_i and w_i are provided in Table 1.

Table 1 Values of p_i, c_i, and w_i corresponding to problem 1 and 3

i	1	2	3	4	5
p_i	1	2	3	4	2
c_i	2.33×10^{-5}	1.45×10^{-5}	0.541×10^{-5}	8.050×10^{-5}	1.950×10^{-5}
w_i	7	8	8	6	9

Table 2 Values of p_i, c_i, and w_i corresponding to problem 2

i	1	2	3	4	5
p_i	2	4	5	8	4
c_i	2.50×10^{-5}	1.45×10^{-5}	0.541×10^{-5}	0.541×10^{-5}	2.10×10^{-5}
w_i	3.5	4.0	4.0	3.5	3.5

2. Series–parallel system

$$\text{Max } R_S = \prod_{i=1}^{5} 1 - (1 - z_1 z_2)[1 - (z_3 + z_4 - z_3 z_4) z_5]$$
$$\text{where } z_i = 1 - (1 - r_i)^{y_i}$$
$$\text{s.t.} \quad \sum_{i=1}^{5} p_i y_i^2 - 180 \le 0$$
$$\sum_{i=1}^{5} c_i \left(\frac{-1000}{\log(r_i)} \right)^{\beta_i} \left[y_i + e^{y_i/4} \right] - 175 \le 0$$
$$\sum_{i=1}^{5} w_i \left[y_i \cdot e^{y_i/4} \right] - 100 \le 0$$
$$\beta_i = 1.5 \quad \text{for all } i$$
$$0.5 \le r_i \le 1 \quad \text{for all } i$$
$$1 \le y_i \le 5, \text{ and } y_i \in Z^+ \text{ for all } i$$

The values of p_i, c_i and w_i are provided in Table 2.

3. Complex bridge system

$$\text{Max } R_S = z_5(1 - q_1 q_3)(1 - q_2 q_4) + q_5[1 - (1 - z_1 z_2)(1 - z_3 z_4)]$$
$$\text{where } z_i = 1 - (1 - r_i)^{y_i}$$
$$\text{s.t.} \quad \sum_{i=1}^{5} p_i y_i^2 - 110 \le 0$$
$$\sum_{i=1}^{5} c_i \left(\frac{-1000}{\log(r_i)} \right)^{\beta_i} \left[y_i + e^{y_i/4} \right] - 175 \le 0$$
$$\sum_{i=1}^{5} w_i \left[y_i \cdot e^{y_i/4} \right] - 200 \le 0$$
$$\beta_i = 1.5 \quad \text{for all } i$$
$$0.5 \le r_i \le 1 \quad \text{for all } i$$
$$1 \le y_i \le 5, \text{ and } y_i \in Z^+ \text{ for all } i$$

The values of p_i, c_i, and w_i are provided in Table 1.

Table 3 Values of p_i, c_i, and w_i corresponding to problem 4

i	1	2	3	4
p_i	1	2	3	2
c_i	1.0×10^{-5}	2.3×10^{-5}	0.30×10^{-5}	2.3×10^{-5}
w_i	6	6	8	7

4. Overspeed protection system

$$\text{Max } R_S = \prod_{i=1}^{4} 1 - (1 - r_i)^{y_i}$$

$$\text{s.t.} \quad \sum_{i=1}^{5} p_i y_i^2 - 250 \leq 0$$

$$\sum_{i=1}^{4} c_i \left(\frac{-1000}{\log(r_i)} \right)^{\beta_i} \left[y_i + e^{y_i/4} \right] - 400 \leq 0$$

$$\sum_{i=1}^{4} w_i \left[y_i \cdot e^{y_i/4} \right] - 500 \leq 0$$

$$\beta_i = 1.5 \text{ for all } i$$

$$0.5 \leq r_i \leq 1 \text{ for all } i$$

$$1 \leq y_i \leq 5, \text{ and } y_i \in Z^+ \text{ for all } i$$

The values of p_i, c_i, and w_i are provided in Table 3.

3.2 Determination of Optimum Cost

In this subsection, two problems of determining the minimum cost in life-support system in a space capsule and complex bridge system are considered. The description of both problems is as follows.

5. Life-support system in a space capsule

This problem is about the minimization of the cost of life-support system with system reliability constraint. This problem is nonlinear but continuous in nature. The line diagram of this problem is depicted in Fig. 2. Originally, this problem is reported by Ravi et al. [14]. This system has four components and r_i is the ith component reliability. Therefore, the reliability of the system is defined as

$$R_S = 1 - r_3[(1 - r_1)(1 - r_4)]^2 - (1 - r_3)[1 - (1 - r_4)(1 - r_1)]^2$$

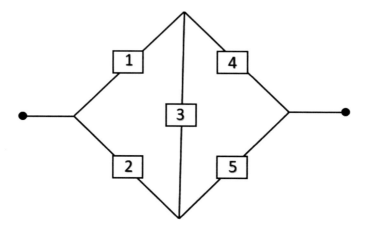

Fig. 2 Line diagram of complex bridge network

Mathematically the problem can be presented as follows:

$$\textbf{Min}\quad C_S = 2\left(K_1 r_1^{\beta_1} + K_2 r_2^{\beta_2}\right) + K_3 r_3^{\beta_3} + 2K_4 r_4^{\beta_4}$$

$$\textbf{s.t.}\quad 0.9 \le R_S \le 1$$

$$0.5 \le r_i \le 1 \quad \text{for all } i$$

$$\text{where } \beta_i = 0.6 \quad \text{for all } i$$

$$K_1 = K_2 = 100, \, K_3 = 200, \, K_4 = 150$$

6. Complex bridge network

The objective of the problem is cost minimization of complex bridge network with constraint defined on the reliability of system. This problem is taken from Mohan and Shanker [22]. The line diagram of this problem is shown in Fig. 3. The complex bridge network system has five components and r_i is the ith component reliability. Therefore, the system reliability can be defined as

$$R_S = r_1 r_4 + r_2 r_5 + r_2 r_3 r_4 + r_1 r_3 r_5 + 2r_1 r_2 r_3 r_4 r_5 - r_2 r_3 r_4 r_5 - r_1 r_3 r_4 r_5 - r_1 r_2 r_4 r_5$$
$$- r_1 r_2 r_3 r_5 - r_1 r_2 r_3 r_4$$

The mathematical form of the problem is given by

$$\textbf{Min}\quad \sum_{i=1}^{5} \alpha_i e^{d_i/(1-r_i)}$$

$$\textbf{s.t.}\quad 0.99 \le R_S \le 1$$

$$0.5 \le r_i \le 1 \quad \forall i = 1, 2, 3, 4, 5.$$

$$\text{where } d_i = 0.0003, \, \alpha_i = 1 \; \forall i = 1, 2, 3, 4, 5.$$

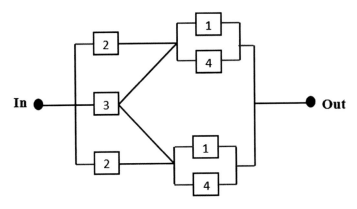

Fig. 3 Line diagram of space capsule

4 Overview of Gray Wolf Optimizer (GWO)

GWO is a recently developed, efficient, and reliable algorithm which mimics the behavior of gray wolves. GWO is proposed by analyzing the dominant leadership characteristic in a gray wolf pack. Gray wolves always try to find an optimal way to find the prey. Gray wolf optimizer mimics a similar search mechanism to determine the efficient solution of the optimization problem. In the hunting mechanism by gray wolves, a leadership hierarchy is followed which is shown graphically in Fig. 4. In a gray wolf pack, wolves have divided four different groups—Alpha (dominant wolf), beta (subordinate wolf to the alpha), delta (caretakers and hunters of the pack), and

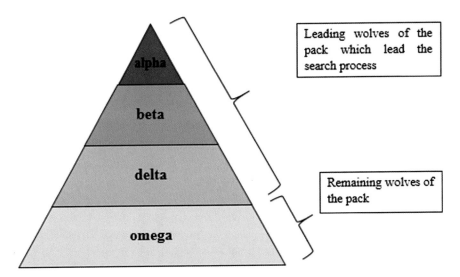

Fig. 4 Leadership hierarchy in a pack of gray wolves

omega (remaining wolves). These wolves follow the three steps to hunt the prey—1. Chasing 2. Encircling 3. Attacking the prey. In the optimization problem, the leading wolves according to their fitness are called as alpha, beta, and delta. To encircle the prey, the following equations are proposed in [7]

$$Y(t + 1) = Y_P(t) - A \cdot D \tag{1}$$

where

$$D = |CY_P(t) - Y(t)| \tag{2}$$

$$A = 2\tau \, \text{rand}_1 - \tau \tag{3}$$

$$C = 2\,\text{rand}_2 \tag{4}$$

$$\tau = 2 - 2\left(\frac{t}{T}\right) \tag{5}$$

where $Y(t + 1)$ and $Y(t)$ denote the locations of wolf at generation t and $t + 1$, respectively. $Y_P(t)$ represents the state of prey at generation t. A and C are the random numbers which are responsible for exploration and exploitation in search process. τ is a parameter that decreases linearly in algorithm. rand_1 and rand_2 are the random numbers which are selected from uniform distribution and lies in the interval $(0, 1)$. To attack a prey, it was assumed that the leading hunters have enough knowledge about the prey. The following mathematical equations are proposed by Mirjalili et al. [7].

$$Y_1 = Y_\alpha(t) - A_1 \cdot D_1 \tag{6}$$

$$Y_2 = Y_\beta(t) - A_2 \cdot D_2 \tag{7}$$

$$Y_3 = Y_\delta(t) - A_3 \cdot D_3 \tag{8}$$

where

$$D_1 = |C_1 Y_\alpha(t) - X_t| \tag{9}$$

$$D_2 = |C_2 Y_\beta(t) - X_t| \tag{10}$$

$$D_3 = |C_3 Y_\delta(t) - X_t| \tag{11}$$

$$Y(t + 1) = \frac{Y_1 + Y_2 + Y_3}{3} \tag{12}$$

Here, the values of A_1, A_2, A_3 and C_1, C_2, C_3 can be obtained with the help of Eqs. (3) and (4), respectively. Random numbers, C and A are liable for the exploitation and exploration of a search space during the search process. But it has been observed that GWO suffers from the stagnation problem and the pack trapped during the search process which results in premature convergence. Therefore in [9, 10], Gupta and Deep have proposed the enhanced version of GWO called RW-GWO by proposing the search equations for leading hunters. In RW-GWO, the mechanism of updating states for omega wolves is kept same and the search equation for the leading hunters are given as

$$Y^{\text{new}} = Y^{\text{old}} + K \times C_{\text{rand}} \tag{13}$$

where Y^{new} denotes the updated position of leader using random walk, Y^{old} is the current positions of leader, K is a parameter which decreases linearly from 2 to 0 as the iterations of algorithm proceeds, and C_{rand} is a random number which has been selected from the Cauchy distribution. The comprehensive details of the algorithm can be found in [9, 10]. The steps of algorithm are shown in Algorithm 1.

5   Constraint Handling Used in RW-GWO

As the reliability optimization problems are complex, constrained, and nonlinear in nature. Therefore, a simple constraint handling based on natural selection of the best candidate is incorporated in GWO. This constraint handling can be defined as

I.   Sort the wolf pack in ascending order of constraint violation.
II. Again sort the feasible wolves (solutions) in decreasing order of fitness. (i.e., for minimization problem sort the feasible solutions to their increasing order of objective function values).
III. Select the top three solutions as leading wolf alpha, beta, and delta.

The constraint violation viol_Y [39] corresponding to the solution Y for general optimization problem

$$\textbf{Min } f(X), \quad X = (x_1, x_2, \ldots, x_d) \in \textbf{R}^d$$
$$\textbf{s.t. } g_j(X) \leq 0, \quad j = 1, 2, \ldots, p.$$
$$h_k(X) = 0, \quad k = 1, 2, \ldots, q.$$
$$a_i \leq x_i \leq b_i, \quad i = 1, 2, \ldots, d.$$

can be evaluated as

$$\text{Cons\_viol}_Y = \sum_{j=1}^{p} \max\big(g_j(Y), 0\big) + \sum_{k=1}^{q} |h_k(Y)|$$

where g_j and h_k are inequality and equality constraints which can be linear or nonlinear. $f : S \subseteq R^d \rightarrow R$ is an objective function which can also be linear or nonlinear. S denotes the feasible space formed by inequality, equality, and bound constraints. The presented constraint handling is very simple to implement in any algorithm and it is a natural way of selecting the best-fitted solutions in the population.

Algorithm 1. Pseudo code of RW-GWO

1.	Initialization of the wolf population using uniform distribution
2.	Calculate the fitness of each wolf using objective function
3.	Initialize the parameters T (maximum number of iterations) and coefficient τ
4.	Select the leading wolves
5.	Initialize the loop counter $t = 0$
6.	**while** $t < T$ do
7.	**for** each leading wolf do
8.	update the location using random walk
9.	**end**
10.	**for** each omega wolf do
11.	Update the location with the help of equations (1) - (12).
12.	**end**
13.	Calculate the fitness of each wolf using objective function
14.	update the leading wolves
15.	update the coefficient τ
16.	$t = t + 1$
17.	**end**

---

6   Experimental Setup and Results

To implement RW-GWO on reliability problems, 30 search agents (size of wolf pack) are considered to lead the search process and the maximum number of iterations is fixed as 1,000. The obtained results on reliability engineering problems are presented in Tables 4, 5, 6, 7, 8, and 9. In these tables, the results are compared with GWO and various other methods proposed by researchers, to investigate the efficiency of random walk gray wolf optimizer in solving the reliability problems. From the presented results, it can be observed that the random walk gray wolf optimizer algorithm outperforms other algorithms in all the problems.

Table 4 Comparison of results for problem 1

Algorithm	Gopal et al. [29]	Kuo et al. [27]	Hsieh et al. [32]	Hikita et al. [28]	GWO [7]	RW-GWO [9, 10]
y	(3, 2, 2, 3, 3)	(3, 3, 2, 3, 2)	(3, 2, 2, 3, 3)	(3, 2, 2, 3, 3)	(3, 2, 2, 3, 3)	(3, 2, 2, 3, 3)
r_1	0.80000	0.77960	0.779427	0.774887	0.782759265	0.777668159
r_2	0.8625	0.80065	0.869482	0.870065	0.871654753	0.872684982
r_3	0.90156	0.90227	0.902674	0.898549	0.903274049	0.903403893
r_4	0.7	0.71044	0.714038	0.716524	0.710063008	0.710195698
r_5	0.80000	0.85947	0.786896	0.791368	0.786339671	0.789463986
R_S	0.930289	0.92975	0.931578	0.931451	0.931656041	0.931662895

Table 5 Comparison of results for problem 2

Algorithm	Chen [34]	Hikita et al. [28]	Wu et al. [37]	Yeh and Hsieh [36]	Hsieh et al. [32]	Kim et al. [35]	GWO [7]	RW-GWO [9, 10]
y	(2, 2, 2, 2, 4)	(3, 3, 1, 2, 3)	(2, 2, 2, 2, 4)	(2, 2, 2, 2, 4)	(2, 2, 2, 2, 4)	(2, 2, 2, 2, 4)	(2, 3, 2, 2, 4)	(3, 2, 2, 2, 4)
r_1	0.812485	0.83819295	0.81918526	0.8197457	0.785452	0.812161	0.840300996	0.77554094
r_2	0.843155	0.85506525	0.8436642	0.8450080	0.842998	0.853346	0.794168271	0.86895059
r_3	0.897385	0.87885933	0.89472992	0.8954581	0.885333	0.897597	0.891499254	0.890614316
r_4	0.894516	0.91140223	0.89537628	0.9009032	0.917958	0.900710	0.89547738	0.89300287
r_5	0.870590	0.85035522	0.86912724	0.8684069	0.870318	0.866316	0.865215359	0.863536231
R_S	0.99997658	0.99996875	0.99997664	0.99997731	0.99997418	0.99997631	0.999984407	0.999986316

Table 6 Comparison of results for problem 3

Algorithm	Kim et al. [35]	Hikita et al. [28]	Yeh and Hsieh [36]	Hsieh et al. [32]	GWO [7]	RW-GWO [9, 10]
y	(3, 3, 3, 3, 1)	(3, 3, 2, 3, 2)	(3, 3, 2, 4, 1)	(3, 3, 3, 3, 1)	(3, 3, 2, 4, 1)	(3, 3, 2, 4, 1)
r_1	0.807263	0.81448	0.828087	0.814090	0.823928677	0.829388102
r_2	0.868116	0.82138	0.857805	0.864614	0.856745698	0.857867267
r_3	0.872862	0.89615	0.704163	0.890291	0.915389329	0.91334922
r_4	0.712673	0.71309	0.648146	0.701190	0.650969587	0.6472368
r_5	0.751034	0.81409	0.914240	0.734731	0.722725037	0.705081068
R_S	0.99988764	0.99978937	0.99948407[a]	0.99987916	0.999889344	0.999889574

[a]Represents the infeasible solution

Table 7 Comparison of results for problem 4

Algorithm	Chen [34]	Dhingra [38]	Kim et al. [35]	Yokota et al. [33]	GWO [7]	RW-GWO [9, 10]
y	(5, 5, 5, 5)	(6, 6, 3, 5)	(5, 5, 5, 5)	(3, 6, 3, 5)	(5, 5, 5, 5)	(5, 5, 5, 5)
r_1	0.903800	0.81604	0.895644	0.965593	0.898610088	0.898435674
r_2	0.874992	0.80309	0.885878	0.760592	0.886050128	0.885577014
r_3	0.919898	0.98364	0.912184	0.972646	0.916062592	0.916392324
r_4	0.890609	0.80373	0.887785	0.804660	0.885377572	0.885874802
R_S	0.999942	0.99961	0.999945	0.999468	0.999946123	0.999946135

Table 8 Comparison of results for problem 5

Algorithm	Tillman et al. [40]	Ravi et al. [13]	Ravi et al. [14]	GWO [7]	RW-GWO [9, 10]
r_1	0.50001	0.50006	0.50000	0.500000	0.500000
r_2	0.84062	0.83887	0.83892	0.838923916	0.838921122
r_3	0.5	0.50001	0.5	0.500000	0.500000
r_4	0.5	0.50002	0.5	0.500000	0.500000
R_S	0.90050	0.90001	0.90000	0.900001061	0.90000028
C_S	642.040	641.8332	641.8240	641.8240535	641.8236938

Table 9 Comparison of results for problem 6

Algorithm	Mohan and Shanker [22]	Ravi et al. [13]	Ravi et al. [14]	GWO [7]	RW-GWO [9, 10]
r_1	0.93924	0.93747	0.93635	0.933708869	0.934353163
r_2	0.93454	0.93291	0.93869	0.935313602	0.935176536
r_3	0.77154	0.78485	0.80615	0.777558341	0.799453798
r_4	0.93938	0.93641	0.93512	0.936446475	0.934747738
r_5	0.92844	0.93342	0.93476	0.935536801	0.934577824
R_S	0.99004	0.99000	0.9905	0.9900011	0.9900001
C_S	5.02001	5.01993	5.02042	5.0199301	5.019920337

7 Conclusion

In the present article, RW-GWO is applied for constrained, nonlinear, and complex reliability problems. The reliability problems are difficult to solve as compared to general nonlinear problems because of NP hardness. In the present work, four redundancy allocation and two reliability problems are attempted to solve using RW-GWO. The results of RW-GWO are compared with classical GWO and with other algorithms which are used in the literature to solve these reliability problems. The demonstrated results show the superior ability of RW-GWO on reliability engineering problems as compared to other algorithms.

References

1. Yang, X.-S.: Nature-inspired optimization algorithms. Elsevier (2014)
2. Wolpert, D.H., Macready, W.G., et al.: No free lunch theorems for search. Technical Report, Technical Report SFI-TR-95–02-010, Santa Fe Institute (1995)
3. Eberhart, R., Kennedy, J.: A new optimizer using particle swarm theory. In: Proceedings of the Sixth International Symposium on Micro Machine and Human Science, 1995, MHS'95, pp. 39–43. IEEE (1995)

4. Dorigo, M., Birattari, M., Stutzle, T.: Ant colony optimization. IEEE Comput. Intell. Mag. **1**(4), 28–39 (2006)
5. Karaboga, D., Basturk, B.: A powerful and efficient algorithm for numerical function optimization: artificial bee colony (abc) algorithm. J. Global Optim. **39**(3), 459–471 (2007)
6. Mirjalili, S., Lewis, A.: The whale optimization algorithm. Adv. Eng. Softw. **95**, 51–67 (2016)
7. Mirjalili, S., Mirjalili, S.M., Lewis, A.: Grey wolf optimizer. Adv. Eng. Softw. **69**, 46–61 (2014)
8. Črepinšek, M., Liu, S.-H., Mernik, M.: Exploration and exploitation in evolutionary algorithms: a survey. ACM Comput. Surv. (CSUR) **45**(3), 35 (2013)
9. Gupta, S., Deep, K.: A novel random walk grey wolf optimizer. Swarm Evol. Comput. (2018a)
10. Gupta, S., Deep, K.: Random walk grey wolf optimizer for constrained engineering optimization problems. Comput. Intell. (2018b)
11. Kumar, A., Singh, S.: Reliability analysis of an n-unit parallel standby system under imperfect switching using copula. Comput. Model. New Technol. **12**(1), 47–55 (2008)
12. Coit, D.W., Smith, A.E.: Reliability optimization of series-parallel systems using a genetic algorithm. IEEE Trans. Reliab. **45**(2), 254–260 (1996)
13. Ravi, V., Murty, B., Reddy, J.: Nonequilibrium simulated-annealing algorithm applied to reliability optimization of complex systems. IEEE Trans. Reliab. **46**(2), 233–239 (1997)
14. Ravi, V., Reddy, P., Zimmermann, H.-J.: Fuzzy global optimization of complex system reliability. IEEE Trans. Fuzzy Syst. **8**(3), 241–248 (2000)
15. Shelokar, P.S., Jayaraman, V., Kulkarni, B.: Ant algorithm for single and multiobjective reliability optimization problems. Qual. Reliab. Eng. Int. **18**(6), 497–514 (2002)
16. Deep, K., Deepti: Reliability optimization of complex systems through C-SOMGA. J. Inf. Comput. Sci. **4**(3), 163–172 (2009)
17. Mutingi, M., Kommula, V.P.: Reliability optimization for the complex bridge system: fuzzy multi-criteria genetic algorithm. In: Proceedings of Fifth International Conference on Soft Computing for Problem Solving, pp. 651–663. Springer
18. Kuo, W., Prasad, V.R.: An annotated overview of system-reliability optimization. IEEE Trans. Reliab. **49**(2), 176–187 (2000)
19. Levitin, G., Lisnianski, A.: A new approach to solving problems of multi-state system reliability optimization. Qual. Reliab. Eng. Int. **17**(2), 93–104 (2001)
20. Kumar, A., Pant, S., Ram, M.: System reliability optimization using gray wolf optimizer algorithm. Qual. Reliab. Eng. Int. **33**(7), 1327–1335 (2017)
21. Majety, S.R.V., Dawande, M., Rajgopal, J.: Optimal reliability allocation with discrete cost-reliability data for components. Oper. Res. **47**(6), 899–906 (1999)
22. Mohan, C., Shanker, K.: Reliability optimization of complex systems using random search technique. Microelectron. Reliab. **28**(4), 513–518 (1988)
23. Hikita, M., Nakagawa, Y., Nakashima, K., Yamato, K.: Application of the surrogate constraints algorithm to optimal reliability design of systems. Microelectron. Reliab. **26**(1), 35–38 (1986)
24. Yalaoui, A., Châtelet, E., Chu, C.: A new dynamic programming method for reliability & redundancy allocation in a parallel-series system. IEEE Trans. Reliab. **54**(2), 254–261 (2005)
25. Liang, Y.-C., Chen, Y.-C.: Redundancy allocation of series-parallel systems using a variable neighborhood search algorithm. Reliab. Eng. Syst. Safety **92**(3), 323–331 (2007)
26. Kulturel-Konak, S., Smith, A.E., Coit, D.W.: Efficiently solving the redundancy allocation problem using tabu search. IIE Trans. **35**(6), 515–526 (2003)
27. Kuo, W., Hwang, C.-L., Tillman, F.A.: A note on heuristic methods in optimal system reliability. IEEE Trans. Reliab. **27**(5), 320–324 (1978)
28. Hikita, M., Nakagawa, Y., Nakashima, K., Narihisa, H.: Reliability optimization of systems by a surrogate-constraints algorithm. IEEE Trans. Reliab. **41**(3), 473–480 (1992)
29. Gopal, K., Aggarwal, K., Gupta, J.: An improved algorithm for reliability optimization. IEEE Trans. Reliab. **27**(5), 325–328 (1978)
30. Aggarwal, K., Gupta, J., Misra, K.: A new heuristic criterion for solving a redundancy optimization problem. IEEE Trans. Reliab. **24**(1), 86–87 (1975)
31. Xu, Z., Kuo, W., Lin, H.-H.: Optimization limits in improving system reliability. IEEE Trans. Reliab. **39**(1), 51–60 (1990)

32. Hsieh, Y.-C., Chen, T.-C., Bricker, D.L.: Genetic algorithms for reliability design problems. Microelectron. Reliab. **38**(10), 1599–1605 (1998)
33. Yokota, T., Gen, M., Li, Y.-X.: Genetic algorithm for non-linear mixed integer programming problems and its applications. Comput. Ind. Eng. **30**(4), 905–917 (1996)
34. Chen, T.-C.: Ias based approach for reliability redundancy allocation problems. Appl. Math. Comput. **182**(2), 1556–1567 (2006)
35. Kim, H.-G., Bae, C.-O., Park, D.-J.: Reliability-redundancy optimization using simulated annealing algorithms. J. Qual. Maintenance Eng. **12**(4), 354–363 (2006)
36. Yeh, W.-C., Hsieh, T.-J.: Solving reliability redundancy allocation problems using an artificial bee colony algorithm. Comput. Oper. Res. **38**(11), 1465–1473 (2011)
37. Wu, P., Gao, L., Zou, D., Li, S.: An improved particle swarm optimization algorithm for reliability problems. ISA Trans. **50**(1), 71–81 (2011)
38. Dhingra, A.K.: Optimal apportionment of reliability and redundancy in series systems under multiple objectives. IEEE Trans. Reliab. **41**(4), 576–582 (1992)
39. Deb, K.: An efficient constraint handling method for genetic algorithms. Comput. Methods Appl. Mech. Eng. **186**(2–4), 311–338 (2000)
40. Tillman, F., Hwang, C., Kuo, W.: Optimization of system reliability. Marecel Dekker (1980)

Optimal Control of Roll Axis of Aircraft Using PID Controller

V. Bagyaveereswaran, Subhashini, Abhilash Sahu and R. Anitha

Abstract In this paper, the Proportional Integral Derivative (PID) controller is tuned using genetic algorithm. The optimally tuned controller is implemented in order to increase the stability and the performance of aircraft. The safety feature of flight system could be enhanced with the tuning of PID parameters of the controller for roll axis of any flight. The design of a mathematical model is necessary for describing the latitudinal roll axis of an aviation aircraft. The PID controller can be employed based on the dynamic as well as mathematical modelling of the aircraft system. The Zeigler Nichols (ZN) Method and Genetic Algorithm (GA) optimization technique are considered to tune the PID controller parameters. The fitness function considered for the optimization algorithm is an Integral Absolute Error (IAE) criterion. The MATLAB simulation result shows that the PID controller tuned by the GA method for aviation aircraft dynamics gives better results.

Keywords Roll axis control · PID controller · Optimization · Modelling · Tuning · Genetic algorithm

1 Introduction

Several aircraft control systems and various other process industries are implementing conventional controllers such as PID controller for improvement in the process dynamic performances [1]. The necessity for a fighter pilot to focus on the target

V. Bagyaveereswaran (✉) · Subhashini · A. Sahu · R. Anitha
School of Electrical Engineering, VIT, Vellore 632014, Tamilnadu, India
e-mail: vbagyaveereswaran@vit.ac.in

Subhashini
e-mail: subhashini.2017@vitstudent.ac.in

A. Sahu
e-mail: abhilash.sahu2017@vitstudent.ac.in

R. Anitha
e-mail: ranitha@vit.ac.in

© Springer Nature Singapore Pte Ltd. 2020
K. N. Das et al. (eds.), *Soft Computing for Problem Solving*,
Advances in Intelligent Systems and Computing 1048,
https://doi.org/10.1007/978-981-15-0035-0_76

creates a base for an autopilot system in flights [6, 7]. Any hardware failure destroys the properties of aircraft and the consequences are critical, which leads to the requirement of optimal control [1, 2]. Generally, an aircraft is controlled by the axes. They are elevator, rudder and the ailerons. There two ailerons, which are connected to each other and both ailerons usually moved in the opposite direction to each other [12]. The ailerons are utilized to bank the aircraft. It generates an unbalanced force, which is a part of larger wing lift force. This force leads to flight path to be curved. The aircraft rolling motion can be controlled by regulating any of the roll angles. In aeroplane modelling stage, the aerodynamic forces such as lift and drag as well the planes inertia are taken into consideration. The flight's pitch control is accomplished by shifting the lift on one of a forward or after control surface. The flapped portion of the tail surface of an aircraft is called an elevator if a flap is used. The yaw control is accomplished by deflecting the flap on vertical tail. It is called rudder and roll control is done by glancing the small flaps placed outboard towards the edges of the wing in a differential manner [1]. The PID controller is the widely used controller for flight control. A PID controller does calculate the differences between the measured actual output values of the process at any set point that is required [8]. The controller minimizes the error by adjusting the inputs of the process controlled but still gives high overshoot and long settling time. The GA optimization technique can be used for achieving better performance.

The remaining sections of the paper are structured as follows. The mathematical modelling of the aircraft roll axis is presented in Sect. 2. In Sect. 3 the basics of PID is discussed and GA fundamentals are deliberated in Sect. 4. The simulation results are discussed in Sect. 5 and the proposed methodology is concluded in Sect. 6.

2   Mathematical Modelling

The orientation of axis is chosen in a way that x-axis is parallel to the velocity vector in the initial equilibrium state, which is the stability axes. The schematic block diagram to control the roll axis of the aircraft is shown in Fig. 1. The aircraft roll axis is given in Fig. 2.

The forces acting on the body are as follows:

(1) Lift—along Z-axis
(2) Drag—along X-axis

$$F = ma \tag{1}$$

$$dF = a * dm \tag{2}$$

The rate of change of coordinates in the inertial frame instantaneously coincident with the body axes.

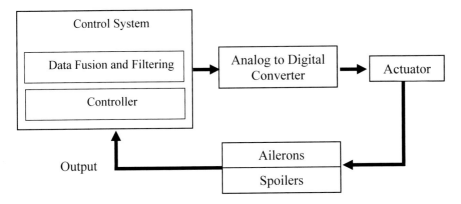

Fig. 1 Schematic block diagram of roll axis control

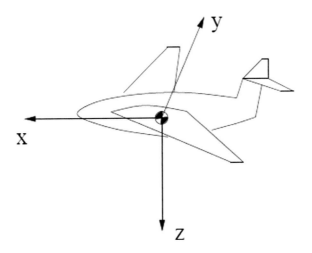

Fig. 2 Diagram of roll axis [2]

$$\dot{X} = u + qZ - rY \qquad (3)$$

$$\dot{Y} = y + rX - pZ \qquad (4)$$

$$\dot{Z} = w + pY - qX \qquad (5)$$

p, q, r = Components of the angular rate of roll, pitch and yawn, respectively
u = Velocity component of roll
v = Velocity components of pitch

- w is the force acting on the body should be equal to all the net external forces acting upon it.

- The forces due to gravitation and propulsion forces are considered as external forces acting on it.
- For an equilibrium state, net external force is equal to force acting on the system.

The system is assumed to be in steady-state cruise at some constant height and pitch angle does not change to any particular speed of the aircraft.

Based on two assumptions, the system can be defined by the following set of equations: [3, 12].

$$Y + mg\cos(\theta) = m(v + ru + pw) \tag{6}$$

$$L = I_x\dot{p} - I_{xz}\dot{r} + qr(I_z - I_y) - I_{xz}pq \tag{7}$$

$$N = -I_{xz}\dot{p} + I_z\dot{r} + pq(I_y - I_x) + I_{xz}qr \tag{8}$$

where

θ = Pitch angle (rotation about the y-axis)
L, m, N = Aerodynamic moment components

The x-axis orientation is selected in a way that the value of product of inertia. As the centre of mass is equal to the origin, $I_{xz} = 0$.

The nonlinear equations are linearized using small disturbance theory and it is given in Eq. (9):

$$\begin{aligned}
u = u_0 + \Delta u, \, Y = Y_0 + \Delta Y, \, L = L_0 + \Delta L \\
M = M_0 + \Delta M, \, \delta = \delta_0 + \Delta\delta, \, v = v_0 + \Delta v \\
w = w_0 + \Delta w, \, p = p_0 + \Delta p, \, r = r_0 + \Delta r
\end{aligned} \tag{9}$$

The propulsive forces acting are assumed to be constant. These assumptions make the system to be considered as linear. Therefore,

$$v_0 = p_0 = q_0 = r_0 = \psi_0 = 0$$

where

Ψ = Heading angle (rotation about z-axis)
$\O$ = Banking angle (rotation about x-axis)

By rearranging the linearized equations shown in Eqs. (10)–(12),

$$\left(\frac{d}{dt} - Y_v\right)\Delta v - Y_p\Delta p + (u_0 - y_0)\Delta r - g\cos(\theta)\Delta\varphi = Y_{\delta r}\Delta\delta_r \tag{10}$$

$$-L_v\Delta v + \left(\frac{d}{dt} - L_p\right) - \left(\frac{I_{xz}}{I_x}\frac{d}{dt} - L_r\right)\Delta r = L_{\delta a}\Delta\delta_a + L_{\delta r}\Delta\delta_r \tag{11}$$

$$-N_v \Delta v + \left(\frac{d}{dt} - N_r\right) - \left(\frac{I_{xz}}{I_z}\frac{d}{dt} - N_p\right)\Delta p = N_{\delta a}\Delta\delta_a + N_{\delta r}\Delta\delta_r \tag{12}$$

Consider $\Delta\beta = \tan^{-1}(\Delta v/u_0)$

where

$\Delta\beta$ = Slide slip angle
Δv = Slide velocity

The state model of the system is given by Eq. (13) [4, 5].

$$
\begin{bmatrix} \Delta\beta_0 \\ \Delta P_0 \\ \Delta\Re_0 \\ \Delta\emptyset_0 \end{bmatrix} = \begin{bmatrix} Y\beta/\upsilon & Yp/\upsilon & -(1-Yr)/\upsilon & s\cos(\theta)/\upsilon \\ L\beta & Lp & Lr & 0 \\ N\beta & Np & Nr & 0 \\ 0 & 1 & 0 & 0 \end{bmatrix}\begin{bmatrix} \Delta\beta \\ \Delta P \\ \Delta\Re \\ \Delta\emptyset \end{bmatrix}
$$
$$
+ \begin{bmatrix} 0 & Y\delta r/\upsilon \\ L\delta a & L\delta r \\ N\delta a & N\delta r \\ 0 & 0 \end{bmatrix}\begin{bmatrix} \Delta\delta a \\ \Delta\delta r \end{bmatrix} \tag{13}
$$

3 PID Controller

Generally, PID generates the controller output value that is proportional to the error value. The error is the difference between the output value of the given system and any desired reference value. The controller reduces the error by means of adjusting it with respect to the pitch control inputs. Since PID has three tuning parameters they are called triple-axis control. The tuning parameters of PID controller represented as gain K_P, K_I and K_D, respectively [10, 11]. The proper tuning of these parameters by a suitable algorithm, the controller could deliver control action that is designed for any specific aircraft requirements. The general transfer function of the PID controller is given in Eq. (14). The closed loop diagram for controlling the aircraft with PID is shown in Fig. 3.

$$G(s) = K_P + \frac{K_I}{s} + K_D s = K_P\left(1 + \frac{1}{\tau_I s} + \tau_D s\right) \tag{14}$$

where

K_P—Proportional gain
K_I—Integral gain
K_D—Derivative gain
τ_I—Integral time constant
τ_D—Derivative time constant.

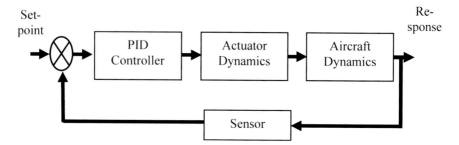

Fig. 3 Block diagram of the aircraft system with a PID controller

4   Genetic Algorithm

Many heuristic algorithms have occurred to find the solution for the optimization problems. One of the best and heuristic algorithms for search used is a Genetic Algorithm. It is based on the theory of selection as well as genetics involved. Genetic Algorithm (GA) is much robust, unlike other algorithms. The main advantage of GA is, a system implemented with GA does not break easily even if there are any slight changes in the inputs (or) in the presence of reasonable noise. So, this GA can be utilized for optimizing the coefficient of the PID controller and the derived controller is referred to PID-GA controller [9].

While implementing GA, it is required to set the values of upper bound as well lower bound for the values to be optimized. The optimized values will be obtained within the prescribed limit on tuning parameters. The selection of these bounds is also an important and critical task. The nearest values of this coefficient are considered as lower bounds and upper bounds. In the next step after the bound selection, fitness function of the PID controller is considered for evaluation and rank wise fitness scaling can be done. The Integral Absolute Error (IAE) is considered as the cost function for the optimization. From scattered to two-point, heuristic, intermediate, custom methods, arithmetic, a scattered method of crossover is considered and the optimization process is called. The factors used in the genetic algorithm are given in Table 1. The above set of parameters that are obtained are considered and GA is implemented for the designed PID controller shown in Fig. 4.

Table 1 GA parameters

Parameter	Type/Value
Maximum generations	200
Population size	10
Encoding	Binary
Selection	Uniform
Crossover	Single point
Mutation	Uniform

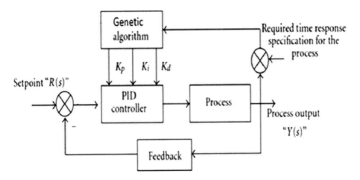

Fig. 4 Block diagram of PID controller tuned with genetic algorithm

5 Simulation and Results

Substituting the values considering general aviation (NAVION), the state model given in Eq. (13) can be transformed as (by ignoring the rudder angle) given in Eq. (15).

$$\begin{bmatrix} \Delta\beta_\circ \\ \Delta P_\circ \\ \Delta\Re_\circ \\ \Delta\emptyset_\circ \end{bmatrix} = \begin{bmatrix} -0.254 & 0 & -1 & 0.183 \\ -15.969 & -8.935 & 2.19 & 0 \\ 4.549 & -0.349 & -0.76 & 0 \\ 0 & 1 & 0 & 0 \end{bmatrix} + \begin{bmatrix} 0 \\ -28.916 \\ -0.224 \\ 0 \end{bmatrix} [\Delta\delta a] \quad (15)$$

The equivalent system transfer function for the state model shown in Eq. (15) can be written as (Fig. 5).

$$\frac{\Delta\varphi}{\Delta\delta_a} = \frac{-28.2s^2 - 29.81s - 140.8}{s^4 + 9.4s^3 + 14.02s^2 + 48.5s + 0.3979}$$

where

$\Delta\delta_a$ = Deflection angle
$\Delta\emptyset$ = Pitch angle

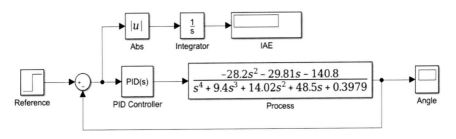

Fig. 5 Simulink diagram for roll axis control

Table 2 PID tuning
parameters

Tuning parameter	Auto tuning	ZN	GA
K_P	−0.0044	−0.001	−0.05
K_I	−0.00072	−0.0003	−0.0045
K_D	0.02555	0.05	0.0345

The tuning parameters obtained through auto tuning, Zeigler Nichols and GA optimization is given in Table 2. The closed loop response for step input with PID is shown in Fig. 6. The performance of closed loop response is compared in terms of various time domain specifications. It is shown in Table 3.

The PID controller that is implemented with these above values gives better performance in terms of any of the required values of the parameters such as rise time, settling time and the peak overshoot. We can find that PID-GA performance is better than the other two methods from Table 3. Therefore, the PID-GA controller is the required efficient controllers modelled to correct/minimize the error effectively.

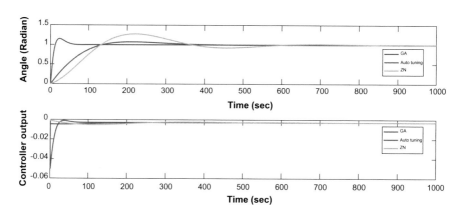

Fig. 6 Closed loop step response with PID controller for different tuning methods

Table 3 Performance
comparison of the closed loop
response

Specifications	PID-Auto tuning	PID-ZN	PID-GA
Rise time (sec)	96.9163	96.1716	9.1416
Settling time (sec)	354.9680	734.9123	55.1097
Peak overshoot (%)	7.3529	28.0674	15.0939
Offset	0	0.0032	0
IAE	62.92	124.4	8.735

6 Conclusion

The genetic algorithm can be used for getting the optimized values of PID controller tuning parameters. The integral absolute error is considered as a cost function for GA optimization. The designed PID controller with the GA has a much efficient response than other tuning methods. The classical tuning methods prove better in giving insight about the starting points for the PID tuning values. However, the GA tuned PID controller proves much better according to the time domain performance parameters such as rise time and settling time. In future other available popular optimization algorithms such as particle swarm optimization, ant colony optimization, simulated annealing and so many can be considered to tune the PID controller for controlling the aircraft roll angle.

References

1. Nair, V., Dileep, M.V., George, V.I.: Aircraft yaw control system using LQR and fuzzy logic controller. Int. J. Comput. Appl. **45**(9), 25–30 (2012)
2. Usta, M., Akyazl, O, Akpmar, A.S.: Aircraft roll control system using LQR and fuzzy logic controller. In: International Symposium on Innovations in Intelligent Systems and Applications (2011)
3. Gouthami, E., Asha Rani, M.: Modelling of an adaptive controller for an aircraft roll control system using PID, fuzzy-PID and genetic algorithm. IOSR J. Electron. Commun. Eng. **11**(1), 15–24 (2016)
4. Akyazi1, O., Usta, M.A., Akpinar, A.S.: A self-tuning fuzzy logic controller for aircraft roll control system. Int. J. Control. Sci. Eng. **2**(6), 181–188 (2012)
5. Fossen, T.I.: Mathematical Models for Control of Aircrafts & Satellites, 2nd edn. (Department of Engineering Cybernetics, NTNU 2011)
6. Murali, S.: Autopilot design for Navion aircraft using intelligent controllers. Int. J. Sci. Eng. Res. **4**(5), 107–110 (2013)
7. Riberio, L.R., Oliveira, N.M.F.: UAV autopilot controllers test platform using matlab/simulink and X-plane. In: 40th ASEE/IEEE Frontiers in Education Conference, Washington, DC (2010)
8. Skarpetis, M., Koumboulis, F.N., Ntellis, A.S.: Longitudinal Flight Multi-Condition Control using Robust PID Controllers. In: IEEE Conference on Emerging Technologies and Factory Automation, Toulouse, France (2011)
9. Jayachitra, A., Vinodha, R.: Genetic algorithm based PID controller tuning approach for continuous stirred tank reactor. Adv. Artif. Intell. (2014). https://doi.org/10.1155/2014/791230
10. Åström, K., Hagglund, T.: PID Controllers: Theory, Design, and Tuning, p. 1994. Instrument Society of America, Research Triangle Park, NC, USA (1994)
11. Dwyer, A.O.: Handbook of PI and PID Controller Tuning Rules, 3rd edn. Imperial College Press (2009)
12. Perrusquía, A., Tovar, C., Soria, A., Martínez, J.C.: Robust controller for aircraft roll control system using data flight parameters. In: 13th International Conference on Electrical Engineering, Computing Science and Automatic Control (CCE) (2016)

Adaptive Noise Cancellation Using Improved LMS Algorithm

Sai Saranya Thunga and Rajesh Kumar Muthu

Abstract In order to attain a higher reduction of the interfering noise, and to improve transmission and reception of the signal-to-noise (SNR) ratio adaptive noise cancellation (ANC) is used. In the application of adaptive noise cancellation most widely used adaptive filtering technique is the least mean square (LMS) algorithm. In this paper an improved least mean square algorithm of flexible step length for adaptive noise cancellation is been used to achieve better noise suppression ability and faster convergence. In MATLAB simulation environment adaptive noise cancellation system is been realized using a speech signal with Gaussian white noise as interfering source. Performance of the LMS algorithms with fixed and variable convergence factor is studied along with performance comparison of designed algorithm against conventional LMS, NLMS and RLS.

Keywords LMS · NLMS · RLS · Adaptive filter · Adaptive noise cancellation · SNR

1 Introduction

An adaptive noise canceller (ANC) is extensively used in echo elimination, fetal heart rate recognition and adaptive antenna system. The purpose of an ANC is to attain an enormous attenuation towards interfering noise for refining signal transmission and reception of the signal-to-noise ratio. In the application of adaptive noise cancellation one of the most popular algorithms is least mean square (LMS). It has widely investigated in the writing, and a substantial amount of outcomes on its unfaltering state misadjustment and its following execution have been acquired [2–8]. The dominant parts of these documents look at the LMS algorithm with a fixed step size.

S. S. Thunga · R. K. Muthu (✉)
Department of Communication Engineering, School of Electronics Engineering, VIT, Vellore, India
e-mail: mrajeshkumar@vit.ac.in

S. S. Thunga
e-mail: saranyathunga@gmail.com

© Springer Nature Singapore Pte Ltd. 2020
K. N. Das et al. (eds.), *Soft Computing for Problem Solving*,
Advances in Intelligent Systems and Computing 1048,
https://doi.org/10.1007/978-981-15-0035-0_77

The choice of the step size reflects a trade-off between misadjustment and the speed of adaptation. In [1], derived expressions show that smaller step size gives small misadjustment at the cost of high convergence time constant. Alternative works have deal with the issue of the optimization of the step size or schemes of varying the step size to get performance-wise efficiency. It seems to us, however, that there is no such detailed analysis or study of a variable step size algorithm that is simple to execute and is capable of giving both fast convergences and minimal misadjustment.

In this work, we proposed an improved LMS algorithm where the square of error is used to adjust the step size. The intention is that a high inaccuracy will cause the step size to rise to provide convergence faster while a minimal slip will result in a reduction in the step size to yield smaller misadjustment. The modification equality can be executed in a simple way, and its form is such that a thorough examination of the algorithm is possible under the usual independent expectations which are frequently made in the works [1] to streamline the investigation of LMS algorithms.

This paper is planned as follows: in Sect. 2 ANC system model is explained in detailed. Section 3 deals with existing and proposed method for adaptive noise cancellation. Section 4 simulations results obtained by traditional LMS, NLMS and RLS along with the proposed method are described. Section 5 gives the conclusion.

2 System Model

As revealed in Fig. 1, an adaptive noise canceller (ANC) has two inputs (a) primary and (b) reference. The primary input obtains a signal's' from the signal source that is degraded by the existence of white noise 'n' which is uncorrelated with the input signal. The reference input receives a white noise n_o which is uncorrelated with the input signal but correlated in some way with the white noise n. The noise n_o passes through an adaptive filter to produce an output $\hat{n}$ that is a close estimate of primary input noise. This noise estimate is deducted from the degraded input signal to generate an estimate of the input signal at $\hat{s}$, the ANC system output.

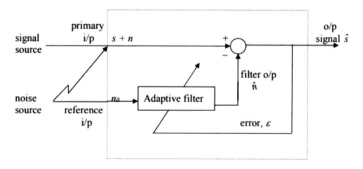

Fig. 1 Adaptive noise canceller

In noise cancelling systems, a real-world objective is to generate a system output $\hat{s} = s + n - \hat{n}$ that is a finest fit in the least-squares sense to the signal 's'. This goal is accomplished by feeding back the system output or the estimated signal to the adaptive filter and tuning the filter weights through an adaptive algorithm to minimalize total system output power.

$$\hat{s} = s + n - \hat{n} \tag{1}$$

Take square on both sides in Eq. (1)

$$\hat{s}^2 = s^2 + (n-\hat{n})^2 + 2s(n-\hat{n}) \tag{2}$$

Take expectation $E[.]$ on both sides and assume s is uncorrelated with n and n_o

$$E[\hat{s}^2] = E[s^2] + E[(n-\hat{n})^2] \tag{3}$$

The signal power $E[s^2]$ will be unaltered as the filter is tuned to minimalize $E[\hat{s}^2]$.

3 Methods

Individually every method establishes an adaptation algorithm that regulates the adaptive filter weights in order to reduce the accompanying error norm.

A. LMS

The most commonly used adaptive filtering technique in ANC is least mean square (LMS) algorithm. Let primary input be $d = s + n$, reference input $x(n)$ be passed through the adaptive filter with weight vector W then the error signal as per Fig. 1 is given by

$$e(n) = d(n) - W^T(n) * x(n) \tag{4}$$

The output error signal is used to update the weight vector W for the next iteration.

$$W(n + 1) = W(n) + \mu x(n)e(n) \tag{5}$$

where μ is the step size parameter which has an important role in the convergence characteristics of the algorithm as well as in its stability condition.

An estimate for the upper bound of this parameter is given in the technical writings [9–11] and may be stated as

$$0 < \mu\text{max} < \frac{2}{\text{trace}[R]} \tag{6}$$

where trace [.] denotes trace operator of matrix; R is the autocorrelation matrix given by

$$R = X^H(n).X(n) \tag{7}$$

B. NLMS

Alternative methods which endeavour to upsurge efficiency at the rate of minimal supplementary computational complication have been projected and are widely deliberated in [3, 4]. One such methodology that has been effectively engaged in circumstances where signal statistics are unknown is the online calculation of the convergence factor which takes part in updating the filter weights [12, 13]. The normalized LMS (NLMS) algorithm can be added in this category [12, 14].

Weight update is given as follows:

$$W(n + 1) = W(n) + \frac{\mu e(n)x(n)}{x^T(n)x(n)} \tag{8}$$

In the above equation if $\mu = 0$, $w(n + 1) = w(n)$ and the weight updating is halted. When $\mu = 1$, the fastest convergence is obtained at the cost of higher misadjustment then the one obtained for $0 < \mu < 1$.

C. RLS

The RLS-type methods have a high convergence rate which is independent of the eigenvalue spread of the input correlation matrix. These schemes are also very useful in applications where the environment is slowly varying. The cost of all these advantages is a considerable increase in the computational complexity of the algorithms belonging to the RLS family.

The adaptive filter weights adaption equations are given by

$$w(n + 1) = w(n) + e(n).k(n) \tag{9}$$

Where $w(n)$ is the filter coefficients and $k(n)$ is the gain vector, $k(n)$ is defined by the following equation:

$$k(n) = \frac{\lambda^{-1}p(n - 1).x(n)}{1 + \lambda^{-1}x^T(n).p(n - 1).x(n)} \tag{10}$$

where λ is the forgetting factor. The value of forgetting factor must be slightly less than 1. In the above equation $p(n)$ is given by

$$p(n) = \lambda^{-1}p(n - 1) - \lambda^{-1}k(n)x^T(n)p(n - 1) \tag{11}$$

D. *Improved LMS*

In the normal LMS μ is a fixed value. Here in this proposed algorithm step size μ is considered as variable μ. Adjustment is given by

$$\mu(n+1) = \alpha.\mu(n) + \gamma.e^2(n) \tag{12}$$

where $0 < \alpha < 1$ and $\gamma > 0$, $\mu_{min} < \mu < \mu_{max}$

Preliminary step size μ_o is typically taken to be μ_{max}, although the algorithm is not sensitive to the choice. From above equation μ is always positive and is controlled by the size of the prediction error and the parameters α and γ. To be more precise, an enormous prediction error increases the step size to provide faster tracking. If the error decreases, the step size will be decreased to reduce the misadjustment. The constant μ_{max} is selected to guarantee that the mean square error of the algorithm remains bounded. A satisfactory criterion for μ_{max} to assurance bounded MSE is

$$\mu max < \frac{2}{3\,trace[R]} \tag{13}$$

4 Simulation and Results Analysis

In this simulation, a speech signal as shown in Fig. 2 is taken as input signal to which a white noise is superimposed. This corrupted speech signal as shown in Fig. 3 which is to be denoised is considered as the primary input to the ANC. An adaptive filter LMS, NLMS, RLS or ILMS is used to denoise the primary input.

Fig. 2 Input speech signal

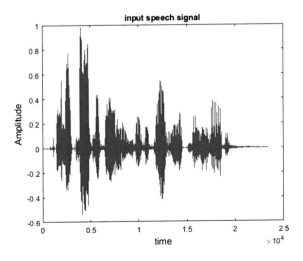

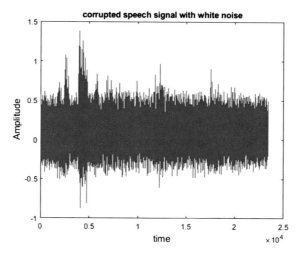

Fig. 3 Corrupted speech signal with white noise

Figure 4 shows the simulation result of the estimated speech signal using the LMS adaptive filter algorithm with step size $\mu = 0.05$. In this case, the filter order is selected as 16.

Figure 5 shows the simulation result of the estimated speech signal using NLMS adaptive filter algorithm with step size $\mu = 0.05$.

Figure 6 shows the simulation result of the estimated speech signal using RLS algorithm. Forgetting factor is chosen as 0.998 which is slightly less than 1.

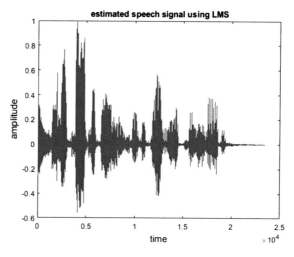

Fig. 4 Estimated signal using LMS

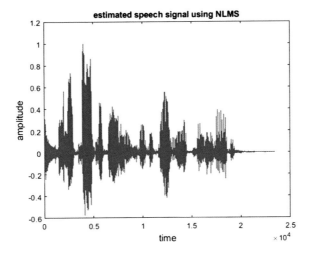

Fig. 5 Estimated signal using NLMS

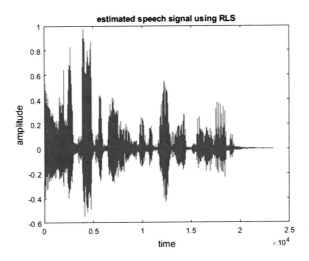

Fig. 6 Estimated signal using RLS

Figure 7 shows the simulation result of the estimated speech signal using proposed algorithm. In this case, the initial step size is considered as $\mu = 0.05$, forgetting factor $\alpha = 0.98$ and $\gamma = 0.01$.

Figure 8 shows the mean square error performance metrics of the LMS, NLMS, RLS and ILMS.

Table 1 shows the SNR and correlation coefficient comparison of the LMS, NLMS, RLS and ILMS algorithms in presence of white noise. ANC input signal SNR is -19.1358 dB. Tabled SNR gives the estimated signal SNR, and it is evident that the ANC improves the SNR by cancelling the noise.

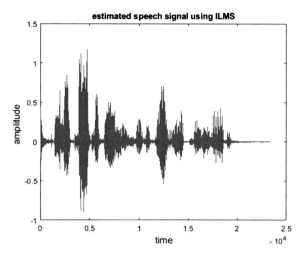

Fig. 7 Estimated speech signal using ILMS

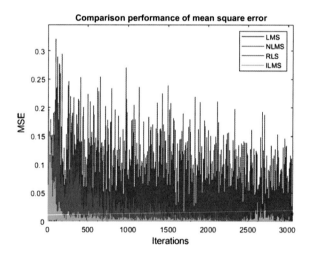

Fig. 8 Mean square error performance of LMS, NLMS, RLS, ILMS

From the table when compared with different algorithms over white noise improved LMS algorithm is found to be more effective. As the algorithm converges at faster rate and has better noise cancellation capability when compared with other algorithms.

Table 1 Performance Comparison

Parameter		White noise			
Step size	Metrics	LMS	NLMS	RLS	ILMS
0.05	SNR	−0.6281	−0.9091	−0.5856	−0.38
	Correlation coefficient	0.9520	0.9103	0.9302	0.9594
	MSE	0.00059	0.0012	0.00084	0.0004
	PSNR	36.8983	34.3049	35.4073	37.8921
0.1	SMR	−0.8833	−0.8619	−0.3340	−0.2514
	Correlation coefficient	0.9539	0.9283	0.957	0.9738
	MSE	0.00062	0.00093	0.00049	0.0003
	PSNR	36.4902	33.6132	37.5663	39.6711
0.15	SMR	−2.1243	−1.006	−0.2401	−0.2060
	Correlation coefficient	0.8555	0.9259	0.9681	0.9789
	MSE	0.0025	0.00099	0.000363	0.00024
	PSMR	38.5132	32.4557	38.6745	40.4208

5 Conclusion

Adaptive noise cancellation for speech signal in noise dominating environment using different adaptive algorithms is implemented in MATLAB simulation environment. Step size is varied in order to study the performance of the algorithms. We can conclude that the performance of the improved least mean square algorithm is more effective when compared with RLS, NLMS and LMS algorithms. The convergence rate of the system using proposed method is highly efficient when compared with LMS, NLMS and RLS.

References

1. Widrow, B., McCool, J.M., Larimore, M.G., Johnson Jr., C.R.: Stationary.and.nonstationary.learning characteristics of the LMS adaptive filter. Proc. IEEE **64**, 1151–1162 (1976)
2. Widrow, B., Glover, Jr, J.R., McCool, J.M., Kaunitz, J., Wiliams, C.S., Hearn, R.H., Zeidler, I. R., Dong, Jr., E., Goodlin, R.C.: Adaptive noise cancelling: principles and applications. Proc. IEEE **63**, 1692–1716 (1975)
3. Widrow, B., Steams, S.D.: Aduprive Signal Processing. Prentice-Hall, Englewood, Cliffs, NJ (1985)
4. Sandhi, M.M., Berkley, D.A.: Silencing echoes on the telephone network. Proc. IEEE **68**, 948–963 (1980)
5. Horowitz, L.L., Senne, K.D.: Performance advantage of complex LMS for controlling narrow-band adaptive arrays. IEEE Trans. Acoust. Speech Signal Process. **29**, 722–736 (1981)
6. Macchi, O., Eweda, E.: Second-order convergence analysis of stochastic adaptive linear filtering. IEEE Trans. Autom. Control **28**, 76–85 (1983)

7. Feuer, A., Weinstein, E.: Convergence analysis of LMS filters with uncorrelated Gaussian data. IEEE Trans. Acoust. Speech Signal Process. **33**, 222–230 (1985)
8. Eweda, E., Macchi, O.: Tracking error bounds of adaptive nonstationary filtering. Automatica. **21**, 293–302 (1985)
9. Chern, S.J., Chang, C.Y.: Adaptive linearly constrained inverse QRD-RLS beam former for moving jammers suppression. IEEE Trans. Antennas Propag. **50**(8), 1138–1150 (2002)
10. Resende, L.S., Romano, J.T., Belanger, M.G.: A fast least-squares algorithm for linearly constrained adaptive filtering. IEEE Trans. Signal Process. **44**(5), 1168–1174 (1996)
11. Johnson, D.H., Dudgeon, D.E.: Array Signal Processing Concepts and Techniques. Prentice-Hall, Englewood Cliffs, NJ, USA (1993)
12. Frost III, O.L.: An algorithm for linearly constraint adaptive array processing. Proc. IEEE **60**(8), 926–935 (1972)
13. Chern, S.J., Chang, C.Y.: Adaptive MC-CDMA receiver with constrained constant modulus IQRD-RLS algorithm for MAI suppression. Signal Process. (Elsevier) **83**(10), 2209–2226 (2003)
14. Schodorf, J.B., Wiliams, D.W.: Array processing techniques for multiuser detection. IEEE Trans. Commun. **45**(11), 1375–1378 (1997)

Variant Roth-Erev Reinforcement Learning Algorithm-Based Smart Generator Bidding as Agents in Electricity Market

P. Kiran and K. R. M. Vijaya Chandrakala

Abstract The dynamically changing deregulated electricity market involves different entities and the aim of each entity is to achieve maximum profit while performing electricity price and power bidding. The agent-based modeling of electricity systems was used to model the market entities under whole sale electricity market operation. This paper discusses about the strategic learning ability of generators in an IEEE 30 bus system using Variant Roth-Erev learning algorithm. It also analyzes the variation in the generator commitments through the implemented learning algorithm during the present day schedule and helps the generator to perform smart bidding in the next electricity market operation. The results presented show that the smart generators are able to bid strategically in the electricity market and which will reflect in its net earnings in a market scheduled on a day-ahead basis.

Keywords Deregulated electricity market · Variant Roth-Erev learning · Agent-based modeling · Independent system operator · Generator company (GenCo)

1 Introduction

The deregulated electricity systems are a market-based system which introduces competition in energy trading and will also provide open access to the infrastructure. Each generating company (GenCo) will attempt to exercise market power and thereby increasing its net profit. The major entities in deregulated electricity market are GenCo, Load Serving Entity (LSE), and the Independent System Operator (ISO) [1]. Various learning algorithms are adopted by the market entities in order to ensure strategic learning behavior [2–4]. The deregulated system being dynamic in nature

P. Kiran (✉) · K. R. M. Vijaya Chandrakala
Department of Electrical and Electronics Engineering,
Amrita School of Engineering, Coimbatore, Amrita Vishwa Vidyapeetham, India
e-mail: kiran7p@gmail.com

K. R. M. Vijaya Chandrakala
e-mail: krm_vijaya@cb.amrita.edu

© Springer Nature Singapore Pte Ltd. 2020
K. N. Das et al. (eds.), *Soft Computing for Problem Solving*,
Advances in Intelligent Systems and Computing 1048,
https://doi.org/10.1007/978-981-15-0035-0_78

suffers with overloading situations during market operation. Under this condition congestion of power flow may occur in transmission lines. This will prevent the system operator to dispatch additional power and this will vary the GenCo commitments. The entities in the market-based system must be intelligent and need to interact with other entities for the dynamic changes happening in the system. Various literature focused on different multi-agent platforms and their modeling [5–8].

In this work the electricity systems are modeled using agents, AMES tool is used to dynamically test the electricity system and to implement the strategic learning capabilities. In the static test case the GenCo's will submit their actual cost and capacities of generators to the ISO but during dynamic testing using AMES, the GenCo's uses Variant Roth-Erev (VRE) reinforcement learning to evaluate the offers and supply. The reinforcement model uses learning strategy and is evolved by Roth and Erev to study the experimental data. Later its modified version MRE reinforced learning and VRE learning algorithm was prepared [9–11].

Different market models available are single buyer model, bilateral model, hybrid model, and pool-co model. Here for the entire analysis the market model considered is a pool-co type and the bidding was done on a day-ahead basis [12–15]. Section 2 deals with the configuration of smart generator agent, Sect. 3 analyzes agent-based approach in the IEEE 30 bus system, and Sect. 4 concludes the paper.

2 Configuration of Smart Generator Agent

The smart generators will have inbuilt learning abilities. Each generator 'i' has a reasonable capacity interval for each hour of a day as given in Eq. 1.

$$\text{Cap}_i^L \leq \text{PG}_i \leq \text{Cap}_i^U \tag{1}$$

where Cap_i^L and Cap_i^U are lower and upper production limit and PG_i is the real power production. The generator cost function varies exponentially with respect to the output power of the generator. The total generator cost function of 'i' (TC_i) is calculated based on Eq. 2.

$$\text{TC}_i = a_i.\text{PG}_i + b_i\text{PG}_i^2 + \text{FCost}_i \tag{2}$$

a_i, b_i, and FCost_i are constants based on the type of generator [15]. The generator marginal cost function of 'i' (MC_i) is calculated based on Eq. 3.

$$\text{MC}_i = a_i + 2.b_i.\text{PG}_i \tag{3}$$

Initially, at the starting of a day 'D' the 'ith' generator chooses an offer supply from action domain AD_i. The AD_i must have a finite cardinality, i.e., $M_i \geq 1$ The AMES framework uses VRE reinforcement learning algorithm which has four learning parameters. They are $q(0)$—initial propensity, C—cooling parameter, r—recency parameter, and e—experimentation parameter. Apart from this generator's action domain is constructed using six action domain parameters. They are {M1, M2, M3, RIMax^L, RIMax^U, SS}, here M1, M2, and M3 are three integer-valued density control parameters, RI is range index parameters and SS is slope parameter [2, 15].

The purpose of initial propensity is to choose a supply offer (m $\in$ AD$_i$) from the action domain and this can be a real number specified by the user. In this analysis, we consider these initial propensities as equal. The choice probabilities that generator 'i' would help to choose the supply offer for a day 'D' is as given in Eq. 4.

$$P_i(D) = \frac{\exp(q_i(D)/C_i)}{\sum_{j=1}^{Mi} \exp(q_{ij}(D)/C_i)} \tag{4}$$

When the cooling parameter $C_i \rightarrow \infty$, then $P_i(D) \rightarrow 1/M_i$, so there is no role for the propensity values and when $C_i \rightarrow 0$, then the choice probabilities becomes high since it is having high propensity values and thereby increasing the probability of choosing that particular supply offer. At the end of day 'D' the initial propensity is updated. Let m^R be the reported supply offer by generator 'i' to the day-ahead market and let the profit attained be Profit$_i(D)$ [2].

$$q_i(D + 1) = [1 - r_i]q_i(D) + \text{Response}_i(D) \tag{5}$$

$$\text{Response}_i(D) = [1 - e_i].\text{Profit}_i(0) \text{ if } m = m^R$$
$$e_i.q_i(D)/[M_i - 1] \text{ if } m \neq m^R \tag{6}$$

The regency parameter on Eq. 5 acts as a hinder to the growth of propensities over time and the experimental parameter on Eq. 6 permits experimentation with other supply offers.

The AMES tool permits synchronized action across all GenCo's. AMES is open source and is capable of dynamic testing with learning traders. The supply offer selected by the GenCo will be based on profit in the past outcomes using a type, namely, Java Reinforcement Learning Algorithm (JReLM) [2, 15]. The algorithm approach to provide maximum utilization of the generator scheduling in the market is shown in Fig. 1.

The AMES is developed in Java language to enable readability and usage. Its open source architecture permits the user to modify the code in order to suit their needs. The dynamic flow with learning implementation is shown in Fig. 2. In a deregulated power market, bids and offers from the traders are collected by ISO as such ISO performs the optimal power flow and calculates the locational marginal price at each node. The inbuilt modules in agent-based analysis are DCOPFJ for power flow solution and JReLM for learning purpose. AMES supports with DCOPF which is used for LMP simulation or forecasting based on production cost model solved using linear programming. The reported cost coefficients (a_i^R, b_i^R) and the reported upper production limit (Cap$_i^{RU}$) determines AMES generator learning. The

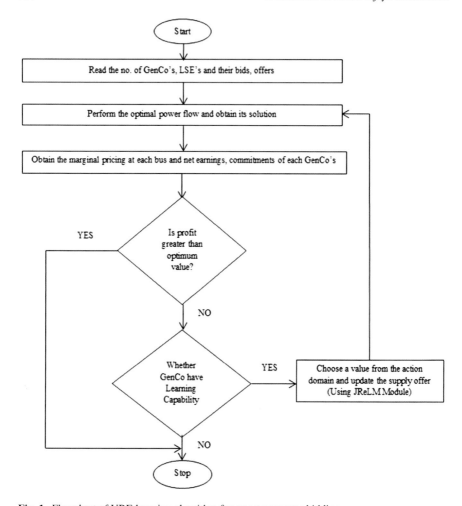

Fig. 1 Flowchart of VRE learning algorithm for smart generator bidding

reported supply offer is represented in the form $S_i^R = (a_i^R, b_i^R, \mathrm{Cap}_i^{RL}, \mathrm{Cap}_i^{RU})$ [15]. The ISO activities for a day are shown in Table 1.

3   Analysis on IEEE 30 Bus System

The 30 bus system comprises 9 GenCo's, 21 LSEs, and 41 branches as shown in Fig. 3. The GenCo's exercise economic capacity withholding and will impose cost greater than the true marginal cost. Table 2 shows the true GenCo cost parameters, and Table 3 shows the learned GenCo cost parameters.

Fig. 2 Dynamic flow
performed by the smart
generator for effective
bidding

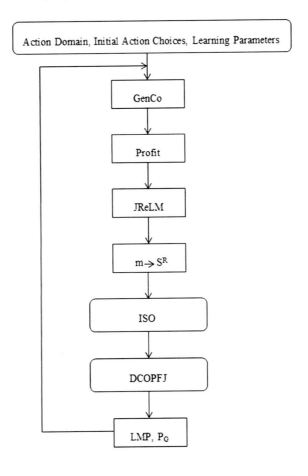

Table 1 ISO activities of a
day

Hour	Settlement for day-ahead market
0 to 10th	ISO collects GenCos and LSE bids/offers
11th to 15th	ISO evaluates bids and offers of demand and suppliers, respectively
16th to 23th	ISO solves the optimal power flow for the next day projects the dispatch and pricing schedule

Figure 4 shows the GenCo commitments without learning for a typical day as scheduled to meet the load as per the market scheduled.

It clearly indicates from Fig. 4, that the low-cost GenCo 1 and 4 is the most utilized in all the hours of a day and the high-cost GenCo 8 and 9 was not at all used. This would imply that the GenCo's scheduled and operated in the market are all low-cost generators but to favor high-cost GenCo is not performed at all. But through generator learning the market strategy would help to favor high-cost GenCo

Fig. 3 IEEE 30 bus system

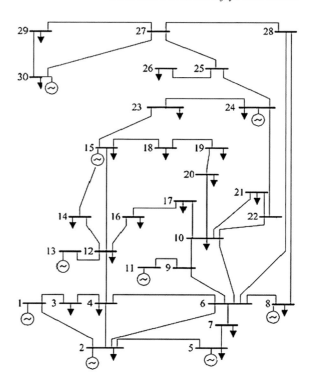

Table 2 GenCo true cost parameters

GenCo	At bus	a ($/MWh)	b ($/Mw^{2}h)	Cap^L (MW)	Cap^U (MW)
GenCo1	1	10.694	0.0046	0	100
GenCo2	2	18.1	0.0061	0	80
GenCo3	5	13.327	0.0087	0	50
GenCo4	8	13.353	0.0089	0	50
GenCo5	11	37.889	0.0143	0	20
GenCo6	13	19.327	0.0103	0	70
GenCo7	15	18.3	0.0071	0	60
GenCo8	24	39.889	0.0163	0	20
GenCo9	30	49.327	0.0243	0	20

to schedule reducing the burden on low-cost GenCo keeping the track of marginal cost fixed by the ISO. Therefore, applying VRE learning algorithm GenCo is trained depending on the capacity and its fuel cost as shown in Table 3.

Figure 5 shows the GenCo commitments with learning. All GenCo's report their offer higher than the true marginal cost and it is clear that the GenCo 6, which is a high-cost GenCo, is utilized only during the peak hour 17. And after learning, GenCo 6 which is a high-cost generator is strategically learnt and now it is being used almost

Table 3 GenCo learned cost parameters

GenCo	At Bus	a^R ($/MWh)	b^R ($/Mw^{2}h)	Cap^L (MW)	Cap^U (MW)
GenCo1	1	11.6662	0.0417	0	100
GenCo2	2	43.44	0.8145	0	80
GenCo3	5	39.981	0.1999	0	50
GenCo4	8	17.804	0.0594	0	50
GenCo5	11	51.556	3.789	0	20
GenCo6	13	23.1924	0.1657	0	70
GenCo7	15	24.4	0.2847	0	60
GenCo8	24	59.8335	0.748	0	20
GenCo9	30	98.654	0.8221	0	20

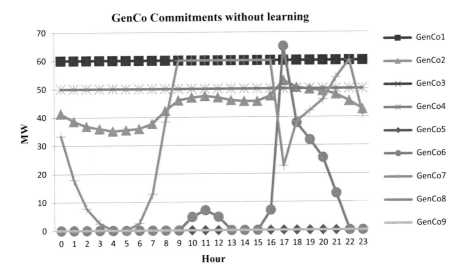

Fig. 4 GenCo commitments without learning

all hours of a day. Also, the strategic supply offer makes GenCo 2 a high-cost GenCo and hence its commitment reduces. This clearly indicates that dynamically changing behavior of the entities makes use of the learning technique. Figure 6 shows the commitment of GenCo's with and without learning in hour 17 of a day.

In this work, all the GenCo's are acting as players since every GenCo is changing its cost parameters to become a winner. The result also shows that irrespective of learning, the location of GenCo and its capacity also plays a major role in the commitments. Also due to the presence of congestion in transmission line the consumer is forced to go for a high-cost GenCo rather than a low-cost one.

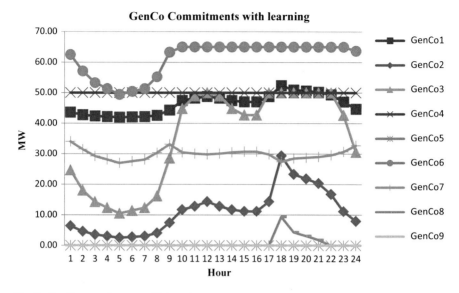

Fig. 5 GenCo commitments with learning

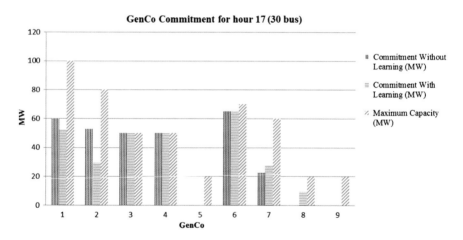

Fig. 6 GenCo commitment with and without learning

4 Conclusion

This paper shows how a GenCo can act as an agent in a deregulated electricity market. The intelligence required for an agent can be provided by the learning module. Here the learning technique applied for the agent-based analysis is VRE learning algorithm. The various learning parameters will be updated based on the past experience, here the profit obtained. A wholesale electricity market having 30 bus was considered for the entire analysis and the results indicate the effectiveness of the agent-based tool in the electricity market. The result shows that GenCo's learn each time by reporting a new marginal cost function based on previous experience.

References

1. Shirmohammadi, D., Wollenberg, B., Vojdani, A., Sandrin, P., Pereira, M., Rahimi, F.: Transmission dispatch and congestion management in the emerging energy market structures. IEEE Trans. Power Syst. **13**, 1466–1476 (1998). https://doi.org/10.1109/59.736292
2. Sun, J.: Dynamic testing of wholesale power market designs: an open-source agent-based framework. Comput. Econ. Springer **30**, 291–327 (2007). https://doi.org/10.1007/s10614-007-9095-1
3. Fang, R.S., David, A.K.: Transmission congestion management in an electricity market. IEEE Trans. Power Syst. **14**, 877–883 (1999). doi:https://doi.org/10.1109/59.780898
4. Conejo, Antonio J.: Electricity markets: analysis & operations. IET Gener. Transm. Distrib. **4**, 123–124 (2010). https://doi.org/10.1049/iet-gtd.2010.9059
5. Foo, Y.S., Gooi, H.B., Chen, S.X.: Multi agent system for distributed management of microgrids. IEEE Trans. Power Syst. **30**, 24–34 (2015). https://doi.org/10.1109/TPWRS.2014.2322622
6. Krishnamurthy, D., Li, W., Tesfatsion, Leigh: An 8-zone test system based on ISO New England data: development and application. IEEE Trans. Power Syst. **31**, 234–246 (2016). https://doi.org/10.1109/TPWRS.2015.2399171
7. Huang, S., Wu, Q., Zhao, H., Li, C.: Distributed optimization based dynamic tariff for congestion management in distribution networks. IEEE Trans. Smart Grid **1**, 1–10 (2017). https://doi.org/10.1109/TSG.2017.2735998
8. Ebrahimian, H., Barmayoon, S., Mohammadi, Mohsen, Ghadimi, Noradin: The price prediction for the energy market based on a new method. J. Econ. Res. **31**, 313–337 (2018). https://doi.org/10.1080/1331677X.2018.1429291
9. Yang, J., Zhao, J., Luo, F., Wen, F., Dong, Z.Y.: Decision-making for electricity retailers: a brief survey. IEEE Trans. Smart Grid **9**, 4140–4153 (2017). https://doi.org/10.1109/TSG.2017.2651499
10. Chen, T., Pourbabak, H., Su, W.: A game theoretic approach to analyze the dynamic interactions of multiple residential prosumers considering power flow constraints. In: Proceedings of the 2016 IEEE power and energy society general meeting, pp. 17–21 (2016). doi:https://doi.org/10.1109/pesgm.2016.7741082
11. Song, M., Amelin, M.: Purchase bidding strategy for a retailer with flexible demands in day-ahead electricity market. IEEE Trans. Power Syst. **32**, 1839–1850 (2017). https://doi.org/10.1109/TPWRS.2016.2608762
12. Balamurugan, S., Lekshmi, R.R.: Control strategy development for multi-source multi area restructured system based on Genco and Transco reserve. Int. J. Electr. Power Energy Syst. **75**, 320–327 (2016). https://doi.org/10.1016/j.ijepes.2015.09.015
13. Solanki, Z., Wani, U., Patel, J.: Demand side management program for balancing load curve for CGPIT College, Bardoli. In: 2017 international conference on energy, communication, data analytics and soft computing (ICECDS) (2017). https://doi.org/10.1109/icecds.2017.8389542
14. Kiran, P., Chandrakala, K.R.M.V., Nambiar, T.N.P.: Multi-agent based systems on microgrid—a review. In: 2017 international conference on intelligent computing and control (I2C2), (2017). doi:https://doi.org/10.1109/i2c2.2017.8321880
15. Kiran, P., Chandrakala, K.R.M.V., Nambiar, T.N.P.: Day ahead market operation with agent based modeling. In: Proceedings of the IEEE international conference on technological advancements in power and energy (Tap Energy 2017), pp. 690–693 (2017). doi:https://doi.org/10.1109/tapenergy.2017.8397302

Standalone Solar Photovoltaic Fed Automatic Voltage Regulator for Voltage Control of Synchronous Generator

Garapati Vinayramsatish, K. R. M. Vijaya Chandrakala
and S. Sampath Kumar

Abstract The aim of this paper is to model standalone solar PV fed Automatic Voltage Regulator (AVR) for controlling the synchronous machine output voltage. The objective of AVR is to sense the output voltage of synchronous machine, alters the field current fed with DC supply and maintain the output terminal voltage constant when load changes. Normally, the DC voltage is given from an additional DC source to the field of the synchronous generator. As solar power is cost free fuel energy; in this work solar power is used to change the field excitation of synchronous machine and excess output power from solar is fed to battery to maintain power balance. Perturb and Observe (P&O) method is used to extract maximum power from solar PV cell through boost converter and buck converter is used to regulate the field current across the generator. The performance of the solar fed AVR to the synchronous machine connected to load is developed and analyzed using MATLAB/Simulink.

Keywords Automatic voltage regulator · Buck converter · Boost converter · Maximum power point tracking (MPPT)

1 Introduction

Renewable energy is becoming a major resource toward modern-day applications because of deficiency of nonrenewable energy sources. Therefore, utilization of renewable energy is carried out through different types of sources like solar, wind, ocean, and biowaste [1]. Solar energy is one of the main renewable resources which has increased drastically in pre years and successively due to extinction of nonrenewable energy sources; it is establishing its pace for the next future applications to be

G. Vinayramsatish (✉) · K. R. M. Vijaya Chandrakala · S. Sampath Kumar
Department of Electrical and Electronics Engineering, Amrita School of Engineering, Amrita
Vishwa Vidyapeetham, Coimbatore, India
e-mail: vinayramgarapati@gmail.com

K. R. M. Vijaya Chandrakala
e-mail: krm_vijaya@cb.amrita.edu

S. Sampath Kumar
e-mail: s_sampathkumar@cb.amrita.edu

© Springer Nature Singapore Pte Ltd. 2020
K. N. Das et al. (eds.), *Soft Computing for Problem Solving*,
Advances in Intelligent Systems and Computing 1048,
https://doi.org/10.1007/978-981-15-0035-0_79

991

driven with. And, synchronous generator being as a source of supply to many applications in the next decades. Normally, to excite the synchronous generator a small DC excitation need to be given and to maintain the required output voltage level the DC excitation is also varied. The DC field excitation is varied by using Automatic Voltage Regulator (AVR), which is again an inverted generator fed separately excited machine, namely, static brushless DC excitation system [1–3]. As the cost of involving machine is higher, to get rid of it solar power is utilized to provide DC excitation to field of synchronous generator. The solar power output is not constant because of continuously varying irradiance. So, for that battery is connected across the solar photovoltaic (PV) system to store the solar power and be utilized to vary the field as on requirement of load to be met during no solar power. When the irradiance is less, solar power will not be sufficient to drive the field of the synchronous machine. In that case, battery will supply the required power and also it is used to store the excess power from solar panel. Recent days, in power system one of the major problems is load side voltage drop. When load increases, the generator output voltage falls down and because of voltage drop on the load, it will draw more current. This excess current leads to thermal instability of the machine and the efficiency drops down [4–6]. Because of these problems, there is a need to maintain output voltage of the generator constant during dynamic change of load. Out of many methods, field supply control is one of the easiest and flexible methods to be followed by which dynamic voltage control is possible. However, depending on the error difference between the required and set DC voltage of the synchronous generator, the steady-state error is improved by the usage of Proportional Integral (PI) controller which is widely used under control applications being highly robust in nature [7–10]. Therefore, in this work PI controller is used to tune the error value in AVR loop of the synchronous generator which regulates the output voltage on dynamic varying demand.

2   Proposed Model of Standalone Solar Photovoltaic and Battery Fed Automatic Voltage Regulator

As availability of solar power and battery as energy storage can play a major role in most of the electrical applications. Instead of having static brushless DC excitation system comprising of two machines to control the output voltage of the synchronous generator, a cost-effective AVR is proposed in this work. In this paper, standalone solar photovoltaic (PV) DC output power with battery support is connected to extract the maximum power depending on the availability of power from either source to control the voltage output of the synchronous generator by varying its field excitation. The complete block diagram of solar power fed AVR to synchronous generator is shown in Fig. 1. This methodology proposed is also much economical to implement when compared to static brushless DC excitation system which is in existence.

From Fig. 1, it shows the complete operation of extracting the maximum power from the PV panel through MPPT tracker, storing in battery during excess solar power and depending on the availability of solar/battery the buck converter connected to the source helps to control the field excitation of the synchronous generator as per

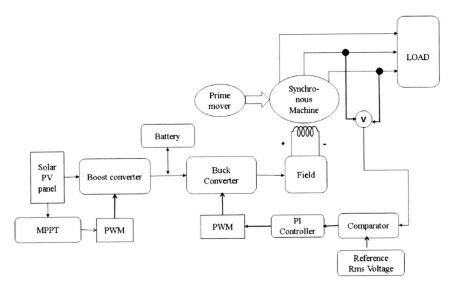

Fig. 1 Standalone solar PV and battery fed Automatic Voltage Regulator (AVR) to control the output voltage of the synchronous generator

the required amount of load. Here, solar PV is selected based on the load. Usually, the solar PV rating is chosen more than rating of the load because of sufficing the load based on the irradiance level. And, practically the solar panel efficiency is less than 20%, the change of irradiance causes more effect on current and change of temperature causes more effect on voltage. Therefore, in this work only irradiance change is considered with a constant temperature. For a load of 8 kW connected across the synchronous generator, the field current of 18.76A (Full load) on requirement the output generated voltage is be controlled to 400 V. For which, solar panel is chosen to be of $V_{max} = 30.7$ V, $I_{max} = 8.15$ A, $P_{max} = 250$ W, number of panels $= 4$, and total power rating $= 4*250$ W $= 1000$ W. In this work, solar panels are connected in parallel, so the output voltages of all panels are same. But output current will be adding up depending on the parallel panel connection.

From solar panel, the maximum power cannot be extracted at normal operating mode because the load impedance and source impedance are not same. As per the maximum power transfer theorem, the maximum power can transfer only when the source and load impedance are same. And, this is met through MPPT algorithm which extracts maximum voltage, current, and power from solar panel. As Perturb and Observe (P&O) method is one of the easiest methods to extract the maximum power from the solar panel is used in this paper. The detailed flowchart of P&O method is shown in Fig. 2.

As the output power extracted by MPPT algorithm is a lower value for which a boost converter is designed and connected along with the lithium–ion battery at a common point as shown in Fig. 3.

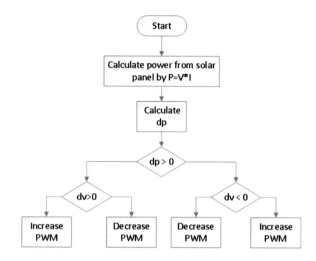

Fig. 2 Pertain and Observe (P&O) method MPPT algorithm for solar PV output

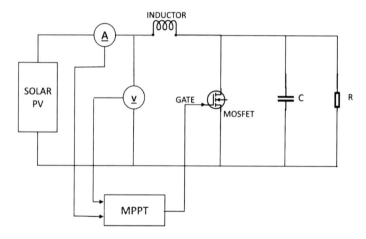

Fig. 3 Circuit diagram of solar PV with MPPT and boost converter

Solar PV cannot supply at constant rate because of continuous variation of sun irradiance. Therefore, for uninterrupted power supply it needs a battery backup. 48 V, 20 Ah lithium–ion battery is connected across the solar PV. Boost converter is operated by MPPT controller and its output terminals are connected to battery.

The maximum power flow through boost converter is 1000 W, and output voltage of boost converter is 48 V. Through MPPT maximum power extraction is possible but the control of output voltage across the boost converter is not possible. The design of inductor and capacitor of boost converter are as follows:

$$C = \frac{D \times I_O}{F \times \Delta V_O} = 500 \ \mu F \tag{1}$$

$$L = \frac{V_{IN} \times (V_O - V_{IN})}{\Delta I_L \times F \times V_O} = 0.85 \ mH \tag{2}$$

Capacitor ripple voltage is 2% and inductor ripple current is 2%.

Therefore, the output voltage of the generator is maintained constant by buck converter which is connected to either the solar panel or the battery, depending up the power requirement to meet the load. The buck converter acts a field voltage regulator of the synchronous generator. As, synchronous machine field winding resistance is 1.208 Ω and the maximum field voltage required to operate during full load is 23 V. Therefore, the field current of 19 A with maximum power of 437 W is required across the synchronous generator. For which the boost converter inductor and capacitor is designed as follows:

$$L = \frac{(V_S - V_O) \times D}{\Delta I_{L1} F} = 1.125 \ mH \tag{3}$$

$$C = \frac{\Delta I_L}{8 \Delta V_O F} = 5.2 \ \mu F \tag{4}$$

Ripple factor for inductor current and capacitor voltage is 2%. The Automatic Voltage Regulator (AVR) maintains the synchronous generator terminal voltage to its rated value through PI controller and buck converter fed by solar/battery power source which compares the output voltage of the machine and reference voltage as required as shown in Fig. 4.

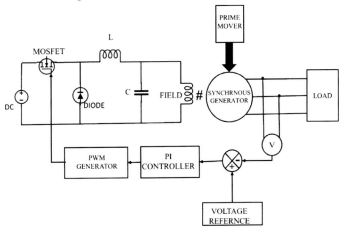

Fig. 4 Schematic diagram of AVR controlled synchronous machine

3 Simulation Results and Discussions

The proposed solar-powered AVR control for the synchronous machine is developed in MATLAB/Simulink® environment as given in Fig. 5. In this simulation, resistive loads are used for loading the generator. The synchronous generator is driven by the prime mover, in this work prime mover is DC motor. Hence, DC motor maintains the constant speed for synchronous machine. Rating of synchronous machine is 8.1 KVA. The output RMS voltage is compared with the rated reference voltage 400 V. The difference between the reference voltage and actual voltage from the synchronous generator is fed to PI controller. The gain values of the PI controller (K_p and K_i) are selected by trial and error method. The PI controller is used for reducing the error value and to generate the control voltage signal for PWM generation. The PWM generator generates the PWM pulses to buck converter

The specification of the synchronous machine used in this work is shown in Table 1.

The solar fed boost converter with MPPT controller is developed as shown in Fig. 1 which is modeled and simulated using MATLAB/Simulink. The output of the boost converter is used to charge the battery and also used as input to the AVR. The solar output voltage, output current, and output power with respect to irradiance

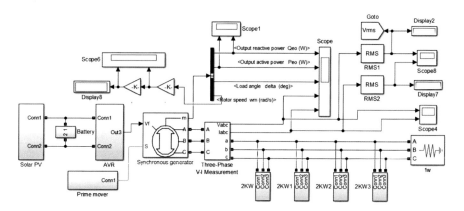

Fig. 5 Proposed model Simulink model

Table 1 Specifications of synchronous machine

Specification	Rating
Frequency	50 Hz
Rated voltage	400 V
Rated power	8.1 KVA
Rated speed	15,000 RPM
Field resistance	1.208 Ω

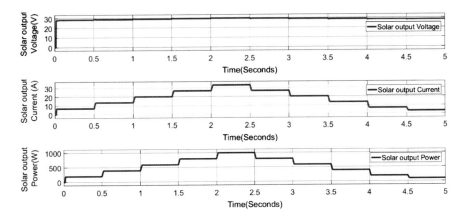

Fig. 6 Output voltage, current, and power at Solar PV

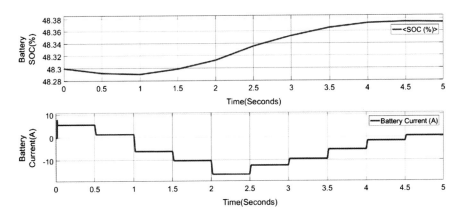

Fig. 7 Battery SOC and current

variation of 200 W/m², 400 W/m², 600 W/m², 800 W/m², 1000 W/m², 800 W/m², 600 W/m², 400 W/m², 200 W/m² are shown in Fig. 6.

The waveform in Fig. 6 clearly depicts that the output voltage is maintained nearly constant and the output power is in phase with output current. The State of Charge (SOC) characteristics of the battery along with battery current are shown in Fig. 7.

From Fig. 7, it clearly states that the negative values in battery current depict the charging period of the battery. Figure 8 shows the battery voltage and power with respect to charging and discharging nature.

As seeing the above figure, whenever the solar output is not enough to maintain the input to the AVR, the excess power is drawn from the battery. When solar power is abundant, it is stored in battery.

The output of the buck converter in AVR is fed to the field of the synchronous generator. The switching signals to the buck converter are from the PI controller.

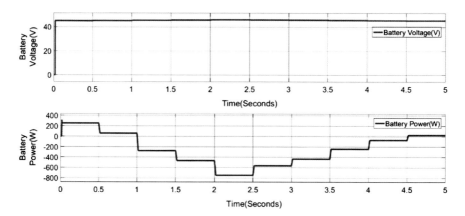

Fig. 8 Battery voltage and power

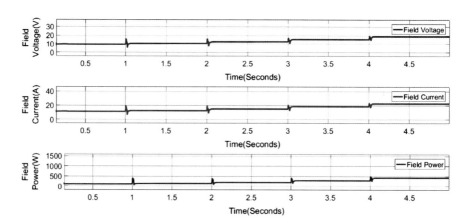

Fig. 9 Field voltage, field current, and field power of the synchronous machine

The responses are shown in Fig. 9 clearly proves that according to the loading of the synchronous machine, the field excitation of the machine varied.

The closed-loop control of the buck converter in the system is given in Fig. 10 with the PI controller. The PI controller gain values are $K_p = 0.007$, $K_i = 1.87$ and switching frequency is 20 kHz. The loading of the machine and the respective terminal voltage and field current are tabulated in Table 2.

The output voltage and output current of the AVR controlled synchronous generator are shown in Fig. 11. From Fig. 11 and Table 2, it is obvious that the terminal voltage is maintained constant irrespective of the load variation.

The active and reactive power variation is displayed in Fig. 12, it shows that the reactive power is almost zero due to the resistive load.

Figure 13 shows the responses of the frequency and speed of the synchronous machine, it proves that the frequency is maintained at constant 50 Hz and the machine runs at the synchronous speed.

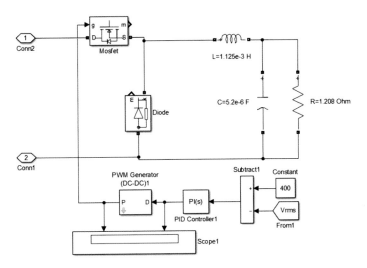

Fig. 10 Closed-loop control of buck converter with PI controller

Table 2 Terminal voltage and field current for applied load

Load in watts	Terminal voltage (V)	Field current (A)
No load	400	9.6
¼ load (2 KW)	400	10.43
½ load (4 KW)	400	12.5
¾ load (6 KW)	400	15.35
Full load (8 KW)	400	18.76

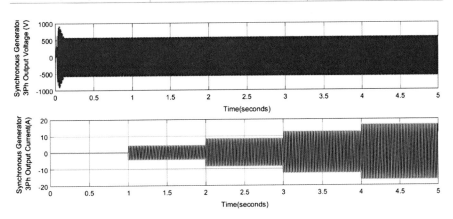

Fig. 11 Voltage and current simulation output responses of AVR controlled synchronous generator

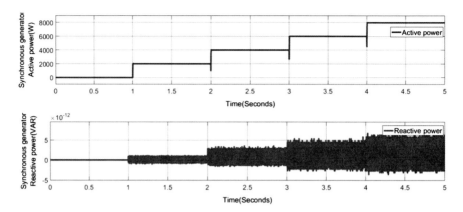

Fig. 12 Active and reactive power output of synchronous generator

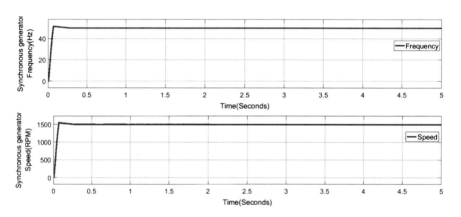

Fig. 13 Output RMS voltage and RMS current of synchronous generator

From the above analysis, it proves that solar fed AVR controls synchronous generator output voltage and it maintains the constant irrespective of irradiance and load variation with low-cost mechanism.

4 Conclusions

Standalone solar PV/battery fed AVR to control the synchronous generator is proposed. The AVR senses the output voltage of the synchronous generator and alters the field voltage as per the requirement of the load. Here, from solar PV the maximum power is extracted using P&O method and the boosted voltage is connected across the point of battery to charge to during the excess solar power output. The field voltage of the synchronous generator is regulated based on the solar power/battery the output voltage of the synchronous generator is controlled under varying load conditions.

References

1. Sudhakar, N., Jain, S., Jyotheeswara Reddy, K.: Solar PV fed standalone excitation system of a synchronous machine for reactive power generation. In: IOP conference series: materials science and engineering vol. 263, (2017). doi:https://doi.org/10.1088/1757-899x/263/5/052017
2. Teja, V.R., Balamurugan, S., Sampath Kumar, S.: Development of ALFC and AVR control loop as a laboratory model using DC-DC Buck Chopper. In: National power engineering conference (NPEC) (2018). doi:https://doi.org/10.1109/npec.2018.8476742
3. Reddy, C.Y., Krishnakanth, V., Sanjay, R., Krishna, V.N.V., Jayabarathi, R.: Laboratory implementation of automatic voltage regulator. In: Biennial international conference on power and energy systems: towards sustainable energy (PESTSE) (2016). doi:https://doi.org/10.1109/pestse.2016.7516477
4. Kutsyk, A., Semeniuk, M.: An application of fuzzy voltage regulator to a static excitation system of a phase compound synchronous generator. In: IEEE international young scientists forum on applied physics and engineering (YSF), pp. 46–49 (2017). doi:https://doi.org/10.1109/ysf.2017.8126590
5. Bayram, M.B.: Lab view based Volt/Hertz controller for synchronous generator excitation systems. In: 58th international scientific conference on power and electrical engineering of riga technical university (RTUCON) (2017). doi:https://doi.org/10.1109/rtucon.2017.8124796
6. Shayeghi, H., Dadashpour, J.: Anarchic society optimization based PID control of an automatic voltage regulator (AVR) system. Electr. Electron. Eng. 2(4), 199–207 (2012). https://doi.org/10.5923/j.eee.20120204.05
7. Park, S.-H., Lee, S.-K., Lee, S.-W., Yu, J.-S., Lee, S.-S., Won, C.-Y.: Output voltage control of a synchronous generator for ships using compound type digital AVR. In: 31st international telecommunications energy conference (2009). doi:https://doi.org/10.1109/intlec.2009.5352002
8. Djagarov, N., Lazarov, T.: Investigation of automatic voltage regulator for a Ship's synchronous generator. Eur. Trans. Electron. Power Eng. 33, 16–21 (2016). doi:https://doi.org/10.7250/pee.2016.003
9. Vijaya Chandrakala, K.R.M., Balamurugan, S.: Simulated annealing based optimal frequency and terminal voltage control of multi-source multi area system. Int. J. Electr. Power Energy Syst. 78, 823–829 (2016). https://doi.org/10.1016/j.ijepes.2015.12.026
10. Adarsh, N.K., Venkatesh, R., Rengarajan, S., Jayabarathi, R.: Simulation and implementation of FPGA controlled distributed solar generation for residential network. IEEE PES Asia-Pacific power and energy engineering conference (APPEEC) (2017). doi:https://doi.org/10.1109/appeec.2017.8308996

Optimizing Vertical Air Gap Location Inside the Wall for Energy Efficient Building Enclosure Design Based on Unsteady Heat Transfer Characteristics

Saboor Shaik, Sunnam Nagaraju, Shaik Mohammed Rizvan and Kiran Kumar Gorantla

Abstract The chief principle of this paper is to optimize the location of vertical air space within composite walls based on thermal unsteady response state parameters that include admittance, transmittance, attenuation factor, and time lag. For computation of these parameters, a MATLAB code has been generated. This code solves 1-D heat flow diffusion equation with convective periodic boundary conditions. Six building construction materials such as laterite stone, burnt brick, mudbrick, reinforced brick, fly ash brick, and concrete block were selected and computations were made for 42 configurations of the composite walls. From this, it is concluded that composite walls with air space located at the outer side of the external wall and the mid-center of the external wall are energy efficient from higher time lag, higher thermal admittance, and lower thermal transmittance perspective and the composite walls with air space located at outer and inner sides of the external walls are the best from the lower decrement factor perspective, among seven studied configurations. The results of the study reduce the air conditioning loads in buildings.

Keywords Optimum air space location · Time lag · Decrement factor · Admittance method

S. Shaik (✉)
School of Mechanical Engineering, Vellore Institute of Technology, Vellore 632014, Tamilnadu, India
e-mail: saboor.nitk@gmail.com

S. Nagaraju
Mechanical Engineering Department, MLR Institute of Technology, Hyderabad 500043, Telangana, India

S. M. Rizvan
Mechanical Engineering Department, SVR Engineering College, Nandyal 518503, Andhra Pradesh, India

K. K. Gorantla
Mechanical Engineering Department, National Institute of Technology Karnataka, Surathkal, Mangalore 575025, Karnataka, India

© Springer Nature Singapore Pte Ltd. 2020
K. N. Das et al. (eds.), *Soft Computing for Problem Solving*,
Advances in Intelligent Systems and Computing 1048,
https://doi.org/10.1007/978-981-15-0035-0_80

1 Introduction

Buildings demand a substantial quantity of power for space conditioning. Thus, it is essential to focus on the building wall parameters defining energy efficiency in the building. The rooms can be maintained at comfortable conditions by walls with a low attenuation factor and higher time lag values [1]. The influence of wall material properties, wall thickness, insulation position within wall/roof, and air gap thickness inside the wall on attenuation factor and it's time lag were reported earlier [2–7]. This study aims at focusing influence of unventilated continuous vertical air space location inside the wall on thermal unsteady response state parameters of composite walls. The admittance method was used and a computer simulation program with graphical user interface was developed in MATLAB to compute thermal unsteady response state parameters of different composite wall materials. This paper recommends the optimum location of air space inside the composite wall based on thermal unsteady response state parameters.

2 Thermal Admittance Methodology for Thermal Unsteady Response State Parameters

In order to compute dynamic thermal response characteristics of multilayer walls with air gaps, a cyclic admittance method was employed. The cyclic admittance procedure solves 1-D diffusion equation with convective periodic boundary conditions. The heat transfer is considered to be 1-D through the thickness of the wall. The diffusion equation for 1-D heat flow is given as [8],

$$\frac{\partial^2 T(X, t)}{\partial X^2} = \frac{\rho Cp}{k} \frac{\partial T(X, t)}{\partial t} \tag{1}$$

The solution of Fourier equation and the arrangement of the terms in matrix form for a single layer solid wall are as follows:

$$\begin{bmatrix} A_1 + jA_2 & (A_3 + jA_4)/c \\ (-A_4 + jA_3).c & A_1 + jA_2 \end{bmatrix} \tag{2}$$

where $\alpha = \sqrt{\pi \rho c_p / \lambda P}$, cyclic thickness $(x) = \sqrt{\pi \rho c_p X^2 / \lambda P} = \sqrt{\pi Cr / P}$ and characteristic admittance of slab $(c) = \sqrt{j2\pi \lambda \rho c_p / P} = \sqrt{j2\pi C / r P}$ and constants $A_1 = \cosh(x)\cos(x)$, $A_2 = \sinh(x)\sin(x)$, $A_3 = [\cosh(x)\sin(x) + \sinh(x)\cos(x)]/\sqrt{2}$, and $A_4 = [\cosh(x)\sin(x) - \sinh(x)\cos(x)]/\sqrt{2}$.

Transmission matrix form of double multilayer wall with an air gap at the mid-center of the wall can be found with given correlation as given below,

$$\begin{bmatrix} T_i \\ q_i \end{bmatrix} = \begin{bmatrix} 1 & -R_{si} \\ 0 & 1 \end{bmatrix} \begin{bmatrix} a_1 & a_2 \\ a_3 & a_1 \end{bmatrix} \begin{bmatrix} 1 & -R_a \\ 0 & 1 \end{bmatrix} \begin{bmatrix} b_1 & b_2 \\ b_3 & b_1 \end{bmatrix} \begin{bmatrix} 1 & -R_{so} \\ 0 & 1 \end{bmatrix} \begin{bmatrix} T_0 \\ q_o \end{bmatrix} \qquad (3)$$

Here, a and b are the multilayer wall quantities.

The reduced transmission matrix form is as follows

$$\begin{bmatrix} T_i \\ q_i \end{bmatrix} = \begin{bmatrix} e_1 & e_2 \\ e_3 & e_4 \end{bmatrix} \begin{bmatrix} T_0 \\ q_0 \end{bmatrix} \qquad (4)$$

For reduced cooling loads, transmittance and decrement factor should be least and admittance and time lag should be more.

The attenuation factor or decrement (f) is the reduction in amplitude of the sine trend of the heat as it penetrates through the thickness of the single/multilayer wall. It is used to measure the thermal mass of any wall. It can be calculated by

$$f = \left| -\frac{1}{U e_2} \right| \qquad (5)$$

The time lag (ϕ) is the lag of time between the scheduling of the highest internal temperature and the highest heat transfer out of the outer single/multilayer wall surface. It is computed by

$$\phi = \frac{12}{\pi} \arctan\left(\frac{\mathrm{Im}\left(-\frac{1}{U e_2} \right)}{\mathrm{Re}\left(-\frac{1}{U e_2} \right)} \right) \qquad (6)$$

The admittance (Y) is the quantity of heat departing the interior side single/multilayer wall surface into the room space per unit degree of temperature fluctuation. It is also used to measure thermal mass of the wall effectively.

$$Y = \left| \left(\frac{q_i}{\theta i} \right)_{T_{o=0}} \right| = \left| -\frac{e_1}{e_2} \right| \qquad (7)$$

Table 1 demonstrates building construction materials with their thermophysical properties and Fig. 1 demonstrates the building materials and composite wall configurations used in the present study. Thermophysical properties of laterite [9] and other building materials are considered as per IS: 3792–1978 [10]. The computer code has been validated with the values of CIBSE standards and found to be accurate. In the current study, outside and inside surface resistances considered are 0.040 m² K/W and 0.130 m² K/W, respectively; and walls are considered as external walls. The air space resistance considered for 25 and 50 mm air space thickness is 0.18 m² K/W as per CIBSE environmental design guide.

Table 1 Thermophysical properties of building materials

Building material	Code	Thermal conductivity k (W/mK)	Density ρ (g/m³)	Specific heat capacity Cp (J/kgK)	Thermal diffusivity α (m²/s)
Laterite Stone	BM1	1.3698	1	1926.1	7.11×10^{-7}
Burnt brick	BM2	0.811	1.82	880	5.06×10^{-7}
Mudbrick	BM3	0.75	1.731	880	4.92×10^{-7}
Reinforced brick	BM4	1.10	1.92	840	6.82×10^{-7}
Fly ash brick	BM5	0.360	1.7	857	2.47×10^{-7}
Concrete block	BM6	1.31	2.24	840	6.96×10^{-7}
Plaster board	P	0.21	0.7	1000	3.0×10^{-7}

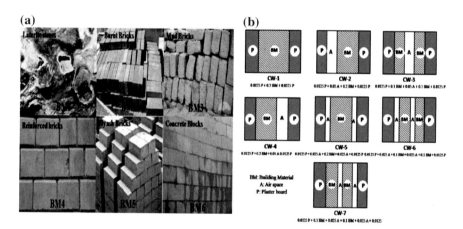

Fig. 1 a Building materials, **b** wall configurations

3 Results and Discussions

3.1 Influence of an Air Space Location Within Wall on Transmittance and Admittance of Composite Walls

Thermal admittance and transmittance are significant parameters to identify thermal mass and thermal insulation of the building walls, respectively. Figure 2 shows the influence of an air space location within a wall on admittance and the transmittance of the composite walls. From the outcomes, it is clear that CW-6 gives highest thermal admittance and lowest thermal transmittance values for all building materials and among all the seven configurations studied. This configuration CW-6 is helpful in reducing cooling loads in the buildings.

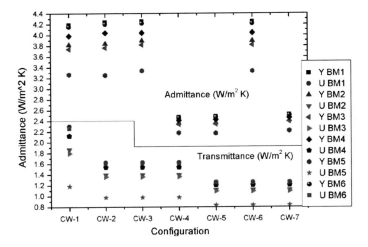

Fig. 2 Admittance and transmittance of composite walls

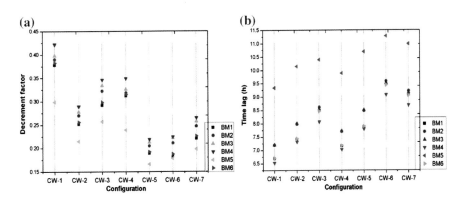

Fig. 3 **a** Attenuation factor of composite walls. **b** Time lag of composite walls

3.2 Influence of An Air Space Location Within a Wall on Attenuation Factor and Time Lags of Composite Walls

Figure 3a, b shows the influence of an air space location within a wall on attenuation factor and it's time lags of composite walls. CW-1 without air space gives the higher decrement factors of 0.378, 0.390, 0.398, 0.422, 0.299, and 0.383 and lower time lags of 6.69 h, 7.2 h, 7.19 h, 6.53 h, 9.34 h, and 6.7 h with composite walls of building materials BM1, BM2, BM3, BM4, BM5, and BM6, respectively. From the outcomes, it is clear that the laterite stone with the air space positioned at the external side and mid-center plane (CW-6) gives the minimum attenuation factor (0.1823) and maximum time lag (9.46 h) among all the configurations of laterite composite walls. It is noted from the results that burnt brick with CW-5 gives the minimum decrement factor (0.205) and burnt brick with CW-6 gives the maximum time lag (9.58 h) among all the composite walls of burnt brick. It is seen from the results that mudbrick with CW-5 gives the minimum decrement factor (0.33) and mudbrick with CW-6 gives the maximum time lag (9.51 h) among all the configurations of mudbrick composite walls. It is inferred from the results that reinforced brick with CW-5 gives the minimum decrement factor (0.21) and reinforced brick with CW-6 gives the maximum time lag (9.07 h) among all the configurations of reinforced brick composite walls. It is also evident that fly ash brick with CW-5 gives the minimum decrement factor (0.16) and fly ash brick with the CW-6 gives the maximum time lag (11.28 h) among all the configurations of all considered building material composite walls. It can also be observed from the results that concrete block with CW-6 gives the minimum attenuation factor (0.187) and maximum time lag (9.44 h) among all the configurations of concrete block composite walls.

4   Conclusions

The composite wall CW-6 is highly recommended configuration for all building materials studied from highest time lag viewpoint. The composite wall CW-5 is recommended for burnt brick, mudbrick, reinforced brick, and fly ash brick composite walls from the lowest attenuation factor viewpoint. The composite wall CW-6 is recommended for laterite and concrete block composite walls. The results of the study are very useful to reduce air conditioning loads in buildings.

References

1. Knowles, T.R.: Proportioning composites for efficient thermal storage walls. Sol. Energy 31(3), 319–326 (1983)
2. Asan, H., Sancaktar, Y.S.: Effects of wall's thermo physical properties on time lag and decrement factor. Energy Build. 28(2), 159–166 (1998)
3. Asan, H.: Effects of wall's insulation thickness and position on time lag and decrement factor. Energy Build. 28(3), 299–305 (1998)
4. Asan, H.: Investigation of wall's optimum insulation position from maximum time lag and minimum decrement factor point of view. Energy Build. 32(2), 197–203 (2000)
5. Shaik, S., Babu, T.P.A.: Optimizing the position of insulating materials in flat roofs exposed to sunshine to gain minimum heat into buildings under periodic heat transfer conditions. Environ. Sci. Pollut. Res. 23(10), 9334–9344 (2016)
6. Shaik, S., Ashok Babu, T.P.: Effect of air space thickness within the external walls on the dynamic thermal behavior of building envelopes for energy efficient building construction. Energy Procedia. 79, 766–771 (2015)
7. Balaji, N.C., Mani, M., Reddy, B.V.V.: Dynamic thermal performance of conventional and alternative building wall envelopes. J. Build. Eng. 21, 373–395 (2018)
8. CIBSE: CIBSE Environmental Design Guide-A, 7th edn. London (2006)

9. Saboor, S., Ashok Babu, T.P.: Influence of ambient air relative humidity and temperature on thermal properties and unsteady thermal response characteristics of laterite wall houses. Build. Environ. **99**, 170–183 (2016)
10. SP: 41. (S&T) Handbook on functional Requirement of Buildings other than industrial buildings. Bureau of Indian Standards, India, pp. 33–40 (1987)

Author Index

A

Absara, A., 737
Adak, Asish, 1
Adarsha, B. S., 353
Aditi, 687
Agarwal, Vernika, 671, 699
Aggarwal, Sugandha, 621
Ahana Priyanka, N., 377
Ajay Kumar, H., 737
Akinmusuru, J. O., 197
Anand, R., 261
Anitha, A., 443
Anitha, R., 961
Anusha, Gorantla, 341
Arock, Michael, 493
Arunkumar, B., 197, 849
Arvind, C. S., 55
Arya, Dhruv, 239
Assad, Assif, 749, 941
Awoyera, P. O., 197, 849

B

Babu, Challa, 311
Baby, Cyril Joe, 365
Bag, Soumen, 455
Bagyaveereswaran, V., 961
Balaji, S., 657
Balamurugan, M., 563
Banati, Hema, 805
Bansal, Jagdish Chand, 749
Bansal, Viraj, 727
Banuselvasaraswathy, B., 329
Behera, Ritanjali, 563
Bhalchandra, Parag, 285
Bhanutej, 907, 925

Bharath, K. P., 657
Bhuyan, R. K., 893
Brisilla, R. M., 711, 727, 773, 879
Busawon, K., 907, 925

C

Chandra, Dhairya, 573
Chaudhary, Reshu, 805
Chaudhuri, Dibyendu Roy, 573
Chinnadurai, T., 329

D

Darbari, Jyoti Dhingra, 671, 687, 699
Debanjan, Kundu, 341
Deep, Kusum, 941
Deheri, G. M., 27
Deheri, Gunamani B., 9
Devi, Salam Shuleenda, 319
Dey, Nilanjan, 405
Dharavath, Ravi, 43
Dutta, Ajoy, 1

G

Gobinath, R., 197, 849
Gorantla, Kiran Kumar, 1003
Govinda Chowdary, V., 879
Gupta, Praveen Kumar, 1
Gupta, Shubham, 941

H

Hambarde, Kailash, 285
Hannah, Esther, 467

© Springer Nature Singapore Pte Ltd. 2020
K. N. Das et al. (eds.), *Soft Computing for Problem Solving*,
Advances in Intelligent Systems and Computing 1048,
https://doi.org/10.1007/978-981-15-0035-0

Harish, Narayana, 353
Hariyale, Neelam, 391
Hemanth Kumar, M. B., 145
Hunagund, P. V., 181
Hussaian Basha, C. H., 711, 727, 773, 879,
     907, 925

I
Itagi, Anirudh, 365

J
Jacob Raglend, I., 43
Jain, Priyansh, 271
Jaladi, Satyendra, 519
Janardhan, Prashanth, 353
Jayabarathi, T., 295
Jayanthi Sree, S., 91
Jha, P. C., 621, 671, 687, 699
Jisha, T. E., 479
Jyothi, 539

K
Kamalanand, K., 405
Kanish, T. C., 251
Kaul, Arshia, 621, 687
Kavitha, G., 377
Khamitkar, Santosh, 285
Khilar, Rashmita, 551
Kiran, P., 981
Kulkarni, Govind, 285
Kumar, Anshuman, 893
Kumari, N. Sandhya, 209
Kumar, S. N., 737
Kumar, Subham, 585
Kumar, Vinod, 431

L
Lakshman, S., 261
Lenin Fred, A., 737

M
Madarkar, Jitendra, 597, 609
Mandal, Sukomal, 353
Mani, Geetha, 629
Mathew, Derick, 907, 925
Mitra, Arkajyoti, 455
Mittal, Ankush, 573, 761
Mohanty, Chinmaya P., 893
Monoth, Thomas, 479

Munshi, Mohmmadraiyan M., 27
Muralikumar, K., 105, 415, 787, 823, 863
Muthu, Rajesh Kumar, 657, 971

N
Naga Jyothi, Grande, 341
Nagaraju, Sunnam, 1003
Nagaraju, T. Vamsi, 795, 839
Nagar, Atulya K., 79
Nair, Shekhar, 505
Narayana Moorthy, N., 251
Natarajan, V. Anantha, 209
Nithish, M. S., 261
Niyaz, Usma, 67

O
Odofin, S., 711, 727, 773, 879

P
Padha, Devanand, 67
Pandit, Purnima, 155
Patel, A. R., 27
Patel, Rakesh M., 9
Patel, Sakshi, 657
Ponnambalam, P., 105, 311, 415, 539, 647,
     787, 823, 863
Prahlada Rao, K., 181
Prasad, Ch. Durga, 795
Prasad Meesaraganda, L. V., 119
Prasad, Ritu, 391
Priya, M., 105, 415

R
Raamesh, Lilly, 467
Rajalaxmi, T. M., 531
Rajinikanth, V., 405
Raju, M. Jagapathi, 795
RamachandranPillai, Resmi, 493
Ramalingam, Senthil Prabu, 505
Raman, Sundaresan, 585
Rana, Ajay, 761
Rani, C., 505, 551, 563, 711, 727, 773, 879,
     907, 925
Rao, T. E., 519
Rathore, Manjari Singh, 391
Rautela, Ankit, 823
Ray, Soumi, 431
Revathi, T., 531
Rizvan, Shaik Mohammed, 1003
Rupini, D. K. D. B., 839

S

Saha, Sriparna, 79
Sahoo, Sarat Kumar, 551, 563
Sahu, Abhilash, 961
Salgotra, Rohit, 79
Samale, Pranita, 285
Sambaiah, Kola Sampangi, 295
Sampath Kumar, S., 991
Sangeetha, G., 197, 849
Saravanan, B., 145
Saravanan, T. Y., 647
Sarkar, Nilarghya, 119
Satapathy, Suresh Chandra, 405
Saurabh, Praneet, 391
Senthilnath, J., 55
Shaha, Aditya, 135, 239
Shaik, Abdul Munaf, 893
Shaikh, Husen, 285
Shaik, Saboor, 1003
Shanmugam, Prabhakar Karthikeyan, 505, 551, 563
Shanthi, T., 261
Sharma, Harish, 749
Sharma, Kritika, 749
Sharma, Nirmala, 749
Sharma, Poonam, 597, 609
Sharma, Rashi, 699
Sharmila Banu, K., 271
Shiva Krishna, A., 197
Silahtaroğlu, Gökhan, 285
Singh, Neha, 225
Singh, Ngangbam Herojit, 319
Singh, Payal, 155
Singh, Rimjhim, 597
Singh, Rimjhim Padam, 609
Singh Sambyal, Abhishek, 67
Singh, Urvinder, 79

Siva Krishna, A., 849
Sivakumar, C., 823
Solanki, Rahul, 671
Soni, Siddharth, 365
Srinath, A., 519
Srinivas, A., 849
Subhashini, 961
Sumathi, 467
Sundar, S., 365
Suresh, V., 737

T

Tamsekar, Pritam, 285
Tarafder, Nilanjan, 119
Thongam, Khelchandra, 319
Thunga, Sai Saranya, 971
Tripathi, Prasun Chandra, 455
Tripathy, B. K., 135, 225, 239, 271

U

Udhay Sankar, V., 907, 925

V

Vani, R. M., 181
Varday, Harshal, 271
Vasanthanayaki, C., 91
Vashi, Yogini D., 9
Vijaya Chandrakala, K. R. M., 981, 991
Vimalathithan, R., 329
Vinayramsatish, Garapati, 991

Y

Yadavalli, Venkata S. S., 699

Printed in the United States
By Bookmasters